# 2012 28th Annual IEEE Semiconductor Thermal Measurement and Management Symposium

# (SEMI-THERM 2012)

San Jose, California, USA
18 – 22 March 2012

| | |
|---|---|
| IEEE Catalog Number: | CFP12SEM-PRT |
| ISBN: | 978-1-4673-1110-6 |

**Copyright © 2012 by the Institute of Electrical and Electronic Engineers, Inc**
**All Rights Reserved**

*Copyright and Reprint Permissions*: Abstracting is permitted with credit to the source. Libraries are permitted to photocopy beyond the limit of U.S. copyright law for private use of patrons those articles in this volume that carry a code at the bottom of the first page, provided the per-copy fee indicated in the code is paid through Copyright Clearance Center, 222 Rosewood Drive, Danvers, MA 01923.

For other copying, reprint or republication permission, write to IEEE Copyrights Manager, IEEE Service Center, 445 Hoes Lane, Piscataway, NJ 08854. All rights reserved.

***This publication is a representation of what appears in the IEEE Digital Libraries. Some format issues inherent in the e-media version may also appear in this print version.*

IEEE Catalog Number:          CFP12SEM-PRT
ISBN 13:                      978-1-4673-1110-6
ISSN:                         1065-2221

**Additional Copies of This Publication Are Available From:**

Curran Associates, Inc
57 Morehouse Lane
Red Hook, NY  12571 USA
Phone:       (845) 758-0400
Fax:         (845) 758-2633
E-mail:      curran@proceedings.com
Web:         www.proceedings.com

# 2012 28th Annual IEEE Semiconductor Thermal Measurement and Management Symposium (SEMI-THERM 2012)

San Jose, California, USA
18-22 March 2012

IEEE Catalog Number:   CFP12SEM-POD
ISBN:   978-1-46731-110-6

# TABLE OF CONTENTS

## SEMICONDUCTOR THERMAL MEASUREMENT AND MANAGEMENT SYMPOSIUM

## Session A1: 3-D Packaging
### Chair: Atila Mertol, Ph.D., LSI Corporation, USA

**Measurement of Microbump Thermal Resistance in 3D Chip Stacks** ........................ 1

Evan Colgan, Paul Andry, Bing Dang, John Magerlein, Joana Maria, Robert Polastre ......................
.................................................................................................. IBM T.J. Watson Research Center, USA
Jamil Wakil ...................................................................................................... IBM Austin, USA

**Experimental Thermal Resistance Evaluation of a Three-Dimensional (3D) Chip Stack,
Including the Transient Measurements** ............................................................... 8

Keiji Matsumoto, Soichiro Ibaraki, Kuniaki Sueoka, Katsuyuki Sakuma, Hidekazu Kikuchi, Yasumitsu
Orii, Fumiaki Yamada ........... ASET (Association of Super-Advanced Electronics Technologies), Japan

**Impact of Die-to-Die Thermal Coupling on the Electrical Characteristics
of 3D Stacked SRAM Cache** ............................................................................. 14

Subho Chatterjee, Minki Cho, Saibal Mukhopadhyay ..................................... Georgia Tech, USA
Rahul Rao .......................................................................................................... IBM, USA

**Dynamic Compact Thermal Model for Stacked-Die Components** ............................ 20

Eric Monier-Vinard, Cheikh Tidiane Dia, Valentin Bissuel, Olivier Daniel ..... Thales Global Svcs, France
Cheikh Tidiane Dia, Najib Laraqi ...................................................... Université Paris Ouest,, France

## Session B1: Liquid Cooling
### Chair: George Meyer, Celsia Technologies, USA

**Two-Phase Flow Control of On-Chip Two-Phase Cooling Systems of Servers** ................... 29

Jackson Marcinichen and John Thome ........................................... LTCM, EPFL, Switzerland

**Numerical Prediction of the Junction-to-Fluid Thermal Resistance of a 2-Phase
Immersion-Cooled IBM Dual Core POWER6 Processor** ......................................... 36

Levi Campbell ................................................................................................. IBM, USA
Phil Tuma ............................................................................................... 3M Company, USA

**Thermal Performance of Sub-atmospheric Loop Thermosyphon
with and without Enhanced Boiling Surface** ...................................................... 45

Shyy Woei Chang ........................................... National Kaohsiung Marine University, Taiwan
Kuei-Feng Chiang, Chuan-Chin Huang, Marine University and Asia Vital Components Co. Ltd, Taiwan

**High-Performance Nickel Wick Development for Loop Heat Pipes** ............................ 52

Vijit Wuttijumnong, Randeep Singh, Masataka Mochizuki, Kazuhiko Goto, Thang Nguyen,
Tien Nguyen, Koichi Mashiko .................................................................... Fujikura, Japan

## Session C1: Sustainable Data Centers
### Chair: Veerendra Mulay, Ph.D., Facebook, USA

**A Comparison Analysis of Air, Liquid, and Two-Phase Cooling of Data Centers** ................ 58

Michael Ohadi, Serguie Dessiatoun, K. Choo, Michael Pecht ................. University of Maryland, USA
John Lawler ............................................................................................... ATEC,Inc., USA

**Design And Management Of Data Center Effectiveness, Risks And Costs** ..................... 64

Mark Seymour, Sherman Ikemoto ...................................... Future Facilities Ltd., UK & USA

**Sustainable Data Centers Powered by Renewable Energy** ...................................... 362

Levente J. Klein, Sergio Bermudez, Hans-Dieter Wehle, Stephan Barabasi,
Hendrik F. Hamann .................................................................................. IBM Corp.

## Session A2: Multidisciplinary Thermal Management
### Chair: Rahima Mohammed, Ph.D., Intel, USA

**Novel 3D Electro-Thermal Robustness Optimization Approach of Super Junction Power MOSFETs under Unclamped Inductive Switching** ........................................................ **69**

Joseph Rhayem, Aarnout Wieers, Andrej Vrbicky, Peter Moens, Jaume Roig, Piet Vanmeerbeek, Marnix Tack ....................................................ONSemiconductor, Belgium
Anna Villamor-Baliarda ............................................ Instituto de Microelectrónica de Barcelona, Spain
Andrea Irace and Michele Riccio ..................................................University of Naples, Italy

**Energy Reduction in Server Cooling Via Real Time Thermal Control** ........................... **74**

Xuefei Han, Yogendra Joshi ... Dept. of Mechanical Engineering, Georgia Institute of Technology, USA

**A Case Study on Impact of Free Air Cooling on Telecom Equipment Performance** ............................. **82**

Jun Dai, Diganta Das, Michael Pecht, Michael Ohadi .................. CALCE, University of Maryland, USA

**Enhancement of Photovoltaic Solar Module Performance for Power Generation in the Middle East** .. **87**

Valerie Eveloy, Peter Rodgers, Shrinivas Bojanampati ............................The Petroleum Institute, UAE

## Session C2: Data Center Modeling and Analysis
### Chair Dave Saums, DS& LLC, USA

**Data Center Cooling Management And Analysis – A Model Based Approach**...................…..............**98**

Rongliang Zhou, Zhikui Wang, Cullen Bash, Alan McReynolds......................Hewlett-Packard Co, USA

**Datacenter Power Savings through High Ambient Datacenter Operation: CFD Modeling Study** ........ **104**

Nishi Ahuja ................................................................................. Intel Corp, USA

**CFD Analysis of Free Cooling of Modular Data Centers**............................................................. **108**

Betsegaw Gebrehiwot, Kushal Aurangabadkar, Naveen Kannan, Dereje Agonafer................................
.......................................................................... University of Texas at Arlington, USA
Deepak Sivanandan, Mark Hendrix ................................................ CommScope, Inc,.USA

## Session A3: Thermal Management in Multi-Core Architectures
### Chair: Attila Aranyosi, Ph.D., Juniper Networks,USA

**Performance Optimization of Multi-Core Processors Using Core Hopping - Thermal and Structural** .**112**

Sunil Lingampalli, Fahad Mirza, Thiagarajan Raman,
Dereje Agonafer................................................................ University of Texas at Arlington, USA

**Thermal System Identification (TSI): A Methodology for Post-silicon Characterization and Prediction of the Transient Thermal Field in Multicore Chips** ........................................ **118**

Minki Cho, William Song, Sudhakar Yalamanchili, Saibal Mukhopadhyay ............................................
.......................................................................... Georgia Institute of Technology, USA

**On-Chip Cooling of Hot-Spots With a Copper Micro-Evaporator** ........................................ **125**

Etienne Costa-Patry and John R. Thome ......................................LTCM, EPFL, Switzerland

**Energy Efficient Liquid-Thermoelectric Hybrid Cooling for Hot-Spot Removal** ....................... **130**

Vivek Sahu, Andrei Fedorov, Yogendra Joshi ............................Georgia Institute of Technology, USA
Kazuaki Yazawa, Amirkoushyar Ziabari, Ali Shakouri ............University of California, Santa Cruz, USA
Ali Shakouri................................................................................ Purdue University, USA

**Test ASIC for Investigation of Thermal Coupling in Many-Core Architectures** ...................... **135**

Michal Szermer, Cezary Maj, Piotr Pietrzak, Marcin Janicki, Piotr Zajac, Andrzej Napieralski ...............
.......................................................................... Technical University of Lodz, Poland

## Session B3: Experimental Methods
Chair: Marta Rencz, Ph.D., Budapest University of Technology and Economics, Hungary

**Socket Thermal Testing Procedure and Correlation** ........................................................ 139
Ted Lee and Michelle Lin ........................................................ Intel Corporation, USA & Taiwan

**A Method to Measure Heat Dissipation from Component on PCB** ........................................... 143
Zhongwei Qi ........................................................................... GE Healthcare, USA

**Thermal Conductivity Measurements of Novel SOI Films Using Submicron Thermography and Transient Thermoreflectance** ........................................................ 150
Mihai Burzo ........................................................... University of North Texas, USA
Peter Raad and Pavel Komarov ..................................... Southern Methodist University, USA
Taehun Lee ......................................................................... INTEL Corporation, USA

**Measurement of Thermal Conductivity of Thin and Thick Films by Steady-state Heat Conduction** ........................................................ 157
Ramesh Shrestha, Tae Choi ........................................... University of North Texas, USA
Wonseok Chang .................................... Korea Institute of Machinery and Materials, Korea

**Application of Thermal Transient Testing for Solar Cell Characterization** ........................... 162
Andras Vass-Varnai, Zoltan Sarkany, Marta Rencz ...................... Mentor Graphics, Hungary
Andras Vass-Varnai, Balazs Plesz, Adrian Malek, Marta Rencz
........................................................ Budapest Univ of Technology and Economics, Hungary

## Session A4: Thermal Issues in GaN Devices
Chair: Ross Wilcoxon, Ph.D., Rockwell Collins, USA

**Integration of a Phase Change Material for Junction-Level Cooling in GaN Devices** .......... 169
Daniel Piedra, Min Sun, Tomás Palacios .......... Massachusetts Institute of Technology, USA
Tapan Desai and Richard Bonner ........................... Advanced Cooling Technologies, Inc., USA

**Thermoreflectance CCD Imaging of Self Heating in AlGaN/GaN High Electron Mobility Power Transistors at High Drain Voltage** ........................................................ 173
Kerry Maize ........................................... University California, Santa Cruz, USA
Eric Heller and Donald Dorsey ................................. Air Force Research Laboratory, USA
Ali Shakouri ................................................................. Purdue University, USA

**Thermal Factors Influencing the Reliability of GaN HEMTs** ............................................. 182
Jason Carter, Jeremy Acord, Andrew Trageser ...................... Pennsylvania State University, USA
Charles Pagel ........................................... Crane Naval Surface Warfare Center, USA

## Session B4: Heat Transfer Fundamentals I
Chair: Wendy Luiten, Philips Research, Netherlands

**A Framework Theory for Dynamic Compact Thermal Models** ........................................... 189
Mohamed-Nabil Sabry ........................................... Mansoura University, Egypt
Mohamed Dessouky ........................................... Mentor Graphics, Egypt

**Heatsink Design Optimization Using the Thermal ShortCut Concept** ............................... 195
Byron Blackmore, John Parry and Robin Bornoff ........................... Mentor Graphics, UK

**A Method to Adapt $Z_{th}$-Junction-to-Ambient Curves to Varying Ambient Conditions** .......... 205
Dirk Schweitzer ........................................... Infineon Technologies AG, Germany

## Session C4:Data Center Liquid Cooling
### Chair: Saket Karajgikar. Ph.D. Future Facilities, USA

**Server Liquid Cooling With Chiller-less Data Center Design
to Enable Significant Energy Savings** ............................................................ 212

Madhusudan Iyengar, Milnes David, Pritish Parida, Vinod Kamath, Bejoy Kochuparambil, David Graybill, Mark Schultz, Michael Gaynes, Robert Simons, Roger Schmidt, Timothy Chainer ... IBM, USA

**Experimental Investigation of Water Cooled Server Microprocessors and Memory Devices
in an Energy Efficient Chiller-less Data Center** ............................................ 224

Pritish R. Parida, Milnes David, Madhusudan Iyengar, Mark Schultz, Michael Gaynes,
Vinod Kamath, Bejoy Kochuparambil and Timothy Chainer ......................... IBM Corp., USA

**Experimental Characterization of an Energy Efficient Chiller-less Data Center Test Facility
with Warm Water Cooled Servers** .................................................................. 232

Milnes David, Madhusudan Iyengar, Pritish Parida, Robert Simons, Mark Schultz,
Michael Gaynes, Roger Schmidt and Timothy Chainer ................................ IBM Corp., USA

## Session A5: Fan Cooling
### Chair: Sharon Adam, Infinera, USA

**A Multiple Vibrating-Fan System Using Interactive Magnetic Force and Piezoelectric Force** ............. 238

H.K. Ma, W. F. Luo, H. C. Su ........................................ National Taiwan University, Taiwan

**Study of a Cooling System with a Piezoelectric Fan** ................................................ 243

H.K. Ma, C. L. Liu, H. C. Su, W. H. Ho ............................. National Taiwan University, Taiwan

**Modeling of Fan Failures in Networking Enclosures** ................................................ 249

Susheela Narasimhan .................................................. Cisco Systems Inc., USA
Gokul Shankaran and Sankar Basak ...................................... Ansys, Inc., USA

**SynJet Augmented Cooling of a 1U Security Chassis** .............................................. 255

Raghav Mahalingam and Brandon Noska ................................... Nuventix, Inc., USA
Susheela Narasimhan .................................................. Cisco Systems, Inc., USA

## Session B5: Heat Transfer Fundamentals II
### Chair: Peter Rogers, Ph.D., Petroleum Institute, UAE

**Volume Averaging Theory (VAT) Based Modeling and Closure Evaluation of
Scale-Roughened Plane Fin Heat Sink** ............................................................ 260

Feng Zhou, David Vasquez, George DeMoulin, David Geb and Ivan Catton .................. UCLA, USA

**Two-Layer Heat Spreading Revisited** ........................................................... 269

Clemens Lasance ......................................... Consultant, SomelikeitCool, the Netherlands

**Heat Spreading From a Small Source on a Thin Plate** ............................................ 275

Wendy Luiten ...................................................... Philips Research, the Netherlands

**Design Considerations for Heat Spreader in High Heat Flux Systems** ............................... 283

Khairul Alam, Xiaoping Shen, and Rahat Taposh ................................. Ohio University, USA

## Session C5: Data Center Cooling Management
### Chair Nishi Ahuja, Intel, USA

**Achieving Energy Efficient Data Centers Using Cooling Path Management
Coupled With ASHRAE Standards** ................................................................ 288

Matthew Green and Saket Karagjikar ........................................ Future Facilities Inc., USA
Phillip Vozza, Nick Gmitter and Dan Dyer ........................ DLB Associates Consulting Engineers, USA

**Use of Ducting to Improve Inlet Conditions for Side-to-Side Airflow Switches in Data Centers** ......... 293

Jim Fleming ................................................................ Panduit, USA
Susheela Narasimhan ...................................................... Cisco Systems, Inc., USA

## Session D1: Thermal-Aware IC Design
### Chair: Xi Wang, Ph.D., Intersil, USA

**Enabling Power Density and Thermal-Aware Floorplanning** ........................................ 302
Ehsan K. Ardestani, Amirkoushyar Ziabari, Ali Shakouri and Jose Renau ............UC Santa Cruz, USA

**Cache Leakage Power Estimation Using Architectural Model
for 32 nm and 16 nm Technology Nodes** ........................................................... 308
Piotr Zajac, Marcin Janicki, Michal Szermer, Cezary Maj, Piotr Pietrzak, Andrzej Napieralski ................
.......................................................................................... Technical University of Lodz, Poland

**New Simulation Approaches Supporting Temperature-Aware Design of Digital ICs** .......................... 313
Gergely Nagy, András Tímár, Albin Szalai, Márta Rencz, András Poppe ..........................................
.......................................... Budapest University of Technology and Economics, Hungary
András Poppe ........................................................ Mentor Graphics, Hungary

## Session D2: LED Thermal Management
### Chair: Dave Saums, DS&A LLC, USA

**An Innovative Passive Cooling Method for High Performance Light-Emitting Diodes** ........................ 319
Angie Fan and Richard Bonner...........................Advanced Cooling Technologies. Inc., USA
Stephen Sharratt and Y. Sungtaek Ju .................................. University of California, Los Angeles, USA

**A Step Forward in Multi-Domain Simulation of Power LEDs** ....................................... 325
András Poppe .......... Budapest University of Technology and Economics & Mentor Graphics, Hungary

**How Thermal Environment Affects OLEDs' Operational Characteristics** ............................. 331
Zsolt Kohári, László Pohl, András Poppe ........ Budapest Univ of Technology and Economics, Hungary

## Session D3: Advances in Materials, Thermal Imaging and Validation Platform
### Chair: Kazuaki Yazawa, Ph.D., UC Santa Cruz, USA

**Development of a Flexible Chip Infrared (IR) Thermal Imaging System for Product Qualification** ...... 337
Chenzhou Lian, Marc Knox, Kamal Sikka, Xiaojin Wei, Alan Weger ......................... IBM Corp., USA

**Side-by-Side Comparison between Infrared and Thermoreflectance Imaging
Using a Thermal Test Chip with Embedded Diode Temperature Sensors** ............................. 344
Dustin Kendig, Kazuaki Yazawa ......................................Microsanj LLC, USA
Amy Marconnet and Mehdi Asheghi.............................. Stanford University, USA
Ali Shakouri........................................................... Purdue University, USA

**High Performance Thermal Interface Materials with Enhanced Reliability** ........................... 348
Sihai Chen and Ning-Cheng Lee ........................................ Indium Corporation, USA

**Performance Improvements of Air-Cooled Thermal Tool with Advanced Technologies** ..................... 354
Rahima Mohammed, Yi Xia, Ridvan Sahan, Ying-feng Pang ........................Intel Corp., USA

---

# SUBMIT A PAPER FOR SEMI-THERM 29!

As you further develop a technique or application, consider documenting it
for the thermal community. **SEMI-THERM 29** will begin accepting
abstracts during the summer (deadline is September 15, 2012).
We welcome your submissions! Visit us at **www.semi-therm.org.**

# Twenty Eighth Annual IEEE
# SEMICONDUCTOR THERMAL MEASUREMENT AND MANAGEMENT SYMPOSIUM

The 2012 Semiconductor Thermal Measurement and Management (SEMI-THERM) Symposium is an annual international forum for the presentation of new developments in and applications relating to generation and removal of heat within semiconductor devices, and measurement of junction temperatures under various application and environmental conditions.

Attendance at the Symposium is limited, to preserve the close interaction among attendees and presenters. The format of the symposium this year couples nine sessions of selected technical papers, a more-intimate Poster Session for one-on-one discussion of results, a luncheon talk, and a series of Workshops focused on products and techniques.

This year, the Symposium is preceded by five Short Courses: **"Thermal Management for LED-based Applications," "Thermal Design of Electronic Systems for Use in Data Centersv" "Data Center Design Cannot Ignore Electronic System Details,", "Thermal Management for LED-based Applications,"** and **"Transitioning from Air to Liquid Cooling: Design Fundamentals for Heat Sinks and Cold Plates."** In addition, an exhibits area offers displays of equipment, software, and other resources within the thermal measurements field.

We trust you will take advantage of the rich array of information and experiences developed by this year's Steering and Program Committees, and consider submitting an abstract for next year's SEMI-THERM.

**General Chair**
Herman Chu,
    Cisco Systems.

**Program Chair**
Zeki Celik,
    LSI Corporation

**Vice Program Chair**
Genevieve Martin,
    Philips Research

**Symposium/Exhibitor Management/Registration**
Tom Tarter,
    +1-408-505-0946,
    ttarter@semi-therm.org

**European Liaisons**
Clemens J. M. Lasance,
    SomeLikeItCool
John Parry, Mentor Graphics

**Asia Liaison**
Prof. Hsiao-Kang Ma,
    National Taiwan University

**Proceedings**
Paul Wesling, Kathe Erickson

The Components, Packaging, & Manufacturing Technology (CPMT) Society, Symposium sponsor, is one of the 35+ technical groups within the Institute of Electrical and Electronics Engineers. See the profile of our Society on the Web: **http://www.cpmt.org**. CPMT is chartered as the focus for the application of technology within the MCM, optoelectronics, MEMS and chip packaging fields, with an emphasis in thermal management, techniques, and characterization. Our journals have a strong focus on thermal management. We also sponsor a number of related conferences:

IEEE COMPONENTS, PACKAGING AND MANUFACTURING TECHNOLOGY SOCIETY

- Intersociety Conference on Thermal Phenomena in Electronic Systems (I-THERM)
- Electronic Components & Technology Conference (ECTC)
- Electronic System-Integration Technology Conference (ESTC)
- International Workshop on Thermal Investigations of ICs and Systems (THERMINIC)
- International Conference on Thermal and Mechanical Simulation in Microelectronics (EuroSim-E)

We invite you to consider membership in our Society, and to participate with us in furthering advancements in these fields. If you already belong to another professional society, then you can affiliate with CPMT at reduced annual dues. Refer to the back cover for additional information, or call 1-800-678-IEEE for membership information.

## March 18-22, 2012
## DoubleTree Hotel
## San Jose, CA   USA

# Welcome to IEEE SEMI-THERM 28

On behalf of the entire SEMI-THERM 28 Organizing Committee, I would like to welcome everyone to the 2012 symposium. We are looking forward to seeing our industry friends and colleagues; and at the same time, meeting and greeting new participants. We would also like to extend special thanks and recognition to all our industry sponsors in bringing us the advanced and emerging technologies conveniently all under one roof. As a fully voluntary non-profit organization, I would like to encourage anyone who is interested in supporting SEMI-THERM (a great opportunity to build up your network of industry colleagues and leaders) to contact me or any of the committee members listed in this year's program and proceedings.

The internet technology revolution that started 25 years ago has brought us the social network landscape of today. It has impacted not just how we interact with each other socially and commercially through social media like Facebook and Twitter, but the unprecedented empowerment of the masses in organizing and communicating around the world. The rate of growth of the connectivity of the global citizens to the worldwide network is extraordinary. In the developed countries, like North America, Western Europe and Japan, they are well over 70% penetration of their populations. However, the world's most populous developing countries, like China, India and Sub-Saharan Africa, have been coming online, and along with the increased bandwidth demands from new technologies and services, the IP networks are growing at a pace that the world has never seen before.

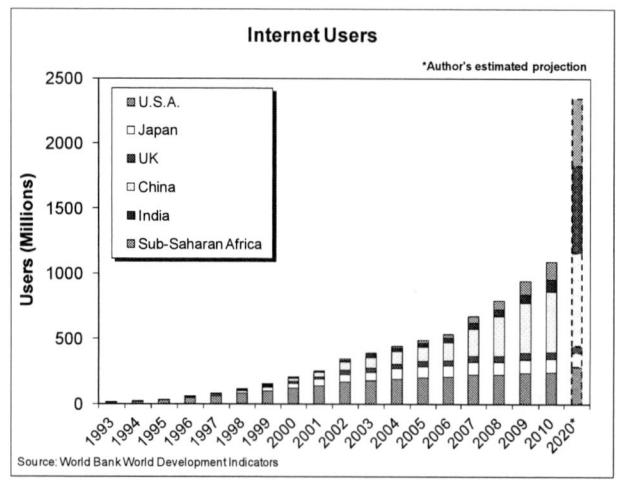

Unfortunately, even with unparalleled advancements in technologies with submicron scaling,

semiconductor materials, optical transport, and SSDs, just to name a few, the power continues to go up dramatically. To deliver the ultra-high bandwidth networking equipment and computing IT equipment, the industry has seen power for a rack increasing from 5kW yester-year, to 10kW today, and in some cases even higher than 20kW. Operating costs have become a dominant portion of the total cost of ownership of the equipment. To address these concerns, this year SEMI-THERM 28 has integrated for the first time a parallel Data Center Track.

It benefits the conference attendees to provide a thermal- and energy-concentrated Data Center Track that brings together leading members of the industry to discuss key industry issues, solutions and tools. This track encompasses technical presentations, two short courses and a panel discussion. The focus of this track ranges from component and system level, all the way to room and facility level.

This year, the keynote is delivered by the world renowned thermal technologist and staunch supporter of SEMI-THERM, Dr. Alfonso Ortega. He is the James R. Birle Professor of Energy Technology, Associate Dean for Graduate Studies and Research, College of Engineering of Villanova University. The topic of his talk is "The Energy Costs of Cooling Electronic Systems: Reflections on Past Lessons and Future Opportunities".

For the auxiliary program, there are:

- Six short courses, include topics such as thermal management of LEDs, harsh environment, data center, air and liquid cooling, and CFD;
- Embedded tutorial on lighting LEDs by Dr. Andras Poppe of Budapest University of Technology and Economics, Hungary;
- Two luncheon talks by Dr. Peter Rogers from The Petroleum Institute, UAE, and Terry Turchie, US FBI (retired);
- Evening tutorial by Dr. Seri Lee, Nanyang Technical University, Singapore, on "Thermal Challenges and Opportunities in Concentrated Photo-voltaics".

Finally, I would like to thank my Program Chair, Dr. Zeki Celik, and Vice-program Chair Genevieve Martin, for leading the Program team of thirty-plus members in putting together this exciting program; Dr. Veerendra Mulay for leading the effort for the Data Center Track; the leadership of the Steering Committee under the direction of Bernie Siegal, Thomas Tarter and George Meyer; Dr. Ross Wilcoxon in leading the Technical Committee for the selections of the keynote and embedded tutorial speakers. Special thanks to Bonnie Crystall, Walter Schuch, John Schuch and Bill Schuch of C/S Communications and Thomas Tarter of Package Science for all their hard work in pulling the logistics together for the conference venue, vendor exhibits and sponsors; and to Paul Wesling and Kathe Erickson of IEEE for the technical manuscript process and reviews. And lastly, we are very fortunate this year to have Bette Cooper from Meptec and the Meptec organization in assisting SEMI-THERM in our marketing campaign.

In summary, I would like to ask all the attendees to take advantage of this year's program. Don't be passive, but actively participate with all the technical discussions, vendor workshops, short courses and the tutorials. Together let's make this another great and memorable event!

Herman Chu
General Chair, SEMI-THERM 28
Cisco Systems, Incorporated

*The 28th Annual IEEE Thermal Measurement, Modeling and Management Symposium*

# ST28 ORGANIZING COMMITTEE

## SEMI-THERM 28 Program Committee

**General Chair**
**Herman Chu** Cisco Systems, Inc.

**Program Chair**
**Zeki Celik** LSI Corporation

**Vice-Program Chair**
**Genevieve Martin** Philips Research

### Program and Paper Selection and Review Committee

| | |
|---|---|
| **Atila Mertol** | LSI Corporation |
| **Attila Aranyosi** | Juniper Networks |
| **Bennett Joiner** | Freescale Semiconductor |
| **Bonnie Mack** | Thermal Specialist |
| **Cathy Biber** | Biber Thermal Design, Ltd. |
| **Chris Aldham** | Future Facilities Ltd |
| **Clemens Lasance** | Consultant@SomelikeitCool |
| **Dave Saums** | DS&A LLC |
| **David Copeland** | Oracle Corporation |
| **Genevieve Martin** | Philips  Research |
| **George Meyer** | Celsia Technologies |
| **Greg Xiong** | NetApp, Inc. |
| **Herman Chu** | Cisco Systems, San Jose |
| **Himanshu Pokharna** | Sheetak |
| **Hsiao-Kang Ma** | National Taiwan University |
| **Kazuaki Yazawa** | University of California, Santa Cruz |
| **Keiji Matsumoto** | ASET |
| **Marta Rencz** | Mentor Graphics |
| **Peter Rodgers** | The Petroleum Institute, UAE |
| **Prasad Tota** | Mentor Graphics |
| **Qian Han** | Huawei |
| **Raghavan Mahalingam** | Nuventix |
| **Rahima Mohammed** | Intel |
| **Ross Wilcoxon** | Rockwell Collins |
| **Sai Ankireddi** | Intersil, Milpitas |
| **Sharon Adam** | Infinera |
| **Shlomo Novotny** | Vette Corporation |
| **Susheela Narasimhan** | Cisco Systems, San Jose |
| **Thomas Tarter** | Package Science Services LLC |
| **Veerendra Mulay** | Facebook Inc. |
| **Wendy Luiten** | Philips |
| **William Maltz** | Electronic Cooling Solutions, Inc. |
| **Xi Wang** | Intersil |
| **Yongguo Chen** | Cisco Systems, Shanghai, China |
| **Younes Shabany** | Flextronics |
| **Zeki Celik** | LSI Corporation |

## SEMI-THERM 28 Steering Committee

**Chair / Operations Subcommittee Chair**
**Bernie Siegal** — Thermal Engineering Associates

**Vice-Chair**
**Clemens Lasance** — Philips Research

**2012 General Conference Chair**
**Herman Chu** — Cisco Systems, Inc.

**Finance / Business Chair**
**Jim Wilson** — Raytheon

**Technical Subcommittee Chair**
**Ross Wilcoxon** — Rockwell Collins

**Technical Editor**
**David Copeland** — Sun Microsystems

**Membership Subcommittee Chair**
**Bill Maltz** — Electronic Cooling Solutions

**Marketing Subcommittee Co-Chair**
**George Meyer** — Celsia Technologies

**Marketing Subcommittee Co-Chair**
**John Parry** — Mentor Graphics

**Secretary**
**Jay Nigen** — Apple

**International Liaisons**

**Europe**
**Clemens J. M. Lasance** — SomelikeitCool
lasance@onsnet.nu

**John Parry** — Mentor Graphics, Flomerics Division
John_Parry@mentor.com

**Asia**
**Prof. Hsiao-Kang Ma** — National Taiwan University
skma@ntu.edu.tw

**Symposium Event and Conference Management**

**Bonnie Crystall** — CS Communications, Inc.
cscomm@earthlink.net — Phone: +1 480-839-8988

**Thomas S. Tarter** — IEEE ST Exhibits
stexhibits@semi-therm.org — Phone: +1 408-642-5170

**Publications**
**Paul Wesling** — IEEE/CPMT Vice-President of Publications

**Marketing Director**
**Bette Cooper** — MEPTEC
bcooper@meptec.org

**Web Master**
**Clark Brown** — Clark Brown Creative
clark@clarkbrowncreative.com

**Members at Large**
**Dereje Agonafer** — University of Texas at Arlington
**Bruce Guenin** — Sun Microsystems.
**Peter Raad** — Southern Methodist University
**Dave Saums** — DS&A LLC
**Tom Tarter** — Package Science Services LLC

**Members Emeritus**
**Kaveh Azar** — Advanced Thermal Solutions
**David Blackburn** — NIST
**Paul Hundt** — Genesis LED Solutions
**Bonnie Mack** — Magma
**Bob Simons** — IBM

**www.SEMI-THERM.org**

*The 28th Annual IEEE Thermal Measurement, Modeling and Management Symposium*

# The 2012 Harvey Rosten Award

## *For Excellence in the Physical Design of Electronics*

**The Harvey Rosten Award For Excellence** has been established by the family and friends of Harvey Rosten. The Award commemorates Harvey's achievements in the field of thermal analysis of electronics equipment, and the thermal modeling of electronics parts and packages. Its aims are to encourage innovation and excellence in these fields. Harvey Rosten was a special friend to all of us at SEMI-THERM. The Harvey Rosten Award is sponsored by Mentor Graphics, Mechanical Analysis Division.

We continue to honor his legacy and great achievements by presenting this prestigious award to:

## Alfonso Ortega, Ph.D.

for the paper

### An Investigation of Multi-Layer Mini-Channel Heat Sinks with Channel Geometric Scale Variation Suggested by Constructal Scaling Principles

*Presented at the 27th Semiconductor Measurement, Modeling and Management Symposium SEMI-THERM, March 2011 San Jose CA USA*

*Dr. Al Ortega* is the James R. Birle Professor of Energy Technology at Villanova University and the Director of the Laboratory for Advanced Thermal and Fluid Systems. For more than 25 years he has been a leading researcher in the area of electronics cooling fundamentals and applications and a teacher of the fundamentals of fluid flow and heat transfer and the design of thermal-fluid systems. He has published and lectured widely on air and liquid-cooling of electronics and experimental methods. He is a former General Chair of SEMI-THERM and received its 2003 "Thermi" Award. He is currently Associate Technical Editor of the ASME Journal of Heat Transfer and a co-Principal Investigator in the new *NSF Industry/University Cooperative Research Center on Energy Efficient Electronic Systems (E3S)*.

**Abstract:** In previous work, we have shown that in single phase flow, stacked multi-layer liquid cooled heat sinks with square or circular channels have advantages over traditional single layer designs with high aspect ratio channels. In particular, it has been found that the thermal performance per unit pressure drop, as characterized by cost effectiveness metric, can be superior when properly optimized. The primary benefits seem to be increased surface area per unit volume available for convective cooling and increased flow area without sacrificing heat conduction paths to the coolant channel surfaces. The principle drawback of stacked multi-layer heat sinks is the difficulty in conducting heat through the metal matrix to the coolant channels farthest from the surface where heat is applied. In previous work we used validated two equation porous media formulations to model the behavior of these "deterministic" porous heat sinks with good success. Porous media formulations reduce the geometric complexity of the problem to two parameters, namely porosity and pore diameter. With this approach, it was shown that geometric scale variation, in which either the characteristic pore diameter of the channels in each layer or the layer porosity was allowed to vary from layer to layer, could result in lower thermal resistance and lower pressure drop, compared to heat sinks in which the pore diameter and porosity were uniform. Furthermore, the behavior of pore-diameter scaled compared to porosity-scaled heat sinks was quite distinct. In the present study, we examine the behavior of deterministic stacked mini-channel heat sinks with parallel channels of square cross section, where the porosity is varied from layer to layer, but the channel diameter is fixed. The scaling rules, developed in the porous media equivalent models, are based on biologically inspired constructal principles. Such scaling principles have led to superior optimal designs in a number of engineering applications. Experimentally validated conjugate CFD simulations were used to characterize the heat sinks. It was found that when the porosity is allowed to increase away from the surface onto which heat is applied, the increased mass flow and advection counteracts the cumulative conduction resistance thereby producing a more isothermal heat sink and a lower overall thermal resistance. Increasing the porosity away from the heat source also increased the flow area thereby producing lower overall pressure drop, compared to a non-scaled heat sink, in which the first layer of channels is the same in both cases. The volumetric thermal performance of the porosity scaled heat sinks exceeded the performance of non scaled heat sinks over a wide range of porosity scaling ratios and the pressure drop was consistently lower.

**www.SEMI-THERM.org**

*The 28th Annual IEEE Thermal Measurement,
Modeling and Management Symposium*

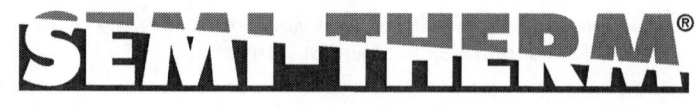

# SEMI-THERM®
## EXHIBITOR LISTING

### ADVANCED COOLING TECHNOLOGIES, INC.     304
1046 New Holland Ave
Lancaster PA 17601 USA
717-295-6080
www.1-ACT.com

Advanced Cooling Technologies Inc. develops innovative thermal technologies and provides technology-based thermal products to customers in the electronics, energy systems, aerospace, military and government sectors. ACT specializes in custom thermal product design and fabrication. Two-phase flow heat transfer, heat pipes and advanced thermal systems are particular areas of expertise.

### AIM SPECIALTY MATERIALS     401
25 Kenney Drive
Cranston RI 02920 USA
401-463-5605
www.aimspecialty.com

AIM offers specialty materials to fulfill the requirements of a broad range of applications. AIM Specialty Materials include indium, bismuth, and gold alloys available in many forms and configurations.

### AI TECHNOLOGIES, INC.     406
70 Washington Road
Princeton Jct. NJ 08550 USA
609-799-9338
aitechnology.com

Since pioneering the use of flexible epoxy technology for microelectronic packaging in 1985, AI Technology has been one of the leading forces in development and patented applications of advanced material and adhesive solution for electronic interconnection and packaging. AI Technology now has one of the high reliability adhesives and underfills for die bonding for the largest dies, stack-chip packaging with dicing die-attach film (DDAF), flip-chip bonding and underfilling and high temperature die bonding for single and multiple-chip modules for applications beyond 230°C. The company continues to provide the adhesive solution for component and substrate bonding for both military and commercial applications. Its thermal interface material solutions of patented phase-change thermal pads, thermal grease and gels and thermal adhesives set many bench marks of performance and reliability for power semiconductor and modules, computer and communication electronics.

### ALPHA NOVATECH, INC.     301
473 Sapena Ct. #12
Santa Clara CA 95054 USA
408 567-8082
www.alphanovatech.com

Since Alpha Company Ltd. was founded in 1972, we have produced cold-forged parts in a wide range of fields, including electronics components and automobile parts. In the process, we constantly strive to develop new concepts for original technology and techniques which have earned our high reputation. In 1989, we began to apply our technology to heat sinks. Since then, we have developed high performance heat sinks previously thought impossible to make and at the same time developed MicroForging, a new technology surpassing any other forging techniques. We specialize in high performance heat sinks and precision cold forging of complex light material parts.

### ANALYSIS TECH INC.     501
6 Whittemore Terrace
Wakefield MA 01880 USA
781-245-7825
www.analysistech.com

SEMICONDUCTOR COMPONENT THERMAL TESTERS for resistance, impedance, and die-attachment using electrical-junction temperature-measurement on any device type (transient and steady state); Thermal-lab test products for component thermal characterization and contract test services. THERMAL INTERFACE MATERIAL TESTERS (TIM Testers) for thermal conductivity measurements of electronic-packaging interface-materials; test services also offered. EVENT DETECTORS for solder joint and connector reliability testing.

### ANSYS, INC.     409
275 Technology Drive
Canonsburg, PA 15317 USA
866-267-9724
www.ansys.com

Today's business environment is rife with competitive challenges, customer requirements and financial pressures. This combination of factors has resulted in the need to find new methods for engineering more innovative products and manufacturing processes — while minimizing costs and time to market. Virtually every industry now recognizes that a key strategy for success is to incorporate computer-based engineering simulation early in the development process, allowing engineers to refine and validate designs at a stage where the cost of making changes is minimal. At ANSYS, we bring clarity and insight to customers' most complex design challenges through fast, accurate and reliable simulation. Our technology enables organizations to predict with confidence that their products will thrive in the real world. They trust our software to help ensure product integrity and drive business success through innovation.

### AOS THERMAL COMPOUNDS     403
22 Meridian Road # 6
Eatontown NJ 07724 USA
732-389-5514
www.aosco.com

AOS Thermal Compounds manufactures unique thermal interface materials including: a variety of non-silicone thermal greases, pre-formed Sure-Form gap filling pads, one part dispensable gap fillers, and patented Micro-Faze thermal pads. This novel thermal solution offers excellent thermal grease performance in a tacky but dry-to-the-touch pad. NOT a phase change material, they require no burn in, work at any temperature with minimal pressure, will never pump-out, and the latest offering has a thermal resistance as low as 0.038 C/W.

**Official Media Sponsor for SEMI-THERM 28**

### Electronics**Cooling**
**MAGAZINE**
EXHIBIT BOOTH 104

**www.SEMI-THERM.org**

**The 28th Annual IEEE Thermal Measurement, Modeling and Management Symposium**

# EXHIBITOR LISTING

## THE BERGQUIST COMPANY          206
18930 W 78th Street
Chanhassen, MN 55317 USA
952-835-2322
www.bergquistcompany.com

The Bergquist Company designs and manufactures high performance thermal management materials used to cool electronic components. Bergquist supplies the world with some of the best-known brands in the business: Sil-Pad® thermally conductive interface materials; Gap Pad® gap fillers; Hi-Flow® phase change grease replacement materials; Bon-Ply® thermally conductive adhesive tapes, and Thermal Clad® insulated metal substrates.

## COLDER PRODUCTS COMPANY          407
1001 Westgate Dr.
St. Paul, MN 55114 USA
651-645-0091
www.colder.com

Colder Products Company is the leading provider of quick disconnect couplings and fittings for life sciences, industrial and chemical handling markets. Colder creates cleaner, faster, safer and smarter connections. Used in a broad range of applications, innovative coupling and connection technologies from Colder allow flexible tubing to be quickly connected and disconnected.

## CPS TECHNOLOGIES          207
111 South Worcester St.
Norton, MA 01880 USA
508-222-0614
www.alsic.com

CPS Technologies Corporation is the worldwide leader in the design and high-volume production of AlSiC (aluminum silicon carbide) for high thermal conductivity and device compatible thermal expansion. AlSiC thermal management components manufactured by CPS include Hermetic electronic packages, Heat sinks, Microprocessor & Flip chip heat spreader lids, Thermal substrates, IGBT base plates, Cooler baseplates, Pin Fin baseplates for Hybrid Electric Vehicles (HEV), Microwave & Optoelectronic Housings and System in Package (SiP) Heat Spreader Lids that address multiple IC's on a PCB with a single lid design.

## DECAGON DEVICES          402
2356 Ne Hopkins Ct.
Pullman, WA 99163 USA
509-332-2756
www.decagon.com

Decagon Devices' KD2 Pro Thermal Properties Analyzer can measure thermal conductivity, diffusivity, and heat capacity of many natural and engineered materials, including liquids. First developed for soil scientists, the KD2 Pro is used in many diverse industries. Stop by booth 402 to get hands-on with the KD2 Pro and learn how Decagon can help measure your materials.

## DEGREE CONTROLS, INC.          507
18 Meadowbrook
Milford, NH 03055 USA
603-672-8900
degreec.com

DegreeC provides intelligent cooling solutions to address the heat generated by densely packed electronics for a variety of industries. We use sophisticated thermal analysis and computer simulation to design and manufacture proprietary fan trays and intelligent controllers critical to thermal management and system reliability. We call this Integrated Thermal Management (ITM). We also offer design validation and reliability testing along with compliance testing to all relevant standards, including UL/CSA, NEBS, and CE requirements.

## ELECTRONICS COOLING MAGAZINE          104
300 Nickerson Road
Marlborough MA 01752 USA
508-870-0714
www.electronics-cooling.com

Electronics Cooling magazine is a unique publication providing practical information and advice exclusively on thermal management of electronics. Over the past 13 years, ElectronicsCooling magazine has established an excellent reputation for high-quality, vendor-independent technical articles and news coverage. The magazine's editors are all highly qualified engineers with extensive experience in thermal management of electronics. ElectronicsCooling magazine is a global publication issued quarterly in February, May, August and November and delivered free of charge by request only to subscribers who complete a subscription form.

## ELECTRONIC COOLING SOLUTIONS          101
Electronic Cooling Solutions, Inc.
2915 Copper Road
Santa Clara, CA 95051 USA
408-738-8331
www.ecooling.com

ECS provides services for companies in a wide variety of industries and applications. Our customers develop products for the avionics, consumer, computing, medical, networking and telecommunications industries. Special needs can also be addressed, such as cooling the electronics for a unique telescope, and the thermal issues in manufacturing processes.

## ELEMENT SIX TECHNOLOGIES US CORP.          503
3901 Burton Drive
Santa Clara, CA 95054 USA
408-986-2400
www.e6.com

Element Six is a global leader and innovator in supermaterials with a history spanning more than 50 years. Our products are used across a wide range of industries. Element Six's activities are organized into four distinct areas – Advanced Materials, Hard Materials, Oil & Gas and Technologies plus a venture capital fund to invest in innovative activities related to our businesses. The company has production sites in China, South Africa, Sweden, The Netherlands, Ukraine, Germany, Ireland and the UK with a global sales organisation, research, technical support and service network.

**www.SEMI-THERM.org**

**The 28th Annual IEEE Thermal Measurement, Modeling and Management Symposium**

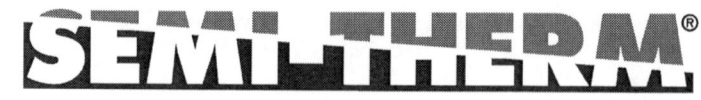

# EXHIBITOR LISTING

## ENZOTECHNOLOGY CORPORATION — 505
14776 Yerba Ct.
Chino, CA 91710 USA
909-993-5140
www.enzotech.com

Enzotechnology is a leading edge material forming and thermal management company. We specialize in providing both standard and customized heatsink solutions. We help our customers in conquering their most challenging thermal dissipating issues by offering engineering support, quick response, short lead time, and quality assurance using diverse leading edge technologies.

## FUJIPOLY AMERICA CORPORATION — 200
900 Milik Street
Carteret NJ 07008 USA
732-969-0100
www.fujipoly.com

Fujipoly is a global company with locations in North America, Europe, Japan, Thailand, Singapore, China and Hong Kong. Our thermal management materials range in performance to suit your application. We specialize in thin films, gold fillers, putty, paste and grease.

## FUTURE FACILITIES INC. — 205
2055 Gateway Place, Suite 110
San Jose CA 95110 USA
408-436-7701
www.futurefacilities.com

Future Facilities announces the release of 6SigmaET, the next generation CFD based software for thermal design in electronics and the first major breakthrough for a thermal design tool in 10+ years! 6SigmaET features a highly efficient, 3rd generation modeling environment that significantly improves analysis-based thermal design from concept to final implementation. In addition, 6SigmaET, as a key module of the 6SigmaDC suite that supports the important trend toward integrated design of electronics and data centers.

## GATEKEEPER LABORATORIES PRIVATE LTD. — 405
4 Engineering Dr. #3,02-12
117576 Singapore
65-9640-9465
www.gatekeeperlabs.com

Based in Singapore, Gatekeeper Laboratories has developed a method to enhance vapor chamber performance by up to three times. Gatekeeper's vapor chamber technology can be incorporated into various forms of air cooled heatsinks for different applications. Currently, Gatekeeper has built in its vapor chamber into custom CPU heatsinks for the high performance server/processor market.

## HENKEL ELECTRONIC MATERIALS LLC — 500
14000 Jamboree Road
Irvine, CA 92606
714-368-8000
www.henkelna.com

Henkel is the world's leading supplier of advanced materials for next-generation electronics assembly and packaging applications. From wafer-level to board-level through to final assembly, Henkel's world-renowned Hysol®, Loctite® and Multicore® product portfolios provide today's electronics specialists with the proven, reliable and compatible material solutions they need to stay competitive.

## IEEE CPMT - SANTA CLARA VALLEY CHAPTER — 106
www.cpmt.org/scv

The IEEE Components, Packaging and Manufacturing Technology (CPMT) Society is the leading international forum for scientists and engineers engaged in the research, design and development of revolutionary advances in microsystems packaging and manufacture. The CPMT Society's objectives are to provide a forum for the dissemination of technical information within its assigned areas. CPMT's fields of interest encompass the materials science, chemical processes, reliability technology, mathematical modeling, education and training utilized in the design and production of discretes, hybrids, and electronic packaging. Also included are fiber optics, connector technology, and semiconductor processing. Manufacturing technology includes systems, concepts, management, and quality as they relate to electronic component manufacturing.

## INDIUM CORPORATION — 201 - 203
34 Robinson Road
Clinton, NY 13323
315-853-4900
www.indium.com/TIM

Indium Corporation is a premiere materials supplier to the global electronics assembly, semiconductor fabrication and packaging, solar photovoltaic, and thermal management markets. Founded in 1934, the company offers a broad range of products, services, and technical support focused on advanced materials science. With facilities in the PRC, Singapore, South Korea, the United Kingdom, and the USA, the company is a four-time Frost & Sullivan Award winner and registered to ISO-9001.

## INTERMARK, INC. USA — 100
1310 Tully Road #117
San Jose, CA 95122 USA
408-971-2055
www.intermark-usa.com

INTERMARK, INC. was established in 1990, as the USA branch operation for KITAGAWA Industries Co., LTD. Based in Nagoya, Japan, KITAGAWA was founded in 1955 (public owned company since 1996) and became soon a leading worldwide supplier for a variety of components for the electronic industry. The product lines comprise a wide range of ferrite products for the EMI suppression, cable shielding materials, grounding fasteners and conductive gaskets. Other product line segments include a complete selection of plastic fasteners and hardware for the electronic packaging, optical fiber inspection equipment and thermal management components.

## JENOPTIK — 303
ESW GmbH - Sensor Systems Business Unit
Pruessingstrasse 41
Jena Thuringia D-07745 Germany
+49 3641 65-3942
www.jenoptik.com/infraredechnology

**www.SEMI-THERM.org**

*The 28th Annual IEEE Thermal Measurement, Modeling and Management Symposium*

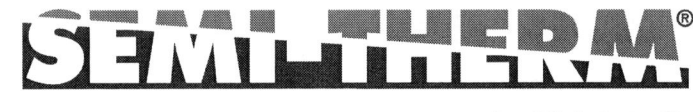

# EXHIBITOR LISTING

## JENOPTIK (CONT'D)     303

State-of-the-art Thermal Imaging Solutions: Thermographic camera systems of the VarioCAM® high resolution family and IR-TCM infrared camera modules by Jenoptik, a leading manufacturer, stand out for image quality, resolution, ergonomic handling and long-time reliability. Especially in electronics R&D and manufacturing processes, the innovative infrared imaging systems are essential tools for thermal design, thermal characterization testing, inspection and quality control of electronic components and systems. Long-time performance in terms of reproducibility and homogeneity + Great variety of IR optics available (wide angle, tele, macro). The thermal camera systems and IR optics are developed and manufactured in Germany by ESW GmbH, a company of JENOPTIK I Defense & Civil Systems (www.jenoptik.com/dcs) and leading manufacturer of thermographic and thermal imaging systems. North America residents might want to contact our United States subsidary: JENOPTIK Optical Systems, Inc., 16490 Innovation Drive Jupiter, FL 33478.

## LONG WIN SCIENCE AND TECHNOLOGY CORPORATION     103

No. 7. Shih 2nd Road, youth Ind. Park
Yangmei
Taoyuan32657 TAIWAN
886 3 464 3221 x 217
www.longwin.com

Long Win specializes in research, design, manufacture and service of scientific instruments for thermal managing, material & fluid mechanic and educational fields, and they hold a leading position on research, measurement and inspection apparatus for the electronic cooling market. Some of their products include thermal-related measurement apparatus for fan performance, TIMs, cooler modules, heat pipes, vapor chambers, IC packages, LEDs, and natural-convection simulation. They have more than 100 types of equipment in their 18,000 sq. ft. lab in Taiwan.

## MENTOR GRAPHICS - MECHANICAL ANALYSIS DIVISION     300-302

300 Nickerson Road
Marlborough, MA 01752 USA
508-303-5395
www.mentor.com/mechanical

Mentor Graphics Corporation is a world leader in thermal testing hardware and software solutions for design, providing products, consulting services and support to meet the thermal challenges of the electronics, semiconductor and systems industries. Engineering productivity in thermal design for applications from electronics board or package level through to data centers is enhanced with the latest developments in Computational Fluid Dynamics (CFD) software with FloTHERM®, FloEFD® and FloVENT® solutions. Thermal testing products including T3Ster® - the thermal transient tester and the TERALED® system for high power LED thermal and radiometric/ photometric characterization provide the highest testing speed and throughput of testing solutions available on the market.

## MICROSANJ LLC     102

3287 Kifer Road
Santa Clara, CA 95051 USA
408-256-1255
www.microsanj.com

Microsanj is a leading provider of high-resolution transient thermal imaging services and solutions for both commercial and research applications. The system is based on optical thermoreflectance characterization; digital signal processing and advanced patented software algorithms that supports electronic and optoelectronic components measurement, thermal design validation of ICs, defects and failure analysis and biological samples.

## MOMENTIVE PERFORMANCE MATERIALS     307

22557 West Lunn Road
Strongsville OH 44149 USA
440-878-5700
www.momentive.com

Momentive Performance Materials offers a family of thermal management products for today's high power electronics, ranging from PolarTherm* boron nitride powders to highly-conductive TPG* thermal pyrolytic graphite materials and composites. PolarTherm boron nitride powders offer a unique combination of thermal, electrical, and mechanical properties that make them excellent candidates for use in a range of thermal management materials, including gap fillers and underfills, potting and molding compounds, silicone and other compliant pads, liquid encapsulants, and compounded thermoplastics. Highly-conductive TPG thermal pyrolytic graphite materials and TC1050* thermal cores offer solutions for high performance heat spreading and dissipation requirements. These passive technologies, exhibiting thermal conductivities from 2.5 to 4 times that of copper, can provide unique solutions for high performance heat spreading and dissipation needs. *PolarTherm, TPG, and TC1050 are trademarks of Momentive Performance Materials, Inc.

## RATHBUN ASSOCIATES     306

48890 Milmont Dr. Suite 109D
Fremont, CA 94538 USA
510-661-0950
www.rathbun.com

Rathbun Associates is a Select 3M Converter specializing in 3M Thermally Conductive Interface Material solutions. 3M solutions include Acrylic Thermal Pads, Silicone Thermal Pads, and High Adhesion Thermally Conductive Acrylic Transfer Tapes and Thermal Grease. Rathbun carries a large selection of Thermal Pads in-stock, ready to ship for JIT applications.

## SHIN-ETSU MICROSI     400

10028 S 51st Street
Phoenix AZ 85044 USA
480-893-8898
www.microsi.com

Shin-Etsu MicroSi- Inc.- together with our parent company Shin-Etsu Chemical Co.- Ltd. represent world-class leadership in the development and manufacture of specialty materials for the semiconductor industries. Our product lines are specifically designed to address today's photolithography- packaging and flexible printed circuit requirements.

**www.SEMI-THERM.org**

# The 28th Annual IEEE Thermal Measurement, Modeling and Management Symposium

# SEMI-THERM®
## EXHIBITOR LISTING

**SOFTWARE CRADLE CO., LTD.** 308
70 Birch Alley Suite 240
Beavercreek, OH 45440 USA
937-912-5798
www.cradle-cfd.com

Software Cradle Co. Ltd. (Osaka, Japan), a leading developer and service provider of CFD software, is delighted to exhibit award-winning CFD software designed for design engineers as well as CFD experts. Established in 1984, Software Cradle offers a suite of CFD products consisting of SC/Tetra (general purpose, unstructured mesh CFD), STREAM (general purpose, structured mesh CFD), HEAT Designer (structured mesh CFD specialized for electronics cooling) and CADthru (CAD to CFD geometry translation). Software Cradle products are distributed and supported worldwide by its offices in Japan and North America (Cradle North America, Dayton, OH), and distributors throughout Asia, North America and Europe. In this conference, Cradle will mainly exhibit HEAT Designer which is used by the majority of Japanese electronics companies that are utilizing CFD in their design process. Its outstanding operationablity, fast and stable computation, ability to handle thousands of parts, flexibility and powerful postprocessor have been contributing to the innovation of Japanese electronics technologies. These traits enable users to import real CAD data and simulate as it is. Some of the users utilize 100 to 200 million meshes on Windows platform. Please come and visit our booth to find out how HEAT Designer can maximize your work.

**SP3 DIAMOND TECHNOLOGIES, INC.** 404
1605 Wyatt Drive
Santa Clara CA 95054 USA
408-492-0630
www.sp3diamondtech.com

sp3 Diamond Technologies provides diamond-based solutions for electronics thermal management and enhanced cutting surfaces to companies worldwide, across a broad spectrum of industries. By supplying wafer-scale diamond-on-substrate product and services utilizing nano- and micro-crystalline diamond morphology sp3 is expanding the commercial reach of CVD diamond. sp3 understands diamond and manufacturing equipment, as well as the cost, reliability and quality needs of its customers. In addition to running CVD diamond manufacturing fabs, sp3 is unique in selling hot filament CVD reactors so customers can manufacture their own CVD diamond in-house. Founded in 1993 and headquartered in Santa Clara, California, USA, sp3 Diamond Technologies is a subsidiary of sp3 Inc., a privately-owned, full service provider of products and services relating to thin-film and thick-film diamond deposition and other polycrystalline diamond materials. sp3 Inc. and its operational units have deposited diamond on over one million cutting tools and completed more than 15,000 successful diamond deposition runs.

**STÄUBLI CORPORATION** 202
201 Parkway West
Duncan, SC 29334 USA
864-433-1980
www.staubli.com

Stäubli is a leading manufacturer of textile machinery, quick release couplings and robotics systems. With a workforce of over 3500 employees Stäubli is present in 25 countries supported by a comprehensive distribution network in 50 countries worldwide. Stäubli's North American headquarters is located in Duncan, South Carolina and has over 200 employees supporting Textile, Connectors and Robotics customers, with a dynamic sales force strategically placed from the East Coast to the West Coast, Canada and Mexico.

**THERMAL ENGINEERING ASSOCIATES, INC.** 101
3785 Kifer Road
Santa Clara, CA 95051
650-961-5900
www.thermengr.com

TEA is a company founded by Bernie Siegal, a 35+-year veteran and recognized technical leader in the semiconductor thermal field. The company's mission is to provide a central source for the products and services necessary for proper semiconductor thermal measurement and modeling and solutions to attendant thermal management problems. Through its own products and services, augmented by an extensive network of technical experts around the world, TEA can assist customers in finding solutions. The Tech Briefs and Hot Links pages provide useful information to those interested in semiconductor and electronics thermal issues. We welcome the opportunity to discuss your thermally-related measurement, modeling and/or management requirements.

**THERMSHIELD, LLC** 502
PO Box 1641
Laconia NH 03247 USA
603-524-3714
www.thermshield.com

Thermal management products for the electronics cooling market including heat sinks, skived and stack fin, heat pipe assemblies, extrusions, stampings, water-cooled plates, fans and other support equipment. Thermal design and design for low cost manufacturing is our specialty.

**WOLVERINE TUBE, INC. - MICROCOOL DIVISION** 408
2100 Market Street, NE
Decatur AL 35601 USA
256-353-1310
www.microcooling.com

Wolverine Tube, Inc. is a world-class manufacturer of copper and copper alloy tubes, fabricated components and assemblies, metal joining products, copper and copper alloy rod and bar products, as well as highly enhanced heat transfer tubes. Micro-Deformation Technology (MDT) is a patented machining process that was originally developed for the production of highly enhanced heat transfer tubes at Wolverines and has been adapted to efficiently process superior quality, low cost micro-channels. The MDT process can achieve a variety of channel depths, channel widths, fin shapes and angularity. MDT provides design engineers a unique, cost-effective alternative to create new components and expand available materials for thermal management hardware applications. With the capability of fabrication and assembly, Wolverine's China facility can build full cooling systems.

**www.SEMI-THERM.org**

*The 28th Annual IEEE Thermal Measurement, Modeling and Management Symposium*

# VENDOR WORKSHOPS

---

## TUESDAY, MARCH 20, 2012

---

### SEMI-THERM 28 VENDOR WORKSHOP PROGRAM

---

**2:00 - 2:45 p.m.**      OAK

### Analysis Tech, Inc.

*Testing of Thermal Interface Materials (TIMs)*

**Dr. John W. Sofia**

This technical presentation deals with testing Thermal Interface Materials (TIMs) using ASTM 5470 testmethod. ASTM 5470 is arguably the most accurate and directly-beneficial method for evaluating the performance of TIMs. But despite is seeming simplicity, this method is often misunderstood and the results misinterpreted. This presentation will cover various topics related to this test method:

- the underlying physics
- practical application for various material types
- sources of error
- meaningful interpretation of the data

The presentation will conclude with a general question and answer period.

---

**2:00 - 2:45 p.m.**      FIR

### Mentor Graphics Mechanical Analysis

FloTHERM V9: New Features and Advanced Integration to EDA Toolsets

**Byron Blackmore**

The Mechanical Analysis Division of Mentor Graphics invites you to learn about the latest release of FloTHERM. Version 9 has a wealth of features and benefits that continue to ensure its position as the #1 CFD-based software tool used for electronics cooling design. The headline feature in this release is the ability to interface with DC IR drop analysis results from HyperLynx Power Integrity. Inclusion of the Joule heating effects in the board improve the accuracy of FloTHERM results and enable the prediction of failure mechanisms in complex power distribution nets.

---

**3:00 - 3:45 p.m.**      OAK

### Future Facilities

*Simulation Techniques: A Modeling Approach to Reduce Operating Costs and Increase Reliability for the Data Centre*

**Chris Aldham**

This presentation will discuss various ways to improve efficiency, reduce cost and increase resilience within data centers. It will show how simulation of the airflows and temperatures with the data center is an essential part of the process; as blindly following "rules of thumb" and "best practice" will not always lead an optimum solution. Using 6SigmaDC simulation software we will demonstrate both the benefits and drawbacks of several popular techniques to improve efficiency.

We will also show that a detailed simulation of a data center down to the server level is necessary to accurately predict the airflow within a data center and how recent developments in 6SigmaET software enable the automatic production of simple, but accurate, models of IT electronic equipment for use in data center simulations.

---

**3:00 - 3:45 p.m.**      FIR

### GE Corporate

Next Generation Electronic Cooling Technologies

**John Vogel, Marni Rutkofsky, and Peter deBock**

Over the past five years, GE Global Research has developed an advanced, dual piezoelectric, thermal cooling jet (DCJ). The DCJ, a 1 mm thick, ultra-compact bellows device provides air flow more efficiently than a traditional cooling fan and in less space. It's thin form factor allows for more design flexibility in electronics applications, such as tablets, where improved thermal solutions are required. These devices produce pulsed jet air flow, which may have velocities 10-20 times greater than average fan velocities. As a result, positioning one or more of these jets closer to the fins can cause high velocity air currents in tightly spaced fin gaps and enhance the surface heat transfer.

GE Licensing's Technology Ventures business is actively engaging thermal cooling companies as technology transfer partners and is working closely with OEMs/ODMs to drive DCJ adoption in a variety of applications. With close ties to GE Global Research, GE Licensing's Technology Ventures business offers its customers access to leading-edge technology, direct support from world-class subject matter experts, and the entrepreneurial spirit to drive commercial adoption of GE's technology.

---

**www.SEMI-THERM.org**

**The 28th Annual IEEE Thermal Measurement, Modeling and Management Symposium**

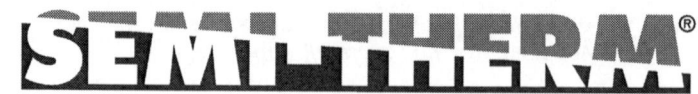

# VENDOR WORKSHOPS

## WEDNESDAY, MARCH 21, 2012

### SEMI-THERM 28 VENDOR WORKSHOP PROGRAM

---

2:00 - 2:45 p.m.                                                    OAK

### Thermal Engineering Associates

#### Thermal Load Boards - Description, Options, and Development

#### Bernie Siegal

Thermal Load Boards (TLB) are system-level thermal management tools used to simulate real application boards before the latter is available. The combination of reasonably short TLB development time and TLB versatility also provides the thermal designer the opportunity to validate various thermal solutions before system enclosure mechanical design is finalized. Heat source simulation can be as simple as the use of strategically placed resistors or as complex as using a multiple heat source semiconductor thermal test vehicle. The heat sources can be powered in a static mode or supplied with transient power to simulate various operating conditions of the actual system. This presentation will describe the various TLB design options - power levels, heat source simulators, and electrical connection variations - and the information requirements for development.

---

2:00 – 2:45 p.m.                                                    FIR

### Mentor Graphics, Mechanical Analysis Division
**MicRed Product Line:** *High power thermal transient testing and reliability characterization*

#### Andris Vass-Varnai

The MicReD hardware products of the Mentor Graphics Mechanical Analysis Division are designed for thermal transient characterization of packaged semiconductor devices and assemblies. Our technology enables the in-situ, non-destructive junction-temperature measurement of LEDs, power transistors and VLSI ICs. The extremely fine temperature resolution (0.01 °C) and fine time resolution (1 µs) of the measurements enables a unique post-processing of the measured results. Due to the high resolution and signal-to-noise ratio of our systems most of these tasks can be carried out at low power levels.

We will cover the following topics:

- Standards and methods for testing transistors (BJTs, MOSFETs and IGBTs)
- Measurement based thermal model generation and detailed numerical model calibration using FloTHERM
- Overview of our booster products and accessories
- The new high current (200A/7V) booster family for reliability testing and high power thermal characterization.

---

3:00 - 3:45 p.m.                                                    OAK

#### Microsanj
Thermal Imaging of Electronic and Optoelectronic Devices
**Dustin Kendig, VP Engineering**

Thermal characterization on a micron scale is an emerging challenge for electronic and optoelectronic device development. Junction temperatures are no longer a meaningful measurement since multiple hotspots exist on a die. Space limitations on the die preclude the use of embedded thermal sensors for thermal characterization. Full field thermal imaging technology on the other hand, provides rich information to enable visual detection of irregularities beyond what can be detected with spot temperature measurements.

Thermoreflectance imaging is a method for quickly obtaining a thermal image on micron-scale devices with excellent spatial and temperature resolution. The uniqueness and benefits of the thermoreflectance technique are introduced with the NT200-Series Thermoreflectance Image Analyzer and the SanjVIEWTM software package. Examples include the thermal imaging of a CMOS thermal test chip from TEA Inc., in a wire-bond package and a flip-chip package (with through-silicon imaging). The transient response of the built-in heaters provides a footprint of the chip's thermal profile. These thermoreflectance images are compared with conventional infrared (IR) images on the same samples. The example further explores other types of devices, such as LEDs, micro coolers, micro vias, and solar cells. The thermal images help to identify potential defects that affect reliability in devices with submicron features.

---

3:00 - 3:45 p.m.                                                    FIR

#### CPMT Santa Clara Valley Chapter

The IEEE Xplore Digital Library: Inside Tips and What's New

#### George Plosker, IEEE Client Services Manager

This talk is an overview of the IEEE Xplore Digital Library and an update on its features. Focus will be on recent significant changes to IEEE Xplore including:

- new "sort by Most Cited" (plus ability to view all cited by papers including IEEE and other publishers) on-going improvements in the area of personalization such as new Search History and Alerting features

The speaker will also cover:

- basic and advanced search options
- constructing successful search strategies for precise results
- using author affiliation to locate information by company or organization name
- how to refine and drill down into searches by multiple criteria
- improved browse options -- browse by document type, most popular or what's new

The speaker will cover the expanded tools available for IEL/Xplore subscribers, as well as the more limited set of features available to the general research community.

---

**www.SEMI-THERM.org**

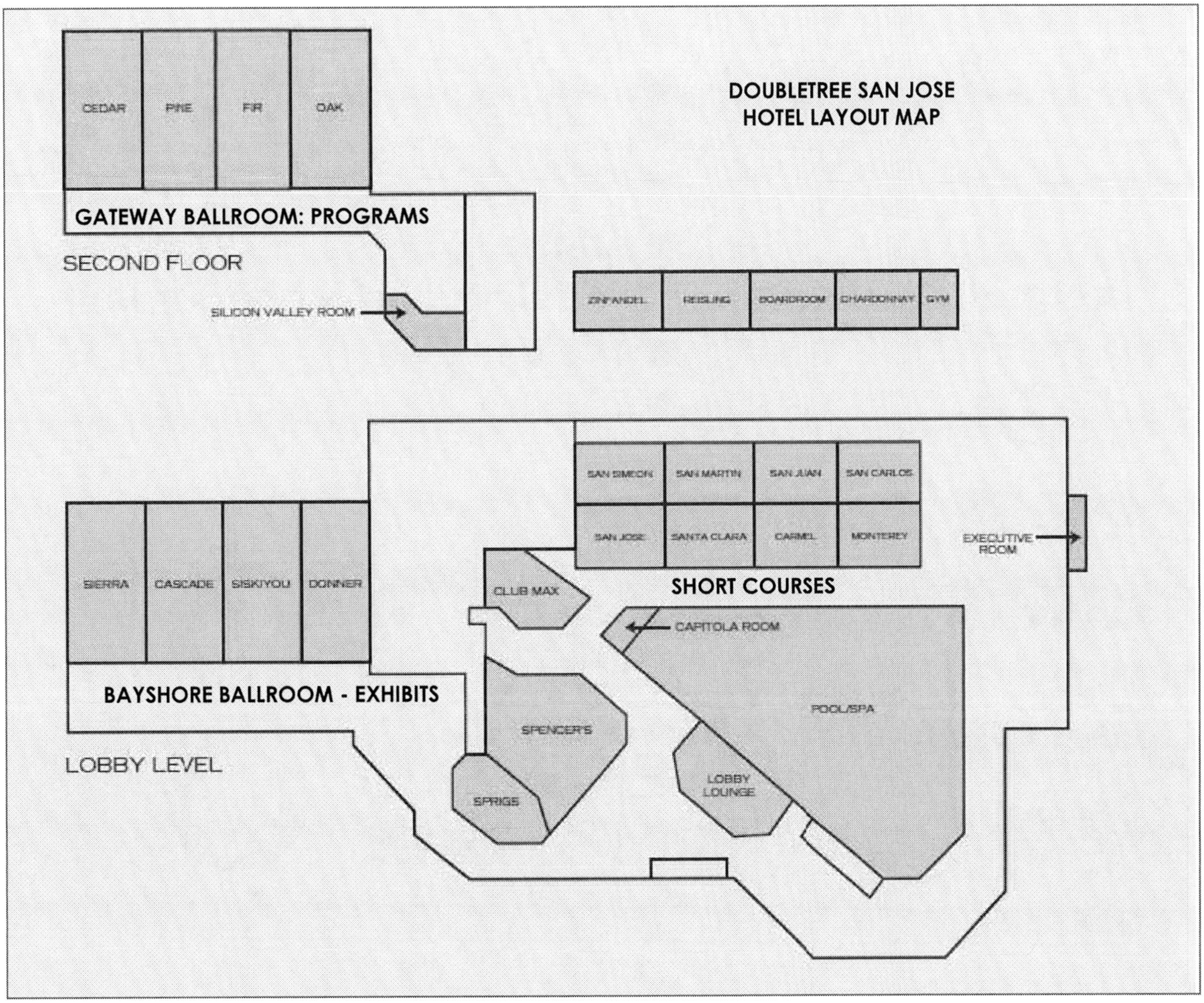

## DOUBLETREE SAN JOSE HOTEL LAYOUT MAP

**Second Floor**

| CEDAR | PINE | FIR | OAK |
|---|---|---|---|

**GATEWAY BALLROOM: PROGRAMS**

SECOND FLOOR

SILICON VALLEY ROOM →

| ZINFANDEL | REISLING | BOARDROOM | CHARDONNAY | GYM |
|---|---|---|---|---|

| SAN SIMEON | SAN MARTIN | SAN JUAN | SAN CARLOS |
|---|---|---|---|
| SAN JOSE | SANTA CLARA | CARMEL | MONTEREY |

EXECUTIVE ROOM →

**SHORT COURSES**

CLUB MAX

CAPITOLA ROOM ←

| SIERRA | CASCADE | SISKIYOU | DONNER |
|---|---|---|---|

**BAYSHORE BALLROOM - EXHIBITS**

LOBBY LEVEL

SPENCER'S

SPRIGS

POOL/SPA

LOBBY LOUNGE

# Join Us for the 29th Annual
### IEEE Thermal Measurement, Modeling and Management Symposium

### THERMAL INNOVATIONS THAT MAKE THE WORLD'S TECHNOLOGY COOL...

## March 17-21, 2013
## Doubletree by Hilton
## San Jose, California, USA

**For more information visit: www.SEMI-THERM.org**

## The 28th Annual IEEE
## Thermal Measurement, Modeling and Management Symposium

Mechanical Analysis Division

Licensing

Electronics**Cooling**

**circuitnet**

Micronews

**ELECTRONICS PROTECTION**

# FINAL PROGRAM
## MARCH 18-22, 2012
## Doubletree by Hilton
## San Jose, California USA

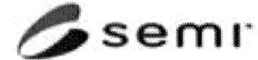

For more information visit: www.SEMI-THERM.org

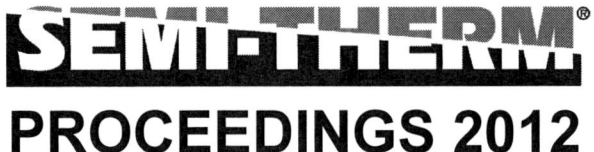

# PROCEEDINGS 2012

The PROCEEDINGS from past SEMI-THERM Symposia are available from the IEEE. For publication ordering information, please contact:

Paul Wesling
IEEE
12250 Saraglen Drive
Saratoga, CA 95070 U.S.A.
email: p.wesling@ieee.org

Curran Associates
57 Morehouse Lane
Red Hook, NY 12571 USA
Phone: +1-845-758-0400
Web: **www.proceedings.com**

The papers from **1988** through **2011** are available on the IEEE's **IEL/XPLORE** on-line system. Any researcher can use full-text search across the three million papers in the XPLORE database and access the abstracts of previous **SEMI-THERM** papers. Subscribers may download the PDFs of any **SEMI-THERM** papers. Non-subscribers may purchase single copies at a reasonable fee. To access this resource, please visit:

## ieeexplore.ieee.org

For information about membership in our Society, or participation in one of the many Technical Committees (including TC-Therm on Thermal Management) or local Chapters within the Society, please contact:

**IEEE COMPONENTS, PACKAGING AND MANUFACTURING TECHNOLOGY SOCIETY**

Marsha Tickman
IEEE Technical Activities
PO Box 1331
Piscataway, NJ 08855 USA
m.tickman@ieee.org

See our Web site: **cpmt.ieee.org**

# Measurement of Microbump Thermal Resistance in 3D Chip Stacks

E. G. Colgan, P. Andry, B. Dang, J. H. Magerlein, J. Maria and R. J. Polastre
IBM T.J. Watson Research Center, 1101 Kitchawan Road, Yorktown Heights, NY 10598

J. Wakil
IBM Austin, 11400 Burnet Road, Austin, TX 78758

## Abstract

The thermal resistance of Pb-free ~25 μm diameter microbumps with pitches of 50, 71, and 100 μm has been measured with and without underfill in four high chip stacks. With underfill, the unit thermal resistance values were 8.0, 15.5, and 19.0 C-mm$^2$/W for 50, 71, and 100 μm pitch microbumps, respectively. The average microbump height was 16.1 microns. For the 50 μm pitch case, the thermal conduction through the underfill is roughly equal to that of the microbumps alone.

## Keywords

Microbump, Chip stack, Thermal resistance.

## 1. Introduction

Significant efforts are underway to develop 3D chip stacks, which improve system performance by increasing the interconnect density and reducing the interconnect length[1]. The stacking of multiple chips with through silicon vias (TSV) and fine pitch microbumps between them not only increases the bandwidth and reduces the latency between the chips, but it also increases the difficulty of adequately cooling the devices during operation.

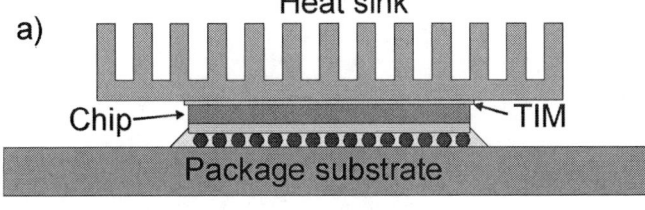

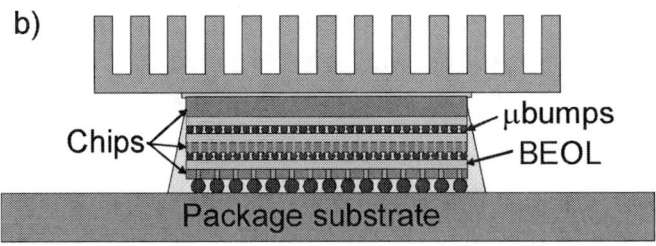

**Figure 1**: Schematic diagram of lid-less chip package (a) and chip stack (b).

With a conventional lid-less chip package, Fig. 1(a), the thermal path from the active circuits is through the silicon substrate, which can act as a heat spreader, and through the thermal interface material (TIM) layer to the heat sink. Typically, only a small fraction of the heat flows through the back-end-of-line (BEOL) wiring layers on the chip and the solder bumps into the package substrate. With a 3D chip stack, the situation is very different as both the BEOL wiring

layers and the microbump layer between chips are now in the thermal path, Fig. 1(b). Additionally, the chips which contain TSVs are thinned so that they are less effective at spreading heat from hot spots.

To be able to design systems based on 3D chip stacks, it is necessary to accurately characterize the additional thermal resistance from the BEOL and the microbump layers between chips. There have been extensive theoretical studies of the thermal resistance in 3D chip stacks[2-6], but only limited experimental measurements have been reported. In work by Matsumoto et al.[7,8], the temperature distribution of 3-layer chip stacks were measured and a corresponding thermal model was built. The chips layers were joined with a low volume Pb-free solder and Cu posts about 20 μm tall and 90 μm in diameter and the measured unit thermal resistance of the interconnect layer, without underfill was 12.9 – 22.0 C-mm$^2$/W. The thermal impact of TSVs and Cu-Cu bonding on the temperature profile of the top and bottom chips in a two chip stack was measured and modeled by Oprins et al.[9].

This paper reports on thermal resistance measurements in 4-layer chip stacks with ~25 μm diameter Pb-free solder microbumps with pitches of 50, 71, and 100 μm both with and without underfill. The design and fabrication of the thermal chip stack is described in section 2 and the thermal measurement procedure in section 3. The results and discussion are presented in section 4 and the conclusions in section 5.

## 2. Thermal Chip Stack Design and Fabrication

The design of the thermal chip stack test vehicle is described in Fig. 2. The test vehicle consists of a 50 x 50 mm ceramic substrate, a silicon carrier, and four thermal chips, shown schematically in Fig. 2(a). The thermal chips are ~ 20.5 x 20.5 mm in size and a large serpentine M1 heater is provided in the center 18 x 18 mm area. Each thermal chip includes nine M2 resistive temperature sensors which are 500 x 500 μm in size and are distributed in a 3 x 3 array. The thermal chips were joined together using ~ 25 μm diameter Pb-free microbumps. The assembly and underfill process has been described previously[10]. The underfill material used had a thermal conductivity of 0.55 W/m-C. This value is reported by the vendor and was verified by independent measurements. The microbump pitch was varied between 50 and 100 μm in different regions of the thermal chips. The region above the heater was divided into nine 6 x 6 mm areas where the microbump pitch was 50 μm in the corner areas, 100 μm in the areas located along the center of each side, and 71 μm in the center area; see Fig. 2(b). The resistive temperature sensors were centered in each of these areas and were aligned above each other in the chip stack. Note that in the thermal chip stack, the thermal chips were stacked with the heaters

978-1-4673-1110-6/12 $31.00 © 2012 IEEE

and sensors facing up. As shown in Fig. 2(a), there were no TSVs through the thermal chips in the central 18 x 18 mm area where the heater was located; the electrical connections between layers were all in the perimeter region.

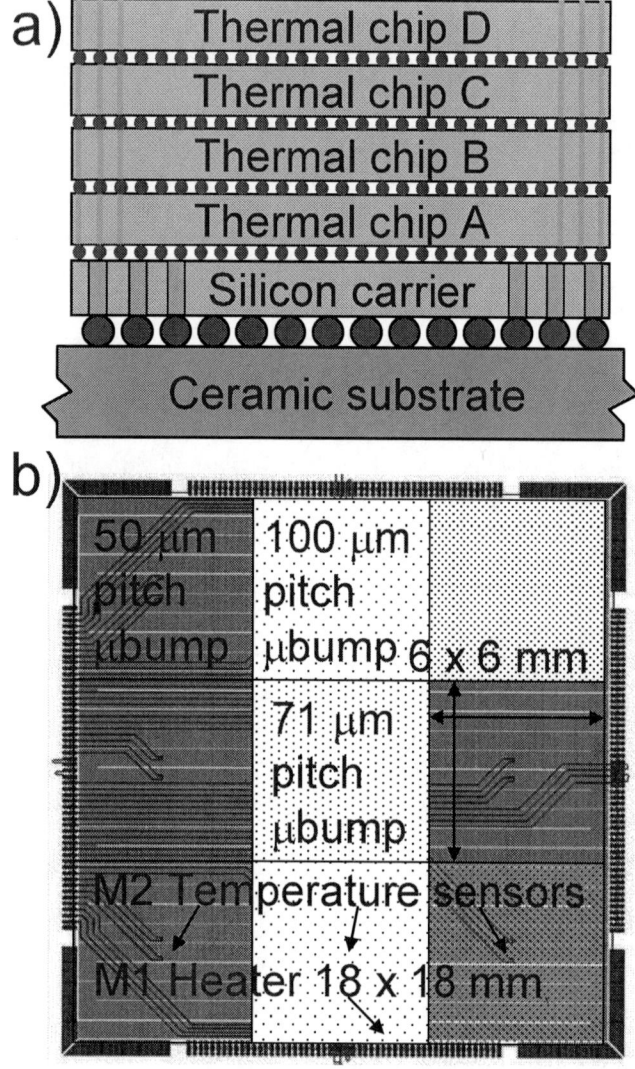

**Figure 2**: Schematic side view of chip stack test vehicle (a) and thermal chip details (b).

Cross-sectional images of an assembled thermal chip stack are shown in Fig. 3(a-c). The overall stack height, from the bottom of the carrier to the top of the uppermost chip was measured by scanning electron microscopy (SEM) at nine locations on two samples, G22 and G24, and the average height was 509 μm with a standard deviation of 4 μm. A typical image from near one edge of a chip stack is shown in Fig. 3(a). The microbump height was also measured by SEM at the same nine locations on G22 and G24 and the average height was 16.1 μm with a standard deviation of 1.3 μm for the top three microbump layers, where the thermal measurements were performed.

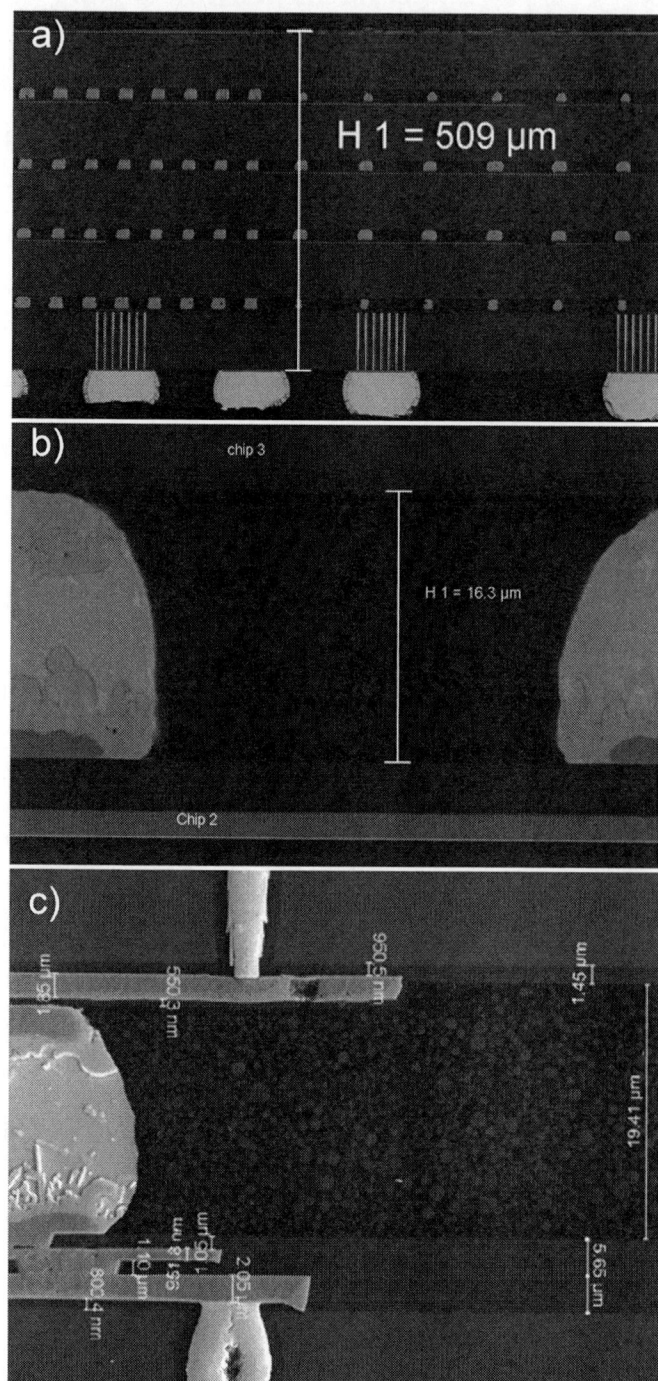

**Figure 3**: Cross sectional SEM images of an assembled chip stack (a), of underfilled microbumps (b) and the insulator and metal layers on the test chips (c).

A typical image of a microbump layer, after underfill was applied, is shown in Fig. 3(b). On the bottom of the thermal chip, the nominal insulator stack was 1 μm of PECVD (plasma enhanced chemical vapor deposited) silicon nitride, and 0.5 μm of PECVD silicon dioxide. On the top of the thermal chip, the nominal thickness of the Copper M1 used to form the heater is 2.1 μm and the nominal thickness of the Copper M2 used to form the sensor resistor is 1 μm. The M2 insulator is PECVD silicon dioxide. The nominal insulator

978-1-4673-1110-6/12 $31.00 © 2012 IEEE

layer thickness between the Si substrate and the M1 is 1 μm of thermal oxide, between M1 and M2 is 0.1 μm silicon nitride/ 1 μm silicon dioxide, and between M2 and the microbump is also 0.1 μm silicon nitride/ 1 μm silicon dioxide; see Fig. 3(c). The total insulator thickness in the thermal path for each chip is approximately 3.5 μm PECVD silicon dioxide, 1 μm thermal oxide, and 1.2 μm PECVD silicon nitride where the contribution from the M1 heater, which covers ~92% of the heated area, can be neglected. From the above measured and nominal values, the average silicon thickness for each thermal chip is about 81 microns. Assuming the thermal conductivity values shown in Table 1, the unit thermal resistance through each thermal chip is about 5.5 C-mm$^2$/W.

| Location | Material | Thickness (μm) | k (W/m-K) | $R_{th}$; C-mm$^2$/W |
|---|---|---|---|---|
| Bottom | PECVD SiO$_2$ | 0.5 | 1.1 | 0.5 |
| | PECVD SiN$_x$ | 1 | 1.5 | 0.7 |
| Si substrate | Si | 81 | 117 | 0.7 |
| Top | Thermal SiO$_2$ | 1 | 1.2 | 0.8 |
| | PECVD SiO$_2$ | 3 | 1.1 | 2.7 |
| | PECVD SiN$_x$ | 0.2 | 1.5 | 0.1 |
| | | | Total | 5.5 |

**Table 1**: Chip insulator layers and thermal resistance values.

## 3. Measurement Procedure

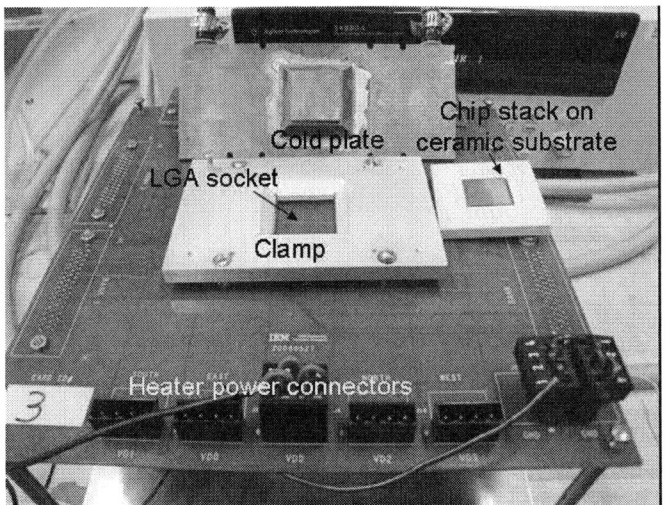

**Figure 4**: Test station used for thermal chip stack characterization.

The thermal chip stack was characterized in a test station where the substrate was clamped to a hybrid LGA (land grid array) socket on a test card and a water cooled cold plate was coupled to the top of the chip stack by using a removable TIM material (see Fig. 4). The temperature of the water circulating through the cold plate was controlled by a recirculation chiller and monitored by the data acquisition system. Four-point resistance measurements were performed on all the resistive temperature sensors and a constant current source was used to power the heater resistors where the voltage drop was measured directly at the heater resistor inputs on each chip.

The sensor resistors were all individually calibrated by varying the water temperature, measuring the sensor resistor values, and performing a least-square-fit to the resulting data. A typical data set for calibrating the nine sensors on one thermal chip is shown in Fig. 5, where a stair-step ramp of the water temperature from 20$^0$ C to 50$^0$ C and back was used. The uncertainty in the measured thermal resistance is ≤ 5% for the 50 μm pitch and ≤ 10% for the 100 μm pitch regions and is mainly due to uncertainty in the sensor calibration and heat flowing into the substrate or spreading laterally rather than through the chip stack to the cold plate.

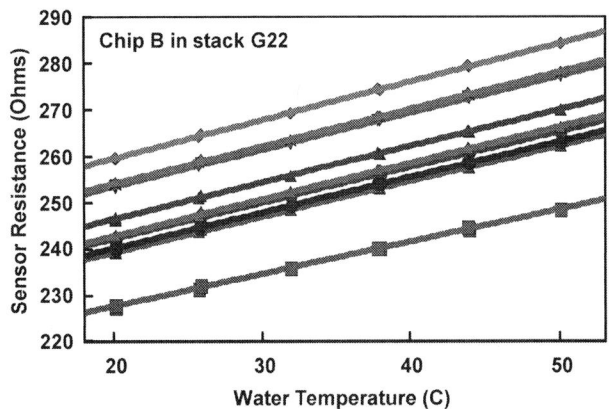

**Figure 5**: Sensor resistance vs. water temperature.

## 4. Results & Discussion

Results are shown in Fig. 6 for the G22 chip stack before underfill (a) and after underfill was applied (b). Power was supplied to the heater of the bottom chip of the stack (chip A, see Fig. 2a) and the temperate values were measured at the various temperature sensors in the chip stack. Note that 31 of the 36 sensors were operational before underfill was applied and only 28 of 36 sensors were operational after underfill was applied. The three sensors which were no longer operational after underfill were all on chip D, which is the top chip and since the chips are face up, the TIM removal processes may have damaged the sensors after the initial testing was performed. In Fig. 6, the sensor locations are indicated on the x-axis and as noted before, the microbump pitch in the corner areas (NW, NE, SE & SW) is 50 μm, in the center of the chip (Ctr) is 71 μm, and in the center of each side (N, E, S & W) is 100 μm. As would be expected, the chips lower down in the stack, i.e. further from the heat sink, are hotter. The variation in the temperature of chip D, which is the top one, is believed to reflect variations in the thickness of the TIM layer between the chip and the cold plate. Note that the temperature difference between adjacent chips increases as the pitch of the microbumps is increased. The temperature difference between chips is not directly comparable between the results without underfill, Fig. 6(a), and with underfill, Fig. 6(b), as the power applied was different, as indicated in the Figures.

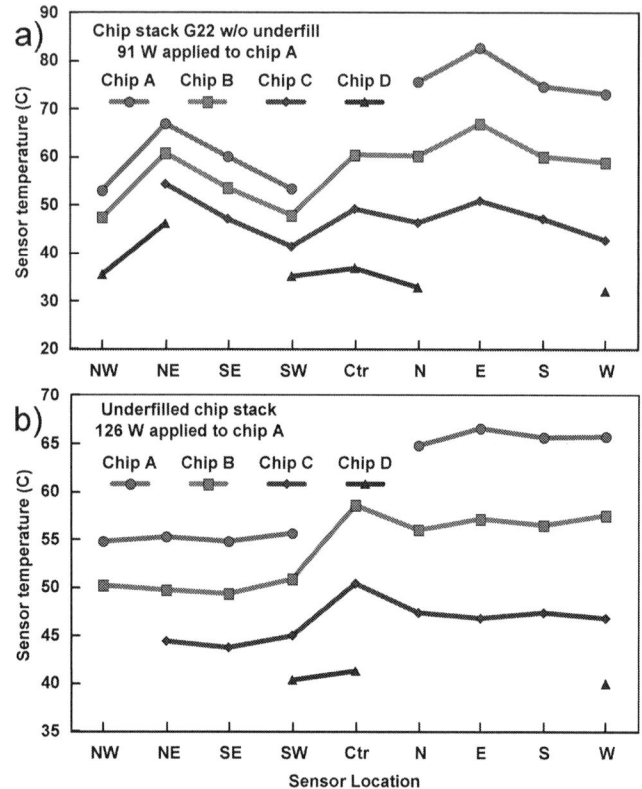

**Figure 6**: Sensor temperature vs. location with bottom chip A powered without underfill (a) and with underfill (b).

Since the microbump pitch varies in different areas of the thermal chip and there is a perimeter region which is not powered, it is necessary to consider the effect of heat spreading in the chip stack along with heat loss into the ceramic substrate and test board. We have examined this experimentally by measuring the thermal resistance across a microbump layer either by powering the chip directly below or by powering chips further down the chip stack. In Fig. 7(a,b), the thermal resistance is plotted on the vertical scale and is simply calculated as the average temperature difference between aligned sensor pairs on the indicated chips divided by the total applied power. The data for the microbump layer with a 50 or 100 μm pitch between the B & C chips are based on the average of 3 or 4 sensor pairs, respectively, whereas the remaining values are based on a single sensor pair. The x-axis in Fig. 7(a,b) indicates the relative location of the powered thermal chip relative to the reported thermal resistance location. For example, for the B-C chip thermal resistance values, direct indicates that chip B was powered and down 1 indicates that chip A was powered. Similarly, for the C-D thermal resistance values, direct indicates that chip C was powered, down 1 that chip B was powered, and down 2 that chip A was powered. The results for chip stack G22 without underfill are plotted in Fig. 7(a) and with underfill in Fig. 7(b).

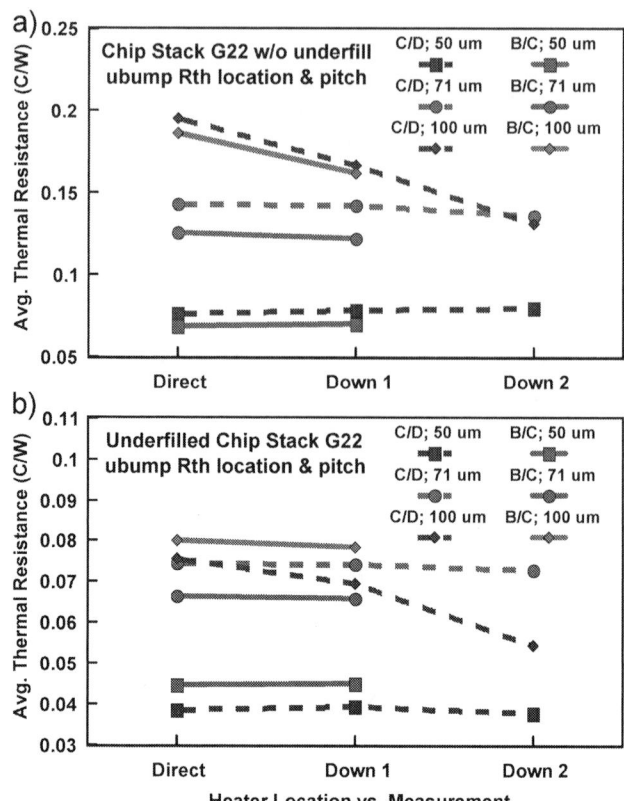

**Figure 7**: Average thermal resistance versus powered chip location and microbump pitch without underfill (a) and with underfill (b).

Comparing the thermal resistance for the direct with the down one or down two values, with a 50 or 71 μm pitch, the thermal resistance values were approximately constant. In general, with the 100 μm pitch, the thermal resistance values decreased significantly for the down one or down two cases compared with the direct case. The decrease was larger for the down two then for the down one measurement.

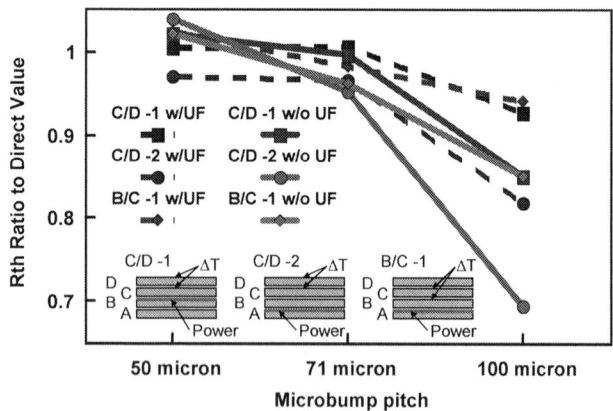

**Figure 8**: Thermal resistance ratio versus microbump pitch with and without underfill.

Similar measurements were performed on chip stacks G24 and G25 and the results for the average of all three chip stacks are plotted in Fig. 8 where the ratio of the down one (-1) or down two (-2) thermal resistance values to the direct thermal resistance value is plotted versus the microbump pitch for the B-C or C-D microbump layer either with, or without, underfill. With the 50 μm pitch microbump layer, the ratio ranged from 0.97 to 1.04 with an average value of 1.01. This suggests that for the 50 μm pitch area, the thermal spreading, or heat flux into the substrate, is minimal. With the 71 μm pitch microbump layer, the ratio ranged from 0.95 to 1.01 with an average value of 0.98. The average down two ratio, 0.96, was smaller then the average down one ratio, 0.99. Also, the average C-D down one ratio was 3% larger then the average B-C down one ratio, which suggest a heat loss into the substrate of ~ 3%. With the 100 μm pitch microbump layer, there were even more pronounced variations in the thermal resistance ratio. The average down one ratio was 0.89 and the down two ratio was 0.76, were the individual values were larger without underfill then with underfill. There was not a significant difference between the C-D down one and B-C down one values, suggesting little or no heat loss into the substrate. This suggests that the large down one and down two ratios are due to heat spreading from the 100 μm pitch regions to adjoining regions which provide a lower thermal resistance path to the cold plate. Note that for the 50 μm pitch region with no underfill, the down two ratio was 1.04 and the down one ratio was 1.02, which would be consistent with increased heat flux due to spreading from the adjacent 100 μm pitch regions. To minimize any spreading effects, results will only be reported when using a direct powered configuration and as noted earlier, the measurement uncertainty is larger for the 100 μm pitch regions then for the 50 μm pitch regions.

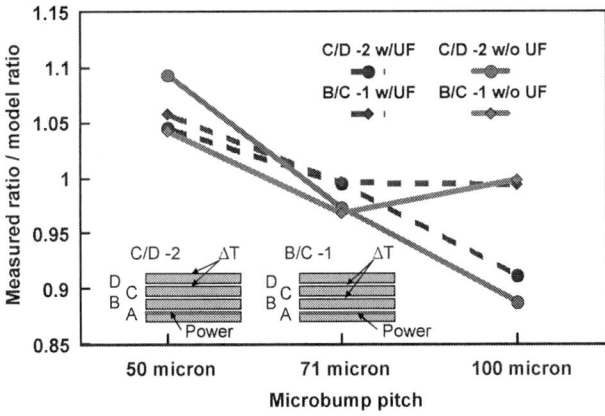

**Figure 9**: Measured ratio divided by modeled ratio versus microbump pitch with and without underfill.

The chip stack was modeled using the commercially available software Flotherm™. A conduction model was used which included the die stack, package substrate, board and mesh cold plate. Heat transfer coefficient boundary conditions were used to simulate the water cooling and any losses through the board. The heat transfer boundary conditions on the bottom of the board were varied from 0 W/m²-K (completely insulated) to 20 W/m²-K to try and determine whether spreading or heat loss through the board

was contributing to different resistance values at different layers. We found that the difference in results from varying the heat loss to the board were insignificant and could not explain the differences in resistances at different layers. The spreading was found to be the dominant factor. The different microbump regions between the die layers were modeled as effective thermal resistances. The ratio of the down one (-1) or down two (-2) thermal resistance values to the direct thermal resistance value was modeled versus the microbump pitch either with, or without, underfill. The measured ratio divided by the modeled ratio is plotted in Fig. 9 as a function of microbump pitch. For the down one case, the variation between the measured and modeled ratio is at most about 5%. The variation between the measured and modeled ratio is larger for the down two case, at most about 10%.

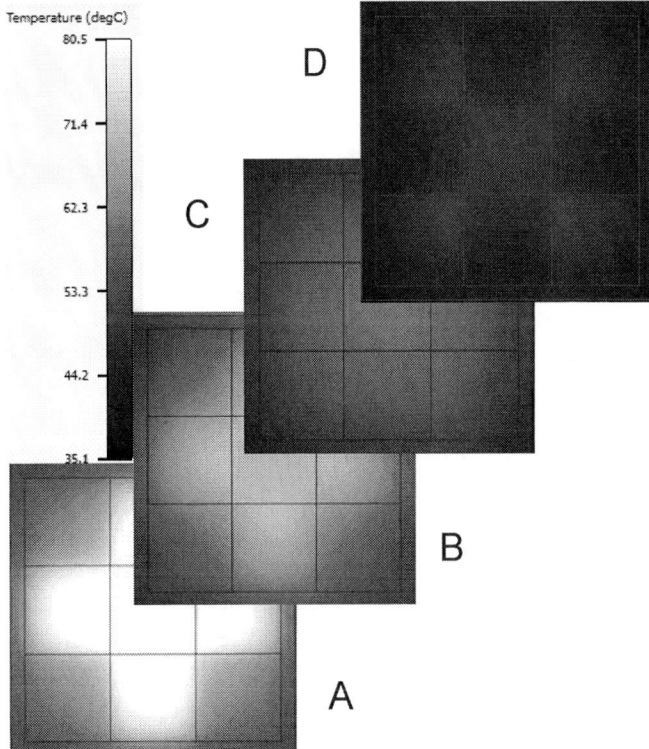

**Figure 10**: Modeled temperature contours with the bottom chip powered in a chip stack without underfill.

The modeled temperature contours, when the bottom chip, A, was powered are shown in Fig. 10 for the four high chip stack without underfill. As noted previously, the microbump pitch in the corner areas (NW, NE, SE & SW) is 50 μm, in the center of the chip (Ctr) is 71 μm, and in the center of each side (N, E, S & W) is 100 μm. The systematic temperature variations in Fig. 10 are roughly consistent with the measured values plotted in Fig. 6.

978-1-4673-1110-6/12 $31.00 © 2012 IEEE

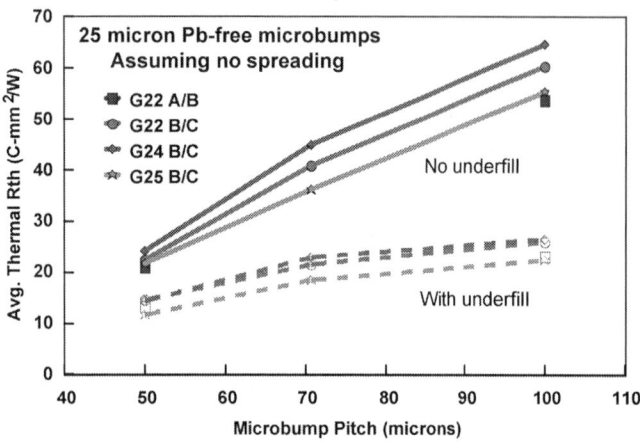

**Figure 11**: Average thermal resistance values versus microbump pitch with and without underfill.

Results are plotted in Fig. 11 for the measured unit thermal resistance of ~25 μm diameter Pb-free microbumps with a pitch of 50 μm, 71 μm, and 100 μm both with and without underfill. These results are from three separate thermal chip stacks, G22, G24 & G25. For the 50 μm and 100 μm pitch areas, the results are only reported when 3 or 4 aligned sensor pairs are available across the specific microbump layer. For the 71 μm pitch area, only one sensor pair is available for each microbump layer and the results from this are included in Fig. 11 for the microbump layers where 50 μm and 100 μm pitch results were included. The unit thermal resistance values were calculated assuming no thermal spreading; i.e. the temperature difference between the aligned sensor pair times the heated area of 18 mm x 18 mm and divided by the applied power. Note that the values plotted in Fig. 11 include the thermal resistance of ~ 5.5 C-mm²/W from the thermal chip; see Table 1.

| C-mm²/W | w/o Underfill | w/Underfill |
|---|---|---|
| 50 μm pitch | 16.9 | 8.0 |
| ~71 μm pitch | 35.3 | 15.5 |
| 100 μm pitch | 53.1 | 19.0 |

**Table 2**: Average thermal resistance values for microbump layers versus pitch with and without underfill.

Table 2 lists the average values of the unit thermal resistances for the microbump layer between chips, and was determined by subtracting 5.5 C-mm²/W from the values plotted in Fig 11. This microbump layer unit thermal resistance includes the contribution from the BLM (ball limiting metallurgy) layers on the chips and intermetallic formation in the Pb-free solder microbumps. The measured values are higher than parallel/series estimates using bulk conductivity values for the solder and underfill. This can be explained by material interfacial resistances, voids, and grain boundary effects. For comparison, with a typical underfilled C4 layer with 200 μm pitch and about 70 μm height, the

comparable thermal resistance value is about 100 C-mm²/W. For the 50 μm pitch case, if the unit thermal resistance of the microbump layer is assumed to be from two thermal resistance terms in parallel, where the value measured without underfill corresponds to the thermal resistance of the microbumps alone, then the thermal conductance of the underfill can be estimated to be roughly equal to that of the microbumps alone. This estimation ignores the spreading resistance term and the conduction through air in the no underfill case.

## 5. Conclusions

The thermal resistance of Pb-free ~25 μm diameter microbumps with pitches of 50, 71, and 100 μm has been measured with and without underfill in four-high chip stacks. With underfill, the unit thermal resistance values were 8.0, 15.5, and 19.0 C-mm²/W for 50, 71, and 100 μm pitch microbumps, respectively. For the 50 μm pitch case, the thermal conduction from the underfill is roughly equal to that of the microbumps alone. In the future, it would be interesting to extend this work to higher thermal conductivity underfill materials and low-volume solder "pancake" microbumps[11].

## Acknowledgments

We are deeply indebted to V. Arena, C. Jahnes, A. Prabhakar, N. Ruiz, C. Tsang and the staff of the IBM Yorktown Microelectronics Research Laboratory for fabrication of the 3D silicon wafers, to M. Gaynes for assistance with the thermal testing, and to M. Interrante at IBM Systems and Technology Division, East Fishkill, NY, for joining the Si carrier to the substrates. We would also like to thank T. Wiggins for the SEM sections, M. Lu for CSAM imaging, and J. Knickerbocker for management support and encouragement.

## References

1. Knickerbocker, J. U., Andry, P. S., Dang, B., Horton, R. R., Interrante, M. J., Patel, C. S., Polastre, R. J., Sakuma, K., Sirdeshmukh, R., Sprogis, E. J., Sri-Jayantha, S. M., Stephens, A. M., Topol, A. W., Tsang, C. K., Webb, B. C., Wright, S. L., "Three-dimensional Silicon Integration", IBM Journal of Research and Development , Vol. 52, No. 6, pp. 553-569, 2008.

2. Black, B., Annavaram, M., Brekelbaum, N., DeVale, J., Jiang, L., Loh, G.H., McCaule, D., Morrow, P., Nelson, D.W., Pantuso, D., Reed, P., Rupley, J., Shankar, S., Shen, J., Webb, C., "Die Stacking (3D) Microarchitecture", Proc. of 39th IEEE/ACM International Symposium on Microarchitecture, MICRO-39, pp. 469-479, 2006.

3. Leduca, P., de Crecy, F., Fayolle, M., Charlet, B., Enot, T., Zussy, M., Jones, B., Barbe, J.-C., Kernevez, N., Sillon, N., Maitrejean, S., Louisa, D., "Challenges for 3D IC Integration: Bonding Quality and Thermal Management", Proc. of IEEE International Interconnect Technology Conference, pp. 210-212, 2007.

4. Agonafer, D., Kaisare, A., Hossain, M., Lee, Y., Dewan-Sandur, B., Dishongh, T., Pekin, S., "Thermo-Mechanical

Challenges in Stacked Packaging", Heat Transfer Engineering, Vol. 29, No. 2, pp. 134-148, 2008.

5. Jain, A., Jones, R.E., Chatterjee, R., Pozder, S., "Analytical and Numerical Modeling of the Thermal Performance of Three-Dimensional Integrated Circuits", IEEE Transactions on Components and Packaging Technologies, Vol. 33, No.1, pp. 56-63, 2010.

6. Wakil, J., Colgan, E. G., Chen, S., "Back-End-of-Line and Micro-C4 Thermal Resistance Contributions to 3-D Stack Packages", IEEE Transactions on Components, Packaging and Manufacturing Technology, Vol. 1, No. 7, pp. 1007-1014, 2011.

7. Matsumoto, K., Ibaraki, S., Sakuma, K., Sueoka, K., Kikuchi, H., Orii, Y., Yamada, F., , "Thermal Resistance Evaluation of a Three-Dimensional (3D) Chip Stack", Proc. 12th Electronics Packaging Technology Conference (EPTC), pp. 614-619, 2010.

8. Matsumoto, K., Ibaraki, S., Sueoka, K., Sakuma, K., Kikuchi, H., Orii, Y., Yamada, F., "Experimental Thermal Resistance Evaluation of a Three-Dimensional (3D) Chip Stack", Proc. of 27th IEEE Semiconductor Thermal Measurement and Management Symposium (SEMI-THERM), pp. 125-130, 2011.

9. Oprins, H., Cherman, V., Vandevelde, B., Torregiani, C., Stucchi, M., Van der Plas, G., Marchal, P., Beyne, E., "Characterization of the Thermal Impact of Cu-Cu Bonds Achieved using TSVs on Hot Spot Dissipation in 3D Stacked ICs", Proc. 61st IEEE Electronic Components and Technology Conference (ECTC), pp. 861-868, 2011.

10. Maria, J., Dang, B., Wright, S. L., Tsang, C. K., Andry, P., Polastre, R., Liu, Y., Wiggins, L., Knickerbocker, J. U., "3D Chip Stacking with 50 μm Pitch Lead-Free Micro-C4 Interconnections", Proc. 61st IEEE Electronic Components and Technology Conference (ECTC), pp. 268-273, 2011.

11. Andry, P., Dang, B., Knickerbocker, J., Tamura, K., Taneichi, N., "Low-Profile 3D Silicon-on-Silicon Multi-Chip Assembly", Proc. 61st IEEE Electronic Components and Technology Conference (ECTC), pp. 553-559, 2011.

978-1-4673-1110-6/12 $31.00 © 2012 IEEE

# Experimental Thermal Resistance Evaluation of a Three-dimensional (3D) Chip Stack, Including the Transient Measurements

Keiji Matsumoto[+1], Soichiro Ibaraki*, Kuniaki Sueoka[+], Katsuyuki Sakuma[+], Hidekazu Kikuchi*,
Yasumitsu Orii[+] and Fumiaki Yamada[+]

: ASET (Association of Super-Advanced Electronics Technologies),
[+]: 1623-14 Shimotsuruma, Yamato-shi, Kanagawa-ken 242-8502, Japan.
*: 550-1, HigashiAsakawa-cho, Hachiouji-shi, Tokyo, 193-8550, Japan.
[+1] Corresponding author's e-mail : keim@jp.ibm.com

## Abstract

For the thermal management of three-dimensional (3D) chip stack, its thermal resistance needs to be clearly understood. In this study, 3D stacked test chips are fabricated, which are implemented with PN junction diodes for temperature sensors and diffused resistors for heating. At SemiTherm2011, the equivalent thermal conductivity of the interconnection, including BEOL(Back-End-Of-the-Line, wiring layer) is experimentally obtained to be 1.6W/mC and this time, we measure the thermal effect of Cu TSVs and it is experimentally supported that as the Cu TSV area ratio increases, the thermal conductivity of chip with TSVs in the vertical direction increases, on the contrary, that in the horizontal direction decreases. Also, the transient thermal measurement is performed and its result is compared with steady state measurement result. Further, the thermal capacitance measurement of 3D stacked test chip with hot spot heating is performed, which is essential to determine the transient thermal performance of 3D chip stack.

## Keywords

Three-dimensional (3D) chip stack, thermal resistance, Through-Silicon-Via (TSV), transient thermal characterization

## 1. Introductiion

Three-dimensional (3D) chip stacks have been widely accepted as a way for system performance enhancements, owing to their higher interconnect density and shorter interconnect length. They are a different approach from further device scaling [1-3]. In 3D chip stacks, the heat density is expected to be much higher than conventional 2D ICs, and cooling solutions are a big concern. However, the guideline of cooling solutions is not yet identified, because the thermal performance of 3D chip stacks is not clearly understood. The experimental data on the total thermal resistances of 3D chip stacks is quite lacking. It is necessary to know the total thermal resistance of 3D chip stacks correctly to propose an appropriate cooling solution for each 3D chip stack, depending on each heat generation.

A 3D chip stack is composed of interconnections (joints), TSVs (Through Silicon Vias), back-end-of-the-line (BEOL) layer, front-end-of-the-line (FEOL) layer, and silicon. Interconnections are regarded as one of the thermal resistance bottlenecks of a 3D chip stack and we have measured the thermal conductivity of SnAg with Cu post to be 37-41 W/mC [4,5], in the previous works. Also, the thermal conductivity of C4 (Pb97Sn3) [6] and copper-tin (CuSn) interconnections [7] were measured previously. We have also fabricated 3D stacked test chips, which are implemented with heaters and temperature sensors, to determine the total thermal resistance of actual 3D chip stack structure. At SemiTherm2011, we have reported the equivalent thermal conductivity of the interconnections between stacked chips including the wiring layer, as 1.6W/mC, based on the experimental temperature distribution of 3D stacked test chips [8]. Other than interconnections between stacked chips, a unique feature of 3D chip stacks is TSV(Through-Silicon-Via). At SemiTherm 2011, Oprins et al. have presented the experimental and simulated transient thermal characterization of chip with TSV[9], but no other reports are found. In this paper, we report the thermal effect of Cu TSVs and the thermal conductivities of chip with Cu TSVs in the vertical direction and that in the horizontal direction are discussed. Also the transient thermal measurement is performed and it is compared with the steady state measurement result. Further the thermal capacitance measurement of 3D stacked test chip with hot spot heating is performed, which is essential

**Table 1.** 4 different configurations of 3D stacked test chips.

| No. | TSV | Interconnection |
|-----|-----|-----------------|
| 1 | Area array | Area array |
| 2 | Peripheral | Area array |
| 3 | Area array | Peripheral |
| 4 | Peripheral | Peripheral |

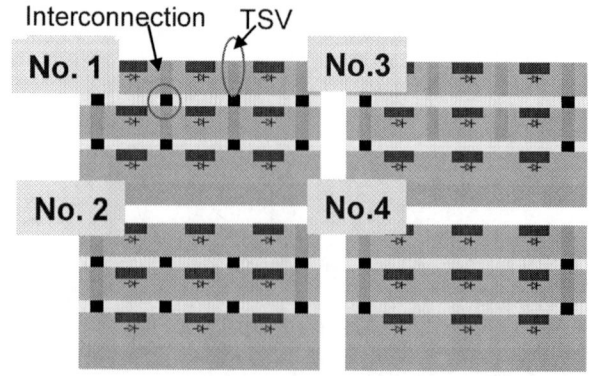

**Figure 1.** 4 different configurations of 3D stacked test chips.

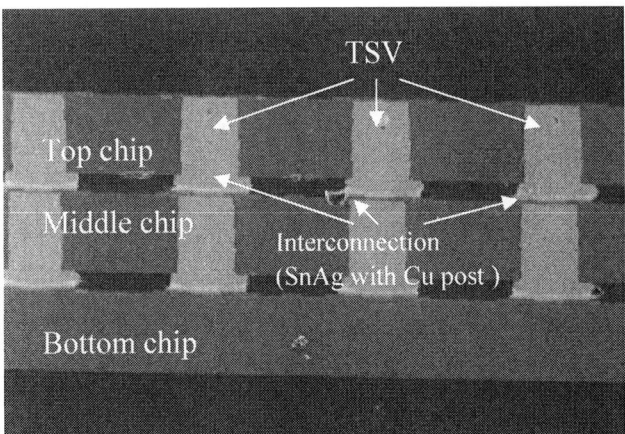

**Figure 2.** Cross section SEM micrograph of 3 layer stacked test chip with the area array TSV and the area array interconnection (Configuration No.1 in Figure 1).

to determine the transient thermal performance of 3D chip stack.

In this paper, the term "interconnections" refers to joints (for example, interconnections between two Si chips).

## 2. Structure of 3D stacked test chip

The detailed information of 3D stacked test chip is shown at SemiTherm2011 [8], and here, it is simply described. The 4 different configurations of 3D stacked test chips are shown in Table 1 and Figure1. For example, No.1 configuration has the area array TSV (Through-Silicon-Via) and the area array interconnection. The interconnection between stacked chips is considered as one of the thermal resistance bottleneck of a 3D chip stack[1], and again, at SemiTherm2011, we have reported the equivalent thermal conductivity of the interconnection between stacked chips, as 1.6W/mC, based on the experimental temperature distribution of configuration No.2 [8]. This year, the thermal effect of TSV is experimentally clarified, similarly, based on the experimental temperature distribution of configuration No.1. The cross section photo of configuration No.1 is shown in Figure 2. The area ratio of TSV (TSV area penalty) for electrical purpose in actual chip is said to be only a few %, but when the TSV concentrated area exists (in one area, TSV is denser than in other area), one should pay attention to the thermal effect of TSV.

## 3. Thermal effect of Cu-TSV in the vertical direction and in the horizontal direction

As a first step, we evaluate the thermal conductivity of a chip with TSV in the horizontal direction (x, y direction) and in the vertical direction (z direction) by using a simple model. Figure 3, and Table 2 show a model and thermal conductivity parameters used in the simulation, respectively. The simulated equivalent thermal conductivity of a chip with TSV, as a function of the TSV area ratio, is shown in Figure 4. The equivalent thermal conductivity is derived by assuming the whole model of a chip with TSV is considered as one unit.

| Material | Thermal conductivity (W/mC) |
|---|---|
| Si | 148 |
| Cu | 398 |
| SiO$_2$ | 1.3 |

**Table 2.** Thermal conductivity parameters to evaluate the thermal conductivity of a chip with TSV [3]:.

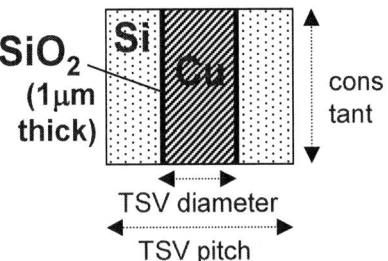

**Vertical direction (z direction)**

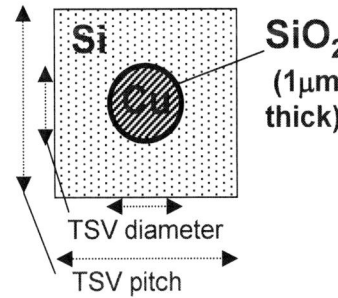

**Horizontal direction (x,y direction)**

**Figure 3.** Model to evaluate the thermal conductivity of a chip with TSV.

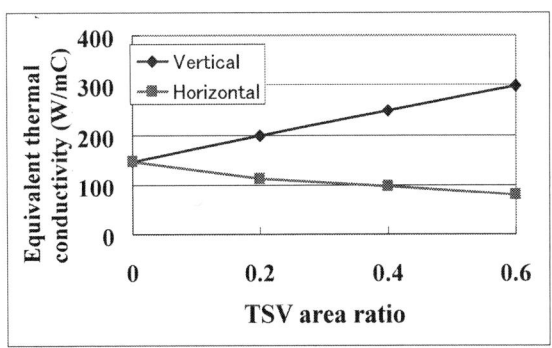

**Figure 4.** Simulated equivalent thermal conductivity of a chip with TSV in the vertical direction (z direction) and in the horizontal direction (x, y direction).

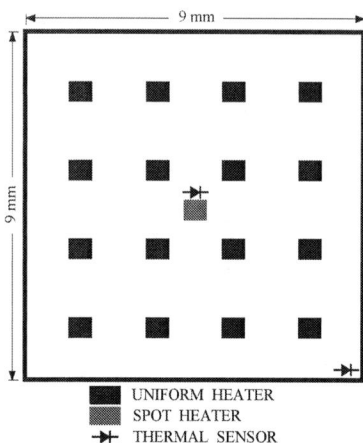

**Figure 5.** Horizontal locations of thermal sensors and heaters (Presented at SemiTherm2011)

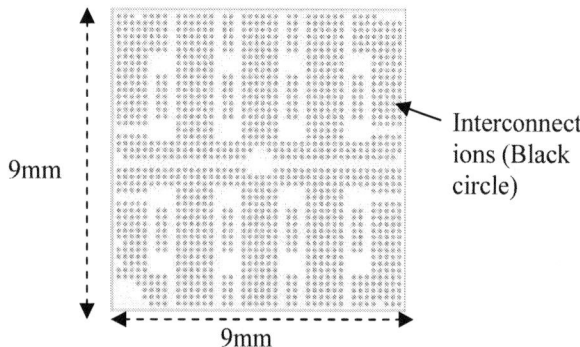

**Figure 6(a).** Layout of interconnections (Presented at SemiTher2010)

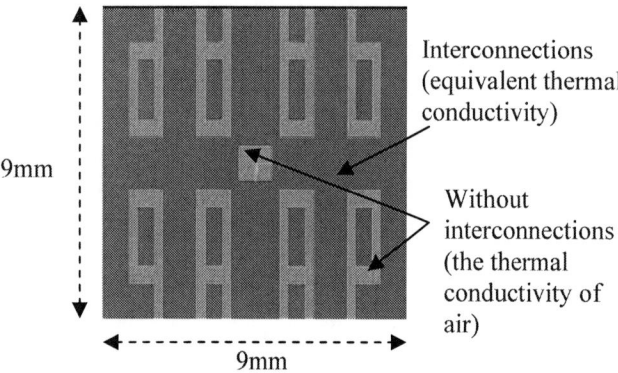

**Figure 6(b).** Simulation model of the interconnection layer (Presented at SemiTherm2010)

Figure 4 indicates as the Cu-TSV area ratio increases, the equivalent thermal conductivity in the vertical direction increases because of the increased Cu volume ratio. On the contrary, the equivalent thermal conductivity in the horizontal direction decreases, as the TSV area ratio increases, because of the $SiO_2$ layer which surrounds Cu. As the TSV area ratio increases, the $SiO_2$ diameter also increases, and as a result, the equivalent thermal conductivity in the horizontal direction decreases. The Cu-TSV area ratio of the 3D stacked test chips

in this study is 0.13 (TSV pitch : 200μm, TSV diameter : 80μm), and it means the equivalent thermal conductivity in the vertical direction is around 180 - 190 W/mC and that in the horizontal direction is around 115 – 125 W/mC. Then, we measure the temperature distribution of 2-stacked test chips with the area array TSV and with the area array interconnection (Configuration No.1).

The horizontal locations of thermal sensors and heaters in each chip are shown in Figure 5 and the interconnection layout is shown in Figure 6(a), respectively. The measured temperature distribution when total 3W is applied to the uniform heaters of the top chip is shown in Figure 7(a) and the measured temperature distribution when 2W is applied to the center hot spot heater of the top chip is shown in Figure 8(a), respectively. The temperature in the center of the bottom chip is set as 0C. The entire simulation model is shown in figure 9 and it simply consists of two components, which are the chip with TSV, and the interconnection. It should be noted that the thermal conductivity of TSV is not parameterized solely, but parameterized as the equivalent thermal conductivity of chip with TSV (vertical direction, horizontal direction). The simulation model of the interconnection layer is shown in Figure 6(b). In the simulation, the equivalent thermal conductivity of a chip with TSV in the vertical direction is set as 185 W/mC, and that in the horizontal direction is set as 120 W/mC. The simulated temperature distribution corresponding to Fig 7(a), is shown in Fig 7(b) and that corresponding to Fig 8(a) is shown in Fig 8(b), respectively. It is found that when the equivalent thermal conductivity of a chip with TSV is set as the above, the measured temperature distributions and the simulated temperature distributions are in reasonable agreement with each other. As a reference, when the equivalent thermal conductivity of a chip with TSV in the vertical direction and that in the horizontal direction, both, are set as 148 W/mC (Si), corresponding to Fig 7(a), is shown in Fig 10(a) and that corresponding to Fig 8(a) is shown in Fig 10(b), respectively. In this case, the agreement with the measured temperature distribution is not so good as Fig 7(b) and Fig 8(b), and it means that the temperature distribution of 3D stacked test chip is influenced by the equivalent thermal conductivity of a chip with TSV, both, in the horizontal direction and in the vertical direction. We will further fabricate 3-stacked test chips with the area array TSV, and examine the preciseness of the obtained equivalent thermal conductivity of a chip with TSV.

Whether TSV is preferable or not is determined, depending on each heat dissipation case. For example, in hot spot heating, horizontal head spreading is important and TSV may not be preferable, especially in thin chip. Whether TSV is preferable or not in hot spot heating case is determined by considering the balance between the horizontal heat spreading and the vertical thermal conductivity. In uniform heating, Cu-TSV may be preferable because of the increase thermal conductivity in the vertical direction. It should be noted that this tendency is determined by TSV structure and material, including dielectric.

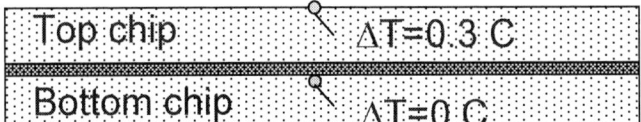

**Figure 7(a).** Measured temperature distribution of 2-stacked chips with the area array TSV and the area array interconnection (Configuration No.1), when total 3W is applied to the uniform heaters of the top chip

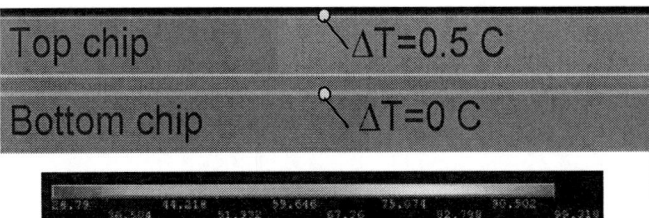

**Figure 7(b).** Simulated temperature distribution of 2-stacked chips with the area array TSV and the area array interconnection (Configuration No.1), when total 3W is applied to the uniform heaters of the top chip

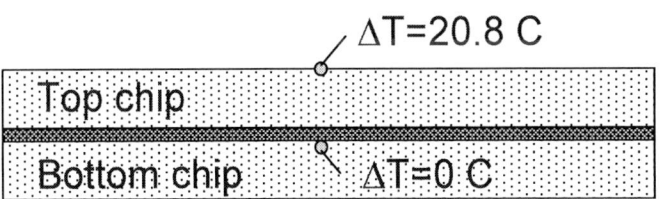

**Figure 8(a).** Measured temperature distribution of 2-stacked chips with the area array TSV and the array interconnection (Configuration No.1), when 2W is applied to the center hot spot heater of the top chip

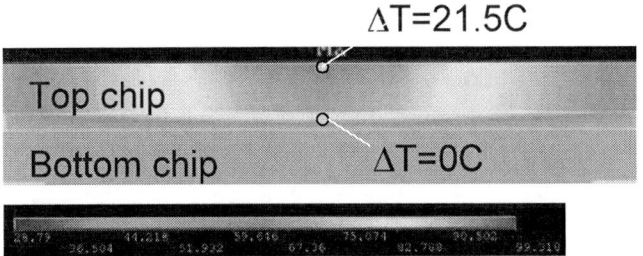

**Figure 8(b).** Simulated temperature distribution of 2-stacked chips with the area array TSV and the area array interconnection, when 2W is applied to the center hot spot heater of the top chip

## 4. Transient thermal measurements and thermal capacitance measurements

The temperature distribution measurement results by steady state method, which is presented at SemiTherm 2011, are compared with the thermal transient measurement by

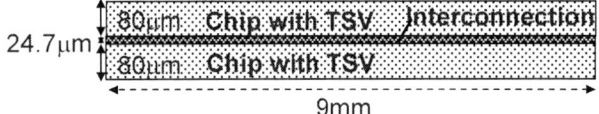

**Figure 9.** Cross section view of the simulation model of a 3D stacked test chip (Configuration No.1 of Figure 1). (the scales of the horizontal direction and the thickness direction are not the same.)

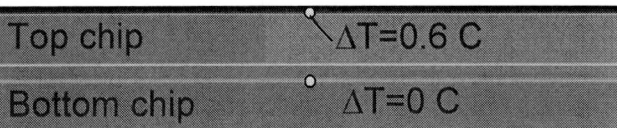

**Figure 10(a).** Simulated temperature distribution of 2-stacked chips with the area array TSV and the area array interconnection (Configuration No.1), when total 3W is applied to the uniform heaters of the top chip ( the equivalent thermal conductivity of a chip with TSV in the vertical direction and that in the horizontal direction,

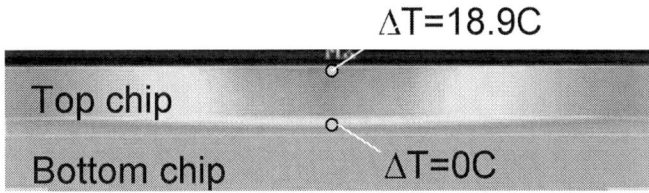

**Figure 10(b).** Simulated temperature distribution of 2-stacked chips with the area array TSV and the area array interconnection (Configuration No.1), when 2W is applied to the center hot spot heater of the top chip ( the equivalent thermal conductivity of a chip with TSV in the vertical direction and that in the horizontal direction, both, are set as 148W/mC (Si))

"T3Ster (offered by Mentor Graphics)." Figure 11 shows the experimental temperature distribution of 3-layer stacked test chips with the peripheral TSV and the area array interconnection (Configuration No2 of Figure 1) by steady state method, when 2W is applied to the center hot spot heater of the top chip. When the temperature of the center of the bottom chip is set as 0C, the temperature of the center of the top chip is 24.0C, and that of the middle chip is 4.4C. The transient temperature measurement result by "T3Ster" is shown in Figure 12. 2W is applied for a while and then at Time=0s, 2W is shut down and the transient temp decrease of each chip are shown here. When the temperature of the center of the bottom chip is set as 0C, the temperature of the center of the top chip is 24.1C and that of the middle chip is 4.6C at Time=10s, as shown in Figure 13. These are in good agreement with the steady state measurement results (the temperature of the center of the top chip is 24.0C, and that of

978-1-4673-1110-6/12 $31.00 © 2012 IEEE

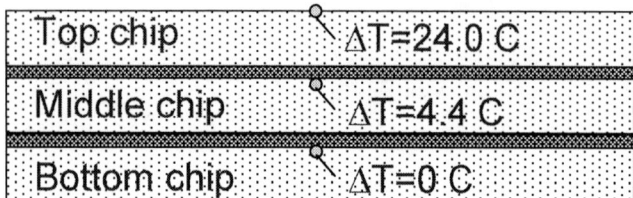

**Figure 11.** Measured temperature distribution of 3-layer stacked test chip with the peripheral TSV and the aarea array interconnection (Configuration No.2 of Figure 1) by steady state method, when 2W is applied to the center hot spot heater of the top chip (Presented at SemiTherm 2011)

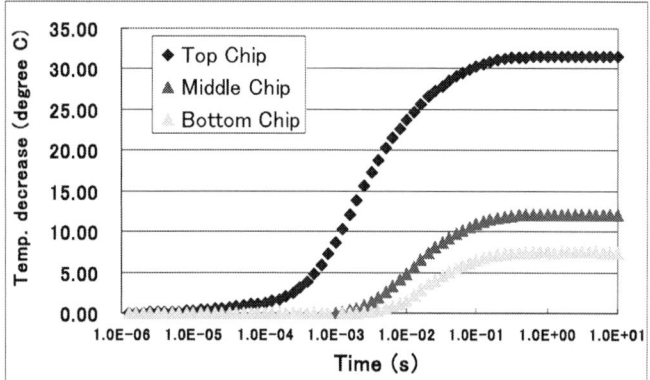

**Figure 12.** Measured transient temperature decrease of each chip by "T3Ster", when 2W is applied to the center hot spot heater of the top chip for a while, then Time=0s, 2W is turned off.

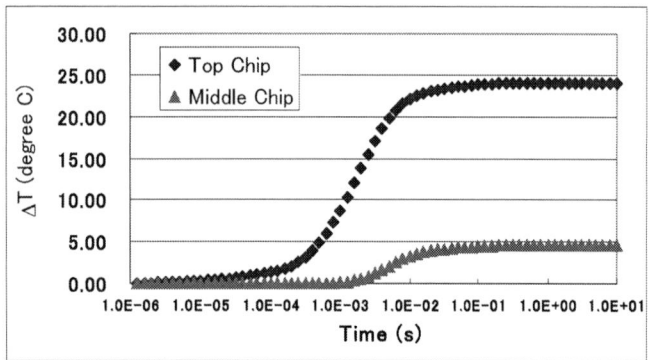

**Figure 13.** Measured transient temperature of the top chip and the middle chip by "T3Ster", when the temperature of the bottom chip is set as 0 .

the middle chip is 4.4C). The transient temperature measurement, like "T3Ster", is more time-efficient and therefore, it will be an appropriate method for the thermal resistance measurement.

We further investigate the thermal capacitance of 3D chip stack by "T3Ster", based on the structure function, and using the same 3-layer stacked test chips (Configuration No.2). The thermal capacitance is a very important property when

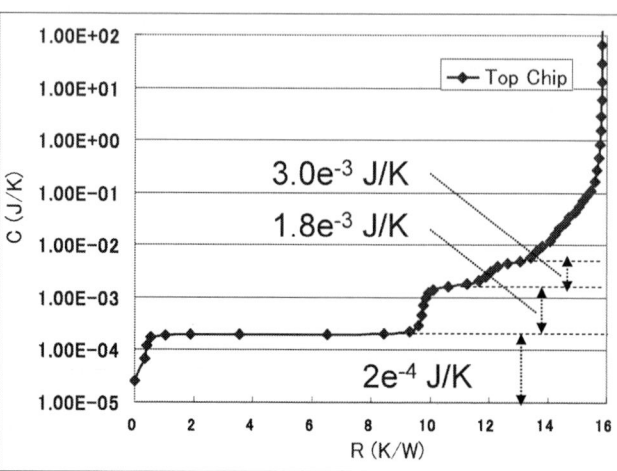

**Figure 14.** Structure function, when 2W is applied to the center hot spot heater of the top chip for a while, then at Time=0s, 2W is turned off.

considering hot spots (small time duration and high heat density). The structure function, when 2W is applied to the center hot spot heater of the top chip, is shown in Figure 14. This plot shows that the thermal capacitance of the top chip is about $2e^{-4}$ J/s, and the thermal capacitance of the middle chip, including the interconnection between the top chip and the middle chip, is about $1.8e^{-3}$ J/s, and the thermal capacitance of the bottom chip, including the interconnection between the middle chip and the bottom chip, is $3.0e^{-3}$ J/s. It should be noted that these values include the heat loss to the ambient. It is further necessary to determine this heat loss, but these thermal capacitance values seem to suggest that the thermal capacitance gradually increases, from the top chip to the middle chip, finally to the bottom chip. This corresponds to the phenomena that the horizontal heat spreading of the hot spot gradually increases, as the heat passes from the top chip to the middle chip, finally to the bottom chip. In other words, the horizontal heat spreading at the top chip is relatively small, because the chip is thin (80μm), and at the middle chip, the horizontal heat spreading increases because the total chip thickness increases to 160μm, and finally at the bottom chip, the horizontal heat spreading increases more because the total chip thickness further increases to 240μm. Compared with the thermal capacitance of a chip (Si, 80μm thick, 9mm x 9mm) of $1e^{-2}$ J/s, the measured values ($2e^{-4}$ J/s, $1.8e^{-3}$ J/s, $3.0e^{-3}$ J/s) are smaller than it ($1e^{-2}$ J/s), which supports that the horizontal heat spreading gradually increases, as the heat go through the top chip, to the middle chip, finally to the bottom chip.

The preciseness of the absolute thermal capacitance values needs to be further examined by additional measurements, but this thermal capacitance measurement will be an effective tool to evaluate the temperature rise by hot spots and propose an appropriate 3D chip stack design, considering hot spots.

## 5. Conclusions

The thermal effect of Cu TSVs in 3D chip stack is experimentally clarified based on the measured results of 3D stacked thermal test chip including Cu TSVs. It is experimentally supported that as the Cu TSV area ratio increases, the thermal conductivity of chip with Cu TSVs in the vertical direction increases, on the contrary, that in the horizontal direction decreases. Theses results can be simply mentioned that, for horizontally uniform heat generation cases, the vertical thermal conductivity is essential to determine the thermal resistance of 3D chip stack and Cu TSV is preferable. For hot-spot cases, the balance between the horizontal thermal conductivity and the vertical thermal conductivity is important and whether Cu TSV is preferable or not is determined, depending on each hot spot case. Also it is experimentally clarified that the transient thermal measurement result is in good agreement with the steady state measurement results. The transient thermal measurement is more time-efficient and it will be useful. Further, the thermal capacitance of 3D stacked test chip is measured to know the transient thermal performance of 3D chip stack. The transient thermal performance is especially important for hot spot heating, and we will collect these transient thermal characterization data in the near future.

## Acknowledgments

This work was entrusted by NEDO "Development of Functionally Innovative 3D-Integrated Circuit (Dream Chip) Technology" Project that is based on the Japanese government's METI "IT Innovation Program".

## References

[1] M. Koyanagi, H. Kurino, K.W. Lee, K. Sakuma, N. Miyakawa, and H. Itani, "Future System-on-Silicon LSI chips," IEEE MICRO, Vol 18, no. 4, pp.17-22, Jul/Aug. (1998)

[2] P. Ramm, A. Klumpp, R. Merkel, J. Weber, R. Wieland, A. Ostmann, and J. Wolf, "3D System Integration Technologies," Materials Research Society Symposium Proceedings, Boston, (2003)

[3] K. Sakuma, P. S. Andry, C. K. Tsang, S. L. Wright, B. Dang, C. S. Patel, B. C. Webb, J. Maria, E.J. Sprogis, S.K. Kang, R.J. Polastre, R.R. Horton, and J. U. Knickerbocker, "3D Chip-Stacking Technology with Through-Silicon Vias and Low-Volume Lead-Free Interconnections," IBM J. Res. & Dev. 52, No. 6, 611–622, (2008)

[4] K. Matsumoto, S. Ibaraki, K. Sakuma and F. Yamada, "Thermal resistance measurements of interconnections for a three-dimensional (3D) chip stack", IEEE International 3D System Integration Conference (3DIC) , 2009

[5] K. Matsumoto, S. Ibaraki, M. Sato, K. Sakuma, Y. Orii, F. Yamada, "Investigations of cooling solutions for three-dimensional (3D) chip stacks", 26th Annual IEEE Semiconductor Thermal Measurement and Maganement Symposium (Semi Therm), p.25, 2010.

[6] K. Matsumoto; Y. Taira, "Thermal resistance measurements of interconnections, for the investigation of the thermal resistance of a three-dimensional (3D) chip stack" , 25th Annual IEEE Semiconductor Thermal Measurement and Maganement Symposium (Semi Therm), p.321, 2009.

[7] K. Matsumoto; K. Sakuma; F. Yamada; Y. Taira, "Investigation of the thermal resistance of a three-dimensional (3D) chip stack from the thermal resistance measurement and modeling of a single-stacked-chip" , Internatinal Conference on Electronics Packaging, p.478, 2008.

[8] K. Matsumoto, S. Ibaraki, K. Sueoka, K. Sakuma, H. Kikuchi, Y. Orii, F. Yamada, "Experimental thermal resistance evaluation of three-dimensional (3D) chip stacks", 27th Annual IEEE Semiconductor Thermal Measurement and Maganement Symposium (Semi Therm), p.125, 2011.

[9] H. Oprins, V. Cherman, M. Stucchi, B. Vandevelde, G. Van der Plas, P. Marchal, E. Beyne, "Steady State and Transient Thermal Analysis of Hot Spots in 3D Stacked ICs using Dedicated Test Chips," SemiTherm, pp.131-138, March 2011.

# Impact of Die-to-Die Thermal Coupling on the Electrical Characteristics of 3D Stacked SRAM Cache

Subho Chatterjee*, Minki Cho*, Rahul Rao** and Saibal Mukhopadhyay*

*School of ECE, Georgia Institute of Technology, Atlanta GA, U.S.A
**IBM Research, T.J.Watson Research Center, Yorktown Heights, NY, USA
Email:{subho.chatterjee, mcho8}@gatech.edu,raorahul@us.ibm.com,saibal@ece.gatech.edu

## Abstract

We study the thermal coupling in a 3D stack with multiple cores in one tier and an SRAM array (cache) in a second tier with face-to-back bonding. For identical statistical distribution of power dissipation in cores, the SRAM sub-arrays experience much higher mean and variance in temperature in a 3D stack compared to a conventional 2D system. The increased variability in temperature increases leakage, degrades performance, and accelerates aging in 3D integrated SRAM. This is studied using 32nm predictive technology. Further, the spatial and temporal variations in performance of SRAM blocks become a strong function of the power variations in cores.

**Keywords:** 3D, SRAM, thermal coupling, performance

## 1. Introduction

3D integration allows partitioning different components of a system and stacking them in separate vertical layers connected by through-silicon-via (TSV). Folding a single 2D design into a 3D die-stack, hereafter referred to as die-folding, can reduce system footprint, interconnect delay, and power [1-2]. Due to its potential advantage, system architecture as well as process technology of 3D integrated SRAM caches are receiving significant attention [3,15]. When multiple power dissipating dies are stacked vertically with a reduced system footprint and hence, lower cooling efficiency, the system temperature can increase. The analysis and mitigation of thermal behavior of 3D die-stacks have been a key area of research [4-6].

Fig. 1 illustrates the system architecture of a 2D and 3D integrated core and cache system. As illustrated in Fig. 1, the heat escape path for 2D and 3D cases are similar, but the heat distribution paths are different in 3D. The heat generated in one tier can flow to the other tier through the die-to-die interface, resulting in the die-to-die thermal coupling [7-8]. In a many-core processor the processor cores generates much higher and time-varying power compared to the SRAM sub-arrays (much lower power) in cache. Hence, it is important to analyze how die-folding and thermal coupling modulates the cache performance in 3D systems. Such analysis is critical to exploit advantage of 3D integration as performance, leakage power, and stability of SRAM are sensitive to temperature [9]. Further, temperature strongly modulates the time-dependent degradation of devices and SRAM cell stability due to bias temperature instability (BTI) [10].

We analyze the implications of 3D die-to-die thermal coupling on power, performance, and aging of SRAM. We show that in 3D system, the worst case temperature of the cache is tightly coupled to core power and temperature. Hence, we observe that a change in the power spread in cores (due to variations in applications and workload) strongly modulate the temperature variation of the SRAM blocks in 3D integrated cache. Consequently, 3D SRAM blocks have higher spatial and temporal variation. The higher average value and spread in temperature significantly degrade the access time (~30% increase maximum access time), increase array leakage (~2X), and accelerate the time-dependent device aging. Further, the spatial and temporal variations in performance of SRAM blocks become a strong function of the power variations in the cores.

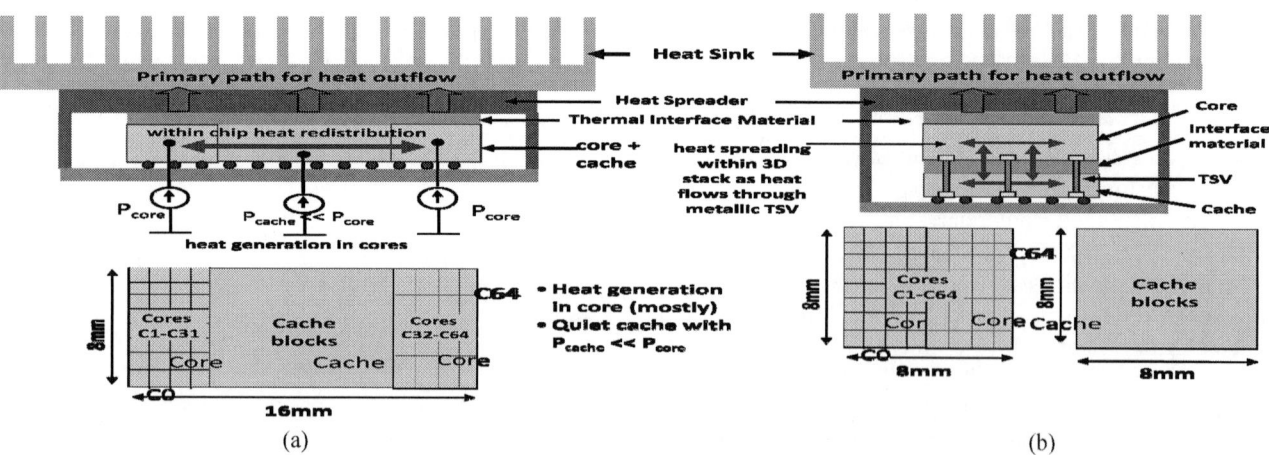

Fig 1: Thermal behavior of a multi-core processor with (a) 2D integration of cores and caches and (b) 3D integration of cores and caches. The figure shows a case of die-folding where the microprocessor in the left is folded into core and cache in the right. The heat sink is represented in blue and provides primary heat escape path. The heat outflow efficiency is higher for 2D due to higher footprint. The generated heat in 3D can be redistributed in both tiers in a 3D system leading to cache performance degradation.

978-1-4673-1110-6/12 $31.00 © 2012 IEEE

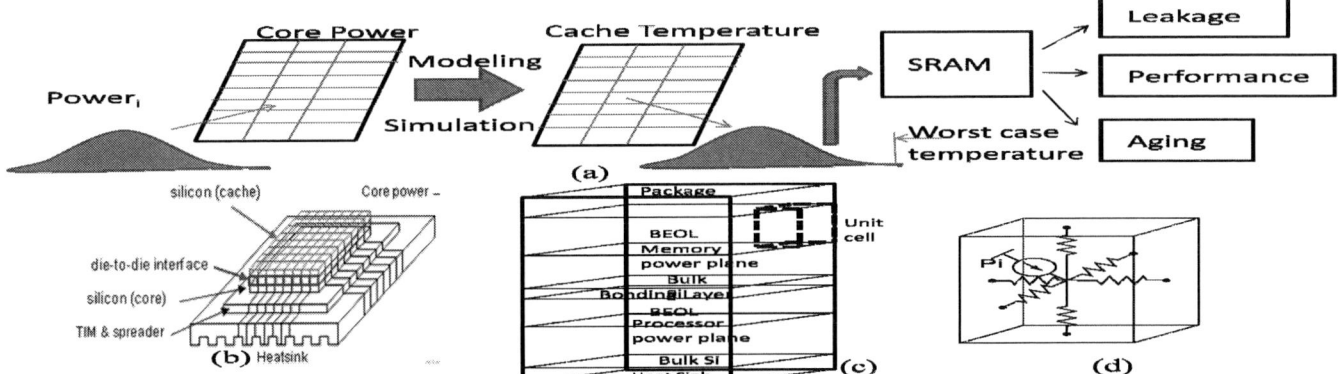

Fig 2: Detailed R-C Thermal Model for 3D analysis. The vertical layers are shown in right figure. The RC values are obtained from material properties. The thermal resistance of the die-to-die interface (BEOL) was obtained considering parallel combination of interface material and copper TSV. (a) The methodology. The core is simulated under power following Gaussian distribution with a combined mean of 1.5 W/core. The temperature distribution across the cache is captured and the worst case temperature for a block is used to simulate SRAM metrics (leakage, performance, aging) (b) Stacked 3D chip on heat sink (c) Modeled vertical layers (d) A unit body centered cell. Pi represents the power. The resistances represent the thermal resistances in a body centered structure. This structure is used for the thermal analysis.

## 2. Related Work

Thermal modeling for 3D system on chip has been an active area of research. The finite element methods [4] and the computationally less complex electrical mesh based methods [11] are normally used for analysis. Efforts have also been directed in finding the sensitivity of temperature distribution to various packaging parameters [12-13]. It has been argued that folding a logic die into multiple stacks can increase the overall chip temperature due to higher power density [14-15]. This work concludes that thermal coupling is a major factor in 3D without exploring the cache performance impacts for core-cache stacking. There are works emphasizing the electro-thermal implications of die stacking [16-18]. However, there is a limited understanding of the electrical impact of the die-to-die thermal coupling between the core and SRAM tiers on the behavior of SRAM. The primary contribution of this paper is the analysis of the above effect considering random variations in the core power symbolizing workloads. To achieve this goal, we adopt a distributed RC based 3D thermal modeling approach suitable to evaluate thermal variations across a plane. Loi et. al [16] have evaluated the effect of thermal cross-talk in a 3D core-cache-memory stack, but they used a lumped model to represent the thermal behavior of each tier. Hence, the effect of within tier power and thermal variations were not captured. Using the distributed analysis, we have shown that power variations among cores and strong thermal coupling lead to much higher spatial and temporal variations in temperature among SRAM sub-blocks in 3D than in 2D. Further, our analysis connects the effect of temperature to behavior of cache electrical parameters that includes not only performance and leakage but also aging and lifetime reliability. To the best of our knowledge, the interaction of 3D stacking and the aging behavior of SRAM have not been reported in earlier literature.

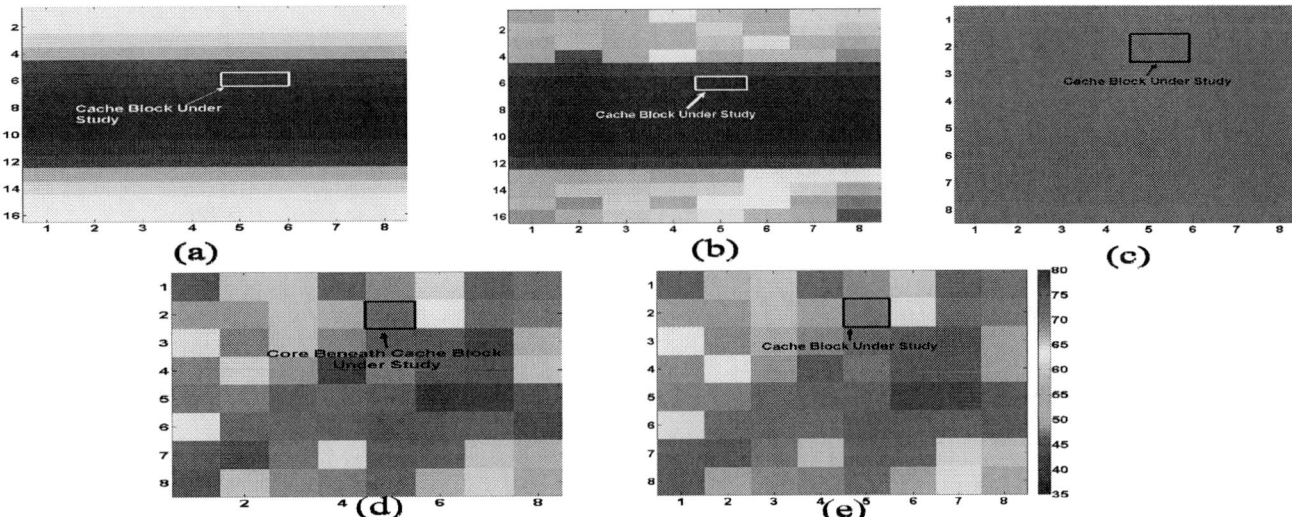

Fig 3: Thermal Distributions for: (a) 2D uniform power (b) 2D non-uniform power (c)3D cache for uniform power across cores (d)3D cores under non-uniform power (e) 3D caches under non-uniform power. All plots use same color bar shown with Fig. (e). While 3(a) and (b) show that for 2D, the cache thermal gradient is fixed in nature and exists even for uniform power distribution, while (c),(d),(e) establish the fact that the cache in case of 3D the cache follows the core thermal distribution closely.

978-1-4673-1110-6/12 $31.00 © 2012 IEEE          15

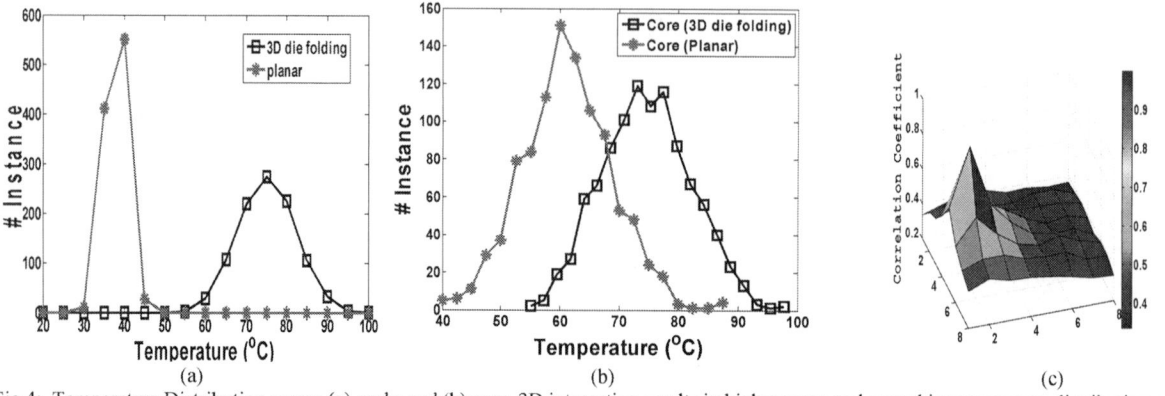

Fig 4: Temperature Distribution across (a) cache and (b) core. 3D integration results in higher mean and spread in temperature distribution of cores and cache blocks considering all random power patterns. (c)Thermal correlation for a cache block (location: 2, 3) with core blocks directly below it. Maximum correlation is observed with the core block directly below and it falls off rapidly with increasing distance. Hence, it shows cache temperature is more strongly correlated to cores vertically below itself.

## 3. System Models and Simulation Environment

We consider a 64 core system as shown in Fig. 1. In 2D system, the cache is placed in between two sets of 32 cores (Fig 1a) to ensure the shared cache acts as a thermal buffer between the cores. In the 3D die-folded system cores and cache are distributed in two different layers.

### 3.1. Thermal Modeling with 3D Distributed RC Grid

The thermal framework used in this study is constructed using distributed RC grid where R represents the thermal resistance and C represents the specific heat. The grid was

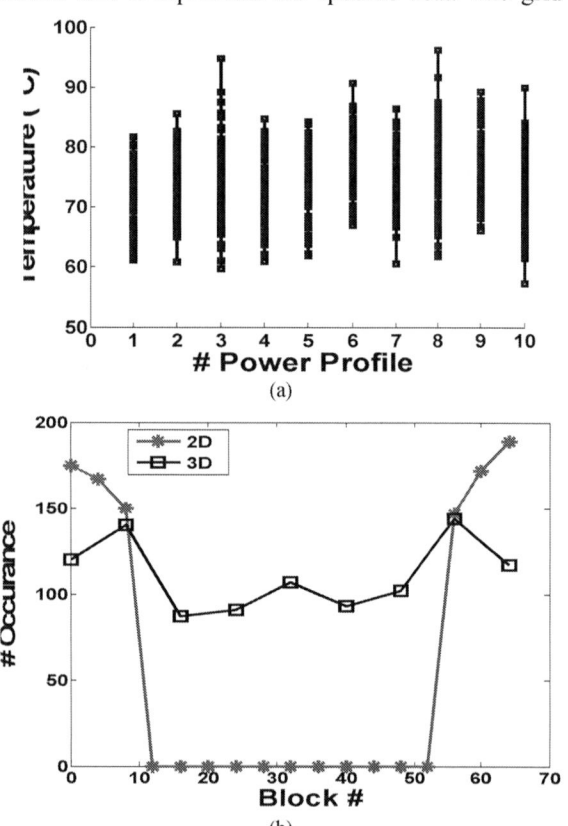

Fig 5: **(a) Distribution of Temperature with Power Profile in 3D and (b) histogram of the identification number of the hottest cache block.** In 3D the spatial difference in temperature is much higher. Moreover, all the blocks have equal probability of being the hottest block in 3D while in 2D location of hottest blocks is deterministic.

structured for the 2D case to include the impact of heat sink, silicon and insulator layers (Fig. 2). Modeling of 3D stacked systems using the electrical equivalent circuit has been extensively used [7]. We use distributed RC models method to analyze the thermal coupling issue. In this paper, we restrict ourselves to the steady state study using the R grid. We take a face-to-back bonded die with processor-memory stacking, as shown in Fig. 2c. The layers consist of the thermal package i.e. heat sink, spreader, and thermal interface material, the bulk silicon, the active silicon (processor), the back end of the line including metal connections (for the processor layer), the die-to-die interface material, the bulk silicon, the active silicon (memory), the back end of the line (for the memory layer), and the electrical substrate/package (i.e. die-to-package interface). We simulated an 8mmX8mm chip having 64 cores. The width of the die considered was 350µm. The values of the bonding material, heat sink and package conductivity was set in accordance with the values reported in [11]. The thickness of the bonding layer was taken to be 10µm. The back end of the line resistance was determined, using oxide to metal ratio of 1:3. The thermal resistivity of the die-to-die interface layer (core and SRAM) was modified to consider the effect of the heat flow through the TSV. HSPICE was used as the RC circuit simulator. Power consumption wise the cache was assumed to consume 10% of the total power. This choice is mainly motivated by the L2 cache power consumption percentage of current microprocessor [19].

### 3.2. Analysis Methodology - Coupling of Thermal and Circuit Simulations to Estimate SRAM Parameters

We first consider that the power of cores (and SRAM) in the 2D and 3D systems are same. To emulate the effect of workload variation on core power, we assume core power follows a Gaussian distribution (mean of 1.5W and a variance of 0.5W) (fig. 2a).

We estimate the temperature patterns in cores and caches in 2D and 3D system considering a large number of random power patterns applied to the cores. Each power pattern is applied for constant period of time in simulation (until steady-state is reached). Each such power pattern applied to the cores results in a particular temperature pattern across the cache blocks for 2D and 3D systems. For each power pattern, we estimate the temperature for each SRAM block and maximum

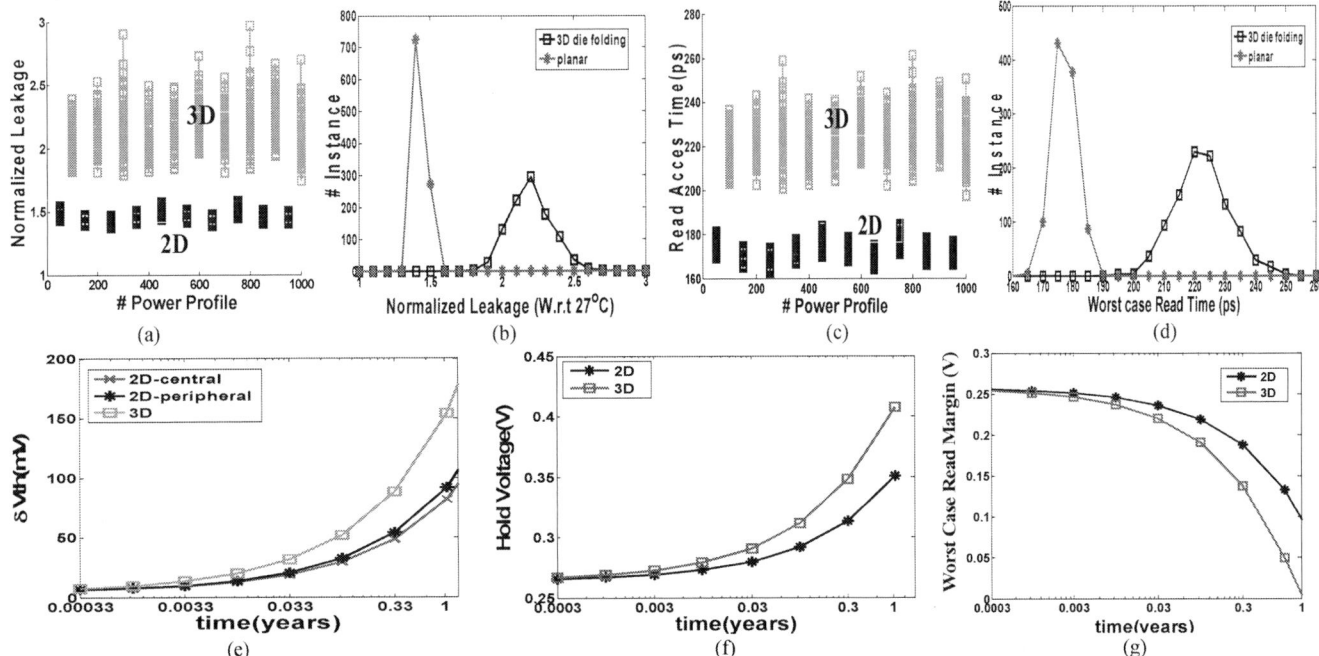

Fig. 6: Effect of die-folding on cache leakage: (a) leakage of different sub-arrays for a core power pattern (spatial spread in leakage) and (b) total SRAM cache leakage which is summation of leakage of all blocks considering temperature & process variation. Effect of die-folding on cache performance (c) block access time ($3\sigma$ read time considering process variation) of sub-arrays for a core power pattern (spatial spread in access time) and (d) variation in cache access time (or worst-case read time over all blocks) considering all random power patterns. (e) Rate of threshold voltage degradation, (f) Hold Voltage, and (g) Read margin. The higher temperature results in faster degradation of temperature for 3D scenario.

and minimum temperature for the entire cache (i.e. maximum and minimum over all blocks at that power pattern). We first study the temperature behavior of SRAM blocks and maximum/minimum temperature of cache over time (i.e. over power patterns). This study indicates how 2D and 3D integration modulates the temperature behavior of SRAM cache over time (*temporal*). We next study how the *spatial* pattern in temperature in the cache is different in 2D and 3D design. The temporal study is critical to understand what ranges of temperature SRAM blocks will experience in 2D and 3D design. The spatial study also helps analyze the how 2D and 3D integration modulates the location of thermal hotspots in 2D and 3D.

Finally, we connect the temperature estimated from thermal simulations to the power, performance, and reliability (stability under PMOS aging) estimated from circuit simulations. The thermal simulations for different random power patterns provide temperature of SRAM blocks. We assume all cells within a block have same temperature. The spatial and temporal patterns of temperature estimated from thermal simulation are used to compute the corresponding patterns in performance, leakage, and reliability.

## 4. Simulation Results and Analyses

### 4.1. Comparison of Thermal Distribution

Fig. 3 shows the spatial variation in temperature for cache blocks considering 2D and 3D designs. Fig. 4 shows the possible temperature variation for a given core and SRAM cache block in the 2D and 3D case considering different power pattern (i.e. at different time point). In Fig. 5(a), given a power pattern (in x-axis), the points in the y-axis represent the temperature of different cache blocks in the 3D design.

We obtained similar plot for 2D as well (not shown). Fig. 5(b) summarizes the spatial observation and plots the histogram (over all power patterns) of identification number of the cache block that becomes the hottest for a power pattern. We summarize the observation as follows:

- ***The core and cache temperature in 3D is higher than 2D.*** We first observe that the cores and cache experience a much lower temperature in 2D case due to the larger spatial area (i.e. lower average power density and larger heat spreader/heat sink contact area) (Fig. 3, 4). In the 3D scenario, the temperature of the cores is higher due to the reduced footprint (higher average power density and smaller heat sink/heat spreader contact area). The elevated core temperature is coupled through the die-to-die interface to the caches on next tier resulting in an increased temperature for the SRAM blocks ($\sim 30\text{-}40^{0}$C higher than 2D SRAM blocks) (Fig. 4).

- ***The temporal variation in temperature of cache blocks is much higher for the 3D system compared to the 2D system.*** Fig. 4(a, b) shows that for the 2D scenario, at any particular location of the cache block, the variance in temperature over time is small. For the 3D system, the individual cache blocks may experience significantly higher average temperature as well as larger temperature spread over time (Fig. 4a, b).

- ***The spatial thermal pattern in 3D is more non-deterministic and more strongly correlated to power patterns of cores compared to 2D.*** Fig. 3(a)-(b) shows that for 2D case, although the power patterns across cores are very different, the cache blocks towards the central region remain the coldest and those towards the periphery are the

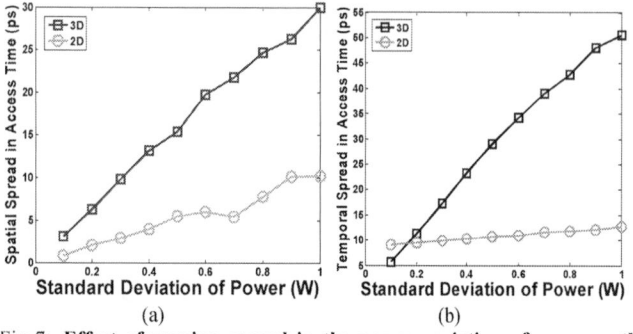

(a)　　　　　(b)

Fig 7: **Effect of varying spread in the power variation of cores on the performance of SRAM cache in 2D processor and 3D die stack.** (a) Spatial Spread (b) Temporal Spread. Increase in the standard deviation of core power variation results in an increase in spatial and temporal spread in access time

hottest. For the 3D scenario we observe that the temperature pattern of the cache follow the temperature pattern across the core distribution very closely. This suggests that as the power pattern across the cores can vary significantly depending on the workload pattern so does the thermal pattern across the cache. This is observed in Fig. 5(a) which shows that in 3D one can have significant spatial variation in temperature across cache blocks. This is further evident from Fig. 4(c) which shows correlation between the temperature of a cache block and all the cores. We observe that the cache temperature is very closely correlated to the temperature of the core directly below it and falls off very sharply (within the order of 1 core distance) to a relatively low value. ***When the cores have spatiotemporally non-uniform and random power profile (as is expected in most applications), temperature can be considered as a source of spatiotemporal random variation for the cache blocks in 3D system.*** Therefore, for the 2D case, even under statistical variation in the power profile of cores, both the temperature of cache blocks and locations of hottest cache blocks are more predictable. But in 3D each block has much higher temperature spread (over time) and all blocks are equally likely to be hottest block. This is illustrated in Fig. 5(b) which shows the location of the hottest cache block for 2D and 3D case.

### 4.2. Implications For Power, Performance And Reliability of SRAM

In this section we evaluate effect of changing temperature distribution on the read access time, leakage power, and reliability of SRAM blocks and cache considering process variation using 32nm predictive technology [20].

- Leakage: With increase in temperature the nominal value of the leakage increases exponentially. Moreover, due to the exponential dependence, the rate of change of leakage due to threshold voltage variation depends exponentially on temperature. For any instance of power distribution across the cores (i.e. at a given time point), the spatial spread in temperature is much higher in 3D [Fig. 6(a)]. This translates to large spatial spread in leakage of different sub-arrays for 3D integrated SRAM than the 2D counterpart. For a given power pattern, we next add the estimated leakage of all blocks to compute the *cache leakage* for that power pattern. The statistical distribution

of the total cache leakage for all power patterns (i.e. over time) is shown in Fig. 6(b). Due to the higher average temperature and spread, both the mean leakage power (~2X) and the spread in the cache leakage over time (~4X) increase significantly as we move to 3D.

- Read Access Time of an SRAM cell is defined as the time required to develop a pre-defined bit-differential (~100mV) during read operation. During reading, the bit line BL discharges due to the read current ($I_{read}$) flowing through the selected cell while bit line complement BR discharges due to bit-line leakage ($I_{bitline\_leakage}$) current. At higher temperature the cell read current reduces due to lower on current for access and pull-down NMOS devices of the selected SRAM cell. Moreover, higher temperature also increases the bit line leakage current through the other access transistors. The overall performance of the cache (*cache access time*) for a power pattern (i.e. at a given time point) is determined by the maximum of the block access time over all SRAM blocks. Fig. 6(c) shows the distribution of *block access time* of SRAM sub-arrays for a given power pattern i.e. the spatial variation in access time across the entire cache. We observe that the access times of the cells in 2D are much predictable (tighter distribution) while in a 3D scenario, they are much widespread. The statistical distribution of the *cache access time* (i.e. maximum of the block access time for a power pattern) for all power patterns (i.e. over time) is shown in Fig. 6(d). We observe that the average *cache access time* increases by nearly 50ps leading to ~28% performance degradation.

- Device Degradation due to NBTI caused with prolonged usage is known to have significant dependence on the voltage and thermal stress [10]. It is known that increase in the Vth of the PMOS devices due to NBTI degrades the cell stability – i.e. reduces read margin and increases minimum data retention voltage (DRV). Using the NBTI models presented in [10], we estimate the change in PMOS Vth for each block considering its *average operating temperature*. The estimated PMOS Vth shift is for an SRAM block used to compute the $3\sigma$ worst-case values of read margin and $V_{min}$ (both under process variation) of that block. As the cache temperature is much higher for a 3D SRAM cache block, the rate of degradation of threshold voltage for PMOS will also be faster in comparison to the 2D case (Fig. 6(e)). But for 2D the blocks near the cores degrade at a faster rate. The faster degradation of PMOS threshold voltage also results in a faster reduction of read margin and increase of minimum DRV for 3D caches (Fig. 6(f) and 6(g)).

### 4.3. Effect of the Non-uniformity of the Power Pattern

We next study the correlation between spatiotemporal variation in the performance of SRAM and the non-uniformity in the power profile of cores. We maintain a constant mean (1.5W) but repeat the Monte-Carlo simulations for different standard deviation (0.1W to 1W). This study is performed to understand the impact of running different workload with varying power variability among the cores (as different threads running on different cores will have varying

978-1-4673-1110-6/12 $31.00 © 2012 IEEE

computational power). For each standard deviation of power profile, we compute the spatial and temporal spread in SRAM properties. For brevity, here we present the analysis of the access time for 2D and 3D SRAM for different non-uniformity in the power pattern – the leakage and device aging follows the same trend. For a given standard deviation of core power, we consider the different instances of power pattern and estimate the spatial variation in block access time for each pattern. We next compute the statistical variation in the *cache access time* considering those 1000 Monte-Carlo instances. The computed spread of the cache access time distribution for a given standard deviation of core power variation is defined as the 'temporal spread' in access time for that standard deviation. The results are shown in Fig 7. We clearly observe that an increase in the standard deviation of core power variation results in an increase in spatial and temporal spread in access time. However, the increase in much higher in 3D integrated SRAM. Therefore, the workload that introduces a significant non-uniformity in power patterns (higher standard deviation) across cores increases the access time spread compared to workload creating uniform power profile.

## 5. Conclusions

3D die folding leads to significant deviations of the cache thermal profile in a core-cache die-stack. Temperature acts an additional source of variation leading to significant degradation for leakage and performance and accelerated aging. 3D integration reduces interconnect latency but increases the access delay and leakage power of the cache. Our analysis points to the fact that the power pattern and workloads are important factors deciding the extent of thermal coupling enabled performance degradation. This suggests a strong need for a cohesive analysis considering the trade-off between the gains in interconnect delay and temperature induced degradation in SRAM power and performance.

## Acknowledgments

This work is supported in part by Semiconductor Research Corporation (#1836.075), National Science Foundation (CCF-0917000, CNS-1054429), Intel Corp, and IBM Faculty Award.

## References

[1] C.C.Liu et al., "Bridging the processor-memory performance gap with 3D IC technology", IEEE Design and Test of Computers, 2005.

[2] K.Banerjee et al., "3-D Heterogeneous ICs: A Technology for the Next Decade and Beyond", 5th IEEE Workshop on Signal Propagation of Interconnects, 2001.

[3] K.Puttaswamy et al., "Implementing caches in a 3D technology for high performance processors", IEEE ICCD 2005.

[4] S.Im et al, "Full chip thermal analysis of planar (2-D) and vertically integrated (3-D) high performance ICs", IEEE IEDM 2000.

[5] K.Puttaswamy et al., "Thermal analysis of a 3D die-stacked high-performance microprocessor", IEEE GLSVLSI 2006.

[6] M. S. Bakir, C. King, D. Sekar, H. Thacker, B. Dang, G. Huang, A. Naeemi, and J. D. Meindl, "3D heterogeneous integrated systems: liquid cooling, power delivery, and implementation," in Proc. IEEE Custom Integrated Circuits Conf., 2008, pp.663-670

[7] T.Chiang et al., "Thermal analysis of heterogeneous 3D ICs with various integration scenarios", IEEE IEDM 2001.

[8] A.Jain et al., "Analytical and Numerical Modeling of the Thermal Performance of Three-Dimensional Integrated Circuits", IEEE Transactions on Components and Packaging Technologies, Vol 33, No 1, 2010.

[9] S.Mukhopadhyay et al., "Modeling of failure probability and statistical design of SRAM array for yield enhancement in nanoscaled CMOS", IEEE TCAD 2005.

[10] R.Vattikonda et al., "Modeling and Minimization of PMOS NBTI Effect for Robust Nanometer Design", IEEE/ACM DAC 2006.

[11] A.Jain et al., " Thermal-electrical co-optimisation of floorplanning of three-dimensional integrated circuits under manufacturing and physical design constraints", IET Computers and Digital Techniques 2011 Vol 5 Issue 3 pp:169-178

[12] G.Sun et al,"Exploration of 3D stacked L2 cache design for high performance and efficient thermal control", 14th ACM/IEEE international symposium on Low power electronics and design ISLPED '09

[13] W.Yun et al, "Thermal-aware energy minimization of 3D-stacked L3 cache with error rate limitation", IEEE ISCAS 2011 pp: 1672 – 1675.

[14] D. Sekar, C. King, B. Dang, T. Spencer, H. Thacker, P. Joseph, M. Bakir, and J. Meindl, "A 3D-IC technology with integrated microchannel cooling," in Proc. IEEE Int. Interconnect echnol. Conf., 2008, pp. 13-15.

[15] TEZZARON SEMICONDUCTORS. 2005. Tezzaron unveils 3d SRAM. http://www.tezzaron.com.

[16] G.Loi et al., "A thermally-aware performance analysis of vertically integrated (3-D) processor-memory hierarchy", ACM/IEEE DAC 2006.

[17] W. Zhao et al, "New generation of Predictive Technology Model for sub-45nm early design exploration," IEEE TED 2006.

[18] H.Hua et al., "Exploring Compromises among Timing, Power and Temperature in Three-Dimensional Integrated Circuits", IEEE/ACM DAC 2006.

[19] V.George et al., "Penryn: 45-nm Next Generation Intel® Core™ 2 Processor", IEEE Asian Solid-State Circuits Conference 2007 pp:14-17.

[20] asu.ptm.edu

# Dynamic Compact Thermal Model For Stacked-Die Components

Eric MONIER-VINARD[1], Cheikh Tidiane DIA[1,2], Valentin BISSUEL[1],
Najib LARAQI[2], Olivier DANIEL[1]

[1] Thales Global Services,
18 Rue du Maréchal Juin, 92 360 Meudon La Forêt
[2] Université Paris Ouest, Laboratoire Thermique Interfaces Environnement, EA 4415, GTE,
50 Rue de Sèvres, F-92410 Ville d'Avray
Email: eric.monier-vinard@thalesgroup.com

## Abstract

The present work proposes an approach to generate Dynamic Compact Thermal Models or "DCTMs" dedicated to electronic components. This one is based on the European project DELPHI, which defined the first comprehensive methodology concerning the generation of thermal behavioral model, Boundary Condition Independent, called Compact Thermal Models or "CTMs". Unfortunately, the scope of "CTMs" was limited to the steady state as well as for single chip packages [1][2][3][4].

But, the latest trend toward higher and higher density packaging using several chips requires henceforth a methodology capable to take into account the transient regime for 3D integration technologies like stacked-die solution.

Following the CTM's modus operandi the DCTMs were conceived to propose a RC network able to predict a set of sensitive component temperatures with a minimized difference during component duty cycle.

This work suggests the use of the genetic algorithms fitting technique that turns out relevant for the realization of DCTM, as well as the conventional DELPHI CTM.

## Keywords

Dynamic compact thermal model, DELPHI methodology, Genetic Algorithm, Thermal simulation, Transient, Stacked-die.

## Nomenclature

| | |
|---|---|
| C | Capacitance, J. $K^{-1}$ |
| Cp | Specific heat capacity, $J.kg^{-1}.K^{-1}$ |
| h | Heat transfer coefficient, $W.m^{-2}.K^{-1}$ |
| k | Thermal conductivity, $W.m^{-1}.K^{-1}$ |
| $P_H$ | Power dissipation, W |
| Q | Heat flow rate, W |
| $\rho$ | Density, $kg.m^{-3}$ |
| R | Thermal resistance, $K.W^{-1}$ |
| T | Temperature, °C |
| t | Time, s |

## Superscripts and subscripts

| | |
|---|---|
| CTM | Compact thermal model |
| DTM | Detailed thermal model |
| DCTM | Dynamic compact thermal model |

## 1 Context

These last decades, the continuous trend towards electronics miniaturization and its densification, to execute more functions in less space, without diminishing the performances, has drastically affected the thermal management of electronic components. Electronic component packages are ceaseless getting smaller and closer to the chip scale which led to a tremendous increase of power density that has to be dissipated from more overpopulated and warmer Printed Circuit Board.

Besides the conventional single chip packages are already unable to perform all the mandatory functions of the new electronic designs. This means an extended demand for always more compact and complex packaging using embedded multiple chips.

The operating temperature limits that are seen today with single-chip components will only be magnified with the development of 3D integration technology, such as the pyramid stacked die package or the recent System In Package devices. To meet the design needs of the upcoming years, new modeling practices have to be investigated in order to better predict the thermal behavior of these electronic devices.

Some years ago, the concept of the Compact Thermal Models appears us to be more and more appropriate for this purpose and a innovative creation process was developed.

This one, based on the European DELPHI methodology, was improved by a genetic algorithm fitting technique and then completed by the superposition principle to address multi-chip components.

CTM was elaborated to replicate the thermal path from the most sensitive elements to a set of external surfaces by the mean of complex thermal resistances network with the aim to keep a high accuracy level and to minimize computation time.

This approach has been validated on a large number of electronic components and is not only powerful but also robust and capable of producing BCI CTM models whose predictions are very close to the Detailed Thermal Models or DTM, generally below 10% of error.

However, the respect of the maximum operating temperature for reliability purpose can not be guaranteed for all the overheated components of a board, in particular for harsh environmental conditions.

A conventional way to reduce the highest operating temperatures to a compliant level consists in optimizing the duty cycle of the chips. In such way, their electronic functions are activated only in a transient manner in order to keep the temperatures below maximum operating limits.

So our new objective is to establish a methodology to create this kind of thermal models, so called Dynamic Compact Thermal Models, or DCTM, which will be able to accurately predict the transient responses of the sensitive parts of the latest electronic devices.

978-1-4673-1110-6/12 $31.00 © 2012 IEEE

## 2 Process of reduction

For generating a boundary condition independent DCTM, our process of reduction is divided into two successive stages.

- Steady State model building

The first step corresponding to CTM generation is according to European DELPHI methodology [5] and allows us to derive the most significant resistances network for a consistent set of boundary conditions.

The simulation can be performed from a none exhaustive number of commercial thermal software [6].

The most appropriate software is used to build a realistic three-dimension numerical model of the component which described in detail its thinnest geometry aspect as well as its thermo-physical properties.

Then the tool permits to apply on all external surfaces of the model the set of 49 uniform convective boundary conditions of Appendix 1, and finally to compute mandatory simulations.

For multi-chip components, according to superposition principle, the initial simulations set is repeated considering that each of its n chips is successively active, then all of them.

The purpose of these numerous simulations is the creation of behavioral models with high degree of Boundary Condition Independence.

Then using a customized statistical process of optimization, its CTM model is derived from the data sets of extracted temperatures and heat flow rates.

Genetic Algorithm fitting technique has been selected to ensure a constant quality of deducted thermal networks.

In addition, using DELPHI CTM style affords us to modeling a large set of component external surfaces as well as their subdivisions to more efficiently match with the behavior of the DTM.

- Dynamic model building

Following the CTM's process, the "DCTMs" are created, this time, from a set of temporal responses of the fully detailed model of the component.

Thus, a new set of boundary conditions is run and followed by a second reduction stage which permits to determine a set of added thermal capacitances dedicated to each node of the CTM network.

Our previous own works [7], according with the results of Christiaens and al [8] analysis showed that the simplest network definition, limited to junction and external surface bonding, is most of the time insufficient to accurately characterize the transient response of the DTM.

In fact, the initial deducted RC network appears to be irrelevant to report the transient heat spreading in the surroundings of the chip, during the first time steps.

Thus to resolve that issue, the thermal resistances of the network that have a significant impact on the chip cooling are split into two or more elements to achieve a convenient agreement. Subdivision factors of the selected resistances are so derived from the transient reduction process using GA technique.

Finally a capacitance value is associated with each node of the network which henceforth includes a set of fictive internal nodes.

The DCTM process reduction is illustrated from relevant test cases in the content of this paper.

## 3 Optimization from genetic algorithm

As mentioned previously, the methodology of resolution is based on the extended use of genetic algorithm [4].

- Steady State model building

Its first step consists in constructing a randomly chosen initial population. The proper fitting population among the random chosen ones is selected to produce the offspring for the next generation, which inherit some characteristics of both parents depending of crossover and mutation rates. After many generations, the evolutionary process gives the best candidate to a criterion.

The starting topology of CTM is composed of the full bonding matrix of all model nodes.

Then, the Genetic Algorithm fitting process is divided into four stages iterating until the criterion is matched:

o "Evaluation" for a $\gamma$ population of the adaptation of each network to a cost function.

o "Selection" of $\mu$ networks having the best scores.

o "Crossing and mutation strategy": creation of $\lambda$ new networks from the $\mu$ parents. Generate two new networks for two random parent networks with rearrangements and evolution of their resistances.

o "Criterion of ruling" based on the "best score"

The setting parameter choice is the population "$\gamma = \mu + \lambda$" for which in every generation we preserve the parents and the children, with $\mu = 50$ and $\lambda = 350$. And the strategy of their evolution is based on a fixed mutation rate of 10% and:

o An initial crossing rate of 50%: the generated children inherit, randomly, about 50% of the both parents.

o A decreasing step of 5% of the initial crossing rate value, if the best individual does not change after 200 generations.

The number of generations is not limited, more than 4000 can be necessary to achieve a good score, of the cost function, as called $F_{CTM}$, given in Eq 1.

$$F_{CTM} = \frac{W}{2}\left(1 - \sum_{bc=1}^{nBC}\sum_{n=1}^{nN}\frac{\Delta T_{n,bc}}{nBC \cdot nN}\right) + \frac{W}{5}\left(1 - \max(\Delta T_{n,bc})\right)$$

$$+ \frac{W}{3}\left(1 - \sum_{i=1}^{nJ}\max(\Delta T_{n=J_i,bc})\right) + \frac{1-W}{2}\left(1 - \sum_{bc=1}^{nBC}\sum_{n=1}^{nN}\frac{\Delta Q_{n,bc}}{nBC \cdot nN}\right)$$

$$+ \frac{1-W}{2}\left(1 - \max\left(\frac{Q_{n,bc}^{CTM} - Q_{n,bc}^{DTM}}{Q_{n,bc}^{DTM}}\right)\right)$$

$$\text{with} \quad \Delta T_{n,bc} = \frac{T_{n,bc}^{CTM} - T_{n,bc}^{DTM}}{T_{n,bc}^{CTM} - T_A}$$

$$\text{and} \quad \Delta Q_{n,bc} = \frac{Q_{n,bc}^{CTM} - Q_{n,bc}^{DTM}}{P_H}$$

*Eq 1: Customized cost Function definition*

where:

- $T_{n=J,bc}$ is the maximal temperature of i junction nodes,
- $T_{n,bc}$ is the average temperature of the other N nodes,
- $Q_{n,bc}$ is represents the heat flow rate leaving the package,
- nBC and nN are respectively scenario and node numbers,
- $T_A$ is the ambient temperature,
- $P_H$ is the power dissipated by the chip(s).

This cost function takes into account the maximal junction temperature, the average temperatures of other nodes as well as the heat flow rates leaving the package.

An important weight W is applied to the temperatures and especially to the junction ones which is our primary concern.

The cost function must be close to 1 that in order to have a good agreement between the fully Detailed Thermal Model, or DTM, and its behavioral CTM model.

- Dynamic model building

The proposal method of reduction dedicated to DCTM model is based on a nodal resolution of heat equation in transient regime.

Unlike the CTM model, the DCTM does not require a large number of scenarios. The set of new boundary conditions can be limited to ten added cases, for instance the Table 1 ones.

As for CTM building, the most appropriate heat Capacitance matrix candidate depends on the score of a cost function, called $F_{DCTM}$.

| BC number | $h_{TOP}$ | $h_{BOTTOM}$ | $h_{SIDES}$ |
|-----------|-----------|--------------|-------------|
| 1 | 15 | 100 | 5 |
| 2 | 50 | 250 | 15 |
| 3 | 800 | 20 | 5 |
| 4 | $1.10^{-8}$ | $1.10^{-8}$ | $1.10^{-8}$ |
| 5 | 40 | 1000 | 15 |
| 6 | $1.10^{-8}$ | 100 | 50 |
| 7 | 1000 | 40 | 100 |
| 8 | 100 | $1.10^{-8}$ | 5 |
| 9 | 15 | 200 | 200 |
| 10 | 30 | 200 | 50 |

**Table 1:** *Set of uniform heat transfer coefficients for DCTM generation*

Eq 2 defines the chosen expression that is wholly based on temperature fitting.

$$F_{DCTM} = 1 - \frac{1}{C} \cdot \sum_{bc=1}^{nBC} \sum_{n=1}^{nN} \sum_{k=1}^{nt} \left[ W \cdot \Delta T(t_k)_{n=J,bc} \right]^2$$
$$- \frac{1}{C} \cdot \sum_{bc=1}^{nBC} \sum_{n=1}^{nN} \sum_{k=1}^{nt} \left[ \frac{(1-W) \cdot \Delta T(t_k)_{n,bc}}{nN - 1} \right]^2$$

*With* $C = nBC \cdot nN \cdot nt$

**Eq 2:** Cost *Function* dedicated to DCTM

where nt characterize the number of time steps and $t_k$ is the current time at the step k.

## 4 State of art: Single chip component

The comparison of thermal performances is focused on the popular plastic Quad Flat-pack No leads (QFN) package. In fact, this package type has a set of 16 buried leads but these ones do not stick out from its body.

The selected QFN16 is a single chip; its main characteristics are detailed in Figure 1.

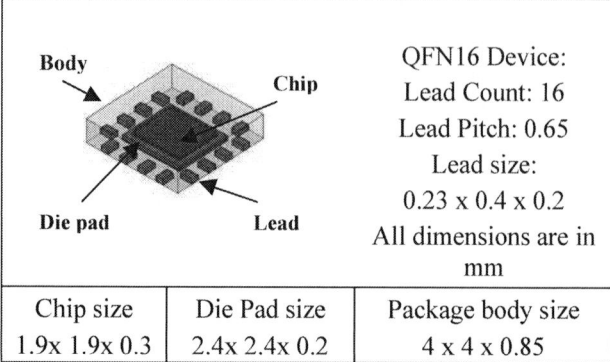

| | QFN16 Device: Lead Count: 16 Lead Pitch: 0.65 Lead size: 0.23 x 0.4 x 0.2 All dimensions are in mm |
|---|---|

| Chip size | Die Pad size | Package body size |
|-----------|--------------|-------------------|
| 1.9x 1.9x 0.3 | 2.4x 2.4x 0.2 | 4 x 4 x 0.85 |

**Figure 1**: *Geometry of the QFN16 package*

More over QFN package includes an exposed thermal pad, which is dedicated to chip housing and to enhance heat dissipation through PCB structure. Exposed pad and lead areas are soldered to PCB and the average solder joint thickness is usually 35µm.

The semiconductor material of the chip is Silicon.

Its thermal conductivity is temperature-dependent in accordance with the following formula:

$$k_{Si} = 148 \times \left( \frac{T+273}{300} \right)^{-1/3}$$

**Eq 3:** *Silicon thermal conductivity profile*

The main thermal properties of the other material of the package are listed in Table 2.

| Constituent | Material name | k | Cp | ρ |
|-------------|---------------|------|------|-----|
| Resin | SUMITOMO | 0.66 | 1900 | 920 |
| Chip | Silicon | Eq 3 | 2330 | 708 |
| Lead-frame | OLIN C194 | 260 | 8900 | 385 |
| Solder | SnPb 63/37 | 51 | 8400 | 167 |
| Die attach | ABLESTIK | 2.1 | 2400 | 728 |

**Table 2**: *Component constituent material*

A uniform power dissipation of 1Watt is applied to a thin layer on the upper surface of the chip that is assimilated to the junction node, called (J).

The CTM definition of QFN16 package considers as external nodes, its four sides (S) and two subdivisions, named inner and outer, of its upper (T) and lower (B) surfaces which provides much more reliable results. Top inner (Ti) and Bottom inner (Bi) external surfaces have respectively the size of the chip and of the die-pad.

978-1-4673-1110-6/12 $31.00 © 2012 IEEE

QFN16's CTM resistor network, connecting its sensitive junction element towards its external side nodes, is described in Table 3.

| Network node | Ti | Bi | To | Bo | S |
|---|---|---|---|---|---|
| J | 160.4 | 4.7 | 217.8 | 213.5 | - |
| Ti | - | 4666.7 | 199487 | 3025.7 | - |
| Bi | - | - | 4312.8 | 718.1 | - |
| To | - | - | - | 171.7 | 206.3 |
| Bo | - | - | - | - | 202.3 |

**Table 3:** *Significant thermal Resistance (R) of QFN16 CTM network*

Thus the full detailed geometry of the component, Figure 1, is reduced under the simplified cubical shape using an embedded thermal network.

The behavior comparison of both DTM and CTM is relevant for the complete range of boundary conditions as shown on Figure 3.

The agreement $\Delta T_{J,BC}$ of the two models, reported on the left Y-axis, is according with the relationship of Eq 1.

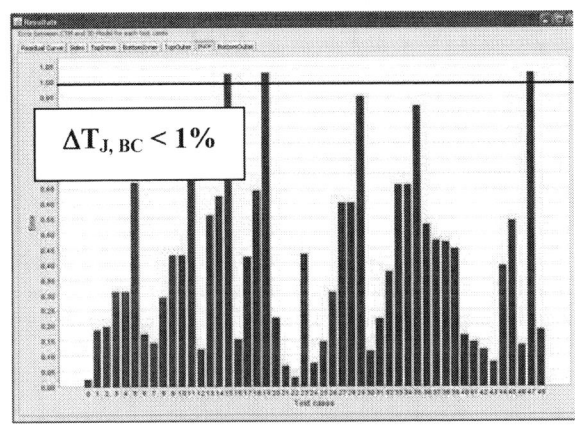

**Figure 2**: *Junction temperature agreement of the DTM and CTM for the 49 scenarios*

Figure 2 demonstrates that the CTM model provided a junction temperature estimation with a discrepancy less than 2%.

Further, the behavior of CTM network has been analyzed by considering different independent Boundary Condition scenario, such as the test case defined in Table 4.

| BC number | $h_{TOP}$ | $h_{BOTTOM}$ | $h_{SIDES}$ |
|---|---|---|---|
| Random | 20 | 800 | 10 |

**Table 4:** *Uniform heat transfer coefficients (h) randomly chosen*

Table 5 presents the prediction of maximal and average temperatures and heat flow rates for all CTM nodes when the ambient temperature is fixed at 0°C.

| Node | $T_{DTM}$ | $T_{CTM}$ | $\Delta TN$ | $Q_{DTM}$ | $Q_{CTM}$ | $\Delta QN$ |
|---|---|---|---|---|---|---|
| J | 109.6 | 109.9 | <1% | -1 | -1 | |
| Ti | 106.1 | 106.1 | <1% | 0.007 | 0.0077 | <1% |
| To | 76.7 | 75.0 | 2.2% | 0.019 | 0.019 | <1% |
| Bi | 105.6 | 107.2 | 1.5% | 0.486 | 0.494 | 1.6% |
| Bo | 59.4 | 57.5 | 3.2% | 0.479 | 0.471 | 1.7% |
| S | 59.6 | 65.2 | **9.6%** | 0.008 | 0.0089 | **9.5%** |

**Table 5**: *Detailed comparison of model thermal behaviors for an applied power of 1W*

$\Delta TN$ and $\Delta QN$ discrepancies (Eq 1) between the DTM results and those of all the nodes of its CTM model have to be below ± 10%.

However, that criterion can be hard to achieve when some nodes are poorly influent on chip cooling, for instance, if less of 1% of its heating power is drained by the node surface.

Besides, the surface temperature profiles of the DTM can be inefficiently represented by the selected set of external isothermal surfaces.

By rule, the pertinence of CTM predictions has to be guaranteed if the power dissipation of the chip is modified.

Table 6 shows the models results if of the uniform heating power of the chip is decreased at 0.5Watt.

| Node | $T_{DTM}$ | $T_{CTM}$ | $\Delta TN$ | $Q_{DTM}$ | $Q_{CTM}$ | $\Delta QN$ |
|---|---|---|---|---|---|---|
| J | 54.8 | 54.9 | <1% | -0.5 | -0.5 | |
| Ti | 52.8 | 53.1 | <1% | 0.0038 | 0.0038 | <1% |
| To | 38.3 | 37.5 | 2.1% | 0.00959 | 0.0092 | 4.1% |
| Bi | 52.6 | 53.6 | 1.9% | 0.243 | 0.247 | 1.6% |
| Bo | 29.6 | 28.7 | 3.0% | 0.240 | 0.235 | 2.1% |
| S | 29.8 | 32.6 | **9.4%** | 0.0041 | 0.0044 | **7.3%** |

**Table 6**: *Detailed comparison of model thermal behaviors for an applied power of 0.5W*

Even if the statistical network given by genetic algorithm has no physical meaning, it duplicates a more physical thermal mapping of the upper and lower surfaces of the component, contrary to simplified network possessing only a single node by surface, such as 2R network

The deduced heat capacitances from the DCTM reduction process are summed up in Table 7.

It occurs that the sum of the heat capacitances is physical and that its estimated value matches the DTM one, which is equal to 25.8 mJ.K$^{-1}$.

In fact, both models must have the same time constant; if the DCTM model fits the curve of the 3D model, thus the thermal mass of the models should be very close.

| Node | J | Ti | Bi | To | Bo | S | $\sum C_n$ |
|---|---|---|---|---|---|---|---|
| C x 10$^{-3}$ | 5.7 | 1.4 | 1.8 | 9.4 | 2.3 | 7.3 | 27.9 |

**Table 7**: *Allocated heat capacitance (C) values to QFN16 CTM nodes*

978-1-4673-1110-6/12 $31.00 © 2012 IEEE

Re-using the random boundary condition of the Table 4, the transient responses of the junction temperature of the DTM and DCTM models were superposed and compared in the Figure 3.

Thus for this boundary condition, the junction temperature of the chip is moving from 0 to a steady state value of 110°C as seen on left Y-axis.

Right Y-axis reports the difference profile between the transient responses of both thermal models.

The agreement is good enough, less ± 4°C for each time-step. In fact, the main error occurs when the heat capacitance of the chip is loaded and that the heat fluxes are spreading into the other constituents of the package such as die-pad part.

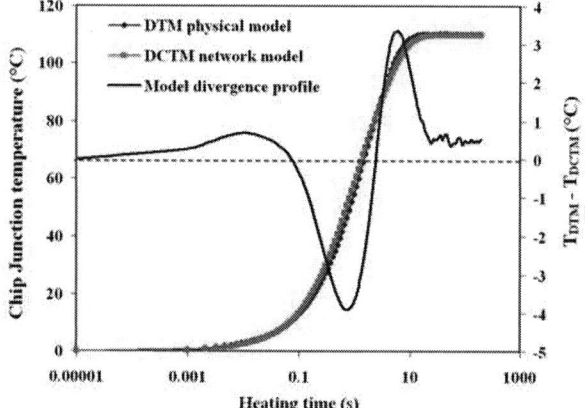

*Figure 3*: *Comparison of the transient responses of junction temperature given by DTM and DCTM models*

However, the temperature response of an electronic component, having several time constants, is more properly represented if its structure is characterized by several ladders of RC values.

Thus the use of n-stage Foster RC network and its transformation to a "physical" Cauer RC model is henceforth a current practice to ground component thermal impedance.

Being inspired by this method, additional internal (I) nodes were integrated into the DELPHI network with the aim of better modeling transient phenomena.

Actually, some existing resistances of the existing CTM are split into two or more elements to get this improvement as pictured in Figure 4.

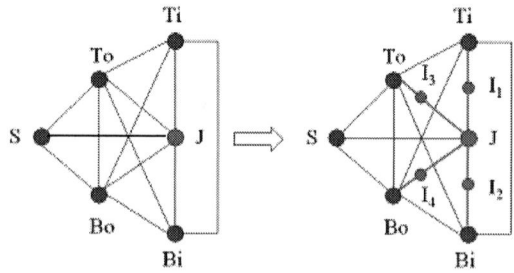

*Figure 4*: *QFN16 revised network topology style*

Resistance subdivision factors and new heat capacitances are derived from the previous transient reduction process.

Updated capacitance identity matrix and added resistance sub-matrix are given Table 8.

The integration of fictitious nodes has a weak influence on the global heat capacitance of the DCTM model that is estimated to 26.3 mJ.K$^{-1}$.

Nevertheless, this last one aims toward the true value. We can also see that the location of fictitious node $I_4$ is very close to the external node Bo.

| C x $10^{-3}$ | | 1.77 | 0.925 | 2.46 | 3.91 | 0.24 |
|---|---|---|---|---|---|---|
| C x $10^{-3}$ | Node | J | Ti | Bi | To | Bo |
| 0.736 | I1 | 56.9 | 103.5 | - | - | - |
| 2.58 | I2 | 4.47 | - | 0.27 | - | - |
| 2.57 | I3 | 99.5 | - | - | 118.3 | - |
| 2.32 | I4 | 214 | - | - | - | 0.0021 |
| 8.79 | S | - | - | - | - | - |

*Table 8*: *Capacitances matrix of QFN16 package using fictitious nodes*

The benefit of the evolution of the network, considering the same boundary condition, is reported in Figure 5.

Left Y-axis shows that the temperature differences of DTM and DCTM predictions through the heating curve are minimized, in particular in first time steps. The n-stage RC modification of the initial network has significantly improved the DCTM performance.

This technique allows us to reach a fair accuracy as demonstrated on right Y-axis. Similar performances are noticed on the other nodes of the DCTM network.

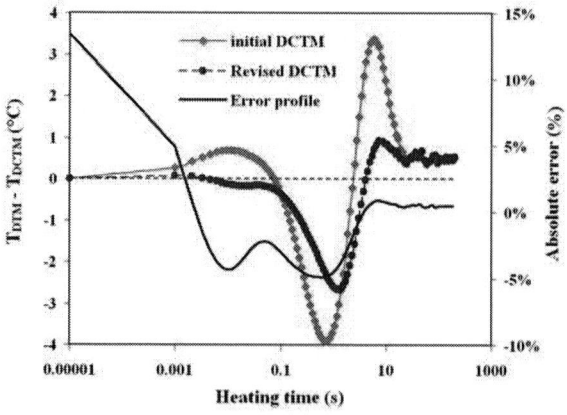

*Figure 5*: *DTM and DCTM of junction temperature divergence between using or not fictitious nodes*

The proposed approach has been tested with success on different sizes of QFN family package, such as QFN32, QFN48 and QFN64.

Thus, the reduction process has proved its pertinence for the DCTM extraction of mono-chip component and its extension to multi-chip packages is discussed farther.

## 5  Reduction of 3D stacked-die component

The stacked-die configuration can take several shapes but

our work, as previous ones [12], focused on the conventional pyramid style. This one consists of die stacking with the decreasing size ones that are overtopped the biggest ones [9].

The characteristics of the pyramid two stacked-die package which has been studied are defined Figure 6.

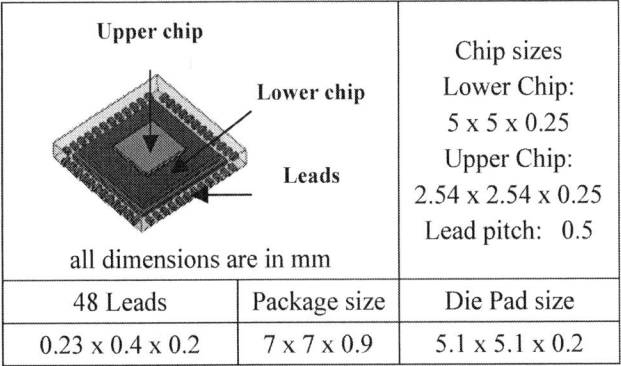

| Upper chip | Chip sizes |
| | Lower Chip: |
| Lower chip | 5 x 5 x 0.25 |
| | Upper Chip: |
| Leads | 2.54 x 2.54 x 0.25 |
| | Lead pitch: 0.5 |
| all dimensions are in mm | |

| 48 Leads | Package size | Die Pad size |
|---|---|---|
| 0.23 x 0.4 x 0.2 | 7 x 7 x 0.9 | 5.1 x 5.1 x 0.2 |

*Figure 6: Geometry of the two pyramid stacked dies*

The thermal properties of the material are similar to Table 2 ones. The thickness of two die attach film is fixed at 10μm.

This time, the transient thermal response of both chips is mainly depending of the resin constituent. Its higher volume compared to the other constituents of the package has a major impact on the transient behavior of the device.

Top inner (Ti) and Bottom inner (Bi) external surfaces have respectively the size of the upper chip and of the die-pad area. The data collection of different scenarios uses the maximal temperature about the upper die source and the mean temperature for the lower die as well as for the other external nodes.

An example of updated network topology is depicted in Figure 7. Three fictitious nodes are added to better represent the fist time steps of chips heating.

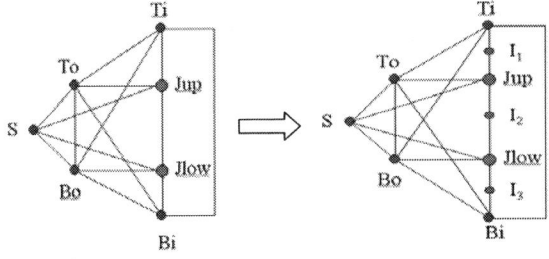

*Figure 7: QFN48 two pyramid-stacked-die network style*

The network given by our process is described in

Appendix 2.

It can be determined that the thermal mass of DCTM network is again close to the one of the 3D model, which is 86.5 mJ.K$^{-1}$. The divergence is lower than 0.3%.

The scenario of previous case has been kept with a heating power applied on each chip of 0.5Watt.

The "junction" temperature plots of the upper die and lower die are respectively displayed in Figure 8 and Figure 9.

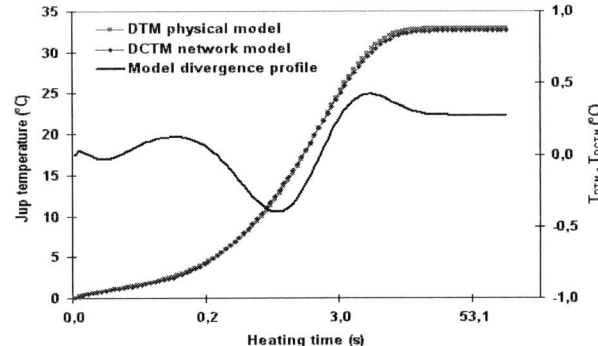

*Figure 8: DTM and DCTM thermal responses of the upper die of QFN48 device*

Right Y-axis shows that DCTM model matches the thermal transient response of both chips of the DTM with a discrepancy of ± 0.5°C, for the complete heating curve.

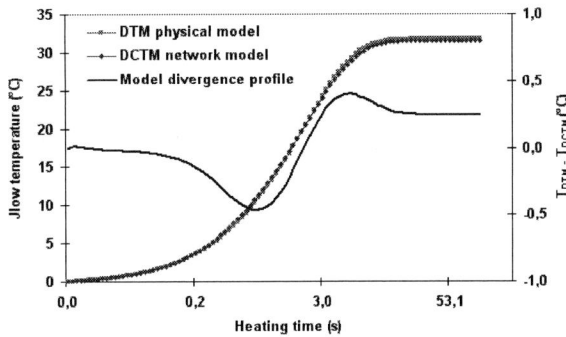

*Figure 9: DTM and DCTM thermal responses of the lower die of QFN48 device*

However, the thermal management of overheated device required to be able to master the activation and deactivation cycle in order to limit the operating temperature of its chips.

So, the comparison of the thermal response of the models performances has been extended a longer duty cycle.

The transient heating powers which are simulated for each chip are reported on Y-Right axis of the curves of Figure 10 and Figure 11. As previously, the left Y-axis displays the "junction" temperature plots of the upper die and lower die.

The left Y-axis of Figure 12 reveals that DCTM model fits the thermal transient response of both chips of the DTM with a discrepancy lower than ± 1°C, for the entire activation-deactivation profile.

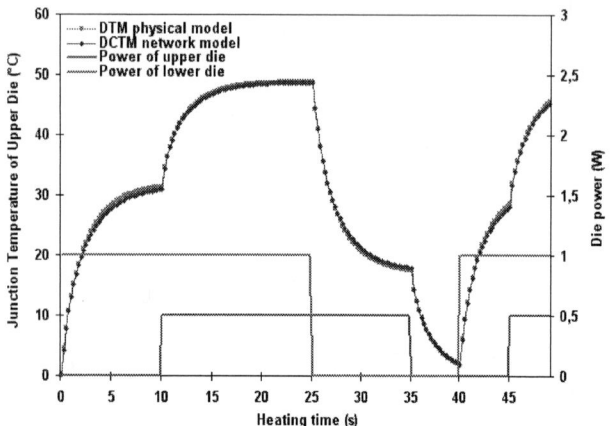

**Figure 10:** *Thermal response of the upper die of QFN48 device submitted to a duty cycle*

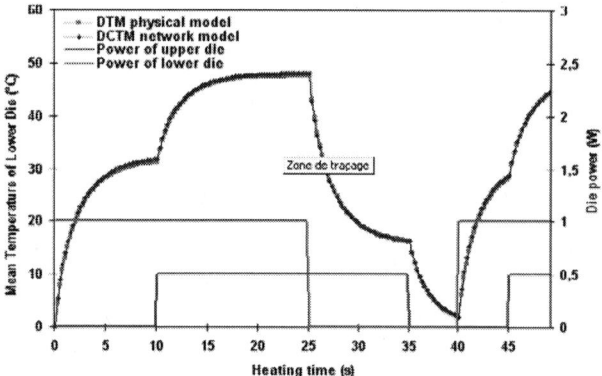

**Figure 11:** *Thermal response of the lower die of QFN48 device submitted to a duty cycle*

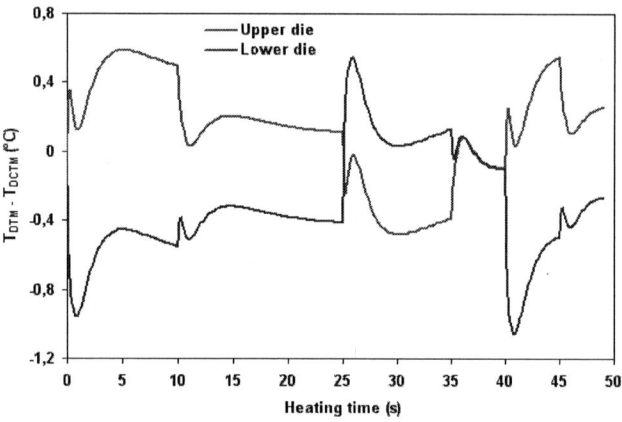

**Figure 12:** *Difference between DTM and DCTM for upper and lower die*

The agreement of DCTM is depending on the activation of both chips. Thus the absolute error decreases for both chips as soon as the upper die is activated, for instance, when the total power dissipated within the device rises from 1 Watt to 1.5 Watt.

So the critical temperatures encountered by the chips will be calculated with higher accuracy.

In the next section, the comparison of models is performed for a more realistic test case.

## 6 DCTM Compliance to realistic boundary conditions

In the preceding cases, the quality of DCTM models has been checked using a trio of uniform heat transfer coefficients fixed on the outside edges of the component.

This time, the comparison is carried out considering the realistic case of a component mounted on an electronic board which is cooled by coupled free convection and radiation phenomena for an ambient temperature of 35°C.

Thus the computed values of the heat transfer coefficients and the heat conduction spreading resistances of the inner and outer surfaces of the DELPHI style network are distinct.

The previous DTM and CTM models of pyramid-stacked-dies QFN48 are located on the upper face of an horizontal board as pictured in Figure 13. The size of the board and the locations of the centers of the components are described in the same figure.

This one is constituted of 2 equidistant full copper layers of 35μm which are buried in a FR4 dielectric substrate.

Nevertheless to simplify the analysis, this layer detailed model is replaced by a lumped PCB model. Its equivalent thermal conductivities of in-plane and through-plane of the board have being fixed respectively to $k_{XY} = 20$ W.m$^{-1}$.K$^{-1}$ and $k_Z = 0.35$ W.m$^{-1}$.K$^{-1}$ and its heat capacity to C=13.53 J.K$^{-1}$.

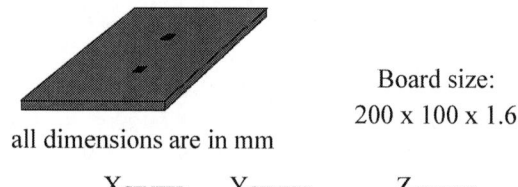

all dimensions are in mm

Board size:
200 x 100 x 1.6

|  | $X_{CENTER}$ | $Y_{CENTER}$ | $Z_{CENTER}$ |
|---|---|---|---|
| DCTM | 50 | 50 | 1.6 |
| DTM | 150 | 50 | 1.6 |

**Figure 13:** Board sizes and locations of components

The heating powers of the upper and lower dies are respectively 0.5W and 1W.

The heating and cooling cycles of both chips are depicted in Figure 14 and Figure 15.

The outcome of numerical simulation is given in Figure 14 through Figure 16.

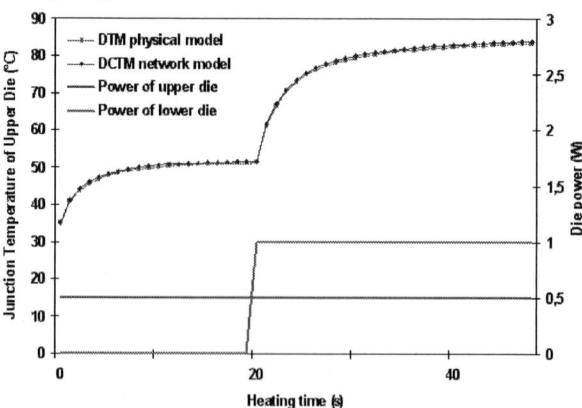

**Figure 14:** *Transient temperature profile of QFN48 upper die on board*

978-1-4673-1110-6/12 $31.00 © 2012 IEEE

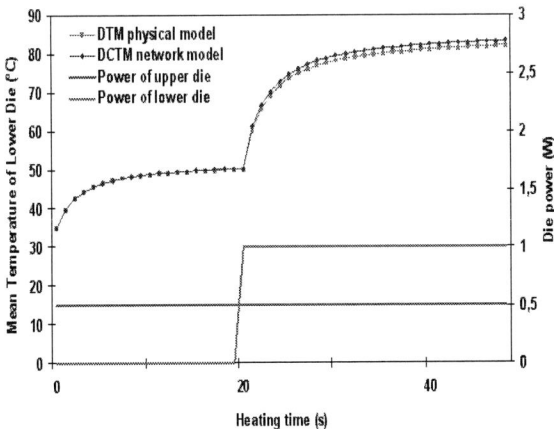

**Figure 15**: *Transient temperature profile of QFN48 Lower die on board*

For the first 20s of duty cycle, when only the upper chip is active, the difference of both models is practically null.

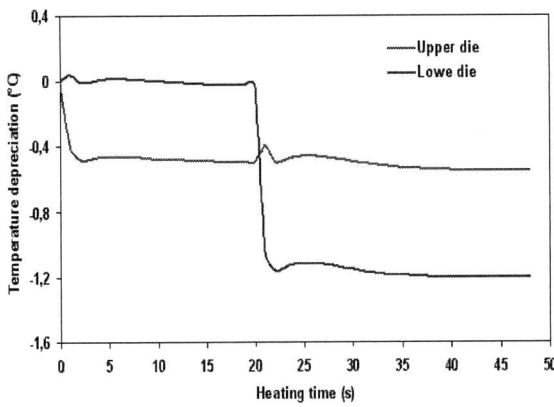

**Figure 16**: *Difference between DTM and DCTM of QFN48 on Jedec board*

Then when lower die is turned on, this one disrupts the accuracy of the DCTM prediction. But the results are still relevant; the maximum discrepancy is limited to 1.2 °C, until the steady state.

## 7 Summary

In this article, a methodology for the generation of Dynamic Compact Thermal Model for mono-chip and two pyramid-stacked-die components is proposed. The hereby-presented methodology is based on the extended use of Genetic Algorithm fitting technique.

This approach seems very powerful and reliable for extracting DELPHI DCTM style for complex devices like stacked-die packages. Fictitious nodes can be added into the network depending on the expected accuracy, especially if the analysis of the first time-steps is the primary concern.

Nevertheless, the manufacturers and the end-users should be aware that DCTM depends strongly on the model information and tool modeling capabilities. The Detailed Thermal Model has to accurately match the physical characteristics of a real device.

Our future works is going on the generation of DCTM models for devices with horizontal integration like side-by-side components, but also for Systems-in-Package components.

*Appendix 1*: *Uniform heat transfer coefficient (h) scenarios*

| Type | Test case | Top face | Bottom face | Sides | Leads |
|---|---|---|---|---|---|
| Free convection conditions | 1 | 5 | 1 | 5 | 1 |
| | 2 | 5 | 10 | 5 | 10 |
| | 3 | 5 | 25 | 5 | 50 |
| | 4 | 5 | 50 | 5 | 50 |
| | 5 | 5 | 50 | 5 | 100 |
| | 6 | 5 | 100 | 5 | 200 |
| | 7 | 15 | 1 | 15 | 1 |
| | 8 | 15 | 10 | 15 | 10 |
| | 9 | 15 | 25 | 15 | 50 |
| | 10 | 15 | 50 | 15 | 50 |
| | 11 | 15 | 50 | 15 | 100 |
| | 12 | 15 | 100 | 15 | 200 |
| Forced convection conditions | 13 | 30 | 5 | 30 | 5 |
| | 14 | 30 | 30 | 30 | 30 |
| | 15 | 30 | 50 | 30 | 50 |
| | 16 | 30 | 200 | 30 | 200 |
| | 17 | 80 | 5 | 80 | 5 |
| | 18 | 80 | 30 | 80 | 30 |
| | 19 | 80 | 50 | 80 | 50 |
| | 20 | 80 | 200 | 80 | 200 |
| | 21 | 200 | 5 | 200 | 5 |
| | 22 | 200 | 30 | 200 | 30 |
| | 23 | 200 | 50 | 200 | 50 |
| | 24 | 200 | 200 | 200 | 200 |
| Free convection and heat-sink conditions | 25 | 25 | 1 | 25 | 1 |
| | 26 | 25 | 10 | 25 | 10 |
| | 27 | 25 | 25 | 25 | 50 |
| | 28 | 25 | 50 | 25 | 50 |
| | 29 | 25 | 50 | 25 | 100 |
| | 30 | 25 | 100 | 25 | 200 |
| | 31 | 75 | 1 | 75 | 1 |
| | 32 | 75 | 10 | 75 | 10 |
| | 33 | 75 | 25 | 75 | 50 |
| | 34 | 75 | 50 | 75 | 50 |
| | 35 | 75 | 50 | 75 | 100 |
| | 36 | 75 | 100 | 75 | 200 |
| Forced convection and heat-sink conditions | 37 | 150 | 5 | 30 | 5 |
| | 38 | 150 | 30 | 30 | 30 |
| | 39 | 150 | 50 | 30 | 50 |
| | 40 | 150 | 200 | 30 | 200 |
| | 41 | 500 | 5 | 200 | 5 |
| | 42 | 500 | 30 | 200 | 30 |
| | 43 | 500 | 50 | 200 | 50 |
| | 44 | 500 | 200 | 200 | 200 |
| Cold plate conditions | 45 | 10 | 50 | 10 | $10^3$ |
| | 46 | 10 | $10^3$ | 10 | $10^4$ |
| | 47 | $10^3$ | 5 | 10 | 50 |
| | 48 | $10^4$ | 50 | 10 | 500 |
| Infinite | 49 | $10^9$ | $10^9$ | $10^9$ | $10^9$ |

*Appendix 2: DCTM network matrix of QFN48 two pyramid-stacked-dies*

| C x10$^{-3}$ | C x10$^{-3}$ | 2.65 | 0.48 | 3.25 | 30.25 | 8.28 | 10.65 |
|---|---|---|---|---|---|---|---|
| | Node | J$_{UP}$ | Ti | Bi | To | Bo | J$_{LOW}$ |
| 3.25 | Bi | 9.55 | 89452 | - | - | - | - |
| 30.25 | To | 522.1 | - | 183.8 | - | - | - |
| 3.25 | Bo | 38657 | - | 3224.3 | 39.8 | - | - |
| 8.65 | S | - | - | - | 347.4 | 149.9 | - |
| 10.65 | J$_{LOW}$ | - | 175.6 | - | 47.1 | 67.2 | - |
| 15.51 | I2 | 3.20 | - | - | - | - | 2.9 |
| 2.65 | I1 | 20.41 | 22.1 | - | - | - | - |
| 4.33 | I3 | - | - | 3.02 | - | - | 1.45 |

## References:

[1] Sabry M.N., **Compact Thermal Models for electronic Systems,** IEEE Transactions on components and packaging technologies, Vol.26, N°1, March 2003

[2] Clemens J.M, Lasance, **Two Benchmarks for the study of Compact Thermal Modelling Phenomena,** Phillips Research Laboratories

[3] Gerstenmaier Y.C., Pape H., Wachutcha G., **Boundary Independent Exact Thermal Model for Electronic Systems,** Nanotech Vol.1 Technical Proceedings of the 2001 International Conference on Modelling and Simulation of Microsystems

[4] Parthiban A., Kankanhally N.S., Ishak A.A., **Determination of Thermal Compact Model via Evolutionary Genetic Optimization Method,** IEEE Transactions on components and packaging technologies June 2005.

[5] Shidore S., Sahrapour A., **Delphi Compact Models Revolutionize Thermal Design,** Flomerics Inc.

[6] **Flotherm® of Mentor Graphics or Icepak® of ANSYS**

[7] Monier-Vinard E., DIA C., Bissuel V., Najib Laraqi N., **Extension of the DELPHI methodology to Dynamic Compact Thermal Model of electronic Component,** Therminic 2011, p p36-41

[8] Filip Christiaens, Bart Vandevelde, Eric Beyne, Robert Mertens, Jan Berghmans, **A Generic Methodology for Deriving Compact Dynamic Thermal Models, Applied to the PSGA Package,** IEEE Transaction on Components, Packaging and Manufacturing Technology, Part A, vol.21, pp. 565–576, 1998.

[9] Annette Teng Cheung, **Dicing Die Attach Films for High Volume Stacked Die Application,** 2006 Electronic Components and Technology Conference

[10] Monier-Vinard E., Bissuel V., Murphy P., Daniel O., Dufrenne J, **Thermal modeling of the emerging Multi-Chip Packages,** EuroSimE XII, Bordeaux, April 2010.

[11] Monier-Vinard E., Bissuel V., Murphy P., Daniel O., Dufrenne **J, Delphi style compact modeling for multi-chip package including its bottom board area based on genetic algorithm optimization,** IEEE Itherm conference, Las Vegas, June 2010.

[12] Marta Rencz, Vladimir Székely, **Structure function evaluation of stacked dies,** 20 th IEEE Semi-therm Symposium, 2004.

# Two-Phase Flow Control of On-Chip Two-Phase Cooling Systems of Servers

Jackson Braz Marcinichen[1] and John Richard Thome
Heat and Mass Transfer Laboratory (LTCM)
EPFL STI IGM LTCM / ME G1 475 / Station 9 CH-1015
Lausanne/Switzerland
[1] jackson.marcinichen@epfl.ch

## Abstract

Thermal designers of data centers and server manufacturers are showing a great concern regarding the cooling of new generation data centers, which are more compact and dissipate more power than is currently possible to cool by conventional air conditioning systems. With very large data centers exceeding 100 000 servers, some consume more than 50 MW [1] of electrical energy to operate, energy which is directly converted to heat and then simply wasted as it is dissipated into the atmosphere. A potentially significantly better solution would be to make use of on-chip two-phase cooling [2], which, besides improving the cooling performance at the chip level, also adds the capability to reuse the waste heat in a convenient manner, since higher evaporating and condensing temperatures of the two-phase cooling system (from 60-95°C) are possible with such a new *green* cooling technology. In the present project, two such two-phase cooling cycles using micro-evaporation technology were experimentally evaluated with specific attention being paid to energy consumption, overall exergetic efficiency and controllability. The main difference between the two cooling cycles is the driver, where both a mini-compressor and a gear pump were considered. The former has the advantage due to its appeal of energy recovery since its exergy potential is higher and the waste heat is exported at a higher temperature for reuse.

## Keywords

Data center, microprocessor, on-chip two-phase cooling cycle, micro-evaporator, controller.

## Nomenclature

Roman

| | |
|---|---|
| $\dot{E}_d$ | rate of exergy destruction due to irreversibilities within the control volume, W |
| $\dot{e}_{fi}, \dot{e}_{fe}$ | inlet and outlet flow exergies, J/kg |
| $\dot{m}_i, \dot{m}_e$ | inlet and outlet mass flow rate, kg/s |
| $\dot{Q}_j$ | heat transfer rate, W |
| $T_0$ | dead state temperature, K |
| $T_j$ | instantaneous temperature, K |
| $\dot{W}_{cv}$ | energy transfer rate by work, W |
| $x_o$ | MEs' outlet vapor quality, - |

Acronyms

| | |
|---|---|
| CPU | central processing unit |
| EEV | electric expansion valve |
| iHEx | internal heat exchanger |
| LA | liquid accumulator |
| LP | liquid pump |
| LPR | low pressure receiver |
| ME | micro-evaporator |
| MP$_{AE}$ | microchannel cold plate applied on auxiliary electronics |
| PCV | pressure control valve |
| TCV | temperature control valve |
| VC | vapor compressor |
| VSC | variable speed compressor |

## 1. Introduction

Under the current efficiency trends, the energy usage of data centers in the US is estimated to become more than 100 billion kWh by 2011, which represents an annual energy cost of approximately $7.4 billion [3]. With the introduction of a proposed carbon tax in the US [4], the annual costs could become as high as $8.8 billion by 2012, increasing annually. With the US having an annual increase of total electrical generation of approximately only 1.5% combined with the current growth rate of electrical energy by data centers being between 10-20% per annum (driven now even more by smart phones), data centers potentially will consume all of the electrical energy produced by 2030 if current growth rates continue! With air cooling of the servers in data centers accounting for most of the non-IT energy usage (up to 45% [5] of the total energy consumption), this is the logical energy consumer that needs to be attacked to reduce its wasteful use.

Nowadays, the most widely used cooling strategy is refrigerated air cooling of the data centers' numerous servers. When making use of this solution, nevertheless, 40% or more of the refrigerated air flow typically by-passes the racks of servers in data centers all together, according to articles presented at ASHRAE Winter Annual Meeting at Dallas (January, 2007), while also "cooling" thousands of servers that are not even in operation. This massive waste of energy motivates the search for a new "green" cooling solution to the future generation of higher performance servers that consume much less energy for their cooling.

Recent publications show the development of primarily four competing technologies for the cooling of chips: microchannel single-phase flow, porous media flow, jet impingement cooling and microchannel two-phase flow [2]. Leonard and Phillips [6] showed that the use of such new technology for cooling of chips could produce savings in energy consumption of over 60%. Agostini et al. [2] highlighted that the most promising of the four technologies was microchannel two-phase, being due to its low thermal resistance, low pumping power requirements and high heat removal capabilities.

The most promising working fluids for these applications appear to be conventional refrigerants, for instance HFC134a and HFO1234ze, as opposed to low pressure dielectric coolants (such as HFC245fa) or water-cooling [7, 8]. The new refrigerant HFO1234ze of Honeywell Inc. is considered as a potential substitute of HFC134a. This fluid has a "Global Warning Potential" of only 6 against 1410 of HFC134a, i.e. it is considered as an immediate/future replacement for HFC134a. Both HFC134a and HFO1234ze are dielectric fluids and thus compatible with electronics. HFC134a is currently the most widely used refrigerant for refrigeration and air conditioning systems.

Based on this background, the main objectives of this paper are to show the overall development and to compare the performance of two different cooling systems using micro-evaporator elements (multi-microchannel evaporators or MEs) for direct cooling of the chips and memories on a blade server board. The specific focus was to work with two-phase cooling using the dielectric refrigerant HFC134a, a liquid pump or a vapor compressor to drive the working fluid, a micro-evaporator for cooling of the chip and, for now, a simple tube-in-tube condenser for heat recovery, which can reduce the demand of cooling energy with respect to air cooling and water cooling by an impressive amount [1, 9]. A multi-purpose test bench was constructed to experimentally evaluate the performance of the cooling systems under various typical blade server operating conditions of transient, steady state, balanced and unbalanced heat loads on the system's two pseudo CPU's, which in turn was directly cooled by means of micro-evaporators. Further, a preliminary exergy analysis was also performed, taking into account experimental results for the two-phase cooling systems operating at steady state conditions.

## 2. Experimental setup

Figure 1 and Figure 2 depict potential two-phase cooling cycles, in which the cycle drivers are a liquid pump and a vapor compressor, respectively [10].

The liquid pumping cycle (Figure 1) consists of a liquid pump, condenser, liquid accumulator, subcooler and stepper motor valves prior to each chip. The condenser is used to remove the latent heat gained from the boiling process in the micro-evaporators (ME) and microchannel cold plates for auxiliary electronics ($MP_{AE}$), with the liquid accumulator (LA) and subcooler ensuring subcooled liquid enters the pump. The stepper motor valves are used to control the flow rate through each ME according to the flow requirements of each.

The vapor compression cycle (Figure 2) consists of a variable speed vapor compressor, condenser, liquid accumulator, internal heat exchanger, low pressure receiver, electronic expansion valve and stepper motor valves prior to each chip. After the variable speed compressor (VSC) the flow passes through three heat exchangers; the condenser, an internal heat exchanger (iHEx) and the low pressure receiver (LPR). This guarantees subcooling and superheating at the inlet of the MEs and the VSC, respectively. The last two heat exchangers also increase the performance of the cooling system [1, 10, 11]. Expansion is ensured by an electric expansion valve (EEV) prior to the fluid flowing through the SMV/ME/$MP_{AE}$.

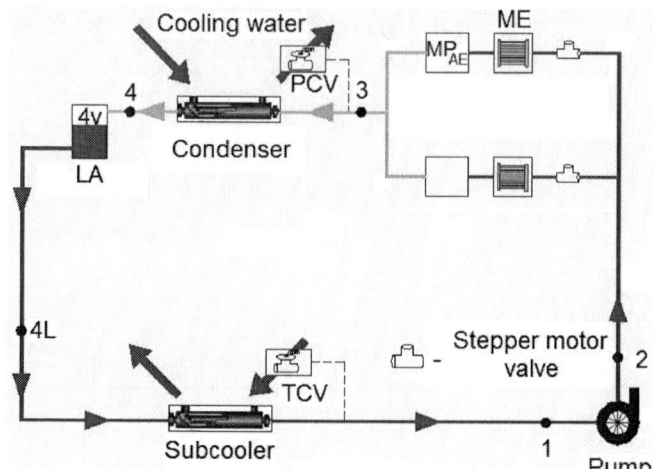

**Figure 1:** Liquid pumping two-phase cooling cycle (LP cycle).

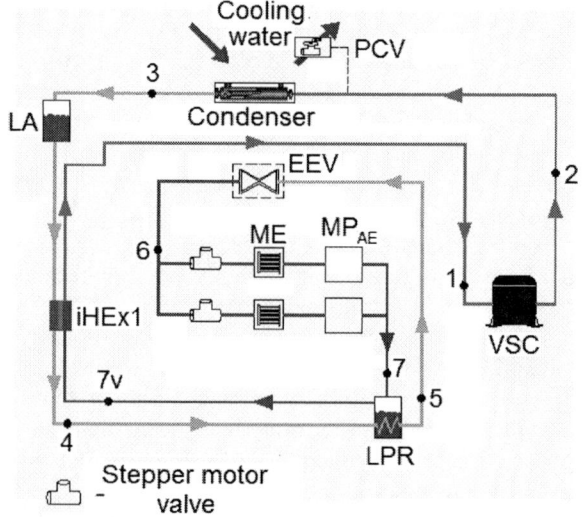

**Figure 2:** Vapor compression two-phase cooling cycle (VC cycle).

The goal is to control the chip temperature to a pre-established level by controlling the inlet conditions of the multi-microchannel cooler (pressure, subcooling and mass flow rate). It is imperative to keep the multi-microchannel cooler outlet vapor quality below that of the critical vapor quality, which is associated with the critical heat flux. Due to this exit vapor quality limitation (it is suggested not to surpass one-half of the critical vapor quality at the evaporator exit as a tentative safety margin), additional latent heat is available for further evaporation, which can be safely done in other low heat flux generating components, such as memory, DC/DC converters, etc. Another parameter that must be controlled is the condensing pressure (condensing temperature). The aim is to recover the energy dissipated by the refrigerant in the condenser to heat buildings, residences, district heating, pre-heating boiler feedwater, etc. when that can be arranged and is viable.

The liquid pumping and vapor compression cooling cycles described above were built and experimentally evaluated in the present study, investigating the cooling system's energy consumption, exergetic efficiency and controllability. For such an evaluation, specific controllers were first designed and tested. The variables to be controlled were the MEs outlet vapor quality, the condensing pressure (LP cycle) and the temperature difference between the water outlet flow and working fluid inlet flow in the condenser (VC cycle). The actuators used were the vapor compressor, the condenser water pump and the stepper motor valve (over-dimensioned to modulate the refrigerant mass flow with a negligible pressure drop). The controllers were achieved by deriving mathematical models capable of representing the dynamic behavior of the system under consideration by means of a system identification process and a PI structure that was used for the controllers since the systems showed low order dynamics. In general, simple SISO strategies were sufficient to attain the requirements of control [12].

Two MEs in parallel (typical for blade server boards with two CPU's) assembled on pseudo chips, each composed of 35 heaters and temperature sensors (2.5 mm by 2.5 mm in size), were used. The MEs' copper microchannel geometry consisted of 53 channels having a height of 1.7 mm and a width of 0.17 mm, with the spacing between channels being 0.17 mm. The effective "footprint" area of the MEs is 12 mm length and 18 mm width. The pseudo chip/ME assembly has been extensively tested to study flow boiling heat transfer, two-phase pressure drops, hot spot cooling with non-uniform heat fluxes, transient cooling, different working fluids etc. by Costa-Patry et al. [13-15]. However, in the present work only uniform heat fluxes were considered. HFC134a was tested as the working fluid and an oil free mini-compressor and a gear pump as drivers. It is important to highlight the characteristic "oil free" operation, which is mandatory for operation of micro-evaporation cooling systems and is considered as an advantage of the new mini-compressor.

Finally, for the present experimental campaign only one SMV was considered for both MEs. The outlet vapor quality used for control was that after the mixture of both flows in the MEs. The condenser is a tube-in-tube type with the secondary fluid being water, where the driver is a controllable speed gear pump.

## 3. Experimental Results

Experiments for set point tracking (for each controller developed), disturbance rejection and non-uniform heat load (last two for all developed controllers integrated / dual SISO, SISO and SIMO strategies) were developed and a short resume is presented below. More details regarding the development of the controllers can be found in [12]. The results presented in items 3.1 and 3.2 are for the LP cycle, however the authors highlight that similar results were obtained with the VC cycle.

### 3.1. Flow distribution for non-uniform heat load

The experimental results showed that for different heat loads applied on the parallel ME's an unbalanced flow exists, which generated a higher temperature on the pseudo chip with higher heat load. Temperatures of 75 °C against 60 °C were obtained when the difference in heat load was 60 W (90 W on ME1 against 30 W on ME2). Despite this, it is important to mention that the temperatures obtained were lower than the typical CPU operating limit of 85 °C and that the difference of

temperatures was reduced when the set point of the outlet vapor quality was reduced (viz. Figures 3 and 4). As can be seen, a total of eight different combinations of heat loads and three outlet vapor qualities were evaluated. Additionally, it is observed that for the same heat load, step 2, the average temperature of both pseudo chips was the same, i.e. about 67 °C. This implies that the distributors (piping) before and after the MEs were well designed and that both MEs have the same mass flow rate.

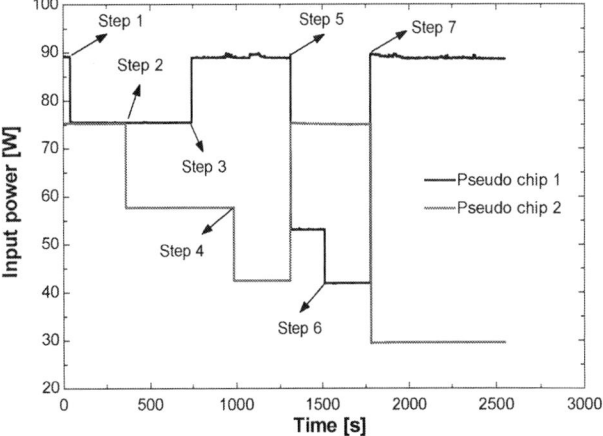

**Figure 3:** Different heat loads on the MEs.

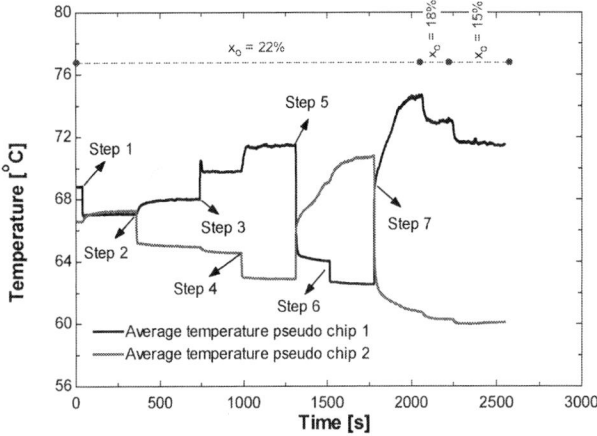

**Figure 4:** Average temperature on the pseudo chips.

The results obtained/presented in this work also proved that since the system is well designed and controlled, a SMV for each ME is not necessary, as was initially proposed by [10] and schematically given in Figures 1 and 2, i.e. only one SMV is sufficient to operate as an actuator for the outlet vapor quality controller. It should be mentioned that this is probably only valid when two MEs are considered (only one blade), with a more general statement only being valid once a complete blade center has been evaluated.

Regarding the controllability, the cooling systems were found to be fast and effective, controlling the condensing pressure or the secondary fluid temperature and the outlet vapor quality at the defined set points under steady state and transient conditions of heat load. For example, *viz.* Figures 5 and 6, controllers of outlet vapor quality and condensing pressure in the LP cycle under evaluation of non-uniform heat load. It can be seen that for all ranges of heat loads investigated, the controllers were able to control and stabilize the condensing pressure and outlet vapor quality at the set point. For the outlet vapor quality, the maximum transient time observed was about 30 s in step 7, all the while maintaining the pseudo chip average temperatures well below 85°C. For the condensing pressure, the maximum disturbance observed was stabilized after 5 min and provoked an overshoot of only 0.1 °C in the condensing temperature.

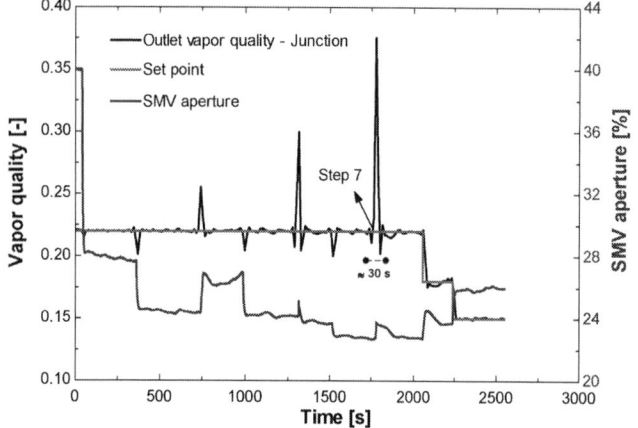

**Figure 5:** Outlet vapor quality and SMV aperture.

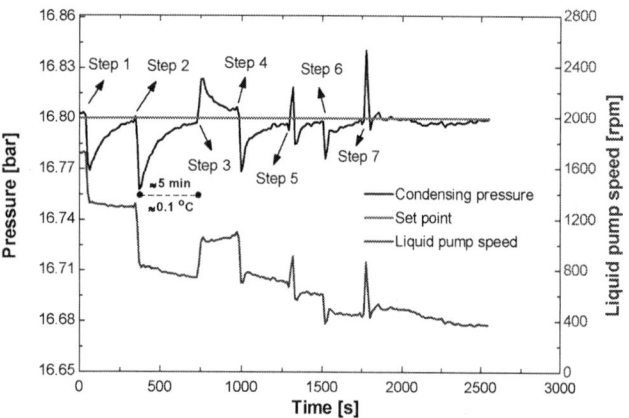

**Figure 6:** Condensing pressure and LPS.

### 3.2. Heat load disturbance rejection

The heat loads on ME1 and ME2 were varied between 90 W and 75 W and 75 W and 60 W, respectively, considering a periodic disturbance time of 1.4 s. Figure 7 shows the input power disturbance on the pseudo chips and the effect on the average temperature of each chip. This temperature was obtained by averaging the temperature from 11 well distributed sensors on each chip. It can be observed that there was a maximum temperature variation of 1.5 °C, which can be considered to be acceptable when compared to the temperature gradient along the chip for on-chip single-phase cooling using water (about 2-3 K [16-18]).

Figure 8 shows the controller's reaction under the situation of a disturbance. It can be seen that the SMV controller was able to maintain the exit vapor quality to within ±5% of the set point. What is important to observe is that the controller was effective, i.e. it showed fast response for the induced disturbance and no instability was observed.

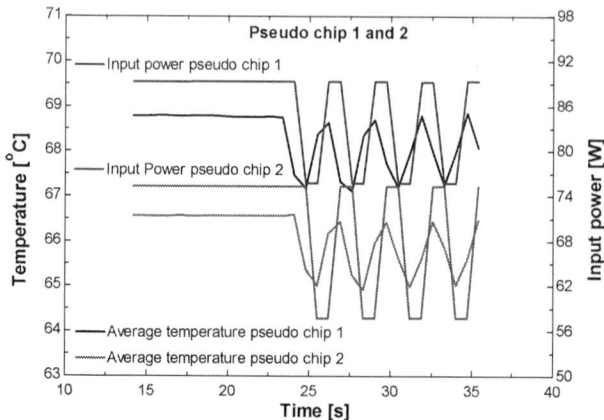

**Figure 7:** Heat load disturbance and pseudo chip temperatures.

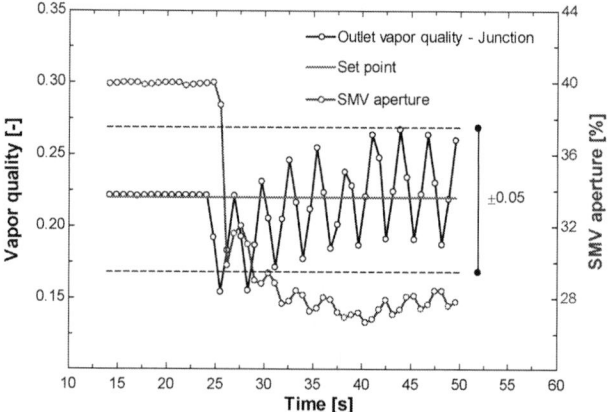

**Figure 8:** Outlet vapor quality and SMV controller.

Finally, it can be highlighted that the control strategies adopted (SISO, dual SISO and SIMO) proved to be a simple and effective way of controlling the specific variables while maintaining the pseudo chips within a safe operating range. The coupling effect between the controllable variables was not strong, in other words the controllers have low interaction effects, implying that it was not necessary to apply a more complex centralized MIMO controller [12].

### 3.3. Energy analysis

To compare the performance of the liquid pumped and vapor compression cooling systems, which were experimentally evaluated and analysed beforehand, a steady state condition was selected from the flow distribution tests. Such a comparison mainly evaluates the difference between the power consumption of the drivers and the available energy and exergy in the condenser.

Table 1 shows the results for the driver's power consumption and overall efficiency, the latter calculated as the ratio between the isentropic pumping or compression and the eletrical input power. It also shows the two systems' input and output energies associated with components and piping and the thermodynamic conditions in the condenser for the main and secondary working fluids. The experimental condition selected for the comparison was that the input powers on pseudo chips 1 and 2 were 90 W (41.7 W/cm$^2$) and 75 W (34.7 W/cm$^2$), respectively. For the VC cycle, a post heater (about 125 W) was considered, which simulated the heat load of auxiliary electronics of servers (memories, DC/DC converters etc.).

| Energy in | LP cycle | VC cycle |
|---|---|---|
| Pump or compressor input power, W | 17.42 | 102.12 |
| Isentropic pumping or compression power, W | 0.048 | 27.77 |
| Driver overall efficiency, % | 0.28 | 27.19 |
| Input power on the pseudo chips, W | 164.47 | 164.51 |
| Input power on the post heater, W | 0 | 125.63 |
| Energy out | LP cycle | VC cycle |
| Heat transfer in the condenser, W | 68.32 | 194.23 |
| Heat loss in the driver, W | 17.37 | 74.35 |
| Heat loss in the piping, W | 96.68 | 124.63 |
| Energy recovery efficiency, % | 37.5 | 49.4 |
| Thermodynamic conditions in the condenser | | |
| Condensing temperature, °C | 59.96 | 80.48 |
| Outlet water temperature, °C | 49.32 | 65.04 |
| Mass flow rate of water, kg/h | 6.28 | 3.31 |

**Table 1:** Energy in and out in the systems and thermodynamic conditions in the condenser.

The results show a higher driver input power for the VC system, about 6 times, which naturally is associated with the energy expended to maintain the difference of pressure between the condenser and micro-evaporator. It is worth observing the drivers' low overall efficiency, which for the pump is mainly a consequence of leakage and slip of HFC134a in the gears. Such a characteristic is due to the low viscosity of the working fluid, being at the lower limit for the specified pump (hence a better pump would be advisable).

Regarding the mini-compressor, despite the high overall efficiency, it is actually considered to be low, especially when compared with conventional household compressors, which have values normally between 50% and 70% [19, 20]. Such a low efficiency is potentially associated with the fact that the mini-compressor is operating at much higher suction/discharge pressures than its actual design conditions.

If one compares the results with a hypothetical air cooling system, considering a COP of 1.22 (45% of the total energy consumption for air cooling system [5]), the energy consumption would be 134.81 W vs. 17.42 W when compared with LP cycle and 237.82 W vs 102.12 W when compared with VC cycle. This represents a reduction of about 87.1% and 57.1% in energy consumption, respectively. The differences in air cooling system energy consumption are due to the input power on the post heater, which was only considered for the VC cycle (*viz.* Table 1).

It can also be seen that 50.6% and 62.5% of the energy out of the VC and LP systems, respectively, are associated with heat losses. It shows that improvements can be done to

improve the overall performance of the system, which would mainly be associated with the reduction of the driver and piping losses and, consequently, to increase the energy recovered in the condenser.

Finally, the results showed a much higher temperature for the secondary fluid at the outlet of the condenser when using the VC system, which is related to the higher condensing temperature. This implies that a higher economic value is obtained for the energy available in the condenser.

### 3.4. Exergy analysis

An exergetic evaluation was developed in both cooling systems considering the experimental results shown in Table 1. The steady state exergy rate balance is defined by Equation 1. The first and second terms in the right side of the equality represent the exergy transfer accompanying heat and work, the third and fourth are the time rate of exergy transfer accompanying mass flow and flow work and, finally, the last term is the rate of exergy destroyed. With some mathematical manipulation it is possible to prove that the last term in Equation 1 [21], i.e. $\dot{E}_d$ is the entropy generation multiplied by the dead state temperature $T_0$.

$$0 = \underbrace{\sum_j \left(1 - \frac{T_0}{T_j}\right)\dot{Q}_j - \dot{W}_{cv} + \sum_i \dot{m}_i e_{fi} - \sum_e \dot{m}_e e_{fe}}_{\text{rate of exergy transfer}} \underbrace{- \dot{E}_d}_{\substack{\text{rate of} \\ \text{exergy} \\ \text{destruction}}} \quad (1)$$

It can be observed that it is necessary that an exergy reference environment be defined. Such an environment represents the state of equilibrium or dead state. This equilibrium state defines the exergy as the maximum theoretical work obtainable when another system in a non-equilibrium state interacts with the environment to the equilibrium. For the present work the reference is defined as 295 K, 100 kPa for water and 295 K, 603.28 kPa and 50% of vapor quality for HFC134a.

The goal of the analysis is to determine, for each system, the exergy supplied, recovered and destroyed for a control volume enclosing the cooling system. With this, the overall exergetic efficiency, defined as the ratio between the recovered and supplied exergies, can be determined. The exergetic efficiency of each component is also evaluated. Table 1 from the previous section and Table 2 show the results obtained regarding energy and exergy, respectively.

Firstly, the total exergy recovered is higher for the VC cooling system, which is a consequence of the higher exergy supplied by the driver and the exergy of the post heater, where the latter was not being considered in the LP system's experimental tests. However, it is highlighted that this high exergy is the subject of interest of the owner of a secondary application of the recovered heat.

Regarding the exergetic efficiency of the components considered in the cooling systems, the driver followed by the condenser showed the lowest values, which implies that to improve the thermodynamic performance of the cooling systems such a components must be optimized in the design. Special attention must also be given to the exergy destroyed in the piping, which represents about 28% and 16% of the overall exergy destroyed in the LP and VC systems, the latter being the same order of magnitude as that in the condenser. This implies that a better insulation of the test unit is required to minimize the heat losses, i.e. exergy lost or destroyed.

| | LP cycle | VC cycle |
|---|---|---|
| Exergy supplied, W | 40.1 | 146.7 |
| | | |
| Exergy destroyed or irreversibility, W | | |
| | | |
| Pump or compressor | 17.4 | 74.4 |
| Condenser | 3.5 | 21.4 |
| ME1 | 3.4 | 3.6 |
| ME2 | 0.9 | 0.8 |
| Post heater | ---- | 9.0 |
| iHEx | ---- | 1.3 |
| LPR | ---- | 0.9 |
| SMV | 0.27 | 3.3 |
| Piping | 9.7 | 21.3 |
| Total | 35.2 | 136.0 |
| Exergy recovered, W | 4.9 | 10.7 |
| | | |
| Exergetic efficiency, % | | |
| | | |
| Pump or compressor | 0.03 | 27.2 |
| Condenser | 58.4 | 33.6 |
| ME1 | 72.6 | 70.7 |
| ME2 | 90.7 | 91.1 |
| Post heater | ---- | 59.3 |
| iHEx | ---- | 78.8 |
| LPR | ---- | 72.7 |
| Overall | 12.3 | 7.3 |

**Table 2:** Exergetic analysis for the VC and LP cooling systems.

It can also be observed that the overall exergetic efficiency was lower for the VC cooling system, with the compressor, condenser and piping being the main culprits. The overall exergetic efficiency also shows that there is a huge need to improve the thermodynamic performance of the cooling systems, since only an average of 10% of the supplied exergy is recovered.

The results and analyses above may lead one to conclude that the LP system is better in terms of exergy and energy. However, such a conclusion is not fair, especially when looking for the potential to improve the components' exergetic efficiency and to reduce the piping's exergy destroyed. It seems both systems can be optimized, i.e. better designed so that improvements will be generated, since the present setups were the first of a kind. It is also important to mention that the results shown here represent only the initial step in a much larger experimental campaign and this more extensive experimental campaign will be used to generalize the results.

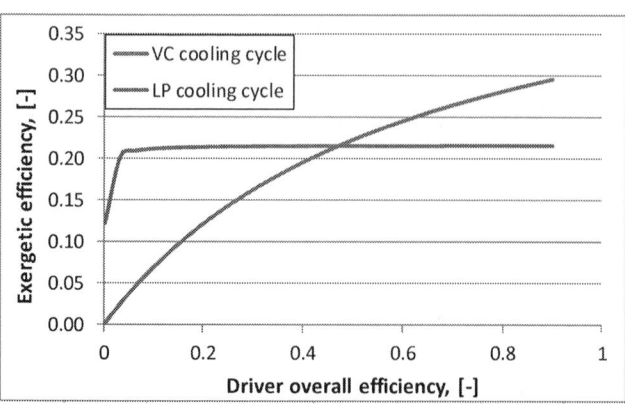

**Figure 9:** Exergetic efficiency versus driver overall efficiency.

Finally, to consider the effect of an improvement in the drivers' overall efficiency (the worst component in terms of exergetic efficiency) on the overall exergetic efficiency of the systems, a thermodynamic simulation was developed considering as inputs the experimental results used in the previous analysis. Additionally, the exergy destroyed by the piping of the VC cycle was considered to be the same as that of the LP cycle, i.e. analogous to an improvement in the insulation material so that both cycles have the same irreversibilities in the piping. Figure 9 shows the results, where it can be seen that the effect of overall driver efficiency on the exergetic efficiency is much greater when using a VC as a driver. There is also a point where the exergetic efficiency of the VC system surpasses that of the LP system (at about 46%). From an exergetic point of view, only after this point does the VC cooling system become competitive with the LP cooling system. It is also important to remember that the other exergy losses must be considered and the matching point of exergetic efficiency in Figure 9 can be changed to higher or lower values in that case.

## 4. Conclusions

Two specific on-chip two-phase cooling cycles described by [10] were built in the LTCM lab and experimentally tested and evaluated. The cycles were differentiated by their drivers, i.e. the first was driven by a liquid pump and the second by a mini-compressor. Aspects such as energy consumption, exergetic efficiency and controllability were investigated.

The controllers designed were evaluated by tracking and disturbance rejection tests, which were shown to be efficient and effective. The average temperatures of the pseudo chips were maintained below the limit of 85 °C for all tests evaluated in steady state and transient conditions. In general, simple SISO strategies were sufficient to attain the requirements of control, i.e. more complex MIMO strategies were not necessary for this application.

Regarding energy and exergy analyses, the experimental results showed that the energy consumption for the VC system was 6 times higher than that of the LP system. However, the appeal of this cycle is in the energy recovery, i.e. higher obtainable condensing temperatures (80 °C - 90 °C), which increases the exergy and economic value of the energy recovered. Second, both systems can be thermodynamically improved since only about 10% of the exergy supplied is in fact recovered in the condenser in the present setup. Finally, the results presented were not generalized since only a limited number of tests were done and a more detailed experimental campaign is necessary to better describe and compare both systems presented.

## Acknowledgments

The Swiss Commission for Technology and Innovation (CTI) contract number 6862.2 DCS-NM entitled "Micro-Evaporation Cooling System for High Performance Micro-Processors: Development of Prototype Units and Performance Testing" directed by the LTCM laboratory sponsored this work along with the project's industrial partners: IBM Zürich Research Laboratory (Switzerland) and Embraco (Brazil). J.B. Marcinichen wishes to thank CAPES ("Coordenação de Aperfeiçoamento de Pessoal de Nível Superior") for a one year fellowship to work at the LTCM laboratory. The authors also wish to thank Vinícius de Oliveira from the Automatic Control Laboratory (EPFL - Lausanne - Switzerland) for support in the development of the control tools.

## References

1. Olivier, J.A., Marcinichen, J.B., and Thome, J.R. *Two-phase Cooling of Datacenters: Reduction in Energy Costs and Improved Efficiencies.* in *13th Brazilian Congress of Thermal Sciences and Engineering – ENCIT2010.* 2010. Uberlandia, MG, Brazil.
2. Agostini, B., Fabbri, M., Park, J., Wojtan, L., and Thome, J.R., *State-of-the-art of High Heat Flux Cooling Technologies.* Heat Transfer Engineering, 2007. Vol. **28**: pp. 258-281.
3. EPA, *Report to Congress on Server and Data Center Energy Efficiency Public Law 109-431.* 2007, U.S. Environmental Protection Agency.
4. Larson, J.B., *America's Energy Security Trust Fund Act of 2009.* H.R. 1337, 2009.
5. Koomey, J.G., *Estimating Regional Power Consumption by Servers: A Technical Note.* 2007, Lawrence Berkeley National Laboratory: Oakland, CA.
6. Leonard, P.L. and Phillips, A.L. *The Thermal Bus Opportunity – A Quantum Leap in Data Center Cooling Potential* in *ASHRAE Transactions.* 2005. Denver, CO.
7. Olivier, J.A., Marcinichen, J.B., Bruch, A., and Thome, J.R., *Green Cooling of High Performance Micro Processors: Parametric Study between Flow Boiling and Water Cooling.* International Journal of Thermal Sciences and Engineering Application, 2011. Vol. **3**: pp. 041003.1-041003.12.
8. Marcinichen, J.B., Olivier, J.A., and Thome, J.R., *Reasons to Use Two-phase Refrigerant Cooling,* in *Electronics Cooling.* March 2011.
9. Marcinichen, J.B. and Thome, J.R. *Refrigerated Cooling of Microprocessors with Micro-Evaporation New Novel Two-Phase Cooling Cycles: A Green Steady-State Simulation Code.* in *13th Brazilian Congress of Thermal Sciences and Engineering - Encit 2010.* 2010. Uberlândia, MG, Brazil.
10. Marcinichen, J.B., Thome, J.R., and Michel, B., *Cooling of Microprocessors with Micro-Evaporation: A Novel Two-Phase Cooling Cycle.* International Journal of Refrigeration, 2010. Vol. **33**(7): pp. 1264-1276.
11. Gosney, W.B., ed. *Principles of Refrigeration.* 1st ed. 1982, Cambridge University Press.
12. Marcinichen, J.B., Olivier, J.A., Oliveira, V., and Thome, J.R., *On-Chip Micro-Evaporation: Experimental Evaluation of Liquid Pumping and Vapor Compression Driven Cooling Systems and Control.* International Journal of Applied Energy, 2011. Vol. **92**: pp. 147-161.
13. Costa-Patry, E., Olivier, J.A., Nichita, B.A., Michel, B., and Thome, J.R., *Two-phase flow of refrigerants in 85µm-wide multi-microchannels: Part I – Pressure drop.* International Journal of Heat and Fluid Flow, 2011. Vol. **32**(2): pp. 451-463.
14. Costa-Patry, E., Olivier, J.A., Michel, B., and Thome, J.R., *Two-phase flow of refrigerants in 85µm-wide multi-microchannels: Part II – Heat transfer with 35 local heaters.* International Journal of Heat and Fluid Flow, 2011. Vol. **32**(2): pp. 464-476.
15. Costa-Patry, E., Olivier, J.A., and Thome, J.R., *Hot-spot effects on two-phase flow of R245fa in 85µm-wide multi-microchannels.,* in *16th International Workshop on Thermal Investigations of IC's and Systems.* 2010: Barcelona, Spain.
16. Ganapati, P. *Water-Cooled Supercomputer Doubles as Dorm Space Heater.* 2009; Available from: http://www.wired.com/gadgetlab/2009/06/ibm-supercomputer/
17. Meijer, G.I., Brunschwiler, T., and Michel, B., *Using Waste Heat from Datacenters to Minimize Carbon Dioxide Emissions,* in *ERCIM News.* 2009. p. 23-24.
18. Brunschwiler, T., Meijer, G.I., Paredes, S., Escher, W., and Michel, B. *Direct Wast Heat Utilization from Liquid-Cooled Supercomputers.* in *14th Int. Heat Transfer Conference.* 2010. Aug. 8-13, Washington, DC, USA.
19. Hermes, C.J.L., Melo, C., Knabben, F.T., and Gonçalves, J.M., *Prediction of the Energy Consumption of Household Refrigerators and Freezers via Steady-State Simulation.* International Journal of Applied Energy, 2009. Vol. **86**: pp. 1311-1319.
20. Gonçalves, J.M., Melo, C., Hermes, C.J.L., and Barbosa, J.R., *Experimental Mapping of the Thermodynamic Losses in Vapor Compression Refrigeration Systems.* Journal of the Brazilian Society of Mechanical Sciences and Engineering, 2011. Vol. **33**(2): pp. 159-165.
21. Moran, M.J., Howard, I., and Shapiro, N., eds. *Fundamentals of Engineering Thermodynamics.* 6th ed. 2010, John Wiley & Sons. 725 pages.

# Numerical Prediction of the Junction-to-Fluid Thermal Resistance of a 2-Phase Immersion-Cooled IBM Dual Core POWER6 Processor

Levi Campbell
IBM Corporation
levic@us.ibm.com
Poughkeepsie, NY, USA

Phillip Tuma
3M Company
petuma@mmm.com
St. Paul, MN, USA

## Abstract

The numerical model used in development of the CPU cold plates for the water-cooled IBM p575 supercomputer is used in this work to predict the junction-to-fluid performance capabilities of passive 2-phase immersion cooling for the same p575 chip module. Experimentally-determined boiling heat transfer coefficients for a porous copper boiling enhancement coating (BEC) were used as a convective boundary condition applied atop the lid in place of the cold plate and secondary thermal interface. The BEC produces 75% and 1500% increases, respectively, in the critical heat flux (CHF) and peak heat transfer coefficients relative to a smooth surface. Lid thicknesses, 3.75mm<t<10mm, were modeled at the peak module power of $Q_m$=158W. A thickness of t=3.75mm eliminated regional dryout of the BEC and yielded the optimal sink-to-fluid thermal resistance based on the lid temperature over the centerline of the chip of $R_{sf}$=0.073°C/W, a value consistent with previous measurements based on electric heaters similar in size to the P6 core. The resultant average junction-to-fluid thermal resistance was $R_{jf}$=0.174°C/W, ~10% lower than the junction-to-water inlet resistance, $R_{jw,i}$ previously modeled for a single water-cooled cold plate used in the production p575.

Immersion system level performance was estimated by assuming that 50% of the volume used for heat sinks in an air-cooled version of the p575 node was available for condensation. The analysis showed roughly equivalent performance to the water-cooled node if the same isolated rack water is used to condense the vapor. If facility water is instead used to condense the vapor directly and at the rack scale, pumps and much of the cooling hardware could be eliminated and the facility water temperature could be raised.

## Keywords

immersion, cooling, datacenter, POWER6, p575, boiling, enhancement

## Nomenclature

| | |
|---|---|
| a | constant |
| A | area [cm$^2$] |
| C | specific heat [J/kg-K] |
| H | heat transfer coefficient [W/cm$^2$-K] |
| I | current [Amperes] |
| ṁ | mass flow rate [kg/s] |
| Q | power [W] |
| Q″ | heat flux [W/cm$^2$] |
| R | thermal resistance [°C/W] |
| R″ | thermal resistivity [°C-mm$^2$/W] |
| t | thickness [mm] |
| T | temperature [°C] |
| V | voltage [V] or volume [cm$^3$] |

## Subscripts

| | |
|---|---|
| a | ambient |
| atm | atmospheric pressure |
| ave | average |
| b | boiling point |
| c | chip |
| CL | centerline of chip |
| cond | condenser |
| f | fluid (2-phase) |
| FW | facility water |
| inc | incipience |
| j | junction or junction layer |
| i | inlet to cold plate or condenser |
| LMTD | log mean temperature difference |
| m | CPU module |
| max | maximum |
| o | outlet |
| s | sink or boiler |
| sat | saturation |
| steady | steady state |
| w | water |

## 1. Introduction

There is growing interest in liquid cooling not only as a means of accommodating higher chip- and system-level power densities but for increasing energy efficiency and making heat available for other purposes. Various approaches are being explored but those that bring the liquid directly to the microprocessor through cold plates or evaporators enable very tight thermal coupling of junction and liquid. The power density in the IBM P7 IH supercomputer reached an unprecedented 250kW per oversize rack with water cooling [1]. In IBM's Aquasar system, tight coupling between CPU junction and the water allows the water coolant temperature to rise to 60°C [2] making it more economical to reject waste heat and more practical to capture and utilize it [3].

This level of performance is not without costs. Cold plates, manifolds, pumps, hoses, heat exchangers, couplings and other components add cost and take precious space within the compute nodes. Controlling the flow and mitigating the loss of coolant through this myriad of components within a node, rack or facility is an engineering challenge exacerbated by the number and variety of devices on each "hot swappable" node. This may be one reason that direct liquid cooling has so far been relegated to the world of supercomputers. The cost and complexity barriers, even in large scale production, appear too high for much of the datacom industry.

Immersion cooling is a possible means of overcoming these barriers. By eliminating the need for node level hardware, immersion cooling systems can, in principle, be much simpler. Companies like Green Revolution Cooling, Hardcore Computer and Iceotope are promoting immersion-cooled systems that employ single-phase passive and pumped immersion. However, the liquids they use, like all dielectric liquids, have much lower specific heat and thermal conductivity than water. The thermal conductivity of a typical fluorochemical fluid, for example, is about 1/9th that of water and the specific heat is about 1/4th that of water. This makes it challenging to produce a liquid phase system that is not inferior in some sense relative to a water-cooled equivalent.

## 1.1. Passive 2-phase Immersion

Dielectric liquids that are sufficiently volatile can transfer heat more effectively by boiling and condensation. Use of passive 2-phase immersion for computing equipment is largely limited to IBM's exploration of the technology for cooling bipolar chips in the 1970s. The liquid encapsulation module or LEM [4], [5], for example, was a $10 \times 10$ array of $4.6 \times 4.6$mm chips immersed in $C_6F_{14}$ liquid that boiled on the bare silicon.

A vulnerability of this technique is the phenomena of incipience overshoot, a large temperature excursion before the inception of boiling that can allow a chip to overheat or stress it mechanically during the sudden temperature drop that follows. This issue was overcome by modifying the silicon surface with sandblasting followed by an aqueous KOH treatment [6]. It was also observed that fluid-borne contaminants could distill out of the fluid onto or under the chip. Under-filling with beeswax kept contaminants away from the C-4 solder bump connections [7].

Boiling $C_6F_{14}$ from silicon produces heat transfer coefficients $\sim 0.6$W/cm$^2$-K resulting in superheats, $T_c-T_f$, exceeding 20°C. LEM chips could dissipate up to 4W and typically operated quite close to the critical heat flux (CHF) of the fluid, a fact that limited the potential of the technology for future generations of higher power chips. The idea of attaching "pegs" or fins to a chip's substrate to spread heat was known [8]. If such fins could be applied to the *silicon* surface and enhanced with Linde's (now Honeywall/UOP's) High Flux™ [9] porous metallic surface coating, which was commercially available at the time, $T_c-T_f$ might have been dramatically reduced and risk of chip damage resulting from incipience overshoot would have been eliminated. However, as one author noted [10] "it is unlikely that chips can be altered to accommodate such surface treatments." There was fear of yield loss and no proven method of attachment.

Though qualified as a shippable product, the LEM was abandoned in favor of thermal conduction module (TCM) technology which provided slightly better performance at the time and could be readily modified to accommodate substantially increased chip heat fluxes. In the years that followed, there were numerous publications [11] addressing the limitations of 2-phase immersion cooling. Much of this work seems predicated on the assumption that 2-phase immersion cooling should be applied to the bare silicon and should be an optimal method, owing to the elimination of a thermal interface, if only the requisite heat fluxes could be managed. Research on subcooling, forced flow and other techniques for enhancing passive 2-phase immersion subsided

with the advent of complementary metal oxide semiconductors (CMOS) which were easily air cooled. However, CMOS quickly evolved to produce system densities that challenged air cooling and chip heat fluxes that far exceeded the CHF of passive boiling. Cray employed 2-phase *spray* cooling as a means to extend the heat flux capability of bare die cooling [12] but there has been little work since on passive techniques.

In the years since IBM's abandonment of the LEM, thermal interface technology has matured. Modern packages that incorporate heat spreaders with high performance interfaces are routine. It is in this context that the notion of attaching an enhanced copper "peg" to a bare silicon surface is being revisited with the peg replaced by a modern lid. By applying a boiling enhancement coating similar to the High Flux surface atop an optimized lid, virtually any chip power can be accommodated, passively, with chip-to-fluid resistances superior to most active techniques, even spray cooling. New approaches to passive 2-phase immersion promise to simplify its adoption [13], [14] and research has shown impressive power density and energy efficiency capabilities [15].

## 1.2. The IBM p575 Supercomputer

The 2008 p575 supercomputer represents a return to IBM's water cooling roots [16] and a good case study for immersion cooling for several reasons. The machined cold plates in the p575 were highly optimized and the water cooling system achieved a rack level power of about 60kW, approximately the practical limitation for forced air cooling if one uses intercoolers as in Cray's ECOphlex system [17]. It therefore represents, in some sense, the entry point for liquid cooling and minimum level of hardware complexity for a practical high performance water cooling system of this scale. Second, the core power density in the P6 chipsets is approximately 80W/cm$^2$. These are more challenging to cool by immersion than many conventional server chipsets. Lastly, a detailed numerical model of the package already existed.

FIGURE 1 shows an overview of the p575 water cooling system. Each hot-swappable compute node or chassis contains 4 parallel circuits of 4 p575 processor modules, each cooled by a dedicated cold plate, resulting in a total CPU heat load of ~2500W. An isolated rack water loop is used so that the water chemistry, pressure and temperature can be controlled to minimize risk of corrosion, leakage and moisture condensation. Heat is transferred to facility water via redundant coolant distribution units (CDUs) within each rack.

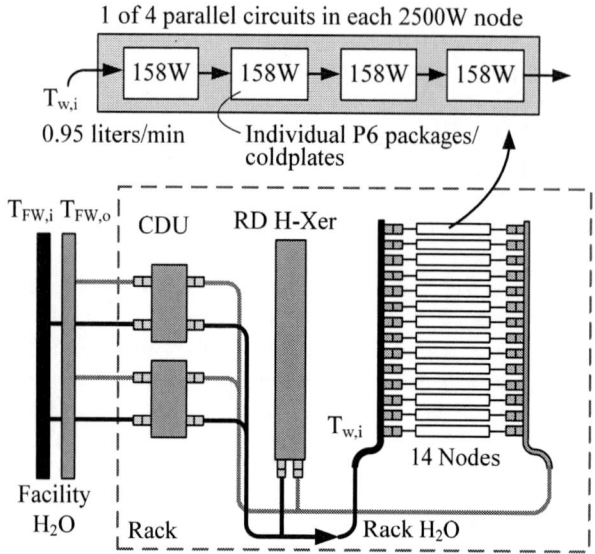

FIGURE 1 - IBM p575 supercomputer water cooling layout.

## 2. Boiling Enhancement Coating (BEC)

### 2.1. Background and Motivation

Various techniques can be used to enhance boiling heat transfer of candidate immersion working fluids like perfluorocarbons, hydrofluoroethers (HFEs) [18], and fluoroketones [19]. Extended surfaces metallic or graphite fins or foams function primarily to spread heat to a lower heat flux thereby reducing the wetted surface superheat. The distinction between extended surfaces and surface coatings can be quite arbitrary. For the purposes of this work, a coating is a porous enhancement with geometric features on the order of the dimensions of a nucleation site, typically 100 micron or less.

Coatings can be organic, in which case the thermal conductivity of the coating is low. An optimal organic coating will be thin and functions primarily to enhance nucleation resulting in significant increases in the sustainable heat flux and heat transfer coefficient. Similarly thin coatings made entirely of high conductivity metals like copper and aluminum produce heat transfer coefficients similar to those of thin organic coatings. At optimum thickness, however, metallic coatings can produce heat transfer coefficients, $H>10$ $W/cm^2$-K at $Q''>30$ $W/cm^2$, performance unmatched by the aforementioned technologies. Webb [20] reviewed several such coatings and later applied one within a thermosyphon [21]. As mentioned previously, such coatings were first developed and patented in the 1970s are used routinely today in distillation towers and reboilers. The most common of these make use of metallic particles bound in a porous matrix to a boiling substrate. Many technologies such as plating [22], [23], sintering, brazing [24], [25] and flame or plasma spraying [26], [27], [28], [29] can be used to make such coatings. However, most suffer inherent limitations that make it challenging to consistently produce coatings on small parts with the proper porosity at the proper length scales (5-10 micron) required to produce optimal heat transfer coefficients.

The metallic BEC that is the subject of this work was developed by 3M for use in the evaporators of discrete thermosyphons such as the one shown in FIGURE 2 that are applied, like a conventional heatsink, to a microprocessor within a desktop computer or rack mounted server. Consistency, cost of manufacturing and performance of a BEC are of utmost importance. Only with a highly optimized, low cost boiling surface can a thermosyphon compete with commodity heat pipe heatsinks or even pumped water cooling systems. Such a coating, if applied to the lid of a modern CPU package, is also ideal for immersion cooling.

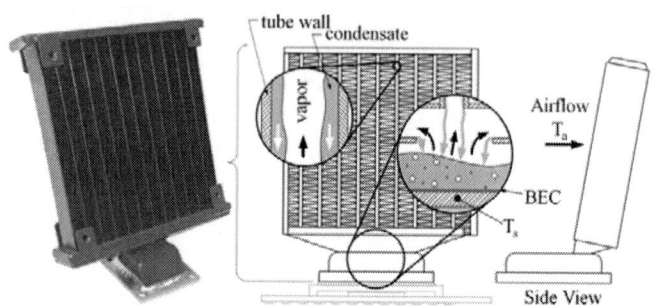

FIGURE 2 - 120×120mm BEC-based thermosyphon CPU cooler produces $R_{sa}<0.08°C/W$ at $1.8m^3/min$ airflow.

### 2.2. BEC Application

The 3M BEC [30], [31] was empirically optimized for HFE fluids starting with optimal values of pore size, porosity and thickness as described by Tehver in his study of plasma-sprayed coatings with chlorofluorocarbon (CFC) 113 fluid [32]. It is made from Chemcopp 1700 FPM copper powder from American Chemet Corporation, a -325 mesh spherical copper powder (mean diameter 10.2 microns) that is plasma coated using a proprietary process with ~0.5wt% silver. The resultant powder is knife coated dry, screen printed or sprayed onto a copper substrate. For screen printing, the powder is mixed with ~13% Dow Corning 704 Diffusion Pump Fluid to create a screen printable mixture that is applied with a Sefar 45-180 mesh polyester screen to the flat copper substrate. The same oil is used for spray coating in a 2-part process useful for 3D surfaces such as pin fin arrays. Dry powder and oil are alternately applied until the proper thickness is achieved by "feel." Screen printing is preferred because substrates are less expensive and the process consistently reproduces the ideal 150 micron thickness after firing.

Coated parts are fired in a 0.01 milliTorr vacuum furnace using the time temperature profile shown in FIGURE 3 to produce the typical coating shown. The ramp and dwell times were largely dictated by the capabilities of the oven and could likely be shortened. It is also possible to substitute a reducing atmosphere for vacuum as a means of suppressing oxidation.

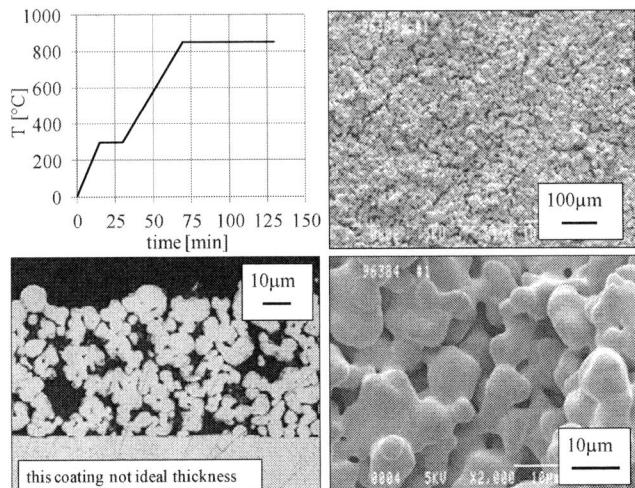

FIGURE 3 - Temperature profile for fusing the BEC and the coating that results.

FIGURE 4 - Standard BEC test disk and schematic of the functional elements of the system used to test disks.

### 2.3. BEC Testing

During the development of the BEC, it was necessary to test hundreds of boiling surfaces. It was impractical to methodically mount thermocouples in each, incorporate them into a thermosyphon and then charge and test each one. Most of the pool boiling apparatus used in academic research comprise large tanks of fluid that do not simulate the application conditions and have to be painstakingly degassed and preheated. A method was needed for quickly swapping and testing surfaces under reproducible conditions, representative of the intended application. A standard test substrate or boiler disk was created. BEC candidates were applied to the central 25mm region (5.07 cm$^2$) on one side (wetted) of a 5cm diameter, 3mm thick 100 series copper disk. The opposite (heated) side of each was lapped and contained a 1x2mm thermocouple groove as shown in FIGURE 4.

The test apparatus (essential features in FIGURE 4) comprises a phenolic platform containing a 25mm diameter copper heater atop 4 thin radial ribs. A thermocouple probe integrated into the platform above the heater is placed so that a greased BEC disk can be placed onto the probe and atop the heater. This probe is bent in such a way that when the disk is locked into the proper x-y position, the probe is gently pressed upward and into the termination of the thermocouple groove to measure the sink temperature $T_s$. The platform moves on z-axis sliders with a lever and spring than engages the BEC disk to a gasketed glass tube into which another thermocouple protrudes to measure $T_f$, the fluid saturation temperature.

Approximately 10cc of fluid is added through a fill port at the top of the apparatus. $C_3F_7OCH_3$[1] was chosen as the standard test fluid in part because its atmospheric pressure boiling point, $T_b$=34°C, minimized parasitic heat losses. Vapor is condensed in an air-cooled condenser and falls back to the pool. The condenser is open at the top so that $P=P_{atm}$ and

$$T_f = T_b = T_{sat}(P_{atm}) . \qquad (1)$$

A typical run begins with a 3 minute warm up at 100W (20 W/cm$^2$) intended to minimize conduction losses from the bottom of the copper heater during subsequent measurements. The power is then lowered to 50W (10W/cm$^2$) and allowed to equilibrate for 2 minutes at which time data are recorded before advancing 10W to the next data point. This continues until $T_s$ exceeds a preset limit, usually about $T_b$+20°C. The data acquisition system queries the DC power supply for the heater voltage, V, and current, I. The heat flux, $Q''$, and heat transfer coefficient, H, are defined as:

$$Q'' = \frac{Q}{A} = \frac{VI}{A} \qquad (2)$$

$$H = \frac{Q''}{(T_s - T_f)} \qquad (3)$$

Typical results for an optimized BEC are shown in FIGURE 5 with results for an uncoated surface. The BEC produces 75% and 1500% increases, respectively, in the critical heat flux (CHF) and peak heat transfer coefficients relative to a smooth surface. Also shown are incipience overshoot measurements,

$$\Delta T_{inc} = T_{s,max} - T_{s,steady} , \qquad (4)$$

where $T_{s,max}$ and $T_{s,steady}$ are the maximum and steady state (following incipience) sink temperatures recorded during a transient experiment in which the power, Q, was applied to a disk that had been resting 10 minutes since the previous measurement.

---

[1] Sold Commercially as 3M™ Novec™ 7000

978-1-4673-1110-6/12 $31.00 © 2012 IEEE

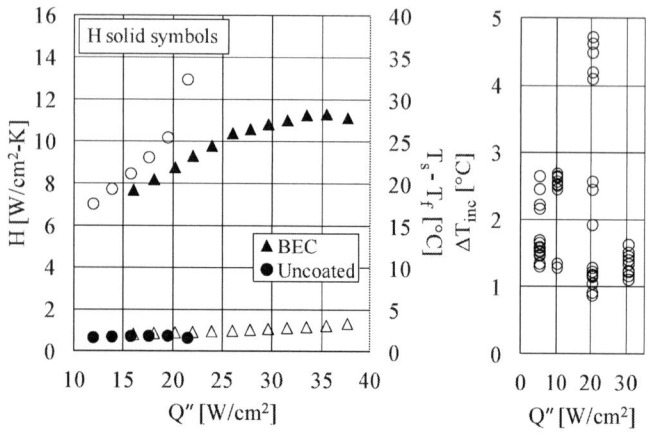

FIGURE 5 - Typical BEC heat transfer coefficient, superheat and incipience overshoot as a function of heat flux with $C_3F_7OCH_3$.

## 2.4. Microprocessor Cooling Predictions

In previous research [33], the BEC was applied to $36\times36\times4$mm copper heat spreaders (boilers) that were applied to heat sources of various surface areas, $A_c$. Boiling $C_3F_7OCH_3$ at pressures near atmospheric, the resultant sink-to-fluid thermal resistance,

$$R_{sf} = \frac{T_s - T_f}{Q}, \quad (5)$$

reached a minimum (FIGURE 6) at a point the author attributed to centerline dryout of the BEC. Numerical conduction modeling predicted this region of film boiling and its growth with subsequent increases in power resulting in increased thermal resistance. Though pin fins can be incorporated onto the boiler and were shown to eliminate this regional dryout throughout the power range of interest, such fins add cost and complexity; they are more difficult to coat with the BEC; and they offer little to no reduction in thermal resistance when compared with a boiler of sufficient thickness to avoid regional dryout.

Experimental $R_{sf}$ data can be regressed as shown in FIGURE 7 to obtain a correlation that is useful for approximating the junction-to-fluid thermal resistance, $R_{jf}$, of a thermosyphon- or immersion-cooled chip package that makes use of the BEC. If the resistivities of the silicon, $R''_{jc}$, and thermal interface, $R''_{cs}$, are known, their sum, $R''_{js}$ can be used, as shown in FIGURE 8, to predict $R_{jf}$ as

$$R_{jf} = R_{sf} + \frac{R''_{cs} + R''_{jc}}{A_c} \quad (6)$$

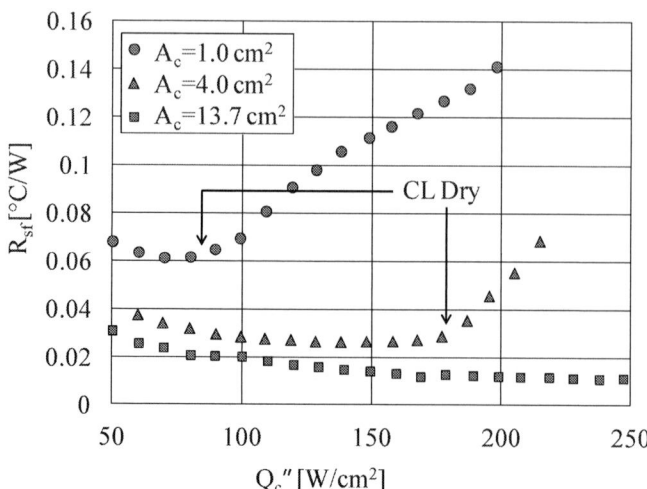

FIGURE 6 - Experimental $R_{sf}$ for 36x36x4mm boiler attached to heat sources of various areas $A_c$. $Q_c''$ is based on $A_c$. From [33].

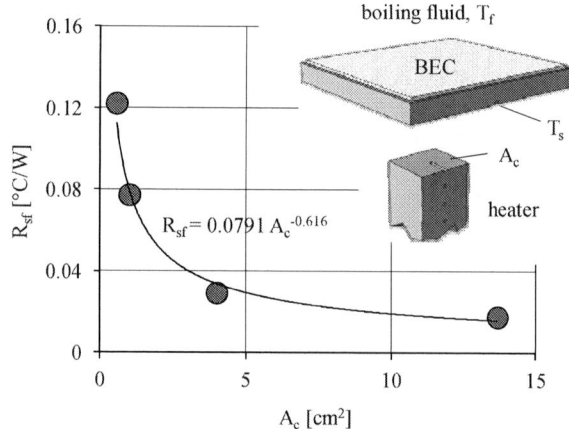

FIGURE 7 - Correlation of experimental data useful for predicting $R_{sf}$ for various die/core sizes.

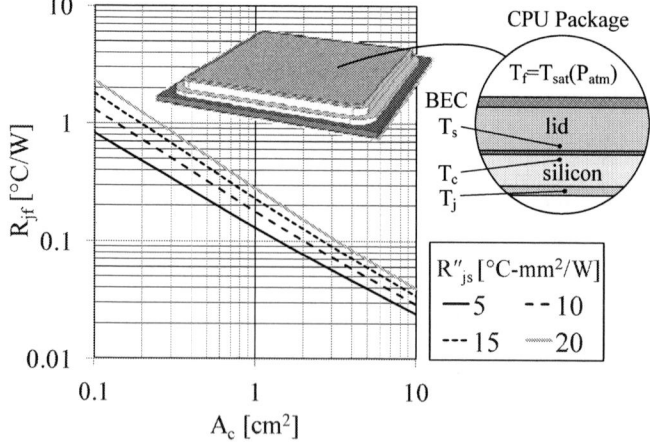

FIGURE 8 - Prediction of $R_{jf}$ based on correlated $R_{sf}$ and known silicon and interface resistivity.

## 2.5. Comparison to Direct-Die, Active Cooling

Though the aforementioned technique is passive and makes use of a thermal interface and heat spreader, thermal performance is superior to most direct-die techniques. Spray cooling, for example, generally produces heat transfer coefficients 1-3W/cm²-K, resulting in significantly higher chip-to-fluid resistances (FIGURE 9) than one can achieve with the passive technique even with a poor thermal interface.

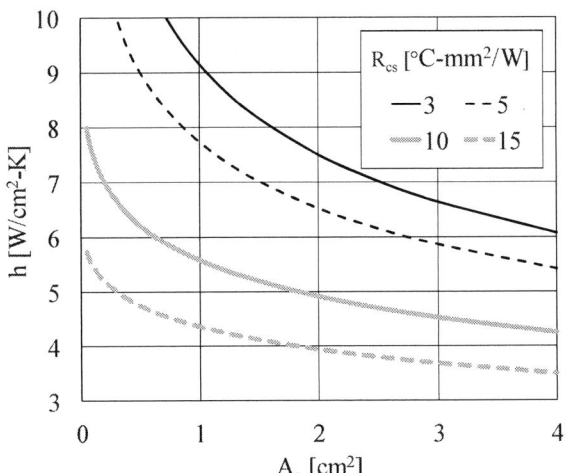

FIGURE 9 - Heat transfer coefficient that a direct-die technology must achieve to match passive 2-phase performance with a given thermal interface resistivity.

## 3. Numerical P6 Model

### 3.1. Production Water-Cooled Machine

The original conjugate CFD/conduction model [34] comprised the 150W dual-core POWER6 or p6 chip (FIGURE 10), 8W memory chip, lid, cold plate, and primary and secondary thermal interfaces (R''=13°C-mm²/W) as shown in FIGURE 11a. The p6 contains two 0.51cm² cores that together represent 54% of the chip's 150W power. The modeled thermal resistance based on the *total* p575 module power, $Q_m$=158W, is defined as

$$R_{jw,i} = \frac{T_{j,ave} - T_{w,i}}{Q_m} \tag{7}$$

where $T_{j,ave}$ is the average p6 *core* temperature and $T_{w,i}$ the cold plate water inlet temperature. $R_{jw,i}$ was found to be ~0.192°C/W at the nominal water flow rate of 0.95 liters/min. Since 4 modules are in series in each flow circuit the last module receives water that has warmed ~7° resulting in an additional advective resistance of 0.045°C/W and a maximum $R_{jw,i}$ ~0.24°C/W based on the water inlet temperature to the series. To maintain the specified junction temperature of $T_j$=65°C for the last p6 in a series, water entering the node, $T_{w,i}$, must be no warmer than 28°C.

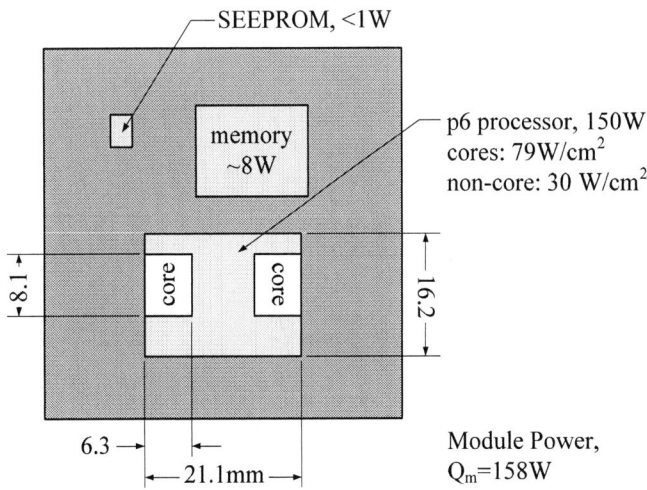

FIGURE 10 - Layout of the IBM p575 dual processor module.

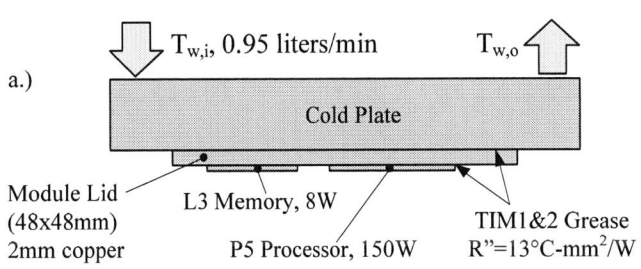

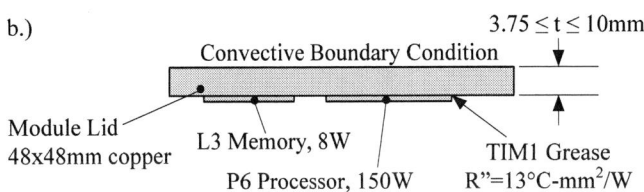

FIGURE 11 - Configurations modeled originally for water-cooled system (a) and in this work for the immersion-cooled system (b).

### 3.2. Immersion-Cooled

The same conduction model was used for the immersion cooling analysis with a convective boundary condition applied atop the lid in place of the cold plate and secondary thermal interface (FIGURE 11b). The experimentally determined BEC convective heat transfer coefficients shown in FIGURE 5 were regressed to express H as a piecewise continuous function of Q'', as illustrated in FIGURE 12. The regressions were utilized in a Fluent user-defined function (UDF) which, at each iteration of the model, assigned a heat transfer coefficient at each boundary face corresponding to the heat flux through the face in the previous iteration.

The p575 module was modeled, first with a uniform chip heat flux to determine the maximum heat load supported by $C_3F_7OCH_3$ in pool boiling with no physical module changes. The results are illustrated in FIGURE 13 and show that a thicker lid (the production lid is 2mm thick) is required to cool a 158W package.

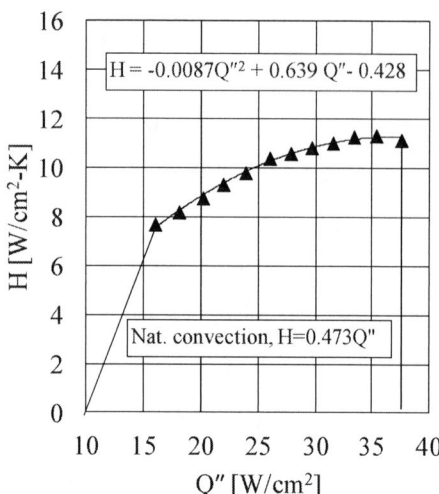

FIGURE 12 - Approximated surface heat transfer coefficients for $C_3F_7OCH_3$ on a horizontal surface with the BEC.

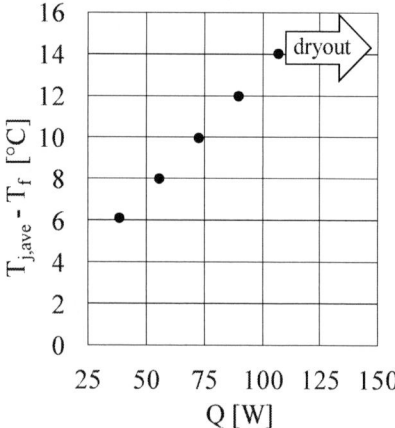

FIGURE 13 - Modeled $T_{j,ave}$ of a production p575 module with a uniform chip heat flux and a horizontal BEC-coated lid.

Lid thicknesses, $3.75<t<10$mm, were next modeled using the *actual* p6 power map at the peak module power of 158W to ascertain the minimum lid thickness required. The results of this analysis are shown in FIGURE 13 and show that the p575 module lid thickness must be at least 3.75mm, with the BEC boiling $C_3F_7OCH_3$, to avoid dryout. A lid thickness of t=3.75mm also corresponds to the optimal sink-to fluid thermal resistance,

$$R_{sf,CL} = \frac{T_{s,CL} - T_f}{Q_m}, \qquad (8)$$

of 0.073°C/W based on the lid temperature over the centerline of the processor chip (FIGURE 14), $T_{s,CL}$. This value is roughly consistent with projections shown in FIGURE 7 if one assumes that the two cores together behave roughly as a 1cm$^2$ heat source. The resultant average junction-to-fluid thermal resistance (based on the average p6 core temperature as in equation 7),

$$R_{jf} = \frac{T_{j,ave} - T_f}{Q_m}, \qquad (9)$$

was 0.174°C/W, meaning that fluid boiling at $T_f$=37°C could maintain the $T_j$=65°C specification.

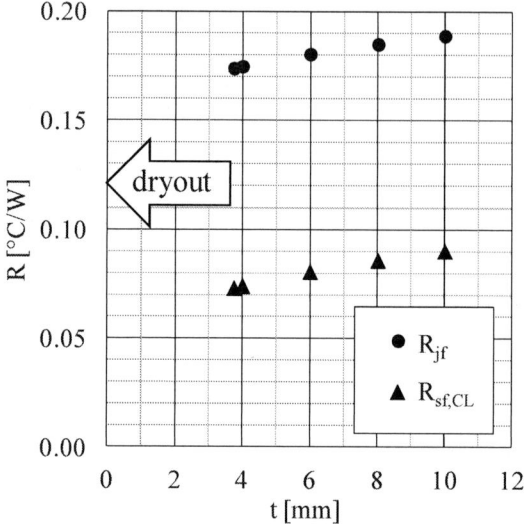

FIGURE 14 - Modeled thermal resistance of the p575 module at $Q_m$=158W as a function of lid thickness for pool boiling of $C_3F_7OCH_3$ on a horizontal BEC-coated lid.

## 4. Immersion System Level Projections

One of the primary advantages of immersion is the elimination of node level cooling hardware and the increased node level power density that results. The liberated volume can be used for condensation. In the analysis to follow, it is assumed that condensation takes place at the node level (2500W) using a condenser that occupies 2 liters or 50% of the 4 liter volume occupied by heat sinks in the air-cooled version of the p575 node.

### 4.1. Condenser Sizing

Previous research has shown that a high density, water-cooled condenser based on a tube-and-fin or enhanced tube technology can achieve a local volume-specific, fluid-to-water inlet resistance,

$$R_{fw} = \frac{V_{cond}(T_f - T_w)}{Q_{cond}}, \qquad (10)$$

of 1.5°C-cm$^3$/W. This number can be used in a log mean temperature difference (LMTD) analysis to predict condenser performance:

$$\Delta T_{LMTD} = \frac{Q_{cond}R_{fw}}{V_{cond}} \qquad (11)$$

$$\Delta T_w = T_{w,o} - T_{w,i} = \frac{Q_{cond}}{\dot{m}C_w} \qquad (12)$$

The resultant inlet water temperature, $T_{w,i}$, is

$$T_{w,i} = T_f + \frac{a\Delta T_w}{1-a} \qquad (13)$$

where

$$a = \exp\left(\frac{\Delta T_w}{\Delta T_{LMTD}}\right). \qquad (14)$$

FIGURE 15a shows results of such calculations for two cases. In the first, the water flow to the condenser is fixed at 3.8 liters/min and the condenser water inlet temperature, $T_{w,i}$, is fixed at 28°C. These are the nominal flow rate and inlet temperature for water supplied to a node in the water-cooled production machine. The aforementioned analysis then dictates $T_f$. $T_j$ is calculated from the modeled $R_{jf}=0.174$°C/W and 158W module power. Also shown is the water temperature that must be supplied to the condenser if $T_j$ is instead fixed at the 65°C design point ($T_f=37$°C). These data show that the immersion-cooled equivalent would produce the same $T_j$ with the same water flow and $T_{w,i}$ as the production machine.

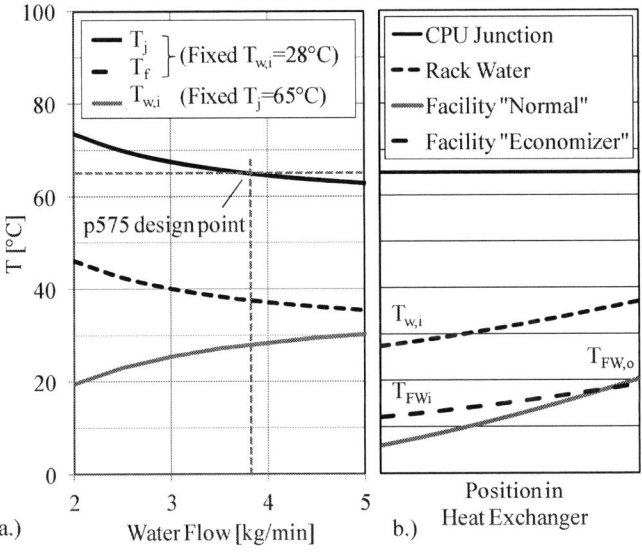

FIGURE 15 - Calculated junction, fluid and condenser water inlet temperatures for the immersion-cooled node (a) compared to production water-cooled system temperatures (b).

### 4.2. Comparison of Immersion and Production p575

The analysis so far has assumed that the condenser is fed by isolated rack water as shown in FIGURE 16a. As already mentioned, the resultant performance is roughly unchanged when compared with the water-cooled machine. Further, this configuration does little to ease complexity as it retains the CDU, manifolds and couplings of the original water-cooled design. While node level plumbing and numerous thermal interfaces can be eliminated, these are replaced by condensers and the challenges of maintaining a hermetic clamshell around each node.

FIGURE 15b shows the design rack and facility water temperatures for the production p575 rack in normal and economizer modes. There is a substantial difference between the facility and rack water temperatures, a difference dependent on the size of the CDUs. This difference can be eliminated in the case of an immersion-cooled system if facility water is flowed directly to the condenser, integrated not at the node, but at the rack level as show in FIGURE 16b. This would simplify the system design considerably by removing all node level cooling hardware.

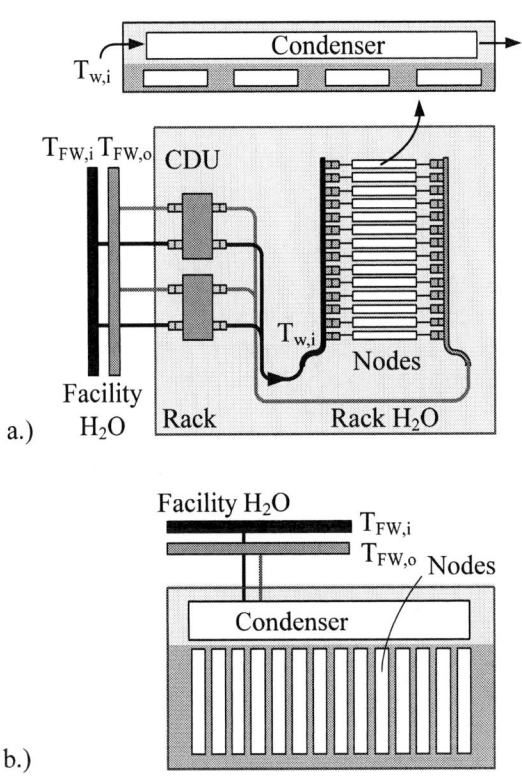

FIGURE 16 - Immersion-cooled system configurations based on an isolated condenser water loop a) and a condenser fed by facility water directly.

### 5. Conclusions

- At the optimal lid thickness of t=3.75mm, the modeled $R_{jf}$ of the immersion-cooled p575 module was 0.174°C/W, ~10% lower than the $R_{jw,i}$ enabled by the production cold plates. This difference rises to 26% if one compares to the 4th module in a series as it receives warmer water than the first.

- If rack water is used to condense the vapor being generated, the $T_j$ in a passive 2-phase immersion-cooled version of the IBM p575 supercomputer would likely be similar to those in the production water-cooled machine.

- If, however, facility water flowed directly to the condenser integrated at the rack scale, a temperature benefit roughly equal to the approach temperature difference between facility and rack water in the production machine could be realized. Such a configuration has the potential to simplify the system thermal design.

978-1-4673-1110-6/12 $31.00 © 2012 IEEE        43

## Acknowledgment

The authors thank Bob Simons for recounting the history of the IBM LEM project.

## References

1. Ellsworth, M.J., "The Water-Cooled Power7 IH Supercomputing Node/System," Presentation IMAPS Advanced Thermal Workshop, Palo Alto, Ca, USA, September 27-30, 2010.

2. Zimmermann, S., et al, "Experimental Investigation of a Hot Water Cooled Heat Sink for Efficient Cooling with High Exergetic Utility," Proc. 2nd European Conf. on Microfluidics, Toulouse, December 8-10, 2010.

3. Brunschwiler, T., et al, "Direct Waste Heat Utilization from Liquid-Cooled Supercomputers," IHTC14-23352, Proc. 14th Int. Heat Transfer Conf., Washington D.C, August 8-13, 2010.

4. Chu, R.C., et al, "Review of Cooling Technologies for Computing Products," IEEE Trans. Device and Material Reliability, 4(4), Dec. 2004.

5. Simons, R.E., "The Evolution of IBM High Performance Cooling Technology," Proc. 11th IEEE SEMI-THERM Symposium, San Jose, CA, Feb., 1995, pp. 102-114.

6. Reeber, M.D. and Frieser, R.G., "Heat Transfer of Modified Silicon Surfaces," IEEE Trans. Components, Hybrids and Manuf. Tech., Vol. CHMT-3(3), Sept. 1980.

7. Personal correspondence with Robert Simons, IBM, December 7, 2011.

8. Simons, R.E. and Moran, K.P., Immersion Cooling Systems for High Density Electronic Packages," IBM TR 00.2837, Feb. 23, 1977.

9. Brochure "High Flux™ Tubing," UOP (a division of Honeywell).

10. Bergles, A.E., Chu, R.C., Seely, J.H., "Survey of Heat Transfer Techniques Applied to Electronic Packages," Proc. NPECON West, Anaheim, CA, March 1977.

11. See for example, 3M Immersion Cooling Reprints Volumes 1-3, for 235 references from this era.

12. Pautsch, G., "Thermal Challenges in the Next Generation of Supercomputers," Presentation, CoolCon 2005, May 16-17.

13. Tuma, P.E., "The Merits of Open Bath Immersion Cooling of Datacom Equipment," Proc. 26th IEEE SEMI-THERM Symposium, Santa Clara, CA, Feb. 21-25, 2010.

14. Tuma, P.E., "A Comparison of Passive 2-phase Immersion and Pumped Water Cooling for Cooling Datacom Equipment," presentation IMAPs ATW on Thermal Management, Palo Alto, CA, USA, Nov. 7-9, 2011.

15. Tuma, P.E., "A Comparison of Passive 2-phase Immersion and Pumped Water Cooling," Presentation IMAPS ATW on Thermal Management, Palo Alto, CA. Nov. 7-9, 2011

16. Ellsworth, M.J., et al, "The Evolution of Water cooling for IBM Large Server Systems: Back to the Future," Proc. 2008 ITherm Conf., Orlando, FL, May 28-31.

17. Gham, D. and Laatsch, M., "Meeting the Demands of Computer Cooling with Superior Efficiency," Cray whitepaper WP-XT01-0709, Copyright Cray, 2009.

18. P. E. Tuma, "Segregated Hydrofluoroethers: Long Term Alternative Heat Transfer Liquids," Proceedings of the 2000 Earth Technologies Forum, Oct.30-Nov.1, Washington D.C., pp. 266-275.

19. P. E. Tuma, "Fluoroketone $C_2F_5C(O)CF(CF_3)_2$ as a Heat Transfer Fluid for Passive and Pumped 2-Phase Applications," Proc. 24th IEEE SEMI-THERM Symposium, San Jose, CA, March 16-20, 2008, pp. 174-181.

20. Webb. R.L. "Principles of Enhanced Heat Transfer," John Wiley and Sons, New York, 1994.

21. Webb, R.L. and Yamauchi, S., "Test Results on a Thermo-Syphon Concept to Cool High-Power Desktop Computers and Servers," Proc. 18th IEEE SEMI-THERM Symposium, San Jose, CA, March 12-14, 2002, pp. 151-158.

22. US Patent 4,129,181

23. US Patent 4,182,412

24. US Patent 3,821,018

25. US Patent 4,064,914

26. US Patent 3,990,862

27. US Patent 4,232,056

28. US Patent 4,354,550

29. US Patent 4,890,669

30. US Patent 7,360,581

31. US Patent 7,695,808

32. Tehver, J., et al, "Heat Transfer and Hysteresis Phenomena in Boiling on Porous Plasma-Sprayed Surface," Experimental Thermal and Fluid Science, Vol. 5, pp. 714-727, 1992.

33. Tuma, P.E., "Evaporator/Boiler Design for Thermosyphons Utilizing Segregated Hydrofluoroether Working Fluids," Proc 22nd Annual IEEE SEMI-THERM Symposium, art. no. 1625209, 2006, pp. 69-77.

34. Campbell, L., Ellsworth, M. and Sinha, A., "Analysis and Design of the IBM Power 575 Supercomputer Node Cold Plate Assembly," Proc. IPACK2009, July 19-23, San Francisco, CA.

# Thermal Performance of Sub-Atmospheric Loop Thermosyphon with and without Enhanced Boiling Surface

S. W. Chang[1], K. F. Chiang[2], C.-C. Huang[3]

1  Corresponding author,
Professor , Thermal Fluids Laboratory, National Kaohsiung Marine University,
No. 142, Hai-Chuan Road, Nan-Tzu District, Kaohsiung, Taiwan, ROC. Post code: 811,
Email: swchang@webmail.nkmu.edu.tw

2  Senior research manager,
CEO Office, Asia Vital Components Co. Ltd.,
7F-3, No. 24, Wu-Chuan 2 Rd., Hsin-Chuang City, Taipei, Taiwan, ROC. Post code: 24892.

3  Engineer,
R&D Office, Asia Vital Components Co. Ltd.,
7F-3, No. 24, Wu-Chuan 2 Rd., Hsin-Chuang City, Taipei, Taiwan, ROC. Post code: 24892.

## Abstract

This experimental study comparatively examines the thermal performances of two-phase loop thermosyphons (TPLP) with and without enhanced boiling surface at sub-atmospheric pressures. The boiling instabilities along with the constituent and total thermal resistances of these TPLPs are analyzed with the aid of boiling flow structures imaged at sub-atmospheric pressures. Boiling heat flux (Q) and thermal resistance of condenser ($R_{th,con}$) are selected as the controlling parameters with their individual and interdependent effects on the thermal performances examined. With the present enhanced boiling surface, the intermittent bursting of large bubbles from liquid pool in the multi-channel evaporator of plain surface is significantly suppressed, leading to the moderate pressure waves agitated by bubble eruptions with reduced boiling instabilities and pressure-drop thermal resistances ($R_{th,\Delta P}$). The effects of TPLP height (H), which affects the driven pressure head for liquid-vapor circulation, on the thermal performances of the enhanced TPLP at various Q and $R_{th,con}$ are subsequently examined. Total thermal resistances ($R_{th}$) measured from the TPLTs with enhanced boiling surface are considerably reduced from the TPLTs with plain boiling surface and reduced to about 0.265 at the test condition of Q=150W, $R_{th,con}$=0.2, H=35.3 tube diameters. A set of $R_{th}$ correlation which permits the evaluation of individual and interdependent Q, $R_{th,con}$ and H impacts on total thermal resistances of the enhanced TPLPs is generated to assist the design activities using this type of enhanced TPLP for cooling of electronic chipsets.

## Keywords

Two Phase Loop Thermosyphon, Enhanced Boiling Surface.

## Nomenclature

### English symbols

$d$  Evaporator channel hydraulic diameter of TPLP (m)

H  Height of TPLP (m)

Q  Boiling heater power (W)

$R_{th}$  Total thermal resistance of TPLT=$(T_c - T_\infty)/Q (KW^{-1})$

$R_{th,BI}$  Boiling instability thermal resistance ($KW^{-1}$)

$R_{th,con}$  Condenser thermal resistance ($KW^{-1}$)

$R_{th,\Delta P}$  Pressure drop thermal resistance ($KW^{-1}$)

$R_{th,ev}$  Evaporator thermal resistance ($KW^{-1}$)

$T_c$  Central temperature of electrical heater foil ($^0C$)

$T_{c,max/min}$  Maximum/minimum $T_c$ over $T_c$ oscillation ($^0C$)

$T_{c,mean}$  Time-mean $T_c$ over one period of oscillations ($^0C$)

$T_{con,in/out}$  Entry/exit temperature of condenser ($^0C$)

$T_{e,in/out}$  Entry/exit temperature of evaporator ($^0C$)

$T_\infty$  Ambient temperature ($^0C$)

### Greek symbols

$\Psi_1, \Psi_2$  Unknown functions

## 1. Introduction

Phase change heat transfer process, which utilizes latent heat at small driven temperature difference by offering the high heat transfer rate in the range of $10^3$-$10^5$ $W/m^2K$ during condensation, boiling or evaporation [1], is of great advantage to cooling of high power density sources where the moderate temperature gradient is required. With heat pipes, liquid from condenser returns to evaporator by means of capillary actions in a variety of wicks. Without wick, a thermosyphon can be designed with less thermal resistance at reduced manufacturing cost but the thermal performances at anti-gravity conditions are considerably degraded. In a Two-Phase Closed Thermosyphon (TPCT), the faster vapor flow interacts with the counteracting liquid flow to incur high drags and flow instabilities, which in turn decreases the maximum cooling capacity as the length to diameter ratio of a heat pipe increases [2]. Particularly, when the vapor plug intermittently holds the liquid slug in a TPCT pipe, the process of vapor-liquid circulation becomes periodic to raise the crisis of partial dry-out on the evaporator wall [3,4]. Taking the form of thin plate, the growth and disruption of vapor bubbles in a TPCT generate significant pressure oscillations, leading to cyclic variations in saturation temperature with amplified wall temperature fluctuations [5]. By separating the liquid and vapor flows to exclude the countercurrent limitation in a TPCT, a Two-Phase Loop Thermosyphon (TPLT) with unidirectional vapor-liquid circulation is capable of long distance heat transmission and generally offering the better thermal performances than a TPCT. The different temperatures between heat source and heat sink lead to differential saturation pressures between evaporator and condense to add the driven pressure head by gravity and therefore promote the vapor-liquid circulation in a TPLT. Interdependent impacts between the thermal resistance of condenser ($R_{th,con}$) and the boiling heat flux (Q) arise from the added pressure potential for vapor-liquid circulation as a result of the prescribed source-to-sink temperature difference, which affects the thermal performances of the TPLT consequently [6]. On balance of the total driven pressure head with the resisting pressure losses, including the friction and form drags as well as the pressure drops in association with flow acceleration/deceleration, contraction/enlargement and bends through the pathway of vapor-liquid circulation [7], the thermal equilibrium state of a TPLT emerges at a particular set of $R_{th,con}$ and Q to maintain the energy conservation between the heating and cooling duties for evaporator and condenser respectively. As local dryness, flow rate of working fluid and boiling/condensation pressures in a TPLT are bound to these balancing conditions for pressure-drop and enthalpy accountancies, the pressure distributions along the flow pathway in a TPLT are interdependent with the various local thermal resistances. Khodabandeh [8] calculated and compared with the pressure drop data for the TPLT through the evaporator and riser with the reported correlation [9]. Within the framework of homogeneous model [9] validated by [8] to approximate the pressure drop through the tubular evaporator, the frictional pressure-drop, which is inversely proportional to the homogeneous fluid density, decreases with the increase of fluid density at the elevated saturation pressure in a TPLT.

Instability phenomena and thermal resistances of a TPLT are the primary design focuses for cooling applications to electronic devices. In a TPLT, the various modes of vapor-liquid interactions such as nucleation, coalescence, fragmentation and interfacial actions trigger a variety of two-phase instabilities. These instabilities are controlled by their characteristic phase change processes and the associated vapor-liquid interactions within the predefined geometries [10-11]. In this respect, Tadrist [11] classified the two-phase instabilities into static and dynamic modes. While the flow excursion, boiling crisis and the transition of flow pattern from one to another steady state are classified as the static instabilities; the oscillation of density wave features the most common dynamic instabilities [11]. With intermittent boiling phase during which the waiting time for new bubbles is much longer than the growth time, Niro and Beretta [12] observed the boiling instabilities with a pulsation character. In the multi-channel evaporator, the flow patterns are subject to large amplitude fluctuations which cause flow reversal in some channels with expanding bubbles pushing the liquid–vapor interface in both upstream and downstream directions [13]. With R134a as the working fluid for unstable operations at low and high heat loads using the evaporators of different channel diameters [14] and with the enhanced boiling surfaces by the copper nano- and micro-porous structures [15], Khodabandeh and Furberg classified the boiling instabilities into Type I and Type II instabilities which occur respectively at low and high heat loads with different frequencies and amplitudes for different evaporator geometries [14-15]. Type I instability at low heat loads is triggered by the instant vapor condensations in the evaporator with the backward liquid flow from the riser; at which the flow through the evaporator instantly turns stagnant with single phase liquid flow at reduced heat transfer rates. Type II instability at high heat load is related to the spatial temperature differences in an evaporator channel within which the inlet heat transfer is more efficient than those in the middle and outlet sections [14-15]. Temperature fluctuations at high heat fluxes are observed due to the unstable boiling activities with the occasional hot dry spots at the upper half of the evaporator [14]. With enhanced boiling surfaces, boiling heat transfer rates are elevated with the instabilities suppressed due to the higher nucleation density and bubble frequency [15]. In general, the higher liquid head in the riser for the evaporators with and without enhanced boiling surfaces could improve the TPLT performance by suppressing the boiling instabilities [14-15]. Nevertheless, there is no detailed analysis reported in [14-15] to examine the impact of liquid-head height on the thermal performances of a TPLP.

It is a common design practice to control the cooling area together with the airflow rate through the condenser, which regulates $R_{th,con}$, to adjust the total thermal resistance of a TPLT at the predefined Q and source-to-sink temperature differences. Although the adjustments of Q can vary the ratio of the constituent thermal resistance to the overall thermal resistance of a TPLT [14-16], the systematic analysis that examined the individual and interdependent impacts of $R_{th,con}$ and Q on each constituent and the total thermal resistances of a TPLT is lacking. Another drawback for envisaging the thermal physics in a TPLT for electronic cooling is the insufficient studies for thermal performances of a TPLT operated at the sub-atmospheric pressures. While the water-copper two-phase heat transfer device, which is a compatible combination for long term TPLT operations, is commonly used by this technical community, the operating pressures corresponding to the working temperatures for electronic cooling systems are usually sub-atmospheric. Most of the reported TPLT data [14-16] were generated at pressures above the atmospheric levels using refrigerants as the working fluids. This experimental study is formulated to devise a highly efficient sub-atmospheric copper-water TPLT using the multi-channel evaporator with enhanced boiling surface. Initially, the boiling structures in the multi-channel evaporator with plain and enhanced boiling surfaces at various $R_{th,con}$ and Q are comparatively examined using a set of snapshots for the flow images taken from a series of flow visualization tests. The relative improvements for reducing the boiling instabilities by deploying the enhanced boiling surface are subsequently analyzed throughout the typical Q range for cooling of electronic chipsets. The individual and interdependent effects of Q, $R_{th}$ and TPLP height (H) on each constituent thermal resistance and the total thermal resistance ($R_{th}$) of the TPLTs with enhanced boiling surface are parametrically studied with the attempt to devise a set of $R_{th}$ correlations which permit the evaluation of individual and interdependent Q, $R_{th,con}$ and H impacts on $R_{th}$ for the TPLP with enhanced boiling surface.

978-1-4673-1110-6/12 $31.00 © 2012 IEEE

## 2. Experimental details

Figure 1 depicts (a) experimental facility of the test thermosyphon loop with the thermocouple locations indicated and the topologies scanned by electronic microscope for (b) plain (c) enhanced boiling surfaces. With plain or the enhanced boiling surfaces, two thermosyphon loops with the identical geometries for flow visualization and thermal performance tests are individually manufactured. All the connected pipes of diameter 3.4mm for each tested TPLT and the fin array attached on the tubular condenser are made of copper. To acquire the sufficient stiffness to tighten the leakage-free transparent endwall on the evaporator for flow visualization tests, the Perspex endwall on evaporator is 20mm thick and sealed with O ring. The electrical heater foil is attached on the back wall of the square-sectioned copper pyramid through which the boiling heater powers (Q) are supplied at the targeting levels. The finned condenser is cooled by an electrical fan with variable speeds to adjust the airflow rate in order to control $R_{th,con}$ at the specific values. Five thermocouples measuring $T_c$, $T_{e,out}$, $T_{con,in}$, $T_{con,out}$, $T_{e,in}$ are sequentially embedded along the thermal network of each tested TPLT, as indicated in Fig. 1(a), to measure raw temperature data at each predefined Q and $R_{th,con}$. In addition to $R_{th,con}$ which is treated as a controlling parameter to adjust the overall thermal performance of a thermosyphon loop, the total thermal resistance of each thermosyphon loop ($R_{th}$) is decomposed into the evaporator thermal resistance ($R_{th,ev}$), the equivalent thermal resistance due to pressure drop between evaporator and condenser ($R_{th,\Delta P}$) and the implicit thermal resistance due to Boiling Instability ($R_{th,BI}$) which raises the entry temperature of evaporator ($T_{e,in}$) from the liquid exit temperature at condenser ($T_{con,out}$). Following the usual definition used by this technical community, the total thermal resistance of the thermosyphon loop ($R_{th}$) is the summation of each constituent thermal resistance evaluating from $T_c$ to $T_\infty$ where $T_\infty$ is the ambient temperature surrounding the test facility. All the thermocouple signals measuring $T_c$, $T_\infty$, $T_{e,o}$, $T_{con,in}$ and $T_{con,o}$ are transmitted to the computer through the Fluke NetDAQ data logger which provides the precision of $0.01\,^0C$. $R_{th}$ and each constituent thermal resistance are determined as equation (1),

$$R_{th} = (T_c - T_\infty) / Q = R_{th,ev} + R_{th,\Delta P} + R_{th,con}$$
$$= (T_c - T_{e,o})/Q + (T_{e,o} - T_{con,in})/Q + (T_{con,in} - T_\infty)/Q \qquad (1)$$

The evaporator deployed for thermal performance tests is produced through a series of brazing, vacuumed and filling process. It has always been a bottleneck in practice that copper can be oxidized into CuO or $CuO_2$. The oxidized copper could lead to the clogged boiling surface so that several treatments for water and copper surface are necessary. The complete assembly of the thermosyphon loop is initially heated in the electrical oven filled with the in-house of AVC $N_2$ and $H_2$ mixture as an attempt to eliminate the oxidized copper through this heating process. Having completed the surface treatment, the dionized water is degassed and the complete loop is vacuumed to 10-5 torr prior to filling the treated water into thermosyphon loop. The material and manufacturing processes follow those used to produce heat pipe with the same criteria for reliability tests to ensure the life time of 20 years. With the similar manufacturing processes, water treatments and component material, it is expected that the present TPLP with the enhanced surface share the similar lifetime of the heat pipe. The treated water is filled to the level about 1/3 channel height of evaporator, giving the filling ratio of 31% when the thermosyphon loop is vertically positioned. As the oscillations of liquid level in the multichannel evaporator is expected as a result of boiling instabilities [6], the backward liquid flow from the evaporator toward the exit of condenser is permissible, leading to the rise of $T_{e,in}$ from $T_{con,out}$. The implicit thermal resistance which considerably affect the total thermal resistance of a TPLT, namely the boiling instability thermal resistance ($R_{th,BI}$), is evaluated as $(T_{e,in} - T_{con,out})/Q$. Variations of $R_{th}$, $R_{th,ev}$, $R_{th,\Delta P}$ and $R_{th,BI}$ responding to the systematic adjustments of Q and $R_{th,con}$ for the TPLT with and without enhanced boiling surfaces at four different TPLP height (H) are examined in details with the aid of the boiling images taken from the flow visualization tests at each set of the controlled Q and $R_{th,con}$ test conditions. The experimental conditions are controlled by maintaining Q at 65, 95, 120 or 150W with $R_{th,con}$ in the range of 0.2-0.42 $^0CW^{-1}$ by adjusting the airflow rate through the condenser.

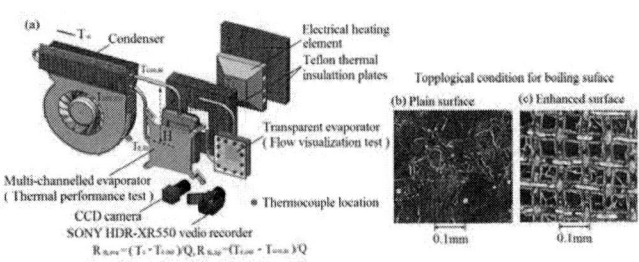

Fig. 1 (a) test facility and topologies of TPLT with (b) plain (c) enhanced boiling surfaces.

## 3. Results and discussion

### 3.1 Boiling flow structures and instabilities

Figure 2 compares the consecutive snapshots of boiling structures over one variation period (T) for the tested TPLTs with and without enhanced boiling surfaces at two sets of Q and $R_{th,con}$. As $T_c$ and $T_{e,o}$ for this set of boiling images falls in the respective ranges of 71.5-85.8 $^0C$ and 63.5-69.6 $^0C$ the saturation pressures in the evaporator remain sub-atmospheric for these flow visualization tests. Similar to the interfacial mechanisms in pulsating heat pipe [17], the bubble agglomeration and pumping action in the interconnected boiling channels come into sight at metastable non-equilibrium conditions, leading to the intermittent slug flows with a pulsation character to permit the vapor-liquid circulation without starting difficulty [6]. With enhanced boiling surface, the degrees of impacts by bubble bursting on liquid pulsation are significantly moderated from the plain TPLT counterparts, which consequently suppresses $R_{th,BI}$.

Capillary actions added by the micro-structures on the enhanced boiling surface improve the wetting performance over the vapor dome in evaporator to allow evaporation processes on the evaporator wall above liquid level.

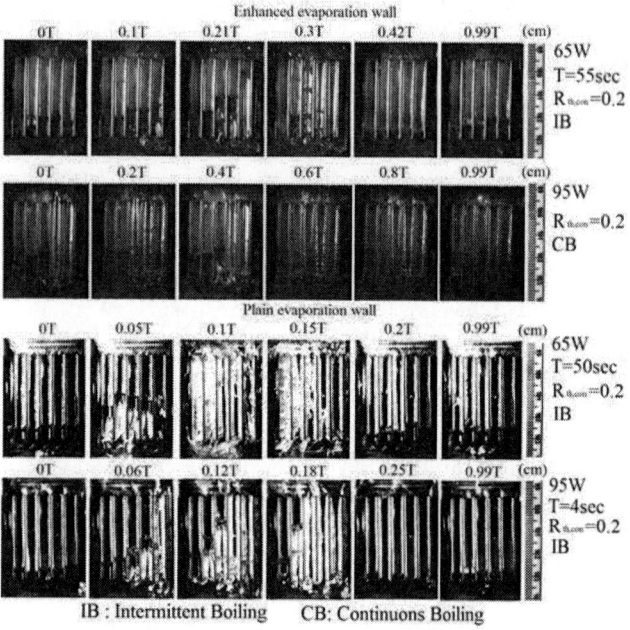

IB : Intermittent Boiling    CB: Continuons Boiling

Fig. 2 Consecutive snapshots of boiling structures over one variation period (T) for TPLTs with and without enhanced boiling surfaces at Q =65, 95W and $R_{th,con}$=0.2 $^0CW^{-1}$.

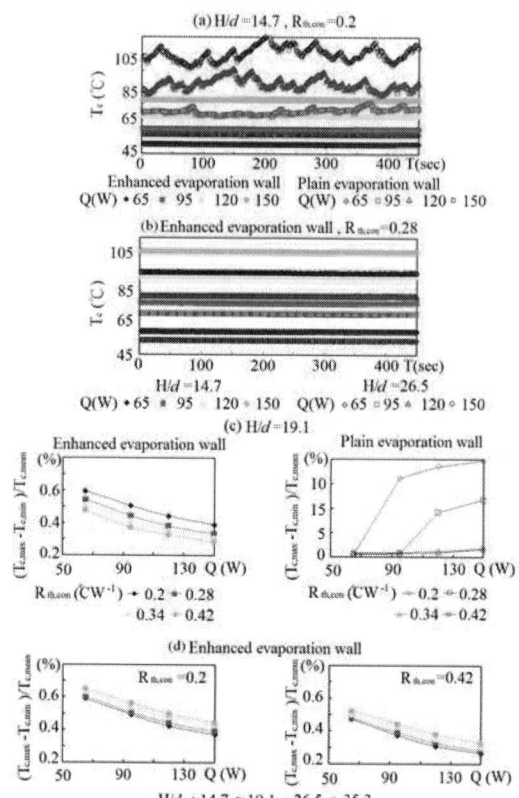

Fig. 3 $T_c$ oscillating amplitudes at various test conditions for TPLTs with and without enhanced boiling surfaces.

With enhanced and plain TPLTs, Fig. 3 compares (a) temporal $T_c$ variations at various Q with $R_{th,con}$=0.2, H/$d$=14.7 between plain and enhanced evaporation walls; (b) temporal $T_c$ variations at various Q with $R_{th,con}$=0.28 with H/$d$=14.7 and 26.5 for the enhanced evaporation wall; (c)variations of normalized mean-amplitudes of $T_c$ oscillations against Q at 19.1 H/$d$ with various $R_{th,con}$; (d) variations of normalized mean-amplitudes of $T_c$ oscillations against Q with various $R_{th,con}$ at H/$d$=14.7, 19.1, 26.5, 35.3. In Figs. 3(c) and 3(d), the normalized amplitude of $T_c$ oscillation is evaluated as $(T_{c,max} - T_{c,min})/T_{c,mean}$ where $T_{c,max}$, $T_{c,min}$, and $T_{c,mean}$ respectively stand for the maximum and minimum $T_c$ over one period of $T_c$ oscillations with the time-mean $T_{c,mean}$. While $T_c$ oscillations for the TPLT with plain boiling surface remain considerable, the $T_c$ oscillations for the TPLT with enhanced boiling surface are negligible. Due to the different boiling flow structures responding to Q variation for the TPLTs with and without the enhanced boiling surface, $(T_{c,max} - T_{c,min})/T_{c,mean}$ respectively decreases and increases as Q increases, Fig. 3(c). Although $(T_{c,max} - T_{c,min})/T_{c,mean}$ are negligible for the enhanced TPLT, the systematic trends of decreased $(T_{c,max} - T_{c,min})/T_{c,mean}$ driven by increasing $R_{th,con}$ at each tested Q are observed, which are generally elevated as H/$d$ increases, Fig. 3(d).

As compared by Fig. 4 for the enhanced and plain TPLT, the spectrum diagrams obtained from the various $T_c$ oscillations collected at ascending Q with fixed H/$d$=19.1 and $R_{th,con}$=0.28 show the similar wide-band patterns. The bubble initiation, growth, departure and agitation for both the enhanced and plain TPLTs take place at wide range of frequencies. The significant events to trigger $T_c$ oscillations appear at low frequencies due to the intermittent Taylor bubble bursting which randomly emerges in each multichannel evaporator at relatively low frequencies. While the amplitudes corresponding to the various fundamental frequencies shown in Fig. 4 are similar at various Q for the TPLT with enhanced boiling surface, the oscillating amplitudes for the plain TPLT increase systematically as Q increases. The more severe bubble bursting develops in the plain evaporator at the high heat loads, whereas the bubble agglomeration and pumping action in the interconnected boiling channels with enhanced boiling surface are less affected by increasing Q to offer the more stable boiling process with less $R_{th,BI}$.

978-1-4673-1110-6/12 $31.00 © 2012 IEEE

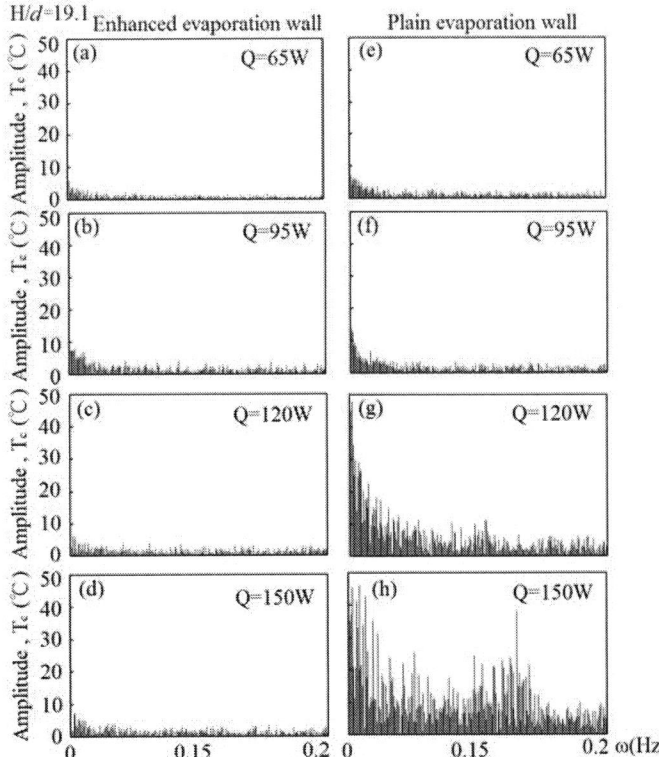

Fig. 4 Spectrum diagrams for various $T_c$ oscillations with ascending Q at H/$d$=19.1 and $R_{th,con}$=0.28 for enhanced and plain TPLTs.

## 3.2 Thermal resistance networks and correlation

Figure 5 comparatively examines the variations of total thermal resistance ($R_{th}$) against Q at fixed $R_{th,con}$ between the TPLTs with and without enhanced boiling surfaces.

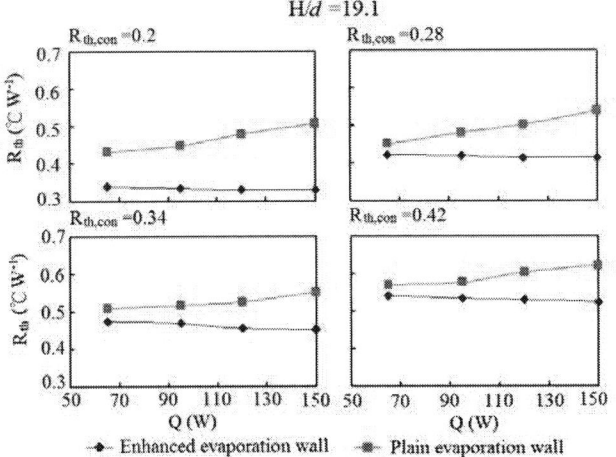

Fig. 5 Variations of $R_{th}$ against Q at fixed $R_{th,con}$ for enhanced and plain TPLTs.

At all tested Q and $R_{th,con}$, the total thermal resistances of enhanced TPLT at any H/$d$ are considerably less than their plain-TPLT counterparts. Two different varying manners of $R_{th}$ against Q for the enhanced and plain TPLTs, which respectively increase and decrease as Q increases, are caused by the contrary Q impacts on the constituent thermal

resistances of $R_{th,\Delta P}$ and $R_{th,BI}$ for the enhanced and plain TPLTs. This will be later demonstrated. Due to the opposing Q-driven $R_{th}$ varying trends for the enhanced and plain TPLTs, the relative thermal performance improvements by the present enhanced boiling surface are amplified by increasing Q.

Figure 6 reveals the H/$d$ impacts on $R_{th}$ for the present enhanced TPLTs by comparing the various Q-driven $R_{th}$ variations obtained at a fixed $R_{th,con}$ with H/$d$=14.7, 19.1, 26.5 and 35.3. For each tested $R_{th,con}$, the increase of H/$d$ from 14.7 to 35.3 incurs consistent downward $R_{th}$ spreads at each tested Q, indicating the reduced total thermal resistance by increasing the gravity driven pressure head at each fixed $R_{th,con}$. The increased driven pressure head to promote coolant circulation by increasing H/$d$ is not offset yet by the according growth of pressure resistances due to the elongated loop length in the range of 14.7≤H/$d$≤35.3 for the enhanced TPLTs.

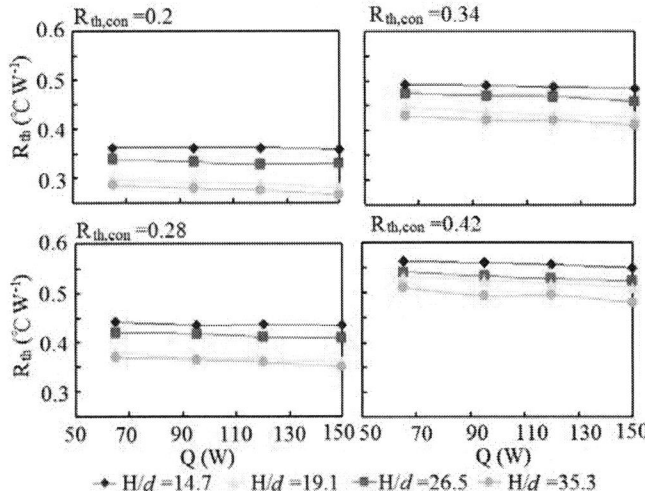

Fig. 6 H/$d$ impacts on $R_{th}$ for the present enhanced TPLTs at fixed $R_{th,con}$ with H/$d$=14.7, 19.1, 26.5 and 35.3.

Each Q-driven $R_{th}$ data series collected in Fig. 6 follows the pattern of linear decrease which can be correlated as:

$$R_{th} = \Psi_1\{R_{th,con}, H/d\} + \Psi_2\{R_{th,con}, H/d\} \times Q \qquad (2)$$

In equation (2), $\Psi_1$ decreases with the increase of H/$d$ but increases as $R_{th,con}$ increases. Justified by the consistent Q-driven $R_{th}$ data trends in Fig. 6, which can be well correlated by equation (2), a set of empirical $R_{th}$ correlation for the present enhanced TPLT can be generated. It is interesting to discover that $\Psi_1$ and $\Psi_2$ in equation (2) also increase linearly with the increase of $R_{th,con}$. Consequently, $\Psi_1$ and $\Psi_2$ can be expressed as the functions of $R_{th,con}$. By way of converting these correlative coefficients in $\Psi_1\{R_{th,con}\}$ and $\Psi_2\{R_{th,con}\}$ functions as functions of H/$d$, the net result of the correlation process is the generation of the empirical $R_{th}$ correlation for the present enhanced TPLT as equation (3) with Q, $R_{th,con}$ and H/$d$ as the controlling variables. Equation (3) is generated to assist the in-house design of the present enhanced TPLT.

978-1-4673-1110-6/12 $31.00 © 2012 IEEE

$$R_{th} = \{(0.253 - 5.56E\text{-}3 \times H/d) + (0.857 + 7.83E\text{-}3 \times H/d) \times R_{th,con}\}$$
$$- \{(-2.67E\text{-}4 + 1.3E\text{-}5 \times H/d) + (7.94E\text{-}4 - 1.46E\text{-}5 \times H/d) \times R_{th,con}\} \times Q \qquad (3)$$

In order to disclose the opposing Q-driven $R_{th}$ variations for the plain and enhanced TPLTs shown by Fig. 5, the variations of each constituent thermal resistances, namely $R_{th,ev}$, $R_{th,\Delta P}$ and $R_{th,BI}$, against Q at fixed $R_{th,con}$ and against $R_{th,con}$ at fixed Q at the representative condition of $H/d$=19.1 are respectively compared in Figs. 7(a) and 7(b).

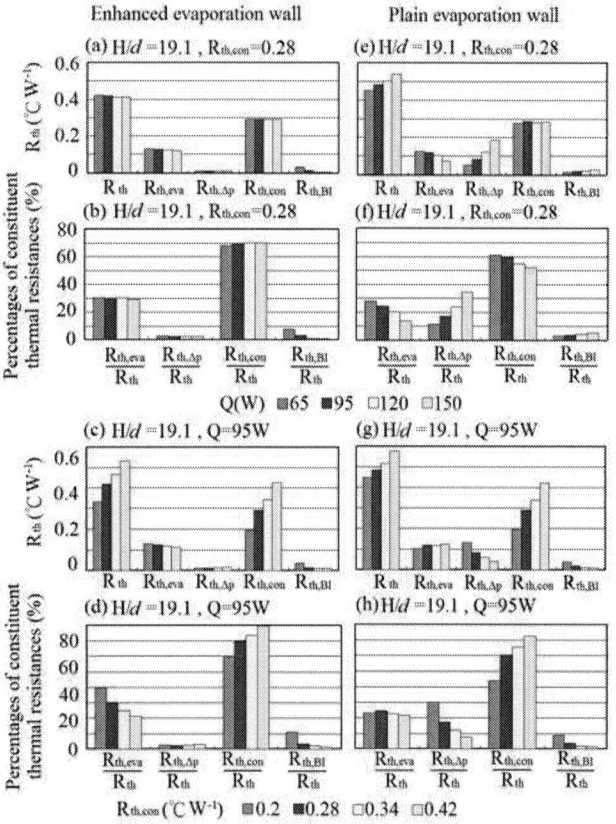

Fig. 7 Variations of $R_{th,ev}$, $R_{th,\Delta P}$ and $R_{th,BI}$ against Q at fixed $R_{th,con}$ and against $R_{th,con}$ at fixed Q with $H/d$=19.1.

As compared by Fig. 7(a), while the $R_{th,\Delta P}$ and $R_{th,BI}$ for the plain TPLT keep increasing with the increase of Q at fixed $H/d$ and $R_{th,con}$, the $R_{th,\Delta P}$ for enhanced TPLT always remain negligible and its $R_{th,BI}$ decreases as Q increases, Fig. 7(a). Our previous works that compared the pressure drops for the single-phase liquid flow and air-water two-phase flows through a vertical swirl tube had demonstrated the significant increases of pressure drops through the swirl tube when the single-phase liquid flow yielded to the two-phase flow [18-19]. The increased $R_{th,\Delta P}$ by increasing Q for the plain TPLT indicates the augmented pressure drops between the evaporator and the condenser due to the reduced dryness factors in this section. With wet vapor, the pressure drops through this section of TPLT inevitably induce temperature drops which add $R_{th,\Delta P}$ by increasing ($T_{e,o}$-$T_{con,in}$). Clearly, with plain TPLT, the sever bubble bursting, which is amplified by increasing Q, tends to reduce the dryness factor of vapor dome in the evaporator of plain wall, leading to the

increased pressure resistances in the tube connecting the evaporator and condenser as a result of the increased wetness. By way of $R_{th,con}$ increase, the pressures in the evaporators of plain and enhanced TPLTs are generally elevated so that $R_{th,\Delta P}$ and $R_{th,BI}$ are generally decreased as $R_{th,con}$ increases for both plain and enhanced TPLTs, Fig. 7(b). Due to the suppressed bubble bursting in the evaporator with enhanced boiling surface, the $R_{th,\Delta P}$ for the enhanced TPLT are also considerably less than the plain-TPLT counterparts, leading to the lower $R_{th}$ for the enhanced TPLT at each $R_{th,con}$ examined. Nevertheless, as $R_{th,ev}$ for plain and enhanced TPLTs at all test conditions are similar, the overall convective heat transfer coefficients for the boiling processes in the evaporators with and without the enhanced surfaces are in close agreements, Fig. 7. Therefore the thermal performance improvements generated by present enhanced boiling surface are mainly attributed to the suppressions of bubble bursting activities, which consequently increases the dryness factor of vapor dome in the evaporator and stabilizes the boiling activities to reduce $R_{th,\Delta P}$ and $R_{th,BI}$ from the plain-TPLT counterparts; rather than the elevated boiling heat transfer rates over the evaporator walls with the enhanced micro-structures.

## 4. Conclusions

This experimental study comparatively examined the thermal performances of TPLT with and without enhanced boiling surfaces. The following concluding remarks emerge from this study.

1. The bubble agglomeration and pumping action in the interconnected boiling channels at metastable non-equilibrium conditions prevail over the plain and enhanced evaporator, leading to the intermittent slug flows with a pulsation character to permit the vapor-liquid circulation without starting difficulty. The impacts of bubble bursting on liquid pulsation are significantly moderated by the enhanced boiling surface, which consequently suppresses $R_{th,BI}$. Capillary actions added by the micro-structures over the enhanced boiling surface improve the wetting performance over the vapor dome in evaporator to permit latent heat conversion above the liquid level in the evaporator.

2. While the $R_{th,\Delta P}$ and $R_{th,BI}$ for the plain TPLT keep increasing with the increase of Q at fixed $H/d$ and $R_{th,con}$, the $R_{th,\Delta P}$ for enhanced TPLT always remains negligible and its $R_{th,BI}$ decreases as Q increases. The increased $R_{th,\Delta P}$ by increasing Q for the plain TPLT is caused by the augmented pressure drops between the evaporator and the condenser due to the reduced dryness factors as a result of the sever bubble bursting in the evaporator, which is amplified by increasing Q. Due to the suppressed bubble bursting in the evaporator by the enhanced boiling surface, the increased dryness factor of vapor dome in the evaporator, which can be further elevated by increasing Q, reduce $R_{th,\Delta P}$. The respective increases and decreases of $R_{th,\Delta P}$ and $R_{th,BI}$ by increasing Q for the plain and enhanced TPLTs have led to the increased and decreased $R_{th}$ for the plain and enhanced TPLTs as Q increases.

3. The empirical $R_{th}$ correlation which permits the evaluations of individual and interdependent Q, $R_{th,con}$ and H/$d$ impacts on $R_{th}$ for the present TPLT with the enhanced boiling surface is generated to assist the in-house design activities.

## Acknowledgments

The research facilities are jointly supported by AVC and National Science Council, Taiwan, under NSC 100-2628-E-022-001MY3 project.

## References

1. L.L. Vasiliev, "Heat pipes in modern heat exchangers", Applied Thermal Engineering, Vol. 25, pp. 1-19, 2005.

2. I. Golobič, B. Gašperšič, "Corresponding states correlation for maximum heat flux in two-phase closed thermosyphon", Int. J. Refrigeration, Vol. 20, pp. 402-410, 1997.

3. T. Fukano, K. Kadoguchi, H. Imuta, "Experimental study on the heat flux at the operating limit of a closed two-phase thermosyphon", JSME Ser. B, Vol. 53, pp. 1065-1071, 1987.

4. Y. Koizumi, T. Yoshinari, T. Ueda, T. Matsuo, T. Miyashita, "Study on dry-out heat flux of two-phase natural circulation", JSME Ser. B, Vol. 60, pp. 545-551, 1994.

5. M. Zhang, Z. Liu, G. Ma, "The experimental investigation on thermal performance of a flat two-phase thermosyphon", Int. J. Thermal Sciences, Vol. 47, pp. 1195-1203, 2008.

6. V. Tsoi, S.W. Chang, K.F. Chiang, C.C Huang, "Thermal performance of plate type loop thermosyphon at sub-atmospheric pressures", J. Applied Thermal Engineering, Vol. 31, pp. 2556-2567, 2011.

7. F. Fantozzi, S. Filippeschi, E.M. Latrofa, "Upward and downward heat and mass transfer with miniature periodically operating loop thermosyphon", Superlattices and Microstructures, Vol.35, pp. 339-351, 2004.

8. R. Khodabandeh, "Pressure drop in riser and evaporator in an advanced two-phase thermosyphon loop", Int. J. Refrigeration, Vol. 28, pp. 725-734, 2005.

9. M.B. Bowers, I. Mudawar, "two-phase electronic cooling using mini-channel and macro-channel heat-sinks — part II, flow rate and pressure drop constraints", ASME J. Electronic Packaging, Vol. 116, pp. 298-305, 1994.

10. A.E. Bergles, V.J.H. Lienhard, G.E. Kendall, P. Griffith, "Boiling and evaporation in small diameter channels", Heat Transfer Engineering, Vol. 24, pp. 18-40, 2003.

11. L. Tadrist, "Review on two-phase flow instabilities in narrow spaces", Int. J. Heat Fluid Flow, Vol. 28, pp. 54-62, 2007.

12. A. Niro, G.P. Beretta, "Boiling regimes in a closed two-phase thermosyphon", Int. J. Heat Mass Transfer, Vol. 33, pp. 2099-2111, 1990.

13. S.G. Kandlikar, "Fundamental issues related to flow boiling in mini-channels and micro-channels", Experimental Thermal Fluid Sciences, Vol. 26, pp. 389-407, 2002.

14. R. Khodabandeh, R. Furberg, "Intability, heat transfer and flow regime in a two-phase flow thermosyphon loop at different diameter evaporator channel", Applied Thermal Engineering, Vol. 30 pp. 1107-1114, 2010.

15. R. Khodabandeh, R. Furberg, "Heat transfer, flow regime and instability of a nano- and micro-porous structure evaporator in a two-phase thermosyphon loop", Int. J. Thermal Sciences, Vol. 49, pp. 1183-1192, 2010.

16. R. Khodabanden, "Thermal performance of a closed advanced two-phase thermosyphon loop for cooling of radio base stations at different operating conditions", Applied Thermal Engineering, Vol. 24, pp. 2643-2655, 2004.

17. S. Khandekar, N. Dollinger, M. Groll, "Understanding operational regimes of closed loop pulsating heat pipes: an experimental study", Applied Thermal Engineering, Vol. 23, pp. 707-719, 2003.

18. S.W. Chang, A.W. Lees, H.-T. Chang., "Influence of spiky twisted tape insert on thermal fluid performances of tubular air-water bubbly flow", Int. J. Thermal Sciences, Vol. 48, pp. 2341-2354, 2009.

19. S.W. Chang, T.L. Yang, "Forced convective flow and heat transfer of upward cocurrent air-water slug flow in vertical plain and swirl tubes", Int. J. Experimental Thermal and Fluids Science, Vol. 33, pp. 1087-1099, 2009.

# High-Performance Nickel Wick Development for Loop Heat Pipes

Vijit Wuttijumnong, Randeep Singh, Masataka Mochizuki, Kazuhiko Goto,
Thang Nguyen, Tien Nguyen, Koichi Mashiko
Fujikura Ltd., Thermal Technology Division, R&D Department,
1-5-1, Kiba, Koto-ku, Tokyo 135-8512, Japan
vijit@fujikura.com, randeep.singh@jp.fujikura.com

## Abstract

In the present investigation, high-performance capillary pump has been developed and evaluated for using inside loop heat pipes with 500 W heat transfer capability up to distance of 250 mm. Wick structure is one of the most critical components of the loop heat pipe that provides the necessary capillary pumping, liquid-vapour phase separation and heat leak barrier from evaporation section to the compensation chamber. In order to fabricate wick structure with appropriate physical characteristics, highly pure (> 99.5%) nickel powders with average particle sizes of 2, 10, 12 and 75 μm were selected for sintering experiment. It was established that nickel powder can be effectively sintered when maintained at 750 – 850 °C temperature for one hour. Out of four nickel powders, NM-12 powder with average particle diameter of 12 μm was able to provide most qualified porous structure, sintered at 850 °C for one hour, with high porosity (> 72 %), high permeability (> 2 x 10⁻¹³ m²), finer pore radius (< 7.2 μm), low shrinkage (< 22%), good axial straightness and acceptable strength. The main issues faced in sintering trials and remedies to avoid them are explained in detail. Center rod extraction was the major problem faced in the sintering experiment which was rectified by replacing carbon rod with high strength, lubricious center rod made from stainless steel with boron nitride coating. The methodologies development in this study can be used for the fabrication of high performance capillary wicks for miniature to large-scale loop heat pipes.

## Keywords

Capillary pump, wick, nickel sintering, loop heat pipe, heat transfer

## 1. Introduction

Loop heat pipes (LHPs) are competitive two-phase thermal management devices due to their ability to transfer high heat loads up to long distances and against gravitational field (i.e. at different orientations and tilts) [1, 2]. LHPs have unique heat transfer characteristics that qualify them as one of the most potential passive thermal control candidates. Loop heat pipe operation is based on the similar principle as conventional heat pipes i.e. cyclic evaporation-condensation process of working fluid circulated by capillary pumping inside wick. However there are number of design-related alterations in LHPs that make its manufacturability and operation very different from conventional capillary heat pipes. LHPs have more developed evaporator section with system of vapour removal channels and with evaporation from wick-wall interface. Unlike conventional heat pipes in which wick is attached to the internal circumference throughout the length of the tube, the porous structure in LHP is confined in the evaporator section only. Constructively, these unique structural modifications in the loop systems provide them with high heat flux transfer capability and ability to transfer heat over longer distances respectively. Nonetheless, on the downside, the design changes in LHP to enhance the entrainment limit (by separating vapour and liquid flow passages) and capillary limit (by making wick stricture fine-pored and confining it inside the evaporator section) poses number of fabrication and operational issues.

It should be noted that dominance of LHPs over conventional heat pipes or electro-mechanically pumped loops is mainly performance and/or cost specific. For the given heat flux range, the operational regimes of these three technologies (heat pipes, LHP & single-phase pumped loop) can be well defined [3]. Heat pipes are competent devices for low heat flux and short distance heat transfer applications, due to high cost of other two technologies and lower performance of LHPs at low heat flux. Liquid cooling (using mechanical pump) performs better than heat pipe but is outperformed by LHP for high heat fluxes due to unprecedentedly high evaporator performance of LHP. In summary, LHPs are potential alternatives over heat pipes for applications where heat pipes cannot meet the performance requirements (performance dominance). On the other hand, when compared to the electrically pumped loops, LHPs provide passive thermal control system without any power demands, lower cost, longer system life, higher runtime reliability and thus sustainable cooling technology for the energy extensive electronic devices (cost/technology dominance).

Manufacturing of heat pipe involves relatively simple processes on metal tube that include cutting, swaging, sinter attachment of powder to tube internal wall, charging working fluid and welding. In loop heat pipe, the evaporator (including evaporator internally grooved portion, compensation chamber and wick), condenser (including flow tubes, fins and connectors) and heat transfer tubes (vapour and liquid lines) are each separately manufactured and then assembled through brazing/welding process. This adds to both cost as well as manufacturing complexity of LHP. Evaporator is the most complex and costly component of the loop heat pipe assembly. Fabrication of evaporator includes machining internal flow channels in the tube (for cylindrical evaporator) or plate (for flat evaporator), sintering of wick structure, integration of wick inside the evaporator, sealing of evaporator container by welding/brazing and attachment of flow tubes (vapour/liquid lines). Any fabrication simplicity or technology development with respect to the abovementioned processes can provide the consequential cost reduction and promotion of the passive loop devices in high flux thermal management applications.

Wick is the most critical component of the loop heat pipe that dictates its runtime reliability and thermal performance [4]. Sintering of the wick structure with the required physical properties needs rigorous evaluation and is one of the most challenging tasks in the fabrication of the LHP. In the present investigation, development of the nickel wick with superior physical properties has been explained in detail. The paper commences with the outline of the design considerations for the LHP wick followed by LHP prototype design. The main discussion agenda includes wick material choice, sintering mould design, optimising sintering conditions, achieving superior wick properties, issues faced in wick sintering and solutions to tackle sintering issues,

## 2. Wick Design Considerations

Wick is the heart of the loop heat pipe and should provide functionalities including: 1) Capillary pumping of the working fluid around the loop (capillary pump), 2) Spread liquid in the evaporator active or heated zone for evaporation (liquid distribution), 3) Minimise heat flow from evaporation zone to liquid reservoir or compensation chamber (heat barrier), and 4) Separate liquid and vapour phases inside the evaporator (internal sealing)

In order to satisfy these roles, the wick structure should possess appropriate physical properties as follow:

Fine pore size → Wick should be able to generate capillary pressure greater than or equal to the total pressure loss by fluid around the loop at maximum heat load. Generally, mean effective pore radius of wick is fixed to possible value, for the given wick material, followed by choice of hydraulic diameter for fluid flow passages. Maximum capillary pressure generated by wick is then equated to total pressure inside loop to iteratively optimise the diameter of flow passages (as per available space) and wick pore size with certain factor of safety for worst case scenario.

High porosity → Bulk of wick structure should have well-connected pores with torturous flow pathways. It provides twofold advantages: 1) effective wick hydration i.e. liquid distribution inside wick volume, and 2) lower effective thermal conductivity of liquid saturated wick (that reduces heat leaks from active zone to liquid reservoir)

High permeability → It should be noted that high wick permeation and smaller pore size are inversely related parameters and therefore need optimization. High permeability implies how easily and effectively the liquid can reach the wick evaporation face. A wick with smaller pore size and high porosity but poor permeability will 1) increase fluid pressure drop through wick, and 2) heat-up wick evaporation face due to lower liquid filtration rate through wick thereby increasing evaporator operating temperature. Generally, wick pore size is fixed by the LHP heat flux and heat transfer distance, and therefore highest possible permeability is achieved for the given pore size by optimising sintering parameters or powder size.

Low effective thermal conductivity → Heat leak from evaporation zone to compensation chamber poses critical performance issues during loop start-up from cold state and steady state operation. Wick with lower effective thermal conductivity (with a given working fluid) is a desirable parameter which can reduce heat leaks and therefore enhance LHP performance.

Small geometrical tolerance → Wick structure should have close geometrical tolerances to provide 1) good thermal contact between wick and vapour flow passages, and 2) internal sealing or separation of liquid phase (on compensation chamber side) and vapour phase (on vapour removal channel side).

The wick lateral thickness (perpendicular to wick evaporation face) is one of the important design parameter which is optimised to minimise heat leak across wick. Favourably, larger wick thickness is an advantage however maximum possible wick thickness is limited by mandatory 1) wick core diameter (for cylindrical evaporator) or compensation chamber thickness (for typical flat evaporator) as dictated by loop internal volume, 2) hydraulic diameter of vapour flow passages as fixed by wick pumping capability, and 3) evaporator container wall thickness as dictated by structural strength requirements. Here, pressure drop across the wick will increase with added thickness therefore design objective should target maximum wick thickness within the pumping capability of the capillary structure. The wick planar dimension (parallel to wick evaporation face) is dependent on the evaporator heated area (or heat source footprint area).

## 3. Loop Heat Pipe Prototype

Figure 1 show the schematic of the loop heat pipe which has been designed to transfer 500 W heat load up to distance of 250 m, in horizontal orientation, while maintaining evaporator temperature below 100 °C with forced convection cooling of condenser using ambient air at 25 °C. In order to support loop operation, cylindrical wick structure, as shown in Figure 2, with dimensions for active zone φ18 mm, thickness 4 mm & total wick length of 100 mm, and with mean pore radius < 12 μm, porosity > 50% and permeability > $10^{-13}$ $m^2$ should be installed inside the loop evaporator.

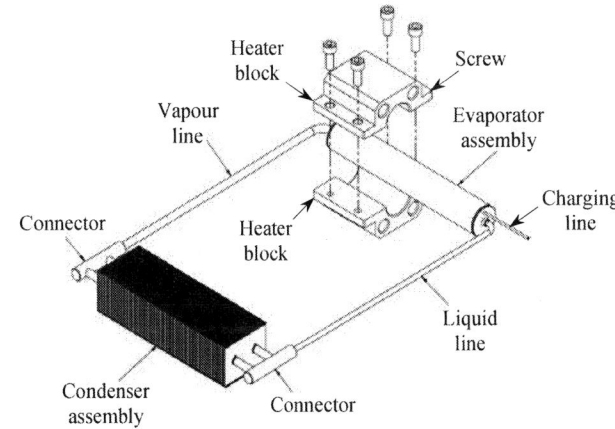

**Figure 1** Schematic of Loop Heat Pipe Prototype

978-1-4673-1110-6/12 $31.00 © 2012 IEEE

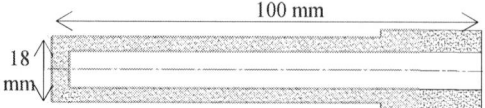

**Figure 2** Cross sectional detail of LHP wick

In this paper, results and discussion on the sintering experiment conducted to develop appropriate wick structure, for the designed LHP, by testing powders with different shape and size are presented in detail.

## 4. Wick Development

### 4.1. Nickel Power Types

In order to develop porous structure with required geometrical dimensions and physical properties, four types of highly pure (> 99.5%) nickel powder that differ in powder size and shape were considered for sintering experiment. Nickel has been considered as wick material for LHP due to its 1) lower thermal conductivity (to avoid heat leaks) 2) sinterable with high porosity and fine pore size and 3) availability in high assay level at reasonable cost and in different powder sizes. Figure 3 presents the magnified view of each powder sample with their average powder size (in μm) denoted by the numeral in their name e.g. T-2 powder has average powder size of 2 μm. It should be noted that the average powder size represents the powder particles present in larger percentage which lies midway between the smallest and largest size powder in the given powder sample.

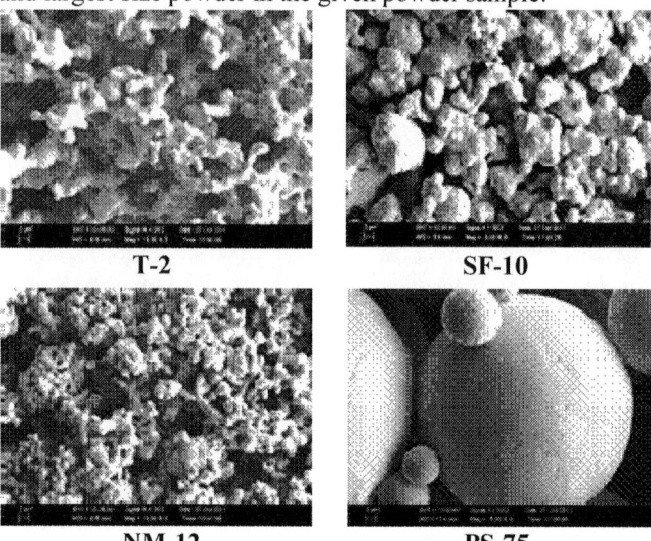

**Figure 3** Different nickel powder types used in the sintering experiment

### 4.2. Sintering Procedure

To conduct the sintering of nickel powder, mould made from carbon was used. Carbon due to its lubrication nature (for extraction of sintered wick from mould) and chemical inertness (to avoid wick contamination) was used in the experiment. Stainless steel (SUS) and ceramics are other potential sintering mould materials which were not used in this investigation due to low lubrication properties of SUS and high cost of ceramic mould. Figure 4 shows the parts and assembly of the carbon mould.

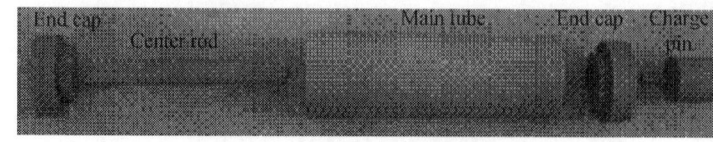

(a) Mould parts

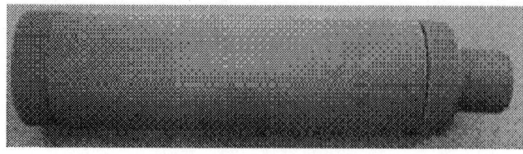

(b) Mould assembly

**Figure 4** Carbon mould used in the nickel sintering experiment

The steps followed for each sintering trial include:
o Assemble the mould
o Charge nickel powder into the mould through charging hole
o Close the charging hole
o Secure mould parts in place using external jig (to avoid mould parts movement or dismantling during transport and sintering)
o Set the sintering temperature and time for the furnace
o Input mould into the furnace
o Remove mould from furnace after sintering process
o Extract sintered wick from the mould
o Measure physical properties of the sintered wick

For sintering, belt-type furnace with three temperature zones (preheating, sintering & cooling zone respectively) was used. Nitrogen (with very low percentage of hydrogen) was used to provide inert atmosphere during high temperature sintering. Hydrogen was used to reduce any oxides that might be formed due to air infiltration. In the experiment, firstly the sintering conditions for nickel powder were optimised and then the effect of sintering temperature (600 – 1000 ºC) on porosity, shrinkage and strength of wick was studied. Pore size and permeability of the sintered wick made from selected powder type was also measured to qualify the wick for designed loop heat pipe. Measurement techniques for porosity (soaking method), pore radius (U-tube bubble point method) and permeability (Darcy's law) are detailed in reference [5]

### 4.3. Sintering Experiment

Sintering conditions for nickel powder were standardised by conducting trials at different temperatures in the range of 600 – 1000 ºC for one hour sintering time. All types of nickel powder were weakly sinterable at 650 ºC. As the sintering temperature was increased beyond 650 ºC, the wick porosity decreases and shrinkage increases, however the sample strength was enhanced with sintering temperature. It should be noted that expected wick strength is very much application dependent e.g. for electronic cooling low strength can be acceptable, however, for automotive application high strength will be a necessity. In these experiments, two simple stress tests were conducted to qualify wick strength 1) compressive stress test: apply radial compression force (on 1 cm axial length) under incremental loading, 2) shock/drop test: drop wick through certain height. The load in the first test and drop

978-1-4673-1110-6/12 $31.00 © 2012 IEEE

height in second test can be altered, as per application, to get the required strength. If the wick sample qualified both tests then it was qualified as strong wick. In the present investigation, the application is stationary electronic cooling therefore test conditions were fixed at 5 kg maximum load and 1 m drop height. On the basis of the sintering runs, 750 – 850 °C temperature and one hour time period was chosen as optimum sintering conditions. The temperature range give tolerance for method designed to integrate wick inside evaporator container i.e. force fitted (need high strength) or sintered attach (low strength acceptable). Figure 5 and 6 plots the porosity and volumetric shrinkage of different powder types sintered at 750 °C respectively. It is evident from the graphs that porosity and shrinkage are inversely related to average powder size (i.e. smaller the powder, higher is the porosity and shrinkage). PS-75 powder has low shrinkage (5.5%) however the porosity was too low (21%). On the other extreme, T-2 powder has very high porosity (80%) but the high shrinkage (21%) makes dimensional/geometry control very difficult. SF-10 & NM-12 powder shows very similar and optimum porosity (76-77%) and acceptable shrinkage (~8%) however SF-10 powder have high impurity contents (estimated by EDX analysis) than NM-12. As a result, NM-12 (1st choice) and T-2 (2nd choice) powders were chosen out of four types.

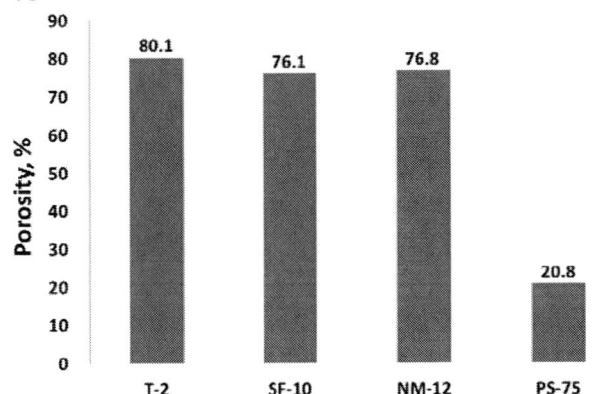

**Figure 5** Porosity of different powders sintered at 750 °C

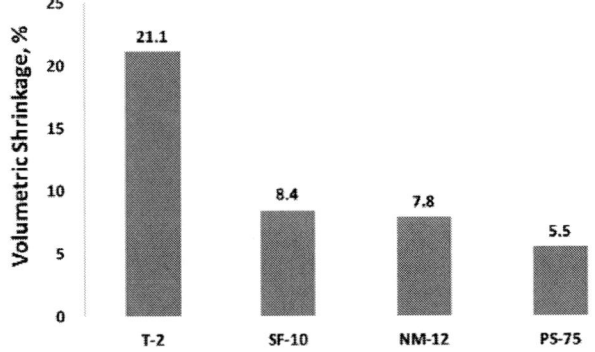

**Figure 6** Volumetric shrinkage of different powders sintered at 750 °C

Figure 7 and 8 further compares the porosity and shrinkage of sintered wick sample made from NM-12 and T-2 nickel powder at different sintering temperatures. It is observed that drop in porosity and increase in shrinkage for T-2 powder is more steep than NM-12 due to its smaller powder size. Beyond 800 °C, porosity crossover is observed between NM-12 & T-2 owing to high volumetric shrinkage of T-2 wick.

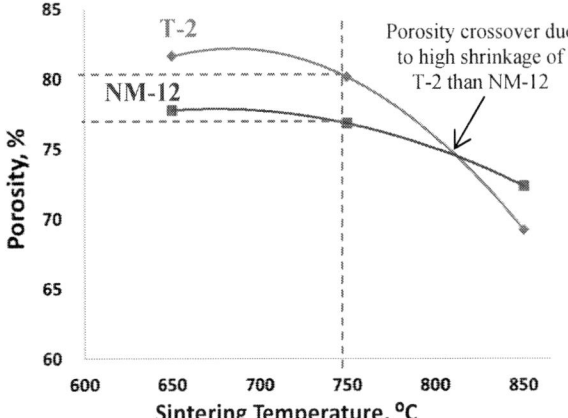

**Figure 7** Dependence of wick porosity on temperature

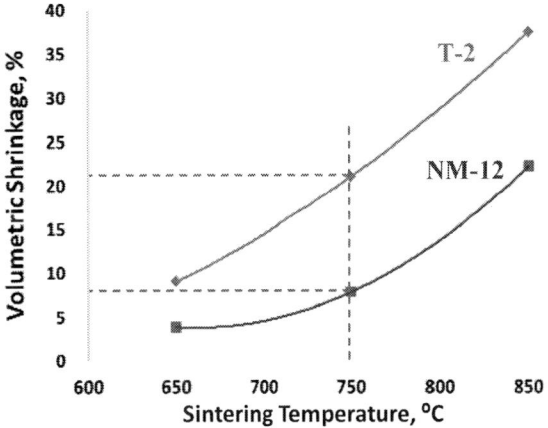

**Figure 8** Dependence of wick shrinkage on temperature

Wick diametric shrinkage was the highest and the wick diameter is the most critical dimensions in the proper integration of wick into evaporator container. Figure 9 plots the diametric shrinkage of the NM-12 wick. The graph clearly shows that shrinkage increases very sharply from 750 to 850 °C that also explains the increased wick strength at high temperatures due to volumetric compaction. For T-2 wick, shrinkage was higher even at lower sintering temperature as shown in Figure 10. Another point evident from Figure 9 & 10 is that straightness (or flatness) of NM-12 wick along the axial direction is very controlled and better than T-2 wick that showed high degree of lengthwise waviness.

For NM-12 wicks, specific permeability was measured to be within $2 – 9$ (x $10^{-13}$) m$^2$ range and largest pore size was below 7.2 μm. It is worth noting that for loop heat pipes, the largest pore size in the given wick sample is very significant parameter because the back flow of vapour from the evaporation zone to compensation chamber will commence through the pore with maximum radius owing to inverse dependence of capillary pressure on pore radius. Therefore, in summary, the wick samples sintered from NM-12 powder exhibited the appropriate physical properties for qualifying wick structure as capillary pump for designed LHP.

978-1-4673-1110-6/12 $31.00 © 2012 IEEE

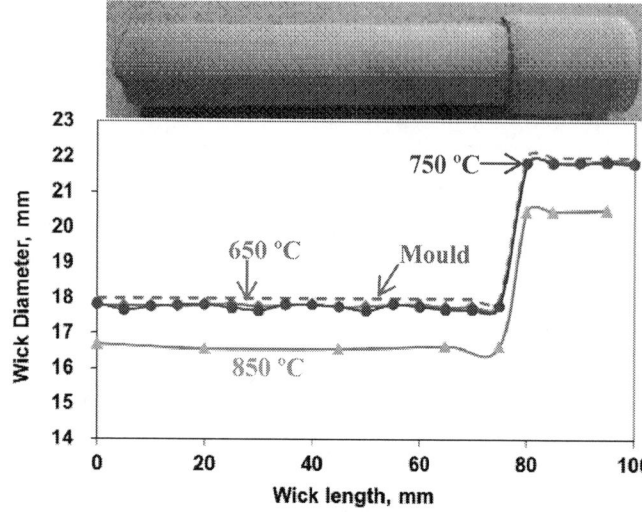

**Figure 9** Diametric variation of NM-12 wick at different temperatures

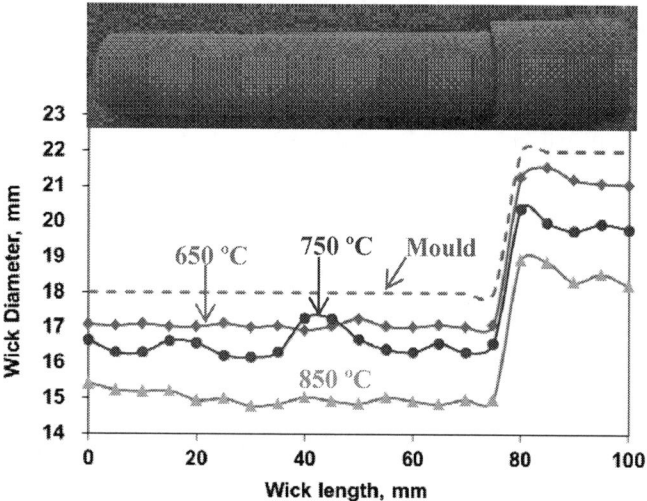

**Figure 10** Diametric variation of T-2 wick at different temperatures

### 4.4. Sintering Issues

The main issues faced in the sintering experiment were related to 1) powder charging related defects, 2) high volumetric shrinkage of the final sintered wick (particularly for smaller powder types) and 3) wick extraction (from mould). To solve these issues, number of developments in the sintering procedure and mould were done.

Improper powder charging of the powder into the mould resulted in improper wick geometry, random cracks on wick outer surface and breakage of wick during extraction. Figure 11 shows the defected and broken wick sample due to inadequate charging. Entrapment of air (air pockets) inside powder volume (i.e. poor powder compaction) and insufficient powder charge at the sharp corners/steps are two most common reasons for charging related defects. To rectify these issues, the powder was charged slowly, in small increments and the mould was agitated at high frequency using ultrasonic vibrator which helped to avoid air entrapment and provide consistent powder charge in the sharp profiles and bulk of mould.

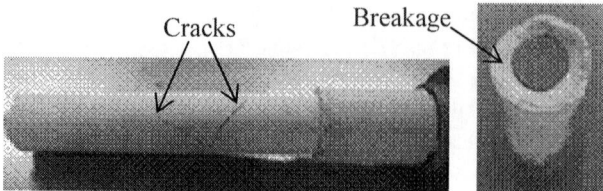

**Figure 11** Powder charging related defects in the sintered wick

Volumetric shrinkage of sintered wick is an unavoidable physical phenomenon that results due to the diffusion of the powder from high concentration to low concentration areas (empty spaces between powder particles) during sintering process. To obtain wick with required volume and geometry, a larger size moulds that were oversized, as compared to original mould, on the basis of volumetric shrinkage data, presented in Figure 8 & 9 were designed and used for sintering experiment. For instance, for sintering at 750 °C or 850 °C, NM-12 mould dimensions should be 2 or 8% oversized respectively. For T-2 wick sintering, corresponding oversizing of dimensions should be 6 and 17% respectively.

The most critical issue in sintering experiment was the extraction of the mould center rod (as shown in Figure 4(a)) from the sintered wick core. Two main reasons for this occurrence were 1) shrinkage stress created by wick on center rod on cooling after sintering process, and 2) wide difference in the coefficient of thermal expansion (CTE) of carbon (2 x $10^{-6}$ K$^{-1}$) and nickel (13 x $10^{-6}$ K$^{-1}$). It was observed that center rod extraction was easier for sintering temperature below 750 °C however for higher temperatures it was increasing problematic due to high shrinkage ratios. To resolve this issue, different approaches as follow were developed:

○ Taper center rod: Replacing cylindrical rod with taper-shape rod increases the workable temperature of carbon center rod, however there were still incidents of rod breakage during extraction, due to low strength of carbon rod.

○ Double sintering technique (rod alteration method): In this method, the final wick structure was obtained by two time sintering; firstly, low temperature sintering (650 °C) with regular cylindrical center rod followed by high temperature sintering (750 ‐ 850 °C) with regular rod replaced by smaller diameter center rod. This method was based on the fact that; center rod was easily extractable after low temperature sintering; and center rod diameter for high temperature sintering was based on the shrinkage tolerance for wick internal diameter at given sintering temperature. With this method, the altered carbon center rod was successfully extractable from the final sintered wick however due to poor support provided by smaller diameter rod during second time sintering process, the axial waviness of the diameter was high (poor straightness).

○ Boron Nitride (BN) coated Stainless Steel (SUS) center rod: This method gives the best result with respect to the easiness of center rod extraction and axial straightness of sintered wick. The method was based on the facts that 1) Boron Nitride can provide very lubricious, non-wetting, chemically inert and anti-oxidation barrier thermal coating for high temperature applications, 2) CTE of SUS ($17 \times 10^{-6}$ K$^{-1}$) is higher than nickel ($13 \times 10^{-6}$ K$^{-1}$), it means that SUS rod will shrink more than nickel wick on cooling after sintering process which will provide easy release of center rod, and 3) BN coating process is very simple and cost effective and it can be easily coated on the SUS (which is cheaper than carbon rod) surface. Using this method, the center rod was easily extractable from the core of wick sintered at 850 °C. Unlike double sintering method, in this case, acceptable geometrical shape and straightness of the wick was achieved due to the proper support at the wick core.

## 5. Conclusions

In summary, wick structure sintered from nickel powder can be qualified as one of the most promising capillary pump for loop heat pipes due to lower thermal conductivity of nickel and its ability to be sintered into required wick shape with high porosity, high permeability and small effective pore radius. Nickel powder can be successfully sintered when heated at 750 - 850 °C temperature for one hour in nitrogen atmosphere. Out of four highly pure (> 99.5%) nickel powder with different average powder sizes, NM-12 powder with average particle diameter of 12 µm was able to provide highly porous (> 72 %) wick, sintered at 850 °C, with high permeability (> $2 \times 10^{-13}$ m$^2$), fine pore radius (< 7.2 µm), low volumetric shrinkage (< 22%), good straightness and acceptable strength for designed LHP, with 500 W heat transfer capability up to distance of 250 mm, in horizontal mode. Center rod extraction was the major problem faced in the sintering experiment which was rectified by replacing carbon rod with high strength, lubricious center rod made from stainless steel with boron nitride coating.

## References

1. Maydanik, Y. F., Vershinin, S. V., Korukov, M. A., Ochterbeck, J. M., "Miniature Loop Heat Pipes — A Promising Means for Cooling Electronics", IEEE Trans. Compon. Packag. Technol., Vol. 28, No. 2, pp.   290–296, 2005

2. Singh, R., Akbarzadeh, A., Dixon, C., Mochizuki, M., Riehl, R.R., "Miniature Loop Heat Pipe with Flat Evaporator for Cooling Computer CPUs", IEEE Trans. Compon. Packag. Technol., Vol. 30, No. 1, pp. 42 – 49, 2007

3. Singh, R., Akbarzadeh, A., Mochizuki, M., "Thermal Potential of Flat Evaporator Miniature Loop Heat Pipes for Notebook Cooling", IEEE Trans. Compon. Packag. Technol., Vol. 33, No. 1, pp. 32 – 45, 2010

4. Singh, R., Akbarzadeh, A., Mochizuki, M., "Effect of Wick Characteristics on the Thermal Performance of the Miniature Loop Heat Pipe", Journal of Heat Transfer, Vol. 131, No. 8, pp. 082601-1-10, 2009

5. Singh, R., Akbarzadeh, A., Mochizuki, M., "Experimental Determination of Wick Properties for Loop Heat Pipe Applications", Journal of Porous Media, Vol. 12, No. 8, pp. 759 – 776, 2009

# A Comparison Analysis of Air, Liquid, and Two-Phase Cooling of Data Centers

M.M. Ohadi, S. V. Dessiatoun, K. Choo, and M. Pecht
Center for Advanced Life Cycle Engineering (CALCE)
University of Maryland, College Park, MD, USA 20742
Advanced Heat Exchangers and Electronics Cooling Laboratory
University of Maryland, College Park, MD, USA 20742
Corresponding Author: ohadi@umd.edu

John V. Lawler
Advanced Thermal and Environmental Concepts, Inc.
7100 Baltimore Ave., Suite 300, College Park, MD 20740

## Abstract

This paper provides analysis of air vs. liquid and two-phase flow cooling for a data center application. A new micro channel-based forced convection evaporation cooling is introduced, and its performance is compared against single phase and conventional phase change cooling systems. The technique offers substantially reduced thermal resistances with associated pumping power requirements significantly below that of conventional systems. It removes the need for compressors in a typical phase change cooling as it relies on a combination of forced convection boiling and thin film evaporation mechanism. Comparison analysis of the three techniques may provide additional incentives for adoption of energy efficient liquid cooling in next generation data centers. Issues remaining with large scale adoption of liquid and phase change cooling in data centers are also addressed in this paper.

## 1. Introduction

Data centers continue to be the backbones of the information economy, crucial to universities and government institutions, and the financial services, medical, media, and high-tech industries. Data centers are the buildings, facilities, and rooms that contain enterprise servers, server communication equipment, and cooling and power equipment. Recently, many companies have built or plan to build new and large data centers to provide diverse services, including cloud-computing services. Collectively, planned spending is expected to be on the scale of billions of US dollars in the coming years.

The continuous growth of communication and data storage industries has caused a rapid increase in the energy consumption of data centers, to the extent that data centers now represent an important share of the national energy consumption and carbon footprint. In the U.S. alone, the energy consumption of data centers is approximately 2% of total U.S. energy consumption and continues to increase steadily [1]. The electricity cost to remove the heat generated from the server racks has continued to rise to the point that the four-year energy costs of operating many data centers exceeds their purchase price.

A typical energy distribution in data centers is shown in Figure 1. The components that consume the most energy are the IT equipment and cooling systems, with combined power consumption of more than 80% of total energy. In general the cooling equipment uses almost the same amount of energy as the IT equipment.

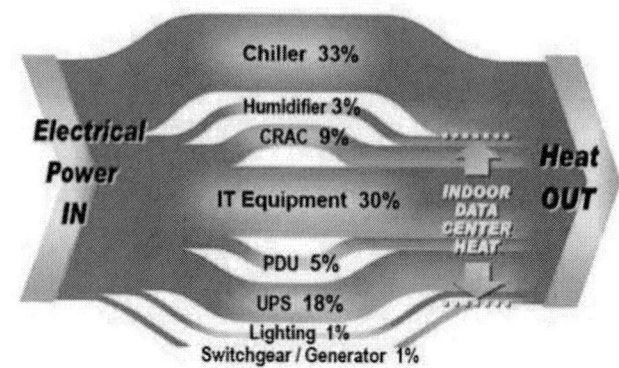

**Figure 1: Typical distribution of electricity consumption in data centers [2]**

A majority of existing data centers use air-cooling systems to maintain desired operating conditions. In typical air-cooling systems, heat generated by the processor is conducted to a heat sink and transferred to the air blowing into the server. The air in turn is cooled by chilled water, which is maintained at a sub-ambient temperature in order to produce sufficient heat transfer, and thus a chiller/refrigeration system is needed to chill this cooling water. Air-cooling has often been preferred due to its wide usage in computer cooling, proven high reliability, and lower initial and maintenance costs. However, the heat dissipation currently associated with some of the processors in new servers is already too high to cool with standard air-cooled heat sinks and fans. These new cooling challenges demand new cooling schemes that not only provide improved cooling, but also reduce energy and equipment costs. However, for sufficiently cold climates where cold external air is available year round, air-cooling may continue to be economical and the most preferred option, as will be discussed in later sections of this paper.

## 2. Analysis of Air, vs. Liquid, vs. Two-Phase Flow Cooling

**Air-cooling.** In typical air-cooling systems, heat generated by the processor is conducted to a heat sink and from there is transferred to the air moving through the server by a fan or several fans located at the back of the server. Some low-profile servers have up to a dozen small fans rotating at speeds up to 10,000 rpm. Cooling air is supplied through an under-floor plenum to the cold aisle in front of the racks and exits on the back to the hot aisle, as shown in Figure 2. Hot air rises and moves to a Computer Room Air-Conditioning unit (CRAC) where it is cooled by chilled water, which must be maintained at a sub-ambient temperature in order to produce a sufficient heat transfer rate, and thus a heating, ventilation, and air conditioning (HVAC) system is needed to chill this cooling water. Heat is usually rejected to ambient at an elevated temperature through a cooling tower or air-cooled condenser for smaller units.

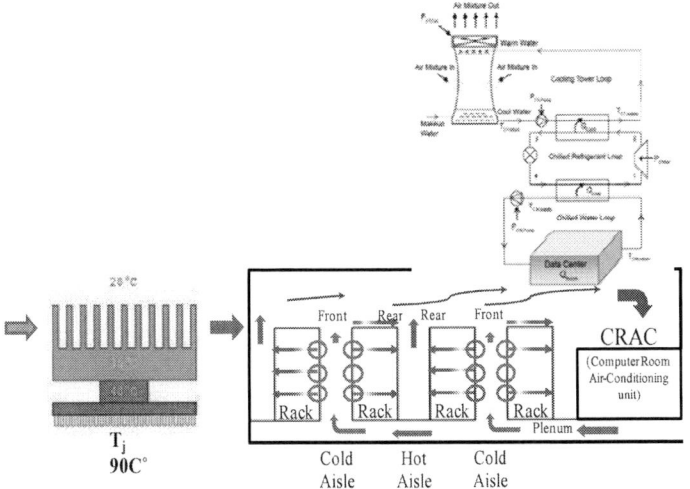

**Figure 2: Traditional air-cooling and the significant resistances between the source and the sink**

Several points of thermal resistance and energy loss are associated with the cooling system depicted in Figure 2. One major point of thermal resistance is between the heat generating processor and the heat sink. This thermal contact resistance is a focus point of research and development. It can be eliminated only by using a chip-integrated heat sink.

Another major point of thermal resistance is between the heat sink and the air. Our CFD simulations of a standard air-cooled heat sink on an 85 W source found that the incoming air had to be cooled to 5°C to keep the temperature of the CPU below 78 °C (air flow rate of 12.7 L/s or 26.7 CFM, $\Delta P = 2.26$ Pa, and pumping power of 29 mW), as shown in Figure 3. Two issues with air-cooling are the significant bypass that occurs and the mixing of warm and cold air downstream of the heat sink, which reduces further the feasibility of harvesting this waste heat. As a result, the heat dissipation of modern CPUs and GPUs now requires the use of quite large finned heat sinks (weighing about 600 g of copper), with multiple fans used to generate the required air flow rate, as shown in Figure 4. This large flow rate of warm air is released into the room and is eventually cooled by an HVAC system. The delivery of the hot air to the CRAC is usually associated with mixing of hot air with cold incoming air and additional heat losses. This issue is significant to the design of newer datacenters. Newer cabinets incorporate chilled water-air heat exchangers, illustrated in Figure 5, so the warm air is cooled inside the cabinet; otherwise, the cabinet is heat neutral to the room. This eliminates mixing losses and significantly reduces the air circulation through CRACs and associated energy losses.

More recently, there has been renewed interest in incorporating two-phase evaporators into the cabinet door, eliminating the intermediate refrigerant-water heat exchanger. Incorporating the two-phase evaporators in the cabinet further

reduces thermal resistance, reduces the cost of equipment, and eliminates the possibility of water damage to the IT equipment. However, the thermal resistance between the heat sink and the ultimate sink (ambient air) still remains high, as do the associated high levels of air circulation, friction losses and noise.

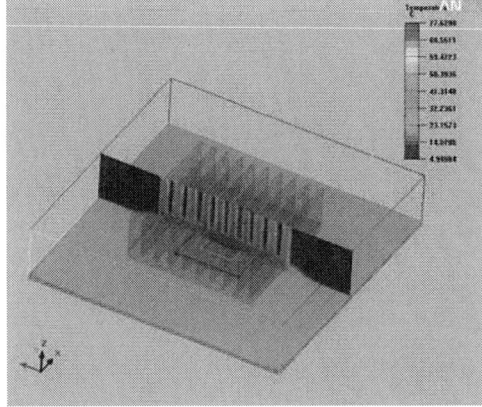

**Figure 3: Temperature contour from CFD simulations of air flowing in a cold plate**

**Figure 4: Photograph of air-cooled heat sink inside a server**

*"Free" Air Cooling.* Free air cooling inherently represents a cooling method that considerably reduces the energy consumption. It uses cool air of temperature and humidity acceptable to the data center electronics as the primary fluid to cool the equipment directly, thus shutting off the chiller equipment and yielding substantial savings in the energy consumption of the air conditioner.

Free air cooling is usually utilized through air-side economizers with a set environmental temperature and humidity range. This environmental range can be based on the recommended operating ranges given by the published standards and/or industry guidelines that may exist. If the outside ambient air conditions are within this range, or if they can be brought within the range by mixing of cold outside air with warm return air, then outside air can be used for data centers for cooling via an air-side economizer fan. If the conditions achievable by economization and mixing of outside air are beyond the set ranges, a chiller and humidity control system could be implemented to internally re-circulate conditioned air instead of outside air for cooling. In some extreme environmental cases, the air side economizer can be isolated in favor of a backup air-conditioning system.

978-1-4673-1110-6/12 $31.00 © 2012 IEEE

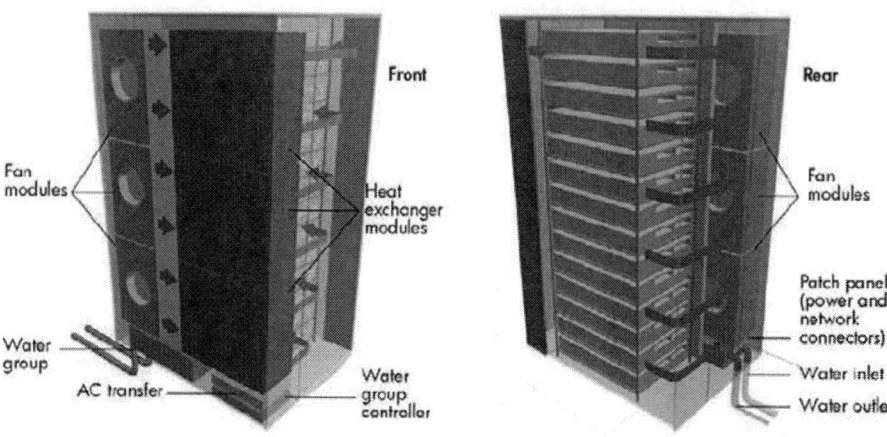

**Figure 5: Chilled water-air heat exchangers incorporated into cabinet door.**

The advantage of free air cooling as an energy saving measure is that it is naturally available and is an inexpensive option. Despite this inherent benefit, it poses several operating environment challenges which need to be addressed by the supply chain of the data center including the operators, regulators, equipment suppliers, and the standards community including ASHRAE. Such challenges include, but not limited to, the following:

- Elevated temperature variation – Storage, server, and communication equipment manufacturers typically specify maximum ambient temperature for operation of their electronic equipment/servers. Long term reliability can be reduced due to thermal cycling and elevated temperatures, which are likely to be higher in a free air cooling system.

- Wider humidity range –Data centers are often maintained between 40 to 60% relative humidity. This controlled environment provides protection against a number of corrosive failure mechanisms, such as corrosion and various forms of electrochemical migration.

- Contaminants (gases and particulate) – With air from outside, various reactive gases can enter the flow within a data center. These gases can interact with metals on the circuit boards and components and can accelerate various failure mechanisms. In the presence of high humidity, some of these mechanisms are accelerated. For example, at 60% RH, a layer of moisture, a few molecules thick, can form on most surfaces. This level of moisture can concentrate the reactive gases from the air, thus gradually damaging the protective oxides on the metals used in data center server electronics. In addition, presence of dust can accelerate several other failure mechanisms, depending on the size and density of the particulate contaminants.

The data center industry is awaiting a set of universal standards/guidelines that can define at least the minimum standards for the use of free air cooling, thus preventing undue compromise in lifetime and/or reliability of the data center electronics.

**Direct liquid-cooling.** This method eliminates two of the least effective heat transfer processes: heat-sink-to-air and air-to-chilled-water. Liquid-cooling improves heat transfer efficiency, decreasing significantly the overall thermal resistance of the heat transfer circuit, energy consumption, and the size and cost of equipment. An IBM study in 2009 [3] found that liquid-cooling can be 3500 times more efficient than air-cooling. Their tests showed a 40% reduction in total energy usage with liquid-cooling. Liquid-cooling also improves working conditions for data center personnel by reducing the noise level, since the multiple fans per server used in air-cooling can be eliminated. However, the choice of cooling liquid is problematic. Water, which has good thermal properties, can damage electronics if leaks occur. Electronic-friendly dielectric liquids, such as certain refrigerants, have poor thermal properties in the single phase and are costly. A recent study by Mandel et al. [4] indicates ammonia may be an optimum working fluid for efficient cooling of high flux electronics; however, the use of ammonia carries safety risks and is therefore regulated.

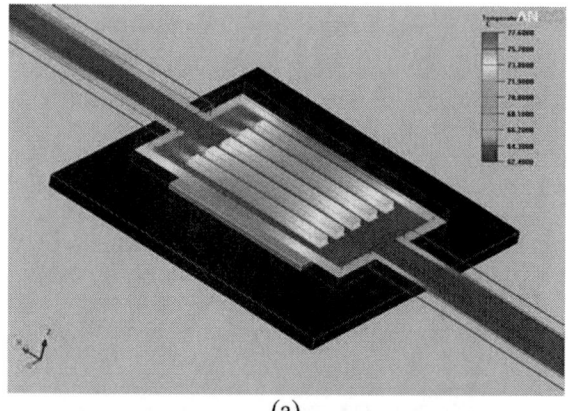

(a)

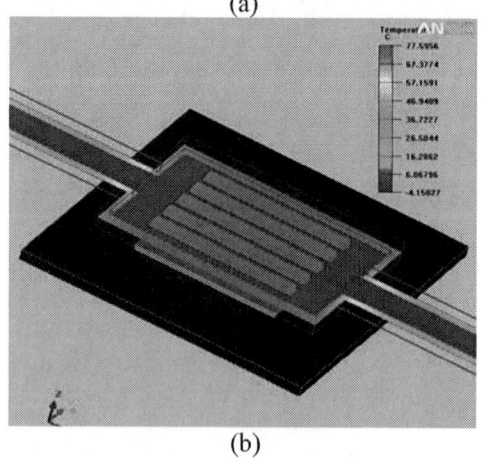

(b)

**Figure 6: Temperature contours from CFD simulations for: (a) water with $T_{in}$ = 62.4 °C (b) dielectric fluid flowing with $T_{in}$ = -4.15 °C.**

Similar to the air-cooled heat sink CFD simulations mentioned above, simulations using a microchannel cold plate on an 85 W source and with both water and FC-72 as cooling fluids showed that incoming temperatures for water could be about 62°C (flow rate of 18 cm$^3$/s or 0.04 CFM, ΔP = 32 kPa, and pumping power of 57 mW). However, the dielectric fluid, such as R-134a, would need an incoming temperature of -4 °C (flow rate of 18 cm$^3$/s or 0.04 CFM, ΔP = 31 kPa, and pumping power of 56 mW). The temperature contours from these simulations are shown in Figure 6. These results indicate that if water (or another high performing fluid such as ammonia) is used as the cooling fluid, the server cooling system for the data center could be operated without compressors, thus saving a significant amount of energy, as illustrated in Figure 7a. However, if the cooling fluid is limited to dielectric fluids, then compressors will be required, as illustrated in Figure 7b.

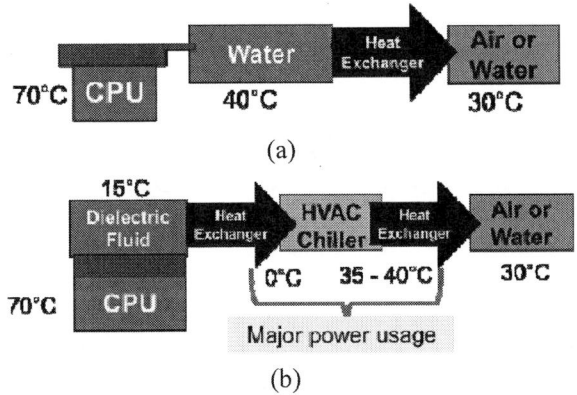

(a)

(b)

**Figure 7: Heat path I data center: (a) water cooling (b) dielectric fluid cooling.**

**Direct two-phase flow cooling.** Direct two-phase refrigerant cooling for data centers, if properly implemented, can eliminate the use of chilled water and HVAC equipment needed to sub-cool the chilled water. In order to eliminate the need for sub-cooling, the thermal resistance of the cold plate in the two-phase heat transfer must be quite low—much lower than currently available cold plates. This is already possible with some of the emerging technologies in embedded liquid cooling. For example, thin film manifolded microchannels, shown in Figure 8, yield remarkably low thermal resistance between the heat source and the cooling fluid [4 and 5]. Thermal resistance of a manifolded microchannel heat sink for chip cooling application can be as low as 0.04 K/W, illustrated in Figure 9, compared to conventional commercially available off-the shelf cold plates thermal resistances of 0.15 to 0.20 K/W for dielectric fluids. In most cases this thermal resistance is low enough that the vapor compression cycle of the HVAC equipment is not required. Heat from the chip can be directly rejected to the ambient of almost any climatic zone using only a pumped refrigerant loop with evaporation on the chip side and condensing of the refrigerant in an air-cooled condenser or a much smaller HVAC system than would otherwise be required by conventional systems.

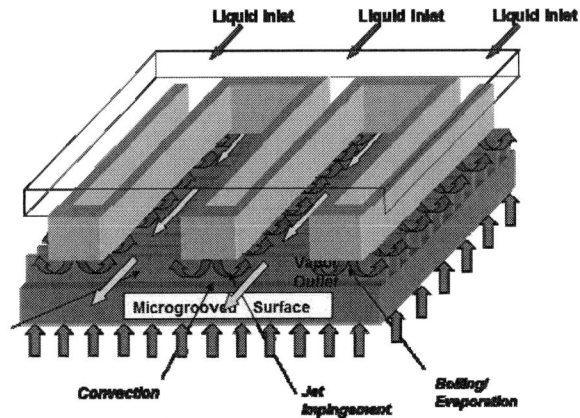

**Figure 8: Thin film manifold microchannel cooling [5].**

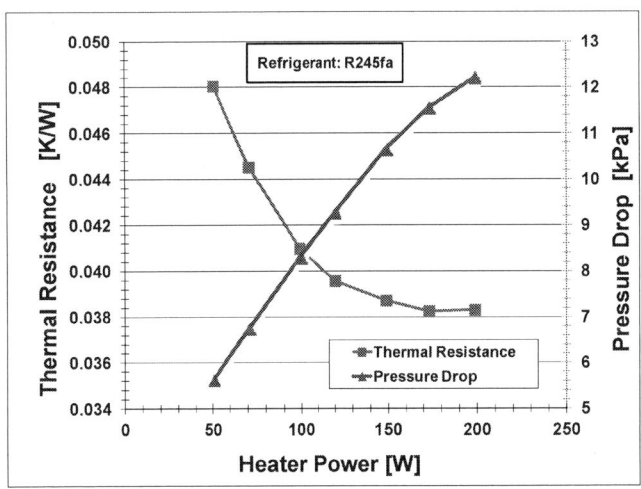

**Figure 9: Thermal resistance and pressure drop of two-phase thin film manifold cooling [5].**

**Figure 10: Temperature contours from CFD simulations of two-phase thin film manifold cooling.**

For the two-phase CFD simulations, a heat transfer coefficient obtained from earlier heat transfer experiments was used. The results showed that an entering liquid temperature of 76.5°C would be sufficient to adequately cool an 85 W CPU (flow rate of 0.54 g/s or 0.46 cm$^3$/s, ΔP = 5 kPa, and pumping power = 2.3 mW), as shown in Figure 10. The fluid exiting these cold plates could easily be cooled using ambient air due to the elevated temperature of the fluid. Thermal energy at such temperatures can be also be used as low-grade thermal energy for heating or cooling through a thermal driven (absorption) heat pump. Other alternatives are also possible, such use of the warm water for district heating and other applications.

Despite its many advantages, there are some challenges in deploying two-phase cooling for servers in data centers. Direct cooling does introduce a significant number of tubing connections. The fluid/vapor handling system must allow individual servers to be swapped in and out of a cabinet. This can be made easier by redesigning server layouts to allow external access to the top surfaces of the CPU's and GPU's. A control system will be needed to detect, isolate, and stop fluid leaks. To date, these issues have delayed deployment of two-phase cooling in commercial systems, thus these issues need to be addressed before any major adoption and full scale technology change takes place.

**A comparison between air, liquid, and two-phase flow cooling.** Table 1 provides a summary of our simulation analysis on air, liquid, and two-phase cooling for the selected application. The thermal resistance of liquid cooling is less than one half of corresponding air-cooling due to the high heat capacity of the liquid. However, the pumping power of the liquid cooling is twice that of air-cooling due to higher pressure drops. In heat transfer design, it is fairly common to trade improved heat transfer performance for increased pressure drop, as long as the associated pressure drop penalties are not excessive.

From the results in Table 1 it is clear that the two-phase flow cooling provides substantially reduced thermal resistance, as much as an order of magnitude less than that of air and significantly below that of liquid cooling. Naturally, the pressure drop associated with conventional two-phase cooling is higher than single-phase liquid cooling, due to presence of vapor acceleration and the inherent pressure drop losses inherent with the phase-change phenomenon [6, 7, 8 ]. As is established in an earlier study [9], the additional pressure drops lead to higher pumping power, thus higher operating cost of the cooling system. Thermal performance increases as pumping power increases, but the operating cost also increases. Therefore, the best cooling method would have high thermal performance with low pumping power. In search of such technique, recently, a thin film manifold cooling which yields remarkably low thermal resistance with low pumping power was introduced by Cetegen [5], with further results available in Mandel et al. [4]. As shown in Figure 11, two-phase cooling by the force-fed manifolded micro channels (Figure 8) has 20 times lower thermal resistance than liquid cooling at 5 times lower pumping power consumption. This is contrary to the generally held view that two-phase cooling flows require higher pumping power than single phase cooling flows. The main reason for this behavior is the fact that the governing regime in the force fed micro channels is a combination of forced convection boiling and thin film evaporation over high aspect ratio micro channels with limited fluid flow running length. An optimized design will aim for dominance of thin film evaporation throughout the process by achieving high vapor quality at the exit while requiring minimum possible fluid flow in circulation. The heat transfer coefficients associated with thin film micro channel cooling are an order of magnitude or higher than what has been reported for two-phase flow for the same application. The magnitude of the heat transfer coefficient is inversely proportional to the film thinness on the wall, thus optimized design of the heat transfer surface and the manifold system can lead to heat transfer coefficients that are on the order of 1000 kW/(m$^2$-K) or higher. Meanwhile, the corresponding pressure drops are significantly below that of liquid cooling or conventional two-phase flow cooling. An economic analysis of use of such systems for data center cooling indicates substantial savings in the cost of infrastructure, including building heights, ducting systems, associated air leaks, and operating costs of the air handling units and the chiller system. More detailed economic feasibility of thin film cooling vs. liquid and air cooling of data centers will be given in a future publication by the same authors.

**Table 1. A comparison between air, liquid, and two-phase cooling**

|  | Air | Water | Dielectric fluid (FC-72) | Two-phase flow R-245fa |
|---|---|---|---|---|
| Generated power | 85 W | | | |
| Fluid inlet Temp. ($T_{in}$) | 5 °C | 62.4 °C | -4 °C | 76.5 °C |
| Thermal Resistance ($R_{th}$) | 0.4 – 0.7 K/W | 0.15 – 0.2 K/W | 0.15 – 0.2 K/W | 0.038 – 0.048 K/W |
| Pumping power ($P_{pump}$) | 29 mW | 57 mW | 56 mW | 2.3 mW |

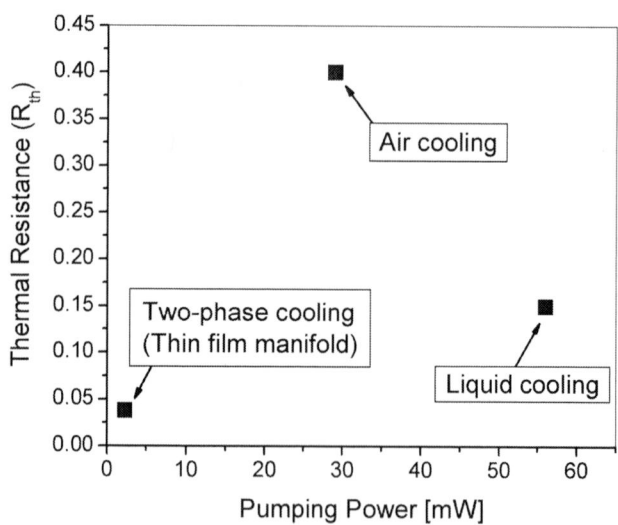

**Figure 11: A comparison of thermal resistance between air, liquid, and two-phase cooling.**

## 3. Emerging Trends

Much of the success of the electronics industry in the past four or so decades is often attributed to three sustained and distinct trends:

I) The increased number of transistors on a chip (Moore's law and the "more than Moore's law") and the resulting increased functionality and speed

II) The reduced feature size, thus enabling continuous and substantial miniaturization

III) The consistent and substantial reduction in cost (more than 1,000,000 times over the past four decades).

Moreover, it appears that for the foreseeable future the CMOS technology will continue to prevail, and thus if these trends are to continue, new challenges will be introduced at all levels of the packaging and implementation. Perhaps the chief challenges are to reduce the energy consumption of individual electronic components and complex systems, such as data centers. The demand for energy efficiency is felt by almost all major industries, considering that (a) more than 80% of total world energy supply by 2030 will still be coming from the limited fossil fuel reserves, and (b) the increased (close to twice as much) energy consumption of developing countries (led by China and India) will increase the world energy demand by more than fifty percent by 2030.

Combining two-phase cooling and advanced cold plates, such as the manifolded microchannels geometry, has the potential to yield significant capital and life cycle cost savings. Such savings include the elimination of need for compressors (since sub-ambient fluids are no longer needed), fans and heat exchangers, the reduction in the amount of cooling fluid being pumped around the system, the reduction in the size of other components, and the reduction in the amount of electricity required to operate the cooling system. Future data centers utilizing advanced phase change cooling systems will have a significant competitive advantage over conventional systems, due to their lower capital costs, operating costs, and energy savings appeal. Moreover, such systems will allow reductions in the sizes of servers by eliminating the need for air flow within the server housing and of cabinets. These space savings can be combined with corresponding reductions in the open space around the datacenter cabinets to generate savings in real estate and building expenses. Even without realizing all of these advantages, the total energy savings from a redesign of an integrated cooling system can lead to potential energy savings of 75% or higher compared to a traditional air-cooled data center.

## 4. Summary

This paper provided performance comparison analysis for cooling of electronics in datacenters by air, liquid and two-phase flow cooling. Advantages and the disadvantages of each of the methods were also discussed. It was demonstrated that air cooling may remain very attractive for select locations where cool air climate is available most of the year. For optimum two-phase cooling, a proprietary manifolded microchannel cooling technology was introduced. Performance comparison with conventional liquid and phase change cooling techniques demonstrates this new cooling technology produces a substantially smaller thermal resistance, with minimal or no additional pressure drop. The associated life cycle cost savings stem from reduced capital and operational cost of cooling systems, savings in reduced

building space and infrastructure, and the total system integration and performance optimization. The important issues holding up a serious investment in liquid and phase change cooling were discussed. One important issue is the lack of industrial standards to regulate use of free air cooling, liquid cooling, and phase change cooling. In the absence of such standards the industry may be slow to adopt technologies that will result in substantial reductions in the usage of energy, capitol, and infrastructure resources.

## References

[1]. Jonathan Koomey. 2011. Growth in data center electricity use 2005 to 2010. Oakland, CA: Analytics Press. July. <http://www.analyticspress.com/data centers.html>

[2]. "The Total Cost Ownership of Data Centers", Qpedia Thermal Magazine, Vol. 5 (8), August 2011, 14-16.

[3]. Schmidt, Roger, "Packaging of New Servers, Energy Efficiency Aspects," presentation given at the Charting a Course to Energy Independence Conference, Aug. 9-12, 2009.

[4]. Mandel R., Dessiatoun S.V., and Ohadi M.M., "Analysis of Choice of Working Fluid for energy efficient cooling of high flux electronics," Progress Report, Electronics cooling consortium, CALCE/S2Ts lab, December 2011.

[5]. Cetegen, Edvin, "Force Fed Microchannel High Heat Flux Cooling Utilizing Microgrooved Surfaces" Ph.D Thesis, University Of Maryland, 2010.

[6]. Choo K., Kim S.J., "Heat transfer and fluid flow characteristics of Nonboiling two-phase flow in microchannels," ASME J. Heat Transfer 133, 102901, 2011

[7]. Ghiaasiaan, "Two-phase flow, boiling, and condensation in conventional and miniature systems," Cambridge University, Cambridge, England, 2008.

[8]. Wang et al., "Characteristics of an Evaporating Thin Film in a Microchannel", International Journal of Heat and Mass Transfer, Vol. 50, pp. 3933 – 3942, 2007

[9]. Choo K.S., Kim S.J., "Heat transfer characteristics of impinging air jets under a fixed pumping power condition," Int. J. Heat and Mass Transfer 53, , 320-326, 2010

# Design and Management of Data Center Effectiveness, Risks and Costs

Mark Seymour, Sherman Ikemoto
Future Facilities
1, Salamanca Street  and  2055 Gateway Place, Suite 110
London, UK                    San Jose, USA
Mark.Seymour@futurefacilities.com, Sherman.Ikemoto@futurefacilities.com

## Abstract

Data centers have been cooled for many years by delivering cool air to the IT equipment via the room. One of the key advantages of this approach is the flexibility that it provides the owner / operator in terms of equipment deployment. In principle it seems that all that is necessary is to determine the maximum power consumption of the equipment and provide an equivalent amount of cooling to the data center. Why then, since we have been building and operating data centers for decades using air cooling, do data centers experience hot spots, operate inefficiently and fail to reach their design expectations for capacity?

This paper explains some of the challenges faced in the search for the perfect data center. In particular it identifies why, given the variability of equipment design and the time varying nature of the data center load, Computational Fluid Dynamics (CFD) should be used for airflow and heat transfer modeling. It shows that although it is not the only tool to be used in design and / or management of a data center, it is an essential tool to avoid lost capacity due to inability to efficiently cool equipment and the resulting potential for overheating.

Important themes include a strategy that encompasses the entire life cycle of the data center, the need for an appropriate level of detail and the critical requirement for model verification and calibration.

## Keywords

Data Center, Cooling, CFD, Simulation, Overheating risk, Design & Management

## 1. Conceptual Design - Is kW enough?

This is not a CFD question at its most basic level. Traditional design simply ensures that the cooling system provides sufficient cooling, on average, in terms of kW per area or per cabinet. When power densities were low, this was sufficient since any re-circulation of air only resulted in moderately warmed air entering the equipment. However the increase in heat density in electronics and the fact that the cooling system does not operate in a purely serial manner means that the assumption "sufficient kW is enough" is fundamentally flawed.

Figure 1 illustrates how over supply of cooling air is likely to result in cool air by-passing the equipment resulting in low air temperatures returning to the cooling units and poor energy efficiency and increased costs. Under supply on the other hand means that some equipment will have to use recirculation of room air resulting in higher inlet temperatures and risk of over-heating. To mitigate the risks, set-points can be reduced on the cooling system if the capacity exists at the

expense of reduced efficiency and higher operational cost or alternatively the less equipment can be installed reducing capacity.

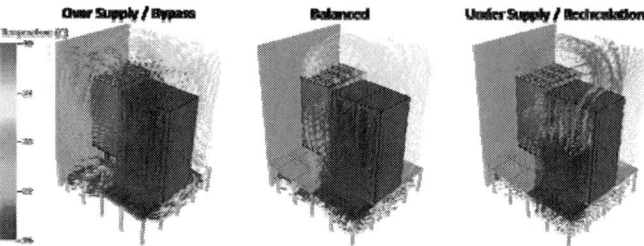

**Figure 1:** The effect of differences in cooling air supply volume and IT air volume.

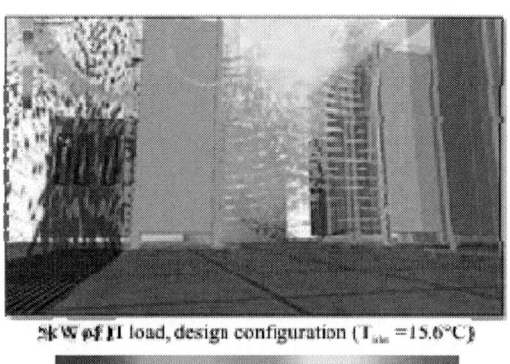

5kW of IT load, design configuration ($T_{sup}$ =15.6°C)

15.6    21    26.4    31.8    37.2
Temperature (C)

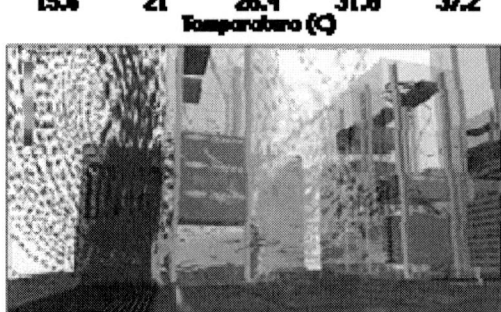

5kW of IT load, design configuration ($T_{sup}$ =37.2°C)

**Figure 2:** Airflow and temperature in a facility with conceptual equipment (top) and realistic equipment (bottom).

It is therefore important to ensure that even a conceptual design considers the airflow requirements of the equipment in the data center with a view to providing enough flow to the right place as well as enough cooling.

Designing a facility from the top down, and more importantly deploying equipment without consideration of the likely impact on the airflow, is potentially a fundamental error as this is an integral part of how the data center 'breathes'. Of

978-1-4673-1110-6/12 $31.00 © 2012 IEEE          64

course at the design stage the end user will probably not be able to identify the exact equipment to be deployed on day 1 let alone the equipment (that has not yet been designed!) for use in later life. The designer should therefore consider sensitivity studies to ensure the design is robust for a range of different equipment deployments.

Figure 2 illustrates how both the airflow patterns and temperature distribution are affected by a change in IT equipment installed even though the both scenarios have the same IT equipment power.

## 2. The impact of real equipment on cooling performance

In a real facility with frequent changes in IT equipment the cooling requirement in terms of temperature and airflow is also an ever changing one. Furthermore, because there are no generally accepted standards, the way the designer of the electronics chooses to package the equipment and cool it internally will change the equipment's demands on its environment; that is the configuration of the rack <u>and</u> the data center. This variability in design (Figure 3.) is most clearly reflected in equipment drawing air in from, and exhausting air from differing locations / faces of the equipment. Further variability comes from the lack of standardization of cooling air volume and consequent temperature rise.

**Figure 3:** Example showing several different equipment airflow regimes. Front to back, upward outflow and mixed side to side and front to back.

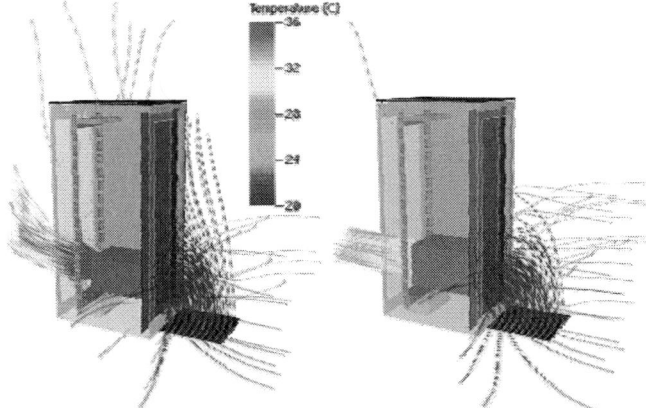

Low CFM, high outlet temperature   High CFM, low outlet temperature

**Figure 4:** Example showing the impact of air volume and temperature rise on the flow from a server.

It is worth recognizing that the variability illustrated in *Figure 4* may be a result of designer choice but also can be a consequence of utilization or environment changes and so this is a time varying requirement. The choice of equipment will, as a result of the equipment design from a cooling standpoint, affect the design, configuration and operation of the facility.

Accepting air is still the dominant vehicle for cooling equipment and that the hardware has variable requirements for cooling, an ideal design will be obliged to have sufficient flexibility in the design for the owner / operator to provide the manufacturer's environmental requirements for any and every piece of equipment, at any point in time.

Lack of consideration of the equipment type and configuration can result in substantially different conditions at both the rack and the room level (Figure 5).

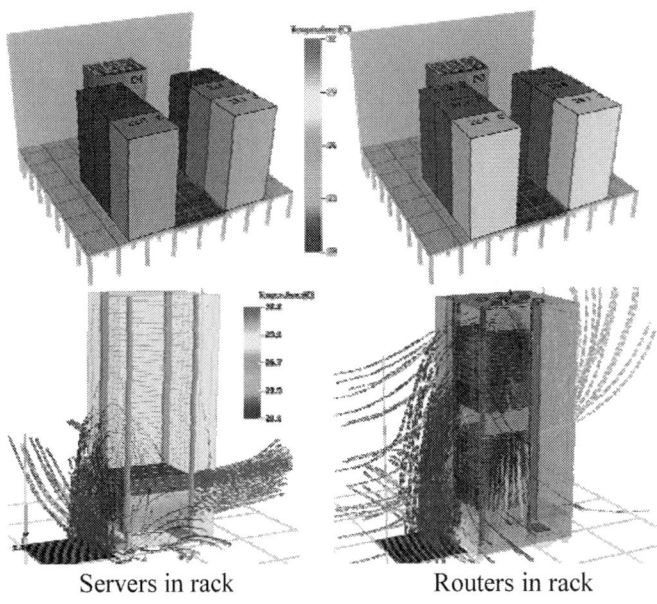

Servers in rack             Routers in rack

**Figure 5:** Example showing front to back servers are cool while routers are hot in the same configuration.

At the conceptual design stage very little will be known about the type of equipment and therefore the airflow demand. As a result the design will be based on nominal values of kW per area or per cabinet. The owner / operator is therefore faced with 2 challenges:

a. How to deploy different types of equipment in the data center side by side which have different demands for airflow and cooling when the data center was intended to provide a relatively uniform cooling capability;

b. How to be prepared for and accommodate future generations of equipment with characteristics yet unknown.

A further challenge is that the cooling system is unlikely to deliver airflow and cooling uniformly throughout the data center. This means that the owner / operator must be aware of the impact of location of deployment in the context of the actual cooling distribution. Figure 6 shows the same equipment deployed in 2 different configurations can completely change whether the cooling system is effective or not.

978-1-4673-1110-6/12 $31.00 © 2012 IEEE          65

Arrangement 1 creates high inlet temperatures

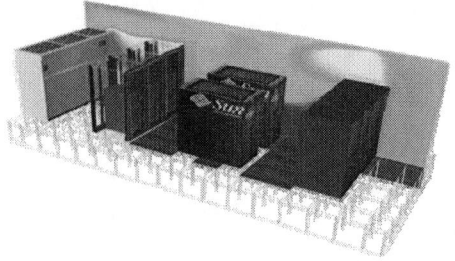

Arrangement 2 creates lower inlet temperatures

**Figure 6.** Example showing 2 different spatial deployments of 2 items of equipment – only option 2 works.

Although success or failure can be a result of the equipment airflow and cooling design, the performance can only be addressed when it is considered in conjunction with center configuration. Figure 7 shows that variation in distribution of air supply resulting from a change in cabling configuration in the rack alone.

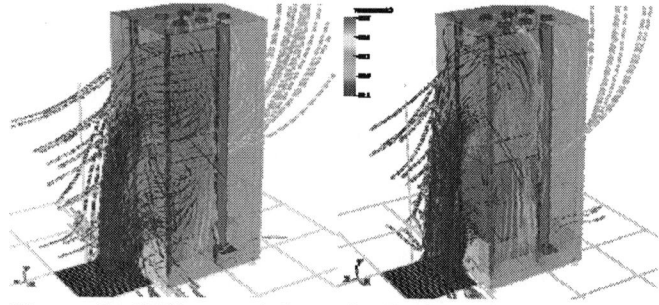

**Figure 7:** Different airflow distribution and temperatures resulting from simple changes to cabling strategy.

Changes to the 3-d configuration of cabling and other obstructions under-floor can be equally critical to the resulting distribution of cool air on a room scale. The resulting variation in temperature will again impact risk unless actions are taken (at the expense of efficiency and cost) to mitigate any increase in air temperature reaching the equipment.

## 3. The evolving data center

It is important to recognize that the data center is an evolving entity in which deployments are easily made but often not easily reversed. The consequences of changes are often not seen until later in the life of the data center. Failure to plan with consideration for these details in the airflow and cooling in respect of deployments of varying IT in a fixed

infrastructure can, and commonly does, result in a gradual deterioration of cooling performance and hence capacity of the data center. In fact it is not unusual for a data center to experience cooling difficulties and hotspots at only 2/3rds design capacity in terms of kW equipment deployed. This apparently unusable capacity is often called 'stranded capacity'. This is appropriate because without further assessment and reconfiguration it may not be possible to use the intended design capacity without creating hotspots. If the business then decides that equipment can no longer be installed, the cost impact of having to build a new data center before time is obvious.

This is not just a data center issue at room level alone but is affected by deployment decisions within the cabinet too. Figure 8 shows the air circulation and temperatures in a cabinet where only 2 IT deployments have been made. First a blade system has been deployed and then 3 1U servers added in the slots immediately above. In this instance the hot air from the blade system is confined to the lower part of the cabinet and recirculates under the cabinet causing high inlet temperatures and reduced resilience.

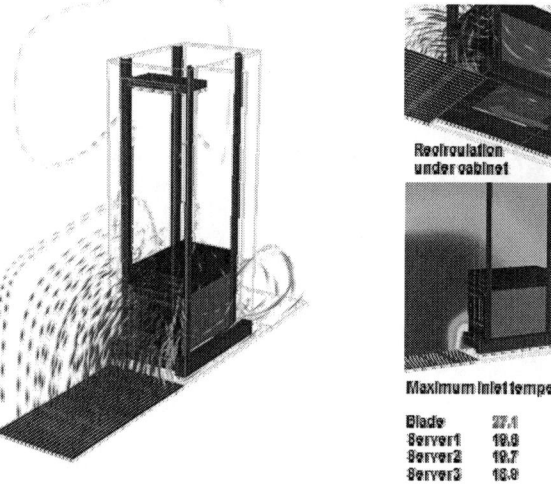

**Figure 8:** Blade system deployed in cabinet below 1U servers.

If the servers are deployed in the opposite sequence however, with the 1U servers in the lower 3 slots and the blade system above then the recirculation is dramatically reduced resulting in lower inlet temperatures and greater resilience.

**Figure 9:** Blade system deployed in cabinet above 1U servers.

The issue of location of equipment in the cabinet affecting the cooling performance is also reflected at room level.

If poor deployment decisions are made this stranded capacity will be reflected by a shortening of the facilities life and a requirement for additional data center capacity ahead of plan.

### 4. Changing cooling strategies.

The increase in power density and soaring energy costs, combined with the growing awareness of a need for environmentally responsible design, has stimulated the fundamental design of data center cooling to be revisited. New approaches, such as aisle containment, are implemented to increase efficiency but the increased efficiency will only be achieved if equipment airflows are appropriately controlled / balanced. If not there may be problems, for example, if recirculation occurs this can cause higher inlet temperatures than in an open scenario. Figure 10 shows example of hot air recirculating into the contained cold aisle causing locally high inlet temperatures for some equipment.

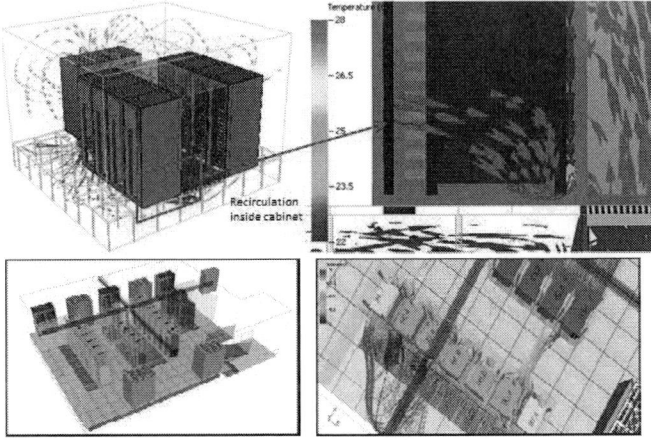

**Figure 10:** Recirculation in poorly contained scenarios.

### 5. Modelling and simulation as a life cycle design and management tool

As you will have understood from the preceding text, many differences in cooling performance depend on relatively small details and characteristics of the equipment configuration and resulting airflow. Computational Fluid Dynamics, or CFD for short, provides a unique tool capable of modeling the data center and equipment installation in conceptual design right through to detailed modeling for operation. Providing a 3-dimensional model of the facility can account for almost any feature and combination of features. In fact this is, in principle, no different than the CFD commonly used in the design of electronic equipment. However, modeling a data center can be somewhat more challenging since the data center is a dynamic changing 'electronics design'. The above illustrations show the potential for CFD to predict the significant impact of these small variations.

Some recommend simplified simulation methodologies [1] to gain computational speed, others [2] provide design or snapshot models using conceptual CFD models. While these can be used for conceptual analysis it has been shown above that details that are not included in simpler models may well dramatically affect the actual performance of the cooling system in the data center. As a result the simpler approaches are insufficient for detailed design or operational management. So, what are the barriers that must be overcome for successful use of CFD? From an academic point of view people may point to the more theoretical aspects of CFD such as turbulence modeling, gridding, solution time and indeed including the full physics, but, in practice, these difficulties are normally far outweighed by the difficulty of capturing the true configuration (such as the features described above) in sufficient detail to predict the resulting environment.

The tool can relatively easily be used for concept design decisions but even here there is need for education about the significant risk of ignoring the type of equipment and their resulting airflow, temperature affects and likely deployment locations. For real facilities it is even more challenging and it is essential that measurements are made alongside the modeling process in order to ascertain that the model reflects reality. Why? Because some details can never be represented precisely (e.g. unstructured cabling, damper settings …) and others may depend on operational factors such as equipment utilization. For the latter it is hard to gather airflow and heat dissipation data for equipment to be able to characterize them fully in deployment. Here the electronics industry can make an important contribution by publishing data more openly and indicate likely trends for planning purposes.

The complexity of the actual data center means therefore that it is important to undertake a careful data gathering exercise. With this data, and with the right tools, it is relatively straightforward to build a CFD model of the data center and check that it reflects reality. Once a 'calibrated' model is achieved it can be used with reasonable certainty to make deployment decisions. It can also be used to undertake other tests such as failure scenarios with confidence.

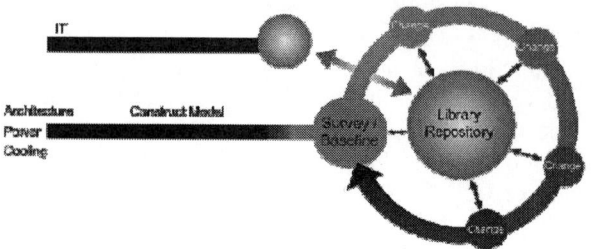

**Figure 11:** Typical process for design, calibration and operation using a virtual facility model.

In the view of the author it is imperative, if the potential for stranded capacity is to be minimized, that simulations be undertaken frequently to understand the implications of deployment decisions and that monitoring and comparison with measured data is made regularly to ensure the model continues to reflect reality. With the increasing availability of measured and live data, maturing CFD tools for data centers and the pressure on energy efficiency whilst maintaining availability, CFD adopted with an appropriate methodology can be a critical tool for design and management.

## 6. Conclusions

Many data centers fail to ever reach their design capacity because hot-spots make it impossible, without a better understanding of the cooling, to safely add additional equipment. This unused capacity, is often termed 'stranded capacity' because the owner / operator is unable to access the ull design capacity. Furthermore the existing data center will almost certainly be operating at a lower Power Usage Effectiveness (PUE) than intended and be at greater risk of overheating. The ultimate outcome, if the situation is not addressed, is that a new data center of data center expansion will be required sooner than planned resulting in substantial additional cost. To manage a data center and avoid the risks and costs associated with stranded capacity it is necessary to understand the airflow accounting for many details in the facility and IT configuration. While a number of alternative tools are available to aid in the design and in particular conceptual configuration of a data center they are not able to capture the critical details that are required for the model to be used for equipment deployment and capacity planning accounting for the likely cooling performance..

## References

1. VanGilder, J, "Real-Time Data Center Cooling Analysis", Electronics Cooling Magazine, September 2011
2. Iyengar, M; Schmidt, R; Caricari, J, "Reducing energy usage in data centers through control of Room Air Conditioning units", 12th IEEE Intersociety Conference on Thermal and Thermomechanical Phenomena in Electronic Systems (ITherm), 2010, pp. 1-11,
3. David J. Cappuccio , "DCIM: Going Beyond IT", Gartner, March 2010

# Novel 3D Electro-Thermal Robustness Optimization Approach of Super Junction Power MOSFETs under Unclamped Inductive Switching

J. Rhayem, A. Wieers, A. Vrbicky, P. Moens, A. Villamor-Baliarda*, J. Roig,

P. Vanmeerbeek, A. Irace**, M. Riccio**, M. Tack.

ON Semiconductor Belgium BVBA, Westerring 15, B-9700 Oudenaarde, Belgium

*Instituto de Microelectrónica de Barcelona (IMB-CNM-CSIC), 08193, Barcelona, Spain

**Department of Biomedical, Electronics and Telecommunication Engineering, University of Naples

Federico II, Via Claudio, 21 80125 Naples, Italy

email :joseph.rhayem@onsemi.com

## Abstract

This paper presents a novel approach to optimize the electro-thermal robustness of a super-junction power MOSFET under unclamped inductive switching (UIS) conditions. The loosely coupled electro-thermal simulation has been used to predict accurately the interaction between the core active device and the termination rings. The simulation results have been validated by the emission microscopy (EMMI) measurements and the transient IR thermography photos.

## Keywords

Super-junction Power MOSFET, UIS, electro-thermal robustness.

## 1. Introduction

Power MOSFET failure due to unclamped inductive switching (UIS) conditions is one of the most prevalent failure modes encountered. The ruggedness, which characterizes the device capability to handle high avalanche currents during the applied stress, is a critical criterion for many applications. In the avalanche regime and during the UIS event, the power MOSFETs generates considerable amount of heat because of huge dissipation of electrical power. In addition to that, the breakdown voltage depends strongly on the junction temperature. Therefore the electro-thermal simulation is necessary to assess the electrical as well as the thermal behavior of the power MOSFETs

The super junction transistor presented in this paper, uses a principle of the local charge balance and is a non-planar structure [1]. Referring to experimental data and TCAD simulations during the UIS pulse [2,3], the avalanche current in the core device flows deeper in the structure, while the avalanche current in the termination rings is located at the top surface. Therefore also the heat dissipation is located at different depth in the silicon which is very specific for this type of power MOSFET and which imposes clear limitations for traditional 1D/2D based finite element analysis and compact thermal models.

In previous works, several approaches of electro-thermal simulations have been presented for power MOSFET. In [4-6] 1D/2D electro-thermal simulation methods were reported to successfully describe the thermal behavior of the power devices. However those 2D methods are not suitable for the type of the power MOSFET (heat sources located at different depth in the silicon) studied in this work, for which the third

dimension is vital to simulate. In [7] a power MOSFET equivalent circuit for the transient thermal impedance is reported for compact modeling purposes. This type of device modeling is relatively simple and offers a solution that can be used in commercial simulators. However this compact type of electro-thermal model does not provide useful information about the device design level. Moreover, for the power MOSFET in avalanche regime (this work), it must be ensured that the maximum permissible junction temperature is not exceeded, which is impossible to solve by conventional means, i.e. using a Zth-diagram.

In more recent work [8], a 3D fully coupled electro-thermal simulation have been reported to evaluate the robustness of power MOSFET in forward condition. The present work focuses on the avalanche regime and uses the loosely coupled method, which represents significant reduction of computation time compared to the fully coupled one.

This paper reports for the first time on a novel 3D approach, for this type of super-junction power MOSFET. The novelty manifests in its capability of predicting accurately the electro-thermal interaction between the core active device and the termination rings. The method presented in this paper, provides good in depth concurrent optimizations of electrical and thermal designs and performance in order to improve the ruggedness of the power MOSFET.

The breakdown voltages as a function of temperature and current have been derived from TCAD simulation in the avalanche regime. An empirical model has been extracted and converted to Spice. The loosely coupled electro-thermal simulations principle has been used. The results of the simulations have been confirmed by the measurements data from the electronic microscopy (EMMI) and from the transient IR thermography photos. The simulation results have been used to optimize the robustness of the power MOSFET.

## 2. Device Description

The device consists of an active area (core device) and a set of termination rings (see Fig. 1a, 1b and 1c). In this device the termination structure does not consist of super junction trenches but a set of rings which provides the freedom to control the voltage window between the breakdown voltage of the termination rings (BVTerm) and the breakdown voltage of the core active device (BVCore). Different scenarios of charge balance and voltage windows between BVTerm and

BVCore have been investigated. This paper reports mainly on the method used by presenting one case of optimal charge balance and BVTerm > BVCore (see Fig. 1d).

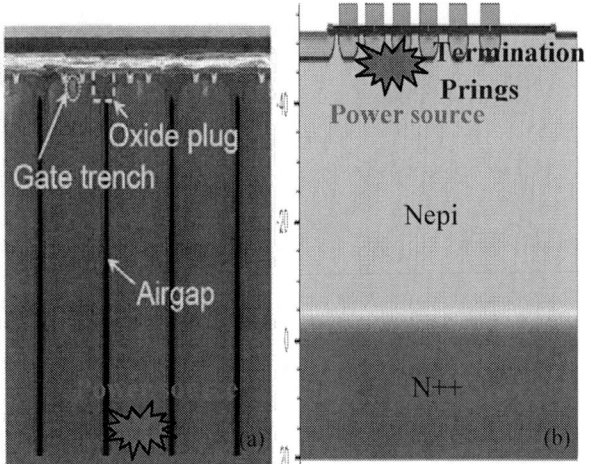

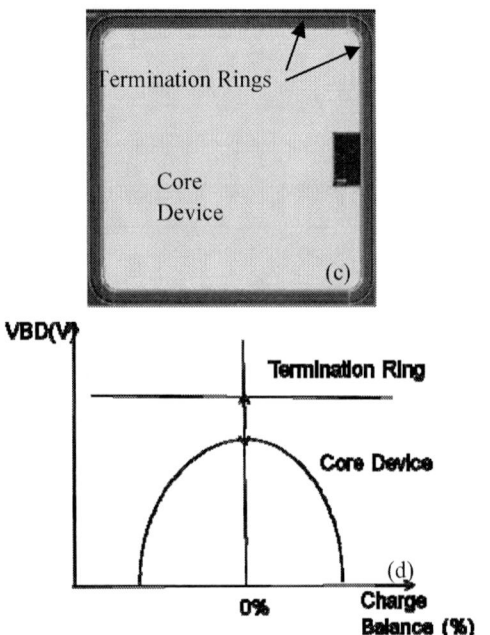

**Fig. 1a**: Schematic cross section of the core device showing the air gap, the super junction and the gate trenches. Under UIS condition the power is dissipated at the bottom of the core active device.

**Fig. 1b**: Schematic cross section of the termination rings. Under UIS conditions the power is dissipated at the surface of the termination rings.

**Fig. 1c**: Top view of the full MOSFETS, core active device and termination rings.

**Fig. 1d**: Illustration of the VBD of the core device and the termination rings versus the charge balance of the super junction trenches.

## 3. Methodology

The novel approach used in the present contribution is described below in three steps:

1/ First TCAD simulations under DC breakdown conditions have been performed on the core device as well as on the termination rings. The TCAD simulations have been calibrated to the measurements data performed in [3]. As an example the resulting I-V characteristics in avalanche regime and for optimal charge balance are presented in Fig. 2. From these figures one can see clearly that BVCore and BVTerm both strongly depend on temperature and injected current.

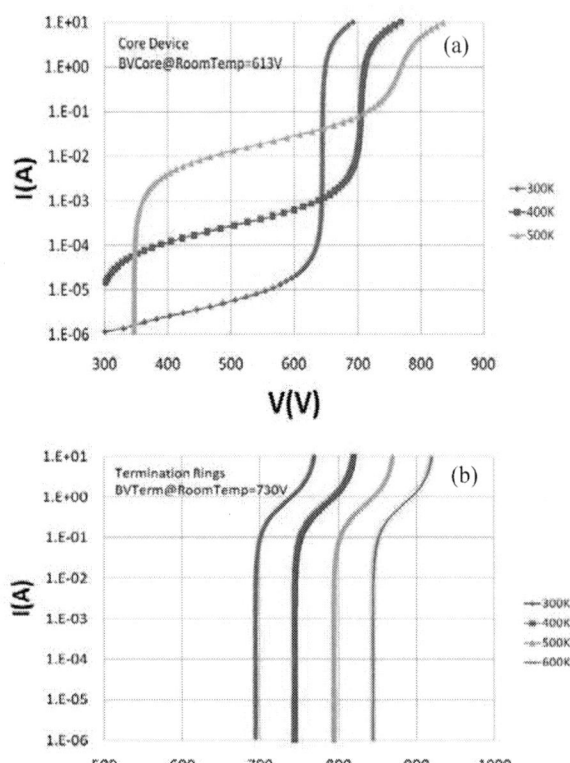

**Fig. 2a**: TCAD simulation of the I-V characteristics of the core device at different temperatures under DC breakdown condition. Simulations are for a 10A rated device. Full line is the behavioral model BVCore (T,ICore).

**Fig. 2b**: TCAD simulation of the I-V characteristics of the termination rings at different temperatures under DC breakdown conditions. Simulations are for a 10A rated device. Full line is the behavioral model BVTerm (T,ITerm).

2/ Second the dependencies of BVCore and BVTerm as a function of temperature and injected current have been empirically modeled. (See the equations of BVCore(T, ICore) and BVTerm(T,Iterm) in Fig. 3). These equations were converted into Spice model.

3/ Third a loosely coupled electro-thermal simulator was used, consisting of HeatWave™ [9], a FEM thermal engine and Spectre as a Spice engine. The simulator uses the relaxation method [10] with separate but synchronized thermal and electrical simulations. The advantages of the method are the relative simplicity of the implementation, the speed up of the simulation time and the capability of full structure analysis. The drawback is the difficulty to achieve convergence in case of fast change of electrical power, which is the case of the studied power MOSFET in this paper. To remedy this problem, a smoothing function was introduced in the behavioral model in order to keep control of the convergence of the simulator and the computation time. To represent non-uniform heat distribution, the core active device as well as the termination rings has been segmented into smaller pieces, represented as well electrical as lumped device in Spice and thermal as power boxes in HeatWave. The number of segments, the size of each segment and the placement in 3D are user defined. This method allows to control the accuracy versus the simulation time. HeatWave reads the layout, the locations as well as the geometry of the power sources and the package properties. The thermal engine solves the heat equations and provides the local increase in temperature, a parameter called (trise), for each segment. This trise (rise in temperature) is passed to the corresponding device segment in Spice. A new updated value of breakdown voltage and thus power is generated and provided back to HeatWave. The principle of the loosely coupled is illustrated in Fig. 3.

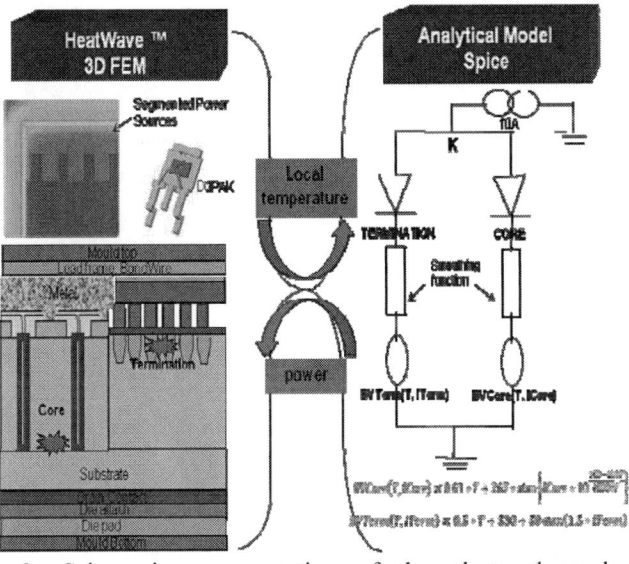

**Fig. 3:** Schematic representation of the electro-thermal concept used. On the left side, the thermal engine, HeatWave, takes into account the layout, location and geometry of dissipated power as well the material and technology properties from the device and the package. On the right the behavioral models of the breakdown voltages of the core device and termination rings are shown. Local temperature and power are exchanged during the electro-thermal simulations under UIS condition.

## 4. Experimental Results

A typical UIS pulse is shown in Fig. 4a. The total avalanche current decays from 10A down to 0A in 125us. In Fig 4a also the simulated currents in the core active device and in the termination rings are shown. One can easily see that a part of the injected current goes into the termination rings. Since the termination rings are much smaller in area compared to the active core, the current needs to be properly balanced. The Emission Microscopy (EMMI) measurements,

shown in Fig. 4b, confirm the current distribution between core active device and termination rings as simulated in Fig. 4a.

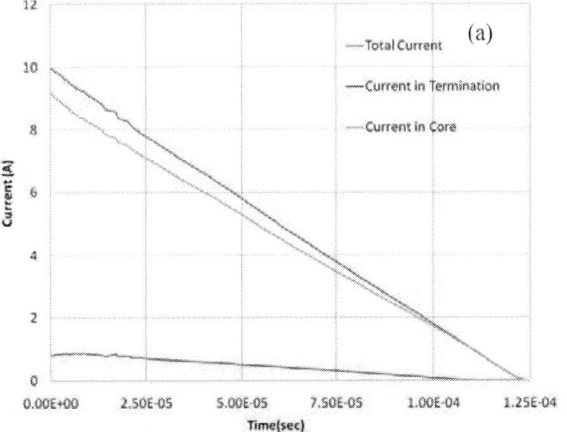

**Fig. 4a:** The simulated total avalanche current, current in the termination rings and in the core device during the UIS pulse.

**Fig. 4b:** EMMI measurements photos show the activation of the termination rings (left) and on the core active device (right) and thus confirms the balance of current between termination rings and core device, for different avalanche current densities.

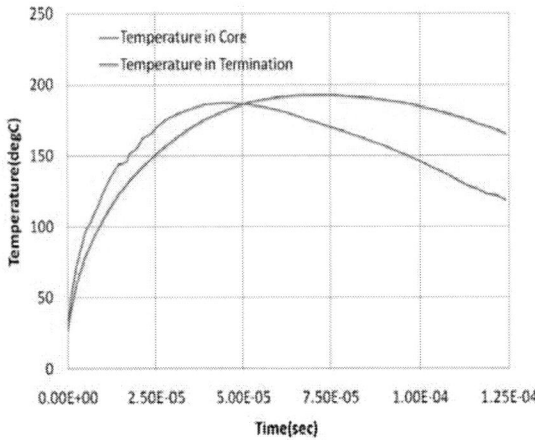

**Fig. 5:** The simulated maximum temperature in the core as well as in the termination rings during the UIS pulse. The termination, being smaller than the core device, heats faster.

The simulated temperatures are depicted in Fig. 5. Since the volume of the heat source in the termination rings is smaller, for the same injected current, the termination rings will heat up faster compared to the core device.

A 3D snap shot of the simulated temperature distribution in the core device as well as in the termination rings is shown in Fig. 6.

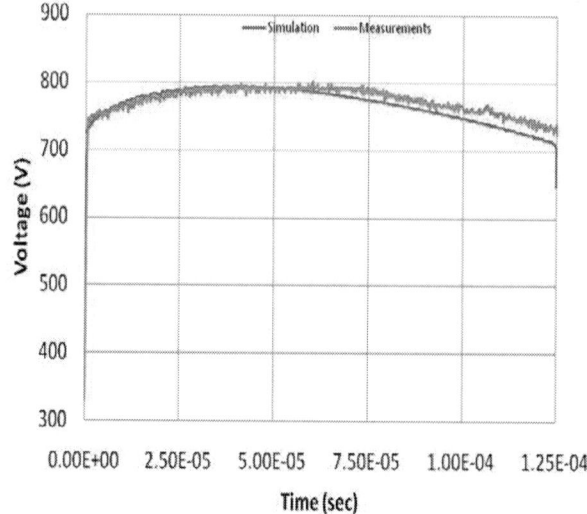

**Fig. 7:** The simulated BV is compared to the measured UIS curve. A good agreement is obtained between the simulation results and the measurements of the UIS curve.

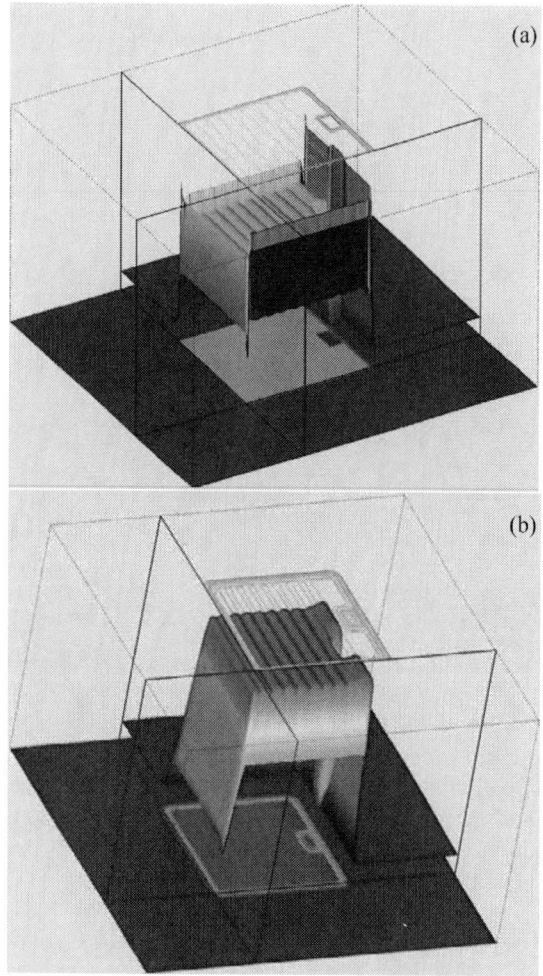

**Fig. 6:** 3-Dimensional temperature profile in the termination rings at 25us (a) and in the core device at 80us (b) during the UIS pulse. The termination rings are heated by the core device, while the core device is not heated by the termination.

In this paper a new approach was presented that can accurately simulate and optimize the electro-thermal robustness of a super junction power MOSFET. This method uses 3D electro-thermal simulation to predict the distribution of UIS current in termination rings and core device, as well for a packaged as for a non-packaged device. Out of these simulations also the optimal voltage window between BVCore and BVTerm can be deducted. In order to validate the new simulation method, the simulated voltage during the UIS pulse has been compared to the measured one and a good agreement has been obtained (see Fig. 7).

**Fig. 8:** First row shows the transient IR thermography photos performed during the UIS pulse ( decaying from 9A down to 0A in 170us) taken at 3 different times and showing the heating-cooling of the core device. The second row shows the corresponding electro-thermal simulation for the same bias condition. The level of temperature is lower than the one reported in Fig. 5, in this case the device is bigger and the power density is lower. A good agreement is obtained between the simulation and the experimental data.

Further validation has been performed by comparing the simulation results to the transient IR thermography [10]. Also good agreement is observed between the simulation and the thermography photos (see Fig. 8).

## 5. Conclusions

By combining the physical aspects of the device taken from TCAD simulations and the geometrical and thermal aspects taken from layout and package, in full 3D, an optimal balance of power between core active device and termination rings is achieved. This method offers a novel and practical approach to analyze and optimize the robustness of super junction MOSFETs for UIS. The EMMI measurements as well as the transient IR thermography photos validate the electro-thermal interaction between core active device and termination rings predicted by the novel approach.

**References**

1. Moens, P., Bogman, F., Ziad, H., De Vlesschouwer, H., Baele. J., Tack, M., Loechelt, G., Grivna, G., Parsey, J., Wu, Y., Quddus, T., Zdebel, P., "UltiMOS: A Local Charge-Balanced Trench-Based 600V Super-Junction Device", Proc of the 23$^{rd}$ International Symposium on Power Semiconductor Devices & ICs, ISPSD, pp. 304-307, 2011.

2. Roig, J., Moens, P., McDonald, J., Vanmeerbeek, P., Bauwens, F., Tack, M., "Energy limits for Unclamped Inductive Switching in Hihg-Voltage Planar and SuperJunction Power MOSFETs", Proc of the 23$^{rd}$ International Symposium on Power Semiconductor Devices & ICs, ISPSD, pp. 312-315, 2011.

3. Villamor-Baliarda, A., Vanmeerbeek, P., Roig, J., Moens, P., Flores, D., "Electric Field Unbalance for Robust Floating Ring Termination", Microelectronics Reliability, 51 (9-11), pp. 1959-1963, 2011.

4. Roig, J., Stefanov, E., Morancho, F., "Thermal Behavior of a Superjunction MOSFET in a High-Current Conduction", Trans on Electron Devices, Vol. 53, No. 7, pp. 1712-1720, 2006.

5. Donoval, D., Vrbicky, A., Marek, J., Chvala, A., Beno, P., "Evaluation of the ruggedness of power DMOS transistor from electro-thermal simulation of UIS behaviour", Solid-State Electronics, 52, pp. 892-898, 2008.

6. Fisher, K., Shenai, K., "Electrothermal Effects During Unclamped Inductive Switching (UIS) of Power MOSFET's", IEEE Trans on Electron Devices, Vol. 44, No. 5, pp. 874-878, 1997.

7. Jakopovic, Z., Sunde, V., Bencic, Z., "Electro-Thermal Modelling and Simulation of a Power-MOSFET", Automatika, 42, pp. 71-77, 2001.

8. de Filippis, S., Kosel, V., Dibra, D., Decker, S., Kock, H., Irace, A., "ANSYS based 3D electro-thermal simulations for the evaluation of power MOSFETs robustness", Microelectronics Reliability, 51, pp. 1954-1958, 2011.

9. HeatWave™ : www.gradient-da.com.

10. Van Petegem, W., Geeraerts, B., Sansen, W., Graindourze, B., "Electrothermal Simulation and Design of Integrated Circuits", IEEE Journal of Solid-State Circuits, Vol. 29, No. 2, pp. 143-146, 1994.

11. Riccio, M., Breglio, G., Irace, A., Spirito, P., "An equivalent time temperature mapping system with a 320x256 pixels full-frame 100kHz sampling rate", Review of Scientific Instruments, 78, pp. 106-108, 2007.

# Energy Reduction in Server Cooling Via Real Time Thermal Control

Xuefei Han, Yogendra Joshi
The George W. Woodruff School of Mechanical Engineering
Georgia Institute of Technology, Atlanta, GA, 30332-0405
Phone: 404.385.2810, Fax: 404.894.8496
Email: yogendra.joshi@me.gatech.edu

## Abstract

Currently, server fans often over-cool the CPUs, contributing to increased use of energy. A thermal model-based real-time fan controller was studied in this paper to reduce the energy consumption in server CPU fans. A reduced order model of the server CPU and its heatsink was developed using proper orthogonal decomposition (POD), which provided 180 times faster simulation times, with reasonable accuracy. This was used in conjunction with a fan control strategy to achieve 27% savings in energy consumed for cooling.

## Introduction

Effective thermal management in modern commercial servers and data centers is critical. Benchmarking studies show that for every 1W used to power servers, an additional 0.5W ~ 1W is required to remove heat from the system [1]. A recent study by the Environmental Protection Agency (EPA) shows that 60 billion kWh, or 1.5% of the total US energy use in 2006, was used to power data centers, and this usage is expected to rise to 100 billion kWh by 2012 [2]. Forced convection cooling of heatsinks using fans is typically employed for thermal management. During usage, with increased activity, as the CPU power density increases, the fans must speed up to deliver more air through the system. The state-of-the-art server fan control schemes often over-provision air flow, resulting in unnecessarily increased energy consumption for cooling fans. Moreover, fans operation at higher voltage can shorten their lifetime and introduce acoustic noise [3]. To address the dynamic thermal management of server CPUs, a reduced-order model based approach is introduced here, which adjusts fan speed based on CPU power dissipation. In the following of this paper, existence of observed over-cooling in servers is presented experimentally, and then simulation and experimental work is presented to demonstrate the concepts of reduced-order model based server fan controller for preventing over-cooling.

## Experimental Observation of Over-Cooling in Servers

The primary energy users in servers are CPUs, cooling fans, memory, and power supply. As an example, the CPU fans in the present study for HP ProLiant DL 360 G3 servers can draw power from 3.3 W to 23.8 W, depending on the operating voltage. Fig. 1 illustrates the configuration of this server, which has temperature sensors built in four regions - the two processors, power supply, and I/O zone. The fans are controlled by software which can adjust their speed in small steps, according to the temperature sensor measurements. If the temperature is below the predefined threshold, the fans run at their normal speed. Once the predefined threshold is exceeded, the fans speed up to cool the CPU below the threshold, and then go back to normal speed. The fans were

observed to be able to adjust speed by small steps, depending on how much the CPU temperature exceeds the threshold. Measurements on CPU temperature, CPU workload and fan speeds were done to confirm the existence of over-cooling in this model.

**Fig. 1** Example Configuration using HP ProLiant DL 360 G3 servers

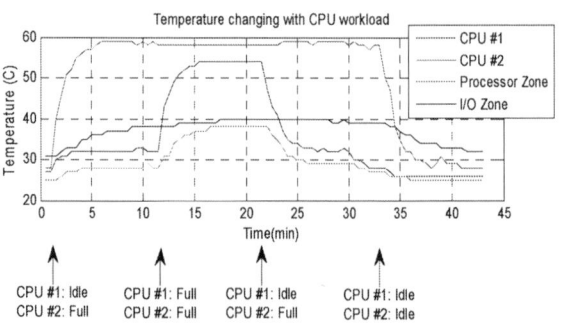

**Fig. 2** CPU Temperature vs. Workload

Initially, both processors were idle. When one was set to 100% workload, its temperature rose to slightly below 60 °C, as shown in Fig. 2. The second CPU was also set to 100% workload after 10 minutes and both CPU temperatures were below the 67 °C threshold predefined for the processors. The test was done with an ambient temperature of 21 °C. During the entire test the fans were running at constant speed, which is the system defined startup fan speed. The CPU temperatures were significantly below the threshold when they were

978-1-4673-1110-6/12 $31.00 © 2012 IEEE

inactive, thus resulting in over-cooling. This can contribute a significant amount to power usage of the server system, due to the fact that fan speed and fan power follow a cubic relation according to fan laws [4]:

$$\frac{rpm_1}{rpm_2} = \left(\frac{Power_1}{Power_2}\right)^3 \qquad \text{Eq. 1}$$

## Reduced-Order Modeling of Server

Overcooling can be prevented if the dynamic heat generation rate in the CPUs is known and a model can quickly determine the mass flow rate needed to cool the CPUs to a pre-defined goal temperature. The CPU power is primarily composed of two parts: the dynamic power and the leakage power [5]. The first term is proportional to $fV^2$, where $f$ is operating frequency and $V$ is supplied Voltage. The leakage power depends, in a complex fashion, on the operating temperature as well as $f$ and $V$. The total power can be estimated by using the Nose-Sakurai Model as discussed in [5], taking the goal chip temperature for calculation. The Nose-Sakurai model (2000) described the relationships between leakage power, dynamic power and total power of any CMOS circuit [5]. The chip power can also be found by monitoring the total current and voltage supplied to the chip, since all the supplied power is finally converted to heat.

The model used to predict the chip temperature has to be fast and achieve good accuracy. The proper orthogonal decomposition (POD) reduced-order modeling method is employed for this objective. The successful applications of this method [6-10] suggest that it is a powerful tool of data analysis, which obtains a low-dimensional approximation of a high-dimensional system. The basic idea of POD reduced-order modeling is to extract one set of orthogonal basis modes from an ensemble of observations of the original high-dimensional system, and linearly combine these POD basis to represent a particular flow or temperature state [6, 11]. The velocity and temperature in POD basis are expressed as

$$u = u_0 + \sum_{i=1}^{k} a_i \phi_i, \qquad \text{Eq. 2}$$

$$T = T_0 + \sum_{j=1}^{l} b_i \psi_i \qquad \text{Eq. 3}$$

where $u_0, T_0$ are source terms, and are usually the mean value of all the velocity and temperature observations; $\phi_i, \psi_i$ are POD basis and $a_i, b_i$ are POD coefficients. $k$ and $l$ in Eq. 2 and Eq. 3 are the numbers of POD basis used construct velocity and temperature, respectively. The POD basis is calculated by singular value decomposition of the observation matrix $X$

$$U\Sigma V^T = svd\left(X^{m \times n}\right) \qquad \text{Eq. 4}$$

where the columns of $U$ are the POD basis. Also, $m$ is the dimension of the system, which is the number of elements or nodes if the observations are generated by numerical calculation, or the number of data points if the observations are generated by measurements, and $n$ is the number of observations, where usually $m \gg n$. In Eq. 4, $\Sigma$ is the singular value matrix with singular value $\sigma_i$ arranged in a descending order. Each POD basis associates with a singular value $\sigma_i$. The singular value can be taken as a measure of how close the corresponding POD basis to the accurate solution. The energy captured by a single POD basis $\phi_i$ is [6]

$$E_i = \frac{\sigma_i}{\sum_{j=1}^{n} \sigma_j} \qquad \text{Eq. 5}$$

Therefore, the more POD basis used in the approximation, the more accurate the result can be. Usually, a small number of POD basis (typically <10 depending on problems) will be sufficient to approximate the original system. As a result, the number of POD basis used in approximation, $k$ and $l$, are usually much smaller than $n$. The POD coefficients can be calculated by several methods, such as Galerkin Projection, linear interpolation and boundary flux matching method [6].

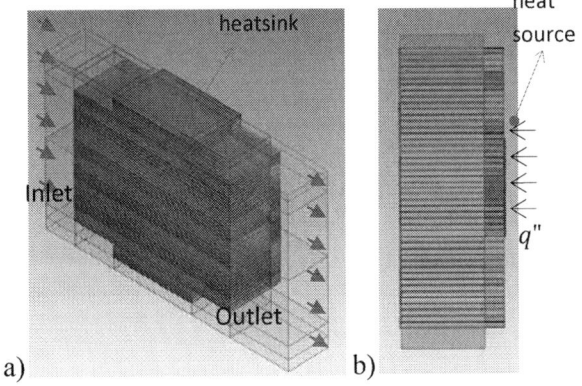

**Fig. 3** Numerical modeling of CPU and heatsink

The observations in the present study were generated by numerical modeling using ANSYS FLUENT. One of the CPU and its heatsink are modeled in detail, as shown in Fig. 3. Uniform heat flux at the CPU region, uniform inlet velocity, zero pressure at the outlet, and symmetry at one side were applied as the boundary conditions for numerical calculation. The near zero pressure at the outlet was verified by experiment, although there were some server components downstream. The symmetry at one side was because the two CPUs inside the server enclosure were placed side by side as in Fig. 1. Before generating numerical observations, a mesh independence study was done to determine the required mesh size. Two models with about 2 million cells and 0.5 million cells respectively, were tested and the results are compared in Table 1.

**Table 1** Mesh Independence Study

| Inlet $\dot{m}$ ($kg/s$) | Heat source power | Number of cells | Maximum temperature ($K$) |
|---|---|---|---|
| 0.004 | 45 | 1 979 930 | 319.62 |
| 0.004 | 45 | 417 858 | 319.22 |
| 0.008 | 45 | 1 979 930 | 312.56 |
| 0.008 | 45 | 417 858 | 312.40 |

The differences in maximum temperature were 0.4 K and 0.16 K for inlet air mass flow rates $\dot{m}$ of 0.004 kg/s and 0.008 kg/s respectively, for heat source power of 45 W of both cases. Considering the fact that POD method requires tens of simulations or more, the model with about 0.5 million cells was used as a compromise between accuracy and speed. The layer of thermal interface material and Kapton film package of the heater were included in the model. In the experiment, the CPU was replaced with film heaters due to the difficulty to vary CPU power in a well-controlled manner.

One set of 42 observations was generated using this model by varying the inlet velocity and heat flux. The observation parameters are attached in Appendix A. The velocity POD coefficients are calculated using linear interpolation and the temperature POD coefficients are calculated using boundary flux matching method (BFM). The BFM method forces the POD approximation to satisfy the boundary heat flux to solve the POD coefficients [6]. The accuracy of BFM highly depends on the number of boundary conditions of the system. In order to achieve better accuracy, the single heat source was divided into four in the model. The singular values of the velocity and temperature observations decrease notably as plotted in Fig. 4. Thus the first few POD basis would be sufficient to capture the physics of the system, as mentioned before. 22 test cases were taken to examine the performance of the POD model predicting the velocity and temperature field with input parameters, as listed in Appendix B, different from observations. The comparison of velocity and temperatures for Test Case 21 using 10 POD basis for both are presented in Fig. 5. The POD predicted velocity and temperature match well with CFD results, with the maximum errors of 0.0096 m/s and 0.60 °C, respectively. The absolute error in POD predicted velocity compared with CFD is always in the range of $10^{-4} \sim 10^{-3}$ m/s. However the energy equation is coupled with velocity, and both velocity and boundary heat fluxes will influence the temperature. As such, the POD temperature prediction sometimes can generate a relatively high error. Fig. 6 presents the maximum temperature errors for all 22 test cases with the highest value of 1.34 °C, still in an acceptable range. The typical calculation time for the CFD model to converge is 30 minutes, compared to 10 seconds using the POD reduced-order model.

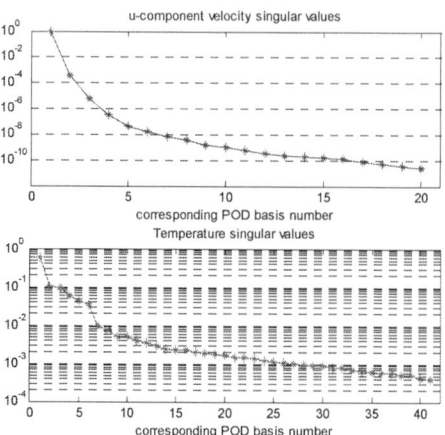

**Fig. 4** Singular Values

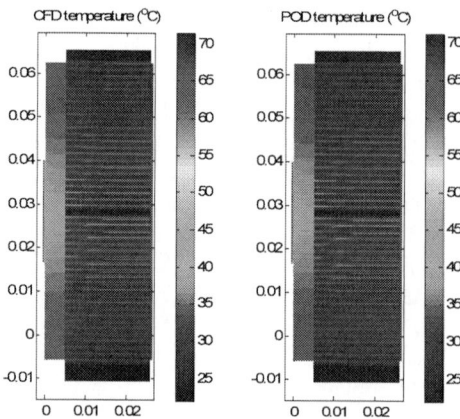

**Fig. 5** a) Temperature at vertical cross-section

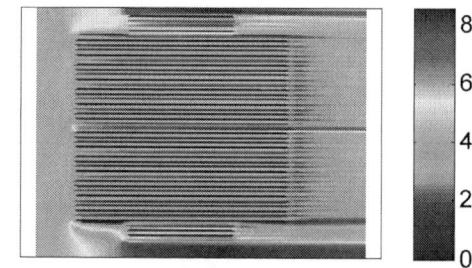

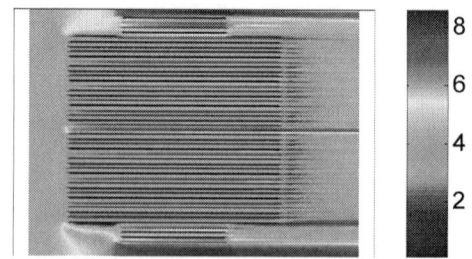

**Fig. 5** b) Velocity magnitude at mid-height

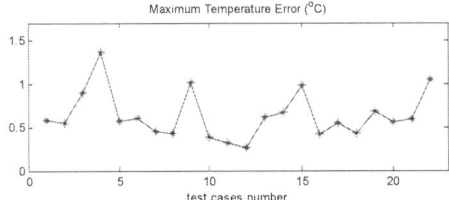

**Fig. 6** POD prediction errors for test cases

**POD Model Based Fan Controller**

Compared to the time-consuming CFD model, the POD is a fast reduced-order model with good accuracy. Based on this property, a POD model based fan controller was designed and implemented to adjust the fan speed according to heat source power.

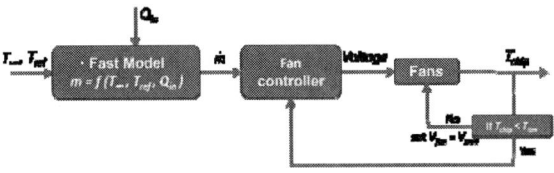

**Fig. 7** Fan control scheme

The fan control scheme is shown in Fig. 7. The fast model takes the ambient temperature $T_x$, the goal CPU temperature $T_{ref}$, and CPU power $Q_{in}$ as input parameters, and determines the mass flow rate $\dot{m}$ needed. Then the fan controller converts the mass flow rate into voltage to adjust fan speed. The goal chip temperature is set at the maximum permissible value, and therefore there is a chance of chip over-heating due to approximations in modeling or the control approach. To address this problem, the chip temperature is monitored constantly. In case that it exceeds the maximum set temperature, the fans are set to run at full speed until it falls below the limit. The limit temperature is 67 °C for the HP server CPUs in this example.

The POD model was used to predict the mass flow rate using the algorithm in Fig. 8. Only a few runs of this loop were needed to find the final mass flow rate.

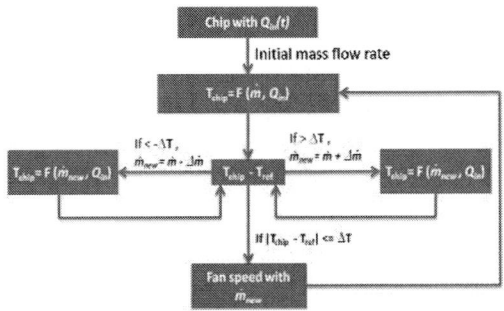

**Fig. 8** Algorithm for determining air flow rate

**Experiment**

The fan curve was tested using a fan tester. Fig. 9 shows the schematic of this equipment [12]. The fan tester measured fan static pressure and differential pressure between the upstream and downstream air flow, passing through a standard nozzle. The air mass flow rate that the fans can deliver under different voltage was found, as shown in Fig. 10, using constant air property at 20 °C and standard atmosphere pressure.

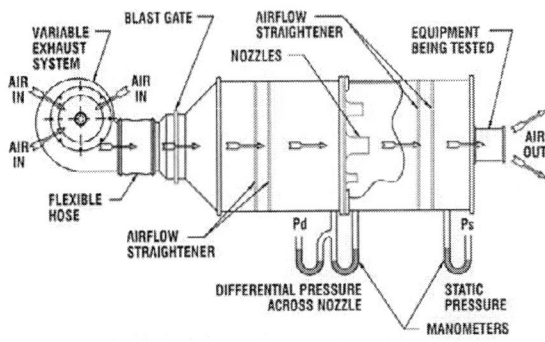

**Fig. 9** Fan tester configuration

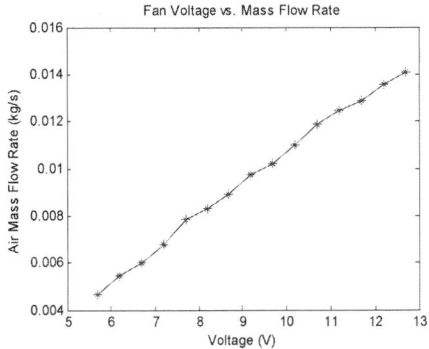

**Fig. 10** Fan Voltage vs. Air Mass Flow Rate

The accuracy of this measurement was tested by combining and comparing the CFD simulation of temperature and experimentally measured temperature as shown in Table 2. The relation of the CFD mass flow rate and fan voltage in experiment follows Fig. 10. The discrepancy between CFD and experimental result tends to increase at high flow rate and high power. Hence, four such test cases are listed in Table 2 to show the worst situation, and those cases with small flow rate and power generate smaller temperature difference and are not tabulated here. As seen in Table 2, the error of maximum temperature between CFD simulation and experiment measurement is within 1~3 °C.

**Table 2** Comparison of CFD and Experiment

| **CFD mass flow rate ($kg/s$)** | 0.012 | 0.012 | 0.014 | 0.014 |
|---|---|---|---|---|
| **Fan voltage in experiment ($V$)** | 11.05 | 11.05 | 12.65 | 12.65 |
| **Heat source power ($W$)** | 35 | 45 | 45 | 40 |
| **CFD $T_{max}$ (°C)** | 55.2 | 65.8 | 63.4 | 58.5 |
| **Expt $T_{max}$ (°C)** | 53.2 | 62.9 | 61.9 | 55.8 |

| $|\Delta T|$ (°C) | | 2 | 2.9 | 1.5 | 2.7 |
| --- | --- | --- | --- | --- | --- |

Due to the limited ability to vary the CPU powers, the CPUs were replaced with film heaters, keeping all other server features unchanged. Six thin wire T-type thermocouples were attached to the backside of each heater close to the upstream of air flow, ensuring the maximum temperature could be read. The backside of heater was well insulated with insulation material. The configuration was shown in Fig. 11. The heater power was increased from 20 W to 45 W in 1 W increment every 20 minutes. The goal was to maintain the heaters within the range of 59~61 °C. The mass flow rate and temperature predicted by the POD model are seen in Fig. 12. The predicted temperature was 2-3 °C off the experimental measurement as in Fig. 13, but the heater temperature was controlled within 58-64 °C, when the heater power rises to the relatively high power range. The uncertainty of the data acquisition equipment is 0.5 °C. At the beginning of experiment, the fans were set to the minimum speed, and the heater temperatures were below the goal range, due to low heat generation rate. When heater power reaches about 30 W, the fan starts to speed up under control of the model-based fan controller. Fan powers were tested as in Appendix C, and the fan energy saved with the controller, compared with starting fans at normal speed (8V), was 27%.

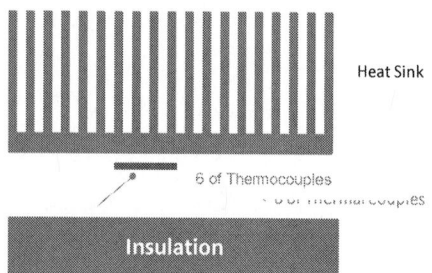

**Fig. 11** Experiment setup with thin film heater replacing chips

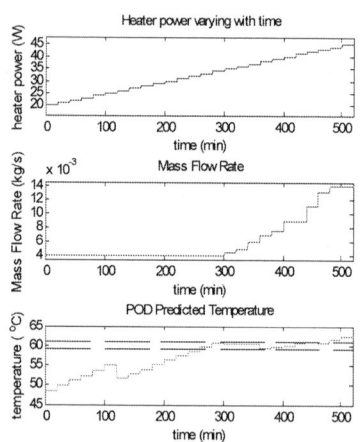

**Fig. 12** POD model predicted $\dot{m}$ and $T$

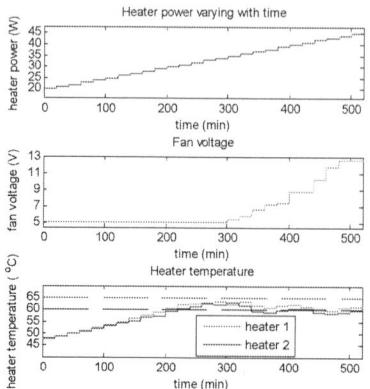

**Fig. 13** Experimental measurement

## Conclusion

A reduced-order model based server fan speed controller was simulated and experimentally implemented in this work. A POD reduced order model of server CPU and heatsink was developed and validated with detailed CFD simulation. The POD model yields good accuracy and generates results significantly faster than detailed CFD simulation. This excellent property of POD modeling enables prediction of chip temperature and finding out the necessary air mass flow rate in a short time. This POD real-time reduced-order model-based fan speed controller approach can reduce energy consumption in server CPU fans, compared to the existing fan speed control scheme. The actual saving depends on the server workload.

## Acknowledgments

This work is supported by the Interconnect Focus Center, one of five research centers funded under the Focus Center Research Program, a Semi-conductor Research Corporation program.

## References

[1] C. D. Patel, *et al.*, "Smart Cooling of Data Centers," *ASME Conference Proceedings,* vol. 2003, pp. 129-137, 2003.

[2] "Report to Congress on Server and Data Center Efficiency: Public Law 109-431," ed: US Environmental Protection Agency ENERGY STAR Program, 2007.

[3] R. H. Lyon and A. E. Bergles, "Noise and Cooling in Electronics Packages," *Components and Packaging Technologies, IEEE Transactions on,* vol. 29, pp. 535-542, 2006.

[4] M. K. Patterson, "The effect of data center temperature on energy efficiency," in *Thermal and Thermomechanical Phenomena in Electronic Systems, 2008. ITHERM 2008. 11th Intersociety Conference on,* 2008, pp. 1167-1174.

[5] D. C. Sekar, "Optimal Signal, Power, Clock and Thermal Interconnect Networks for High-Performance 2D and 3D Integrated Circuits," PhD Dissertation, School of Electrical and Computer Engineering, Georgia Institute of Technology, Atlanta, 2008.

[6] J. D. Rambo, "Reduced-order modeling of multiscale turbulent convection: Application to data center thermal management," PhD Dissertation, Georgia Institute of Technology, Atlanta, 2006.

[7] E. Samadiani, *et al.*, "Reduced Order Thermal Modeling of Data Centers via Distributed Sensor Data," *ASME Conference Proceedings,* vol. 2009, pp. 807-814, 2009.

[8] Q. Nie and Y. Joshi, "Reduced order modeling and experimental validation of steady turbulent convection in connected domains," *International Journal of Heat and Mass Transfer,* vol. 51, pp. 6063-6076, 2008.

[9] J. Rambo and Y. Joshi, "Reduced-order modeling of turbulent forced convection with parametric conditions," *International Journal of Heat and Mass Transfer,* vol. 50, pp. 539-551, 2007.

[10] E. Samadiani and Y. Joshi, "Multi-parameter model reduction in multi-scale convective systems," *International Journal of Heat and Mass Transfer,* vol. 53, pp. 2193-2205, 2010.

[11] P. Holmes, *Turbulence, Coherent Structures, Dynamical Systems and Symmetry*: Cambridge University Press, 1996.

[12] http://www.fantester.com/.

## Appendix A Observation Parameters

| Obs No. | Mass Flow Rate (kg/s) | Heat Source Power (W) | | | | Total power (W) | Obs No. | Mass Flow Rate (kg/s) | Heat Source Power (W) | | | | Total power (W) |
|---|---|---|---|---|---|---|---|---|---|---|---|---|---|
| | | No. 1 | No. 2 | No. 3 | No. 4 | | | | No. 1 | No. 2 | No. 3 | No. 4 | |
| 1 | 0.004 | 2.5 | 2.5 | 2.5 | 2.5 | 10 | 22 | 0.004 | 10.5 | 9.5 | 9.75 | 11.25 | 41 |
| 2 | 0.0045 | 3.75 | 3.75 | 3.75 | 3.75 | 15 | 23 | 0.0045 | 9.25 | 10 | 8.25 | 9 | 36.5 |
| 3 | 0.005 | 5 | 5 | 5 | 5 | 20 | 24 | 0.005 | 7.5 | 9.5 | 8.25 | 6.25 | 31.5 |
| 4 | 0.0055 | 6.25 | 6.25 | 6.25 | 6.25 | 25 | 25 | 0.0055 | 1.75 | 2.5 | 2 | 2.5 | 8.75 |
| 5 | 0.006 | 7.5 | 7.5 | 7.5 | 7.5 | 30 | 26 | 0.006 | 10.5 | 11.25 | 12 | 12.5 | 46.25 |
| 6 | 0.0065 | 8.75 | 8.75 | 8.75 | 8.75 | 35 | 27 | 0.0065 | 3.75 | 3.75 | 3.75 | 3.75 | 15 |
| 7 | 0.007 | 10 | 10 | 10 | 10 | 40 | 28 | 0.007 | 4.5 | 5 | 5 | 5.5 | 20 |
| 8 | 0.0075 | 11.25 | 11.25 | 11.25 | 11.25 | 45 | 29 | 0.0075 | 3.75 | 4.25 | 4 | 3.5 | 15.5 |
| 9 | 0.008 | 11.25 | 7.5 | 11.25 | 11.25 | 41.25 | 30 | 0.008 | 8 | 7.5 | 6.75 | 6.75 | 29 |
| 10 | 0.0085 | 8.75 | 6.25 | 12.5 | 10 | 37.5 | 31 | 0.0085 | 11.25 | 2.5 | 5 | 2.5 | 21.25 |
| 11 | 0.009 | 8.75 | 8.75 | 6.25 | 8.75 | 32.5 | 32 | 0.009 | 2.5 | 2.5 | 2.5 | 2.5 | 10 |
| 12 | 0.0095 | 5 | 7.5 | 7.5 | 7.5 | 27.5 | 33 | 0.0095 | 4.5 | 11.25 | 3.75 | 3 | 22.5 |
| 13 | 0.01 | 6.25 | 6.25 | 5 | 3.75 | 21.25 | 34 | 0.01 | 10.5 | 10.75 | 11 | 10.75 | 43 |
| 14 | 0.0105 | 2.5 | 3 | 3 | 2.5 | 11 | 35 | 0.0105 | 9.5 | 9.5 | 9.5 | 9.5 | 38 |
| 15 | 0.011 | 3.75 | 4.25 | 4.25 | 3.75 | 16 | 36 | 0.011 | 7.5 | 7.5 | 7.5 | 7.5 | 30 |
| 16 | 0.0115 | 5 | 5.5 | 5.5 | 5 | 21 | 37 | 0.0115 | 11.25 | 11.75 | 11.25 | 11.75 | 46 |
| 17 | 0.012 | 6.25 | 6.75 | 6.25 | 7 | 26.25 | 38 | 0.012 | 5.25 | 4.5 | 5.5 | 6.25 | 21.5 |
| 18 | 0.0125 | 7.5 | 8 | 7.75 | 8.25 | 31.5 | 39 | 0.0125 | 4.5 | 3.75 | 3.75 | 4.25 | 16.25 |
| 19 | 0.013 | 8.75 | 9.25 | 10.75 | 9.5 | 38.25 | 40 | 0.013 | 11.25 | 11.25 | 12 | 11.25 | 45.75 |
| 20 | 0.0135 | 11.25 | 10.5 | 12 | 10.75 | 44.5 | 41 | 0.0135 | 3 | 3.75 | 1.75 | 2.75 | 11.25 |
| 21 | 0.014 | 3.75 | 3.75 | 3.75 | 3.75 | 15 | 42 | 0.014 | 11.25 | 12 | 11.25 | 12 | 46.5 |

## Appendix B Test Cases Parameters

| Obs No. | Mass Flow Rate (kg/s) | Heat Source Power (W) | | | | Total power (W) | Obs No. | Mass Flow Rate (kg/s) | Heat Source Power (W) | | | | Total power (W) |
|---|---|---|---|---|---|---|---|---|---|---|---|---|---|
| | | No. 1 | No. 2 | No. 3 | No. 4 | | | | No. 1 | No. 2 | No. 3 | No. 4 | |
| 1 | 0.0113 | 5 | 5 | 5 | 5 | 20 | 12 | 0.0077 | 11.25 | 11.25 | 11.25 | 11.25 | 45 |
| 2 | 0.0113 | 10 | 10 | 10 | 10 | 40 | 13 | 0.0077 | 8.25 | 8.25 | 8.25 | 8.25 | 33 |
| 3 | 0.0113 | 15.5 | 15.5 | 15.5 | 15.5 | 62 | 14 | 0.0052 | 9.25 | 9.25 | 9.25 | 9.25 | 37 |
| 4 | 0.0126 | 1.25 | 1.25 | 1.25 | 1.25 | 5 | 15 | 0.0052 | 11 | 11 | 11 | 11 | 44 |
| 5 | 0.0126 | 5.5 | 5.5 | 5.5 | 5.5 | 22 | 16 | 0.0052 | 3.25 | 3.25 | 3.25 | 3.25 | 13 |
| 6 | 0.0126 | 9 | 9 | 9 | 9 | 36 | 17 | 0.0092 | 9.5 | 9.5 | 9.5 | 9.5 | 38 |
| 7 | 0.0138 | 2.5 | 2.5 | 2.5 | 2.5 | 10 | 18 | 0.0092 | 2 | 2 | 2 | 2 | 8 |
| 8 | 0.0138 | 8.25 | 8.25 | 8.25 | 8.25 | 33 | 19 | 0.0092 | 6.25 | 6.25 | 6.25 | 6.25 | 25 |
| 9 | 0.0138 | 6 | 6 | 6 | 6 | 24 | 20 | 0.0068 | 8.75 | 8.75 | 8.75 | 8.75 | 35 |
| 10 | 0.0138 | 11.25 | 11.25 | 11.25 | 11.25 | 45 | 21 | 0.0068 | 12.5 | 12.5 | 12.5 | 12.5 | 50 |
| 11 | 0.0077 | 4.5 | 4.5 | 4.5 | 4.5 | 18 | 22 | 0.0068 | 3 | 3 | 3 | 3 | 12 |

## Appendix C Fan Power

| Voltage (V) | 5 | 5.5 | 6 | 6.5 | 7 | 7.5 | 8 | 8.5 | 9 | 9.5 | 10 | 10.5 | 11 | 11.5 | 12 |
|---|---|---|---|---|---|---|---|---|---|---|---|---|---|---|---|
| Currant (A) | 0.64 | 0.71 | 0.77 | 0.84 | 0.91 | 0.98 | 1.05 | 1.13 | 1.22 | 1.28 | 1.37 | 1.45 | 1.5 | 1.58 | 1.73 |

# A Case Study on the Impact of Free Air Cooling on Telecom Equipment Performance

Jun Dai, Diganta Das, Michael Pecht, and Michael Ohadi
Center for Advanced Life Cycle Engineering (CALCE)
University of Maryland, College Park, MD, USA 20742
Corresponding Author: pecht@calce.umd.edu

## Abstract

Energy consumption and its environmental impacts have become key concerns in the telecommunications industry and its data centers. As an energy-efficient approach for cooling, some data centers are adopting free air cooling, which uses ambient air outside the data centers, rather than air conditioning, to cool the electronic equipment. Traditionally, telecom equipment qualifications are based on passing a set of tests of industry standards that assume pre-defined environmental conditions. However, free air cooling changes the operating conditions and may go beyond those pre-defined conditions, which may affect the reliability and performance of the telecom equipment. This paper evaluates impact of free air cooling on the performance of telecom equipment. It compares the performance variations of telecom equipment under free air cooling with those under the traditional air conditioning to identify the impact of free air cooling on the telecom equipment performance.

**Keywords:** telecom equipment; data center; energy saving; free air cooling; performance variations.

## 1. Introduction

Data centers for digital information management are now both common in and essential to a safe and secure economy in almost every sector of today's society, including financial services, media, communications, academia, and government institutions. Data centers are all buildings, facilities, and rooms that contain enterprise servers, server communication equipment, and cooling and power equipment.

Energy consumption is one of the leading data center operating expenses. The worldwide energy consumption of data centers increased about 56% between 2005 and 2010, reaching approximately 237 terawatt hours (TWh)[1] in 2010, accounting for about 1.3% of the world's electricity use [1]. In the U.S., data center energy consumption increased roughly 36% from 2005 to 2010, reaching 76 TWh and accounting for approximately 2% of the total U.S. electricity consumption in 2010 [1]. Computer room air conditioning (CRAC) systems maintain data center operating conditions (temperature and humidity) in narrow ranges, and as a result, the energy consumption for air conditioning is a key cost driver in data centers 0. In 2009, about 40% of the energy consumed by data centers was devoted to the cooling that support data center equipment 0 [3], as shown in Figure 1.

Free air cooling is an approach to saving cooling energy that uses ambient air to cool data center equipment, thereby reducing the need for air conditioning and, hence, cooling expenditures. It is usually implemented with airside economizers[2], and a typical example is shown in the American

---

[1] This value is the midpoint of the low bound and high bound of the energy consumption estimation.

[2] There are generally two kinds of economizers: airside economizers and waterside economizers. Waterside economizers are not included in this paper

---

Society of Heating, Refrigerating and Air-Conditioning Engineers (ASHRAE) standard [4]. This cooling method is being increasingly used in industry. Intel used ambient external air to cool its equipment at a 10-megawatt (MW) data center, saving US$2.87 million annually and reducing energy use by 67% in the data center [5]. In 2009, Microsoft and Google used free air cooling to replace traditional chillers in some of their new data centers in Europe [6] [7]. Due to the potential energy savings from free air cooling, regulations and standards are being updated to facilitate its adoption. For example, the "EU Code of Conduct on Data Centers" [8] and the ASHRAE Standard 90.1 [9] recommend free air cooling as the preferred cooling method in data centers.

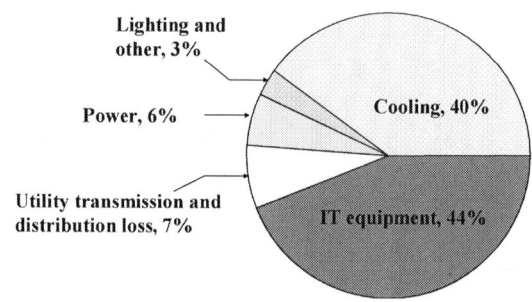

**Figure 1: Energy consumption breakup in data centers [3]**

Despite its endorsements, free air cooling presents risks to data center equipment, including servers, storage, and telecom equipment (routers and switches), due to variations in operating conditions. Some companies have performed research on the reliability of their equipment and obtained preliminary results. For example, Intel has claimed that there is only a small difference between the 4.46% failure rate of equipment with free air cooling and the 3.83% failure rate of equipment with traditional air-conditioned cooling over a 10-month period [5]. Dell has also reported there is only a small difference in the number of hard failures under free air cooling conditions compared to the number of failures in a control cell operating within the ASHRAE recommended range [10]. However, both of these cases focused mainly on server hardware, with communication equipment, such as routers and switches, not included in the tests. Furthermore, they focused on the impact of free air cooling on the reliability of equipment, but ignored the impact on the performance of the equipment.

Accordingly, there is a need for studies on the effect of free air cooling on telecom equipment performance, particularly for mission critical data centers that require high-quality performance on a continuous basis. This paper analyzes the potential performance risks arising from free air cooling, based on the expected operating conditions, and presents a case study to compare the performance variations of telecom equipment under free air cooling conditions and traditional air

---

because they re-circulate water instead of air.

conditioning (A/C), which can identify the impact of free air cooling on the performance of the equipment.

## 2. Potential Performance Risks Associated with Free Air Cooling

Free air cooling usually changes the operating conditions in data centers, compared to data centers in traditional air-conditioner (A/C) conditions. The design rules, test conditions, acceptance conditions, and overall operating cost estimates for data center equipment are impacted by operating conditions. This section analyzes the potential performance risks associated with the implementation of free air cooling.

### 2.1 Operating Conditions

In 2004, ASHRAE published "Thermal Guidelines for Data Centers and Other Data Processing Environments" for data center operating conditions [4]. These guidelines provide recommended limits of 20°C to 25°C and 40% RH to 55% RH, and allowable limits of 15°C to 32°C and 20% RH to 80% RH. The 2008 revised specifications expanded the recommended limits to 18°C to 27°C, which allows for more operating hours of airside economizers in the free air cooling mode [11]. ASHRAE revised the thermal guidelines in 2011 and the new version has more data center classes to "accommodate different applications and priorities of IT equipment operation". In the new version, the recommended limits are the same as the previous version, and the allowable limits are expanded based on the data center classifications [17]. Telcordia Generic Requirements GR-63-CORE [12] and GR-3028-CORE [13] also provide operating condition specifications for telecom equipment. The recommended limits are 18°C to 27°C and 5% RH to 55% RH, and the allowable limits are 5°C to 40°C[3] and 5% RH to 85% RH, which are slightly different from those of ASHRAE.

Currently, most data centers operate at temperatures between 20°C and 25°C. A survey of fourteen of Sun Microsystem (now Oracle) data centers in 2007 showed that eight had the temperature set at 20°C, five at 22°C, and one at 23°C [14]. There are also some reports of extremely low temperatures, around 12°C, for inlet air. Typically, humidity levels in data centers are maintained between 35% RH and 55% RH [15]. With the implementation of free air cooling, the operating conditions may be significantly changed. For example, the inlet temperature range was 18°C to 32°C, and the relative humidity range was 4% RH to more than 90% RH in the Intel case [5]. Depending on the energy saving goals and the local climate, data center equipment may experience great operating condition variations under free air cooling. These operating condition changes may pose risks to the performance of equipment, as will be discussed in the following sections.

### 2.2 Performance Risks Associated with Free Air Cooling

A traditional data center usually use multiple A/C units to "fine tune" data center temperature in addition to the use of air flow, however, in a free air cooled data center, the temperature across the data center is controlled largely by air flow. Thus, additional designs and air flow simulations, as well as measurement and control during the operation must be considered for the proper implementation of free air cooling. When a data center accepts free air cooling, more air flow and

temperature optimizations are required to eliminate unwanted hotspots compared with those in a traditional data center. This kind of optimizations may even have to be performed on selected rack configurations, which helps make sure that all the equipment within the rack are working within the required operating condition envelope. The temperature and air flow optimization also help identify the "weak link" hardware, i.e., that has a lower thermal margin than the other pieces of equipment and then limits the ability of the data center to function at its target maximum temperature.

For data centers that are already in operation, the "weak link" hardware are limitation for the free air cooling implementation since major design modifications are no longer feasible. One solution is to modify the rack configuration or provide more air flow to cool the equipment back into the required operating conditions. However, in some cases, the only option is to replace the "weak link" hardware with equipment that has a wider thermal margin. This can be very costly and time consuming and could interrupt the services of data centers. In these cases, data center operators will need to find the operating condition limits of the data centers based on that of each piece of equipment.

Changes in temperature and humidity levels can impact the electrical parameters and life of components and systems [16]. As a result, the performance may intermittently fail to meet specifications, particularly if those are located at hotspots. Possible reductions in performance will arise if it is necessary to apply uprating methods, such as stress balancing for components or equipment, which trade off temperature for performance (e.g., reduced operating frequency or speed with increases in operating temperature). This approach is common in computer systems (e.g., reduced access speed for memory or lower frequency of operation for microprocessors) and can be applied to telecommunication systems and data centers, but any decreased performance must be assessed against customer expectations regarding service quality and availability.

With free air cooling, some system-level operational performance changes are expected, due to the dependence of the electrical parameters of the system components (parts and materials) on temperature and humidity. In general, good design practices that account for component performance based on their temperature dependence (e.g., worst case analysis), will ensure acceptable operation, as long as the temperature and humidity levels are within those stated by the recommended operating conditions of the datasheets. However, if the free air cooling results in the operation of components at environmental conditions outside the datasheet recommended operating conditions (either on the low or high side), then additional analysis will be necessary to quantify the associated electrical operating parameters' characteristics and assess their impact on the system-level performance.

## 3. Impacts of Free Air Cooling on the Performance of Telecom Equipment

The network architecture in a data center consists of a set of routers and switches. In order to identify the impacts of free air cooling on telecom equipment performance, we conducted a performance monitoring and recording experiment on a typical switch. Since most data centers currently use A/C units as the cooling method, the performance of telecom equipment under A/C conditions is used as the baseline for measuring the impacts of other cooling methods. This section compares the performance of equipment under free air cooling to the performance of equipment under traditional A/C conditions

---

[3] The temperature in the ASHRAE standard is the inlet temperature of telecom equipment; however, the temperature in the Telcordia standard is the ambient temperature of telecom equipment.

978-1-4673-1110-6/12 $31.00 © 2012 IEEE

using experimental data. The performance variations are analyzed to identify the impact of free air cooling on the performance of telecom equipment.

## 3.1 Setup of Experiment

The network equipment selected for this case study was a Zonet zfs 3015P switch. It is widely used in offices and small enterprises, and its primary function is to send packets to their intended destinations. Its rated operating conditions are 0°C to 40°C and 10% RH to 90% RH, which are typical ranges for routers and switches. A photo of the switch is shown in Figure 2.

**Figure 2: Zonet zfs 3015P switch**
**(1) IC+ IP 178C chip; (2) Magnetic components; (3) MC 34063A;(4) Capacitor; (5) Coil**

In the experiment, data packets were sent from one computer to another computer throughout the switch, which was put inside an environmental chamber to simulate the different conditions, as shown in Figure 3. To monitor the performance of the switches in the experiment, we used NetIQ Chariot, a network testing software package. This software sent large files with sizes up to $10^9$ bits through the switch continuously, and then calculates its performance parameters. An advantage of NetIQ Chariot is that the files are generated by the software itself, rather than by the hard drive of computer, so it is not dependent on the limitations of hard drive access speeds.

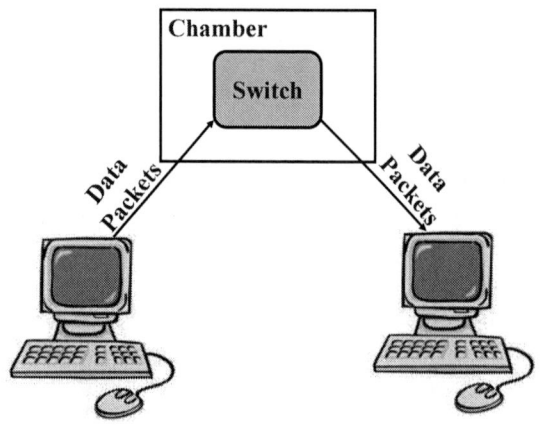

**Figure 3: Overview of switch experiment system**

There were two test conditions in this experiment: free air cooling and A/C. The operating conditions (temperature/humidity) of free air cooling are usually decided

by data center operators based on the local climates, industry standard requirements, equipment specifications, hotspot identification, and other analyses. In this case, the free air cooling condition used 10°C ~ 50°C, and 15% to 85% RH, as shown in Figure 4. Based on the survey of Sun Microsystem data centers described in section 2.1, the A/C condition in the experiment used 20°C and 50% RH. There were three samples under each test condition, and the experiment ran for four months[4].

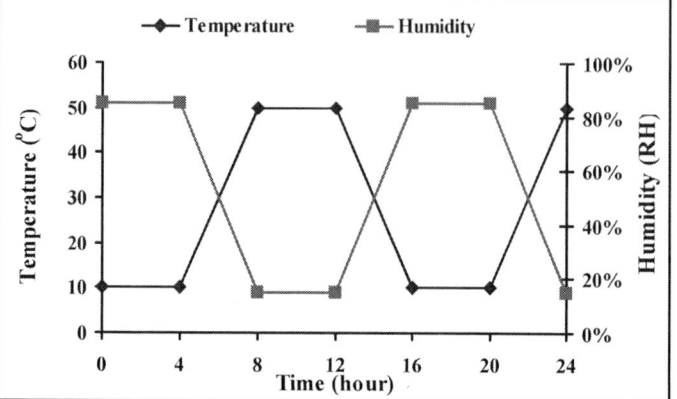

**Figure 4: Free air cooling profile**

## 3.2 Test Result Analysis

Three critical performance parameters of the switches were monitored in the experiment: throughput[5], response time[6], and transaction rates[7]. The three parameters are related and can be derived from each other, so only throughput was considered in this case. Examples of the monitored switch throughput for one day under the A/C and free air cooling conditions are shown in Figure 5 and Figure 6, respectively. Based on the one-day examples, the switch throughputs under the free air cooling condition have larger variations than those under A/C conditions.

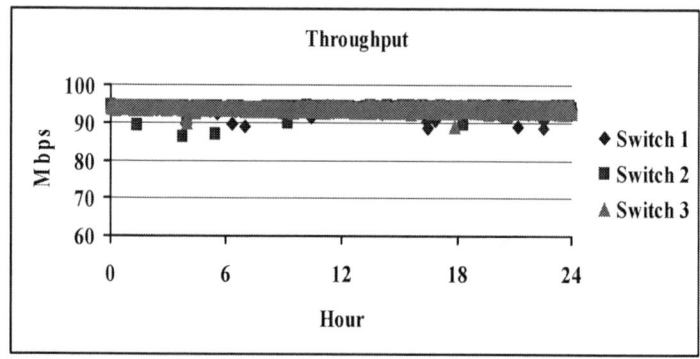

**Figure 5: Monitored throughput of switches under A/C condition in one day**

---

[4] Regarding the four months, two months (about eight weeks) were free air cooling experiment, and the other two months were A/C experiment.

[5] Throughput is the amount of data transferred from one place to another or processed in a specified amount of time. Data transfer rates for disk drives and networks are measured in terms of throughput.

[6] Response time is the elapsed time between the end of an inquiry or demand on a computer system and the beginning of a response.

[7] Transaction rate is the speed of transferring data.

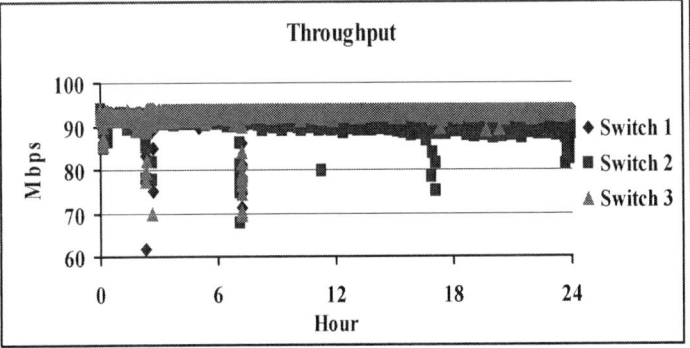

Figure 6: Monitored throughput of switches under free air cooling condition in one day

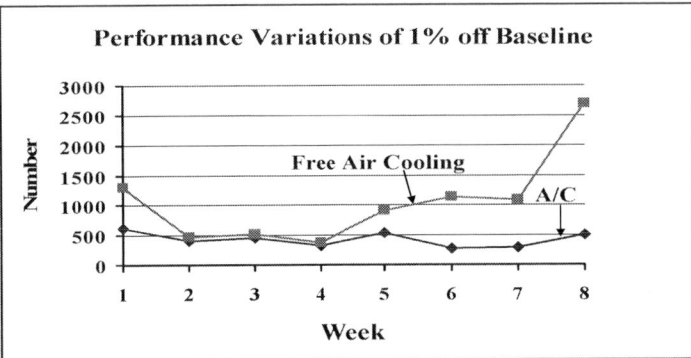

Figure 7: Performance variations of 1% off baseline

In order to evaluate the throughput variations of the switches, throughput baselines had to be created for comparison under the two conditions. In this case, the baseline of every switch was considered to be its average throughput for the first 10,000 data packets (approximately one day). The throughput baselines of the six switches are shown in Table 1. The throughput baselines under the free air cooling condition were slightly lower than the baselines under the A/C condition, since there were more variations, decreasing the throughput ability.

Table 1: Throughput baselines of switches

| Sample | | Throughput Baseline (Mbps) |
|---|---|---|
| A/C Condition | Switch 1 | 93.58 |
| | Switch 2 | 93.74 |
| | Switch 3 | 93.73 |
| Free Air Cooling Condition | Switch 1 | 92.97 |
| | Switch 2 | 91.32 |
| | Switch 3 | 92.78 |

Some metrics are required to compare and evaluate the performance variations of the switches under the two conditions. In this paper, we consider the number/percentage of throughput off the baseline as the possible metrics: 1% off baseline, 2% off baseline, 5% off baseline, 10% off baseline, and 20% off baseline. Comparisons of the throughput variations (average of the three samples) under the A/C and free air cooling conditions with different metrics are shown in Figure 7- Figure 9 [8]. Generally, the average throughput variation frequencies under the free air cooling condition were larger than those under the A/C condition. In addition, the average throughput variation frequencies decreased in the first weeks under both test conditions.

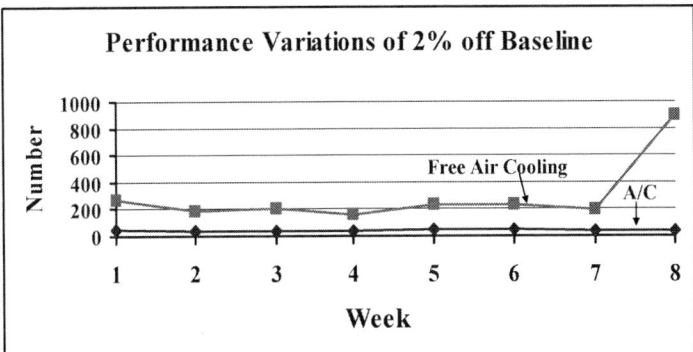

Figure 8: Performance variations of 2% off baseline

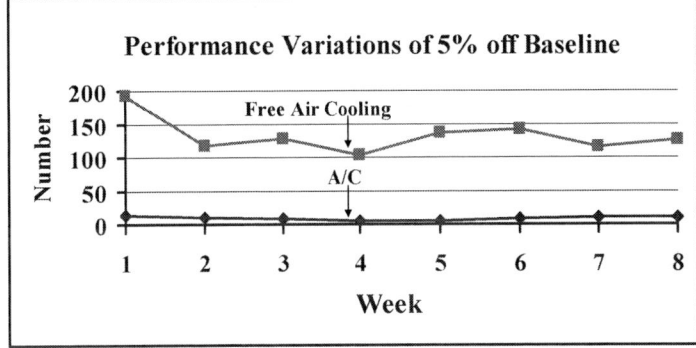

Figure 9: Performance variations of 5% off baseline

The average increases over the eight weeks of throughput variation frequencies between the two conditions are shown in Table 2 Under simulated free air cooling conditions, the increases of throughput variation frequencies compared with those under air conditioning are various when the different metrics are used as the comparison baseline.

Table 2: Increases of Throughput Variation Frequencies

| | A/C | Free Air Cooling | Increase |
|---|---|---|---|
| 1% off baseline | 424 | 1059 | 149% |
| 2% off baseline | 40 | 297 | 643% |
| 5% off baseline | 9 | 133 | 1444% |
| 10% off baseline | 0 | 105 | |
| 20% off baseline | 0 | 55 | |

Based on these metrics, the throughput variations of the switches under A/C conditions are shown in Table 3. Since

---

[8] The variations of 10% off baseline and 20% off baseline under free air cooling are listed in Table 2, but there were no these kinds of variations under the A/C condition, so we don't compare them.

978-1-4673-1110-6/12 $31.00 © 2012 IEEE

there are no throughput variations beyond 10% off baseline and 20% off baseline under A/C condition, the table shows the variation frequencies of 1% off baseline, 2% off baseline, and 5% off baseline. Among the three metrics, the throughput variations with 2% off baseline are the most stable with the minimum standard deviation over mean.

**Table 3: Throughput Variation Frequency under A/C Condition**

| Week | Sample 1 | | | Sample 2 | | | Sample 3 | | |
|------|------|------|------|------|------|------|------|------|------|
| | 1% | 2% | 5% | 1% | 2% | 5% | 1% | 2% | 5% |
| 1 | 476 | 60 | 16 | 713 | 34 | 14 | 660 | 39 | 13 |
| 2 | 233 | 57 | 13 | 469 | 21 | 12 | 416 | 32 | 8 |
| 3 | 374 | 59 | 11 | 515 | 30 | 9 | 449 | 29 | 7 |
| 4 | 264 | 50 | 5 | 372 | 30 | 5 | 341 | 28 | 7 |
| 5 | 434 | 67 | 7 | 608 | 30 | 6 | 554 | 32 | 7 |
| 6 | 239 | 58 | 6 | 310 | 32 | 9 | 278 | 40 | 11 |
| 7 | 244 | 48 | 13 | 326 | 26 | 7 | 319 | 33 | 9 |
| 8 | 404 | 51 | 6 | 552 | 39 | 11 | 547 | 35 | 12 |
| Mean | 346 | 56 | 9 | 483 | 30 | 9 | 445 | 33 | 9 |
| Std[9]/Mean | 0.25 | **0.10** | 0.46 | 0.27 | **0.16** | 0.32 | 0.28 | **0.12** | 0.25 |

## 4. Conclusions

The performance variations vary using different metrics, and thus an appropriate metric is critical to identify the impact of free air cooling on the performance of telecom equipment. Since performance variations exist even under the A/C condition, and 2% off baseline was the most stable metric under the A/C condition in our case, the 2% off baseline can be selected as the metric for the performance variation comparisons of this telecom equipment

When free air cooling is implemented in data centers to maximize the energy savings, the operating conditions may increase beyond the recommended operating conditions (RoC) of telecom equipment. Furthermore, the telecom equipment with hotspots is more likely to run beyond its RoC than equipment cooled by A/C units, which may cause performance variations of telecom equipment. The performance variation is one of the key concerns for free air cooling implementation.

The frequency of the performance variations increases under free air cooling compared with the variations under A/C. The performance variations under free air cooling are also larger than those under the A/C. For example, the performance variations are never beyond 10% off baseline under the A/C; however, they exceed even 20% off baseline under free air cooling conditions. This kind of large performance variation decreases the quality of data center services and may be unacceptable to the customers of data centers. The future research may include identifying the impact of temperature and humidity on the performance of telecom equipment, respectively, which can help data center operators set proper operating condition ranges without affecting the equipment performance significantly during the implementation of free air cooling.

## 5. Acknowledgment

The authors would like to thank the more than 100 companies and organizations that support research activities at the Center for Advanced Life Cycle Engineering (CALCE) at the University of Maryland annually. The authors would like to thank the members of the Prognostics and Health Management Consortium at CALCE for their support of this work. We also thank Dr. John Fitch of Dell for his key comments and suggestions.

## REFERENCES

[1] J. G. Koomey, "Growth in Data Center Electricity Use 2005 to 2010", Analytics Press, Oakland, CA, Aug.2011.

[2] A. Almoli, A. Thompson, N. Kapur, J. Summers, H. Thompson, and G. Hannah, "Computational Fluid Dynamic Investigation of Liquid Rack Cooling in Data Centres", Applied Energy, Vol. 89, Issue 1, pages 150-155, Jan., 2012.

[3] P. Johnson, and T. Marker, "Data Center Energy Efficiency Product Profile", Pitt & Sherry, Report to Equipment Energy Efficiency Committee (E3) of The Australian Government Department of the Environment, Water, Heritage and the Arts (DEWHA), Apr., 2009.

[4] ASHRAE TC 9.9 2004, "Thermal Guidelines for Data Processing Environments", 2004.

[5] Intel Information Technology, "Reducing Data Center Cost with an Air Economizer", IT@Intel Brief; Computer Manufacturing; Energy Efficiency; August 2008.

[6] R. Miller, "Microsoft's Chiller-less Data Center", Data Center Knowledge, Sep. 2009.

[7] R. Miller, "Google's Chiller-less Data Center", Data Center Knowledge, Jul. 2009.

[8] European Commission, "Code of Conduct on Data Centres Energy Efficiency—Version 2.0", Nov. 2009.

[9] American Society of Heating, Refrigerating and Air-Conditioning Engineers (ASHRAE), "Energy Standard for Buildings Except Low-Rise Residential Buildings", Atlanta, GA, Oct., 2010.

[10] T. Homorodi, and J. Fitch, "Fresh Air Cooling Research", Dell Techcenter, Aug. 2011.

[11] ASHRAE, "2008 ASHRAE Environmental Guidelines for Datacom Equipment", Atlanta, 2008.

[12] Bell Communications Research Inc., Generic Requirements GR-63-CORE, "Network Equipment-Building System (NEBS) Requirements: Physical Protection", Piscataway, NJ, Mar., 2006.

[13] Bell Communications Research Inc., Generic Requirements GR-3028-CORE, "Thermal Management in Telecommunications Central Offices", Piscataway, NJ, Dec. 2001.

[14] R. Miller, "Data Center Cooling Set Points Debated", Data Center Knowledge, Sep, 2007.

[15] A. Shehabi, W. Tschudi, and A. Gadgil, "Data Center Economizer Contamination and Humidity Study", Emerging Technologies Program Application Assessment Report to Pacific Gas and Electric Company, Mar., 2007.

[16] P. Lall, M. Pecht, and E. Hakim, "Influence of Temperature on Microelectronics and System Reliability: A Physics of Failure Approach", CRC Press, New York, ISBN: 0-8493-9450-3, May, 1996.

[17] ASHRAE Technical Committee (TC) 9.9, "2011 Thermal Guidelines for Data Processing Environments – Expanded Data Center Classes and Usage Guidance", Atlanta, 2011.

---

[9] Std is "standard deviation" here.

# Enhancement of Photovoltaic Solar Module Performance for Power Generation in the Middle East

Valerie Eveloy, Peter Rodgers and Shrinivas Bojanampati
Department of Mechanical Engineering
The Petroleum Institute
Abu Dhabi, United Arab Emirates
veveloy@pi.ac.ae

## Abstract

Despite the abundance of solar energy in the Middle East, the efficiency and reliability of photovoltaic (PV) modules is severely affected by elevated cell operating temperature, which reduces the effectiveness of sun tracking techniques. In this study the potential of water-cooling to improve the electrical performance of stationary and sun-tracked PV modules is experimentally investigated for application at off-shore oil and gas facilities in the Persian Gulf. In parallel with measurements of PV module electrical characteristics and operating temperature, global solar irradiation, ambient air and cooling water temperatures, wind velocity and relative humidity are also recorded. In autumn conditions in the United Arab Emirates (24.43°N, 54.45°E), water-cooling is found to reduce PV module operating temperature by up to 30°C relative to passive cooling conditions, depending on operating conditions. Using water-cooling, power output is enhanced for a significant portion of the day for fixed geographical South facing modules (e.g., 22% at solar noon, and 13% at 2 p.m.) relative to passive cooling, and even more significantly for sun-tracked modules (e.g., 20% at 2 p.m.). Sun tracking enhances power output by 16% and 24% for passively- and water-cooled modules, respectively, at for example 2 p.m. The incorporation of water-cooling and sun tracking is the most effective, with enhancements in output power of on order 40% at for example 2 p.m. relative to passively-cooled, fixed South facing operation. In winter conditions (e.g., December), modest reductions in operating temperature and improvements in electrical performance are obtained using water-cooling and/or sun tracking, which may not justify the associated capital and operating costs.

## Keywords

Photovoltaic, thermal management, water-cooling.

## 1. Introduction

Due to space constraints and to eliminate the technical and cost issues associated with the transportation and storage of fossil fuels at off-shore oil and gas production and processing facilities, photovoltaic (PV) solar modules have recently been integrated for on-site power generation. This technology is particularly attractive in the Middle East, where solar energy is highly available, and offers environmental benefits compared to the use of fossil fuels. The present study was undertaken in the United Arab Emirates (UAE), which experiences among the highest yearly solar irradiation levels in the world due to its latitude and high insolation clearness index. In Abu Dhabi (24.43°N, 54.45°E), the yearly average daily energy input is typically of 18.48 MJ/m², with one-minute average solar radiation levels culminating at 1,041 W/m². Daily and monthly mean solar radiation typically peak

at 369 and 290 W/m², respectively. In addition, the yearly average insolation clearness index is of approximately 0.58 [1,2].

In the past few years, the UAE government has actively promoted the use of solar energy in efforts to reduce the nation's carbon footprint, which is one of the highest per capita in the world [3]. Solar energy has been introduced in the form of both grid-connected and standalone PV modules, for industrial and housing power generation, respectively. The MASDAR city initiative [4] in Abu Dhabi, for example, is in the process of constructing a 10 MW array of PV modules to power the largest net zero-carbon-dioxide-emissions and zero-waste urban area in the world[1]. With regard to the oil and gas industry, which is the application sector considered in the present study, key off-shore oil and gas applications that presently utilize PV technology include cathodic protection, telemetry and valve control. It is anticipated that a growing number of oil and gas applications will rely on solar power generation in the future, particularly if the technological issues considered in the present work are addressed.

While PV technology has been successfully integrated in many regions of the world, with impressive yearly growth rates reported [5], its implementation in the Middle East faces unique technological challenges, which have been marginally considered to date. These challenges essentially arise from harsh environmental conditions. Typically, commercially-available PV modules only convert 4 to 18% of the solar irradiation to electrical energy, depending upon cell type and operating conditions, with the remainder being primarily converted to thermal energy, and to a lesser extent lost by reflection (20%) [6,7]. The portion of the irradiation absorbed and converted to thermal energy increases cell operating temperature, thereby severely reducing electrical efficiency and potentially causing thermo-mechanical reliability issues. The efficiency of PV cells (i.e., percent of incident solar irradiance that is converted into electricity) typically degrades by 0.2% to 0.5% per °C rise in ambient temperature, depending upon the PV technology[2] [6,8]. This effect is essentially attributable to a reduction of the semiconductor material band gap [6]. The significance of the temperature dependency of cell electrical performance has been highlighted experimentally in Saudi Arabia facilities, where

---

[1] MASDAR city is to cover a desert area of 6 km² in the UAE, and represents a $15 billion investment.

[2] Monocrystalline technology, which dominates the PV market because of its performance and availability, exhibits the highest efficiencies (14 - 18%), but a pronounced temperature sensitivity coefficient. Thin film technologies have lower efficiencies (~10%) than monocrystalline PVs, but lower temperature coefficients. Certain thin film PVs have been claimed to almost match, or potentially even outperform, monocrystalline technology in hot climatic conditions.

PV electrical efficiency has been observed to decrease by up to 30% in summer, relative to standard PV operating conditions [7]. In summer in the UAE, PV modules surface temperatures of up to 80°C have been recorded [4]. Such elevated cell temperatures result in unacceptable power losses, and over a prolonged period of time can cause permanent structural damage [7]. Ambient relative humidity and fouling (e.g., dust, sand) can further degrade electrical performance [9] and reliability. Thus, module soiling conditions (i.e., location, type of fouling particles, frequency of rainfall) have been found to strongly influence the efficiency of PV systems in Saudi Arabia [10]. Moisture ingression-induced failures are of concern in high ambient humidity and temperature conditions [11]. As the introduction of PV technology and related development/research activities in the Middle East are still relatively recent, insufficient data presently exists on the long term performance and reliability of various PV technologies in such conditions. Enhancing heat dissipation to the environment in order to prevent electrical and thermo-mechanical performance degradation, poses challenges at elevated ambient temperatures, which yearly average 29°C in Abu Dhabi, with daily maximum and monthly average temperatures peaking at approximately 51°C and 36°C, respectively [1].

Apart from reducing PV electrical efficiency, the environmental conditions encountered in the Middle East also impair the benefits of sun tracking techniques, that control the orientation of a PV module throughout the day to maximize the absorbed solar irradiation. Sun tracking can be implemented along either one or two axes, with the latter providing a higher power output. The energy yield can thus be increased by approximately 20% to 30%, depending on geographical location and environmental conditions [12]. It has been projected that in standard PV applications installed from 2009 to 2012, trackers will be integrated in approximately 85% of commercial installations greater than 1 MW [13]. In environmental conditions that do not depart significantly from standard PV operational conditions, the improvements in power conversion efficiency obtained with sun tracking relative to stationary operation often outweigh the additional capital and operating expenses. However, in extremely hot climate conditions, the increased absorbed solar radiation constitutes an additional thermal load. The integration of an active PV module cooling solution and sun tracking technique for application in hot climate conditions has not been explicitly addressed in the literature, and is one of the objectives of this study. This is achieved by experimentally investigating the potential of water-cooling to improve the electrical performance of stationary and sun-tracked PV modules operated in the Persian Gulf. To assist the development of a candidate cooling design in this study, previous studies related to the thermal analysis of PV modules are reviewed.

## 2. PV Module Thermal Management

The majority of PV cooling analyzes previously reported have been for hybrid photovoltaic/thermal (PVT) systems, in which cooling results from passive solar energy collection by the cooling fluid and thus is not the primary objective, but contributes to improve system electrical efficiency. Although the PVT literature is a primary source of information on candidate PV module cooling strategies, few PVT actual applications are yet installed [7,14]. In addition, air-type collectors are essentially the only commercial systems built to date [14]. As will be confirmed in this study, air-cooling is unlikely to meet the cooling requirements of PV systems operated in elevated ambient temperatures encountered over a major part of the year in the Middle East. Furthermore, due to a lack of field tests and long term monitoring, insufficient experience presently exists on the performance, life time and reliability of PVT systems, hence PV modules [14] as a function of PV technology, cooling configuration and climatic conditions.

PVT cooling has been based on either natural, forced air or forced liquid convection. Most air-cooled designs have consisted of single- or double-pass cooling ducts placed on either the active or non-active module surface, or both [7]. Surface extension of the module non-active surface has been investigated using fins, ribs, holes, dimples, or Teflon surface roughening, while V-shaped grooves have been evaluated on the absorber surface [14]. Insertion of a blackened metal plate at half height along the air channel, and air impingement on the module non-active surface have also been considered [14].

Liquid-cooled PVT configurations have consisted of ducts placed either above or below the PV cell array, finned or unfinned sheet-and-tubes (e.g., single-pass straight tubes, multi-pass U-shaped serpentine tube), or rectangular-shaped box channels (e.g., extruded or formed using ribs) [7]. Minimization of thermal contact resistance requires consideration of module-cooling channel interface planarity, which may be difficult to achieve in sheet-and-tube configurations, and channel attachment material/process (e.g., clamping, glue, silicone, soldering) [14]. Apart from thermal and hydraulic performance optimization, reliability, manufacturing, assembly, and economic viability are key design challenges, all of which are greatly influenced by materials selection.

The design of a PV cooling system differs from that of a PVT in essentially two aspects. Whereas the cooling fluid of a PVT requires to be thermally insulated from the surroundings to permit utilization of the thermal energy collected for applications (e.g., space heating, hot water production), in a PV system the heat advected by the fluid can be directly dissipated to the surroundings. In addition, the economic payback period of a PV liquid cooling system is likely to exceed that of a PVT, if electrical power is the only energy output. Unlike PV systems, PVTs also collect heat for applications, thereby yielding higher economical benefits. From a comparison of typical PV/PVT electrical efficiencies (i.e., up to 18%) and PVT thermal efficiencies (up to 70% for liquid PVTs [7]), it is evident that re-use of the PV cooling water waste heat in applications typically yields higher energy savings than the increase in PV power output obtained via PV cooling. This has been experimentally observed by Furushima and Nawata [15]. For the later reason, and since most climatic conditions would not justify the capital and operating expenses associated with active PV cooling, few efforts (e.g., [16]) have attempted to develop thermal management solutions for standalone PV modules, operated independently of applications. Recovery of the cooling water waste heat for domestic or industrial applications has either been an integral part of the system constructed (e.g., water desalination [17]) or presumed as an added modular capability for economic viability (e.g., domestic water heating [15]).

Both PV cell-level and module-level thermal management solutions have been considered. Passive cell-level thermal management approaches have included enhancement of the ethylene vinyl acetate (EVA) layer using aluminum particles [18,19], use of a structured glass layer to enhance convection from the PV active surface [19], liquid indium-gallium alloy for a PV-powered desalination unit in Saudi Arabia [20], and phase-change material (PCM) on the non-active surface of a V-through concentrated PV in India [21]. However, the associated material costs are often prohibitive. Also, structured surfaces are prone to fouling and pose difficulties for cleaning [22].

Module-level strategies have been based on either air- or liquid-cooling. In buoyant flow conditions, cooling ducts ([e.g., 16]), aluminum heat spreaders [23], folded [19] or casted [24] aluminum fins attached to the non-active module surface, have been investigated. In addition, filtering of the irradiation [25] and heat pipe integration have been proposed. Most of these enhancements have been for concentrated PVs, in which the enhanced irradiation can significantly elevate module operating temperature, rather than flat-plate PVs, which are the focus of this study.

While forced liquid cooling yields more effective heat transfer rates than free convection, pump efficiency and cooling flow rate should be optimized, as pumping power reduces the net electricity gain. Alternatively, gravity-assisted fluid circulation, thermosyphoning, or integration of the cooling unit with a PV application or separate system incorporating a water source (e.g., city water [15], desalination unit [17], irrigation system [26]) can eliminate the need for a pumping device specifically for module cooling.

Water cooling on the module active surface has been investigated by either trickling [26] in Australian and Jordanian desert conditions, or spraying in Brazil [22] and Iran [27,26]. In the case of [26], the PV cooling unit was integrated with an irrigation system that supplied bypass water pumped from wells. The potential advantages of active surface water cooling include increased irradiation due to refraction by water, increased heat transfer rate due to evaporation, and fouling removal [22,26]. However, this approach is not feasible if sea, process- or waste water serves as the cooling fluid due to scale deposition on the active surface. Depending upon the water source, water filtering or purification may be required [27,28]. Cooling by submerging PVs in shallow water has also been explored experimentally in Italy [29].

Recently, liquid cooling designs have been proposed for the module non-active surface [15,17,30]. Furushima and Nawata [15] built an automated aluminum box-channel liquid cooling system attached to the non-active PV module surface for operation in Japan summer conditions. The device utilized siphonage for water circulation between hot and cold water storage tanks and the PV module. The use of city water and siphonage eliminated the need for a pumping device. It was highlighted that re-use of the cooling water for domestic consumption would yield significantly higher energy production than the increase in PV power output obtained via PV cooling, and that higher energy output enhancements would be anticipated in higher ambient air temperatures and irradiation levels.

Wilson [30] used a gravity-fed water flow on the non-active surface of an 8-cell module in Caribbean conditions. The concept was however limited to regions having natural water supply, such as river or rain water.

Kelley et al. [17] assembled a flat plate heat exchanger to the non-active surface of a PV module used for powering a reverse-osmosis desalination unit, which desalinated the cooling sea water. Water cooling enhanced both PV electrical output and the production rate of desalinated water. A motor-driven pump was required to circulate and pressurize the cooling water for desalination. The reduction in PV module operating temperature due to cooling permitted the incorporation of low-cost flat-plate concentrating mirrors. The system was automated to control both water flow rate and temperature. It was also suggested that a portion of the physical energy of the pressurized desalinated water could be recuperated using an additional recovery device (e.g., turbine). The reported system characterization was however limited to mid-November in Cambridge, MA, USA. Furthermore, as enhancement of desalination water production capacity was the focus of the study, PV temperature reduction and output enhancement due to cooling were not quantified.

Other studies (e.g., [31,32]) have analyzed the heat transfer characteristics of passively-cooled PV modules and developed predictive thermal models, but the design of a cooling scheme was not considered.

Sun tracking requires control electronics, and represents an additional capital and operating expense for PV installations. Although a few cooling investigations undertaken for flat-type PV modules report the use of sun tracking [17,22,28], the impact of tracking and cooling, either individually or combined, on total and net electrical output, and economic viability, were not analyzed.

In summary, the above overview highlights that few studies have attempted to integrate liquid cooling schemes in flat-type PVs. Cooling designs in which the coolant flows on the module active surface would preclude the use of sea water as cooling fluid, which would be the ideal water supply in off-shore oil and gas application environments, as considered in this study. Furthermore, elevated relative humidity such as in Persian Gulf facilities would impede evaporation of the cooling fluid from the active PV surface. Both considerations limit the generation of potential liquid cooling designs to non-active PV surface cooling. In addition, the electrical and thermal performance of passively- and liquid-cooled PV modules in hot climatic conditions has been insufficiently investigated, particularly for non-active surface cooling. Finally, the interaction of liquid cooling with sun tracking has not been explicitly investigated, in terms of electrical power output and economic feasibility.

The objectives of the present study are to experimentally quantify PV module performance losses associated with environmental effects (i.e., solar irradiation, ambient air temperature) in the Persian Gulf, and to explore the potential of liquid cooling in conjunction with sun tracking to improve module power output. Considering the capital and operating expenditure associated with liquid cooling, the performance improvements obtained are also compared to those for forced air-cooling. The application environments considered are off-shore oil and gas production facilities in the Persian Gulf. The present design problem therefore involves unique customer-specific requirements, arising from climatic conditions, the remoteness of off-shore oil and gas facilities, space constraints, and environmental emission regulations. Unlike for other industry or application sectors, economic payback period may not be a primary design consideration.

## 3. Experimentation

The PV modules electrically and thermally characterized consisted of 100 W and 140 W modules referred as Module A and B, which contained 72 and 32 monocrystalline PV cells, respectively. Both module specifications are listed in Table 1.

The PV modules were characterized outdoors between 9 a.m. and 5 p.m., at an altitude of 15 m above the ground level, with their surface inclined at a tilt angle of 24.5° with the horizontal as illustrated in Figure 1. This inclination corresponds to the approximate optimum tilt angle for the latitude of Abu Dhabi (24.43°N).

Each PV module was characterized in both passive and active cooling conditions to assess the power output improvements obtained with active cooling. In passive cooling conditions, the module is cooled by free air convection and radiation to the surroundings.

Owing to the availability of sea water at off-shore oil and gas facilities, water was selected as the cooling fluid for the liquid-cooled application. However as a first step, all measurements were undertaken using fresh water supplied from a city water distribution system. Sea water in the Persian Gulf has a typical salinity of approximately 40 g/kg [33], which does not significantly impact water thermophysical properties [34] for heat transfer. The PV module water-cooling design is illustrated in Figure 2, and consists of a Plexiglas serpentine channel of internal depth 5 mm, attached to the non-active surface of the module. Unlike

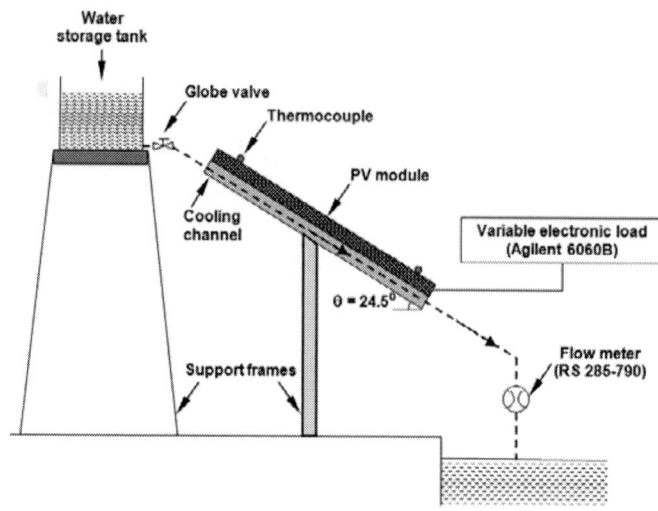

**Figure 1:** Schematic representation of experimental test set-up for water-cooled PV module.

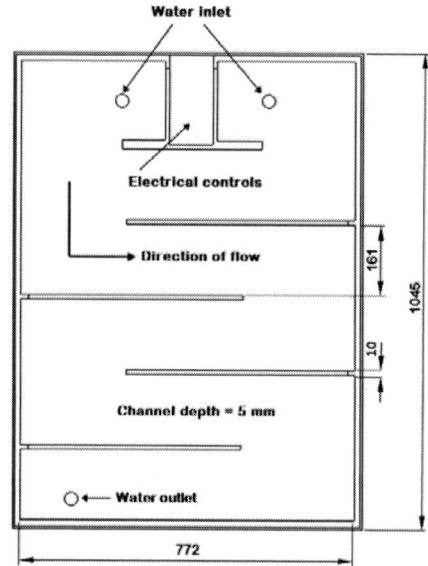

Note: All dimensions are in mm.
**Figure 2:** Schematic representation of PV module water-cooling channel geometry.

in flat-plate PVT devices, where the absorber plate separates the PV cell array from the cooling tube or channel (e.g., [35]), in the present design the fluid directly wets the non-active PV module surface. This eliminates any thermal contact resistance. Water was discharged from a 120 gallon reservoir at a mean velocity of 0.5 m/s through the cooling channel via two inlets at the upper end of the module and discharged from one outlet at the lower end of the module. As the reservoir water itself was supplied from a building-scale reservoir located above ground, the inlet water temperature varied from 25 to 36°C depending upon its usage rate in the building complex, solar loading on the reservoir, and ambient air temperature. Although refrigerating the cooling water was considered, this would increase the complexity, footprint and cost of the cooling system. Consequently all measurements were undertaken for a worst-case, uncooled water temperature of 25 to 36°C, which is representative of sea water surface

**Table 1:** Manufacturer specifications of PV modules.

| Parameter | Value | |
|---|---|---|
| | Module A | Module B |
| Dimensions (mm) | 1,080 x 808 x 35 | 1,485 x 655 x 50 |
| Weight (kg) | 11 | 12 |
| Peak power, $P_{max}$ (W) | 100 | 140 |
| Rated power tolerance | 3% | ± 2.5 W |
| Maximum power current, $I_{mp}$ (A) | 5.56 | 7.64 |
| Maximum power voltage, $V_{mp}$ (V) | 18.00 | 18.32 |
| Short circuit current, $I_{sc}$ (A) | 6.43 | 8.29 |
| Short circuit voltage, $V_{oc}$ (V) | 21.60 | 21.71 |
| Maximum system voltage, $V_{max}$ (V) | 1,000 | 1,000 |

Note: Standard test conditions defined as: Spectral distribution of AM1.5; solar irradiation, G = 1,000 W/m², cell temperature, $T_c$ = 25°C, Wind speed < 1 m/s.

temperatures in the Persian Gulf, that seasonally range from 15 to 32°C [33]. This study being exploratory, the cooling channel design and water flow rate were not formally optimized, but permitted an effective assessment of the technical feasibility of water cooling.

To demonstrate the need for water cooling, a forced air-cooled PV prototype was also constructed. This design comprises an aluminum duct of cross sectional area 811 mm x 160 mm, which is attached to the non-active surface of PV Module A. In addition, eight 120 x 120 x 38 mm 22/21 Watts axial AC Watts fans (SUNON DP200A) are mounted at the duct inlet and outlet, that operate in a push-pull mode.

The measurements were undertaken between October 29 and December 22, 2011 in Abu Dhabi. During this period ambient air temperatures and solar irradiation would be representative of those in summer at for example mid European latitudes.

In both passive and active cooling conditions, the module was characterized in either a fixed geographical South facing orientation[3], or with vertical single-axis sun tracking. The fixed South facing orientation yielded maximum power output at solar noon[4]. For the present geographical location and time period, solar noon varied from 12:14 p.m. to 12:21 p.m. In vertical single-axis tracking, the axis of rotation is vertical with respect to the ground, with the module rotating from East to West over the course of the day. At solar noon, the fixed and sun-tracked module orientations are the same.

To permit direct comparison of module electrical performance in the same environmental conditions for different cooling conditions (i.e., passive, forced-air, forced-water cooling), at least two cooling designs were simultaneously characterized.

PV module surface temperature was measured using four type-K thermocouples having an accuracy of ±1°C, two of which were attached to the module active surface, and the remaining two to its non-active surface. The *quasi* isothermality of the module active surface was assessed using infrared thermography, which revealed less than 2°C spatial temperature variation in worst-case thermal loading conditions, i.e. at solar noon. All module surface temperature measurements were made by ensuring that steady-state conditions were reached. The module current-voltage characteristics were recorded using a 300 W Agilent 6060B DC electronic load. In parallel with module surface and cooling water temperature, and current-voltage measurements, global solar (i.e., total hemispherical) irradiation, ambient air temperature, wind velocity, and relative humidity were recorded using a Virtual Weather Hawk station having accuracies of ±2%, ±0.5°C, starting threshold of 0.78 m/s, and accuracies of ±5% for 90–100% RH and ±3% for 10–95% RH, respectively for these variables. The meteorological data were collected every minute, at the module location. The test matrix is summarized in Table 2.

---

[3] Geographical south differs from magnetic south due to magnetic declination.

[4] Solar noon is the point in time at which the sun is at its daily zenith. This time typically differs from 12 p.m. because of national time settings, and varies on a daily basis.

**Table 2:** PV module test matrix.

| Module | Date | Orientation | Cooling Configuration |
|--------|------|-------------|----------------------|
| A | September 27, 2011 | Tracked | Passive |
| | September 28, 2011 | Fixed | Passive |
| | November 3, 2011 | Fixed | Passive<br>Forced air-cooled<br>Water-cooled |
| B | October 29, 2011 | Fixed | Passive<br>Water-cooled |
| | November 6, 2011 | Tracked | Passive<br>Forced air-cooled<br>Water-cooled |
| | December 14, 2011 | Tracked | Passive<br>Passive-insulated<br>Water-cooled |
| | December 22, 2011 | Fixed | Passive<br>Passive-insulated<br>Water-cooled |

## 4. Results

The measured current-voltage (I-V) and output power characteristics of Module A operated in passive, forced air and water-cooling conditions are presented in Figure 3. Forced air- and water-cooling result in 8.7°C and 26.5°C reduction in module operating temperature relative to passive cooling, respectively, with power output increased by approximately 6% and 19%, respectively. This demonstrates the effectiveness of water cooling in improving PV module output power, while forced air-cooling only results in marginal performance improvements. Forced air-cooling would clearly be ineffective in summer conditions in the Persian Gulf, where ambient air temperatures typically reach 45°C.

Figure 4 shows the variation in measured I-V and output power characteristics of the passively-cooled PV Module A, both in fixed and sun-tracked orientations, as a function of time between 11 a.m. and 5 p.m. on September 27 and 28, 2011. Figures 4a and 4b highlight the strong dependence of $I_{sc}$ to irradiation, G. $I_{sc}$ typically increases linearly with increasing irradiance, and culminates at solar noon (i.e., 12:13 p.m.). The short circuit voltage, $V_{oc}$, increases weakly (i.e., logarithmically) with increasing irradiance, such that this effect is typically ignored. Due to a small variation in environmental conditions between September 27 and 28, that is global irradiance (Figure 5a), ambient air temperature and relative humidity (Figure 5b), the PV module characteristics in Figure 4 slightly differ at solar noon between the two days. Despite this variation, sun tracking results in a maximum output power that is 7% to 120% higher than for a fixed module orientation between 2 p.m. and 4:30 p.m., respectively (Figure 4c). The effect of incorporating a water cooling scheme in conjunction with sun tracking on module electrical performance is assessed in Figures 6, 8 and 10 for Module B.

Figure 6 compares the measured electrical characteristics of Module B for passive and water-cooling, in both fixed South facing (October 29, 2011) and sun-tracked module orientations (November 6, 2011). Due to partly cloudy conditions after 2 p.m. on November 6, which resulted in scattering in the global irradiation data (Figure 7a), the results presented in Figure 6b are limited to clear sky conditions between 11:20 a.m. to 2 p.m. The measured ambient air, module and cooling water temperatures, as well as global irradiation and peak power output are compiled in Table 3.

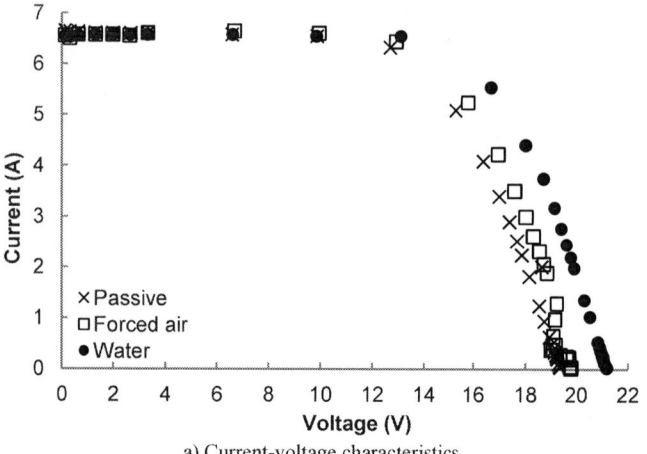

a) Current-voltage characteristics

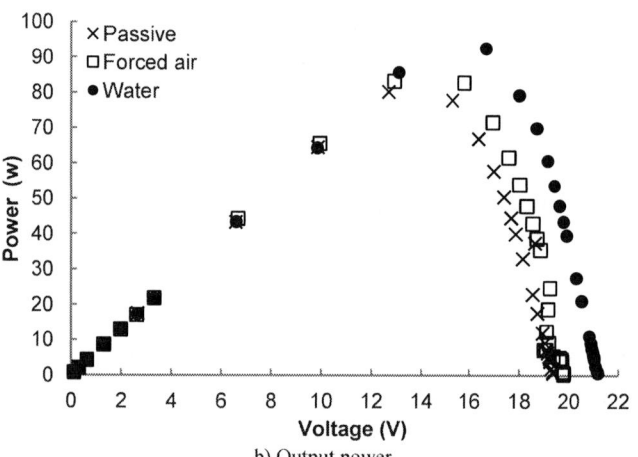

b) Output power

| Cooling Configuration | $T_p$ (°C) | $T_\infty$ (°C) | $T_w$ (°C) | G (W/m²) |
|---|---|---|---|---|
| Passive | 61.4 | 32.5 | --- | |
| Forced air-cooled | 52.7 | 32.5 | --- | 769 |
| Water-cooled | 34.9 | 32.5 | 31.5 | |

Note: $T_p$, $T_\infty$, and $T_w$ refer to module, ambient air and water temperature, respectively. G refers to global solar (i.e., total hemispherical) irradiation.

**Figure 3:** Comparison of measured electrical characteristics for passively cooled, forced air-cooled and water-cooled PV Module A in a fixed orientation. Measurements were taken from 12:06 p.m. to 12:55 p.m. on November 3, 2011, in Abu Dhabi (24.43°N, 54.45°E).

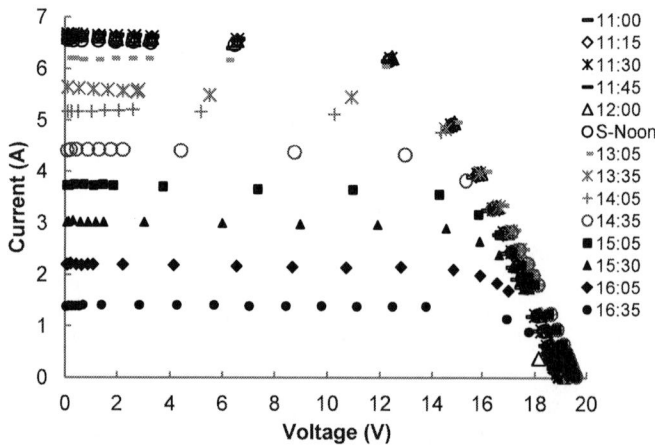

a) Current-voltage characteristics for fixed module on September 28

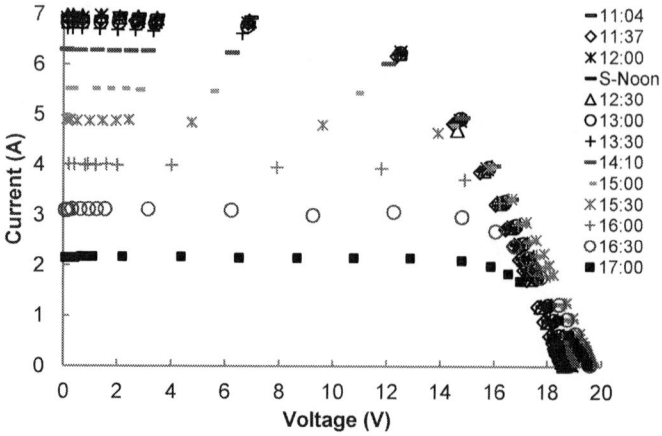

b) Current-voltage characteristics for sun-tracked module on September 27

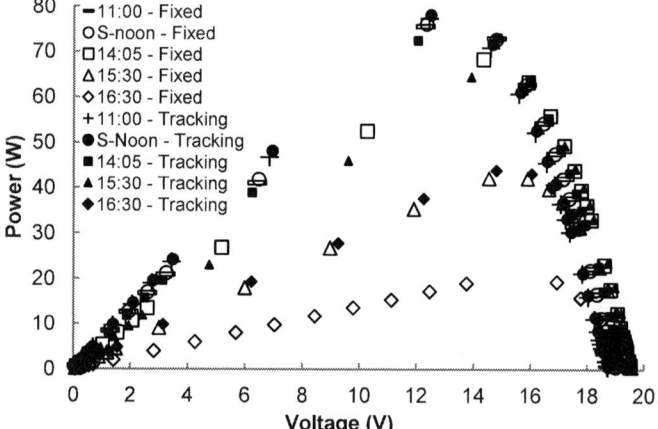

c) Comparison of output power for fixed and sun-tracked module

**Figure 4:** Comparison of measured electrical characteristics of passively cooled PV Module A in fixed and sun-tracked orientations, as a function of time from 11 a.m. to 5 p.m. on September 27 and 28, 2011, in Abu Dhabi (24.43°N, 54.45°E).

**Table 3:** Measured ambient air and cooling water temperatures, and global irradiation, for PV Module B's measurements in Figure 6.

| Date, Time | Module Orientation, Cooling Configuration | $T_p$ (°C) | $T_\infty$ (°C) | $T_w$ (°C) | G (W/m²) | Peak Electrical Power (W) |
|---|---|---|---|---|---|---|
| Oct. 29, S-Noon | Fixed Passive | 61.5 | 31.5 | --- | 734 | 84.0 |
| Oct. 29, S-Noon | Fixed Water | 33.1 | 31.5 | 25.5 | 734 | 102.7 |
| Nov. 6, S-Noon | Tracking Passive | 62.0 | 33.5 | --- | 747 | 82.8 |
| Nov. 6, S-Noon | Tracking Water | 32.1 | 33.5 | 28.5 | 747 | 100.5 |
| Oct. 29, 14:00 | Fixed Passive | 53.5 | 32.2 | --- | 605 | 76.0 |
| Oct. 29, 14:00 | Fixed Water | 33.0 | 32.2 | 28.0 | 605 | 86.0 |
| Nov. 6, 14:00 | Tracking Passive | 55.5 | 34.1 | --- | 538 | 88.5 |
| Nov. 6, 14:00 | Tracking Water | 31.0 | 34.1 | 29.5 | 538 | 106.2 |

Note: S-Noon refers to solar noon. $T_p$, $T_\infty$, and $T_w$ refer to module, ambient air and water temperature, respectively. G refers to global solar (i.e., total hemispherical) irradiation.

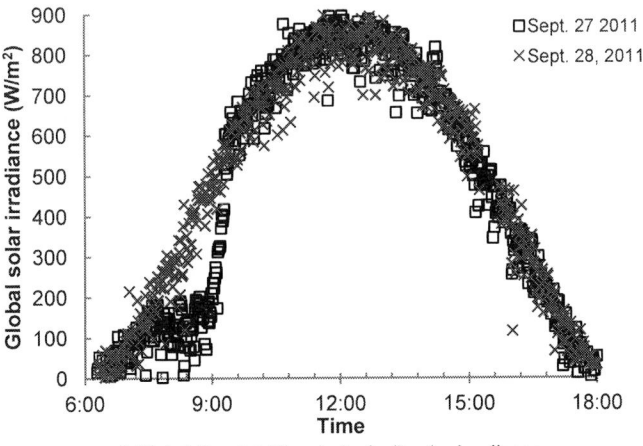

a) Global (i.e., total hemispherical) solar irradiance

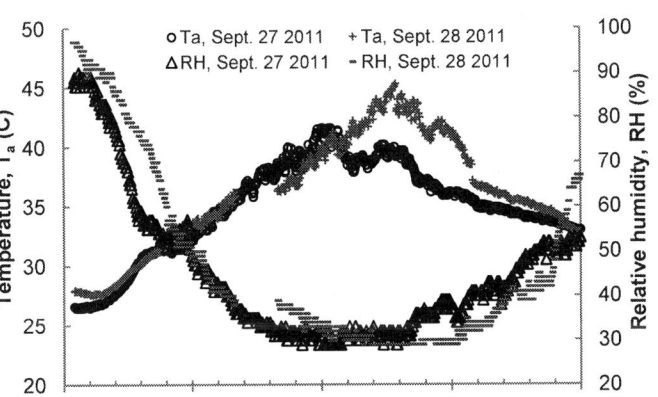

b) Ambient air temperature and relative humidity

**Figure 5:** Measured environmental data on September 27 and 28, 2011, in Abu Dhabi (24.43°N, 54.45°E).

**Table 4:** Measured ambient air and cooling water temperatures, and global irradiation, for PV Module B's measurements in Figure 9.

| Date, Time | Module Orientation, Cooling Configuration | $T_p$ (°C) | $T_\infty$ (°C) | $T_w$ (°C) | G (W/m²) | Wind Speed (km/hr) | RH (%) | Peak Electr. Power (W) |
|---|---|---|---|---|---|---|---|---|
| Dec. 22, S-Noon | Fixed Passive | 42.0 | 25.8 | --- | 746 | 8.5 | 31 | 111.8 |
| Dec. 22, S-Noon | Fixed Water | 30.5 | 25.8 | 28.3 | 746 | 8.5 | 31 | 119.9 |
| Dec. 14, S-Noon | Tracking Passive | 42.2 | 27.6 | --- | 671 | 16.4 | 35 | 107.5 |
| Dec. 14, S-Noon | Tracking Water | 34.0 | 27.6 | 28.9 | 671 | 16.4 | 35 | 112.6 |
| Dec. 22, 14:00 | Fixed Passive | 37.6 | 25.1 | --- | 541 | 13.5 | 32 | 91.9 |
| Dec. 22, 14:00 | Fixed Water | 30.8 | 25.1 | 32.1 | 541 | 13.5 | 32 | 94.1 |
| Dec. 14, 14:00 | Tracking Passive | 42.5 | 31.5 | --- | 583 | 17.4 | 35 | 95.6 |
| Dec. 14, 14:00 | Tracking Water | 37.6 | 31.5 | 36.4 | 583 | 17.4 | 35 | 99.7 |

Note: S-Noon refers to solar noon. $T_p$, $T_\infty$, and $T_w$ refer to module, ambient air and water temperature, respectively. G refers to global solar (i.e., total hemispherical) irradiation. RH refers to relative humidity.

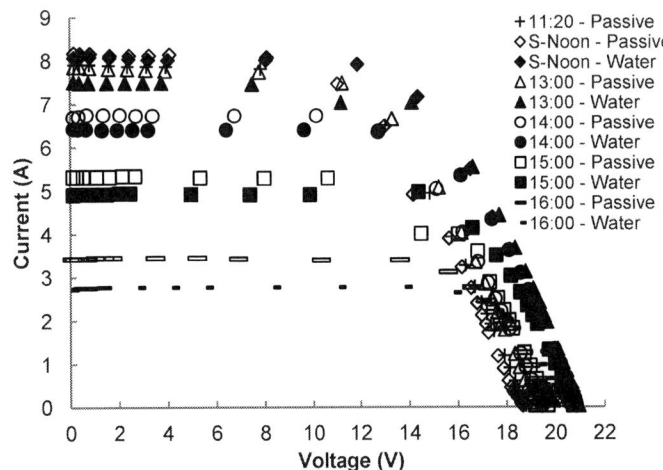

a) Current-voltage characteristics for fixed module on October 29

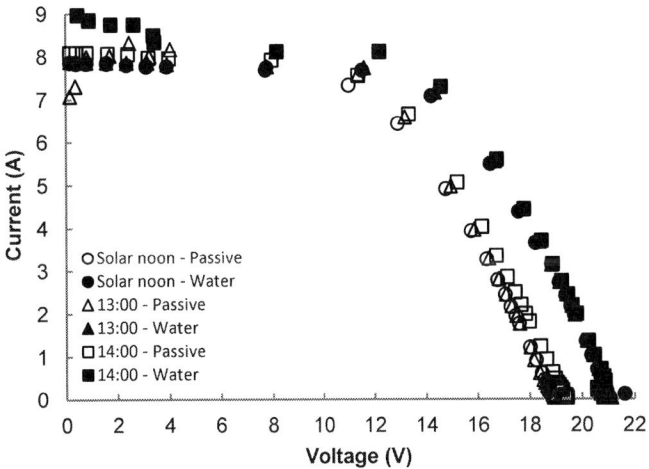

b) Current-voltage characteristics for sun-tracked module on November 6

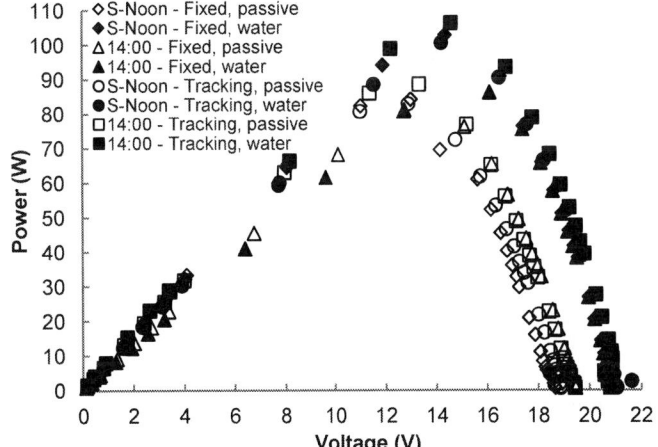

c) Comparison of output power for passively- and water-cooled module, both for fixed and sun-tracked orientations

**Figure 6:** Measured electrical characteristics of passively and water-cooled PV Module B, in fixed geographical South facing (October 29, 2011) and sun-tracked (November 6, 2011) orientations, as a function of time from 11:20 a.m. to 4:00 p.m., in Abu Dhabi (24.43°N, 54.45°E).

978-1-4673-1110-6/12 $31.00 © 2012 IEEE

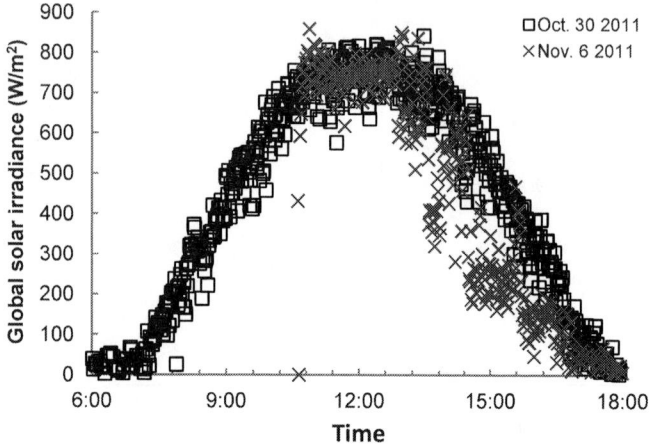

a) Global (i.e., total hemispherical) solar irradiance

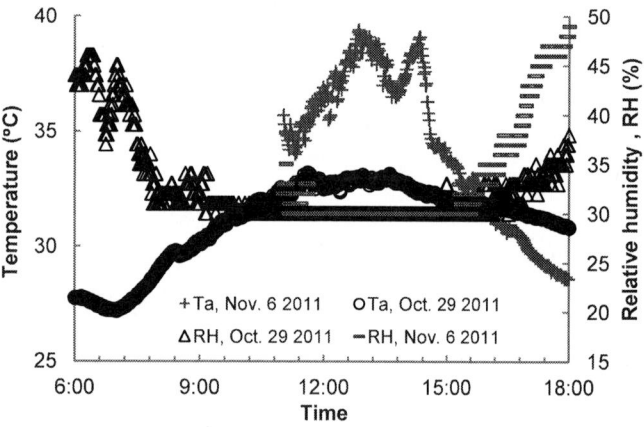

b) Ambient air temperature and relative humidity

**Figure 7:** Measured environmental data on October 29 and November 6, 2011, in Abu Dhabi (24.43°N, 54.45°E).

Module electrical performance in passive and water-cooling conditions can be directly compared as the measurements were undertaken simultaneously for two modules. Water-cooling reduces module operating temperature by approximately 29°C and 22.5°C at solar noon and 2 p.m., respectively, for both the fixed and sun-tracked orientations (Table 3). This results in a significant increase in $V_{oc}$ (Figures 6a and 6b), hence power output (on order 20%, Figure 6c). The interactive effects of sun tracking and water-cooling on operating temperature and electrical performance can be observed from the 2 p.m. data. Since the fixed and sun-tracked orientations for a given cooling configuration were characterized on two different days, the slight variability in ambient air, module and/or cooling water temperature, global irradiation and wind speed needs to be accounted for when assessing the impact of module orientation on electrical performance. At solar noon, module operating temperature for the fixed and sun-tracked orientations differ by a worst-case 1°C (Table 3), which is within the bounds of experimental error. However for the 2 p.m. water-cooling configuration, module temperature is 2°C lower in sun-tracked conditions relative to fixed ones (Table 3). This suggests that the effects of increased irradiation absorption

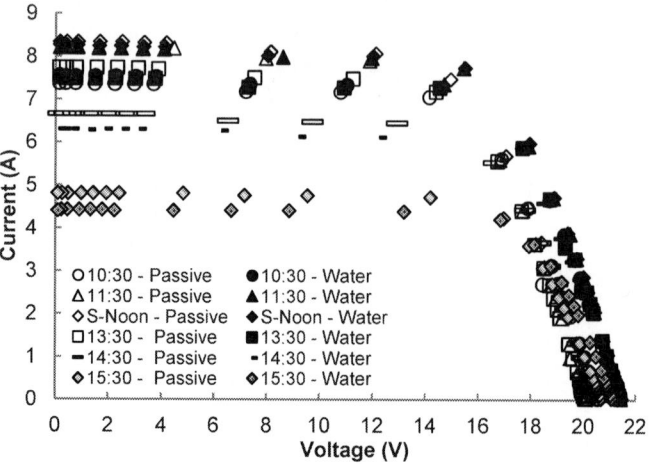

a) Current-voltage characteristics for fixed module on December 22

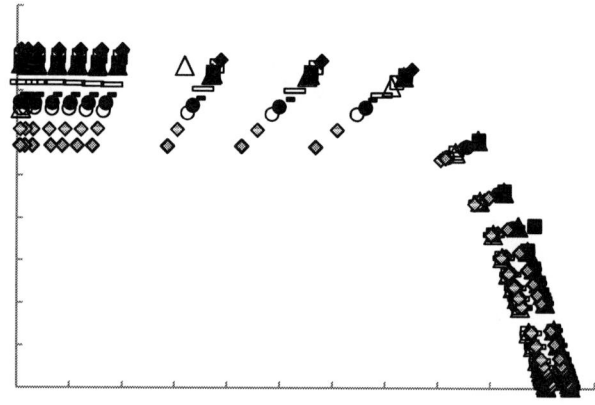

b) Current-voltage characteristics for sun-tracked module on December 14

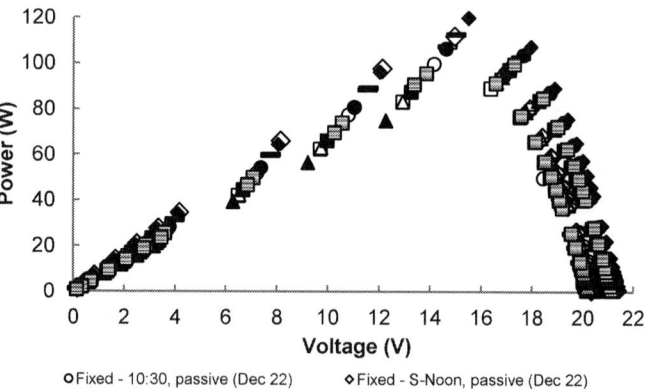

c) Comparison of output power for passively- and water-cooled module, both for fixed and sun-tracked orientations

**Figure 8:** Comparison of measured electrical characteristics of passively and water-cooled PV module B in a fixed geographical South facing (December 22, 2011) and sun-tracked orientation (December 14, 2011) as a function of time from 10:30 a.m. to 3:30 p.m., in Abu Dhabi (24.43°N, 54.45°E).

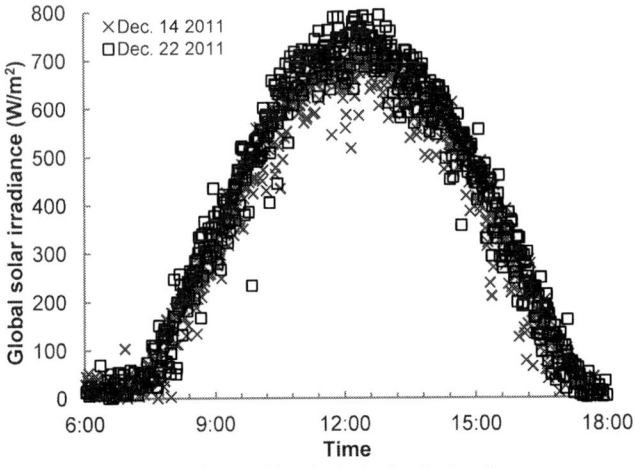

a) Global (i.e., total hemispherical) solar irradiance

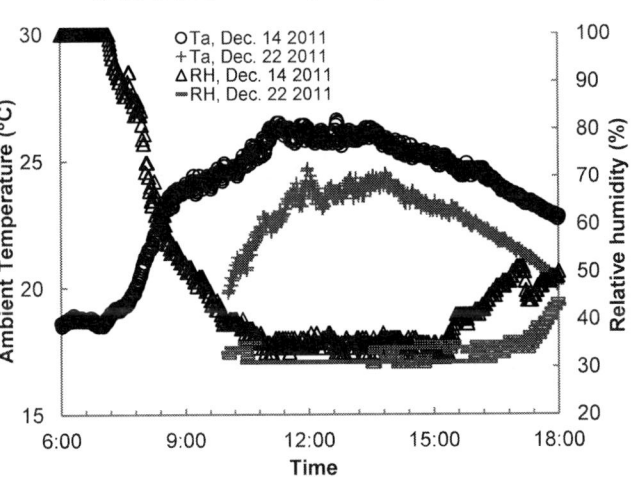

b) Ambient air temperature and relative humidity

**Figure 9:** Measured environmental data on December 14 and 22, 2011, in Abu Dhabi (24.43°N, 54.45°E).

due to tracking, higher water and ambient air temperatures, may be offset by a combination of lower global irradiation and possibly wind speed. On the other hand, for the 2 p.m. passively-cooled configuration, module temperature is 2°C higher in sun-tracked conditions relative to fixed ones (Table 3). In this case, the effects of increased irradiation absorption due to sun tracking and higher ambient air temperature, are not offset by lower global irradiation and possibly lower wind speed. Despite its slightly higher operating temperature (2°C), the passively-cooled module power output is 12.5 W (16%) higher with sun tracking than for the fixed South facing orientation. In spite of the slight variability in operating conditions between the two measurement dates, the results in Figure 6 provide preliminary guidelines for the operation of PV modules in autumn conditions in the Persian Gulf, which are comparable to summer conditions at for example mid European latitudes:

i) Sun tracking does not significantly elevate module operating temperature in passively-cooled conditions, such that tracking without active cooling is beneficial in terms of enhanced power output compared to operation in a fixed, geographical South facing orientation (e.g., +12.5 W or 16% at 2 p.m.).

ii) When used in conjunction with water-cooling, sun tracking increases output power by 20.2 W (24%) at for example 2 p.m. relative to fixed, water-cooled conditions.

iii) Water-cooling enhances the power output of fixed South facing modules for a significant portion of the day (+10 W or 13% at 2 p.m., versus 18.7 W or 22% at solar noon), and even more significantly for sun-tracked modules (+17.7 W or 20% at 2 p.m.).

iv) The integration of water-cooling and sun tracking increases power output by 30 W (40%) at for example 2 p.m. relative to a fixed, passively-cooled module.

The measurements conducted late October to early November in Figure 6 were repeated late December, to evaluate the need for water cooling at the coolest and lowest irradiation period of the year in the Persian Gulf. These results are shown in Figure 8. The variation in global irradiation, ambient air temperature and humidity throughout the day are depicted in Figure 9. Ambient air, module and cooling water temperatures, global irradiation, wind speed, relative humidity and peak power output, are also recorded in Table 4 for both solar noon and 2 p.m. Water-cooling reduces module operating temperature by approximately 11.5°C and 6.8°C at solar noon and 2 p.m., respectively for the fixed orientation, and 8.2°C and 4.9°C, respectively for the sun-tracked orientation (Table 4). The discrepancy in temperature reduction between the fixed and sun-tracked orientations at solar noon are attributable to a variation in ambient conditions between December 14 and 22. The following observations may be made from Table 4 regarding the operation of PV modules during the coolest part of the year in the Persian Gulf:

i) Sun tracking without active cooling marginally enhances power output (e. g., +3.7 W or 4% at 2 p.m.) compared to operation in a fixed, geographical South facing orientation.

ii) When used in conjunction with water-cooling, sun tracking increases output power by 5.6 W (6%) at for example 2 p.m. relative to fixed water-cooled conditions.

iii) Water-cooling enhances the power output of fixed South facing modules by +2.2 W (2%) at for example 2 p.m., versus +8.1 W (7%) at solar noon, and of sun-tracked modules by +4.1 W (4%) at 2 p.m.

iv) The integration of water-cooling and sun tracking increases power output by +7.8 W (9%) at for example 2 p.m. relative to a fixed, passively-cooled module.

The above results suggest that the modest reductions in operating temperature and improvements in electrical performance obtained using either water-cooling or sun tracking, or a combination of both, may not justify the costs associated with cooling and tracking during the winter. Measurements of cumulative daily energy output for a complete year, combined with an economic analysis, would permit the optimum daily and yearly window of operation of water-cooling and sun tracking systems to be determined.

The measured peak power outputs of passively and water-cooled modules throughout the day between October 29 and December 22 are compiled in Figure 10 for fixed and sun-

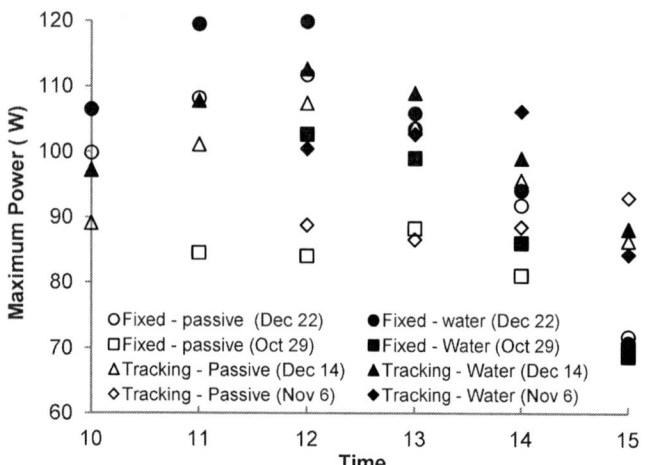

**Figure 10:** Comparison of measured peak output power of PV Module B in fixed and sun-tracked orientations, both for passive and water-cooling, as a function of time from 10 a.m. to 3 p.m. on October 29 and November 6, 2011, in Abu Dhabi (24.43°N, 54.45°E).

tracked orientations. The results for late October to early November clearly show that water-cooling combined with sun tracking results in significantly higher electrical output than a water-cooled module in a fixed orientation. The maximum benefits from water cooling are obtained at solar noon, but are still significant three hours apart.

## 5. Conclusions

In this study, the thermal and electrical performance of water- and air-cooled PV modules was experimentally characterized in autumn to winter conditions in the Persian Gulf. Between late October and early November, water-cooling is found to reduce PV module operating temperature by up to 30°C relative to passive-cooling conditions. Using water-cooling, power output is enhanced for a significant portion of the day for fixed geographical South facing modules (e.g., 22% at solar noon, and 13% at 2 p.m.), and even more significantly for sun-tracked modules (e.g., 20% at 2 p.m.) relative to passive cooling. Sun tracking enhances power output by 16% and 24% for passively- and water-cooled modules, respectively, at for example 2 p.m. The incorporation of water-cooling and sun tracking is the most effective, with enhancements in output power of on order 40% at for example 2 p.m. relative to passively-cooled, fixed South facing operation. In winter conditions (e.g., December), the modest reductions in operating temperature and improvements in electrical performance obtained using water-cooling and/or sun tracking may not justify the associated capital and operating costs. Recovery of the cooling water waste heat for applications such as pre-heating of process streams could be considered to aid the return on investment.

Future work will focus on design optimization and extend the experimental measurements to both summer conditions in the Persian Gulf, and the use of sea water as cooling fluid. In addition, the benefits of water-cooling and sun tracking will be analyzed for a complete year in terms of cumulative daily energy output and economic feasibility to determine the optimum daily and yearly window of operation of water-cooling and sun tracking systems.

Ultimately, long-term monitoring of PV module performance, reliability and life in harsh environments are required to fully assess the techno-economic feasibility of water-cooling, which is highly application-dependent. Although for off-shore oil and gas applications, constraints on the transportation and storage of fossil fuels, and environmental regulations, may outweigh the capital and operating expenses of a dual cooling and sun-tracking solution, the economic payback period may be more critical for other applications.

## References

1  Islam, M.D., Kubo, I, Ohadi, M., Alili, A.A., 2009, "Measurement of Solar Energy Radiation in Abu Dhabi, UAE," Applied Energy, Vol. 86, pp. 511–515.

2  NASA, Surface Meteorology and Solar Energy, A Renewable Energy Resource Website (release 6.0), http://eosweb.larc.nasa.gov/sse/, last accessed December 23, 2011.

3  WWF, 2010, Living Planet Report, http://www.worldwildlife.org/sites/living-planet-report/#, last accessed December 23, 2011.

4  Bullis, K., 2009, "A Zero-emissions City in the Desert," Technology Reviews, Vol. 112, No. 56, www.technologyreview.com/energy/22121/, last accessed December23, 2011.

5  Celik, A.N., Muneer, T., and Clarke, P., 2009, "A Review of Installed Solar Photovoltaic and Thermal Collector Capacities in Relation to Solar Potential for the EU-15," Renewable Energy, Vol. 34, No. 3, pp. 849-856.

6  Anderson, T.N., Duke, M., Morrison, G.L., and Carson, J.K., 2009, "Performance of a Building Integrated photovoltaic/thermal (BIPVT) Solar Collector," Solar Energy, Vol. 83, No. 4, pp. 445-455.

7  Chow, T.T., 2010,"A Review on Photovoltaic/thermal Hybrid Solar Technology," Applied Energy, Vol. 87, No. 2, pp. 365–379.

8  Agrawal, B., and Tiwari, G.N., 2010, "Optimizing the Energy and Exergy of Building Integrated Photovoltaic Thermal (BIPVT) Systems under Cold Climatic Conditions," Applied Energy, Vol. 87, No. 2, pp. 417-426.

9  Hepbasli, A., and Alsuhaibani, Z., 2011, "A Key Review on Present Status and Future Directions of Solar Energy Studies and Applications in Saudi Arabia," Renewable and Sustainable Energy Reviews, Vol. 15, pp. 5021-5050.

10  Bilton, A.M., Wiesman, R., Arif, A.F.M., Zubair, S.M., and Dubowsky, S., 2011, "On the Feasibility of Community-scale Photovoltaic-powered Reverse Osmosis Desalination Systems for Remote Locations," Renewable Energy, Vol. 36, pp. 3246-3256.

11. Jorgensen, G., Terwilliger, K., Glick, S., Pern, J., and McMahon, T., 2003, "Materials Testing for PV Module Encapsulation," NREL/CP-520-33578, presented at the National Center for Photovoltaics and Solar Program Review Meeting, March 24-26, Denver, Co, USA, http://www.osti.gov/bridge, last accessed December 23, 2011.

12 Karimov, K.S., Chattha, J.A., and Ahmed, M.M., 2002, Journal of References, Academy of Sciences of Tajikistan, Vol. XLV, No. 9, pp. 75-83.

13 Ko, J., 2009, "PV Tracking Applications Gather Momentum," March 31, www.renewableenergyworld.com, last accessed on December 23, 2011.

14 Zondag, H.A., 2008, "Flat-plate PV-Thermal Collectors and Systems: a Review," Renewable and Sustainable Energy Reviews, Vol. 12, pp. 891–959.

15 Furushima, K. and Nawata, Y., 2006, "Performance Evaluation of Photovoltaic Power-Generation System Equipped With a Cooling Device Utilizing Siphonage," Transactions of the ASME, Journal of Solar Energy, Vol. 128, pp. 146-151.

16 Brinkworth, B.J., and Sandberg, M., 2006, "Design Procedure for Cooling Ducts to Minimise Efficiency Loss due to Temperature Rise in PV Arrays," Solar Energy, Vol. 80, pp. 89–103.

17 Kelley, L., Bilton, A.M., and Dubowsky, S., 2011, "Enhancing the Performance of Photovoltaic Powered Reverse Osmosis Desalination Systems by Active Thermal Management," 2011, Proceedings of the ASME 2011 International Mechanical Engineering Congress & Exposition (IMECE2011), November 11-17, Denver, Co., USA, Paper No. IMECE2011-62717.

18 Lee, B., Liu, J.Z., Sun, B., Shen, C.Y., Dai, G.C., 2008, "Thermally Conductive and Electrically Insulating EVA Composite Encapsulants for Solar Photovoltaic (PV) Cell," eXPRESS Polymer Letters, Vol.2, No.5, pp. 357–363, www.expresspolymlett.com, last accessed December 23, 2011.

19 Chintapalli, M., Diskin, M., and Guha, I., 2010,"Improving Solar Cell Efficiency: a Cooler Approach," Massachusetts Institute of Technology, Course 3.042, Spring 2010, Technical Poster, http://web.mit.edu/course/3/3.042/team5_10/OurApproach.html,last accessed December 23, 2011.

20 Patel, P., 2010, "Solar-Powered Desalination - Saudi Arabia's Newest Purification Plant will Use State-of-the-Art Solar Technology," Technology Review, Published by MIT, April 8, http://www.technologyreview.com/energy/25010/, last accessed December 23, 2011.

21 Maiti, S., Banerjee, S., Vyas, K., Patel, P., and Ghosh P.K., 2011, "Self Regulation of Photovoltaic Module Temperature in V-trough using a Metal–wax Composite Phase Change Matrix," Solar Energy, Vol. 85, pp. 1805–1816.

22 Krauter, S., 2004, "Increased Electrical Yield via Water Flow over the Front of Photovoltaic Panels," Solar Energy Materials & Solar Cells, Vol. 82, pp. 131–137.

23 Araki, K, Uozumi, H, and Yamaguchi, M., 2002, "A Simple Passive Cooling Structure and its Heat Analysis for 500x Concentrator PV Module," Proceedings of the 29th IEEE Photovoltaic Specialist Conference, New Orleans, LA, USA, May 19-24, pp. 1568-1571.

24 Skyline Solar, Inc., http://www.skyline-solar.com/architecture.htm, last accessed December 23, 2011.

25 Maiti, S., Vyas, K., Ghosh, P.K., 2010, "Performance of a Silicon Photovoltaic Module under Enhanced Illumination and Selective Filtration of Incoming Radiation with Simultaneous Cooling," Solar Energy, Vol. 84, pp. 1439-1444.

26 Behnia, M., and Odeh, S., 2009, "Improving Photovoltaic Module Efficiency Using Water Cooling," Heat Transfer Engineering, Vol. 30, No. 6, pp. 499–505.

27 Abdolzadeh, M., and Ameri, M., 2009, "Improving the Effectiveness of a Photovoltaic Water Pumping System by Spraying Water over the Front of Photovoltaic Cells," Renewable Energy, Vol. 34, pp. 91–96.

28 Kordzadeh, A., 2010, "The Effects of Nominal Power of Array and System Head on the Operation of Photovoltaic Water Pumping Set with Array Surface Covered by a Film of Water," Renewable Energy, Vol. 35, pp. 1098–1102.

29 Tina, G.M., Rosa-Clot, M., Rosa-Clot, P., and Scandura, P.F., 2012, "Optical and Thermal Behavior of Submerged Photovoltaic Solar Panel: SP2," Energy, in press.

30 Wilson, E., 2009, "Theoretical and operational Thermal Performance of a 'Wet' Crystalline Silicon PV Module under Jamaican Conditions," Renewable Energy, Vol. 34, pp. 1655–1660.

31 Jones, A.D., and Underwood, C.P., 2001, A Thermal Model for Photovoltaic Systems," Solar Energy, Vol. 4, pp. 349–359.

32 Armstrong, S., and Hurley, W.G., 2010, "A thermal model for photovoltaic panels under varying atmospheric conditions," Applied Thermal Engineering, Vol. 30, pp. 1488-1495.

33 Kampf, J., and Sadrinasab, M., 2006, "The Circulation of the Persian Gulf: A Numerical Study," Ocean Science, Vol. 2, pp. 27–41, http://www.ocean-sci.net/2/27/2006, last accessed December 23, 2011.

34 Sharqawy, M.H., Lienhard V, J.H., and Zubair, S.M., 2010, "Thermophysical Properties of Seawater: A Review of Existing Correlations and Data," Desalination and Water Treatment, Vol.16, pp. 354–380.

35 Chow, T.T., He, W., Ji, J., 2006, "Hybrid Photovoltaic-Thermosyphon Water Heating System for Residential Application," Solar Energy, Vol. 80. No. 3, pp. 298–306.

# Data Center Cooling Management and Analysis – A Model-Based Approach

Rongliang Zhou, Zhikui Wang, Cullen E. Bash, Alan McReynolds

*Abstract*— As the hub of information aggregation, processing, and dissemination, today's data centers consume significant amount of energy. The data center electricity consumption mainly comes from the IT equipment and the supporting cooling facility that manages the thermal status of the IT equipment. The traditional data center cooling facility usually consists of chilled water cooled computer room air conditioning (CRAC) units and chillers that provide chilled water to the CRAC units. Electricity used to power the cooling facility could take up to a half of the total data center electricity consumption, and is a major contributor to the data center total cost of ownership. While the data center industry has established the best practice to improve the cooling efficiency, the majority of it is rule of thumbs providing only qualitative guidance. In order to provide on demand cooling and achieve improved cooling efficiency, a model based description of the data center thermal environment is indispensable. In this paper, a computationally efficient multivariable model capturing the effects of CRAC units blower speed and supply air temperature (SAT) on rack inlet temperatures is introduced, and model identification and reduction procedures are discussed. Using the model developed, data center cooling system design and analysis such as thermal zone mapping, CRAC units load balancing, and hot spot detection are investigated.

## I. INTRODUCTION

Due to the ever-increasing computing and hence power density of the IT equipment, today's data centers require tremendous amount of cooling power to maintain the desired thermal status. According to [1], [2], about a third to a half of data center total power consumption goes to the cooling system. Highly efficient cooling systems are thus indispensable to reduce the total cost of ownership and environmental footprint of data centers.

Figure 1 shows a typical raised-floor air-cooled data center with hot aisles and cold aisles separated by rows of IT equipment racks. The thermal requirements of IT equipment are usually specified in terms of the inlet air temperatures of the equipment [3]. The equipment temperature thresholds are not necessarily uniform across the entire data center but are dependent on the different functions, such as computing, storage, and networking, which the IT equipment serves. Service contracts of the IT workload hosted in the IT equipment can also affect the temperature threshold.

The blowers of the Computer Room Air Conditioner (CRAC) units pressurize the under-floor plenum with cool air, which in turn is drawn through the vent tiles located in front of the racks in the cold aisles. Hot air carrying the waste heat from the IT equipment is rejected into the hot aisles. Depending on its design, the CRAC unit internal control can regulate the chilled water valve opening to track the given

Sustainable Ecosystems Research Group, HP Labs, Hewlett-Packard Company, 1501 Page Mill Road, Palo Alto, CA 94304-1126. {firstname.lastname}@hp.com

reference of Supply Air Temperature (SAT) or Return Air Temperature (RAT). The flow rate of the cool air supply can also be tuned continuously if a Variable Frequency Drive (VFD) is installed for each CRAC unit to vary the speed of its blowers.

For the particular configuration shown in Fig. 1, neither the cold aisles nor the hot aisles are contained and hence air streams are free to mix. Most of the hot air in the hot aisles returns to the CRAC units, but a small portion of it might escape into the cold aisles from the top, the sides, or even the bottom of the racks and causes recirculation. Recirculation can be also due to the reverse flows with certain IT equipment (some network switches, for example) of which the internal fans blow the hot exhaust air from the hot aisle into the cold aisle. The inlet air flow of the IT equipment is thus a mixture of cool air from the vent tiles in its vicinity and the recirculated hot air [4]. The recirculation of hot air into the cold aisle generates entropy and lowers the data center cooling efficiency.

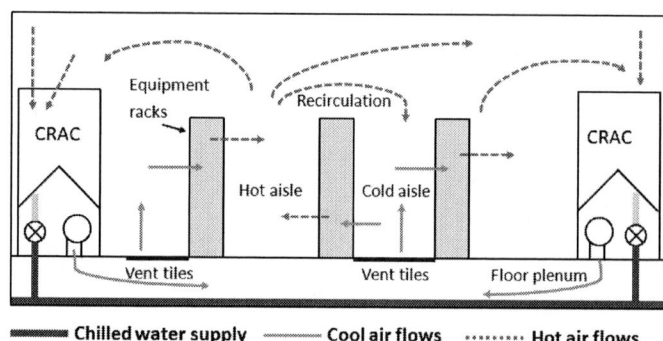

Fig. 1. Typical Raised Floor Data Center

The challenge of data center cooling management is the coordination of CRAC units blower speeds and supply air temperature (SAT) tuning to minimize the power consumption of both CRAC units and chiller plants, while maintaining hundreds or even thousands of rack inlet temperatures below their respective thresholds. The major hurdle to overcome this challenge is the lack of simplified and computationally efficient models that are capable of capturing the complex energy and mass flows within the data centers and performing transient cooling analysis. Computational Fluid Dynamics (CFD) have been used extensively for data center cooling system design, but most of the applications are built upon steady-state system analysis. The few CFD based transient cooling system performance analyses reported so far have been focused on predicting system responses to cooling failures. Beitelmal and Patel [5], for example, use transient CFD simulation to investigate data center temperature distribution

978-1-4673-1110-6/12 $31.00 © 2012 IEEE

change caused by a malfunctioning CRAC unit, and show that acceptable rack inlet temperatures can still be maintained if the IT load and available cooling resources can be appropriately re-organized. In another application, CFD simulation is used to analyze the various failure scenarios of IBM's cooling infrastructure design for the water cooled cluster of 11 racks [6]. While transient CFD analysis provides valuable insights in data center cooling system design as well as measures to handle different failure modes, it is impractical to use it for real-time data center cooling management. This is partly because data center IT load could change from minute to minute, and transient CFD analysis is normally time consuming and could take hours or even longer to finish. In addition, as pointed out in [7], [8],it is usually difficult to capture the true IT and cooling configurations in sufficient details to predict the resulting environment with desired accuracy. Targeting higher computationally efficiency, some alternative approaches have been utilized or developed by researchers for transient data center cooling performances analysis. Khankari [9] uses a simple energy balance model to investigate the availability of data center thermal mass in various configurations during power shutdown, Kummert [10] studies the effects of chiller failure and cooling system thermal inertia on room temperature variations, and Zhang together with VanGilder develops data center transient thermal models. These alternative approaches, however, achieve the improved computational efficiency by sacrificing the spatial non-uniformity witnessed in most data centers, and hence are not suitable for real-time data center thermal management either.

In an attempt to bridge this gap, the authors' recent work [11], [12] develops physics based state-space models that describe the air flow transport and distribution within the data centers. The parameters of the models are obtained from measurement data of system identification experiments, and hence are ensured to reflect the data center reality emphasized in [8]. The physics based data center cooling model is utilized in [11] to coordinate zonal (CRAC units blower speeds and SAT) and local cooling actuation (adaptive vent tiles), and validation on a small portion of a research data center shows significant cooling power savings. Using the same but simplified physics based model, the authors present in [12] a decentralized model predictive control (MPC) design approach for CRAC units SAT and blower speed regulation targeting large scale data centers. Compared with the commonly used CFD modeling, the model employed in this paper is computationally light without losing the data center spatial non-uniformity, and is suitable for both off-line analysis and online dynamic control. In this paper, the computationally efficient model developed is used to perform critical cooling system design and analysis tasks such as thermal zone mapping, CRAC units load balancing, and hot spot detection.

The other sections of this paper are organized as follows. Section II first briefly introduces the dynamic rack inlet temperature model we developed previously using the energy and mass balance principles, followed by a discussion on the model identification and reduction procedures. In Section III,

we demonstrate how model based grouping of rack inlet temperatures and CRAC units can be used for improved thermal zone mapping and coordination of decentralized cooling controllers. Section IV defines Hot Spot Index (HSI) using the model parameters and shows that it is powerful in data center hot spot detection. Finally, Section V concludes the paper with a summary of the work presented.

## II. DYNAMIC COOLING SYSTEM MODELING AND MODEL PARAMETER IDENTIFICATION

### A. Dynamic Cooling System Modeling

In this subsection, we briefly introduce simplified models from the basic mass and energy balance principles to characterize the complex mass and energy flows within the raised-floor air-cooled data centers. Since only the modeling results are presented, the interested readers can refer to [11], [12] for more detailed description on how these modes are derived.

In the open environment, air flow coming into the IT equipment inlet is a mixture of the cool air from the CRAC units (through the vent tiles) and the recirculated hot (exhaust) air that escapes into the cold aisle. In hot aisle contained environment, although significantly reduced, recirculation could still exist because of imperfect containment, or the reverse flows from some network switches that draw hot air from the hot aisle for cooling and reject the even hotter air (up to $40°C$) into the cold aisle.

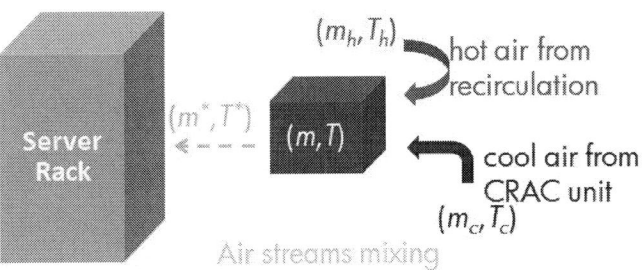

Fig. 2.   Air Mixing at the Rack Inlet

Consider a small control volume in the proximity of the rack inlet with mass $m$ and temperature $T$, as shown in Fig. 2. Cool and recirculated hot air flows with mass and temperature $(m_c, T_c)$ and $(m_h, T_h)$ enter the control volume, mix well with the air $(m, T)$ already in the volume, leave the control volume altogether and enter the rack inlet with total mass $m^*$ and temperature $T^*$. It can be found that the temperature change $\Delta T$ of the air within the control volume before and after the mixing is:

$$\Delta T \triangleq T^* - T = \frac{m_c(T_c - T)}{m + m_c + m_h} + \frac{m_h(T_h - T)}{m + m_c + m_h}, \quad (1)$$

which reveals that the influence of cool and recirculated hot air on rack inlet temperature can be mainly captured by $m_c(T_c - T)$ and $m_h(T_h - T)$, respectively.

In raised-floor data centers, all the CRAC units pressurize the under floor plenum by blowing the cool air into it. The

978-1-4673-1110-6/12 $31.00 © 2012 IEEE

cool air $\dot{m}_c$ flowing into a rack inlet could come from all the CRAC units and hence:

$$\dot{m}_c = \sum_{j=1}^{N_{CRAC}} b_j \cdot VFD_j, \qquad (2)$$

in which $N_{CRAC}$ is the number of CRAC units, $b_j$ quantifies the cooling air contribution from the $j^{th}$ CRAC unit to a specific rack inlet, and $VFD$ stands for the speed of the blower in the percentage of its maximum.

It can be seen from Eqn. (1) that both cool and recirculated hot air contribute to the rack inlet temperature change $\Delta T$. Since the recirculated hot air flow is beyond direct control, we can lump its effect into a time-varying term $C$ and simplify Eqn. (1) as:

$$T^* - T = \frac{\dot{m}_c \Delta t (T_c - T)}{m + \dot{m}_c \Delta t + m_h} + C, \qquad (3)$$

in which $\Delta t$ is the length of the sampling interval.

The discrete form of Eqn. (3) is:

$$
\begin{aligned}
T(k+1) = & \, T(k) \\
& + \Big\{ \sum_{j=1}^{N_{CRAC}} g_j \cdot [SAT_j(k) - T(k)] \cdot VFD_j(k) \Big\} \\
& + C(k),
\end{aligned}
$$
$$(4)$$

in which $T(k+1)$ and $T(k)$ are rack inlet temperatures at time steps $k+1$ and $k$, respectively. In Eqn. (4), $g_j$ quantifies the combined influences of VFD and SAT tuning of the $j^{th}$ CRAC unit, and also lumps the effects of parameters $b_j$, $\Delta t$ together with the nonlinearity associated with $\dot{m}_c$.

The vector form of Eqn. (4) for multiple rack inlet temperatures is:

$$\overline{T}(k+1) = \overline{T}(k) + \overline{F} + \overline{C}, \qquad (5)$$

in which

$$\overline{T} = [T_1, T_2, \cdots, T_{N_T}]^T,$$
$$\overline{F} = [F_1, F_2, \cdots, F_{N_T}]^T,$$
$$F_i = \sum_{j=1}^{N_{CRAC}} g_{i,j}[SAT_j(k) - T_i(k)]VFD_j(k), \quad 1 \le i \le N_T,$$
$$\overline{C} = [C_1, C_2, \cdots, C_{N_T}]^T,$$

and $N_T$ is the number of rack inlet temperatures of interest.

### B. Model Parameter Identification and Model Reduction

Parameters of the dynamic rack inlet temperature model described in Eqn. (5), including $g_{i,j}$ $(1 \le i \le N_T, 1 \le j \le N_{CRAC})$ and $\overline{C}$ can be obtained through model identification experiments. During the experiments, the various available cooling actuation, such as CRAC unit blower speed and SAT, is perturbed through sequential/simultaneous step changes or other specifically designed identification sequences. Both the cooling actuation signals and the corresponding temperature response at the rack inlets are collected.

In order to find the model parameters $g_{i,j}(1 \le i \le N_T, 1 \le j \le N_{CRAC})$ and $\overline{C}$, a nonlinear optimization can

be performed on part of the experimental data collected such that the parameterized model minimizes the error between model prediction and rack inlet temperature measurements. The remaining part of the experimental data can then be used to validate the model identified. Note that after the model identification data is collected, the model parameter identification can be performed per rack inlet temperature, since different rack inlet temperatures only have input coupling. The associated parallelization can be exploited to speed up the model identification process and is extremely useful for large scale data centers with thousands of rack inlet temperatures.

From the physical perspective, every rack inlet temperature of interest is influenced by all the CRAC units, and correspondingly each CRAC unit affects all the rack inlet temperatures within the data center. As a result, the matrix $G = [g_{i,j}]$ $(1 \le i \le N_T, 1 \le j \le N_{CRAC})$ obtained from the model identification process is fully populated. However, it is observed that most rack inlet temperatures are usually affected by a selected small number of CRAC units, and that this subset of CRAC units varies with the location of the specific rack inlet temperature relative to the layout of the cooling facilities. Reflected on the dynamic rack inlet temperature model identified, this observation manifests through the fact that for any rack inlet temperature $T_i$ $(1 \le i \le N_T)$ some of the $g_{i,j}$ $(1 \le j \le N_{CRAC})$ items are dominant with noticeably larger values than the rest, suggesting that the non-dominant items may be ignored without significant effects on modeling accuracy.

In order to obtain a sparse $G$ matrix and hence take the associated advantages such as system decoupling and parallelization of sub-problem solving, we can perform model reduction for each rack inlet temperature. For rack inlet temperature $T_i$, define

$$g_{i,max} = max(g_{i,j}), \qquad 1 \le j \le N_{CRAC}.$$

In the reduced matrix $G^r$, item $g_{i,j}^r$ is set to the corresponding $g_{i,j}$ if and only if

$$g_{i,j} \ge \lambda \cdot g_{i,max},$$

and otherwise $g_{i,j}^r = 0$. In the inequality above $\lambda$ is a adjustable threshold between 0 and 1. The physical interpretation of this model reduction process is that in order to regulate a specific rack inlet temperature, we can only consider the CRAC units that have significant influence over it and simply ignore the CRAC units that just marginally affect this rack inlet temperature of interest.

## III. MODEL BASED GROUPING OF CRAC UNITS AND RACK INLET TEMPERATURES

The matrix $G_r$ obtained after the model reduction process trims the weak relationships between CRAC units and rack inlet temperatures, maintaining only the strong ones. The sparse structure of $G_r$ implies the natural grouping of CRAC units and rack inlet temperatures, which can be used for both static system analysis and dynamic system control.

## A. Thermal Zone Mapping

Previously, CRAC units thermal zone mapping has been based on the absolute values of thermal correlation index (TCI) defined in [13] as:

$$TCI_{i,j} = \frac{\Delta T_i}{\Delta SAT_{CRAC,j}}, \qquad (6)$$

which quantifies the steady-state response of the $i^{th}$ rack inlet temperature to a step change in the SAT of the $j^{th}$ CRAC unit. This steady-state system information based method, however, has a drawback since the TCI values obtained through the system commissioning process is dependent on the blower speeds settings of the CRAC units. The thermal zone of a particular CRAC unit established using TCI could expand or shrink when its blower speed increases or decreases, leading to a family of data center thermal zone mappings under different CRAC units blower speed settings.

The drawback of TCI based thermal zone mapping method outlined above comes from the fact that only steady-state system information is utilized and the rich dynamic system information embedded in system parameter such as the $G_r$ matrix is left out. From the CRAC's perspective, CRAC $\#j$ effectively affects rack inlet temperature $T_i$ ($1 \leq i \leq N_T$) only if the corresponding item $g_{i,j}^r$ of matrix $G_r$ is nonzero. The nonzero items of the $j^{th}$ column of $G_r$ defines the zone of influence of the $j^{th}$ CRAC units, with the value of the corresponding item denoting the exact intensity of the influence of CRAC $\#j$ on a particular rack inlet temperature.

Using this model based approach, Fig. 3 shows the thermal zones of a research data center with 8 CRAC units and 10 rows of racks, and the thermal zone of each CRAC unit is approximated by a balloon. The significant overlapping between the thermal zones of CRAC $\#3$ and $\#4$, and CRAC $\#5$ and $\#6$ are clearly indicated in the figure. Since CRAC units with overlapping of thermal zones all affect the rack inlet temperatures in the overlapping area, some coordination might be necessary between them such that they may have balanced loads when managing the shared rack inlet temperatures.

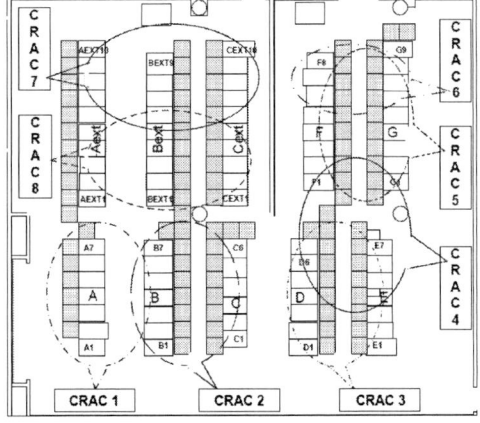

Fig. 3.   Model Based Thermal Zone Mapping of a Research Data Center

Compared with thermal correlation index (TCI) based method, the thermal zone identified using the model based method for each CRAC unit does not vary with its blower speed, since $G_r$ captures the combined effects of CRAC unit SAT and blower speed on rack inlet temperatures. Because of the consistency of this model based thermal zone mapping method, it is valuable for distributed controller design of large scale data centers. In the authors' previous work [12], a decentralized controller is designed for each CRAC unit to regulate the rack inlet temperatures within its established zone of influence.

## B. Load Balancing Based on CRAC Units Grouping

In raised floor air cooled data centers, there might be strong interactions between neighboring CRAC units, thus causing load balancing problems such as load piggybacking and load swapping, which is not uncommon in decentralized or decoupled CRAC unit controller design.

In load piggybacking, one of the CRAC units may keep increasing its cooling provisioning in order to drive a temporary rack inlet temperature violation below the specified threshold, and part of the cool air from this CRAC unit may also be routed to the racks intended to be cooled by the neighboring CRAC units because of the shared underfloor plenum. Observing the rack temperature decrease in its thermal zone, the neighboring CRAC units may piggyback on the CRAC unit with high load and decrease its own load provisioning and in turn cause the high load CRAC unit to reach an even higher load. Load piggybacking often manifests itself as a high load CRAC unit with low SAT and high blower speed, with its neighbor(s) working at the opposite extreme with high SAT and low blower speed. The significant load imbalance between CRAC units from load piggybacking may shorten the CRAC units' life span, and the mixing of supply air streams with big temperature difference also generates entropy and lowers the overall data center cooling efficiency.

Apart from load piggybacking, load swapping can also be observed for CRAC units that are individually controlled by different controllers. When stuck in load swapping, the high and low load status could switch back and forth between neighboring CRAC units, resulting in oscillation in the SAT and blower speeds. Load swapping could be triggered by the temperature disturbances such as a sudden and temporary load increase from a server rack, opening of a cold aisle vent tile due to maintenance, or introduction of free cooling from the outside air, and may not stop without intervention of the operator.

The root cause of both load piggybacking and load swapping is lack of coordination in decentralized or decoupled CRAC unit controllers, in which the physical input coupling between neighboring thermal zones are neglected. The local controller for each thermal zone tries to maintain the thermal status within the zone using the least efforts (usually through minimization of the cooling power required), and uncoordinated local optimization could easily lead to global suboptimal solution (load piggybacking) or instability (load swapping). In order to address these problems, simple and yet effective load balancing mechanism needs to be established between neighboring thermal zones to coordinate the outputs of the CRAC units.

The first step toward CRAC load balancing is to identify for each thermal zone the neighboring thermal zones that it needs to coordinate with. For each rack inlet temperature $T_i$, the corresponding temperature violation $T_{v,i}$ over its reference temperature $T_{ref,i}$ is defined as:

$$T_{v,i} = T_i - T_{ref,i}.$$

In the $j^{th}$ thermal zone, the rack inlet temperature with the highest $T_{v,i}$ is called its master sensor, and is denoted as $T_{m,j}$. At each control interval, the local controller of each thermal zone broadcasts the index and reference temperature of its master sensor, together with its CRAC unit's current SAT and blower speed setting. Each local controller also receives broadcast messages from all other thermal zones, and compares its master sensor with its neighbors. After this information exchange, each local controller has a most recent snapshot of the settings of all the cooling controllers, and CRAC units sharing same master sensor can be grouped for load balancing purpose.

In the case that several thermal zones share the same master sensor, it is desirable that all the CRAC units in this group coordinate to the same SAT since mixing of supply air streams at different temperatures generates entropy and lowers cooling efficiency. In order to coordinate to the same SAT, the thermal zone with the highest SAT of the group adds to the SAT setting of its local controller an additional load balancing term:

$$SAT_c = k \cdot (SAT_{min} - SAT_{max}),$$

in which $SAT_{min}$ and $SAT_{max}$ are the lowest and highest CRAC unit SAT settings of the group, and $k$ is appropriate feedback gain for SAT coordination. Note that in this co-ordination mechanism, load balancing is only performed on the low load CRAC unit to increase its provisioning, and the reason is to minimize the chance of rack inlet temperature violation during the load balancing process.

Figure 4 shows the experimental results in a research data center as shown in Fig. 3 with 8 CRAC units. Before load balancing is enabled shortly after time $t = 1hr$, the data center has already entered a relatively steady state. CRAC #3/4 share the same master sensor, and CRAC #5/6 share another master sensor. Due to the lack of coordination between the decentralized controller which each controls a CRAC unit, the supply air temperature difference is as large as 2.2 $^\circ C$ between CRAC #3 and #4, and 1.8 $^\circ C$ between CRAC #5 and #6. After load balancing is enforced at time $t = 1hr$, CRAC #3 and #4 quickly converge to the same SAT, and the same is true for CRAC #5 and #6. The blower speeds of these four CRAC units, denoted by the percentage of the maximum of variable frequency drive (VFD) output, also reach their new steady-state settings after load balancing is enforced as shown in Fig. 4(b). The trajectories of other CRAC units are not shown here since their settings do not change for the duration of the experiment, either because they do not share master sensors (CRAC #1 and #2) or load balancing has already been achieved (CRAC #7 and #8).

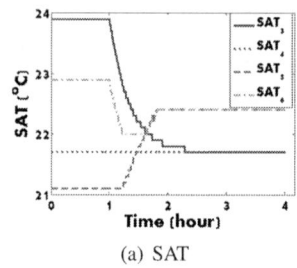
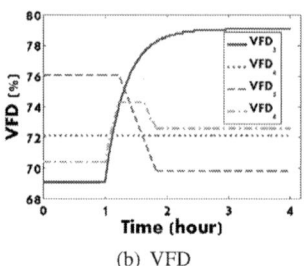

(a) SAT  (b) VFD

Fig. 4. Load Balancing

## IV. MODEL BASED HOT SPOT DETECTION

In data center cooling management, ability to identify the hot spots is essential. While examining the snapshot or temporal trends of the data center rack inlet temperature distribution is helpful, a systematic approach is needed to automate the process.

The detection of hot spots can not be accomplished without the definition of an appropriate metric or measure. While most people tend to believe that the highest rack inlet temperature within the entire data center or a thermal zone indicates a hot spot and thus temperature seems to be the right measure, it is not always the case. First, location where the highest rack inlet temperature observed within the data center or a thermal zone could easily change as the settings of the CRAC units vary. A hot spot previously identified could disappear as the thermal zone it belongs to is over provisioned while the neighboring thermal zones are configured to be insufficiently provisioned. Second, the hot spot detection results using temperature measure might be subject to various disturbances and dependent on whether the detection is performed when the data center has reached a relatively steady state. These drawbacks indicate that using temperature as the measure for data center hot spot detection lacks consistency. New and improved metric for data center hot spot detection needs to be developed.

In order to address this problem, we define the model based Hot Spot Index (HSI) for data center hot spot detection. For a rack inlet temperature $T_i$ with reduced order model:

$$T_i(k+1) = T_i(k)$$
$$+ \left\{ \sum_{j=1}^{N_{CRAC}} g_{i,j}^r \cdot [SAT_j(k) - T_i(k)] \cdot VFD_j(k) \right\}$$
$$+ C_i,$$

HSI is defined as:

$$HSI \triangleq \frac{C_i}{\| g_i^r \|_1}, \tag{7}$$

in which vector $g_i^r = [g_{i,1}^r \ g_{i,2}^r \ \cdots \ g_{i,N_{CRAC}}^r]$, and $\| \cdot \|_1$ stands for the vector 1-norm. From a physical perspective, HSI measures the ratio between effects of hot air recirculation and CRAC units tuning on a specific rack inlet temperature. A high HSI value means that hot air recirculation is severe while the cooling effects from all the CRAC units are weak at the location of interest, and hence indicates a potential hot spot.

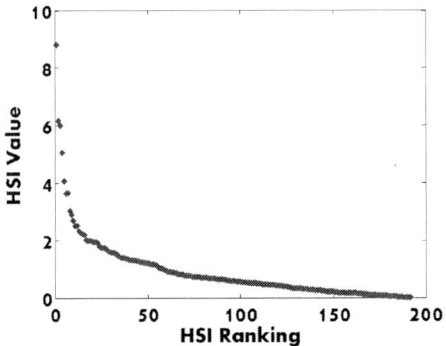

Fig. 5. HSI Values for Rack Inlet Locations of a Research Data Center in Descending Order

Figure 5 shows the HSI values in descending order for the 192 rack inlet temperature locations in the research data center shown in Fig. 3. A closer look at the locations with the leading HSI values points to rack inlet temperatures at rack B6, D5, D6, and G9. Among these hot spots, rack B6, D5, and D6 are affected by severe reverse flow from the network switches mentioned earlier, while G9 is at the end of the row G and is most affected by the hot air escaped from the hot aisle. The locations of these hot spots agree with the observations of the research data center. Furthermore, the severity of these hot spots can also be indicated by their corresponding HSI values, which can not be easily detected by simply observing the data center temperature distribution. Rack B6, for example, has higher HSI values, and it is found that the reverse flow is much more severe than that of rack D5, and D6, which follow B6 in HSI ranking.

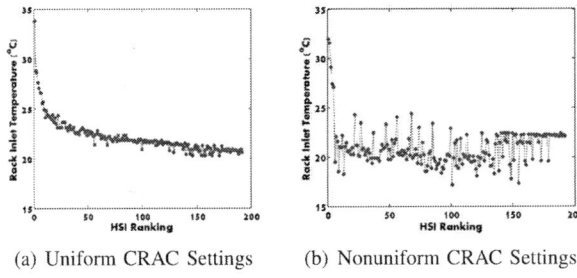

(a) Uniform CRAC Settings     (b) Nonuniform CRAC Settings

Fig. 6. Rack Inlet Temperatures (in Descending Order of HSI)

Compared with HSI based method, relying solely on data center temperature distribution for hot spot detection could lead to misleading results. Figure 6, for example, shows the temperature distribution of the aforementioned research data center in descending order of HSI. In Fig. 6(a), all the 8 CRAC units are configured with the same SAT and blower speed. The trend of rack inlet temperatures roughly follows that of HSI values as shown in Fig. 5, meaning that in this particular data center cooling configuration the rack inlet temperatures can be used as a reference for hot spot detection. The rack inlet temperatures as shown in Fig. 6(b), however, vary significantly in both amplitudes and ranking relative to each other with a nonuniform setting of CRAC units. Although the five rack inlet locations with the highest HSI values still have the highest temperatures, other hot spots

identified by HSI based method, such as those in rack D5 and G9, now have relatively low temperatures among all the rack inlet locations and hence could be left out in hot spot detection.

## V. CONCLUSIONS

In this paper, a data center management and analysis scheme based on dynamic rack inlet temperature model is introduced. The model parameter identification and model reduction procedures are both discussed. Improved thermal zone mapping approach is introduced through model based rack inlet temperature grouping, and load balancing mechanism is investigated through grouping of CRAC units. In order to detect the hot spots within data centers, Hot Spot Index (HSI) is defined using the model parameters and proves to be effective in data center hot spot detection.

### REFERENCES

[1] Steve Greenberg, Evan Mills, Bill Tschudi, Peter Rumsey, and Bruce Myatt. Best practices for data centers: Results from benchmarking 22 data centers. In *2006 ACEEE Summer Study on Energy Efficiency in Buildings*.

[2] Chandrakant D. Patel, Cullen E. Bash, Ratnesh K. Sharma, Monem H. Beitelmal, and Rich J. Friedrich. Smart cooling of data centers. In *IPACK03, The Pacific Rim/ASME International Electronic Packaging Technical Conference and Exhibitions*.

[3] ASHRAE. Datacom equipment power trends and cooling applications. Atlanta, GA, 2005.

[4] Cullen E. Bash, Chandrakant D. Patel, Ratnesh K. Sharma. *Efficient thermal management of data centers – Immediate and long-term research needs*, volume 9, no. 2. HVAC&R Research, Apr 2003.

[5] Monem H. Beitelmal and Chandrakant D. Patel. Thermo-fluids provisioning of a high performance high density data center. *Distributed and Parallel Databases*, 21(2):227–238, 2007.

[6] Roger Schmidt, Mike Ellsworth, Madhu Iyengar, and Gary New. IBM's power6 high performance water cooled cluster at NCAR: Infrastructure design. In *ASME 2009 InterPACK Conference collocated with the ASME 2009 Summer Heat Transfer Conference and the ASME 2009 3rd International Conference on Energy Sustainability (InterPACK2009)*, San Francisco, California, USA, July 19–23 2009.

[7] Jim VanGilder. Real-Time data center cooling analysis. *Electronics Cooling*, September, 2011.

[8] Mark Seymour, Christopher Aldham, Matthew Warner, and Hassan Moezzi. The increasing challenge of data center design and management: Is CFD a must? *Electronics Cooling*, December, 2011.

[9] Kishor Khankari. Thermal mass availability for cooling data centers during power shutdown. *ASHRAE Transactions*, 116(2):205–217, 2010.

[10] Michael Kummert, William Dempster, and Ken McLean. Thermal analysis of a data centre cooling system under fault conditions. In *11th International Building Performance Simulation Association Conference and Exhibition, Building Simulation 2009*, Glasgow, Scotland, July 27–30 2009.

[11] Rongliang Zhou, Zhikui Wang, Cullen E. Bash, Christopher Hoover, Rocky Shih, Alan McReynolds, Niru Kumari, and Ratnesh K. Sharma. A holistic and optimal approach for data center cooling management. In *American Control Conference (ACC2011)*, pages 1346–1351. IEEE, 2011.

[12] Rongliang Zhou, Zhikui Wang, Cullen E. Bash, and Alan McReynolds. Modeling and control for cooling management of data centers with hot aisle containment. In *ASME 2011 International Mechanical Engineering Congress & Exposition*, Denver, USA, November 11-17 2011.

[13] Cullen E. Bash, Chandrakant D. Patel, and Ratnesh K. Sharma. Dynamic thermal management of air cooled data centers. In *Thermal and Thermomechanical Phenomena in Electronics Systems, 2006. ITHERM'06. The Tenth Intersociety Conference on*, pages 445–452. IEEE, 2006.

# Datacenter Power Savings through High Ambient Datacenter Operation: CFD Modeling Study

## Nishi Ahuja
### Senior Data Center Architect
### Datacenter and Connected Systems Group, Intel Corporation

## Abstract

In a typical datacenter, almost 40% of the total power consumption is spent on datacenter cooling. In addition, the capital expenditure costs for the cooling infrastructure are also significant. Large Internet Portal Datacenters are looking at every possible way to reduce the cooling cost. One of the emerging trends in the industry is to move to higher ambient datacenter operation. Some datacenter operators are even wanting to operate the datacenters at ambient as high as 40C. It is shown that both server power increase and facility level cooling power savings must be considered to determine net power savings at the datacenter level. CFD modeling is used to demonstrate that following best practices in airflow management: using blanking panels, floor layout, eliminating cable obstructions, hot/cold aisle containment and bypass and re-circulation reduction are first important steps to get ready for high ambient datacenter operation. It is shown that without these practices, there is large variation in inlet air temperature from one server to other creating hotspots forcing thermostats on CRAC units to be set low. It is shown that CFD modeling can be used to quantify how cooling path management improves with hot/cold aisle containment. The study shows that significant datacenter level power savings can be achieved by operating the datacenter up to 35C with the use of Economizers.

## Keywords

Data Center, Efficiency, Higher Ambient DC operations (HTA), Cooling Control, Computational Fluid Dynamics, Cooling Path Management.

## 1. Introduction

It is well known that leading Cloud Data Center customers have a keen focus on reducing operating costs and emissions generated by power consumption, and many are raising the ambient operating temperature in their data centers to improve their financial results and environmental impact.

Optimizing a data center's total cost of ownership (TCO) is one main objective of new datacenter designs. Because the largest non-IT expenditure in a typical data center is cooling cost, raising data center temperature can be a key consideration in lowering TCO.

As the ambient temperature increases the server power goes up and facility power goes down as the requirements for the cooling infrastructure goes down (The number of CRAC units needed in a datacenter reduces) resulting in improved datacenter cooling efficiency, reduces Opex and Capex cost. See figure 1.

- Operating cost drivers
  - IT equipment ~50%
  - Power, cooling, & related infrastructure ~45%
  - Labor, other misc ~5%
- Power consumption drivers
  - IT equipment ~50%
  - HVAC ~35%
  - UPS ~10%
  - Misc ~5%
- PUE metric
  - Typical at 1.5 – 2.0
  - Best in class at 1.2, working toward 1.125

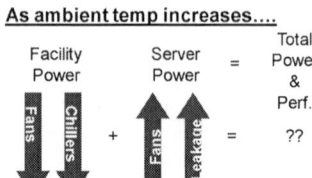

- Need for integrated test that comprehends power, thermal and performance implication of HTA

**Figure 1:** Power and Cost Drivers in Data Center. Impact to facility power vs. Server power with increase in ambient temperature
*PUE [3]

Trade-offs between Facility Power and Server Power must be understood to embrace High Ambient DC Operation

## 2. Server Power vs. Ambient Temperature

Investigating power and performance impact when servers are operated at higher ambient temperature with workloads and then optimizing the system for power is first step to embracing High Ambient DC operation.

The study in this section includes data on how the server power increases at the higher ambient operations. The study covers spread core systems. The server power increase is studied for a number of workloads such as SpecInt, SpecPower, Hadoop, Sandra and Prime95. Server power increase ranges from 0% to 6% in going from 25C to 40C see figure 2. All component temperatures were within their spec limits; No CPU and memory throttling was observed up to 40 C in the systems tested. No degradation of performance for the systems tested as long as specs are met.

**Table 1: Workload vs Ambient Temperature**

| System Ambient, C | Hadoop (Completion time, Sec) | SpecInt Rate Base | SpecPower SSJ_ops | HMMER | LINPACK (GFLOPS) |
|---|---|---|---|---|---|
| 25 | 217 | 185 | 416,152 | 153 | 76 |
| 35 | 217 | 185 | 411,438 | 153 | 76 |
| 40 | 219 | 185 | 411,228 | 153 | 76 |

Table 1 shows the performance for various workloads. Performance did not degrade and was well within run to run variation.

The conclusion from the internal studies is system power increases and system optimization is critical to minimizing system power increase.

Designing servers to comprehend elevated ambient temperature and operating datacenters beyond ASHRAE guidelines requires careful planning and up front diligence (**ASHRAE [1]**). However, elevated air temperatures need not be equated to lower server performance, degradation of component reliability, or even substantial increases in server level power.

Building systems to support high temp operations one needs to take into consideration the following

- Server wall power should be minimized by evaluating wall power at different fan speeds while ensuring that all component specifications are met

- Use the latest generation of Intel CPUs to achieve best energy efficient performance

- Spread core board layout is power optimized

- Copper or high performance heatsinks

Once the server power addressed it is now time to look at how to achieve power savings at datacenter level

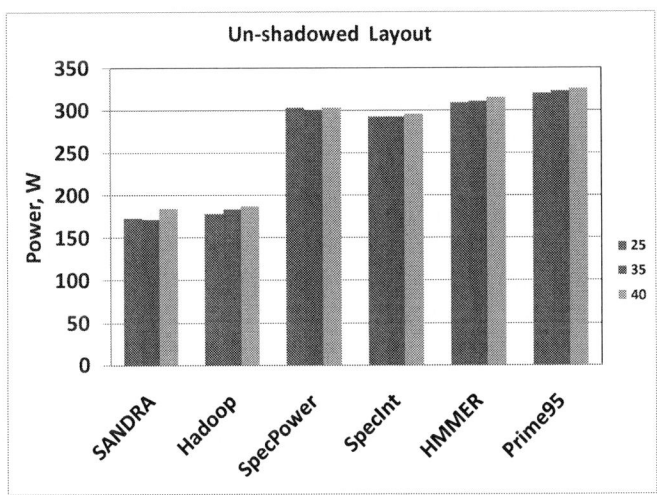

**Figure 2:** Server Power increases with increase in ambient temperature

*Software and workloads used in performance tests may have been optimized for performance only on Intel microprocessors. Performance tests, such as SYSmark and MobileMark, are measured using specific computer systems, components, software, operations and functions. Any change to any of those factors may cause the results to vary. You should consult other information and performance tests to assist you in fully evaluating your contemplated purchases, including the performance of that product when combined with other products.*

* *Results have been estimated based on internal Intel analysis and are provided for informational purposes only. Any difference in system hardware or software design or configuration may affect actual performance. Intel does not recommend running its processors outside their thermal specifications*

### 3. Challenges for going to High Ambient Data Center operations

To achieve efficiencies, running the datacenters warmer isn't as simple as raising the thermostat in a house. Here are some of the key challenges that need to understood and eliminated.

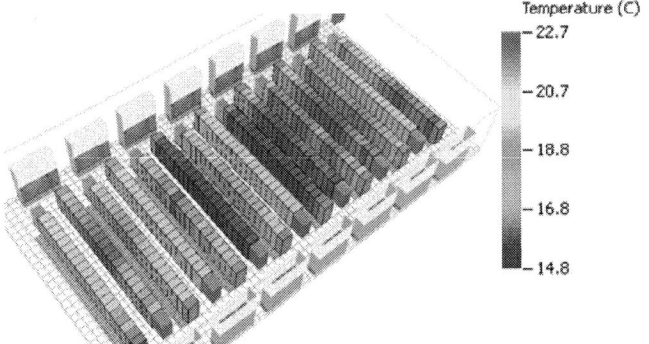

**Figure 3:** Challenges going to High ambient operations in DC

Figure 3 shows Computational Fluid Dynamics simulation model (**Simulation model [2]**) of 2.416MW datacenter with return temperature set point as 21degree C and IT load of 1515kW. This scenario has Hot-Aisle/Cold Aisle layout for datacenter racks with no containment implemented. The temperatures within the datacenter are not uniform. The hottest server dictates the thermostat setting on the CRAC. One of the servers has maximum Inlet temperature of 22.7 degree C, this server will dictate how much temperature in the datacenter can be raised. Inefficiencies in the cooling path management results in hotspots and also the temperature that is supplied by CRAC units is not same as server inlet temperature. If one can make temperatures in the datacenters uniform, one can raise the thermostat settings of CRAC. Implementing the Data Center Best practices is important step to eliminate inefficiencies within cooling path and move to high ambient operations.

### 4. Data Center Cooling Path

Airflow/cooling path in the datacenter needs to be understood to embrace High Ambient DC Operation. Temperature is inseparable from airflow management; datacenter operators must understand how the air gets around, into, and through their server racks. Computational fluid dynamics (CFDs) can help by analyzing airflow on the datacenter floor.

Datacenters with excess cooling are prime environments to raise the temperature set point. Those with hotspots or insufficient cooling can start with implementing simple datacenter best practices for example blanking panels, removing obstructions in the path of cooling, grommets etc. Close-coupled cooling and containment strategies are especially relevant, as server exhaust air, so often the cause of thermal challenges, is isolated and prohibited from entering the cold aisle.

With airflow addressed, users can focus on finding their "sweet spot" - the ideal temperature setting which aligns with business requirements and improves energy efficiency.

Figure 4 Example illustrates cooling path for datacenter no containment implemented and the return temperature set point = 21C. This example is a hot aisle/ cold aisle layout with no containment. In theory, hot aisle / cold aisle layouts increase energy efficiency by allowing for higher temperature set points by directing cooler air closer to equipment inlets. By concentrating the cooling air where it is needed most, CRAC / ACU units can run a few degrees higher than would otherwise be necessary. While modest effectiveness gains are realized by delivering cool air to the proper side of the equipment, hot aisle / cold aisle layouts still present challenges in today's high-density rack environment. Foremost being: "bypass air", where cool air fails to enter the IT equipment; and "recirculation", where heated exhaust air flows back into the cold aisle over the tops of the rack or through open rack space. See figure 4: Not all the air from ACU /CRACs hits the IT Equipment. Some of it bypasses back to ACUs / CRACs. Not all the air leaving the IT equipment makes it back to ACUs. Recirculation brings some of the air back to the inlet of servers

Both of these conditions may lead to data center hot spots and the need for lower temperature set points and significant cooling over-capacity.

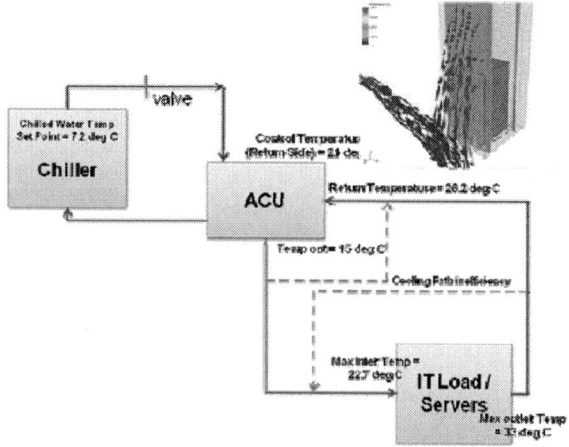

**Figure 4:** Data Center Cooling Path

Due to the impact of bypass and re-circulation, air entering the IT equipment is at a higher temperature than what is supplied by the CRAC units and air entering the ACU/CRAC is at a lower temperature than the exhaust temperature of IT equipment (see Figure 4). The lower return temperature leads to lower cooling capacity of the ACUs. Re-circulation can be fixed by hot & cold isle containment.

Adopting Data Center Best Practices is first step towards High Ambient Temperatures.

## 5. Containment Strategies

In response to the challenges of bypass and re-circulation, data center operators need to contain either their hot or cold aisles. In the figure below we compare case of hot aisle containment implemented and return temperature set point = 27C versus no containment and return temperature set point = 27C. The goal of hot aisle containment is to capture the hot exhaust from IT equipment and direct it to the CRAC/ACU or CRAH units as quickly as possible using the ceiling plenum space for returning the air back to the CRAC units.

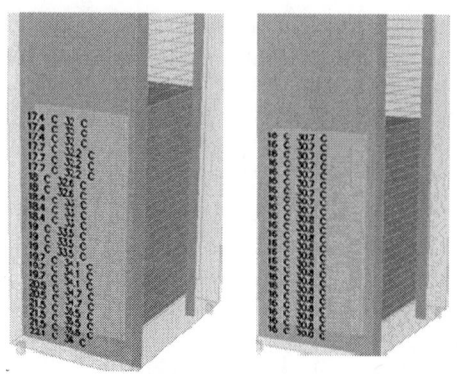

27 Deg C No-Containment      27 Deg C Containment

**Figure 5:** Containment vs. Non Containment.

Figure5 shows containment results in:

- Uniform server inlet temperature throughout Data Center.
- Highest temperature seen at the server inlet reduces with containment
- Containment facilitates the use of ACU supply temperature set points

Hot and Cold Aisle containment is important step to move to High temperature data center operation

## 6. Results

This section covers the case study results for four Computational Fluid Dynamics simulation models of a 2.416MW datacenter. The various scenarios included

- The base case is Hot-Aisle/Cold Aisle Layout for Data Center Racks. No Containment Implemented and the Return temperature set point = 21C. Baseline IT Equipment power is 1515kW

- The second case is Hot-Aisle/Cold Aisle Layout for Data Center Racks. No containment implemented and the return temperature set point = 27C. IT Equipment power at 27 degree C is 1519kW

- Third case is Return Temperature set point = 27 C and Hot Aisle Containment Implemented.

- Fourth case is implemented as return temperature set point = 35C with Containment + Economizer. IT Equipment power at 35 degree C is 1525kW

All the four scenarios listed above incorporate increase in power at server level. Potential power saving shown in table 2 and table 3 takes into account increase in power at the server level.

Table 2 shows the Baseline chilled water plant power at 21C is 541.5KW. Chilled water plant power at 27C is 483.9KW; i.e there is 10.6% power reduction between 21 C and 27C without containment. Chilled water plant power at 27C with containment is 475.3KW; i.e. there is 12.2% power reduction between 21C and 27C containment. Going from 21C to 27C saves 2.39% power at the datacenter level.

Table 2 also shows that the server inlet temperature is lower when containment is implemented, as result additional power savings can be realized if max server inlet temperature is allowed to go to 35C.

In table 3 we see HTA savings if worst server inlet temperature is allowed to increase to 35C.

**Table 2: Potential Saving due to Chilled water temperature increase**

| ACU $T_{return}$ Set point, C | $T_{return,}$ C | Max Server Inlet Temp., C | Chilled water plant power @ $T_{return}$ set point, kW | % Reduction in chiller plant power | % Power savings due to HTA |
|---|---|---|---|---|---|
| 21 | 26.2 | 22.7 | 541.5 | 0.0 | 0.0* |
| 27 | 28 | 26 | 483.9 | 10.6 % | 2.39 % |
| 27 + Cont | 28 | 17.1 | 475.3 | 12.2 % | 2.74 % |

*Baseline datacenter power of 2.416MW used for calculating the last column showing % power reduction due to High Ambient DC operations (HTA)

*Note - Intel Internal estimate, based on Simulation data for a 2416kW baseline data center power , baseline IT load 1515 kW, 5 KW rack and 50 % utilization.*

**Table 3: Power savings when datacenter ambient is raised to achieve Tsys = 35C**

| ACU $T_{return}$ Set point, C | $T_{return,}$ C | Chilled water plant power after raising set point, kW | Data Center Power, kW | % Power savings due to HTA |
|---|---|---|---|---|
| 27 | 28 | 396.8 | 2275 | 5.8% |
| 27 + Cont | 28 | 337.8 | 2216 | 8.3% |
| 35 + Cont | 35 | 0 | 1884.5 | 22.0% |

*Note - Intel Internal estimate, based on Simulation data for a 2416kW baseline data center power , baseline IT load 1515 kW, 5 KW rack and 50 % utilization.*

Table2 shows with baseline of 21C set at ACU return side, Baseline datacenter power at 21C= 2416kW. In both cases chilled water temperature set point is increased to achieve 35C at the server inlet. Datacenter level power savings with 35C server inlet are 5.8% without containment and 8.3% with containment. With air side economizer, the chilled water plant can be eliminated, resulting in 22% Datacenter level power savings.

**Table 4: Effectiveness Metrics**

| | 21 deg C | 27 deg C | 27 deg C + Cont | 35 deg C + Econ + Cont |
|---|---|---|---|---|
| ACU Supply Effectiveness % | 51.6 | 51.4 | 63.6 | N/A |
| ACU Return Effectiveness % | 59.1 | 59.1 | 72.8 | N/A |
| Equipment Supply Effectiveness % | 78.8 | 78.2 | 96.7 | N/A |
| Equipment Return Effectiveness % | 77.9 | 77.6 | 95.4 | N/A |

Table 4 shows Effectiveness Metrics for the four scenarios. These metrics are output from the simulation model. Effectiveness is a measure of how effectively the cold air travels from the cooling systems to the equipment and how effectively warm air is scavenged from the equipment by the cooling systems. It can be viewed in several ways.

- ACU Supply Effectiveness (%) - How much of the air supplied by ACU goes into IT equipment
- ACU Return Effectiveness (%) - How much of the hot air from IT Equipment is scavenged directly by the cooling system
- Equipment Supply Effectiveness - How much of the air entering the equipment inlets is directly from a cooling supply
- Equipment Return Effectiveness - How much of the air leaving the equipment out flows travels directly to a cooling system

Table 4 shows Equipment effectiveness metrics improve with containment; it gets close to 100%. ACU effectiveness metrics did not get close to 100% since ACUs provide more airflow that what is required by the IT equipment. By aggregating server flow (real time sensor data) and tuning ACU to provide the needed airflow will improve the ACU effectiveness.

Variable speed CRAC/ACU units and optimizing cooling with real time data provide additional savings

## 7. Conclusion

Operating Datacenter at High Ambient can result in significant power saving but before embracing High Ambient DC operations trade-offs between facility power and server power must be understood. One needs to also understand and eliminate inefficiencies in cooling path management.

The power optimization needs to start at the system level by use of latest generation of Intel CPUs and further server wall power should be minimized by evaluating wall power at different fan speeds. All component specifications must be met.

## 8. References

1. ASHRAE TC 9.9 2011 Thermal Guidelines for Data Processing Environments– Expanded Data Center Classes and Usage Guidance

2. Simulation Model. 6SigmaRoom From Future Facilities

3. Power Usage Effectiveness = Total datacenter energy use to total IT equipment energy use. Recommendations For Measuring and Reporting Overall Data Center Efficiency Version 2 - Measuring PUE for Data Centers (May 2011)

# CFD Analysis of Free Cooling of Modular Data Centers

Betsegaw Gebrehiwot[1*], Kushal Aurangabadkar[1], Naveen Kannan[1], and Dereje Agonafer[1]
Deepak Sivanandan[2*] and Mark Hendrix[2]
[1]University of Texas at Arlington, 500 W. First Street, Box 19023, Arlington, TX 76019
[2]CommScope, Inc., 1300 E Lookout Dr # 150, Richardson, TX 75082
[1*]Betsegaw.gebrehiwot@mavs.uta.edu and [2*]DSivanandan@commscope.com

## Abstract

Air-side economizers use outside air for cooling information technology (IT) equipment completely or part of the time (coupled with traditional computer room air conditioning (CRAC) units). This system is often used in colder climates and requires less energy since it doesn't use compressors for cooling incoming air. In this paper, thermal performance of an air side economizer for a modular data center is studied using computational fluid dynamics (CFD). The modular data center is comprised of two 12 ft x 40 ft (3.66 m x 12.19 m) IT containers, one 12 ft x 40 ft (3.66 m x 12.19 m) power/cooling module, and a 12 ft x 36 ft (3.66 m x 10.97 m) plenum. Blowers in the power/cooling module draw outside air, pass it through filters, and distribute the air between the two IT modules. The IT modules are arranged in a hot/cold aisle arrangement with a total heat load of 1.2 MW. The two IT modules contain more than 5000 servers, actuator controlled louvers, and sensors needed for cooling system control. CFD analysis of the modular data center showed the importance of louver angles at the inlet of the IT containers. Based on the simulation results improvements have been made to the modular data center by adjusting the angle of the louvers to improve airflow distribution in the servers. This change resulted in significant improvement in the mean exhaust temperature at the servers.

## Keywords

Computational Fluid Dynamics (CFD), FloVENT, Modular Data Center, Containerized Data Center, Cold/Hot Aisle

## Nomenclature

Primitives: Basic building blocks from which all FloVENT geometry is built, either specifically or inherently using assemblies or SmartParts. [1]

SmartParts: FloVENTs geometric tools that are designed to represent actual parts such as fans, filters, hollow boxes, etc. They are "complicated assemblages of primitives[2] which can be quickly generated parametrically, with different levels of modeling to choose the level of detail required." [1]

## 1. Introduction

Data centers comprised of racks and corresponding servers form the backbone of today's cloud computing. Continuous operation of large numbers of servers can generate large amounts of heat which in turn requires high capacity cooling systems. These cooling systems can consume a significant portion of the energy required to run a data center and can negatively impact data center efficiency.

Significant research has been conducted into alternative methods that reduce cooling system energy consumption and improve overall efficiency. Free cooling is one of these alternative methods; in this paper, it is studied for a particular configuration of power/cooling and IT modules in a modular data center.

Free cooling is the most efficient cooling alternative for data center cooling. In this method, ambient air is introduced into the data center through filters, and is then forced by fans that provide the required flow rate to directly cool the server equipment. The heated server exhaust air is then vented back out to the ambient. This cooling method is highly effective at reducing energy consumption for data centers in cooler climates where it can be utilized for a significant portion of the year.

This paper discusses free cooling of a modular data center using CFD analysis. Compared to traditional data centers, modular data centers have the primary advantage in their potential for ease and speed of deployment, possible lower capital and operating costs [2]. The modular data center under study is comprised of one 12 ft x 40 ft (3.66 m x 12.19 m) power/cooling module, two 12 ft x 40 ft (3.66 m x 12.19 m) IT modules, and one 12 ft x 36 ft (3.66 m x 10.97 m) plenum. The power/cooling module contains all of the primary free cooling components (fans, intake filters, actuator-controlled louvers, and environmental controller), while the two IT modules contain actuator controlled louvers and house over 5000 servers along with sensors needed for cooling system control.

Figure 1 shows ambient air passing through the power/cooling module, dividing between the two IT modules and entering into the cold aisles of each IT containers. Each IT module contains two sets of 21 racks placed to the left and right hand side of the cold aisle. The two hot aisles in each IT container are located opposite the cold aisle across each set of 21 racks.

Figure 2 shows details of the power/cooling module. In addition to the cooling components, the power/cooling module contains input power panels, transformers, power distribution, and UPS backup components.

## 2. Modeling and Simulation

### 2.1. Assumptions

Steady state conditions are assumed for all simulations. Ambient air is assumed to be dry air at 27°C. Radiation heat transfer is not considered in the simulation.

### 2.2. Modeling

Containers and louvers of the IT modules and the power/cooling module were first imported from a Pro/E model of the modular data center using FloMCAD into FloVENT. The imported model is then simplified by removing parts and changing geometries that do not play

978-1-4673-1110-6/12 $31.00 © 2012 IEEE

significant role in thermal analysis. Doing so decreases the number of cell count needed to obtain successful CFD simulation thereby reducing computational time. Internal components of the power cooling and IT modules are then represented using the CFD tool's SmartParts. Parts that are small compared to the size of the modular data center were simplified to 2D models. These parts include walls of the containers, filters, blowers, and louvers. Sets of servers in a rack are represented using a Rack SmartPart which captures what happens only at the inlet and exhaust of a set of servers. This SmartPart does not contain any grid cell inside it which further reduces the number of cell count and the computational time.

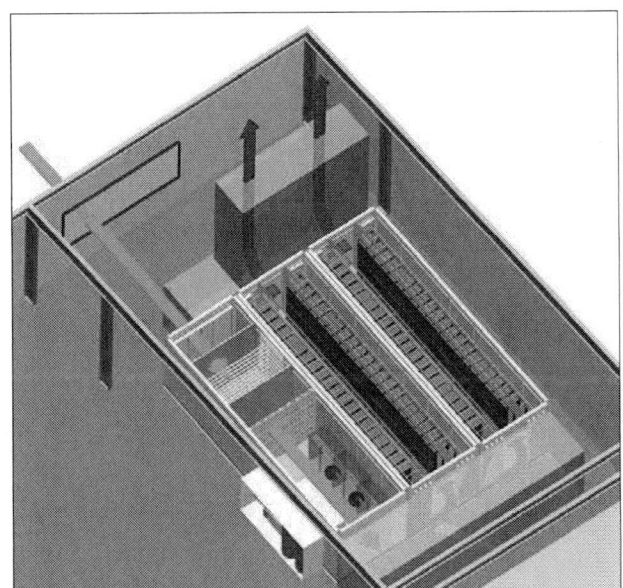

Figure 1 Schematics showing airflow path: Green arrow shows cold ambient air and red arrows show hot exhaust air

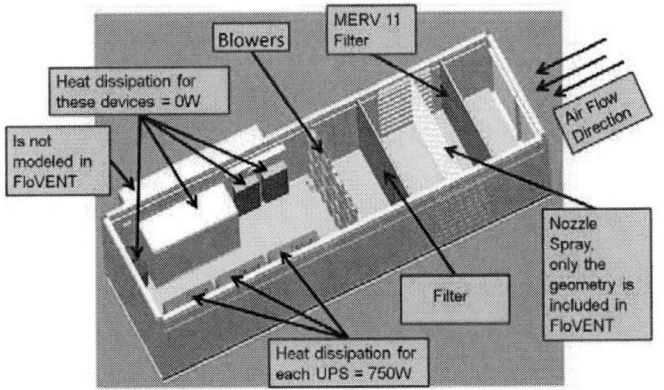

Figure 2 Description of the power/cooling module

Filters, intended to be used in the modular data center were tested in an airflow bench and the resulting system resistance curve is used to create resistance SmartParts which represented the filters.

Figure 3 shows a baseline model of the entire modular data center with roofs removed ready for thermal analysis. Summary of heat dissipation values in the data center is given in Table 1.

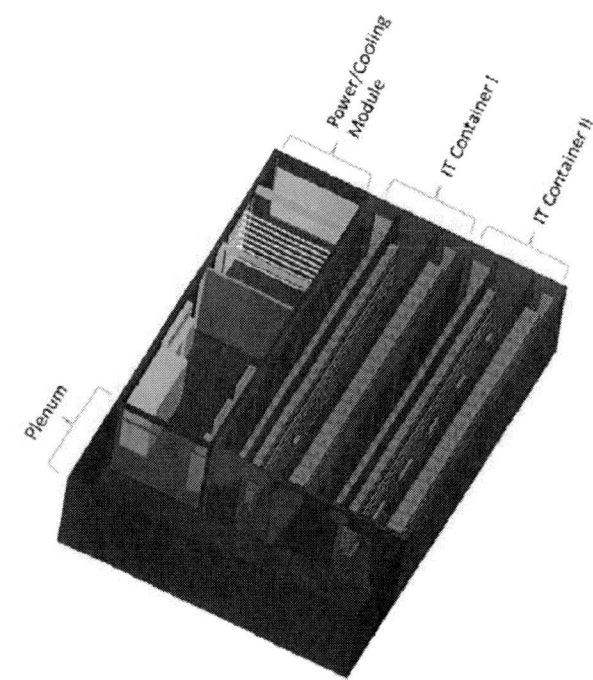

Figure 3 Isometric view of entire system with roofs removed, thermal model

Table 1 Heat load description

| Number of Item(s) | Item | Heat Load (W) |
|---|---|---|
| 1 | Rack | 14226 (at 40% load) |
| 1 | IT Container | 597,475 |
| 2 | IT Containers | 1,194,950 |
| 1 | Power Cooling Module | 2250 |
| The Entire System (i.e. 1 Power Cooling Module + 2 IT Containers) | | 1,197,200 |

In addition to the baseline case, other cases have been run to investigate effect of different components in the modular data center. Case 1 is the same as the baseline case except here all blowers in power/cooling module have been removed. In Case 2 effect of number of servers on the operating point of server fans is shown by running two different simulations on a single IT container. In Case 2a, Figure 4, simulation is run on one IT container with only one set of servers in it. In Case 2b, Figure 5, a total of 42 set of servers, 21 set of servers placed on each side of the cold aisle, are used in the simulation.

978-1-4673-1110-6/12 $31.00 © 2012 IEEE       109

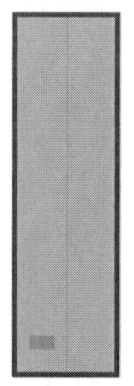

Figure 4 Case 2a: A single set of server mounted in one IT container

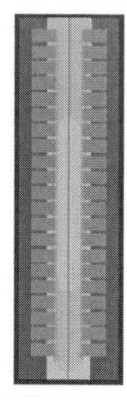

Figure 5 Case 2b: All sets of servers as in the baseline case mounted in one IT container

### 2.3. Meshing and Simulation

Structured Cartesian mesh is used to generate a total of 3.4 million grid cells. K-ε turbulence model is used in the CFD simulation since it is the preferred model for cases where turbulence is to be modeled over large empty volumes within an enclosure [1].

## 3. Results

The thermal profile on a horizontal cut plane passing through the modular data center is shown in Figure 6. The maximum temperature in the modular data center is 16°C above the inlet temperature, 27°C.

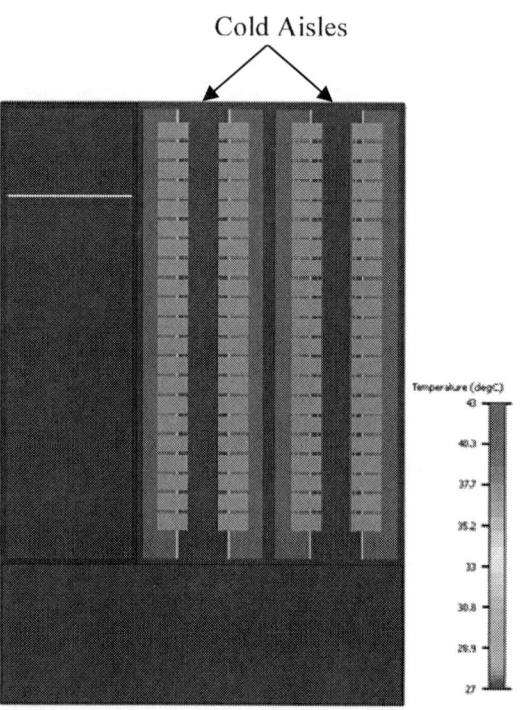

Figure 6 Baseline Case - Temperature Profile (Top View)

Figures 7 and 8 show airflow patterns in the cold aisle of IT module I with IT container inlet louvers inclined at 45° and 0°, respectively. Louvers at these two degrees of inclination are shown in Figures 9 and 10. Figure 7 shows strong air

circulation around mid-length of the cold aisle where as no circulation is visible at this location in Figure 8. Comparison of Figures 7 and 8 clearly show the effect of inlet louvers on airflow distribution in the cold aisle.

Air circulation forms a low pressure region at its center. Thus server fans which have to draw air from regions of strong circulation in the cold aisle will have to overcome higher differential pressure as long as the pressure in the hot aisle is relatively uniform. Overcoming higher pressure differential means less air flow passing through the fans which in turn results in high temperature at the servers. This observation is seen in Figure 11 which shows mean exhaust temperatures at the exhaust of sets of servers numbered in increasing order starting from the rack closest to the inlet louvers.

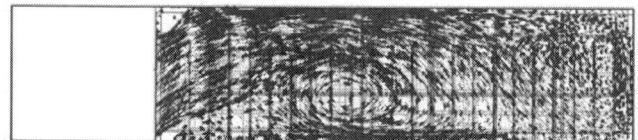

Figure 7 Louvers at 45°: Cold aisle of IT container I, side view

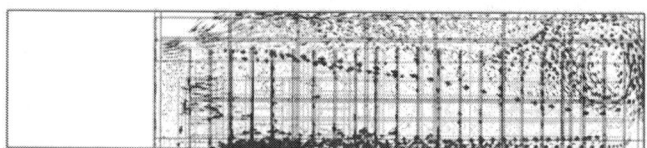

Figure 8 Louvers at 0°: Cold aisle of IT container I, side view

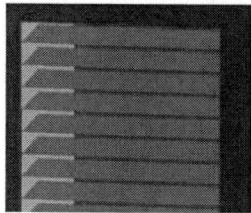

Figure 9 Louvers inclined at 45°

Figure 10 Louvers inclined at 0°

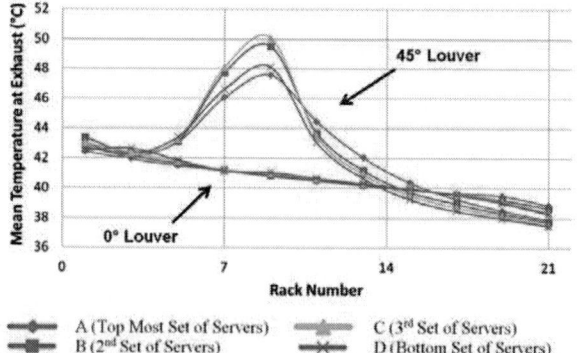

Figure 11 Comparison of effect of 0° and 45° inclined louvers on the mean exhaust temperature of the top set of servers in IT container I

Figure 12 and Figure 13 show a representative server fan operating point for the baseline case and Case 1, respectively. Although the operating point of server fans for Case 1 show

978-1-4673-1110-6/12 $31.00 © 2012 IEEE

the need for overcoming higher pressure differential when compared to the baseline case, the difference is small (less than 1 in $H_2O$ (249 Pa)).

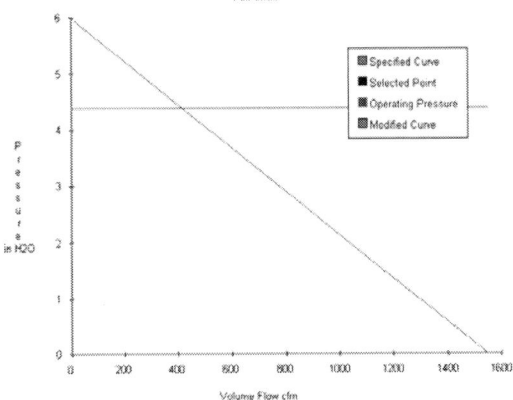

Figure 12 Baseline Case: Fan operating point for one of the top set of servers

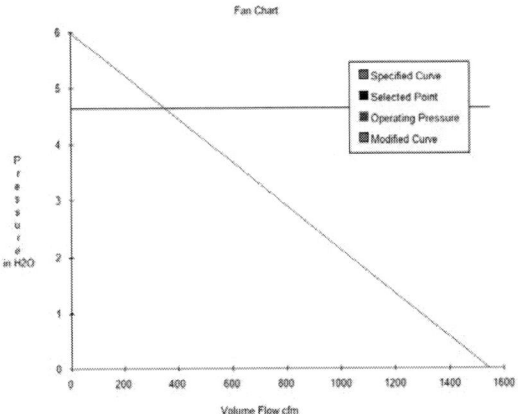

Figure 13 Case 1: A single set of server mounted in one IT container and it fan operating point

Simulation results for Case 2a and Case 2b are shown in Figures 14 and 15, respectively. For the case with only one set of servers in the IT container, the fan on the set of servers has to overcome small resistance when compared to the operating points of server fans in Case 2b. This is due to the additional resistance created at the inlet and exhaust of the IT container when large volume of air is required to pass in Case 2b.

## 4. Conclusion

Airflow pattern of modular data centers is important to understand why temperature and volume flow rate variations occur within rows of racks and sets of servers within a rack. For the particular configuration of power/cooling module, plenum, and two IT containers, CFD analysis has shown maldistribution of volume flow rate in racks and sets of servers caused by circulation of air in the cold aisle. It has been shown that to improve the airflow distribution in the sets of servers, setting the appropriate louver angle is important.

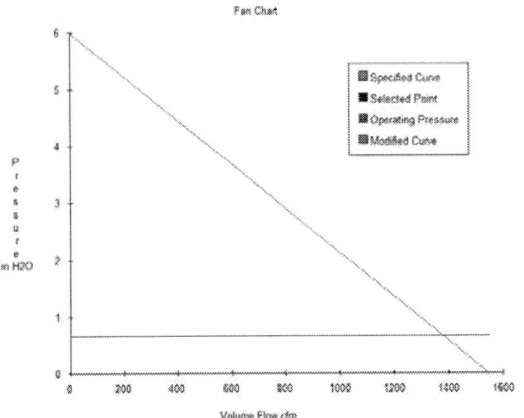

Figure 14 Case 2a: Fan operating point for a single set of server mounted in one IT container

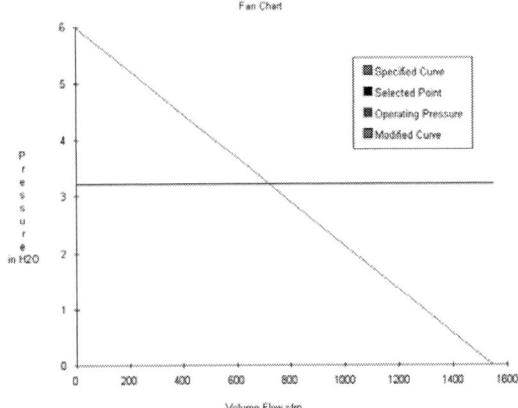

Figure 15 Case 2b: Fan operating point for a representative set of server mounted in one IT container

Results from Case 1 suggest that it could be possible to operate the data center without the need for blowers found in the power/cooling module as long as there are strong server fans in use.

Case 2 showed operating point of a set of server fans changes based on the number of servers in a given IT container. Thus in the design phase of IT containers it is important to consider the number of servers to be placed in a modular data center with respect to the desired server fan operating point.

## Acknowledgments

This work is supported by the National Science Foundation.

## References

1. FloVENT User Guide. Software Version 9.2.
2. Bramfitt, M., and H. Coles. Modular/Container Data Centers Procurement Guide: Optimizing for Energy Efficiency and Quick Deployment. Lawrence Berkeley National Laboratory. February 02, 2011.

# Performance Optimization of Multi-Core Processors using Core Hopping - Thermal and Structural

Sunil Lingampalli, Fahad Mirza, Thiagarajan Raman, Dereje Agonafer

University Of Texas at Arlington, Arlington, TX, USA

## Abstract

As the work load on the single core processor increases, its power density and the die temperature increases as well. The increase in the die temperature results in decreased performance, reliability and increased leakage currents and cooling cost. Also, the non-uniform power distribution across the die results in hot spots. In order to decrease the work load and the cooling cost on the single core processor, multi-core processors have been implemented. Multicore Processors also known as Chip Multi Processors (CMP's). CMPs are processors which contain two or more independent cores on a chip. In CMPs, if one core reaches its critical temperature, the workload is transferred to the other. This phenomenon is termed as core hopping. Core hopping facilitates uniform distribution of the work load among the many cores and leads to improvements in the performance and reliability. The demand for greater performance in applications involving high levels of computing has resulted in many cores being put on a single chip. Every succeeding processor is predicted to hold double the number of cores than the previous one. In this study, core hopping for CMPs is analyzed and the thermal analysis of the chip with core hopping is performed using ANSYS Fluent. The hop sequence is analyzed as a function of chip temperature distribution and a numerical methodology to analyze the coupled thermal and structural integrity of the CMPs is demonstrated.

**KeyWords** – core, CMPs, Moore's law

## Introduction

According to Moore's law the number of transistors on a chip doubles every 18 months [1]. CMPs are the processors which contain two or more independent cores on a chip. This kind of architecture was introduced by Intel with Intel core Duo and has continued to AMD NVIDA and so on. As the transistor count keeps increasing in order to integrate the billions of transistors resulting from the continued scaling of technology, CPMs is one of the effective strategies to integrate such ICs [2]. Figure 1 shows the Moore's law projection.

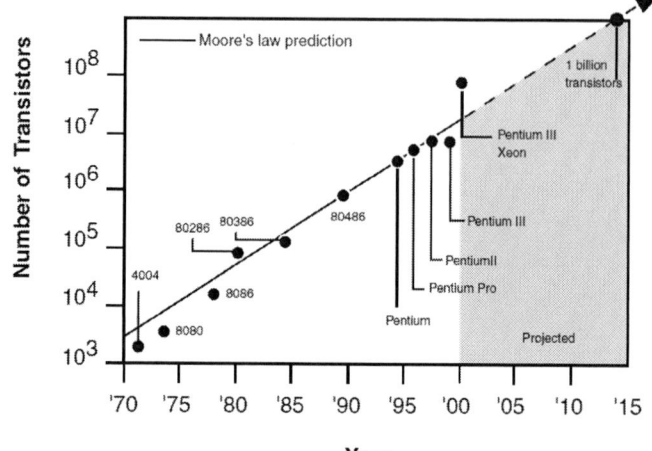

**Figure 1** Moore's law

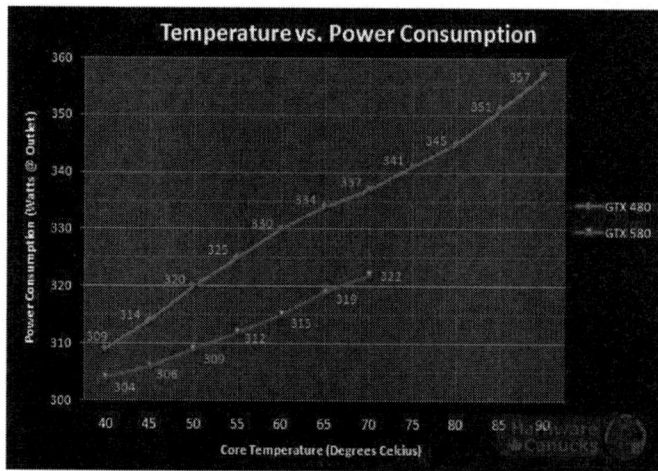

**Figure 2** Power Consumption vs Temperature [3]

As the work load on the single core processor increases, there is an increase in the power densities and die temperature. The increase in die temperature results in decreased performance and reliability and increased leakage currents and cooling costs. In order to decrease the work load and cooling cost on the single core processor, multi-core processors have been implemented. As the technology scaling on CMPs occurs i.e., shrinking of chip geometry in the order of sub- 100nm realm, this results in increase of transistor density and also increases the leakage current leading to excessive power consumption and heat generation which is a major challenge for future CMPs. The scaling down of silicon technology leads to significant thermal coupling between neighboring cores [2].

978-1-4673-1110-6/12 $31.00 © 2012 IEEE

Figure 2 shows two different processors from AMD depicting temperature vs power [3]. With the shrinkage of chip geometry and transitioning to billion transistor microprocessors, the power budget of CMPs must be addressed at the design level. The current and future processor power dissipation increases with increase in clock frequency and transistor count [4, 5]. Figure 3 shows the power consumption of the Intel processors from 1970 to 2005 [6]. Huang et al. [7] derived the expression for power consumed by a core and power density with fixed architecture from across generations, as shown in equation 1 and 2, respectively.

Power consumed

$$1) \quad P_{n+1} = \left(\frac{Vdd_{n+1}}{Vdd_n}\right)^2 P_n$$

Power density

$$2) \quad PD_{n+1} = \left(\frac{1}{s}\right)^2 \left(\frac{Vdd_{n+1}}{Vdd_n}\right)^2 PD_n$$

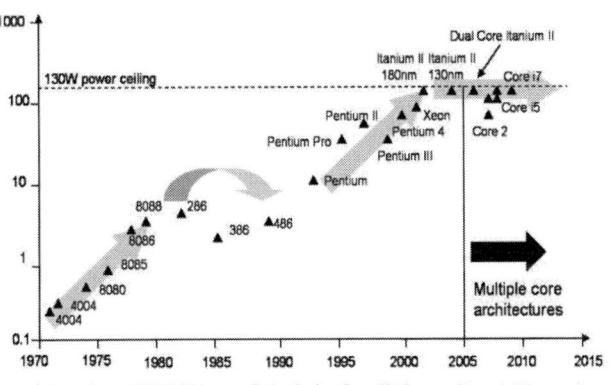

**Figure 3** Intel Power Map

In equations 1 and 2, $V_{dd}$ is the supply voltage, n and n+1 denotes technology generations, and s is the scaling factor. Many cores pose a thermal problem i.e., primary cores consume more power than the other simple cores which results in localized hot spots [7]. Cho et al. [8] demonstrated that at any given point of time not all cores in CMP's will be functioning, i.e., different cores at different locations are active at different times resulting in non-uniformity in power consumption. He used a proactive spatiotemporal power multiplexing method in his study to achieve a lower peak temperature and uniform thermal field on the chip. The spatiotemporal power multiplexing is based on time i.e., it changes the location of power dissipation after a fixed interval of time while maintaining the throughput during the redistribution. Borkar et al. [9] discussed the fine grain power management and system design for the many core system. It is observed through his work that multiprocessors have several benefits such as individual cores can be turned "ON" or "OFF", thereby saving power, lower die temperature can be maintained thus improving reliability. Tasks can be distributed among the many cores such that an overall lower temperature is achieved. The governing rule by which the performance increases by micro-architecture alone is given by Pollack's rule. Pollack's rule states that performance increase is roughly

proportional to square root of increase in complexity as illustrated in the figure 4.

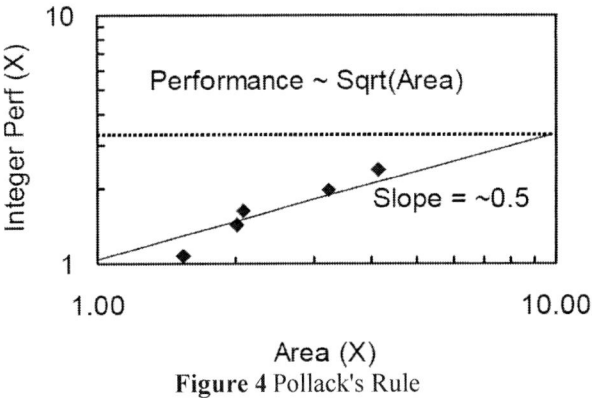

**Figure 4** Pollack's Rule

Huang et al. [10] states that an asymmetric architecture with many core creates a huge thermal problem where in the primary or more complex cores create localized hot spots due to higher power consumption. It also states that due to thermo-spatial low pass filtering effect in smaller cores the equivalent thermal resistance reduces i.e., for same power density small cores produce less heat than large cores. He also mentioned that some techniques used by Intel to improve performance such as Intel "turbo mode" used for boosting processing speed by increasing supply voltage and frequency to those cores that are active, results in increasing hot spots. Shayesteh et al. [11] studied the core swapping technique on a dual micro core architecture triggered thermally, swapping is done with the use of helper engine that reduces the overhead of swapping by buffering the core state during the swapping processes. The author came to a conclusion that core swapping leads to maintaining temperatures below the threshold. In this paper, core hopping for CMPs is analyzed and thermal analysis of the chip is performed using ANSYS. The hop sequence is studied as a function of chip temperature distribution and thermo-mechanical analysis of the chip will be carried out to estimate its structural integrity.

**Modeling and Methodology**

A typical flip chip package is represented in figure 5. The test vehicle (TV) consists of a heatsink, thermal interface material-(TIM-2), heat spreader, TIM-1, die, C4, underfill, substrate, copper pads, solder balls and the printed circuit board (PCB). Thermal analysis was performed with boundary condition of natural convection at the PCB (with a heat transfer coefficient of $10W/m^2K$) and forced convection ($1200W/m^2K$) being applied at the top surface of the heat sink. Such high value of heat transfer coefficient was used to compensate for the heat sink fins (heat sink modeled as a block). The dimensions and thermal properties of the components used in the study are listed in table 1. Mechanical properties are given in Table 2. The geometry is based on Intel Pentium processor used in [12]. Figure 5 shows the full model of the module in ANSYS Workbench.

978-1-4673-1110-6/12 $31.00 © 2012 IEEE

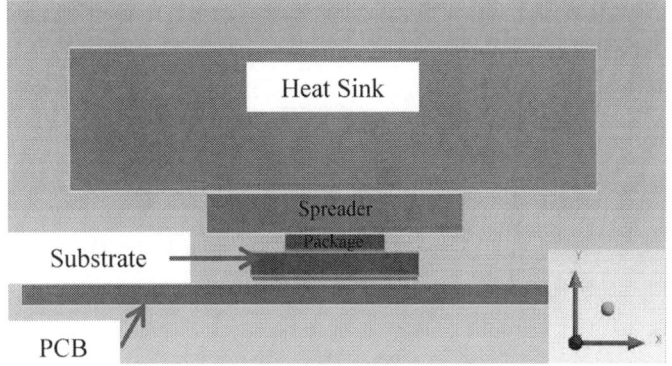

**Figure 5** Full Model

**Table 1 Dimensions and thermal properties of the Components in the model**

| Components | Dimension (mm) | Thickness (mm) | Thermal Conductivity (W/mK) |
|---|---|---|---|
| Heat sink | 64 x 64 | 6.35 | 247 |
| TIM-2 | 31 x 31 | .075 | 6 |
| Heat Spreader | 31 x 31 | 18 | 390 |
| TIM-1 | 12 x 12 | .025 | 50 |
| Die | 12 x 12 | .75 | 140 |
| Copper pads | .3 x .3 | .03 | 390 |
| Solder Balls | .4 x .8 | .28 | 57 |
| PCB | 76 x 76 | 1 | 13 |

**Table 2 - Mechanical properties of various components**

| Material | E(GPa) | CTE(ppm) | Poisson's Ratio (v) |
|---|---|---|---|
| PCB | 21.9( X or Z); 9.99(Y) | 17e-6(X or Y); 70e-6(Y) | 0.28 |
| Solder Bump | 38 | 2.21e-05 | 0.36 |
| Substrate | 25.99(X or Z); 11(Y) | 17e-6(X or Z); 52e-6(Y) | 0.39 & 0.11 |
| Copper Pad | 82.7 | 1.27e-05 | 0.34 |
| C4/Underfill | 14.5 | 2e-05 | 0.28 |
| Die | 150 | 3e-06 | 0.3 |
| TIM 1 | 4e-04 | 1.75e-04 | 0.28 |
| Heat Spreader | 121 | 1.73e-05 | 0.3 |
| TIM 2 | 4e-04 | 1.75e-04 | 0.28 |
| Heat Sink | 68 | 2.4e-05 | 0.3 |

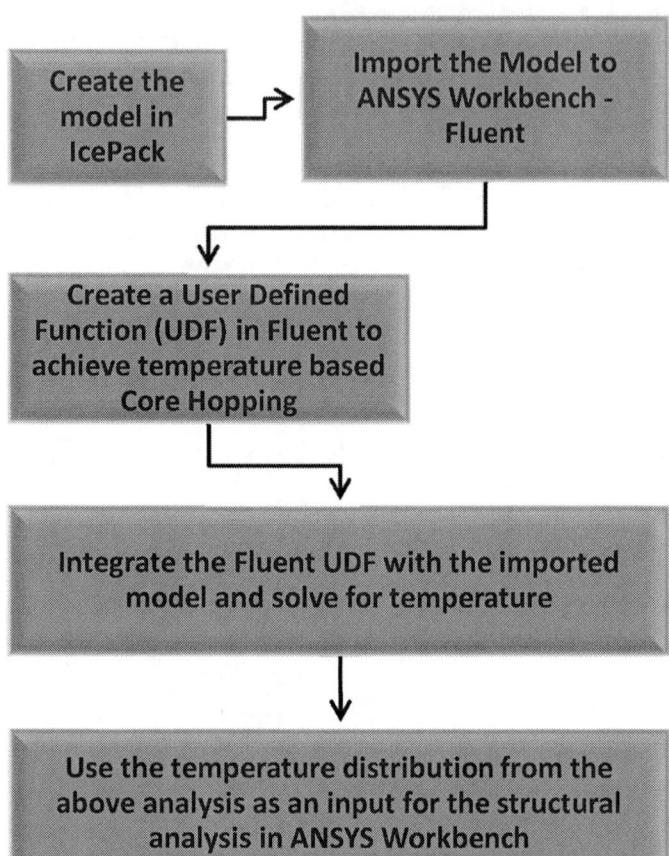

**Figure 6** Analysis Flowchart

The package shown in Figure 5 was first modeled in Icepak 13.0.2 in form of simple blocks. The heat sink is modeled as a block without fins and to compensate for the fins a higher value of heat transfer coefficient is used. The surface of the die was divided into 16 equal areas modeled as heats source each representing a core. The copper pads on the top and bottom of the solder ball and the solder ball were modeled as three separate blocks in order to decrease the computing time. The isometric view of the model created in Icepak is shown in Figure 7.

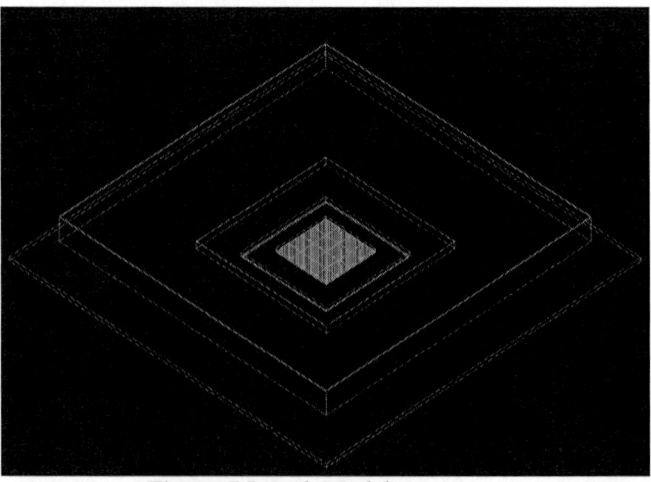

**Figure 7** Icepak Model

978-1-4673-1110-6/12 $31.00 © 2012 IEEE

After creating the model, it is meshed in Icepak 13.0.2; a Fluent case file was written using Icepak solver after that the case file was imported to Fluent 13.0. Using the User Defined Functions in fluent, a UDF code was written to assign the boundary conditions i.e., the maximum and minimum temperature at which the cores starts hopping and the conditions were integrated to the respective zones in the model. Figure 8 shows the arrangement of all the cores on the chip.

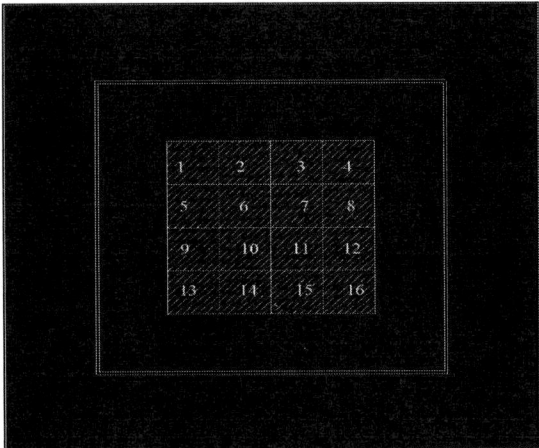

**Figure 8** Arrangements and Numbering of Cores

The UDF code was written in such a way that at any given point of time four out of sixteen cores are active and generating heat. The sequence in which the cores start hopping is given in the Table 3 where $T_{min}$ is the minimum threshold temperature i.e., the core gets activated only when it cools down to this temperature and $T_{max}$ is the maximum threshold temperature i.e., the temperature where the core has to be deactivated. Table 3 shows the hop sequence simulated through the UDF.

**Table 3 Sequence in which the Cores Start Hopping**

| Core | ON | OFF |
|---|---|---|
| 1 | Tcore1<=Tmin | Tcore1>=Tmax |
| 2 | Tcore1>=Tmax and Tcore6>=Tmax and Tcore5>=Tmax and Tcore2<=Tmin | Tcore2>=Tmax |
| 3 | Tcore4>=Tmax and Tcore7>=Tmax and Tcore8>=Tmax and Tcore3<=Tmin | Tcore3>=Tmax |
| 4 | Tcore4<=Tmin | Tcore4>=Tmax |
| 5 | Tcore1>=Tmax and Tcore6>=Tmax and Tcore5<=Tmin | Tcore5>=Tmax |
| 6 | Tcore1>=Tmax and Tcore6<=Tmin | Tcore6>=Tmax |
| 7 | Tcore4>=Tmax and Tcore7<=Tmin | Tcore7>=Tmax |
| 8 | Tcore4>=Tmax and Tcore7>=Tmax and Tcore8<=Tmin | Tcore8>=Tmax |
| 9 | Tcore3>=Tmax and Tcore10>=Tmax and Tcore9<=Tmin | Tcore9>=Tmax |
| 10 | Tcore13>=Tmax and Tcore10<=Tmin | Tcore10>=Tmax |
| 11 | Tcore16>=Tmax and Tcore11<=Tmin | Tcore11>=Tmax |
| 12 | Tcore16>=Tmax and Tcore11>=Tmax and Tcore12<=Tmin | Tcore12>=Tmax |
| 13 | Tcore13<=Tmin | Tcore13>=Tmax |
| 14 | Tcore13>=Tmax and Tcore10>=Tmax and Tcore9>=Tmax and Tcore14<=Tmin | Tcore14>=Tmax |
| 15 | Tcore16>=Tmax and Tcore11>=Tmax and Tcore12>=Tmax and Tcore15<=Tmin | Tcore15>=Tmax |
| 16 | Tcore16<=Tmin | Tcore16>=Tmax |

# Results

## Thermal Analysis

Thermal analysis of the various cores is completed based on temperature. The hopping occurs among the four cores depending on temperature conditions given in the UDF code. At any instant four cores are active and if the temperature of any core reaches the threshold (~305K), that particular core is switched "OFF" and the operation is transferred to a different core. Among all the core hopping sequence investigated, the cases shown in table 3 had uniform temperature distribution. In this work, heat flux of $1e^6$ W/m$^2$ is applied on each core and solution was run for 10 time steps with a step size of 0.5 seconds each. For example, if core 1 reaches the threshold, it was replaced by the 6th core; similarly core hopping takes place between 4 and 7, 13 and 10, 16 and 11, respectively. The hopping always took place among the above mentioned 8 cores and that the cores 2, 3, 5, 8, 9, 12, 14, and 15 never came "ON" till 5 seconds. However when the heat flux is increased from $1e^6$W/m$^2$ to $6e^6$W/m$^2$ core hopping occurred among all the 16 cores because the condition for activation was satisfied at some point of time within the 5 second interval. Figure 9a and 9b show the hop sequence among the 8 cores mentioned above.

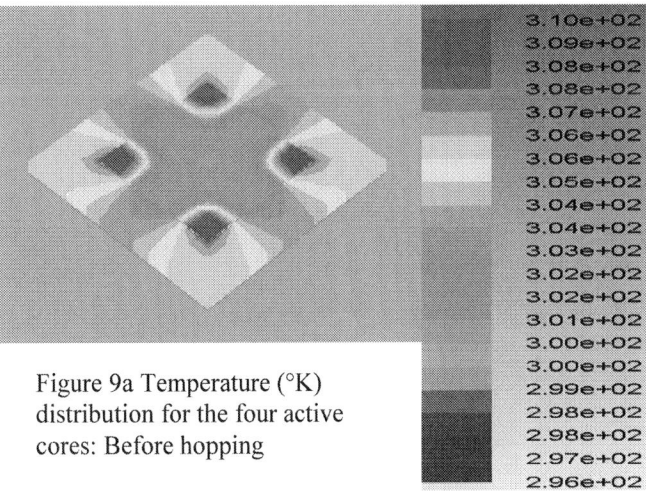

Figure 9a Temperature (°K) distribution for the four active cores: Before hopping

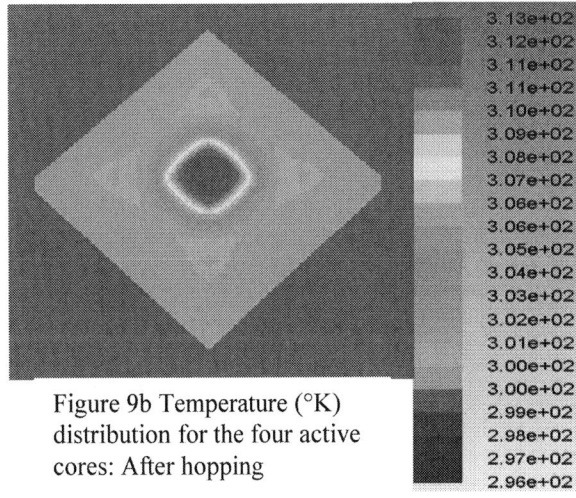

Figure 9b Temperature (°K) distribution for the four active cores: After hopping

## Structural Analysis

Thermo mechanical analysis is carried out using ANSYS Work Bench 13.0, the package geometry is imported to work bench in .IGES format which was written using ANSYS Icepak. The model has been imported to transient structural analysis in work bench. Transient structural analysis is a time based analysis where the loading changes with time. In this study loads were thermal (temperature distribution). Results from the thermal analysis done in Fluent were imported as input loads to Workbench. Figure 10 shows the transfer of thermal loads from fluent to Work bench transient structural analysis. The whole idea was to investigate the thermal-mechanical stresses in the chip region induced due to the thermal gradients within the chip. Compared to the chip region the other components were pretty much stresses free at all instants. The von misses stress results for the entire package are shown in the Figure 11. Maximum stress is seen in the chip region with a magnitude of 47.9 MPa. The stress distribution on the chip changed with the core hopping and the maximum stress was always observed at the chip edges. Maximum warpage in the chip was 13-μm, as shown in figure 12.

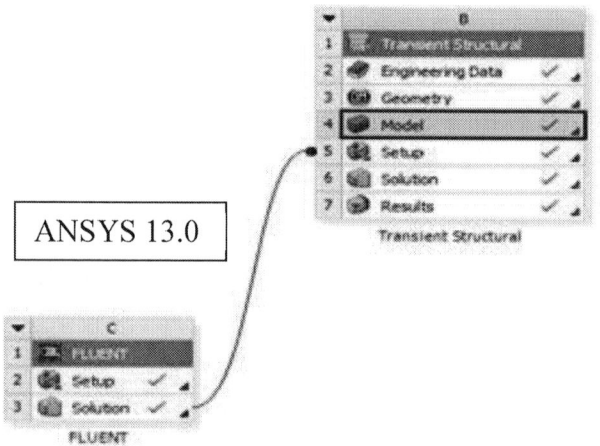

**Figure 10** Load transfer from Fluent to Workbench

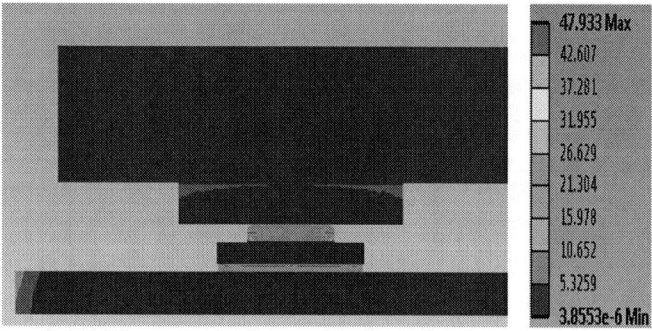

**Figure 11** Stress distribution for the entire package

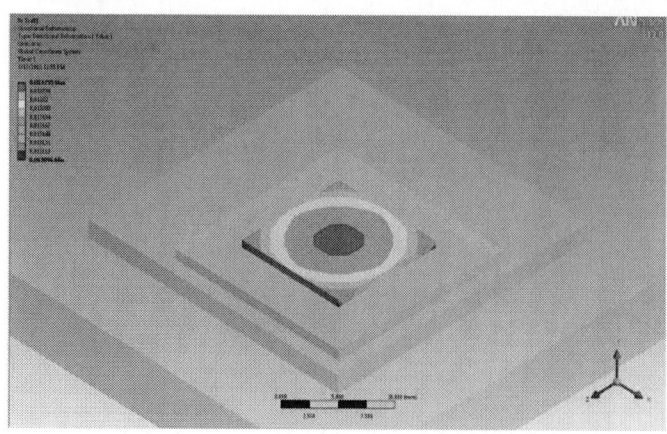

**Figure 12** Warpage in the chip (Y deformation)

## Conclusion

Core Hopping has been studied based on the chip temperature distribution. Temperature based core hopping is more realistic than the time based counter-part because in the time based core hopping, the core is activated for a certain period of time irrespective of the temperature, which could result in exceeding the safe temperature limit for which the chip was designed for. This may lead to premature chip failures. A methodology has been demonstrated to analyze the coupled thermo-mechanical integrity of CMPs by integrating ANSYS Fluent with Workbench. For the stress analysis, the temperature based hopping results were imported to Work Bench and transient structural analysis was performed. Maximum stress, as expected was seen in the chip region (die and the C4) and other components show very little stress throughout the hop sequence. The maximum stress in the chip region is around 50MPa which is almost 2 orders of magnitude less than the yield strength of silicon. This concludes that in the core hopping phenomenon, the key parameter is temperature management only and that mechanical stresses do not play much role in causing failures.

## References

[1]. Fundamentals of Microsystems Packaging - Rao R. Tummala, McGraw Hill

[2]. Hot Spots and core-to-core Thermal Coupling in Future Multi-core ArchitecturesM.janicki, J. H. Collet, A. Louri and A. Napieralsk

[3]. www.amdzone.com

[4]. Dynamic Thermal Management for High-Performance Microprocessors David Brooks and Margaret Martonosi (Department of electrical engineering Princeton University), in Proc of 7th international Symposium on Performance computer Architecture 2001

[5]. Managing the Impact of increasing Microprocessor Power Consumption, Stephen H. Gunther, Frank Binns, Douglas M. Carmean, Jonathan. C. Hall, Intel Technology Journal, 2001.

[6]. http://thewarrencentre.blogspot.com/2010/08/what-will-you-do-with-100-cores.html

[7]. Interaction of scaling Trends in Processor Architecture and cooling", Wei Huang, Mircea R. Stan, Sudhanva

Gurumurthi, Robert.J. Ribiando and Kevin Skadron, 26th IEEE SEMI-THERM Symposium, 2010

[8]. "Proactive Power Migration to reduce maximum value and spatiotemporal non- uniformity of On-Chip temperature distribution in homogenous many-core processors", M.Cho, N.Sathe, M.Gupta, S.Yalamanchalli, and S.Mukhopadhyay,In Proc. Of SEMITHERM 2010.

[9]. "Thousand core chips - A Technology perspective", S.Borkar, DAC 2007.

[10]. "Exploring the Thermal Impact on Manycore Processor Performance", Wei Huang, Kevin Skadron, Sudhanva Gurumurthi, Robert.J. Ribando and Mircea R. Stan, 26[th]IEEE SEMI-THERM symposium, 2010.

[11]. Reducing the Latency and Area Cost of Core Swapping Through Shared Helper Engines, A.Shayesteh, E.Kursun, T.Sherwood, S.Sair and G.Reinman, In Proc of International Conferanc on Computer Design, 2005.

[12]. Multi-Objective Optimization to Improve Both Thermal and Device Performance of a Nonuniformly Powered Micro-Architecture, Saket Karajgikar, Dereje Agonafer, Kanad Ghose, Bahgat Sammakia, Cristina Amon, Gamal Refai-Ahmed, Journal of Electronic Packaging JUNE 2010, Vol. 132 / 021008-1

[13]. Characterization of Microprocessor Chip Stress Distributions During Component Packaging and Thermal Cycling, Jordan Roberts, SafinaHussain, M. Kaysar Rahim, Mohammad Motalab, Jeffrey C. Suhling, Richard C. Jaeger, PradeepLall, 2010 IEEE, 2010 Electronic Components and Technology Conference.

# Thermal System Identification (TSI): A Methodology for Post-silicon Characterization and Prediction of the Transient Thermal Field in Multicore Chips

Minki Cho, William Song, Sudhakar Yalamanchili, and Saibal Mukhopadhyay
School of ECE, Georgia Institute of Technology
266 Ferst Drive, Klaus Advanced Computing Building
Atlanta, GA, USA
mcho8@gatech.edu

## Abstract

This paper presents a methodology for *post-silicon thermal prediction* to predict the transient thermal field a multicore package for various workload considering chip-to-chip variations in electrical and thermal properties. We use time-frequency duality to represent thermal system in frequency domain as a low-pass filter augmented with a positive feedback path for leakage-temperature interaction. This thermal system is identified through power/thermal measurements on a packaged IC and is used for post-silicon thermal prediction. The effectiveness of the proposed effort is presented considering a 64 core processor in predictive 22nm node and SPEC2006 benchmark applications.

## Keywords

Thermal prediction, multi-core, system identification, leakage, and process variations

## 1. Introduction

Characterization of the spatiotemporal variation of the on-chip junction temperature (*the transient thermal field*) is crucial for thermal-aware design, assembly, and management for reliable in-field operation of a chip (die and package) [1, 2]. The thermal field is generated by the interaction of time-varying power pattern and the thermal properties (resistivity and heat capacity) of die and package materials. Further, the thermal properties of the die/package assembly [e.g. conductivity of thermal interface materials (TIM)] can vary between different instances of same IC (chip-to-chip variation) or over time (e.g. delamination in TIM [3]). Moreover, imperfections in the manufacturing process leads to die-to-die and within-die process variations in transistor leakage [1]. The leakage and temperature are positively correlated – a higher temperature results in higher leakage which further increase the temperature. Hence, for same dynamic power, chip-to-chip leakage variation leads to variation in on-chip temperature [4]. Fig. 1 illustrates the impact of process variation and leakage-temperature interaction on thermal behavior of a chip using example simulations in predictive 22nm node. As the die-to-die process variation increases with technology scaling, the post-silicon chip-to-chip variation in transient thermal field is also expected to increase. This challenge is further enhanced by many-core processor architectures running increasingly data intensive and unstructured workloads. As the power, performance, and lifetime reliability of processors depends on the transient temperature, in-field reliable operation of many-

core processors needs the accurate characterization of the interaction of workload variation and chip-to-chip/package-to-package variations in thermal/electrical properties. This leads to a new challenge - ***post-silicon prediction of the transient thermal field***. The objective of *post-silicon thermal prediction* is to predict the transient temperature of a particular instance of a packaged IC for various workload and considering chip-to-chip and package-to-package variations in electrical (leakage) and thermal properties.

The existing transient thermal simulation methods (finite element/volume or distributed RC), suitable for fine-grain *design time* transient thermal analysis, require accurate estimation of thermal resistivity and heat capacity of all materials [5-7]. Many works have studied on how to measure the thermal resistance and capacitance of thermal interface material (TIM), heat sink, convective, and heat spreader [8-11]. Many steady-state method works are modeled after ASTM D5470 [8]. A. Poppe et. al presented dynamic electrical temperature measurement [9] and R. Campbell et. al presented the flash diffusivity method for accurate measurement of thermophysical property data [10]. The measurements of thermal resistance and capacitance suffer from repeatability, contamination, pressure, and inaccuracy problems. Even if we measure accurately the thermal resistances of TIM, heat sinks, and interface, in stacking

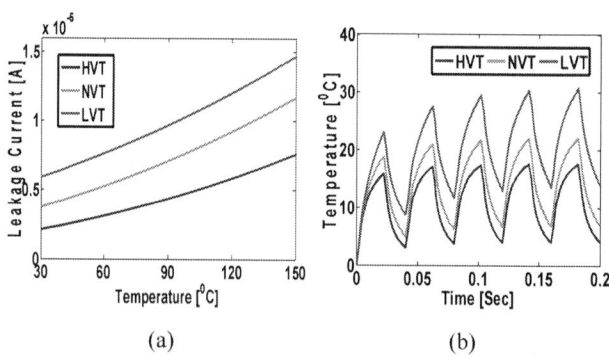

(a)                  (b)

**Figure 1:** Illustration of the need for post-silicon transient thermal analysis considering process variation: (a) the interaction of leakage (average for all input condition) and temperature in a NAND2 gate considering different process corners (HVT – High threshold voltage, NVT – nominal threshold voltage, and LVT – low threshold voltage corner). (b) the effect of such interaction for an example self-consistent thermal simulation (using distributed RC network) considering a square wave dynamic power profile (e.g. turning on and off a the chip after a time-interval) and leakage of 10million NAND2 gate.

condition those values are changed due to imperfect attachment and manufacturing. K. Kurabayashi et. al. presents that the die attach resistance differs substantially from the value predicted using the bulk thermal conductivity of the attachment material because of partial voiding and delamination [12]. Consequently, the fine-grain distributed RC based thermal simulators used during design time are difficult to adopt for post-silicon thermal analysis.

## 2. Contributions and Novelty

This paper presents a unique approach for transient thermal analysis that addresses the specific requirements of *post-silicon thermal prediction*. The proposed approach, referred to as Thermal System Identification or TSI, is based on principles of *system identification, frequency domain signal analysis*, and *positive feedback system*. We develop the mathematical principles of the proposed approach and demonstrate its effectiveness in post-silicon thermal analysis of a 64 core processor at predictive 22nm node [13]. The each core is modeled as close to Intel Nehalem [14] architecture running at 3.0GHz. This post silicon characterization of a multicore chips can be used by operating systems to schedule workloads since the identification of the chip thermal system enables schedulers to reason about the thermal consequences of scheduling a specific workload on a target chip. This understanding can also be exploited in configuring large system (e.g. data centers) via thermally compatible aggregations of multicore packages. Fig. 2 shows the overall flow of the proposed post-silicon thermal prediction approach. This paper makes the following contributions:

- *High-level Transfer Function of the Thermal System including Leakage-Temperature Interaction*: We provide a high-level abstraction of the thermal behavior of a chip as a multi-input multi-output (MIMO) system where power sources are system inputs and observed temperature values at different locations are the system outputs. The interaction of leakage and temperature is used as an integral part of this high-level MIMO system. We show that this thermal system can be represented in frequency domain as a filter matrix. In time domain heat diffusion equation represents a distributed RC network which behaves as a low-pass filter in frequency domain. This is augmented with a positive feedback path representing leakage-temperature interaction.

- *Thermal System Identification - Post-silicon Extraction of Transfer Function of the Thermal System and Fast Prediction of Transient Thermal Field:* We present methodologies that can identify this thermal system (i.e. the thermal filters) after fabrication and packaging using sequences of on-chip power and temperature measurements. These methods allow one to construct a unique thermal system for each chip (thermal system identification or TSI).

We present methods to accurately predict the chip-specific transient thermal fields for varying workloads using the corresponding thermal filter matrix [$\mathbf{H}(\omega)$]. The

frequency response of the temperature variation over a time interval is computed from the Fourier transform of power pattern in that interval and the filter matrix [$\mathbf{T}(\omega)=\mathbf{H}(\omega)\times\mathbf{P}(\omega)$]. The time-domain temperature is obtained from the temperature spectra.

Several methods have been proposed in recent years for fast steady-state spatial thermal map (e.g. power blurring method in [15] and discrete cosine transform (DCT) based method in [16]), fast transient temperature simulations (e.g. [17-18]), and fast spatiotemporal analysis considering multilayers of power and materials (e.g. ThermalScope [19]). The TSI based approach provides important advantages in post-silicon thermal analysis over the above mentioned approaches used in fine-grain design-time thermal analysis. First, the proposed approach performs temperature prediction using the thermal transfer function extracted from the full thermal system (i.e. stacks of heat sink, spreader, TIM, and chip), instead of computing thermal resistance and capacitances of individual materials in isolation. Therefore, the effects of any *non-uniformity and/or uncertainty in the thermal properties of the materials are captured in the extracted transfer function*. Moreover, as the leakage temperature interaction is considered as a part of the MIMO system, the effect of process variation of individual chips is also automatically considered. Second, the *fast simulators mentioned earlier do not consider leakage-temperature interaction*. Currently, the transient temperature estimation considering leakage-temperature interaction is performed using distributed RC based simulators (e.g. Hotspot [21]) where leakage power is updated in each time-step based on the current thermal map [19-22]. Therefore, higher accuracy of the temperature estimation requires fine-grain time-step which in turn increases simulation time. In the proposed approach the leakage temperature interaction is incorporated in the system transfer function and temperature estimation is performed in the frequency domain. Consequently, the accuracy of the proposed method is less sensitive to time-step allowing fast estimation of transient temperature.

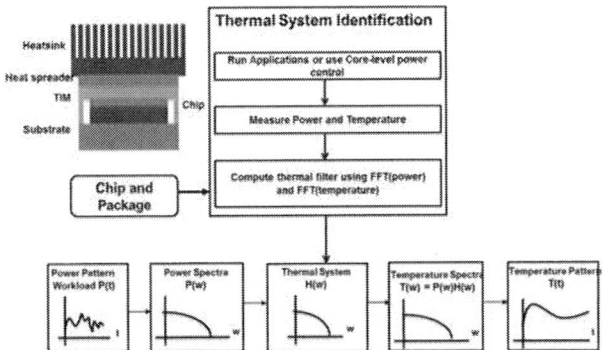

**Figure 2:** Overall methodology of post-silicon prediction of the transient thermal field. The method uses the time-frequency duality to extract thermal system in frequency domain using post-silicon measurement and use that to predict transient temperature profile.

## 3. Mathematical Approach

### 3.1. Modeling the MIMO Thermal System with Leakage-Temperature Interaction

In a MIMO system, the temperature of an observation point is affected by the multiple input power sources. Since a distributed RC network is a linear system, superposition principle can be applied here i.e. the temperature at one location is the additive response of all power sources in the system. Assume that there are M power sources organized into m×m 2D grids. We further assume that there are L numbers of observation points organized in l×l grids. The temperature at the observation point $(i, j)$ in frequency domain can be estimated as:

$$T_{ij}(k)=P_{11}(k)H_{11\to ij}(k)+P_{12}(k)H_{12\to ij}(k)+\cdots+P_{mm}(k)H_{mm\to ij}(k)=\sum_{pq=1}^{m}P_{pq}(k)H_{pq\to ij}(k) \quad (1)$$

Note $H_{ij\to ij}(\omega)$ is defined as the self-transfer function of a location (i.e. the transfer function connecting power and temperature of a location ($H_{self}$)). Likewise $H_{pq\to ij}(\omega)$ ($\forall p,q \neq i,j$) is defined as the cross transfer function ($H_{cross}$) that connects power of one location and temperature of another. The above formulation leads to the 2D filter matrix for the MIMO system (Fig. 3):

$$\begin{pmatrix} T_{11}(\omega) \\ M \\ T_{ll}(\omega) \end{pmatrix} = \begin{bmatrix} H_{11\to 11}(\omega) & \dots & H_{mm\to 11}(\omega) \\ M & M & M \\ H_{11\to ll}(\omega) & \dots & H_{mm\to ll}(\omega) \end{bmatrix} \begin{pmatrix} P_{11}(\omega) \\ M \\ P_{nm}(\omega) \end{pmatrix} \quad (2)$$

We now estimate the self and cross transfer functions considering the leakage feedback. Without loss of generality, we explain this considering two sources and two locations. Consider leakage current ($P_L$) depends on temperature as:

$$P_L(T) = P_L(T_0) + f(T) \quad (3)$$

where $P_L(T_0)$ is the leakage power at room temperature, and the function f(T) represents sensitivity of leakage power to temperature. First, we consider $H_{self}$ i.e. the temperature of location $i$ due to the power source of at location $i$. We obtain:

$$T_i(\omega)=\left[P_{D_i}(\omega)+P_{L0_i}(\omega)+F\left(f(T_i(t))\right)\right]H_{i\to i}(\omega)=\left[\underbrace{\frac{P_{D_i}(\omega)+P_{L0_i}(\omega)}{P_i(\omega)}+\alpha T_i(\omega)}\right]H_{i\to i}(\omega) \quad (4)$$

The last approximation assumes a linear interaction between leakage and temperature to improve analytical tractability. Both the room temperature leakage ($P_{L0}$) and the coefficient ($\alpha$) depends on leakage-temperature interaction. Note $P_i(\omega)=P_D(\omega)+P_L(\omega)$ is the spectral response of power without leakage-temperature feedback (can be estimated from the workload). Now the thermal system model can be represented as (Fig. 3):

$$T_i(\omega) = \underbrace{\left[\frac{H_{i\to i}(\omega)}{1-\alpha H_{i\to i}(\omega)}\right]}_{H_{self}} P_i(\omega) \quad (5)$$

We now evaluate the temperature of location $i$ due to power source at location $k$. We apply superposition principle during this evaluation estimate $T_i(\omega)$ assume $P_i(\omega)=0$. But the heat generated in location $k$ propagates to location $i$ which increases the temperature of location $i$. Increase in temperature at location $i$ triggers the leakage feedback loop at location $i$. This results in leakage power at location $i$ and hence, increase temperature of location $i$. The temperature increase in core $i$ due to power of core $k$ is therefore estimated as (Fig. 2):

$$T_i(\omega)=P_k(\omega)H_{k\to i}(\omega)+\alpha T_i(\omega)H_{self}(\omega)=\underbrace{\left[\frac{H_{k\to i}(\omega)}{1-\alpha H_{self}(\omega)}\right]}_{H_{Cross}}P_k(\omega) \quad (6)$$

### 3.2. Methods for Thermal System Identification

The principle discussed above requires frequency response of the self and cross transfer functions for each chip (i.e. TSI). To perform TSI on the MIMO system, one input power source is excited at a time and temperature is measured at all observation points considered. Hence, the equation (2) transforms to:

$$\forall i\ \&j:\ T_{ij}(\omega)=P_{pq}(\omega)H_{pq\to ij}(\omega)\Rightarrow H_{pq\to ij}(\omega)=T_{ij}(\omega)/P_{pq}(\omega) \quad (7)$$

The above equation can be used to estimate the thermal filter from all inputs power sources to all temperature observation points. As equation (7) is division of two complex numbers, both magnitude and phase of the filter response are extracted. For better accuracy it will be efficient to minimize the leakage of unselected locations.

## 4. Applications of TSI to Thermal Modeling of Many-Core Processors

In this section we apply the TSI based approach to the post-silicon thermal prediction of many-core processor. We consider one temperature sensor is present in each core.

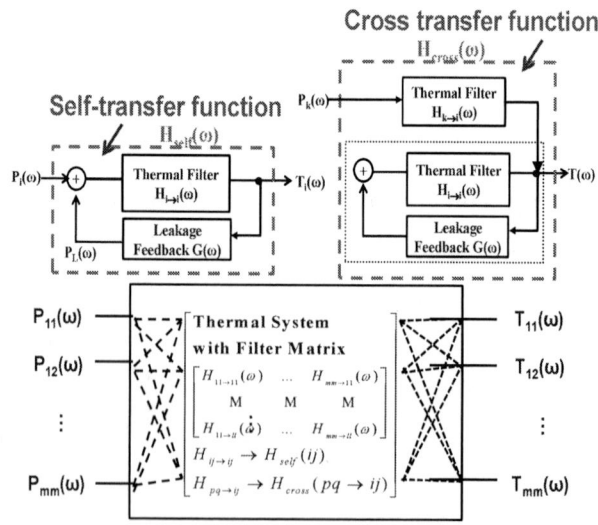

**Figure 3:** Mathematical principle of the proposed approach. The thermal system is considered as a MIMO system.

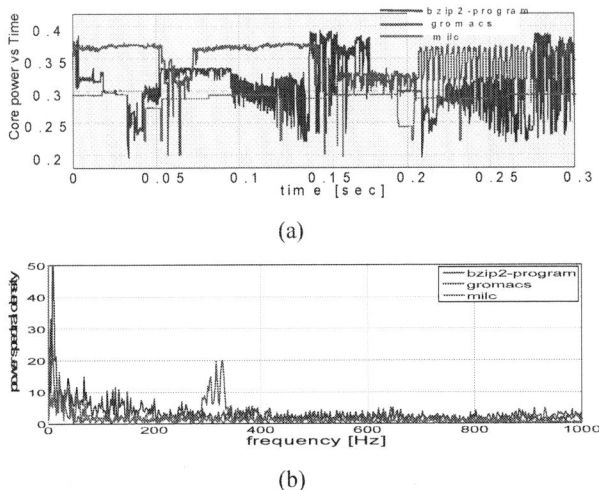

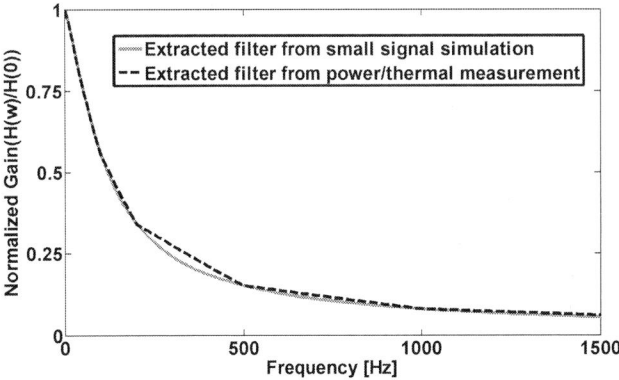

**Figure 6:** The thermal filter extraction through small signal simulation (ideal approach for filter extraction) and sleep control based power/thermal measurement.

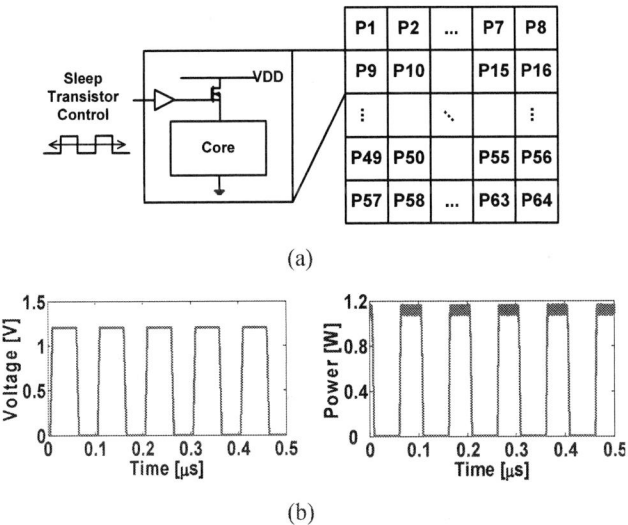

**Figure 4:** (a) Transient power traces of exemplary benchmarks for SPEC2006 applications (b) The frequency response of the power traces.

**Figure 5:** Core gating based approach to power spectra generation. (a) Sleep transistor signal (5 MHz) and (b) power pattern (5 MHz).

Therefore, the MIMO thermal system for many-core has power of each core as an input and temperature of each core as an output.

### 4.1. Baseline Thermal Simulator used for Verification of the Proposed Approach

We first describe the baseline thermal simulation platform used to verify accuracy of the TSI based approach. We consider 3D model of the thermal system including chip, TIM, heat spreader, and heat sink. 3D distributed RC grid is generated for the different regions of the system. We use circuit simulator, HSPICE, for solving the distributed RC grid in time-domain. The power profiles are applied as current sources. The chip is modeled as a homogenous 64 core processor with private cache designed in predictive 22nm technology (total chip area 400mm$^2$, each core and private

cache ~6.25mm$^2$). Each core was modeled as close to Intel Nehalem architecture [14] running at 3.0GHz. We generate power traces of SPEC 2006 benchmark suites using cycle-accurate architecture simulation for timing (Zesto [23]) and power (McPAT [24]) considering x86 architecture. Each benchmark was run or repeated for 0.5 seconds in real time. The above environment considers architectural inputs (e.g, cache sizes, instruction decode width, number of execution units, etc.) and device parameters at various technology nodes to estimate the physical features of the processor. The example power traces obtained from the simulation are shown in Fig. 4.

### 4.2. Thermal System Identification for Many-Core

The practical challenge in TSI of many-core processors is the generation of power spectra in equation (7). The accurate approach is to apply sinusoidal power waveforms of different frequency (small signal analysis). However, generating sinusoidal power waveform in hardware (in a chip) is challenging. We propose two alternative approaches. *Power Spectra Generation with Core-Gating Control:* First, we propose to control the core level power and clock gating (i.e. core-gating available in current processors [25-26]) to generate power pattern of desired frequency spectra. To illustrate this approach we perform SPICE simulation considering core gating (Fig. 5). We consider the core as hundreds of 15-stage ring oscillators to emulate dynamic power. Each core is controlled with a periodic sleep control signal of a given frequency which generates periodic power pattern of same frequency. Hence, by controlling the period of the sleep control signal we can modulate the spectral behavior of the generated power patterns. The on-chip power monitors can be used to sense the core level power [27]. *Application Driven Power Spectra*: The second approach is to run multiple test applications in individual cores and measure power and temperature to compute the filter response. As power profile generated by each application may not contain significant spectral power at all frequency, we consider average of the filter responses computed using different applications as the extracted filter.

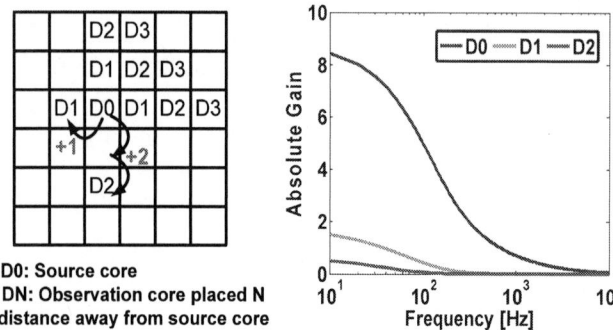

D0: Source core
DN: Observation core placed N
distance away from source core

**Figure 7:** Filter behavior of thermal system: distance between source core and observation node.

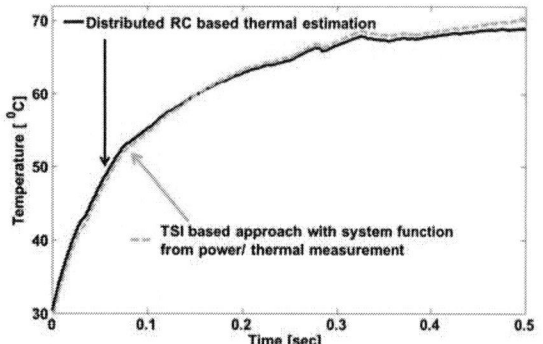

**Figure 8:** Estimation error in transient variation of temperature for a typical core in the 64 core system. The simulations were performed considering random workloads created for all 64 cores using random assignments of benchmark applications for SPEC2006 suites.

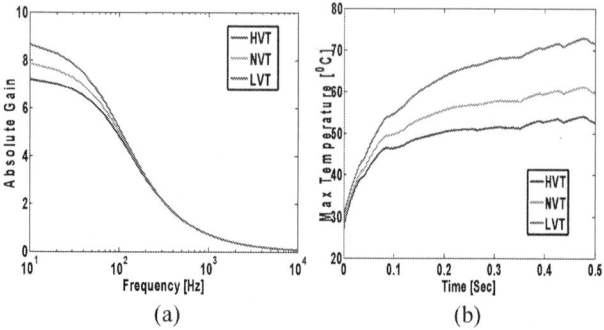

(a)      (b)

**Figure 9:** The application of TSI based approach on the prediction of impact of process variation on transient temperature: (a) the effect of leakage-temperature interaction and (b) time-domain temperature variation

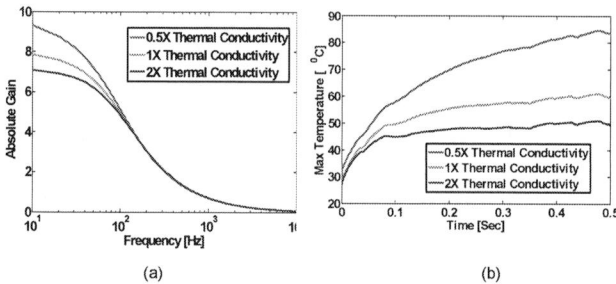

(a)      (b)

**Figure 10:** The application of TSI based approach on the prediction of impact of the conductivity of thermal stack (TIM, spreader, and heat sink) variation on transient temperature: (a) the effect of thermal conductivity in the extracted filter (b) time-domain temperature variation

Figure 6 shows the thermal filter extracted using the practical core-level control closely follows the one from the theoretically ideal small-signal analysis. We observe that the thermal systems behave as the 1st order low-pass filter. The cutoff frequency is located in the low frequency range. Hence, fast time-varying power input has less impact on the temperature while low frequency power variations are more critical. We next study the behavior of the extracted core-to-core cross thermal filters. Fig. 7 shows frequency responses for different location of interest when a power source is applied at core D0. We observe that both self (D1) and cross (D1, D2) transfer functions behave as a low-pass filter. The strength of the cross transfer function reduces significantly with distance i.e. power spread in the distant cores will have minimal impact on the temperature of a core. We also see that the effect of cross transfer function is even less pronounced at higher frequency. We observe that gain at the observation point continues to decrease in all frequency range as it moves away from the source. The decrease in gain due to spatial effect is larger at higher frequencies i.e. fast varying power source has less impact on neighboring regions. We further note that the filter response between a source and an observation node depends on the physical property of the material system that determines the heat flow. It is independent of the magnitude of the generated power, floorplan of the chip, and architecture. The latter factors modulate the power profile and hence, temperature profile but

not the filter response.

### 4.3. Accuracy of TSI based Thermal Prediction

We verify the accuracy of the post-silicon TSI based thermal models against the distributed RC based thermal simulator described in section 4.1. We first create several (60) workloads by randomly assigning the power trace of different application (0.5s of real time data) to different cores and use them for thermal analysis. The same patterns were also run through the baseline distributed RC based thermal simulator. Figure 8 compares the transient temperature variations for a typical core generated from distributed RC based simulation and the proposed approach (with power/thermal measurement driven filter). It can be observed that transient variation is well captured.

### 5. Application to Post-Silicon Thermal Prediction

#### 5.1. Capturing the effect of Process Variations and TIM Conductivity on Thermal Prediction

After verifying the accuracy of TSI based thermal prediction, we next study its effectiveness in post-silicon thermal prediction. We study the ability of TSI in predicting the effect of variations in process corners and thermal conductivity. In this analysis, low-Vt implies a negative 100mV Vth shifts for all devices in a chip while high-Vt implies positive 100mV Vth shifts. The low-Vth dies have

978-1-4673-1110-6/12 $31.00 © 2012 IEEE      122

much higher leakage and stronger leakage temperature interaction. Fig. 9(a) and 10(a) shows that proposed method captures the effect of chip-to-chip variations in leakage and thermal conductivity of the thermal stack consisting of TIM, heat spreader, and heat sink on the extracted thermal filters. We observe that low-Vt die and lower conductivity thermal stack increase the gain in the low-frequency range of the filter transfer function. To illustrate the impact of these variations in filter response, we consider Normal random die-to-die variation of Vth. Each Vth point generated from this Normal distribution represents a unique die for the same many-core processor. For each of such die we consider three different thermal conductivities. TSI is next used to extract the thermal system for all of these die/package condition. The extracted filters for each such instance of the packaged dies are unique. The same workload pattern is applied to all such unique thermal systems to study the effect of process and thermal conductivity variation on chip temperature. Fig. 9(b) and 10(b) show time-domain temperature variation for a typical core for a chip running the same workload but moved to different Vth and thermal conductivity corners.

## 6. Conclusion

We have presented a methodology or post-silicon thermal prediction. The proposed method first identifies the frequency domain response of the thermal system of a packaged die. The extracted filter is used that for fast chip-specific analysis of transient thermal field considering leakage-temperature feedback. The capabilities of post-silicon characterization of the thermal system can benefit thermal design and management at chip as well as large system level.

## Acknowledgment

This work is supported in part by Semiconductor Research Corp (under grant 2084.001), IBM Faculty Award, and Intel Corp. Authors would like to thank Dr. R. Rao, Dr. E. Kursun, Dr. W. Huang, and Dr. P. Bose from IBM Corp. for many helpful discussions.

## References

[1] S. Borkar, "Thousand Core Chips – A Technology Perspective," DAC, 2007.

[2] D. Brooks et. al, "Dynamic thermal management for high-performance microprocessors," HPCA 2001.

[3] T. R. Conrad et. al, "Impact of moisture/reflow induced delaminations on integrated circuit thermal performance," ECTC, 1994.

[4] M. Cho et. al, "Optimization of burn-in test for many-core processors through adaptive spatiotemporal power migration," ITC 2011.

[5] Y. Cheng et. al, "Electrothermal analysis of VLSI system," Kluwer Academic Publishers 2000.

[6] Y. Zhan et. al, "High-Efficiency Green Function-Based Thermal Simulation Algorithms," IEEE TCAD, 2007.

[7] R. Cochran et. al, "Spectral Techniques for High-Resolution Thermal Characterization with Limited Sensor Data," DAC 2009.

[8] ASTM: Standard Test Method for Thermal Transmission Properties of Thermally Conductive Electrical Insulation Materials, Designation D 5470-06, ASTM International, 2006.

[9] A. Poppe and V. Szekely, "Dynamic Temperature Measurement: Tools Providing a Look into packaging and mount structures", Electronic cooling, 2000.

[10] Robert Campbell, "Flash diffusivity method: A survey of capabilities", Electronics cooling magazine, 2002.

[11] S. Y. Kim and R. L. Webb, "Analysis of convective thermal resistance in ducted fan-heat sinks," IEEE Transactions on components, packaging and manufacturing technology, 2006.

[12] K. Kurabayashi and K. E. Goodson, "Precision measurement and mapping of die-attach thermal resistance," IEEE Transactions on components, packaging and manufacturing technology, 1998.

[13] Predictive Technology Model (PTM): http://ptm.asu.edu/

[14] Intel 64 and IA-32 Architecture Optimization Reference Manual, Intel Corp. Nov, 2009, pp. 49-61.

[15] Y. K. Cheng et. al, "An efficient method for hotspot identification in ULSI circuits," IEEE ICCAD, 1999.

[16] Abdullah Nazma Nowroz, Ryan Cochran, and Sherief Reda. 2010. Thermal monitoring of real processors: techniques for sensor allocation and full characterization. In Proceedings of the 47th Design Automation Conference (DAC '10)

[17] T. Kemper et. al, "Ultrafast temperature profile calculation in IC chips," International Workshop on Thermal investigations of ICs, 2006.

[18] D. Schweitzer, "A fast algorithm for thermal transient multisource simulation using interpolated Zth functions," IEEE Transactions on components, packaging and manufacturing technology, Vol. 32, June 2009.

[19] N. Allec et. al, " ThermalScope: Multi-scale thermal analysis for nanometer-scale integrated circuits ", ICCAD 2008.

[20] H. Wang et. al, "Composable thermal modeling and characterization for fast temperature estimation," IEEE EPEPS, 2010.

[21] W. Huang, K. Sankaranarayanan, R. J. Ribando, M. R. Stan, and K. Skadron. "Accurate, Pre-RTL Temperature-Aware Processor Design Using a Parameterized, Geometric Thermal Model" *IEEE Transactions on Computers*, 57(9):1277-88, Sept. 2008.

[22] P. Zhou et. al., "Thermal effects with leakage power considered in 2D/3D floorplanning," IEEE international conference on computer-aided design and computer graphics, 2007.

[23] G. Loh, et. al. "Zesto: A cycle-Level Simulator for Highly Detailed Microarchitecture Exploration," ISPASS 2009.

[24] S. Li, et.al., "McPAT: An Integrated Power, Area, and Timing Modeling Framework for Multicore and Manycore Architecture," IEEE/ACM MICRO, Dec. 2009.

[25] S. R. Vangal, et. al. An 80-Tile Sub-100-W TeraFLOPS Processor in 65-nm CMOS, IEEE JSSC, vol. 43, no. 1,

Jan. 2008, pp. 29-41.

[26] N. A. Kurd, et. al. "Westmere: A family of 32nm IA processors," ISSCC 2010.

[27] N. Mehta, et. al, "In-Situ Power Monitoring Scheme and Its Application in Dynamic Voltage and Threshold Scaling for Digital CMOS Integrated Circuits," ISLPED 2010.

[28] A. Coskun et. al., "Utilizing Predictors for Efficient Thermal Management in Multiprocessor SoCs," IEEE TCAD, 2009.

[29] I. Yeo et. al., "Predictive Dynamic Thermal Management for Multicore Systems," DAC 2008.

# On-Chip Cooling of Hot-Spots with a Copper Micro-Evaporator

Etienne Costa-Patry and John Richard Thome
Heat and Mass  Transfer Laboratory (LTCM)
Ecole Polytechnique Fédérale de Lausanne (EPFL)
CH-1015, Lausanne, Switzerland
john.thome@epfl.ch

## Abstract

Hot-spots are present in micro-electronics and are challenging to cool effectively. Using a copper micro-evaporator mounted on a pseudo-chip, on-chip two-phase cooling was found to very effectively cool the hot-spots without inducing flow instabilities. Building on a flow pattern-based prediction method developed for uniform heat flux conditions, the experimental results could be well predicted.

## Nomenclature

*Latin*

| | |
|---|---|
| c | Constant |
| f | Bubble frequency (Hz) |
| G | Mass flux, based on channel area (kg/m$^2$s) |
| q | Heat flux (W/m$^2$) |
| T | Temperature(°C) |
| x | Vapor quality |

*Greek*

| | |
|---|---|
| α | Heat transfer coefficient (W/m$^2$K) |

*Subscripts*

| | |
|---|---|
| AF | Annular flow regime |
| b | Base |
| ftp | Footprint |
| R1 | Row 1 |
| R2 | Row 2 |
| ref | Reference |
| sat | Saturation |
| w | Channel wall |
| 3z | Three-zone model |

## 1.  Introduction

Many electronic devices, such as CPUs or IGBTs, generate a non-uniform heat flux, with hot-spots several times greater than their background heat flux and high transients in these heat flux. A micro-evaporator mounted on such a chip needs to be able to handle these conditions, because micro-coolers will be positioned as close as possible to the heat source to decrease the overall thermal resistance of the package.

This work presents the results of an experimental investigation of the behavior of two-phase flow cooling in a copper micro-evaporator using a test setup mimicking the actual thermal behavior of a CPU. The objectives were to assess the response of a two-phase multi-microchannel copper evaporator to non-uniform and transient heat fluxes and to analyze the interactions between different power maps and the two-phase cooling system.

## 2.  Experimental setup and data reduction

The test section used for the experiments can be divided into four main layers: the thermal chip, the thermal interface material (TIM), the copper evaporator and manifold. A schematic of the test section is shown in Fig 1. The thermal chip was designed to be able to mimic the behavior of a computer chip and make local temperature and heat transfer coefficient measurements. It was composed of 35 independently controlled heaters arranged in a 5×7 array. Each heater was 2.54mm × 2.54mm in size, had its own temperature sensor and its electrical resistance was of about 25Ω. Thus both a junction temperature map and a base power map were simultaneously recorded.

Between the thermal chip and the  evaporator, a liquid metal thermal interface material (TIM) with a thermal conductivity of about 20W/mK was used. The copper evaporator itself was composed of 52 channels, each 163μm wide and 1560μm high, with fins of 178μm thick, fabricated in collaboration with Wolverine Tube Inc for these tests. The hydraulic diameter was measured to be 246μm by image processing of the channel cut view.

To study the response of two-phase flow to non-uniform heat flux, the conditions found at the root of the fins ("footprint") are needed. The heat flux found at "footprint"

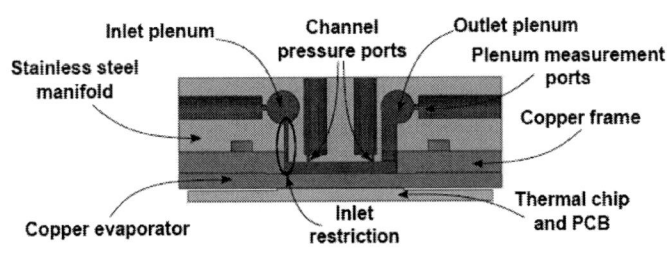

(a) Cut view along the channel axis. Blue: Sections filled with refrigerant.

(b) Cut view of the microchannels

**Figure 1 Description of the test section**

978-1-4673-1110-6/12 $31.00 © 2012 IEEE

level in the evaporator ($q_{ftp}$) differed importantly from what was found at the base of the package ($q_b$), due to thermal conduction ``spreading'' effects inside the package. A multi-dimensional thermal conduction scheme was used to calculate the "footprint" heat flux and temperatures. This scheme has been already described in detail in Costa-Patry et al.[1]. Before proceeding to two-phase flow experiments, liquid single-phase experiments were performed on the facility and the results were in excellent agreement with accepted prediction methods, as detailed in Costa-Patry et al.[2].

## 3. Experimental results

Three fluids (R-134a, R-1234ze(E) and R-245fa) were tested at two saturation temperatures, 30°C and 50°C and at mass fluxes from 205 to 569kg/m²s. The heat flux was incrementally increased until the base temperature reached 85°C and were set to values going from 450 to 4400 kW/m². No attempt was made to reach the critical heat flux, which was avoided to prolong the life of the test section. Seven different hot-spot configurations were studied and are presented in Fig 2, where the flow direction is indicated. They can be divided in three groups: single heater hot-spots, row hot-spots (same heat flux for all heaters in one row) and column hot-spots (same heat flux for all heaters in one column). To present the results, the information about the test conditions are given as they would be needed to define the boundary condition in a numerical simulation of the present thermal package, i.e. $\alpha_{ftp}$ will be plotted for different $T_f$ and $q_b$. A typical example of the impact of a non-uniform heat flux on the base of the chip is given in Fig 3. For a peak heat flux over 3000 kW/m², the base temperature remained everywhere below 70°C. Note that even for such heat flux levels, losses remained below 3% of the total heat flux.

**Heater XY**

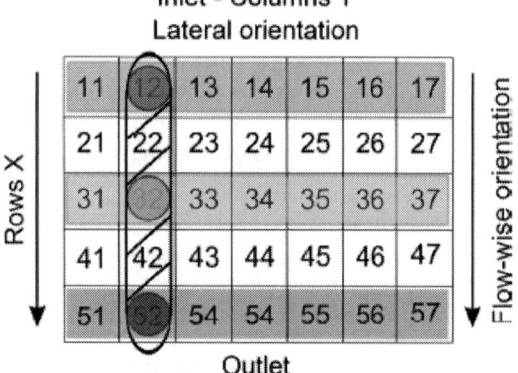

**Inlet - Columns Y**
**Lateral orientation**

**Figure 2 Power map configurations**

Each component of the pressure drop in the test section was directly measured, as explained in Costa-Patry et al.[2]. In non-uniform heat flux situations, the evaporation rate changes over the hot-spot and a modified version of the annular flow model of Cioncolini et al.[3], combined to that of Lockhart and Martinelli[4], was used to calculate the local fluid pressure drops and hence the local saturation temperatures, since it was found to be the most accurate in

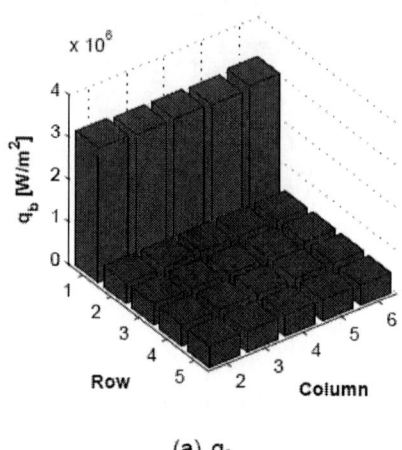

(a) $q_b$

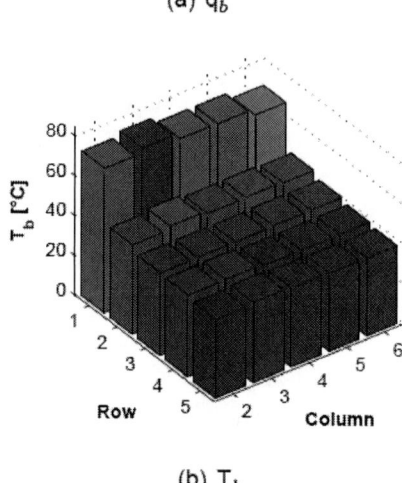

(b) $T_b$

**Figure 3 Power and temperature map for R-1234ze(E), $T_{sat}$=30°C, G=350kg/m²s**

Costa-Patry et al.[2]. Overall the pressure drop across the evaporator was less than 10kPa, which corresponds to a fluid saturation temperature fall of around 0.5°C.

In the case of a row hot-spot, the increase in pressure drop due to the hot-spot will be uniform in all channels and the mass flux will be uniform across all channels since the flow path is normal to the row hot-spot. In the other power configurations, the different evaporation rates and vapor quality profiles will force the mass flux to vary in the parallel channels in order maintain the same differential pressure drop over the test section. Thus in channels running over column and point hot-spot conditions, the mass flux will be lower due to the higher resistance to two-phase, accounted for in the data reduction process. The fluid-side calculations were repeated over each column by changing the local mass flux until the total pressure drop over all columns varied by less than 2.5%.

978-1-4673-1110-6/12 $31.00 © 2012 IEEE        126

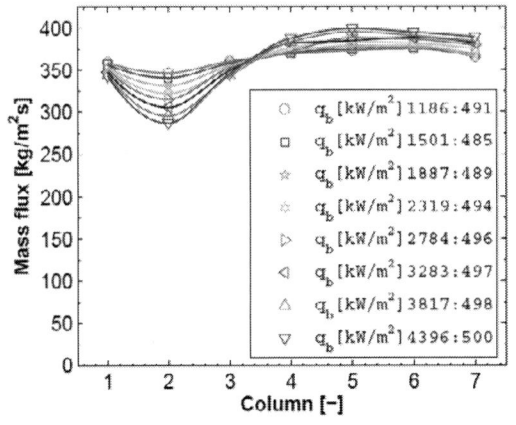 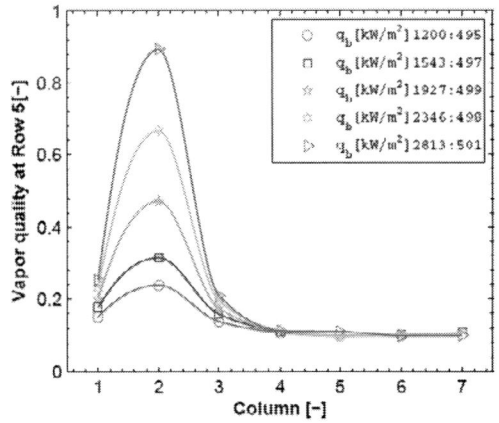

(a) Local mass flux for a hot-spot at Heater 12, G=367kg/m²s.

(b) Local vapor quality taken at Row 5 for a hot-spot at Column 2, G=373kg/m²s.

**Figure 4 Variations due to hot-spots for R-134a at $T_{sat}$=30°C**

The variation of the mass flux across the two evaporators is shown in Fig 4. The mass flux over the hot-spot decreases as its heat flux increases. For a hot-spot at heater 12, $q_b$[kW/m²]4300:500 (i.e. peak $q_b$=4300 kW/m² and background $q_b$=500 kW/m²), the local mass flux represents 78% of the nominal value. This situation could reach dry-out in the channels where the hot-spot is situated. Such cases were encountered in the column hot-spot configuration, as shown in Fig 4, where the calculated exit vapor qualities were close 1, but a temperature run-off, denoting critical heat flux, was not recorded. This means that the heat spreading within the evaporators delayed the onset of the critical heat flux.

The heat transfer coefficients for row hot-spots in the copper test section are plotted for R-1234ze(E) in Fig 5 for different combinations of hot-spot and background heat fluxes at G=210kg/m²s. The footprint heat transfer coefficients vary between 80'000 and 190'000W/m²K. Each datapoint corresponds to a row position. In Row 1 hot-spots (left graph), the heat transfer coefficients are also higher in the row next to the hot-spots. Hot-spots modify the local flow regime compared to what would happen if it were formed only by background heat flux. Close to the inlet, the hot-spot will be cooled by a slug flow, which reacts positively to an increase in heat flux. The increase in heat transfer over Row 2 can be tentatively explained by looking at the mechanism described by the three-zone model of Thome et al.[5]. The bubbles are generated at the inlet where the hot spot is, so that the frequency is higher than if it was generated by the background heat flux. By the time the liquid film has reached the minimum film thickness, the heat flux is lower and the dry-out zone does not grow as fast. Thus on average, the local heat transfer coefficient is augmented because the bubble frequency is higher than under uniform heat flux conditions.

In Row 5 hot-spots (right graph) the heat transfer coefficient is continously increasing and reaches 180'000 W/m²K for $q_b$[ kW/m²] 3249:478. In this configuration, the peak levels are almost the same for all hot-spot heat fluxes. However, the heat transfer coefficients are higher for lower background heat fluxes. Towards the outlet, the flow is probably annular and this flow regime is a predominantly a function of the liquid film. As it becomes thinner, the heat

transfer coefficient increases. In such case the heat flux only has an indirect influence on the heat transfer process when it influences the rate of liquid film thinning.

In some cases, heat spreading from the Row 3 hot-spots creates an almost uniform footprint heat flux condition. It was then possible to make in Fig 6 a comparison with the uniform heat flux results presented by Costa-Patry et al.[2]. For each fluid, the heat transfer coefficients for the hot-spot and uniform heat flux situation are close together, often within the margin of experimental uncertainty (about 10%). The largest difference in heat transfer is found for R-134a at the inlet (R3H: 167'900W/m²K, UHF: 145'700W/m²K), a 14% difference.

Thus, no important distinction was observed for local flow boiling heat transfer coefficients in microchannels between uniform and non-uniform heat flux. In both situations, the pressure drop and the local heat transfer coefficient are a combined function of the local mass flux, footprint heat flux, vapor quality and flow pattern. Fig 6 shows that when the conditions at the footprint surface are the same, the heat transfer coefficients for uniform and non-uniform base heat flux are similar. This is an important conclusion since it means that uniform heat flux heat transfer methods can be applied to non-uniform conditions.

The difficulty with non-uniform heat flux lies in calculating the local footprint conditions. In the case of uniform heat flux, the computations are relatively simple because the mass flux can be assumed to be uniform across the evaporator and the footprint heat flux to be equal to the base heat flux (for a one-dimensional thermal conduction approach). The vapor quality is easily determined from an energy balance and the equations for flow pattern transition are often based on the vapor quality. For non-uniform heat flux conditions, a few assumptions need to be made, since in most cases neither the local mass flux nor the footprint heat flux can be directly known (but only backed out by appropriate calculations). In the present experimental setup, iterative calculations were avoided in determining the footprint heat flux by using the direct multi-dimensional thermal conduction scheme, which greatly speeds up the

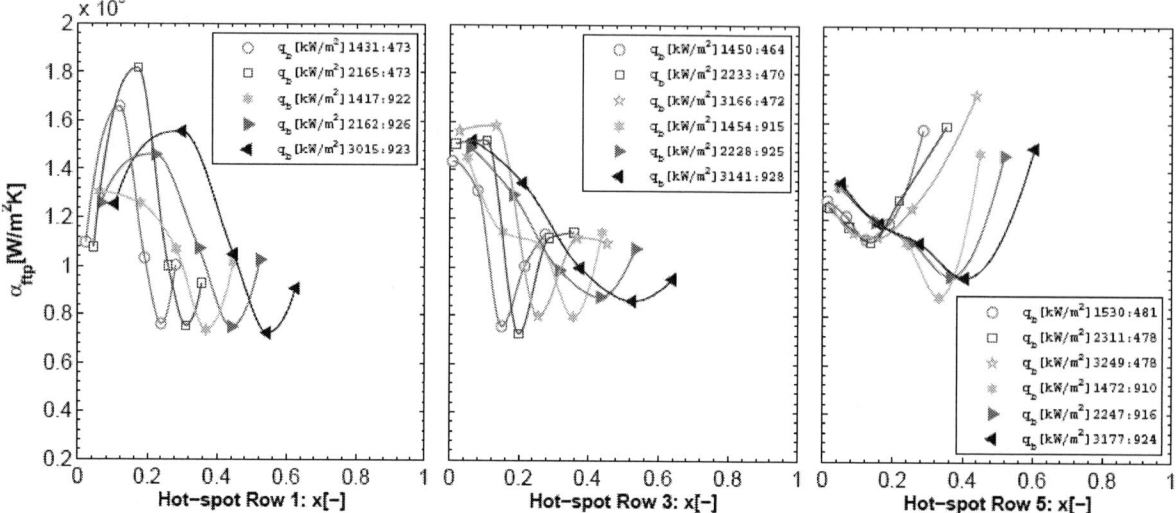

**Figure 5** $\alpha_{ftp}$ for row hot-spots with R-1234ze(E) at $T_{sat}=30°C$

iterative calculation process with respect to a fully numerical analysis.

Although the footprint heat transfer coefficients can be derived from uniform heat flux results, the hot-spot position and orientation remain important factors. It is clearly better to orient ``long" hot-spots perpendicular to the flow direction to diminish the mass flux non-uniformity.

Placing the hot-spot near the inlet or the outlet brings different advantages. Positioned near the inlet, the average heat transfer coefficient will be higher along the entire channel and, as it was discussed by Revellin et al.[6], the critical heat flux value will be higher, but for point hot-spots, the mass flux in the channels running over the hot-spot will be smaller. On the other hand, when the hot-spot is placed near the outlet, the pressure drop will be lower and the mass flux will be more uniform.

## 4. Comparison with prediction methods

Since the characteristics of flow boiling in microchannels do not change locally between uniform and non-uniform heat flux conditions, uniform heat flux prediction methods can extended to non-uniform heat flux situations, although the computation must imperatively be done will the local mass flux and footprint heat flux accounting for the influence of heat spreading. For uniform heat flux (Costa-Patry et al.[2]), two methods were reasonably precise for all fluids and both test sections: the new flow pattern-based method (Costa-Patry et al.[2]) and the Chen-like method of Bertsch et al.[7]. The comparison will be made for the current experimental database and another database of non-uniform heat flux heat transfer coefficients acquired on a silicon micro-evaporator, presented in Costa-Patry et al.[1].

Two modifications were made here to the flow pattern-based method of Costa-Patry et al.[2]. Firstly, since a clear link between the vapor quality and the CB--AF flow pattern transition could not be seen for non-uniform heat flux results, and thus cannot be used, and the criteria for changing from the three-zone to annular flow model was to use the greater of the two values, i.e.:

$$\alpha\_w = max(\alpha\_3Z, \alpha\_AF) \qquad (1)$$

Second, to bring in the increase in heat transfer coefficient over Row 2 when the hot-spot is placed over Row 1, the bubble frequency used in the three-zone model of Thome et al.[5] calculated for Row 1 was used for these power configurations in the downstream calculation for Row 2:

$$f_{R2} = (1-c)\left(\frac{q_w}{q_{ref}}\right)^{1.74} + cf_{R1} \qquad (2)$$

As c increases, the heat transfer over the whole evaporator increases. In the silicon test section, n=0.25 was found to improve the prediction accuracy, whereas in the copper test section, it was better to set n=0, because for high wall heat fluxes, the three-zone model overpredicted the heat transfer coefficients.

As it is listed in Table 1, the flow pattern-based method has a smaller mean average error and places more data within $\pm30\%$ for both database. In the case of the silicon test section, the method of Bertsch et al.[7] was not precise. Overall, the flow pattern method predicts all the data with a mean average error of less than 30%, which is quite good when taking into account the relatively large experimental uncertainty found in non-uniform heat flux situations and the wide range of hot spot heat fluxes and configurations.

**Table 1: Accuracy of prediction methods for the local heat transfer coefficients under non-uniform heat flux.**

|  | Silicon test section | Copper test section |
|---|---|---|
| Berstch et al.[7] | MAE 46% Within ±30%: 18.2% | MAE 27.9% Within ±30%: 55.6% |
| Flow pattern method | MAE 41% Within ±30%: 49.7% | MAE 28.8% Within ±30%: 59.8% |

## 5. Conclusions

In conclusion, the overall cooling was found to be the highest where the hot-spot was positioned. The heat spreading inside the evaporator was found to be important. Its influence was included in the data reduction procedure using a multi-dimensional thermal conduction scheme. Using the local conditions found at the wall, it was possible to show that two-phase flow of refrigerant reacted in the same way to uniform and non-uniform heat fluxes.

The biggest difficulty with a non-uniform heat flux operating condition is that several assumptions have to be made for the flow distribution, such that the local mass flux, wall heat flux and vapor qualities must all be iteratively determined. Once the wall conditions were defined, the flow pattern-based method predicted well the heat transfer coefficients calculated for non-uniform heat flux conditions and is suitable for design of micro-evaporator under uniform and non-uniform heat fluxes.

## References

1. E. Costa-Patry, S. Nebuloni, J. Olivier, and J.R. Thome. On-chip two-phase cooling with refrigerant 85um-wide multimicrochannel evaporator under hot-spot conditions. *Accepted by IEEE Trans. Components and Packaging Tech.*, -:1–10, 2011.

2. E. Costa-Patry, J. Olivier, and J.R. Thome. Heat transfer characteristics in a copper micro-evaporator and flow pattern-based prediction method for flow boiling in microchannels. *Submitted to Frontier Heat Mass Transfer*, -:1–14, 2012.

3. A. Cioncolini, J.R. Thome, and C. Lombardi. Unified macro-to-microscale method to predict two-phase frictional pressure drops of annular flows. *Int. J. Multiphase Flow*, 35:1138–1148, 2009.

4. R.W. Lockhart and R.C. Martinelli. Proposed correlation of data for isothermal two-phase, two-component flow in pipes. *Chem. Eng. Prog.*, 45:39–48, 1949.

5. J.R. Thome, V. Dupont, and A.M. Jacobi. Heat transfer model for evaporation in microchannels. Part I: presentation of the model. *Int. J. Heat Mass Transfer*, 47:3387–3401, 2004.

6. R. Revellin, J.M. Quiben, J. Bonjour, and J.R. Thome. Effect of local hot spots on the maximum dissipation rates during flow boiling in a microchannel. *IEEE Trans. Components and Packaging Tech.*, 31(2):407–416, 2008.

7. S.S. Bertsch, E. Groll, and S.V. Garimella. A composite heat transfer correlation for saturated flow boiling in small channels. *Int. J. Heat Mass Transfer*, 52:2110–2118, 2009.

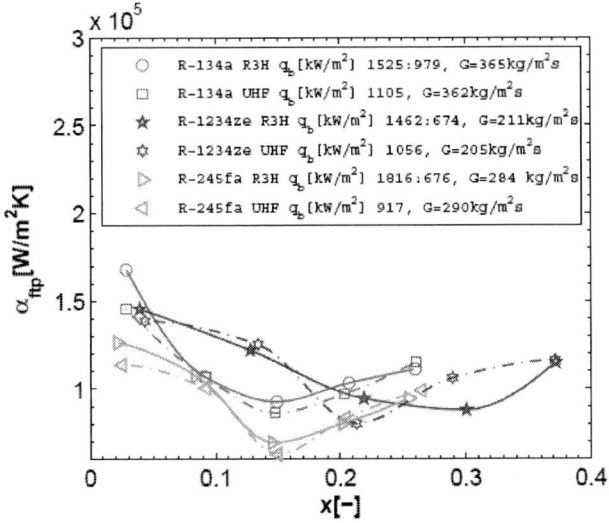

(a) Wall heat transfer coefficient

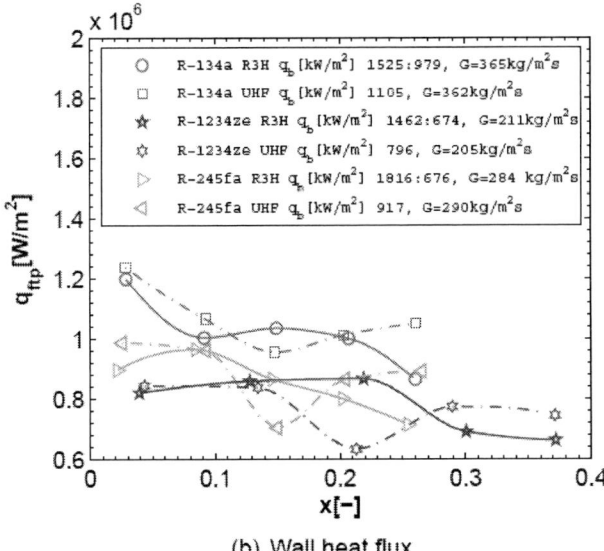

(b) Wall heat flux

**Figure 6 Comparison between some results for Row 3 hot-spots und uniform heat flux results at $T_{sat}$=30°C.**

# Energy Efficient Liquid-Thermoelectric Hybrid Cooling for Hot-Spot Removal

Vivek Sahu[1], Andrei G Fedorov[1], Yogendra Joshi[1]
Kazuaki Yazawa[2*], Amirkoushyar Ziabari[2], Ali Shakouri[2,3]
[1] George W. Woodruff School of Mechanical Engineering, Georgia Institute of Technology
[2] Baskin School of Engineering, University of California Santa Cruz
1156 High St. M/S SOE2, Santa Cruz, CA 95064, USA
[3] Birck Nanotechnology Center, Purdue University
*kaz@soe.ucsc.edu

## Abstract

We report a study on a liquid-thermoelectric hybrid cooling that allows a multiple larger heat flux (>600 W/m$^2$) hotspots on a chip that is never achievable with a reasonable pump power for a microchannel with single phase liquid cooling. Thermoelectric effect is realized in this study by embedding to the silicon chip in superlattice microcooler which has been studied in our previous work. We went through an analytic modeling including spreading resistance through the substrate and modeled the fluid dynamic characteristic of microchannel so that we were able to find the pump power and cooling power of superlattice cooler. We also verified the performance with 3D numerical simulation. The results show that the hybrid system allows much higher heat flux for a hotspot while superlattice cooler locates correctly. As an example, if we have a ZT=0.5 material, a 500μm x 500μm hotspot can be maintained at 85°C (ambient 35°C) with around 850W/cm$^2$ while a simple liquid cooling reaches 620W/cm$^2$ for the same 12W/cm$^2$ of overall cooling power.

## Keywords

liquid cooling, superlattice microcooler, hotspot, energy efficient

## 1. Introduction

With continuous growing of electronics for Information Technologies (IT), hot-spots on a chip are getting more serious challenges for thermal management not only due to the advancement of process technologies called 'More Moore', but also due to the integration in multiple-chip packages. Making enough thermal paths to the heat sink is getting harder and harder by stacking up chips with interconnect enabling technology Through-Silicon-Via (TSV), while this three dimensional exploring approach is called as 'More Than Moore'. Pushing the heat sink thermal resistance lower and lower results a significant increasing of required pump power either for air cooling or liquid cooling. Due to the flow resistance nature, the pump power increase as nearly cube to the cooling power. To adapt this cooling requirement with lower flow rate, we propose following hybrid cooling scheme to maintain the maximum junction temperature below 85°C.

## 2. Hybrid cooling scheme

Figure 1 shows the schematic of the hybrid cooling scheme. Hybrid cooling scheme combines localized solid-state cooling with global liquid cooling to exploit their unique advantages and to overcome the challenges associated with each method [1]. This scheme consists of a microchannel heat sink to remove the background heat flux. Array of superlattice coolers are fabricated at the back of this microchannel heat sink to dissipate the heat from a single or multiple localized hotspot(s).

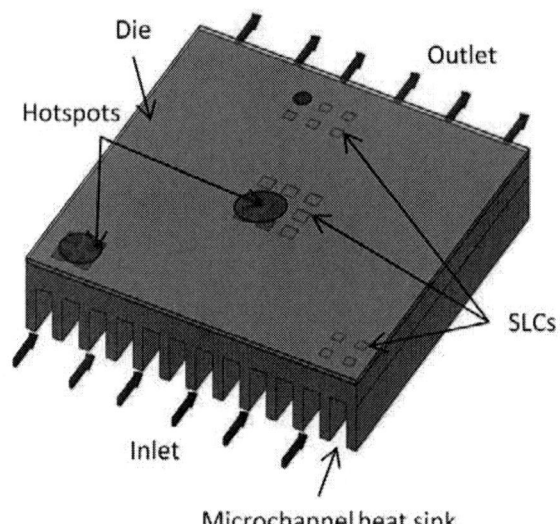

*Figure 1: Schematic of hybrid cooling scheme. SLCs underneath the hotspots are activated. Die is above the microchannel heat sink but shown transparent in the schematic to show the superlattice cooler underneath it.*

Superlattice coolers (SLCs) [2, 3] are solid-state active devices which consist of alternate layers of the epitaxial grow Si and Si/Ge. Each layer is few nanometers thick, and the combined structure is few micrometers thick. Superlattice layer acts as a barrier for electrons flowing from cathode to anode. When electric current is applied to the superlattice, only hot electrons from the cathode, which have sufficient energy to cross the barrier, reach the anode. This creates deficiency of hot electrons in the cathode layer, resulting in cooling at the cathode junction, which is located on the same side as the superlattice cooler. SLCs are silicon micro-fabrication compatible and hence can be directly fabricated on the back of the silicon microchannel heat sink. They are placed at the expected location of hotspots. An array of superlattice coolers can be utilized to manage multiple

978-1-4673-1110-6/12 $31.00 © 2012 IEEE

hotspots, or dynamically shifting hotspots. When a hotspot is detected, SLCs located in the proximity of hotspots are activated providing localized cooling, thus eliminating the hotspots. Heat generated at the ground electrode is rejected to the microchannel heat sink. Thus both the background heat flux and the hotspot heat flux can be managed efficiently.

## 3. Analytic 1-D modeling

Fig. 2 shows the resistance network model to compute the power required to operate the superlattice cooler.

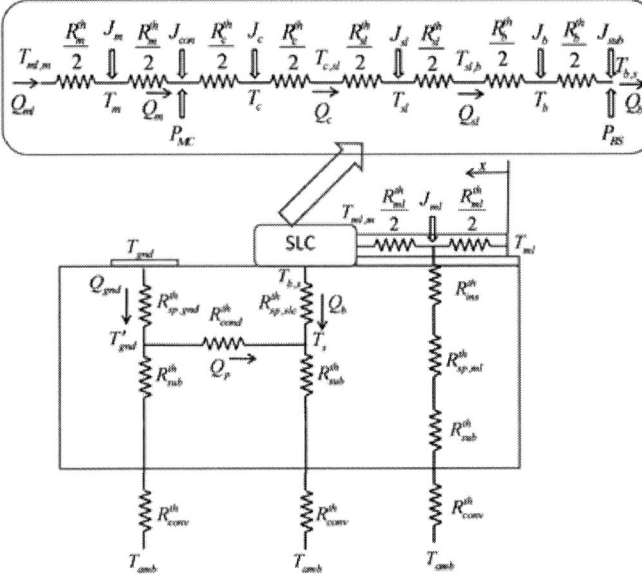

*Figure 2: Resistance network diagram for hybrid cooling scheme.*

It accounts for Peltier cooling, Peltier heating as well as Joule heating in all the layers. Joule heating in the substrate is assumed to be negligible since the substrate has much larger volume compared to SLC device. Joule heating is a volumetric phenomena and which can be modeled in diffusion equation. However, this equation is not suitable to fit in to the thermal resistance network analysis. Hence Joule heating for a particular layer is included by bisecting the layer into two parts and adding a source term at the middle node. Peltier cooling and heating terms are included at the intersection of the layers where they occur.

## Modeling of superlattice cooler

Current and heat flow in the superlattice is assumed to be perpendicular to the surface and spatially uniform. Top metal contact of the superlattice is assumed to be adiabatic, which is justified due to small size of superlattice cooler. Superlattice structure consists of four layers: metal, cap, superlattice, and buffer layer. Temperature difference across any layer is

$$T_{in} - T_{out} = Q_{in} \times R_l\big/2 + Q_{out} \times R_l\big/2 \qquad (1)$$

where $T_{in}$ and $T_{out}$ are the temperature at either side of the layer, $R_l$ is the thermal resistance of the layer, $Q_{in}$ and $Q_{out}$ are the heat going in and going out of the layer, which are related as

$$Q_{out} = Q_{in} + J_l \qquad (2)$$

and

$$J_l = I^2 R_l^e \qquad (3)$$

Here $J_l$ is the Joule heating inside the layer, $I$ is the current flowing through the layer, and $R_l^e$ is the electrical resistance of the layer. Electrical resistance and thermal resistance are defined as:

$$R_l^{th} = \frac{t}{kA} \qquad (4)$$

$$R_l^e = \frac{t}{\sigma A} \qquad (5)$$

In Eqs. (4) and (5) $t$ is the thickness, $A$ is the cross section area perpendicular to heat flow, $k$ is the thermal conductivity and $\sigma$ is the electrical conductivity of the layer. Peltier cooling at the metal-cap and buffer-substrate interface are respectively defined as:

$$\begin{aligned} P_{MC} &= (S_m - S_c)IT_{mc} \\ P_{BS} &= (S_b - S_s)IT_{bs} \end{aligned} \qquad (6)$$

where $S$ is the Seebeck coefficient, and subscripts "m", "c", "b" and "s" stand for metal, cap, buffer and substrate, respectively. $T_{MC}$ and $T_{BS}$ are the temperatures at the metal-cap and buffer-substrate interface, respectively.

Eq (1) can be expressed for each layer in the superlattice structure as seen on Fig.2 and solved for temperature at the top of the superlattice microcooler. Here $T_{ml,m}$ and $T_s$ are the temperatures at the top of the superlattice cooler, and substrate-cooler interface, respectively, $Q_{ml}$ is the heat transport from the metal lead to the top of the superlattice cooler; $J_i$ and $R_i^{th}$ are Joule heating and thermal resistance of layer (subscript $i = m, c, sl, b$ stands for metal, cap, superlattice and buffer layer respectively). $J_{con}$ and $J_{sub}$ are Joule heating source due to finite electrical contact resistance at the metal-cap interface and electrical spreading resistance at buffer-substrate interface, respectively. $R_{con}^e$ is the electrical contact resistance. This analysis assumes electrical contact resistance of $5\times10^{-11}$ $\Omega m^2$. Approximating the superlattice cooler as a cylindrical disk, electrical spreading resistance can be written as:

$$R_{sp}^e = \frac{8}{3\pi^2 \sigma_{sub}} \sqrt{\frac{\pi}{A_{SLC}}} \tag{7}$$

where $A_{slc}$ is the cross sectional area of the superlattice. Thermal spreading resistance from superlattice cooler to the silicon substrate underneath it can be calculated using the expression derived by Yovanovich et al. [4].

$$R_{sp,slc}^{th} = \frac{1}{2a^2 cdk} \sum_{m=1}^{\infty} \frac{\sin^2(a\delta_m)}{\delta_m^3} \varphi(\delta_m) + \frac{1}{2b^2 cdk} \sum_{n=1}^{\infty} \frac{\sin^2(b\lambda_n)}{\lambda_n^3} \varphi(\lambda_n) \tag{8}$$
$$+ \frac{1}{a^2 b^2 cdk} \sum_{m=1}^{\infty} \sum_{n=1}^{\infty} \frac{\sin^2(a\delta_m)\sin^2(b\lambda_n)}{\delta_m^2 \lambda_n^2 \beta_{m,n}} \varphi(\beta_{m,n})$$

where,

$$\varphi(\zeta) = \frac{(e^{2\zeta t_{sub}} + 1)\zeta - (1 - e^{2\zeta t_{sub}})h/k}{(e^{2\zeta t_{sub}} - 1)\zeta + (1 + e^{2\zeta t_{sub}})h/k} \tag{9}$$

and, $a$ and $b$ are length and width of the substrate, respectively $c$ and $d$ are length and width of the heat source (superlattice), respectively, $t_{sub}$ is the thickness of the substrate, $k$ is the thermal conductivity of chip, and $h$ is the heat transfer coefficient.

### Modeling of heat transport in metal lead

Power to superlattice cooler is delivered through metal lead. Even though, metal had two orders of magnitude higher electrical conductivity than silicon ($10^7$ $\Omega^{-1}$m$^{-1}$ as compared to $5 \times 10^4 \Omega^{-1}$m$^{-1}$) due to very high ratio of $l/A_c$, Joule heating in the metal lead is quite significant and cannot be ignored. A large portion of the heat generated in the lead diffuses to the top of superlattice cooler, reducing the cooling obtained at SLC. Heat transport in the metal lead can be calculated by treating it as a fin with constant volumetric heat generation term along with convective heat transfer from the side. Writing down the energy balance for the metal lead yields following governing equation:

$$\frac{\partial^2 T}{\partial x^2} - \frac{h_{eff} P}{kA_c}(T - T_\infty) + \frac{q'''}{k} = 0 \tag{10}$$

where, $k$ is the thermal conductivity, $A_c = t_m w_m$ is the cross sectional area of metal lead with $t_m$ and $w_m$ being the thickness and width of metal lead respectively, $h_{eff}$ is the effective heat transfer coefficient between metal lead and substrate underneath it, $P = w_m$ is the perimeter (through which heat transfer is taking place), $q'''$ is the volumetric heat generation defined as:

$$q''' = \frac{I^2 R_{ml}^e}{A_c l_m} \tag{11}$$

Here, $R_{ml}^e = \frac{l_m}{\sigma_m t_m w_m}$ is the electrical resistance of metal lead, and $l_m$ being the length of metal lead. Yielding the expression for temperature along the metal lead.

$$T(x) = T_{amb} + \left(T_{ml,m} - T_{amb} - m^2 \frac{q'''}{k}\right) \frac{\cosh(mx)}{\cosh(ml_m)}$$
$$m = \sqrt{h_{eff}P \Big/ kA_c} \tag{12}$$

Heat transfer from the metal lead to the superlattice cooler can now be calculated using the following expression:

$$Q_{ml} = -kA_c \frac{\partial T}{\partial x}\Big|_{x=l_m}$$
$$= -\sqrt{h_{eff}PkA_c}\left(T - T_{amb} - m^2 \frac{q'''}{k}\right)\tanh(ml_m) \tag{13}$$

### Modeling of Joule heating in top metal layer

In order to accurately computer the volumetric heat generation in cooling device due to ohmic heating, one needs an accurate estimate of the equivalent 1-D electric resistance of the top metal layer. The current flow in the top metal layer is not strictly one-dimensional, It is flowing along the metal layer as well as perpendicular to its surface into the substrate. Moreover, the magnitude of current is decreasing along the length of the layer due to transport of charge to the superlattice structure underneath it. Since charge transport equation and heat transport equation are analogous, electrical resistance of the layer can be calculated if we can find thermal resistance of metal layer under identical boundary condition. Metal layer can be treated as a fin and heat transport along the layer is given by:

$$\frac{\partial^2 \theta}{\partial x^2} - m_{th}^2 \theta = 0 \; ; \; m_{th} = \sqrt{\frac{h_{eff}^{th} P}{kA_c}} \tag{14}$$

Here $\theta = T - T_\infty$ is non-dimensional temperature (voltage), $h_{eff}^{th}$ is the effective heat transfer coefficient (inverse of leakage resistance). Electrical boundary condition can be converted to analogous thermal boundary conditions to solve Eq. (14). Constant current input at the base of metal layer can be treated as constant heat input to the metal layer. Similarly, no current exists the tip of the metal layer, which can be treated as adiabatic boundary condition.

Solution to Eq. (14) subject to above boundary condition is given by

$$\theta = Q_{ml} \frac{1}{m_{th}}\left(\frac{\sinh\{m_{th}(l_m + l_{ml} - x)\}}{\cosh(m_{th}l_m)}\right) \tag{15}$$

978-1-4673-1110-6/12 $31.00 © 2012 IEEE

where $Q_{ml}$ is the heat input (current) to the metal layer. An equivalent thermal resistance of the fin is defined as:

Using an equivalent thermal resistance of a fin, by the analogy between charge and heat transport we can define the electrical resistance as

$$R_m^e = \frac{l_m}{\sigma_m A_c}\left(\frac{\tanh(m_e l_m)}{m_e l_m}\right) \qquad (16)$$

with

$$m_e = \sqrt{\frac{h_{eff}^e P}{\sigma_m A_c}} \qquad (17)$$

Here $\sigma_m$ is the electrical conductivity, and $h_{eff}^e$ is defined in terms of electrical resistances:

$$h_{eff}^e = \frac{1}{(R_c^e + R_{sl}^e + R_b^e)l_m w_m} \qquad (18)$$

Now, Joule heating in the top metal layer can be expressed as:

$$J_m = I^2 R_m^e \qquad (19)$$

**Parasitic heat transfer from ground electrode**

Portion of heat dissipated at ground electrode is transferred to the superlattice, reducing the cooling at superlattice. The reduction in temperature due to parasitic heat transfer from ground electrode is considered to calculate by solving for the resistance network between superlattice and ground.

**Pumping power**

Power required to pump the coolant in the microchannel is given by

$$\dot{P} = \Delta p \times A_c \times u_m \qquad (20)$$

where, pressure drop is defined as

$$\Delta p = \lambda \frac{\rho u_m^2}{2}\frac{L}{D_h} \qquad (21)$$

where, $\lambda$ is the friction factor, $L$ is the length of channel and $D_h$ is the hydraulic diameter.

## 4. Analysis results

The overall power required to remove the background heat flux in presence of a hotspot using the single phase microchannel heat sink is computed using the resistance network analysis described by Sahu et al. [5]. For the hybrid cooling scheme, the total power is the sum of the pumping power (required for background heat flux removal) and power delivered to SLC (required for hotspot removal). We set the maximum die temperature to 85 °C while the coolant inlet temperature is 35°C. Temperature at the hotspot can be significantly higher than the average temperature, hence to maintain maximum junction temperature below 85°C, total thermal budget (maximum allowable thermal resistance) has to be reduced. The SLC reduces the largest temperature rise at the hotspot and thus relax the thermal budget required for the microchannel heat sink.

Fig. 3 compares the power required for hybrid cooling scheme and that for single phase microchannel heat sink as a function of hotspot heat flux. FC72 is used as a coolant and the microchannel heat sink has the dimensions of 330µm x 55µm x 1cm. Die size is 1cm x 1cm and hotspot area is 500µm x 500µm. The dimension of the microchannel heat sink is optimized for minimum pumping power by using the analysis described in the literature [6, 7]. Background heat flux is fixed to 100W/cm². For lower power density of hotspot, the single phase cooling requires less power than hybrid cooling scheme since the SLC requires additional cooling power and dissipates this additional heat. As hotspot power density increases, the required pump power increases rapidly for single phase convective cooling due to increase in the volumetric flow rate and the SLC helps to shift the heat flux this happen to higher range. Thus, total power consumption of the hybrid cooling scheme shows less power consumption at higher heat flux. Superlattice coolers are assumed to be present at the location of hotspot. Hence all the power dissipated by the hotspot is been dissipated by the superlattice cooler. However, if the power dissipated at the hotspot is more than what superlattice cooler can remove, it is dissipated by the microchannel heat sink. Since COP of the superlattice cooler is low (typically in the range of 0.3-0.5), for lower hotspot heat flux hybrid cooling scheme requires more power compared to single phase. However, there is a transition hotspot heat flux after which single phase requires more power compared to hybrid cooling scheme.

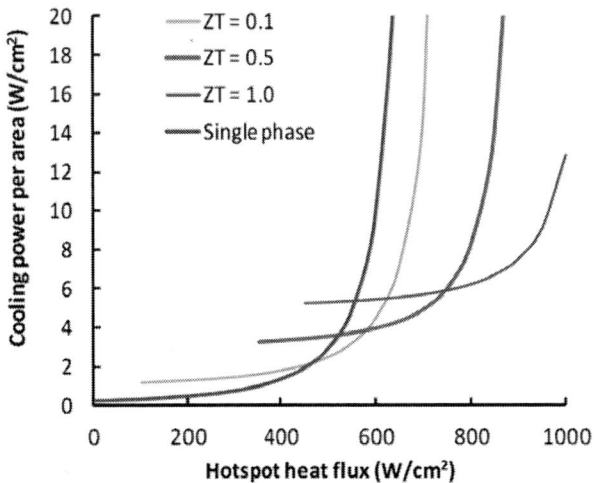

*Figure 3: Comparison of power requirement for hybrid scheme and single phase scheme as a function of hotspot heat flux. Hotspot size is 500 µm x 500 µm.*

978-1-4673-1110-6/12 $31.00 © 2012 IEEE        133

Moreover, hybrid cooling scheme is able to dissipate more power as compared to single phase cooling. For example, the single phase cooling alone removes the maximum hotspot heat flux of approximately 600W/cm² when we limit the pump power to 10W (10% of chip power). Under the same conditions, hybrid cooling scheme (with ZT =2.0) consumes the pump power only around 25% of the single phase cooling. As the dimensionless figure-of-merit (ZT) of the SLC increases, the hybrid cooling scheme increases the allowable hotspot heat flux. If the ZT=2.0 of the superlattice device is available, the hybrid cooling scheme dissipates more than 1000W/cm² hotspot for same cooling power.

Fig. 4 shows a 3D numerical analysis results at 600W/cm² on a hotspot. The superlattice cooler locates just underneath the hotspot. To make the correct spreading resistance to the hotspot, the 500μm silicon substrate is modeled as a flat rectangular shape. The temperature contour shows that heat pumping by superlattice cooler makes a cooling effect on the hotspot surface (top surface of the picture). Due to the heat pump effect, the chip side substrate temperature exceeds 85°C. This mechanism works until reached to the maximum heat pumping condition, which is depending on the ZT of the superlattice cooler.

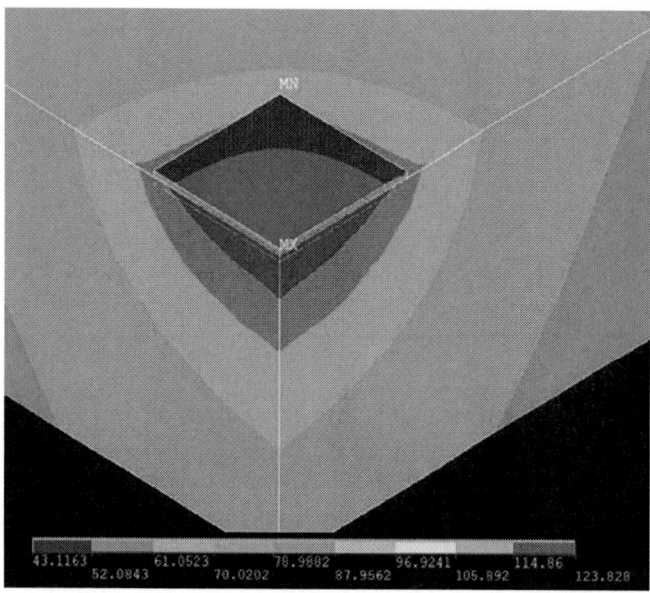

*Figure 4. Temperature contour plot of a quarter section of the model showing the numerical analysis result.*

## 5. Conclusions

We demonstrated the performance of this hybrid scheme for a hotspot size in comparison of the single phase background cooling alone. We observe that the larger advantage of the hybrid liquid-thermoelectric hybrid cooling is found at the higher heat flux of hot spots. There is a breakeven point of heat flux for operating SLC and it can be found analytically. Until hitting the point, the chip cooling can be performed only by liquid cooling. Then beyond the point, operating SLC provides the better COP for cooling. Moreover, hybrid cooling scheme can dissipate much higher heat flux as compared to single phase microchannel heat sink. By improving the ZT of the superlattice material, hybrid cooling scheme can be designed to remove heat flux in excess of 1000W/cm² from 500μm x 500μm hotspot.

## Acknowledgment

The authors acknowledge the support of the Interconnect Focus Center, one of five research centers funded under the Focus Center Research Program, a Semiconductor Research Corporation program.

The authors also acknowledge the contribution of Dr. Xi Wang and Dr. Je-Hyeong Bahk on their efforts on the superlattice cooler design, analysis, fabrication, and tests. These are the fundamental of this work.

## References

1. Sahu, V., Y.K. Joshi, and A.G. Fedorov. *Experimental investigation of hotspot removal using superlattice cooler.* in *12th IEEE Intersociety Conference on Thermal and Thermomechanical Phenomena in Electronic Systems (ITherm).* 2010. Las Veags, NV, United States: IEEE.

2. Shakouri, A., *Nanoscale thermal transport and microrefrigerators on a chip.* Proceedings of the IEEE, 2006. **94**(8): p. 1613-1638.

3. Shakouri, A. and J.E. Bowers, *Heterostructure integrated thermionic coolers.* Applied Physics Letters, 1997. **71**(9): p. 1234-6.

4. Yovanovich, M. M., Muztchaka, Y.S. and Culham, J.R., *Spreading Resistance of Isoflux Rectangles and Strips on Compound Flux Channes,* Journal of Thermophysics and Heat Transfer, 1999, **13**(4), p. 495-500

5. Sahu, V., Y.K. Joshi, and A.G. Fedorov, *Hybrid solid state/fluidic cooling for hot spot removal.* Nanoscale and Microscale Thermophysical Engineering, 2009. **13**(3): p. 135-150.

6. Kleiner, M.B., S.A. Kuhn, and K. Haberger, *High performance forced air cooling scheme employing microchannel heat exchangers.* IEEE Transactions on Components, Packaging, and Manufacturing Technology, Part A, 1995. **18**(4): p. 795-804.

7. Knight, R.W., J.S. Goodling, and B.E. Gross, *Optimal thermal design of air cooled forced convection finned heat sinks-experimental verification.* IEEE Transactions on Components, Hybrids, and Manufacturing Technology, 1992. **15**(5): p. 754-60.

# Test ASIC for Investigation of Thermal Coupling in Many-Core Architectures

Michal SZERMER, Cezary MAJ, Piotr PIETRZAK, Marcin JANICKI, Piotr ZAJAC, Andrzej NAPIERALSKI
Department of Microelectronics and Computer Science, Technical University of Lodz
Wolczanska 221/223, building B18, 90-924 Lodz, Poland
E-mail: michal.szermer@p.lodz.pl; tel: +48 42 631 2722; fax +48 42 636 0327

### Abstract

This paper presents the design of a test ASIC, which was intended for the investigation of thermal coupling in Many-Core Architectures. Particular sections of the paper describe in detail the design concept and simulated operation of the ASIC which is currently sent for manufacturing.

### Keywords

Many-Core Architectures, Thermal Coupling, Test ASIC.

## 1. Introduction

Today processors are designed as multi-core architectures containing 4 or 6 cores. It is predicted that in near future the number of cores integrated in a single chip might exceed 1000 [1]. Such a tendency leads to constant increase of total power dissipated in a chip. Moreover, due to small spacing between cores the thermal coupling among them also increases [2] significantly. In order to investigate this phenomenon in more detail, the authors decided to design a test ASIC containing numerous heat sources, which are placed regularly in a single semiconductor chip.

The main idea behind the project was to investigate how particular cores interact with each other during the realization of some complex calculations. However, the design of a real multicore architecture is difficult and expensive to realize, thus the authors developed a test ASIC in an older technology, in which large heat sources represent particular functional processor blocks. In this way, it is possible to emulate the thermal behavior of state-of the multicore processors based on the analysis of a chip manufactured in a much less expensive technology.

## 2. Design concept

The main part of the design occupies a regular heat source matrix. Each source should be independently controlled and its temperature can be read out at a relatively high frequency. Consequently an extremely flexible architecture is obtained where individual heat sources can be assigned specific power dissipation values. These values could be changed in real time during the operation of the circuit. Moreover, owing to the use of temperature sensors integrated with heat source it should be possible to record detailed dynamic temperature maps of the entire chip.

This versatile structure allows thermal modeling of any VLSI chip provided that power dissipation in particular chip components is known. For processors, as shown in Fig. 1, this data could be obtained from power simulators, which generate power trace files with detailed information on instantaneous power dissipation. Such a power trace file can be used then in the proposed ASIC to generate a dynamic temperature map.

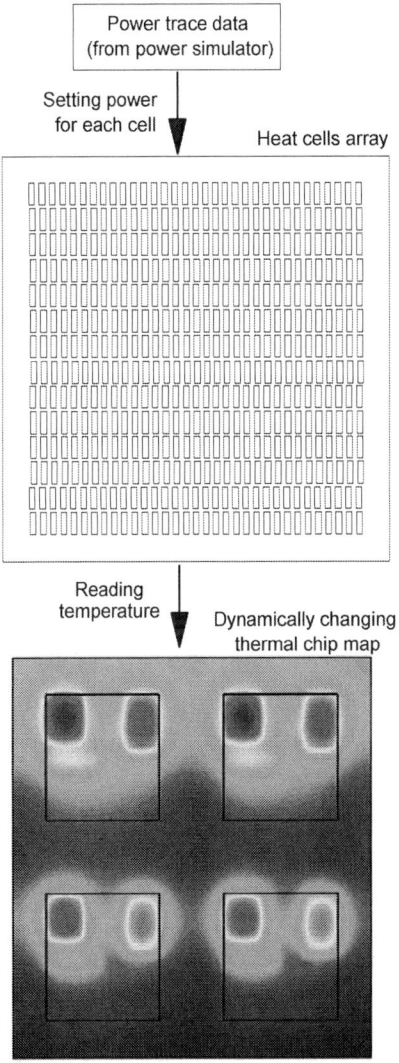

**Figure 1:** Idea of thermal simulations with a configurable heat source matrix.

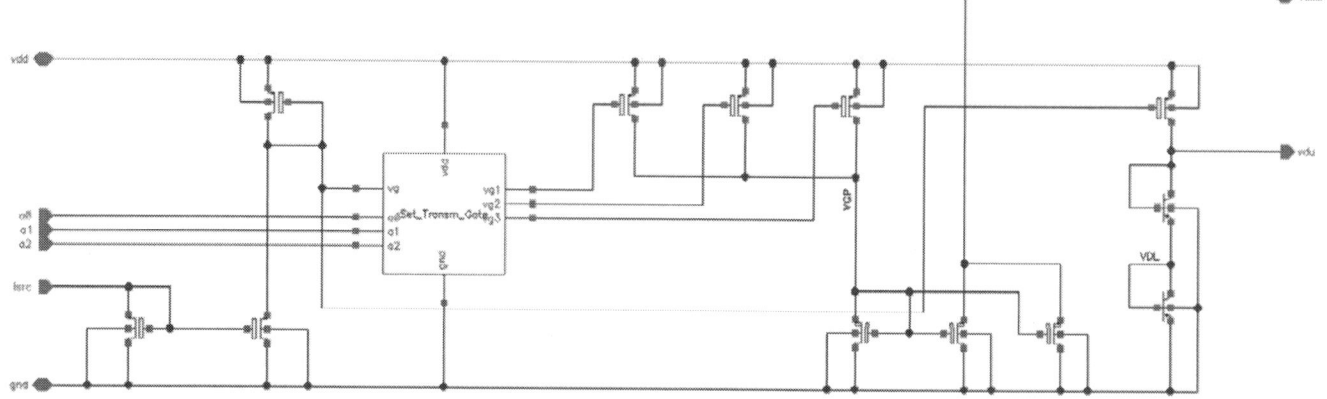

**Figure 2:** Detailed schematic of a heat cell.

## 3. Practical realization

The idea of this ASIC is based on the previous experience, when the authors designed the circuit containing overlapping matrices of 9 heat sources and 25 diode temperature sensors. The detailed description of this work can be found in [3]-[5]. The current ASIC was designed in the 0.35 μm CMOS high voltage technology provided by *austriamicrosystems*® (AMS). This chip contains an array of 384 heat cells organized in the 16 x 24 matrix. The cell dimensions are 320 μm × 160 μm. The total active chip area covered by these heat cells is equal to 6 mm × 6 mm, which is comparable to the area of a real processor. Owing to this approach, it is possible to emulate temperature distribution in multicore chips using the proposed ASIC.

The heat cell, whose detailed schematic is shown in Fig. 2, consists of two large power NMOS transistors, a few current mirrors and diodes. The transistors form the actual heat source whereas the diodes, located in the middle of the transistor layout serve as temperature sensors. Owing to this solution, the diode sensor measures chip temperature very close to the heat source. The current mirrors switch the power transistor according to the state of cell control bits, thus allowing the selection of eight current levels in the range of 0÷8 mA. Moreover, the transistor power supply voltage can be varied between 20 and 50 V. As a result, the power dissipated in the transistor can be adjusted almost fluently from 100 μW in the idle state up to 0.4 W. Moreover, taking into account that the entire chip contains 384 heat cells, very high flexibility of the power dissipation pattern can be achieved.

The temperature is measured, as mentioned previously, by diode sensors associated with each source. However, because of the limited number of available pins, the cells are organized and multiplexed as groups of four, forming 96 cells as the one shown in Fig. 3. Owing to this solution, it is possible to read the output signals from temperature sensing diodes using the dedicated measurement card equipped with only 96 analog channels. This measurement card renders possible recording of thermal transients at the rate of 50 ksps in each channel.

The 3-bit registers of each heat cell in a group form one 12-bit FIFO register. Their values can be set by sequentially clocking the data in. Each 3-bit register is connected to a 3-bit latch. After loading data to all the registers, the digital signals from the D-FlipFlops are copied to the D-Latches at a time instant determined by an external signal common to all heat cells. The particular values written into the D-Latches control the current mirrors switching the heat cells off or on at desired power levels.

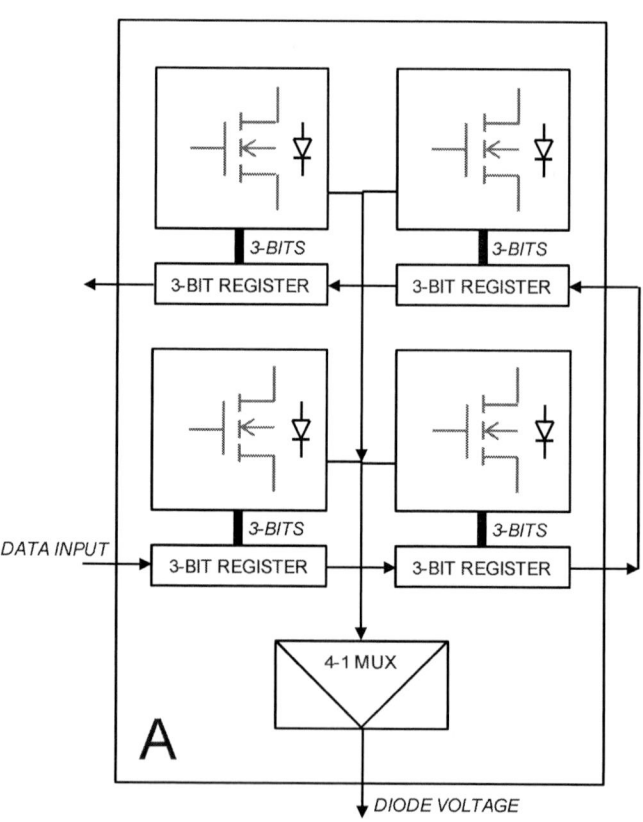

**Figure 3:** Block diagram of heat cell grouping.

978-1-4673-1110-6/12 $31.00 © 2012 IEEE

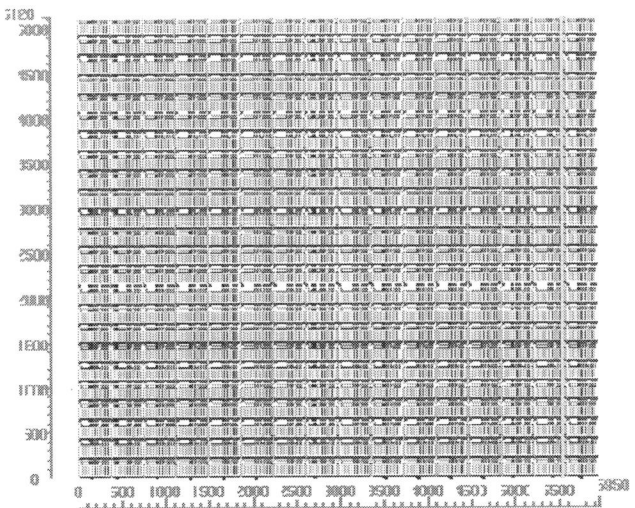

**Figure 4:** Heat cell array layout.

An additional advantage of the proposed solution is the complete elimination of digital clocking signals which could be an important source of interferences during measurements. Moreover, only one input pin is used to set the data in all 384 cells, thus reducing significantly the number of required pins. Obviously, the sequential transmission of bits could be very time consuming, however the authors estimated that for the considered purposes the practically achievable bit rate is more than sufficient.

### 4. Simulated performance

The selected post layout simulations presented in Figs. 4-5 fully proved the correct operation of the circuit. The below figure shows the power transistor drain currents for different values of the heat cell control bits. As can be seen, the widths of transistor in the current mirrors were properly chosen and the current levels are evenly spread in the considered range of transistor currents.

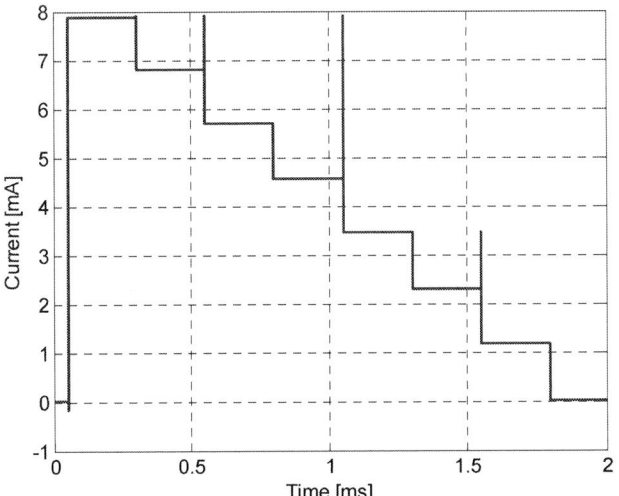

**Figure 5:** Simulated power transistor current levels.

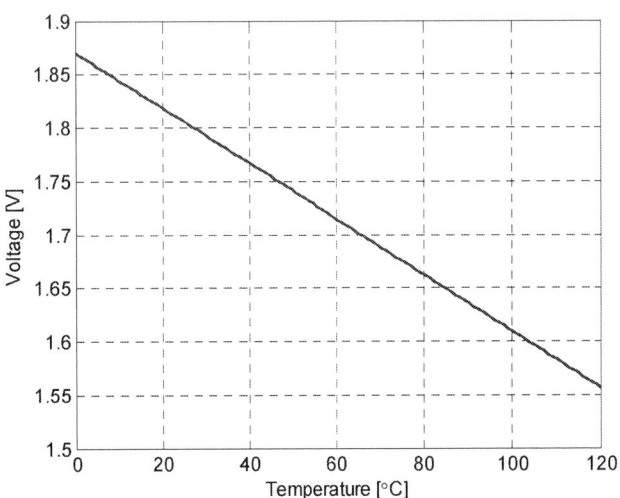

**Figure 6:** Simulated temperature sensor characteristics.

The simulated temperature sensor characteristic is shown in Fig. 5. As can be seen, in the range of temperatures from 0 °C to 120 °C the sensor output voltage gradually decreases from 1.86 V to 1.56 V with the negative slope -2.58 mV/K, which is close to theoretical predictions.

### 5. Experimental verification

Because the main application of the designed ASIC is the investigation of thermal coupling in many-core architectures realized in various technologies, the dimensions of heat cells were chosen expressly to match the sizes of various functional blocks in real processor chips. An example illustrating this idea is given in the below figure presenting the core locations in the 32 nm technology and the future 16 nm technology.

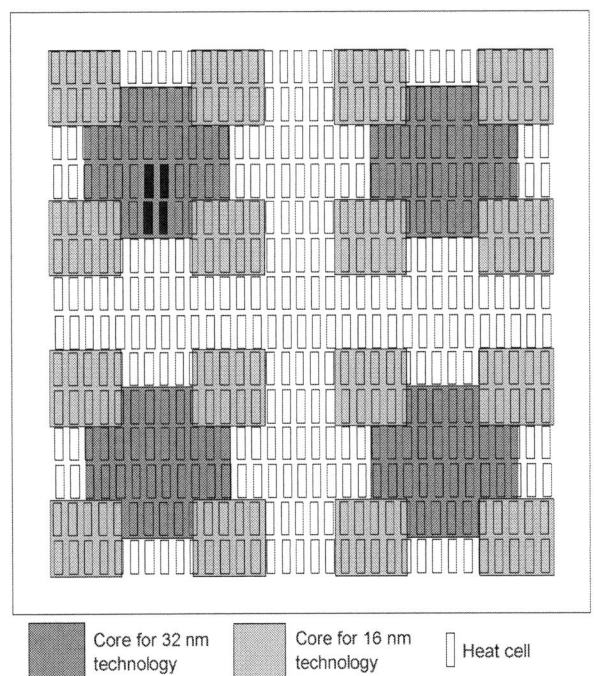

**Figure 7:** Sample core mapping in ASIC layout.

Therefore, as can be seen, real processor floorplans have to be divided into basic functional blocks and mapped onto the test ASIC layout and assigned to its particular heat cells. Then, for the 4 core architecture manufactured in the 32 nm technology each core is mapped on the rectangle composed of 4 x 10 heat cells, which could correspond to individual core components such as ALUs, L1 cache, register file or branch predictors. The same core realized in the 16 nm technology will be assigned only to 10 heat cells and the number of cores in the chip will increase to 16. The area between the cores can be considered as occupied by the shared L2 cache memory.

Once the topology mapping is done, the power trace data obtained from different benchmarks run on simulators could be assigned to individual heating cells. The simulators may be configured in such a way that they provide the power data for each processor block every 100,000 instructions, thus allowing the preparation of the input data sequence for the ASIC.

Let us describe an example for clarity. Let the four black rectangles in Fig. 7 correspond to the ALU of the first core in the quad-core chip. Using the power simulator, we obtain the power dissipated in this ALU in a given time period, for example 2 W. Consequently, in our test chip we set the four heat cells to dissipate exactly 2 W. The same method is used for all processor blocks and for the entire simulation time . Simultaneously, the temperature data from our chip is read, which allows obtaining the dynamic thermal maps of the processor during the execution of the benchmark. Such thermal simulations can be repeated for different configurations, technologies, cooling conditions or benchmarks. It will certainly allow investigating the potential occurrence of hotspots and help us research the problem of thermal coupling, which in our opinion will be especially important in future 16 nm technology.

## 6. Conclusions

The paper presented of a dedicated test ASIC intended for the investigation of core thermal coupling in many-core chips. Owing to the adopted block approach, the same ASIC will allow the investigation of technology migration on the core thermal coupling, which is undoubtedly an innovative feature of the test ASIC.

The thermal measurements of the ASIC will be performed on the special measurement stand consisting of the dedicated multichannel data acquisition and the dual cold plate cooling assembly with Peltier thermo-electric modules allowing active control of cooling conditions [6].

## Acknowledgments

The research is supported by the grant of the National Center of Science No. N515 509140.

## References

1. SKADRON K., STAN M.R., HUANG W., VELUSAMY, TARJAN D., SANKARANARAYANAN, "Temperature-Aware Microarchitecture", Proc. of 30th Intl. Symposium on Computer Architecture, June 2003.
2. JANICKI M, COLLET J.H., LOURI A., NAPIERALSKI A.: "Hot Spots and Core-to-Core Thermal Coupling In Future Multi-Core Architecture", Semiconductor Thermal Measurement and Management Symposium (SEMI-THERM), 2010, 26th Annual IEEE, pp. 205-209
3. SZERMER M., KULESZA Z., JANICKI K., NAPIERALSKI A., "Test ASIC for Real Time Estimation of Chip Temperature", NSTI Nanotech 2008, Hynes Convention Center, Boston, Massachusetts, USA, Vol.3, Jun. 1-5, 2008, pp. 529-532
4. SZERMER M., KULESZA Z., JANICKI M., NAPIERALSKI A., "Design of the Test ASIC for on-line Temperature Monitoring and Thermal Structure Analysis", 15th Int. Conf. Mixed Design of Integrated Circuits and Systems (MIXDES), 2008, Poznan, Poland, Jun. 19-21, 2008, pp. 317-320
5. JANICKI M., SZERMER M., KLAB S., KULESZA Z., NAPIERALSKI A., "Practical Study of Temperature Distribution in a Thermal Test Integrated Circuit", 15th International Workshop on Thermal Investigations of ICs and Systems THERMINIC, 7-9 October 2009, Leuven, Belgium, pp. 136-13.
6. JANICKI M., KULESZA Z., TORZEWICZ T., NAPIERALSKI A., "Automated Stand for Thermal Characterization of Electronic Packages", Semiconductor Thermal Measurement and Management Symposium (SEMI-THERM), 2011, 27th Annual IEEE, pp. 199-202

978-1-4673-1110-6/12 $31.00 © 2012 IEEE

# Socket Thermal Testing Procedure and Correlation

Ted Lee
Intel Corporation
Dupont, WA USA
ted.lee@intel.com

Michelle Lin
Intel Corporation
Taipei, Taiwan
michelle.c.lin@intel.com

## Abstract

CPU socket temperatures have recently been projected to approach, and even exceed, their thermal reliability limits. Up until now, the prediction of socket temperatures was performed only through modeling and simulation. The major drawback of modeling is that it requires a simplification of the motherboard, CPU, and socket construction, estimates of motherboard and CPU substrate power losses, estimates of motherboard and substrate thermal conductivities, and use of pre-silicon CPU powermaps. Ideally modeling methodologies would be supported with test data, however there has been no test data available to correlate with the modeling results. This paper describes the development of a testing methodology that allows for the direct measurement of socket contact temperatures, as well as determining the impact of the motherboard, CPU, and power delivery component temperatures. The test results were then compared to modeling results in order to validate and improve future predictive efforts.

## Keywords

Socket, Thermal, Test, Modeling

## Nomenclature

CFM – Cubic Feet per Minute
CPU – Central Processing Unit
DIMM – Dual Inline Memory Module
FET – Field Effect Transistor
IHS – Integrated Heat Spreader
TTV – Thermal Test Vehicle

## 1. Introduction

The CPU socket temperature has become an increasing area of concern for thermal engineers working on current and future server platforms. While the server CPU silicon temperature is almost entirely dependent on the performance of the heat sink that is attached to the CPU and the CPU power, socket temperatures are dependent on several variables that vary widely within the same platform. Socket temperatures are highly dependent on the CPU packaging, the layout and number of power pins within the socket, the motherboard metal traces, the current passing through the pins and traces, and the proximity and temperature of various power delivery components such as FETs and inductors. Within each server platform, there are several different motherboard layouts and thermal boundary conditions. Due to this large number of variations, thermal modeling has been the only method to determine if the socket contacts meet their thermal reliability requirements. The use of thermal modeling [1] for socket temperatures has recently predicted that the socket contacts may exceed their reliability limits for the next generation of server CPU's.

Typically, thermal modeling requires experimental test data in order to validate the methodology. However, up until now there has been no experimental data available in order to validate the socket thermal modeling methodology. For CPU socket contact temperatures, a new test methodology has been developed to validate the thermal modeling that is currently being performed. The addition of the testing methodology adds to the present and future capability for predicting socket temperatures more accurately to determine if design changes are needed. A new test procedure was designed to utilize a functional server motherboard with thermal test vehicles (TTV). While previous studies have only involved modeling, this is the first time that extensive testing has been utilized to determine socket temperatures and correlate the test results to model predictions.

## 2. Test Setup

The motherboard utilized was a functional Intel server board as shown in Figure 1. This motherboard has two CPU sockets and sixteen DIMM slots. The CPU sockets and DIMM slots are arranged in a shadowed configuration that is slightly staggered. The power delivery components for the CPU's are located mainly in front of the first socket, and behind the second socket. In this configuration, the hottest motherboard and component temperatures will occur around the second CPU socket and the rear set of power delivery components.

Figure 1: Intel Server Motherboard

The TTV's [2] used for this testing were specifically designed to mimic the power and thermal behavior of the Sandy Bridge generation of Intel processors. These TTV's are built to match the same dimensions as the Sandy Bridge server processors, match the power pin locations, and have the capability to power the silicon die using either high current or low current heaters. In the test, the high current heating through the lands was utilized to match the current that the real processors draw.

978-1-4673-1110-6/12 $31.00 © 2012 IEEE

In addition to the TTV silicon temperature, the temperature of one of the TTV contacts to the socket pin was measured. A miniature type-K thermocouple was attached directly to one of the TTV contacts using an epoxy bead on the solder resist, as shown in Figure 2. This thermocouple had a wire diameter of 1 mil. The contact that was chosen was based on preliminary simulation data. The simulation data had shown that the selected power pin was residing in an area having the highest socket contact temperature.

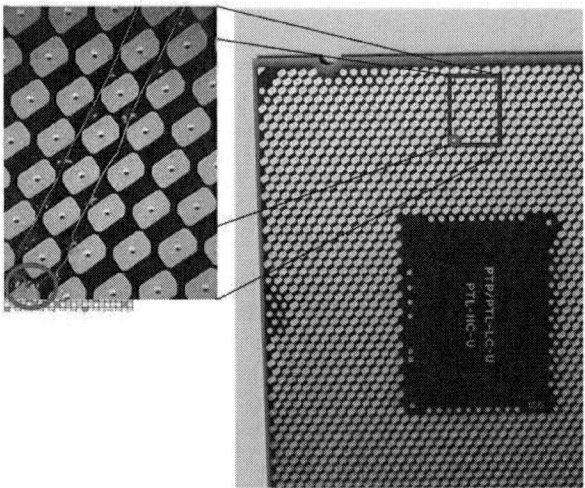

Figure 2: Miniature Thermocouple Attached to TTV Contact

A unique wind tunnel duct was designed and built to accommodate the motherboard. The duct was designed so that the air would flow through only the CPU and DIMM areas of the motherboard. The rest of the motherboard would actually sit outside of the wind tunnel as shown in Figure 3. All areas of potential air leakage were blocked with tape and foam, to ensure that the CPU's and DIMMs received the required air flow.

Figure 3: Wind Tunnel Duct with Motherboard

## 3. Measurements

The test measurements focused on thermocouple temperatures, voltage drops, and infrared imaging. The voltage drop measurements were utilized to accurately determine the current provided by the power supply. With known resistances through the motherboard as well as the TTV, the voltage drop could be correlated with the voltage load line to determine the current and power.

For temperature measurements, thermocouples were placed at various locations on the motherboard, CPU sockets, and power delivery components. Four locations around the

rear CPU socket were chosen for topside motherboard temperatures. The CPU case and junction temperatures were also recorded. The CPU case temperature was measured with a thermocouple and the junction temperature recorded by using the TTV thermal sensor. In addition, the TTV contact temperature was also recorded measured with a miniature thermocouple.

An infrared image was taken of the motherboard backside of the rear CPU socket. The purpose of the infrared image was to obtain a temperature map of a large portion of the motherboard socket region. This provided a more complete picture of how the motherboard temperature was influencing the socket temperature, providing a more complete thermal gradient image than possible by using a single point thermocouple.

## 4. Test Results

Temperature measurements were recorded for multiple thermocouples and TTV sensors at air flow conditions ranging from 60 CFM to 140 CFM within the wind tunnel. The Intel enabled thermal solution design point exists at approximately 90 CFM based on heat sink performance measurements. Figure 4 shows the locations of the various thermocouples and sensors located on the motherboard and socket regions.

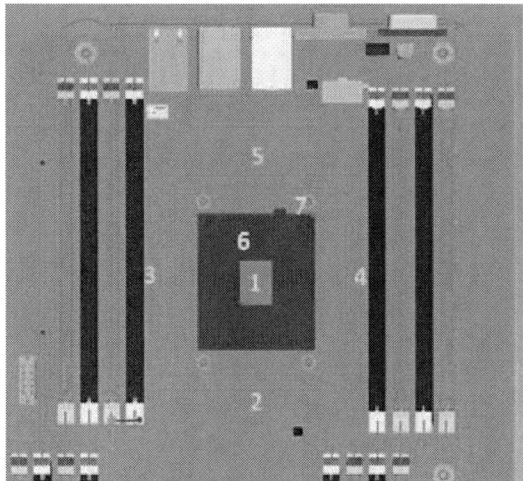

Figure 4: Thermocouple Locations on Motherboard (Numbered for Reference)

The list of thermocouples shown in Figure 4 is as follows;
1. TTV sensor at the center of the die
2. Top surface of the motherboard in front of the socket
3. Top surface of the motherboard at the left side of the socket
4. Top surface of the motherboard at the right side of the socket
5. Top surface of the motherboard behind the socket
6. Socket contact
7. Socket seating plane

A summary of the temperature results for a high current 135 A case is shown in Table 1. The numbered columns on Table 1 correspond to the thermocouple locations listed in Figure 4. The testing was performed at typical room temperatures. In a real data center, the motherboard could see

978-1-4673-1110-6/12 $31.00 © 2012 IEEE          140

elevated temperatures due to higher local air ambient temperatures and also due to preheat from other components not included in this testing such as hard drives and the front CPU. As a result, the temperatures shown in the table are lower than what is expected in a real data center situation.

| CFM | 1 | 2 | 3 | 4 | 5 | 6 | 7 |
|---|---|---|---|---|---|---|---|
| 140 | 52.4 | 32.0 | 32.6 | 30.6 | 91.7 | 57.6 | 56.6 |
| 120 | 53.4 | 32.9 | 33.9 | 31.4 | 94.5 | 59.1 | 58.6 |
| 100 | 55.5 | 34.5 | 36.2 | 33.0 | 101.7 | 61.9 | 61.4 |
| 80 | 57.9 | 36.4 | 38.9 | 34.9 | 110.0 | 64.8 | |
| 70 | 59.5 | 37.9 | 40.9 | 36.3 | 115.0 | 66.7 | |
| 60 | 61.3 | 39.2 | 42.7 | 37.6 | 119.0 | 68.7 | |

Table 1: Temperature Measurements (°C) at Varying Air Flow Rates

Figure 5: Rotated Wind Tunnel Duct with Infrared Camera

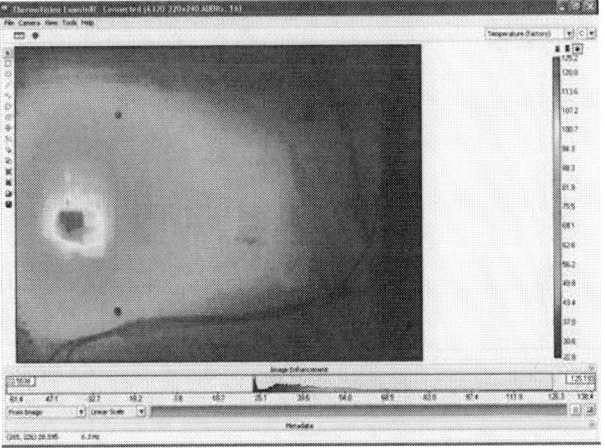

Figure 6: Infrared Image of Motherboard Backside at 100 CFM

While the thermocouple and sensors are able to measure temperatures at specific locations, an overall temperature map was also desired in order to determine the thermal gradients that exist within the motherboard. An infrared camera was used to produce a thermal map of the backside of the motherboard. In this situation the wind tunnel duct was rotated 90 degrees so that the backside could be exposed. Figure 5 shows how the wind tunnel has been rotated to enable an infrared picture to be taken of the motherboard backside. Figure 6 shows the infrared pictures taken at 100 CFM at the high current power condition.

## 5. Modeling Inputs and Assumptions

For the system and board modeling, the motherboard board file was cut down to contain the core region of only the second CPU socket in order to reduce the complexity of the large computational fluid dynamics (CFD) analysis. The cut-down board file was then imported into Icepak CFD [3] using the Cadence CAD tool [4], and then board thermal conductivity is obtained from the imported cooper traces as shown in Figure 7. CPU and memory VR FET, inductors, capacitors, and connectors are modeled as non-conductive blocks that obstruct air flow with smaller-size parts removed from the model. To represent the heat sources, fixed temperatures using downstream board temperatures are assigned to the top of the board underneath the processor VR FET, and the VR heat sink is not modeled. A heat transfer coefficient of 10 W/K-m² was assumed on the bottom side of the board for natural convection with some flushing flow.

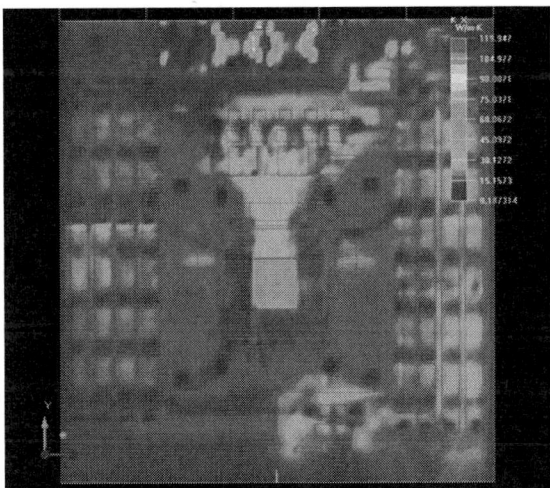

Figure 7: In-Plane Board Thermal Conductivity

Joule heating within the board was determined through a detailed analysis of the 8 layer board stackup and anticipated high current case of 135 A delivered within the traces. The power loss maps are then imported as planar heat sources in the board at the power-weighted average heights, for volumetric heat sources pose certain complexity in Icepak CFD calculations. As for the planar directions, the grid for the power loss maps representing Joule heating are limited to be approximately 5 mm to 10 mm in size, and are 8 mm by 8 mm for the model as shown in Figure 8.

978-1-4673-1110-6/12 $31.00 © 2012 IEEE

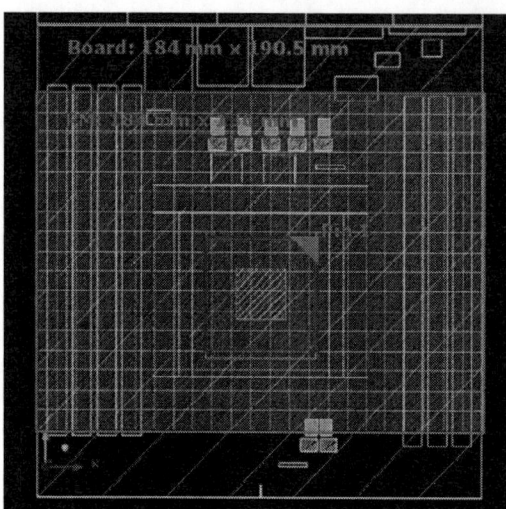

Figure 8: Board Power Map Size and Alignment

For the compact package thermal model, it is assumed that 50% of the measured TTV substrate power goes to the substrate top component and 50% goes to the bottom of the model. The actual measured and calculated die powers from the TTV are used in the CFD model for each case.

## 6. Correlation with Modeling Results

The model board temperature results shown in contour plots for high current 135 A cases with 100 CFM are shown in Figure 9 with the corresponding infrared images of the motherboard backside next to them shown again for visual comparison. The board topside temperatures at the midpoints on the boundary edges 50 mm from the socket center are also listed and compared to the thermocouple temperature readings from location #2, #3, #4, and #5 on the board that were selected to be as close to the defined boundary edges as possible in the test condition. As one can see, the model (shown in red on bottom) and test (shown in black on top) temperatures are very close on three sides, with the average differences been less than 2 °C for 100 CFM case on the front and the both memory connector edges. This indicates a very good correlation between the test and the modeling results. Additionally, the temperature contour plots also demonstrate great correlation of both results on three sides when compared to the infrared images.

It was noticed that the CPU VR temperatures seen in the test from the infrared images were only higher at one or two single VR FET components instead of all eight as shown in the simulation results. This result was due to the fact that the current to the CPU was not supplied evenly through all of the VR phases as expected. This explains the large gap in board temperatures at the VR and power delivery edge between the model and the test on all cases studied. The discrepancy is reasonable for the fact that it had been seen and learned that the board temperatures are most highly influenced by the Joule heating from power losses in the traces and the heat sources from elevated temperatures of the VR components. One can expect higher board temperature to result if the current is more concentrated at fewer points, as in the case of what was seen in the test.

## 7. Conclusions

This paper has described the completed testing and modeling correlation on a motherboard containing two CPU sockets. The results include temperatures for the socket contact, CPU die, IHS, multiple motherboard temperatures around the socket, as well as the socket seating plane. These tests were conducted through a range of air flow velocities which gives a representation of the impact of air flow boundary conditions to the temperatures. In addition, infrared images were taken of the socket backside showing the temperature distribution of the motherboard surrounding the socket.

The socket thermal model of the motherboard with Sandy Bridge CPU package has been updated to match the test conditions. Board and substrate conductivities are still estimated, however, the power losses of the board and substrate have been validated with the measured data. VR FET temperatures in the model are adjusted to the test results. The contact, die, IHS, board, and seating plane temperatures at different air flow velocities have been compared between model and testing to determine the correlation. As a result, there is now increased confidence in the socket temperature modeling capabilities that can be applied to future motherboard layouts.

### Acknowledgments

The authors would like to acknowledge the contributions of Chheang Chay, Tammie Bard, Tom Allyn, Sandy Guo, Kelly Lofgreen, Na Chen, Ed Payton, Dan Stuart, Farzaneh Yahyaei-Moayyed, Neal Ulen, Roger D. Flynn, and Susan Smith.

### References

1. Saeidi, S. M., et al, "Integrated Methodology for Socket Thermal Modeling and Analysis", Intel Design & Test Technology Conference, 2008
2. LGA2011 Thermal Test Vehicle User's Guide, Rev 0.7, Intel Document Number 440179
3. http://www.ansys.com/products/icepak
4. http://www.cadence.com

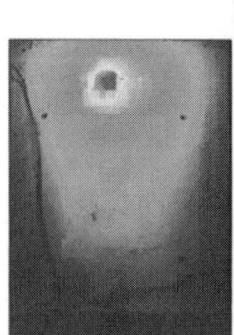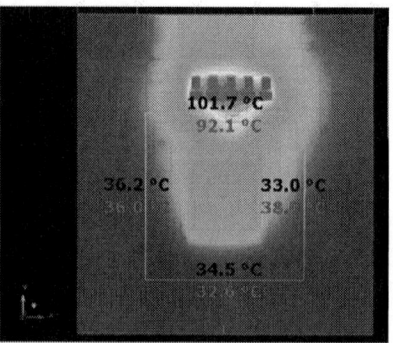

Figure 9: Test (black) and Model (red) Board
Temperatures for 135 A at 100 CFM

# A Method to Measure Heat Dissipation from Component on PCB

Zhongwei Qi

General Electric Technology Infrastructure Healthcare

3000 N. Grandview Blvd

Waukesha, WI, USA, 53188

Zhongwei.Qi@ge.com

## Abstract

Knowledge of heat dissipation from electronic components is critical input for thermal design in the product development phase. Unfortunately heat loss is not always available to design engineers even when a prototype is available. This paper presents a new method to measure and calculate the component heat dissipation on printed circuit board from a thermal standpoint. Based on the error analysis of the thermal resistance network, an error indicator is defined to account for the interference from nearby heat sources. Different cases are simulated to confirm the effectiveness of the measurement method and the error indicator. Using the error indicator in compensation, measurement error can be reduced greatly. Uncertainty analysis is conducted to assess the error sources from measurements and their impacts to result. Preliminary experiments have been conducted on a mock-up board to prove the concept of this method.

## .Keywords

Heat Loss Estimation, Heat Dissipation Measurement Uncertainty Analysis, Thermal Resistance Network, Error Indicator

## 1. Motivation

Heat dissipation from electronic components is a critical input for thermal designers doing product development. Unfortunately heat load can't always be estimated accurately in a straight forward manner, even when a prototype is available. Sometimes it is due to the inability to measure the current through the component, measurement errors associated with AC systems, Electromagnetic Interference noise or due to the fact that the heat load is operating frequency and leakage current dependent. When heat information is not easily available, suppliers' support can be a great resource but should be used with caution. Sometimes using the maximum rating on a data sheet or power-estimation provided by manufacturers can lead to over-engineering or unknown risk in thermal design.

There are some methods found in literature for experimentally measuring heat dissipation. Calorimetric methods [1], [2], [3] have been used to make loss measurements of electric machines or standalone components such as power electronics device, transformer, ferrite and capacitor, etc. Heat flux sensors and thermoelectric modules were used in an apparatus for loss measurements of Integrated Power Electronic Module [4] in which temperature balance is required to minimize heat leakage from all sides other than the heat flux sensor measurement side. Junction temperature (Tj) measurement by Temperature Sensitive Parameter is

another method used if Tj is measurable and the component's thermal model is accurate. The concern with this method lies in the fact that the component's thermal model environmental dependency, material property variation among different lots, and variations in component characterization done by the supplier and Tj measurement done by the end user. Inverse thermal calculation is another way to determine heat dissipation from temperature measurement, since intuitively temperature is induced by the heat dissipation under a certain boundary condition. However with many heat sources, and poorly defined material properties and cooling conditions, it is hard to get good heat dissipation estimates with inverse thermal calculation methods.

This paper proposes a new method to measure heat dissipation of components on a printed circuit board (PCB) under operating conditions by utilizing a heat flux sensor and thermistors, and calculating heat dissipation from a thermal standpoint. Error analysis follows the proposed measurement method to assess the error due to the interference of nearby heat sources. An error indicator is developed to be used for error compensation. Experiments are planned with a mock-up board and heaters. Uncertainty analysis is conducted to identify possible measurement error sources and their impacts to final result with the selected sensors, instrumentations, and testing setup. Finally preliminary tests prove the effectiveness of the proposed method.

## 2. Measurement Method and Error Indicator

It starts with a 2-node thermal resistance network [5] as shown in Fig. 1. There are heat sources at node A and B dissipating $Q_0$ and $Q_1$ heat respectively. Node A and B are connected to each other, and they are connected to ambient sink, as well. $T_0$ represents temperature at node A, and $T_a$ represents ambient temperature. $Q_{ab}$ is the heat exchange between node A and B. $Q_{a0}$ and $Q_{a1}$ are the heat dissipated from A and B to ambient sink, respectively.

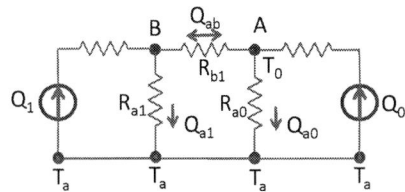

**Fig. 1 – 2-Node Thermal Resistance Network**

There exist Equations (1) to (3) for this 2-node network. Equation (2) shows that $(T_0 - T_a)$ is proportional to the sum of heat dissipation at A and a part of heat from node B. Equation (3) shows that, the proportional ratio $1/R_1$ depends on the thermal resistances only. During testing, if the ambient cooling rate remains unchanged with temperature, which is possible with high cooling rate, thermal resistance elements remain unchanged. Meanwhile, if heat source $Q_1$ remains unchanged, the second term on the left side of Equation (2) is constant too.

$$Q_{a0} = \frac{R_{a1}}{R_{a0} + R_{b1} + R_{a1}} \times Q_1 + \frac{R_{b1} + R_{a1}}{R_{a0} + R_{b1} + R_{a1}} \times Q_0; \quad (1)$$

$$Q_0 + \frac{R_{a1}}{R_{b1} + R_{a1}} \times Q_1 = \frac{1}{R_1} \times (T_0 - T_a); \quad (2)$$

$$\frac{1}{R_1} = \frac{1}{R_{a0}} + \frac{1}{R_{b1} + R_{a1}}; \quad (3)$$

For a general case as shown in Fig. 2, a thermal resistance network represents a certain number of electronic components on a PCB, some dissipating heat ( node A, B,C,D,E, etc. ) and others not. Node A represents the Device Under Test (DUT), whose heat dissipation is to be measured.

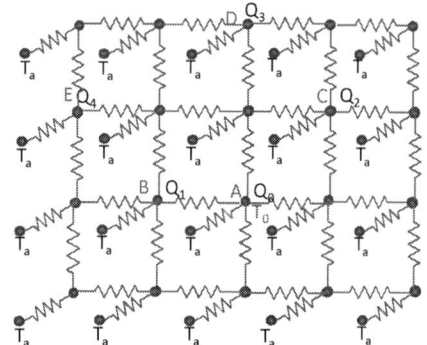

**Fig. 2 – Thermal Resistance Network for General Case**

Using the superposition attribute of a linear network, n-nodes of heat sources can be included in Equation (4). Similarly, $(T_0 - T_a)$ is proportional to the sum of heat to network at node A ($Q_0$) and a part of other heat (node B, C, D, E) in the network. The proportional ratio is a function of thermal resistances in the network, which can remain unchanged at high ambient cooling rate. Let Q be the total heat dissipation from component at node A, including $Q_m$ - a part of Q to be drawn away through heat path other than the network, i.e. component's case top. Then $Q-Q_m$ is the heat spreading to network from node A, which is $Q_0$. Rewrite the Equation (4) into Equation (5) to (7).

$$Q_0 + \sum_{i=1}^{n} f_i(Q_i) = g(R_{b1}, R_{b2}, ..., R_{bn}, R_{a1}, R_{a2,...,}R_{an}) \times (T_0 - T_a); \quad (4)$$

$$Q + E = Q_m + K \times (T_0 - T_a); \quad (5)$$

$$E = \sum_{i=1}^{n} f_i(Q_i); \quad (6)$$

$$K = g(R_{b1}, R_{b2}, ..., R_{bn}, R_{a1}, R_{a2,...,}R_{an}); \quad (7)$$

Equation (5) is a linear equation with two unknowns and two measurable parameters. Under a constant high cooling rate, K is an unknown constant. It is the reciprocal of thermal resistance from the point where $T_0$ is monitored to ambient

reference $T_a$. Q+E is another constant unknown, in which Q is the total heat dissipation from component at node A, and E is the noise from other heat sources on the PCB. $Q_m$ can be modulated with a cooling device, i.e., a thermoelectric module drawing different amounts of heat from the component case top. $Q_m$ is monitored. When $Q_m$ changes, $(T_0 - T_a)$ will change accordingly and be monitored too. Theoretically it takes two sets of measurement data, $Q_m$ and $(T_0 - T_a)$, to calculate the two unknowns, Q+E and K. Fig.3 depicts the above calculation. Q+E is the predicted heat dissipation for DUT, which includes the true heat dissipation Q from DUT, and the artificial part E.

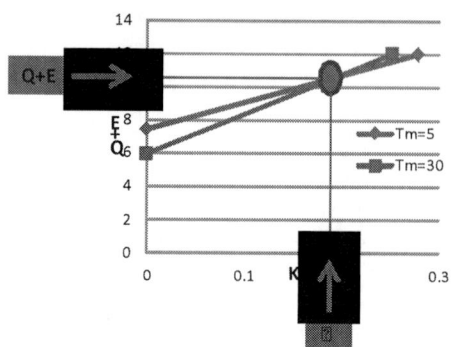

**Fig. 3 – Calculation Graph**

Fig. 4 shows a testing apparatus to implement the measurement method. The entire test board assembly is set in a wind tunnel with high cooling air flow rate, or captured in cold plates except for DUT. A heat flux sensor is placed between DUT and Thermoelectric Cooling module (TEC) with active heat sink. One precise thermistor is attached to the middle area underneath DUT from the other side of PCB, and four other thermistors are attached to the mid-points of the four edges of DUT, respectively, also on the other side of PCB. An insulator covers the temperature sensors to minimize the heat leakage through sensor wire leads. During testing, TEC is modulated to draw different amounts of heat from DUT. It is required to maintain the average temperature from the four points surrounding DUT at a temperature lower than or equal to that in the middle of DUT, so that average heat flow is spreading from center to sides.

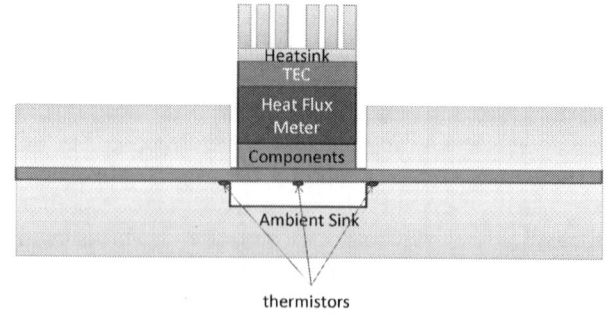

**Fig. 4 – Testing Apparatus**

Equation (4) to (6) show that, when DUT is the only heat source on the PCA, E is zero – no nearby heat sources induced error, theoretically. When heat sources from other than DUT on PCB are present, error is inevitable with this measurement method. Using a 2-node network in Fig. 2 as an example, from Equation (2),

$$E = \frac{R_{a1}}{R_{b1} + R_{a1}} \times Q_1 ; \quad (8)$$

One can see that when $Q_1$ (heat flow from nearby source) is smaller, the error becomes smaller; when $R_{b1}$ is larger (nearby heat is further away from DUT, or board in-plane conductivity is lower), error becomes smaller; and when $R_{a1}$ is smaller (higher cooling rate near $Q_1$), E becomes smaller. These are the three factors affecting the theoretical measurement error. However during testing, except for ambient cooling rate which can be improved to a certain extent, the heat sources on PCB and board conductivity are out of measurement control. Therefore, we need an error indicator during testing to reveal the measurement method error from the above mentioned noise, so that we can compensate it to achieve high accuracy.

In order to define an error indicator, a concept of isotherms hierarchy is used. Fig.5 shows a topology of heat transfer from DUT and other n heat sources flowing through a hierarchy of isotherms to ambient sink. Fig. 6 shows the isotherms hierarchy on PCB plane and thermal resistance network. Let $T_b$ locates at the center of DUT. Let $T_n$ be the 1st isotherm surrounding the DUT only, $T_{n+1}$ be the 2nd isotherm surrounding the DUT and first closest neighbor heat source, $T_{n+2}$ be the 3rd isotherm surrounding the DUT and the first two closest neighbor heat sources, so on and so forth, till $T_{n+n}$ circles all the n neighbor heat sources. It is not necessary for $T_{n+i}$ to be higher than $T_{n+i+1}$. Heat flow $Q_{bi}$ (i=1, ... , n) will alter direction to comply with the law of heat flowing from a high temperature node to a lower temperature node. However, it is required in testing there is always a certain amount of heat, $Q-Q_m$, spreading from DUT into PCB. Therefore, $T_b$ should always be kept larger than $T_n$.

Using network equivalence for all the heat sources except for DUT, Fig. 6 transforms into Fig. 7, in which all nearby heat sources are represented as $Q_{ext}$. Reusing relations derived from Fig. 2 and including $T_b$ into the equation, now we have Equation (9) as,

$$Q + \frac{\frac{R_{e0}}{R_{e0} + R_e}}{1 + R_b \times \left(\frac{1}{R_{a0}} + \frac{1}{R_{e0} + R_e}\right)} \times Q_{ext}$$
$$= Q_m + \frac{\left(\frac{1}{R_{a0}} + \frac{1}{R_{e0} + R_e}\right)}{1 + R_b \times \left(\frac{1}{R_{a0}} + \frac{1}{R_{e0} + R_e}\right)} \times (T_b - T_a); \quad (9)$$

The error percentage due to neighbor heat sources $Q_{ext}$ is,

$$Error\% = \frac{\frac{R_{e0}}{R_{e0} + R_e} \times Q_{ext}}{1 + R_b \times \left(\frac{1}{R_{a0}} + \frac{1}{R_{e0} + R_e}\right) \times Q}$$
$$= \frac{R_{a0} \times \frac{R_{e0}}{R_{a0} + R_{e0} + R_e} \times Q_{ext}}{\left[R_b + R_{a0} \times \frac{(R_{e0} + R_e)}{R_{a0} + R_{e0} + R_e}\right] \times Q}; \quad (10)$$

When TEC draws almost all the heat through the case top, $Q-Q_m$ is close to zero. Therefore,

$$R_{a0} \times \frac{R_{e0}}{R_{a0} + R_{e0} + R_e} \times Q_{ext} = T_n - T_a \ (when \ Q_m = Q); \quad (11)$$

When TEC draws almost none of the heat through case top, Q is close to $(Q-Q_m)$. Therefore,

$$\left[R_b + R_{a0} \times \frac{(R_{e0} + R_e)}{R_{a0} + R_{e0} + R_e}\right] \times Q$$
$$= (T_b - T_a) \ (when \ Q_m = 0)$$
$$- (T_n - T_a) \ (when \ Q_m = Q); \quad (12)$$

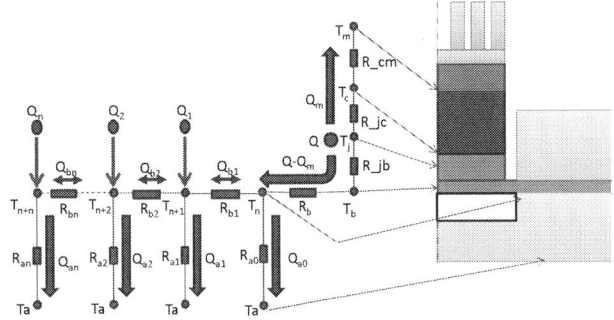

**Fig. 5 – A Topology of Heat Flow to Ambient**

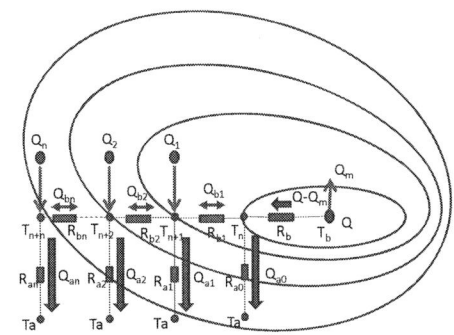

**Fig. 6 – n-Node Thermal Resistance Network**

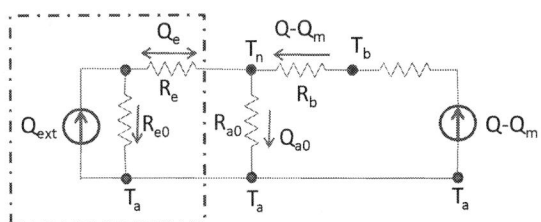

**Fig. 7 – Equivalent Thermal Resistance Network**

In experiment implementation, $T_b$ comes from the thermistor underneath DUT on the other side of PCB. $T_n$ is from the average temperature of four thermistors surrounding the DUT. During measurement, the TEC initially blocks heat dissipating from case top (as when $Q_m=0$), followed by TEC increasingly drawing more heat from case top until $T_b$ is close to $T_n$ (as when $Q_m=Q$). Error indicator is updated with Equation (10), (11), and (12) each time when TEC draws more heat from case top until $T_b$ approaches $T_n$. The error percentage calculated then is the measurement method error due to the nearby heat sources interference.

## 3. Case Simulations and Error Compensation

Case simulations prove the effectiveness of the measurement method and its error indicator. Ten cases of different power levels, neighbor heat sources' distances to

978-1-4673-1110-6/12 $31.00 © 2012 IEEE          145

DUT, PCB conductivities, and ambient cooling rates are simulated with ANSYS\ICEPAK. Fig.8, 9, 10 and 11 show four simulated cases results with different nearby heat sources on PCB.

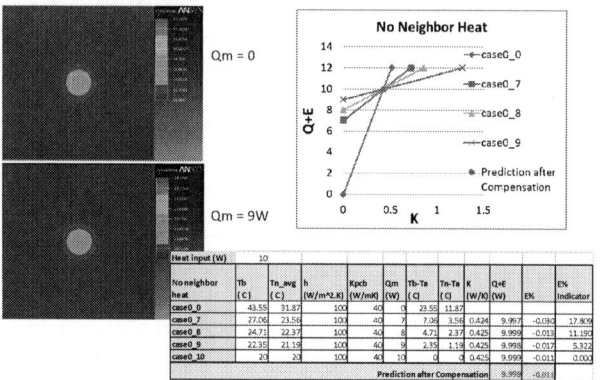

| Heat input (W) | 10 | | | | | | | | | | |
|---|---|---|---|---|---|---|---|---|---|---|---|
| No neighbor heat | Tb (C) | Tn_avg (C) | h (W/m^2.K) | Kpcb (W/mK) | Qm (W) | Tb-Ta (C) | Tn-Ta (C) | K (W/K) | Q+E (W) | E% | E% Indicator |
| case0_0 | 43.55 | 31.87 | 100 | 40 | 0 | 23.55 | 11.87 | | | | |
| case0_7 | 27.06 | 23.56 | 100 | 40 | 7 | 7.06 | 3.56 | 0.424 | 9.997 | -0.080 | 17.809 |
| case0_8 | 24.71 | 22.37 | 100 | 40 | 8 | 4.71 | 2.37 | 0.425 | 9.999 | -0.013 | 11.190 |
| case0_9 | 22.35 | 21.19 | 100 | 40 | 9 | 2.35 | 1.19 | 0.425 | 9.998 | -0.017 | 5.322 |
| case0_10 | 23 | 20 | 100 | 40 | 10 | 3 | 0 | 0.425 | 9.999 | -0.011 | 0.003 |
| Prediction after Compensation | | | | | | | | | 9.999 | -0.011 | |

**Fig. 8 – Case 0 - No Neighbor Heat Source**

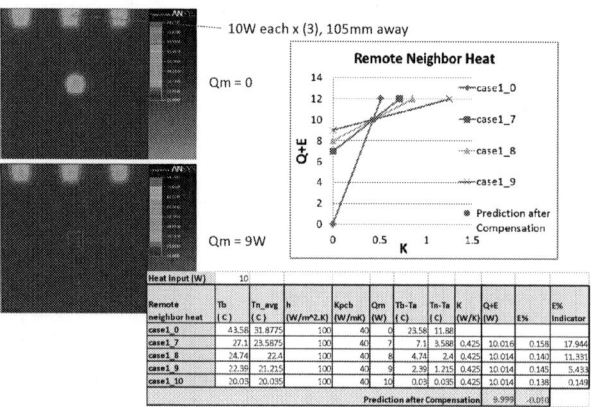

| Heat input (W) | 10 | | | | | | | | | | |
|---|---|---|---|---|---|---|---|---|---|---|---|
| Remote neighbor heat | Tb (C) | Tn_avg (C) | h (W/m^2.K) | Kpcb (W/mK) | Qm (W) | Tb-Ta (C) | Tn-Ta (C) | K (W/K) | Q+E (W) | E% | E% Indicator |
| case1_0 | 43.58 | 31.8775 | 100 | 40 | 0 | 23.58 | 11.88 | | | | |
| case1_7 | 27.1 | 23.5875 | 100 | 40 | 7 | 7.1 | 3.588 | 0.425 | 10.016 | 0.158 | 17.944 |
| case1_8 | 24.74 | 22.4 | 100 | 40 | 8 | 4.74 | 2.4 | 0.425 | 10.014 | 0.140 | 11.331 |
| case1_9 | 22.39 | 21.215 | 100 | 40 | 9 | 2.39 | 1.215 | 0.425 | 10.014 | 0.145 | 5.433 |
| case1_10 | 20.03 | 20.035 | 100 | 40 | 10 | 0.03 | 0.035 | 0.425 | 10.014 | 0.138 | 0.149 |
| Prediction after Compensation | | | | | | | | | 9.999 | -0.080 | |

**Fig. 9 – Case 1 - Remote Neighbor Heat Sources**

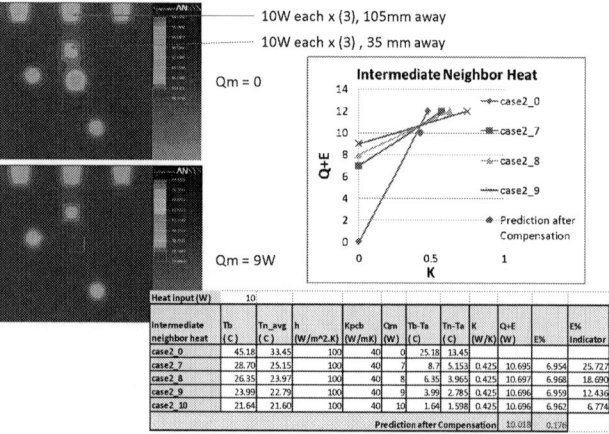

| Heat input (W) | 10 | | | | | | | | | | |
|---|---|---|---|---|---|---|---|---|---|---|---|
| Intermediate neighbor heat | Tb (C) | Tn_avg (C) | h (W/m^2.K) | Kpcb (W/mK) | Qm (W) | Tb-Ta (C) | Tn-Ta (C) | K (W/K) | Q+E (W) | E% | E% Indicator |
| case2_0 | 45.18 | 33.45 | 100 | 40 | 0 | 25.18 | 13.45 | | | | |
| case2_7 | 28.70 | 25.15 | 100 | 40 | 7 | 8.7 | 5.153 | 0.425 | 10.695 | 6.954 | 25.727 |
| case2_8 | 26.35 | 23.97 | 100 | 40 | 8 | 6.35 | 3.965 | 0.425 | 10.697 | 6.968 | 18.690 |
| case2_9 | 23.99 | 22.79 | 100 | 40 | 9 | 3.99 | 2.785 | 0.425 | 10.696 | 6.959 | 12.436 |
| case2_10 | 21.64 | 21.60 | 100 | 40 | 10 | 1.64 | 1.598 | 0.425 | 10.696 | 6.962 | 6.774 |
| Prediction after Compensation | | | | | | | | | 10.018 | 0.176 | |

**Fig. 10 – Case 2 - Intermediate Neighbor Heat Sources**

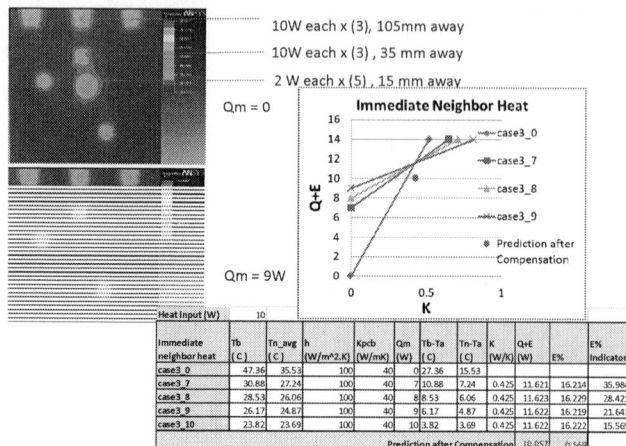

| Immediate neighbor heat | Tb (C) | Tn_avg (C) | h (W/m^2.K) | Kpcb (W/mK) | Qm (W) | Tb-Ta (C) | Tn-Ta (C) | K (W/K) | Q+E (W) | E% | E% Indicator |
|---|---|---|---|---|---|---|---|---|---|---|---|
| case3_0 | 47.36 | 35.53 | 100 | 40 | 0 | 27.36 | 15.53 | | | | |
| case3_7 | 30.88 | 27.24 | 100 | 40 | 7 | 10.88 | 7.24 | 0.425 | 11.621 | 16.214 | 35.984 |
| case3_8 | 28.53 | 26.06 | 100 | 40 | 8 | 8.53 | 6.06 | 0.425 | 11.623 | 16.229 | 28.421 |
| case3_9 | 26.17 | 24.87 | 100 | 40 | 9 | 6.17 | 4.87 | 0.425 | 11.622 | 16.219 | 21.641 |
| case3_10 | 23.82 | 23.69 | 100 | 40 | 10 | 3.82 | 3.69 | 0.425 | 11.622 | 16.222 | 15.565 |
| Prediction after Compensation | | | | | | | | | 10.057 | 0.568 | |

**Fig. 11 – Case 3 - Immediate Neighbor Heat Sources**

As indicated by Equation (10), the model error before compensation is affected by nearby heat sources and their distances to DUT, PCB conductivity, and ambient cooling rate. The errors before compensation are plotted in Fig. 12 from case simulation results.

Error indicator formula's effectiveness is demonstrated by the plot comparing model error and error indicator in Fig. 13. Using error indicator in compensation, most of the error percentages after compensation are reduced greatly, as can be seen in Fig. 13.

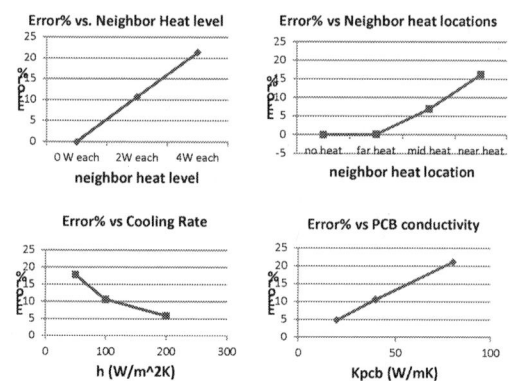

**Fig. 12 – Error (before compensation) Sensitivity**

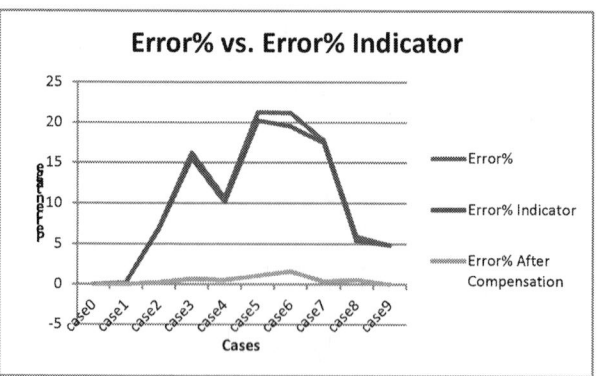

**Fig. 13 – Error% Indicator Effectiveness**

## 4. Uncertainty Analysis

In the proposed experiment method to measure total heat dissipation from component on PCB, Equations (9) to (12) are used to calculate with 5 points temperature measurement on PCB, ambient temperature, and heat flux. In reality, every measurement has error due to the system/sensor interaction, instrumentation error, and operation error [6]. Uncertainty analysis identifies and keeps track of uncertainties in measurements, and further assesses their effects on the accuracy of final result.

There are 3 groups of measurements need to examine the errors,

- Heat measurement
- Ambient temperature measurement
- Surface temperature measurements at 5 points

The measurement of heat drawn from DUT case top is implemented by the product of heat flux and area. Heat flux sensor used in this experiment has 1.7e-6 V/(W/m$^2$) sensitivity according to its manufacturer's calibration report. Data acquisition system has +/-88e-6 V in accuracy. Without knowing the confidence level from manufacturers, assume 95% confidence level regarding to their accuracy. Therefore, the overall variation is +/-52 W/m2 at odds of 20/1, and one sigma is 26 W/m2. Heat flux sensor film is very thin (0.18mm thick) - the same amount of heat entering the sensor from DUT case top side will pass through the sensor on the other side. Heat flux is not assumed to cross the area uniformly, because sensor may not be perfectly aligned with heat source, or heat may not be drawn unevenly across sensor area. A variation in area is used to count for their effect. In this experiment, nominal area is 25.4mm by 25.4mm. Assuming heat flux sensor's linear alignment error is +/-0.5 mm at 95% confidence level. The sensing area variation is 2*25.4e-6 m^2 at odds of 20/1, and one sigma is 25.4e-6 m^2.

Ambient temperature is measured with the same type of thermistor for surface temperature measurement. The interchangeability tolerance of the selected thermistor is +/-0.2C. In calculation of the total DUT heat dissipation, the difference of surface temperature and ambient temperature is used, whose variation is still within +/- 0.2C since any two of them should be interchangeable within +/- 0.2C. A testing record of 6 thermistors placed closely to each is shown in Fig.14. From the graph, we can see at any sampling point the overall variation from thermistors, Wheatstone bridges, and data acquisition system error is proved to be within 0.2C. Therefore it is safe to use the overall variation of +/- 0.2 C at odds of 20/1, one sigma of 0.1C in uncertainty analysis. The ambient temperature is measured at the inlet of wind tunnel, assuming the air is well mixed at the fan inlet with no temperature difference.

Besides the sensor and instrumentation error for temperature measurement discussed above, surface temperature measurement error is strongly influenced by the variations in attachment location, attachment material properties and its thickness. It is an operation-type error. Different setups conducted by difference people will have different errors. However, unlike the sensor and instrumentation errors that occur in every single sampling,

once the thermistors are attached, in the same set up, errors are fixed. Simulations help to identify the variations range because of sensors' disturbance, sensors locations and epoxy thickness that is used to attach thermistors. Fig.15 shows an example of temperature distribution on PCB surface where sensors could be off-centered by +/- 1mm. Fig.16 shows an example of temperature distribution on a surface that cuts through sensors and their attachment epoxy layers to PCB. Fig. 17 shows a graph comparing the undisturbed true temperature, the achieved temperature with 0.5mm epoxy attachment, and the achieved temperature with 0.25mm epoxy attachment at one of surface temperature measurement point. Notice that the undisturbed true surface temperature is higher than the achieved temperature that sensor picks up. Also notice the temperature difference becomes smaller when more heat drawn away from case top.

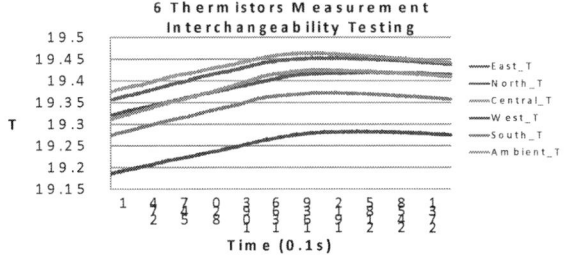

**Fig. 14 Thermistors interchangeability Testing Record**

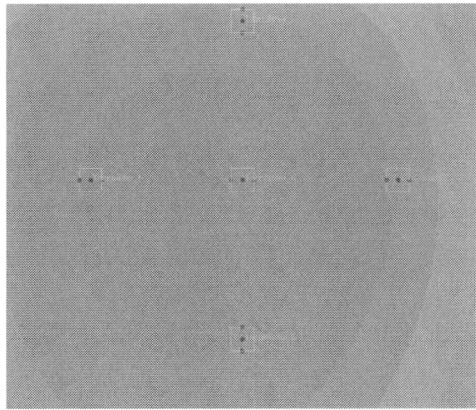

**Fig. 15 Surface Temperature Distribution with**

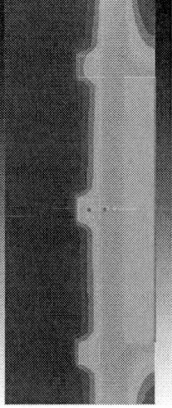

**Fig. 16 Achieved Temperature at Thermistor center**

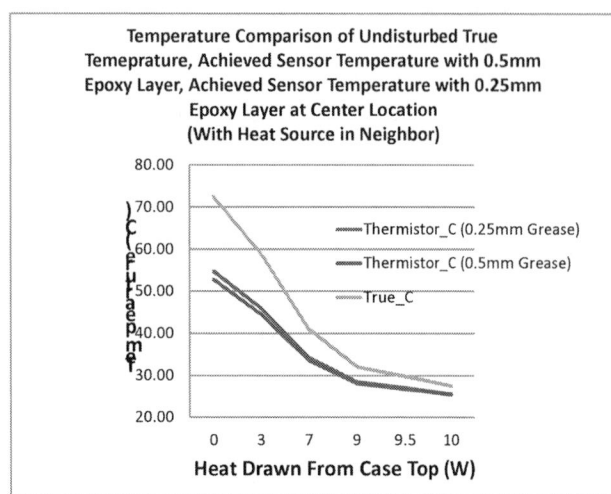

**Fig. 17 Temperature Difference Caused by Sensor Attachment**

There will be random combinations of the center thermistor and four neighbor thermistors locations and epoxy thicknesses, which leads to a random distribution of center and neighbor temperatures. To assess its maximum impact, the combinations of extreme center and neighbor temperatures are selected to use in the following uncertainty analysis. Undisturbed true temperatures are also included as a theoretical case. Fig.18 shows the difference of the temperatures in one case.

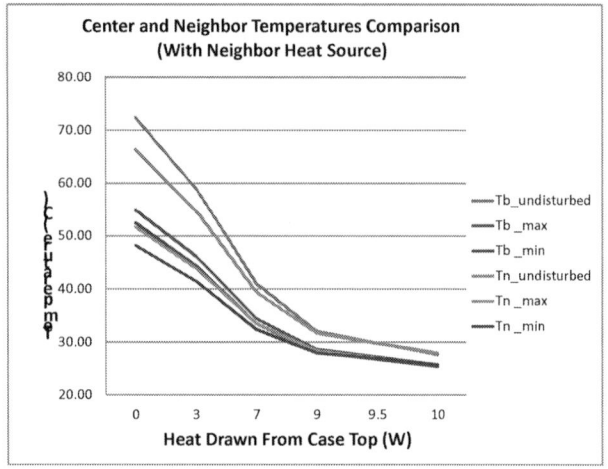

**Fig. 18 Undisturbed True Temperature and Extreme Center and Neighbor Temperatures**

With the error sources identified above, uncertainty analysis is conducted to assess their impacts on the heat prediction result accuracy. Two cases are studied – one is 10W DUT without heat source in neighborhood, the other is 10W DUT with 3.6W heat source 17mm away (edge-to-edge distance). In each case, 5 sets of surface temperature combinations are put into the calculation for DUT heat dissipation, namely,

- Max center with max neighbor temperature;
- Max center with min neighbor temperature;
- Min center with max neighbor temperature;
- Min center with min neighbor temperature;
- Undisturbed center and neighbor temperature;

For errors from sensor and instrumentation that occur in each reading, assume them normally distributed with one sigma values discussed above. Crystal Ball is used as a tool to simulate their effects to each case with Monte Carlo method.

Fig.19 and Fig. 20 show DUT heat prediction error forecasts for cases without and with neighbor heat source respectively. Without neighbor heat source, prediction error varies within +/- 8%. With neighbor heat source, prediction error varies between -38% and 10%.

Uncertainty analysis result also shows, without heat source in neighbor, sensor attachment error does not make difference in result. With heat source in neighbor, thermistor's attachment causes the major variation. Thermistor attachment is the vital factor that affecting heat prediction variation when there are nearby heat sources. Improvements are made to reduce the variations in thermistor attachment material property, and location. They are,

- Using thermal grease with conductivity of 2 W/mK between thermistor and measured surface, with the same thickness variation within 0.25mm to 0.5mm.
- Limiting sensors locations' variation to +/- 0.5mm on PCB.

With the improvements, maximum heat prediction error will reduce from 38% to 14% according to uncertainty analysis, as shown in Fig. 20 and Fig.21.

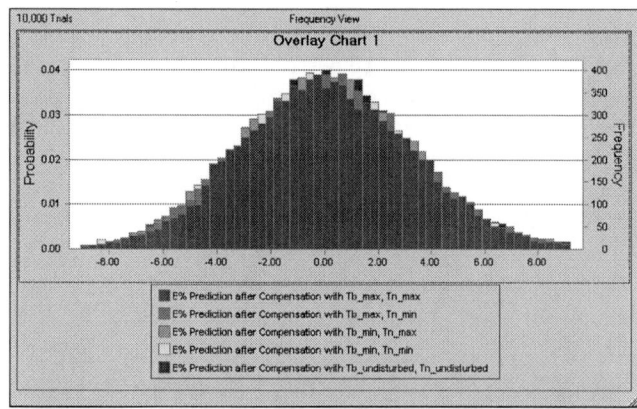

**Fig.19 DUT Heat Prediction Error Forecast without Neighbor Heat Source**

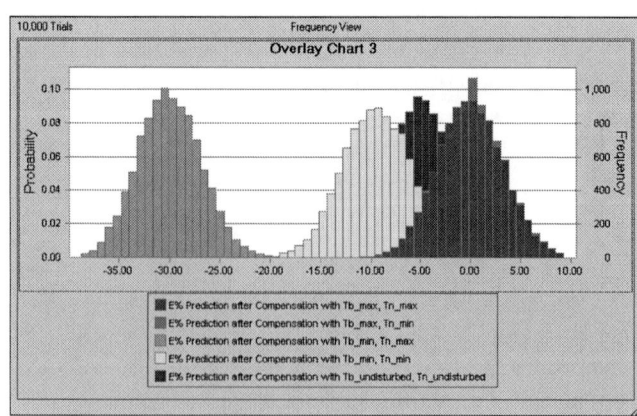

**Fig.20 DUT Heat Prediction Error Forecast with Neighbor Heat Source**

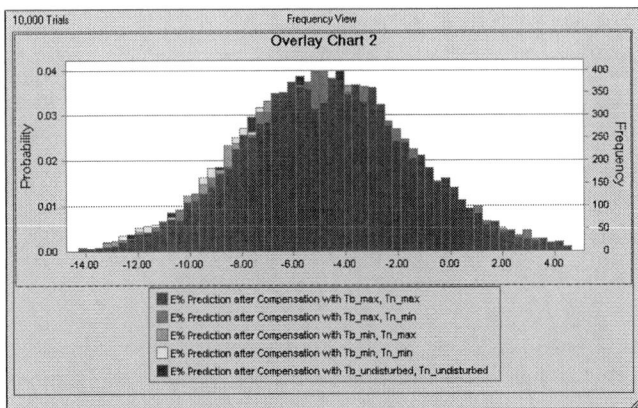

**Fig.21 Improved DUT Heat Prediction Error Forecast with Neighbor Heat Source**

## 5. Preliminary Testing

Preliminary tests have been conducted with a mock-up PCB and heaters mimicking DUT and nearby heat sources. Fig. 22 shows a test apparatus, Fig. 23 and Fig. 24 show two test results, whose errors are within the uncertainty analysis predicted variation range.

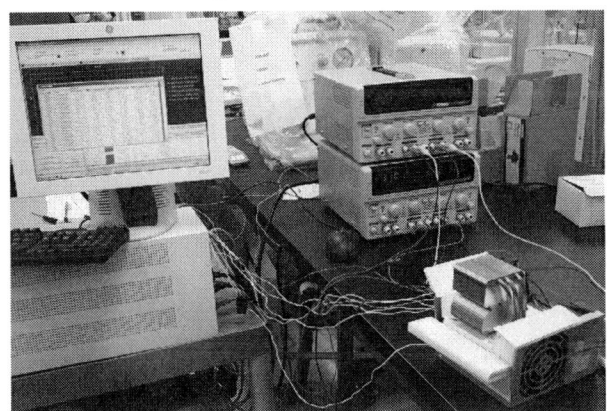

**Fig. 22 – Bench Top Testing and Data Acquisition**

| Heat input (W) | 5.325 | | | | | | |
|---|---|---|---|---|---|---|---|
| 5W with no neighbor heat | Qm (W) | Tb-Ta | Tn-Ta | K | Q+E | E% | E%_Indicator |
| test1_0 | -6.273E-04 | 26.76 | 20.81 | | | | |
| test1_1 | 3.309E+00 | 12.20 | 9.11 | 0.2272377 | 6.08057233 | 14.1891517 | 51.65 |
| test1_2 | 4.890E+00 | 4.53 | 3.01 | 0.2209061 | 5.93475916 | 11.45087623 | 12.67 |
| test1_3 | 5.030E+00 | 3.75 | 2.38 | 0.2190975 | 5.89425618 | 10.69025689 | 9.77 |
| test1_4 | 5.166E+00 | 3.36 | 2.12 | 0.2192498 | 5.8975951 | 10.7529597 | 8.59 |
| test1_5 | 5.231E+00 | 2.91 | 1.73 | 0.2188615 | 5.8893687 | 10.59847333 | 6.93 |
| test1_6 | 5.290E+00 | 2.33 | 1.24 | 0.2178122 | 5.86827115 | 10.20227513 | 4.85 |
| test1_7 | 5.360E+00 | 1.96 | 0.95 | 0.21697 | 5.85197176 | 9.896018326 | 3.67 |
| test1_8 | 5.426E+00 | 1.80 | 0.86 | 0.2166559 | 5.84600223 | 9.784079361 | 3.34 |
| test1_9 | 5.434E+00 | 1.63 | 0.72 | 0.2161936 | 5.83740842 | 9.622693404 | 2.76 |
| Prediction after Compensation | | | | | 5.68051026 | 6.676249077 | |

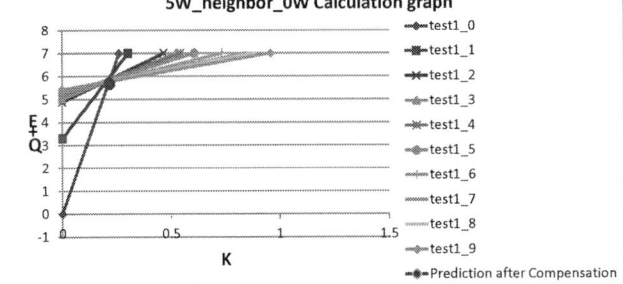

**Fig. 23 – Testing Case 1: 5W DUT with 0W Neighbor Heat**

| Heat input (W) | 5.325 | | | | | | |
|---|---|---|---|---|---|---|---|
| 5W with 3.6W neighbor heat | Qm (W) | Tb-Ta | Tn-Ta | K | Q+E | E% | E%_Indicator |
| test2_0 | -2.147E-05 | 29.18 | 23.52 | | | | |
| test2_1 | 3.409E+00 | 14.02 | 11.34 | 0.224879 | 6.56275992 | 23.24431776 | 63.53 |
| test2_2 | 4.612E+00 | 6.33 | 5.19 | 0.2050327 | 6.05934098 | 13.79044088 | 21.65 |
| test2_3 | 4.818E+00 | 5.11 | 4.20 | 0.2005277 | 5.94934484 | 11.72478569 | 16.83 |
| test2_4 | 5.026E+00 | 3.95 | 3.30 | 0.1982305 | 5.89620234 | 10.72680459 | 12.73 |
| test2_5 | 5.108E+00 | 3.39 | 2.84 | 0.1966204 | 5.86019191 | 10.05055237 | 10.78 |
| test2_6 | 5.202E+00 | 2.30 | 1.94 | 0.1940869 | 5.80765031 | 9.063855541 | 7.13 |
| test2_7 | 5.313E+00 | 1.65 | 1.43 | 0.1921925 | 5.77026679 | 8.361817643 | 5.15 |
| Prediction after Compensation | | | | | 5.4874466 | 3.050640305 | |

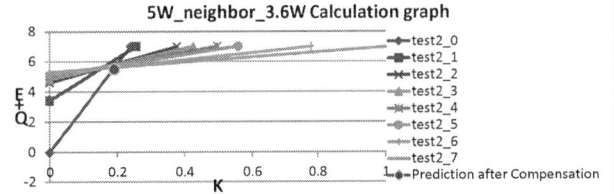

**Fig. 24 – Testing Case 2: 5W DUT with 3.6W Neighbor Heat**

## 6. Conclusions

This paper presents a new method to measure component heat dissipation directly on a PCB in its working condition. Compared to the calorimetric method, it takes less effort to setup tests by allowing heat spreading through PCB and modulating heat from the case top. When DUT is the only heat source, error from the measurement method is minimal. When other heat sources exist on the PCB, the interferences are revealed by an error indicator found in the thermal resistance network analysis of this work. The error indicator's effectiveness was proved by simulations of 10 cases. The error indicator can then be used as compensation to reduce measurement method error to minimum. Uncertainty analysis predicts errors with initial and improved experiment setups. Preliminary tests prove the concept of the proposed method. In summary, the proposed heat loss measurement method provides an innovative and practical way to estimate component heat loss with reasonable error.

## References

1. D. Christem, U. Badstuebner, J. Biela, and J. W. Kolar, "Calorimeter Power Loss Measurement for Highly Efficient Converters", The 2010 international Power Electronics Conference.

2. Chucheng Xiao, Gang Chen, and W.G. Odendaal, "Overview of Power Loss Measurement Techniques in Power Electronics Systems", 2007, IEEE Transactions on Industrial Applications, Vol. 43, Issue:3 pp.657 – 664.

3. B. Seguin, J.P. Gosse, A. Sylvestre, P. Fouassier, and J. P. Ferrieux, "Calorimetric Apparatus for Measurement of Power Losses in Capacitors", IEEE instrumentation and Measurement Technology Conference, St. Paul, Minnesota, may 18-21, 1998, pp. 602-607.

4. Gang Chen, Chucheng Xiao, and W.G. Odendaal, "An Apparatus for Loss Measurement if Integrated Power Electronics Modules: Design and Analysis", 2002, IEEE Industrial Application Conference, pp.222 - 226

5. John R. O'Malley, Circuit Analysis, Prentice-Hall, Inc., 1980.

6. Azar, K, Thermal Measurements in Electronics Cooling, CRC Press LLC, 1997.

# Thermal Conductivity Measurements of Novel SOI Films Using Submicron Thermography and Transient Thermoreflectance

Mihai G. Burzo[1], Peter E. Raad[2], Taehun Lee[3], and Pavel L. Komarov[2]

[1]Mechanical and Energy Engineering, UNT, 3940 N Elm, Denton, TX 76207
[2]Mechanical Engineering, SMU, Dallas, TX 75275
[3]INTEL Corporation, 2501 NW 229th Ave., Hillsboro, OR 97124
Mihai.Burzo@unt.edu

## Abstract

An approach is shown for extracting the thermal conductivity of several buried oxide (BOX) materials from the thermal map of activated devices. First, the surface temperature of ICs that were built on top of the candidate Silicon-on-Insulator structure were determined experimentally and then numerical models of each of the structures were constructed and a numerical simulation tool was finally used to solve the inverse heat transfer problem and extract the effective thermal conductivity of each dielectric layer. The thermal conductivity and the interface thermal resistance of the dielectric films were also measured directly, in-situ, using a laser based non-contact, non-invasive time-domain thermoreflectance approach. The effective thermal conductivity was then calculated based on the measured values of the intrinsic thermal conductivity and interface thermal resistance and the results were compared with the first approach.

## Keywords

Thermal mapping, thermography, thermoreflectance, BOX, buried oxide materials, silicon dioxide, aluminum oxide, aluminum nitride, silicon nitride, diamond like carbon, diamond.

## 1. Introduction

Present-day efforts directed at designing and developing faster and more powerful microelectronic devices that will maintain the exponential trend predicted by Moore's law must address challenges beyond those associated with lithography and miniaturization. An important such challenge is finding ways of improving the capabilities of the microelectronic structure to remove unwanted heat. Enhancing the heat transfer from the active region of the microelectronic device to the surroundings translates into reducing the thermal resistance along the heat path from the source to the sink, which means either decreasing the thickness of the layers in the heat path or increasing the thermal conductivity of the materials making up the device. The effort described in this article focuses on the latter, namely, finding ways to increase the thermal conductivity of the materials used in devices employing the Silicon-on-Insulator (SOI) technology [1,2].

Limitations of the bulk-silicon technology have pushed for new endeavors in using the SOI technology. There are many advantages of the SOI technology such as being suitable for high-energy radiation environments, much smaller parasitic capacitances, no latch-up, ease of making shallow junctions, improved trans-conductance and sharper sub-threshold slope, higher working temperature, etc. Among the disadvantages of SOI technology and one of the major reasons that has hampered wider SOI applications, particularly in the high-performance, high-speed analog and mixed signal ICs, is the *self- heating* effect caused by the low thermal conductivity of conventional buried oxide. Self-heating leads to degraded carrier mobility, reduced drain current, and increased power consumption. Therefore, finding good replacements for the current buried oxide layers (BOX) used in the SOI technology can be an important way of creating faster and more powerful electronics. Several alternative dielectrics were considered in this work and their thermal characteristics were determined as described next.

A direct and telling way of estimating the heat removal efficiency of various SOI structures candidates is to build identical ICs on each SOI structure and compare the temperature that they reach during normal operation. The coolest SOI, the one that leads to the lowest IC temperature, will obviously be the one that will remove the heat in the most efficient way, and therefore be the best candidate. This logic makes sense when considering that the BOX layer is always a significant part of the total resistance of the device under test. However, even though temperatures can be compared arithmetically, the process is still qualitative, and the quantitative aspects of the comparison might or might not be applicable to other devices that employ a different SOI structure or configuration. A better way of comparing the candidate dielectrics is to have an actual value of the effective thermal properties of each SOI material. Another clear advantage of having the actual thermal properties of the dielectrics is that they can be used in the design phase for more accurate calculations, numerical simulations, and analyses of practically any IC in any desired configuration or form factor.

In this work, the thermal conductivities of several new SOI dielectric candidates were determined using two methods:

1. Extracting the thermal conductivity of the dielectric layer from the surface thermal map of the ICs. First, the surface temperature of ICs that were built on top of the candidate SOI structure was determined experimentally and then numerical models of each of the structures were constructed and a numerical simulation tool was finally used to solve the inverse heat transfer problem and extract the effective thermal conductivity of each dielectric layer.

2. Measuring directly both the film (intrinsic) thermal conductivity and the interface thermal resistance of the dielectric films in-situ using a laser based, non-contact, non-invasive thermoreflectance approach. The effective

978-1-4673-1110-6/12 $31.00 © 2012 IEEE

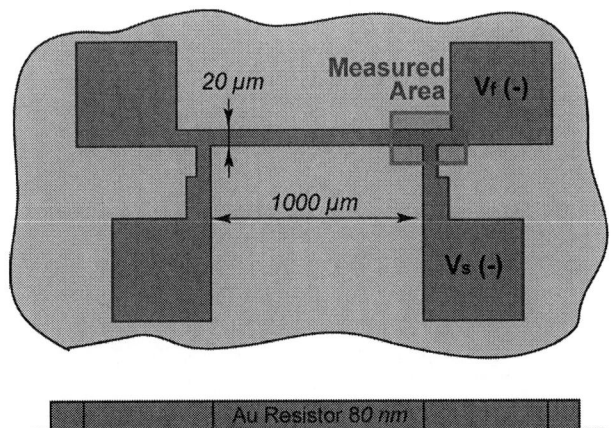

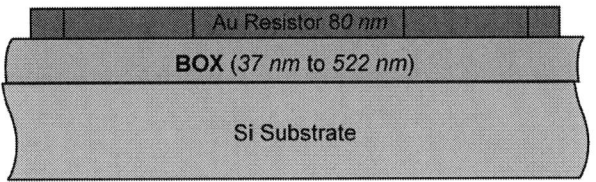

**Figure 1:** Top and cross-sectional views of the DUT (20 μm by 1000 μm microresistors). The thickness of BOX films varies from 37nm to 522nm.

thermal conductivity was then calculated based on the measured values of the intrinsic thermal conductivity and interface thermal resistance.

## 2. Extracting thermal conductivity of dielectric films from the experimental surface temperature map

This approach requires two key steps: a method and system that is capable of measuring the temperature of activated ICs with submicron resolution *and* a numerical simulation that allows ultrafast computation of the heat transfer process. We have created both components and the details are presented next.

Determining the temperature of an active microelectronic device requires the use of a technique that is preferably non-contact (and thus does not influence the temperature reading) and can offer the spatial and thermal resolution required for measuring submicron features that are typical in modern microelectronic devices. Contact methods are another option but present the difficulties of having to access features of interest with an external probe, or in the case of embedded features, fabricate a measuring probe into the device, and then having to isolate and exclude the influence of the probe itself. Non-contact methods are preferable because they can provide

surface temperature profiles or, since most devices use a transparent passivation and field oxide layer, can also provide direct temperature measurement of the channel area of a device with deep submicron spatial resolution.

### 2.1. Samples under test

The geometry and other features of the samples under test for this step of the investigation are shown schematically in Fig. 1 and consist of identical microresistors built on silicon dioxide, diamond-like-carbon, aluminum oxide, aluminum nitride, diamond and silicon nitride BOX layers prepared on 4-inch diameter Si (1-3Ω) wafers. The thickness of the resistors is 80 nm and the length and width of each resistor strip are 1000 μm and 20 μm, respectively. Two pads were constructed at each end of the resistor strip to enable the use of a 4-wire activation scheme (measure force voltage, $V_f$, and sense voltage, $V_s$), which makes it possible to accurately determine the heating power generated by each resistor. The BOX candidate materials were prepared by LPCVD, PECVD, furnace (thermally grown), and RF sputtering systems, as specified in the second row of Table 2. The dielectric properties were also measured and it was found that all BOX materials evaluated here have good dielectric properties, as the values in the last two rows of Table 2 would indicate.

The AlN films were deposited on Si (100) by a radio frequency (rf) reactive sputtering system without heating the stage. A pure Al target (99.999%) with 2.5-in diameter was used. Si wafers were dipped into 1% HF solution for 1 minute to remove the native oxide prior to loading samples in the chamber. The base pressure was $2 \times 10\text{-}8$ 5 mTtorr. During the deposition, the chamber pressure was maintained at 4 mTorr with RF power of 250 W with total N2/Ar mixture gas of 22 sccm. In order to examine the effect of the ratio of N2/Ar gas flow on thermal properties and microstructure of the film, three different mixture gas ratios, $N_2$/Ar, of 37.5%, 57%, and 83.3% were applied. The resulting films are referred to throughout this article as AlN#1, AlN#2 and AlN#3, respectively.

### 2.2. Obtain temperature map

The approach used here for measuring the surface temperature of activated microresistors built on top of various SOI structures is based on the thermoreflectance (TR) method [3, 4, 5], where the change in the surface temperature is found by measuring the change in the reflectivity of the sample. The measurement procedure is as follows: First, the

| BOX material | SiO₂ | DLC | Al₂O₃ | Si₃N₄ | AlN | Diamond |
|---|---|---|---|---|---|---|
| **Deposition method** | Thermally grown | LPCVD/ PECVD | RF sputter | LPCVD | RF Sputter | DoSi™ CVD Diamond on Silicon |
| **Dielectric strength (MV/cm)** | 8-10 | 5 | 5.5-6.5 | 8.6 | 5.5-6.5 | N/A |
| **Dielectric constant** | 3.9 | 5.4 | 7.5-12 | 8.9 | 7.5-12 | N/A |

**Table 2**: Deposition method and dielectric properties for all BOX materials considered.

thermoreflectance coefficient, $C_{TR}$, is determined for each of the surface materials to be scanned. Second, the changes in the surface reflectivity as a function of changes in temperature are measured at each location with submicron spatial resolution. Finally, the resulting reflectivity data are combined to obtain a transient temperature field over the scanned area. In order to maximize the signal to noise ratio, the wavelength of the probing light is chosen such that it produces the maximum value of $C_{TR}$ for the top gold layer.

A functional schematic of the thermography system is shown in Fig. 2. The probing light reflects from the heated surface back along the optical path to the sensitive element of a CCD camera. The intensity of the reflected light depends on the reflectivity (temperature) of the sample's surface. The frames containing the change in surface reflectivity induced by the temperature variations of the DUT are acquired, averaged, and scaled according to the calibration data.

The calibration approach consists of determining the relationship between the changes in reflectance and surface temperature. To do so, the device is held at prescribed low and high temperature levels. Dividing the measured change in reflectance by the prescribed change in temperature yields $C_{TR}$ for a given material at a given wavelength of light.

The temperature maps for several candidate materials are shown in Fig. 3. The maps were obtained by the use of a T°Imager™ Q16 system [6]. An identical power of 4 W was applied to all the microresistors made from a gold alloy. The heat generated by the microresistors, due to self-heating joule effect, produces a surface temperature rise, $\Delta T$, which is proportional to the thermal resistance of the BOX structure. Thus, by comparing the thermal map of the microresistors one can find which DUT runs cooler and hence find the BOX material that produces the lowest effective thermal resistance. The TRTG measurements were made with a light wavelength of 485 nm, which maximizes the value of $C_{TR}$ for the gold alloy. The value of the thermoreflectance coefficient was measured for each sample and found to vary between $1.51 \div 3.05 \times 10^{-4}$ for the gold microresistors considered here.

The positive value of the $C_{TR}$ (@485nm) indicates that an increase in temperature produces an increase in surface

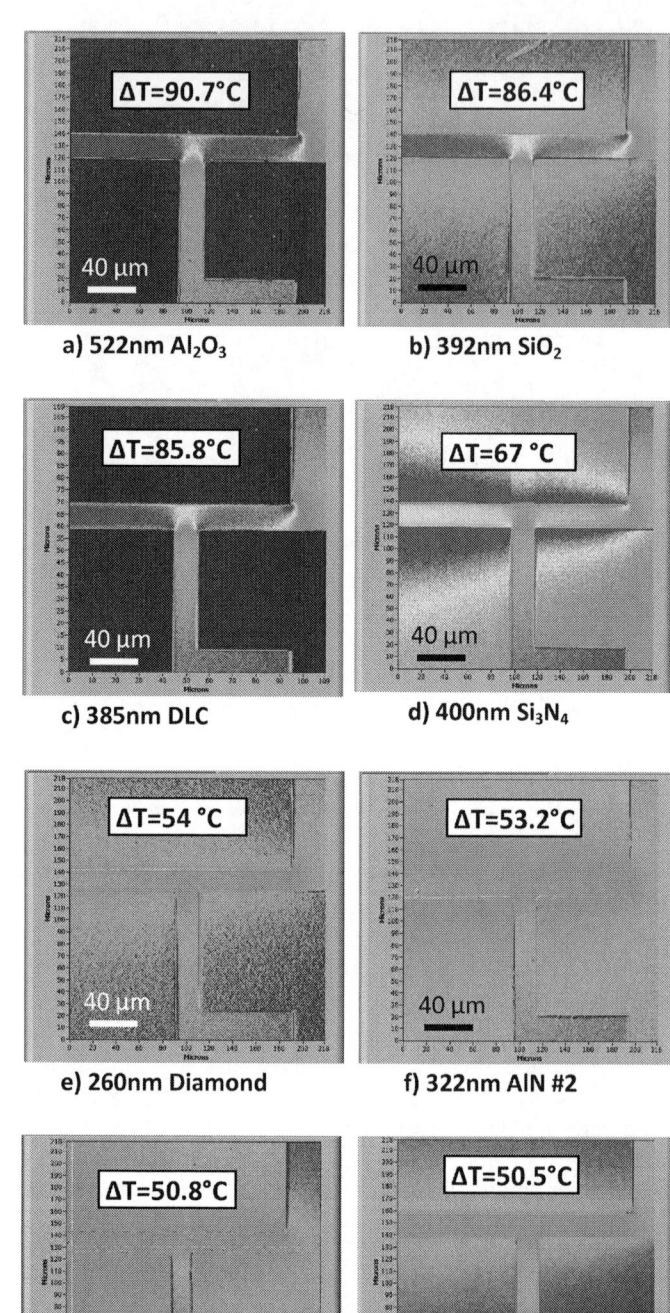

**Figure 3**: Surface temperature of the microresistors constructed on top of various BOX materials. The temperature contours are correct only within the microresistors area (the thermoreflectance coefficient was calibrated only for the Au microresistors area).

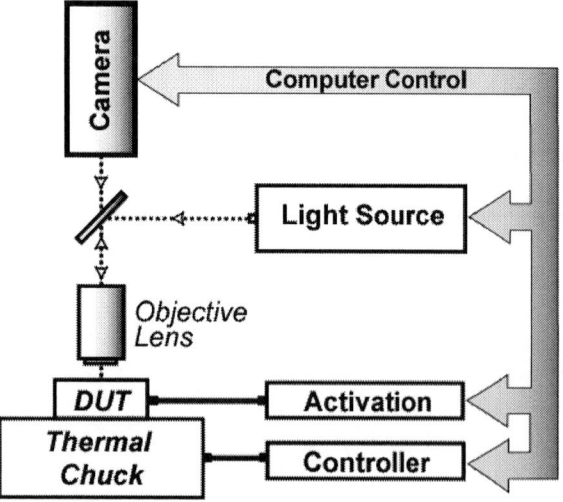

**Figure 2:** Schematic of the thermoreflectance thermography system

reflectivity. The temperature was averaged in a consistent area for all the microresistors as indicated by a box in Fig. 5. The calculated average values are shown in text boxes for each material in Fig. 3. The highest temperature rise of 90.7 °C was measured on aluminum oxide films, and the lowest temperature rise of 50.5°C was obtained for the aluminum nitride BOX sample. It is worth noting that since the thickness of the BOX layer is not the same for all the samples it is not trivial to state which material will provide the lowest effective thermal conductivity among the considered materials. This more conclusive determination can be attained by using a numerical simulation solver as shown next.

## 2.3. Solve inverse problem and obtain thermal conductivity

Once the surface thermal map is determined non-invasively and with submicron spatial resolution, the thermal conductivity of each thin-film layer composing the device is measured and a numerical model is built using measured and known values. The temperature distribution map is then used as input for an ultrafast inverse computational engine, T°Solver™ [1] to extract the thermal conductivity of the BOX layers. Since the heat transfer problem is parabolic in nature, a unique temperature field can be computed in the entire structure by solving the corresponding heat transfer problem. The novel numerical approach [7] begins by solving the corresponding steady-state problem by the use of a grid nesting technique. The nesting technique defines a template that is then used to solve the transient problem in a multi-scale fashion. The advantage to solving with multiple grids in time is that the majority of the problem domain can be resolved in time at lower nest levels (with a coarser grid), while the finer grid resolution is reserved for the parts of the problem that demand

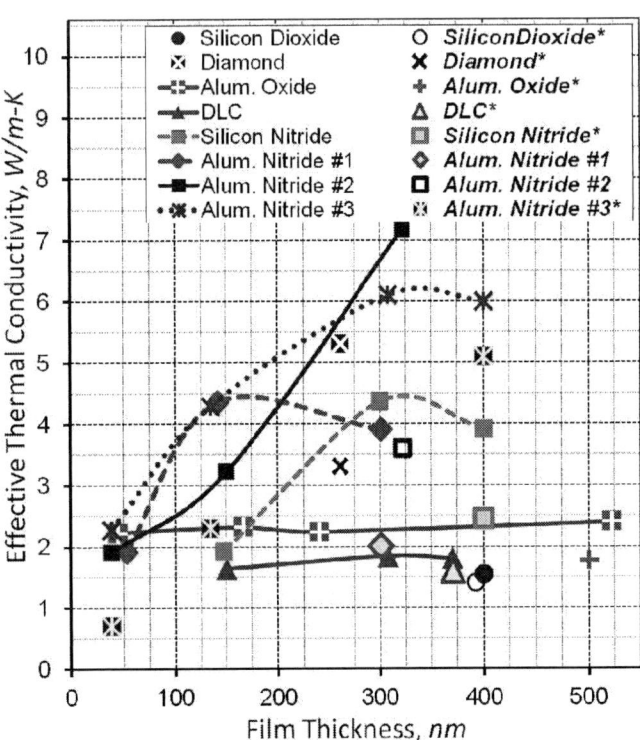

**Figure 4:** Measured effective thermal conductivity for several dielectric films using the two approaches describes in sections 2 and 3: (i) the transient thermoreflectance technique and (ii) the coupled thermoreflectance thermography-numerical simulation approach – data obtained using the coupled approach are marked with an asterisk (*).

the finer resolutions in space and time. The end results of this step are the thermal conductivities of each BOX film. The approach described above was used with several BOX samples and the results are tabulated in Table 2 and also

| BOX Material | Direct Transient Thermoreflectance Measurements | | | | Indirect Thermography-Simulation Results | | | Additional Data | |
|---|---|---|---|---|---|---|---|---|---|
| | Thickness BOX Film _nm_ | $K_{film}$ _W/m·K_ | $R_{th} \times 10^8$ _m²K/W_ | $K_{eff}$ (TTR) _W/m·K_ | Thickness BOX Film _nm_ | $\Delta T_{avg}$ _°C_ | $K_{eff}$ (Fitted) _W/m·K_ | $\rho \cdot Cp \times 10^{-6}$ _J/m³·K_ | $K_{bulk}$ (Bulk Mat.) _W/m·K_ |
| SiO₂ | 400 | 1.59 | 0.78 | 1.54 ± 0.03 | 392 | 86.4 | 1.41 | 2.27 | 1.3-1.5 [8] |
| DLC | 370 | 1.8 | <0.1 | 1.80 ± 0.02 | 385 | 85.8 | 1.6 | 1.998 | 1700-2100 [9] |
| AlN #3 (83% N₂/Ar) | 400 | 10.37 | 2.81 | 6.00 ± 0.05 | 400 | 50.8 | 5.1 | 2.705 | 150-200 [10] |
| | 134 | 7.63 | 1.36 | 4.29 ± 0.04 | 134 | 50.5 | 2.3 | 2.705 | 150-200 [10] |
| | 37.7 | 3.63 | 0.58 | 2.28 ± 0.04 | 38 | 48.8 | 0.7 | 2.705 | 150-200 [10] |
| AlN #2 (57% N₂/Ar) | 322 | 13.72 | 2.12 | 7.17 ±0.07 | 322 | 53.2 | 3.6 | 2.705 | 150-200 [10] |
| AlN #1 (38% N₂/Ar) | 301 | 6.66 | 3.15 | 3.91 ± 0.02 | 300 | 65.1 | 2 | 2.705 | 150-200 [10] |
| Al₂O₃ | 522 | 4.19 | 9.3 | 2.40 ± 0.03 | 500 | 90.7 | 1.76 | 3.423 | 25-35 [11] |
| Si₃N₄ | 400 | 4.88 | 4.85 | 3.06 ± 0.03 | 400 | 67.0 | 2.45 | 2.275 | 30 [12] |
| Diamond | 260 | 9.67 | 2.19 | 5.31 ± 0.05 | 260 | 54.0 | 3.3 | 1.787 | 900-2320 [13] |

**Table 2:** Effective thermal conductivity measured using the laser based thermoreflectance technique and the thermal conductivity extracted using the coupled thermography-numerical simulation procedure as described in section 2.3. Results are shown for silicon dioxide (SiO₂), diamond like carbon (DLC), aluminum nitride (AlN) obtained with various rations of nitrogen (N₂) and argon (Ar) gases, aluminum oxide (Al₂O₃), silicon nitride (Si₃N₄), and diamond.

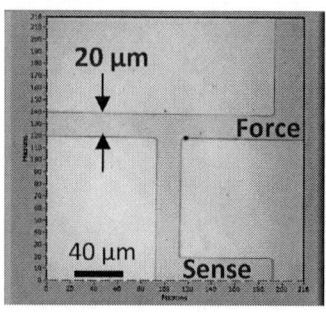
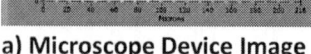

**a) Microscope Device Image**

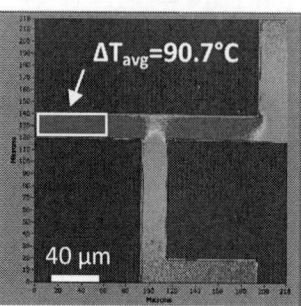

**b) Measured Surface Thermal Image**

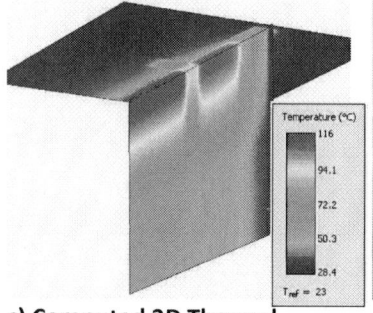

**c) Computed <u>3D</u> Thermal Image (horizontal and vertical slices shown through the center of the microheater)**

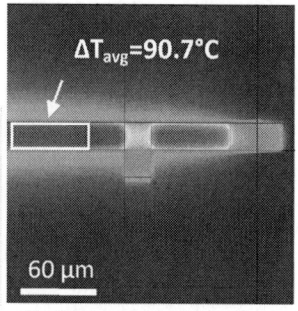

**d) Computed <u>2D</u> Surface Temperature Contours – Fitted to Experimental Thermal Map**

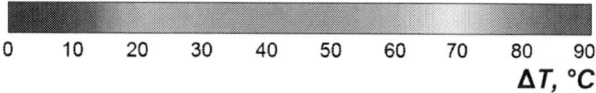

**Figure 5**: Measured and computed temperature contours for the Au microresistor device. Activation power is 4mW ±1%. The DUT is built on a 522nm layer of $Al_2O_3$ that was deposited on a silicon wafer.

plotted in Figure 4. Columns six to eight of Table 2 include the details of the results obtained using this technique. The thickness of the film, shown in column six of Table 2, was measured by a spectroscopic ellipsometer (SENTECH 800), which was also confirmed by cross-sectional transmission electron microscopy (TEM). The thickness of the Au layers was measured using a Veeco profilometer. The temperature rise shown in the table was obtained by averaging over a specific region of the thermal maps measured using the TRTG approach. Other data that were used to build the model are shown in Table 2. The conductivity of the Au was 315 W/m-K and the $\rho \cdot C_p$ value was taken as $2.45 \times 10^6$. Once the computational model was built, the value of the thermal conductivity of the BOX films was varied until the computed temperature rise matched the measured temperature rise. The resulting values are plotted in column eight of Table 2. A comparison between the measured and the computed surface temperature map is shown in Figs. 5b and 5d, respectively. The area that was used to average the temperature is indicated in the picture. The temperature map for the numerical map is

shown as an absolute value while the experimental map is shown as a temperature rise but the averaged temperature in the corresponding areas (boxes) is identical since the numerical data was fitted to the experimental data. An additional benefit of this approach is the fact that the 3D thermal contours are also obtained for the entire structure.

The data obtained using the thermography-simulation approach is compared with the thermal conductivity measurements obtained using a time domain thermoreflectance system as described next.

## 3. Direct thermal conductivity measurements of dielectric films using the transient thermoreflectance technique

The thermal conductivity of several BOX materials was measured directly using the Transient Thermoreflectance method that was described elsewhere [14, 15]. Since the method can measure both the intrinsic thermal conductivity and also the thermal interface, the effective thermal conductivity can be estimated as follows:

$$K_{eff} = \frac{1}{\dfrac{1}{K_{film}} + \dfrac{R_{th}}{h}}$$

(1)

where h is the thickness of the film, $R_{th}$ is the interface thermal resistance, and $K_{film}$ is the intrinsic thermal conductivity of the film.

The results of the measurements are shown in columns two to five of Table 2 and the thermal conductivity data are also plotted in Fig. 4. The sample consists of BOX films that are deposited on silicon wafers and then covered with a layer of gold. The layer of gold was optimized to maximize the responsivity of the transient thermoreflectance method as described elsewhere [14, 16]. As observed in Table 2 the thickness of the BOX films for same samples did not match exactly the thickness that was measured for the microresistors samples. The intrinsic thermal conductivity of the BOX film and the interface thermal resistance between the film and the Au layer is shown in columns three and four in Table 2, respectively. The effective thermal conductivity was then estimated using eqn. (1) and the data is shown in column five of Table 2. The bulk thermal conductivity value of all of the materials considered here is also included, for reference purpose, in the last column of Table 2. More thermal conductivity data is available in Figure 4.

## 4. Discussion

Analyzing the thermal conductivity data shown in Fig. 4 and Table 2, one can conclude that for all amorphous materials, such as $Al_2O_3$, DLC, and $Si_3N_4$ (also valid for $SiO_2$ – variation with thickness not shown here but was carried out in a separate study [17]), the conductivity does not change significantly with the thickness, while for polycrystalline AlN films a strong thickness effect on its thermal conductivity is observed. For all thin-film BOX materials investigated here the film thermal conductivity was much smaller than their corresponding bulk thermal conductivity values (shown in the last column of Table 2). Overall, the AlN films exhibited the highest thermal conductivity among all candidate materials evaluated. A more detailed investigation of the structure of

978-1-4673-1110-6/12 $31.00 © 2012 IEEE          154

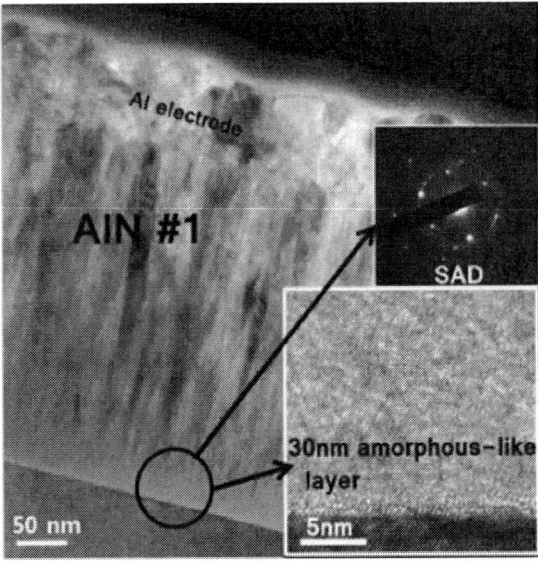

**Figure 6:** Cross-sectional TEM images of as-grown (a) ~320 nm-thick AlN #1 ( 37.5% $N_2$/Ar), and (b)~400 nm-thick AlN #3 (83.3% $N_2$/Ar).

the film was further carried out to determine the reason for the observed thermal conductivity values for the AlN films.

Cross-sectional bright-field TEM image were carried out as shown in Fig. 6. From the pictures, it is evident that the AlN thin-film exhibits a c-axis oriented columnar structure. The grain size is about ~36 nm on average close to its surface. Figure 6(a) shows an ~30 nm thick amorphous-like layer above the interface in the AlN#1 film while the AlN#3 (higher N2 content than AlN#1) clearly shows highly oriented crystalline, as depicted Fig. 6(b).

The phase purity and the crystal orientation of the films were examined by x-ray diffraction (XRD). XRD results confirmed that the AlN#1 film exhibits (10-10) preferential

orientation mixed with (0002) and (10-11) orientation while AlN#2 and AlN#3 films show strong (0002) c-axis-preferred orientation. The area near the interface between the AlN film and the Si substrate was also examined by using scanning transmission electron microscopy (STEM) energy dispersive spectrometer (EDS). The STEM EDS analysis confirmed that ~30 nm amorphous interfacial layer in AlN#1 film shows a gradual increase in Al while AlN#3 showed an abrupt increase in Al. Therefore, it can be concluded that differences of phase purity and microstructure lead to the different thermal conductivity values reported in Fig. 4. As one might expect, during heat transport, materials that exhibit a highly oriented crystalline structure induce less phonon scattering than those materials with weakly oriented crystalline structures.

The thermal conductivity values obtained for the thin films BOX materials is in very good agreement with the thin film thermal conductivity data observed by other investigators, using various measurement methodologies, as discuses next.

The measured thermal conductivity values of 1.8 and 2.4 W/m-K for $Al_2O_3$ films are well within the $1 \div 3$ W/m-K values reported by Bai Su-Yuan et al. [18] and Kato et al. [19].

Zhao et al. [20] reported thermal conductivity values from 1.4 to 4.5 W/m-K for sputtered AlN films that were up to 1050 nm thick, using the photothermal reflectance method. Kato et al. [19] used the micro ac calorimetry approach and reported thermal conductivity values of 5.6 to 8.4 W/mK for AlN films with thickness ranging from 100 to 300 nm. The values measured in this work are similar, ranging from $0.7 \div 6$ W/m-K and the were found to vary with the $N_2$/Ar gas ratio used during the film deposition as well as the film thickness.

For the up to 0.4 μm thick $Si_3N_4$ films the measured thermal conductivity values was found to be around $2 \div 4.4$ W/m-K which is similar to the $2.1 \div 2.3$ data obtained by Von Arx et al.[21] and Stojanovic et al. [22] for a 0.21 μm $Si_3N_4$ film, but the values obtained here are smaller than the 9 W/m-K values reported by Zhang et al. [23], which was obtained for a thicker 1.4 μm film.

## 5. Conclusions

The thermal conductivity results show that all amorphous materials, e.g, $Al_2O_3$ DLC, $Si_3N_4$, and SiO2, experience a smaller thickness effect while polycrystalline AlN film exhibits strong thickness effect on their thermal conductivity values. As expected, all thin-film BOX materials presented much smaller thermal conductivity values as compared to their bulk values (shown in the last column of Table 2). Nonetheless, AlN films exhibited the highest thermal conductivity among all candidate materials evaluated and as expected also produced the smallest increase in the temperature of the built-in microresistors.

The thermal conductivity values determined using the two approaches compare well to each other and as mentioned earlier the values are much lower than the bulk values. The measured effective thermal conductivity values confirm the

978-1-4673-1110-6/12 $31.00 © 2012 IEEE

qualitative assessment made using the results of the thermoreflectance thermography.

Since many thermal analysts and investigators might not have direct access to tools that can measure the thermal conductivity of thin films but have access to thermography imaging systems it was shown that the proposed approach could be successfully used to determine the thermal conductivity of thin films using the thermal map of the device coupled with the results of a numerical simulation. The responsivity (sensitivity) of the approach is quite good. As an example, for the AlN#3 sample with 134nm in thickness, a change of $1°C$ in the surface temperature is induces by a change of 5% in the thermal conductivity.

## Acknowledgments

The authors wish to thank Professor Kim Moon of the University of Texas at Dallas for his support with this project.

## References

1. Aubain, M. S. and P. R. Bandaru, "Determination of diminished thermal conductivity in silicon thin films using scanning thermoreflectance thermometry", Appl. Phys. Lett., 97, p. 97-99, 2010.

2. Liu, W. and M. Asheghi, "Thermal Conductivity Measurements of Ultra-Thin Single Crystal Silicon Layers", J. Heat Transfer, Vol. 128, p. 75-84, 2006.
   3a. 2005 Published

3. Burzo, M. G., Komarov, P. L., Raad, P. E., "Non-Contact Transient Temperature Mapping Of Active Electronic Devices Using The Thermoreflectance Method", IEEE Transactions on Components and Packaging Technologies, Vol. 28, pp. 637 – 643, 2005.

4. Christofferson, J. and A. Shakouri, "Thermoreflectance based thermal microscope", Review of Scientific Instruments, Vol. 76 ( 2), pp. 024903-024903-6. (2005).

5. Tessier,G., S. Pavageau , B. Charlot , C. Filloy , D. Fournier , B. Cretin , S. Dilhaire , S. Gomes , N. Trannoy , P. Vairac and S. Voltz  "Quantitative thermoreflectance imaging: Calibration method and validation on a dedicated integrated circuit",  IEEE Trans. Compon. Packag. Technol., vol. 30, p.604 , 2007.

6. T°Imager ™ and T°Solver™ are registered trademarks and products of TMX Scientific, Inc.

7. Raad, P. E., J. S. Wilson, and D. C. Price, "System and Method for Predicting the Behavior of a Component," U.S. Patent No. 6,064,810 (2000), Korean Patent No. 0501053 (2005), Japanese Patent No. 3,841,833 (2006).

8. Burzo, M. G., Komarov, P. L., Raad, P. E., "Thermal transport properties of gold-covered thin-film silicon dioxide",  IEEE Transactions on Components and Packaging Technologies, p. 80 - 88,  Vol. 26(1), 2003.

9. Jukka Rantala, "Diamonds are  a thermal designer's best friend", Electronics Cooling, 2002.

10. P. S. de Baranda, "The effect of Calcia and Sílica on the Thermal Conductivity of Aluminum Nitride Ceramics, PhD. Thesis, Rutgers, The State University of New Jersey, 1991.

11. Bansal, N. P.; Zhu, D., "Thermal conductivity of zirconia-alumina composites", Ceramics International, Volume 31, Issue 7, Pages 911-916, 2005.

12. James F . Shackelford and William Alexander, CRC Materials Science and Engineering Handbook, p. 281, CRC Press, 2000.

13. Berman, R., P. R. W. Hudson, and M. Martinez, "Nitrogen in diamond: evidence from thermal conductivity", J. Phys. C. 8, 21 (1975) L430-L434.

14. Burzo, M. G., P. L. Komarov, and P. E. Raad, "A Study of the Effect of Surface Metalization on Thermal Conductivity Measurements by the Transient Thermo-Reflectance Method," ASME J. Heat Transfer, Vol. 124, No. 6, pp. 1009-1019, 2002.

15. Komarov, P. L., M. G. Burzo, G. Kaytaz, and P. E. Raad, "Transient Thermo-Reflectance Measurements of the Thermal Conductivity and Interface Resistance of Metallized Natural and Isotopically-Pure Silicon," Microelectronics J. , Vol. 34, pp. 1115-1118, 2003.

16. Burzo, M. G., P. L. Komarov, and P. E. Raad, "Minimizing the Uncertainties Associated with the Measurement of Thermal Properties by the Transient Thermo-Reflectance Method, IEEE Transactions on Components and Packaging Technologies, Vol. 39, pp. 39-44, 2005.

17. Burzo, M. G., P. L. Komarov, and P. E. Raad, "Thermal Transport Properties of Gold-Covered Thin-Film Silicon Dioxide", IEEE Transactions on Components and Packaging Technologies, Vol. 26, No. 1, pp. 80-88, 2003.

18. Yuan, B.S., Tang Zhen-An, Huang Zheng-Xing, Yu Jun and Wang Jia-Qi, "Thermal Conductivity Measurement of Submicron-Thick Aluminium Oxide Thin Films by a Transient Thermo-Reflectance Technique", Chinese Phys. Letters, Vol. 25, pp. 593-596, 2008.

19. Kato,R., A. Maesono, and R. P. Tye, "Thermal Conductivity Measurement of Submicron-Thick Films Deposited on Substrates by Modified ac Calorimetry (Laser-Heating Angstrom Method)", International Journal of Thermophysics, Vol. 22, No. 2, pp. 617-629, 2001

20. Zhao, Y., C. Zhu, S. Wang, J. Z. Tian, D. J. Yang, and C. K. Chen, "Pulsed photothermal reflectance measurement of the thermal conductivity of sputtered aluminum nitride thin films", J. Appl. Phys., 96, 8, pp. 4563 – 4568, 2004.

21. M. von Arx, O. Paul, and H. Baltes, "Process dependent thin film thermal conductivities for thermal CMOS MEMS," J. Microelectromech. Syst., vol. 9, no. 1, pp. 136–145, 2000.

22. Stojanovic, N., Y. Jongsin, E.B.K Washington, J/M. Berg, M. W.Holtz, and H. Temkin, "Thin-Film Thermal Conductivity Measurement Using Microelectrothermal Test Structures and Finite-Element-Model-Based Data Analysis", Journal Of Microelectromechanical Systems, Vol. 16, No. 5, pp. 1269-1275, 2007.

23. Zhang, X. and C. P. Grigoropoulos, "Thermal conductivity and diffusivity of free-standing silicon nitride thin films", Review of Scientific Instruments, Volume 66 (2), pp.1115-1120, 1995.

# Measurement of Thermal Conductivity of Thin and Thick Films by Steady-state Heat Conduction

R. Shrestha

Department of Mechanical and Energy Engineering, University of North Texas,
3940 N. Elm St. Denton TX 76203 USA
rameshstha@yahoo.com

T. Y. Choi

Department of Mechanical and Energy Engineering, University of North Texas,
3940 N. Elm St. Denton TX 76203 USA
choi@egw.unt.edu

W. S. Chang

Korea Institute of Machinery and Materials,
171 Jang-dong Yuseong-gu, Daejeon 305-343, Korea
paul@kimm.re.kr

## Abstract

This paper describes novel glass micropipette thermal sensor fabricated in a cost-effective manner and thermal conductivity measurement of carbon nanotubes (CNT) thin film using the developed sensor. Various micrometer-sized sensors, which range from 2 $\mu$m to 30 $\mu$m, were produced and tested. The capability of the sensor in measuring thermal fluctuation at micro level with an accuracy of $\pm 0.01^{\circ}$C is demonstrated. We have obtained reproducible thermal conductivity data using various micropipette temperature sensors. The average thermal conductivity of the CNT film at room temperature was determined at 73.4 W/m$^{\circ}$C.

## Keywords

Thermal conductivity, CNT films, micropipette

## Nomenclature

A area, m$^2$
E electron energy
e electronic charge
K Boltzmann constant
k thermal conductivity, W/mK
Q absorbed laser power, W
r radial position, m
S Seebeck coefficient, V/K
T Temperature, K
t thickness, m

Greek symbols
$\zeta$ Fermi energy, eV
$\lambda$ mean free path of the conduction electron, m

## 1. Introduction

The study of physical phenomena at a micro scale is made possible today with the development of atomic force microscopy, scanning probe microscopy and ultrafast light sources which have enabled the measurement of thermal, mechanical, chemical, electronic, optical, and acoustic properties of materials [1]. However thermal conductivity of a very thin conducting film with thickness, t < 1 $\mu$m has been difficult to be measured due to the complexity of experimental setup. According to Volklein [2] various plate methods for determination of thermal conductivity of thin

film are not useful as the thermal resistance of the interfaces between film and plates or heat sinks rather than the thermal conductivity of the film is the dominating measuring quantity. The other problem in the measurement of the thermal conductivity is aroused when large temperature sensors are used where the relative loss of heat by sensor may be larger than the heat flow in thin films. The solution to this problem is to employ a laser point source for heating instead of using bulky heating strips that are much wider than the film thickness and avoid thermal resistance of the interfaces between heating strip and film. In addition, the use of a micro scale thermal sensor with high sensitivity and resolution should be employed for measuring the temperature change accurately and avoid the loss of heat from the sensing tip. Such thermocouple probes based on a glass micropipette for the measurement of thermal responses has been demonstrated [3,4]. However, they were unable to provide meaningful temperature data that would validate the accuracy and resolution of the sensor [4]. In addition, the fabrication technique is intricate, requiring several processes [4] or drawing of a micro platinum wire [3] inside a micro-capillary tube. Gold and platinum as thermocouple materials were used [3, 5]. These materials are expensive and generate relatively small thermo power (i.e., a low level of thermoelectric voltage signals).

In this paper, fabrication of a novel glass micropipette thermal sensor is described and a new method for determining the thermal conductivity of thin CNT film at room temperature is presented. The micropipette thermal sensors were fabricated such that different tip sizes were produced with varying coating conditions during physical vapor deposition. Then the micropipette thermal sensors were calibrated individually using a constant-temperature water bath chamber and were used to measure the temperature difference at two spots, which were 10 to 20 $\mu$m apart, on a 100 nm thick CNT film.

## 2. Methodology and Materials

A thick-wall borosilicate glass micropipette with outer diameter of 1.5 mm and inner diameter of 0.86 mm that is used in various biological applications for injecting biological solutions into tissues was used to produce a sensing tip in

978-1-4673-1110-6/12 $31.00 © 2012 IEEE

micrometer size. The pipette puller (P-97, Sutter Instrument) was programmed according to the recipe to create patch pipettes with a tip size of 1 μm and approximately 5 to 7 mm long taper length (Fig. 1a). The pulled pipette was filled with a lead-free soldering alloy mainly composed of tin (Sn) by an injection molding process in conjunction with localized heating of material. The injection molding was accomplished by mechanical pressurization (or pushing) of molten metal at the upper part of the pipette while heating the lower part near the tip with an electronic soldering gun maintained at 300 °C as shown in Fig. 1b.

The next step was the beveling of the tip in order to remove unwanted metal extruded outside the pipette. This step is particularly important to assure there is a smooth and continuous contact between two metals after sputtering of nickel. Therefore, BV-10 micropipette beveler (Sutter Instrument) that was designed for beveling micropipettes with tip diameters between 0.1 and 50 μm was used to sharpen and smoothen the tip. A 2-axis micromanipulator consisting of an angle plate to clamp the pipette was used to adjust the bevel angle between 25–30°. Controlled advancement of the pipette to an ultrafine grinding surface with 0.3 μm alumina abrasive was performed with use of coarse and fine control knobs mounted on the manipulator.

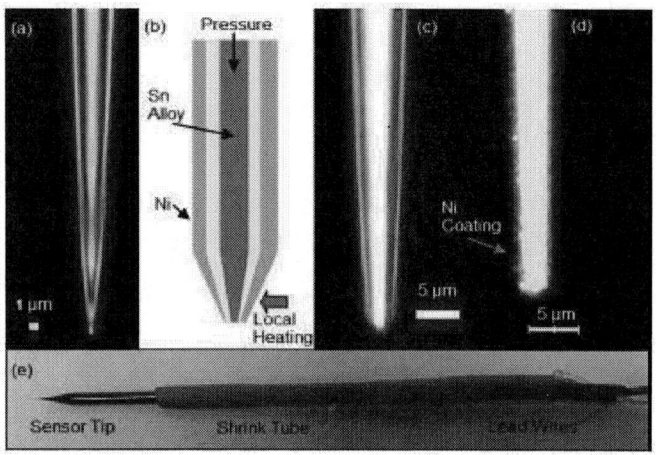

Figure 1 Fabrication steps of pipette thermal sensor: (a) an empty pipette with less than 1 μm opening; (b) a schematic outline of the filled pipette with a solder alloy and coated nickel; (c) after filling with a solder alloy and beveling; (d) after PVD thin film coating of nickel and silver; (e) a prototype micropipette thermal sensor.

In order to investigate the effect of the tip size on the thermoelectric power of the sensor, the pipettes were beveled to produce tip sizes varying from 2 to 30μm. Fig. 1c shows an image of a pipette, which was taken with a high-magnification optical microscope (Nikon Eclipse ME600) after beveling. The pipette was cleaned with ethyl alcohol using an ultrasonic cleaner (SHARPERTEK). Then a sputtering technique was used to coat thin films of nickel on the outer surface of glass and thus forming a Ni-Sn alloy junction at the beveled tip (Fig. 1d). To analyze the effect on the sensitivity of the sensors due to the coating condition, a batch of sensors were produced by differing the heating power to create the plasma

of nickel target inside the sputtering machine and also varying the deposition time. During the sputtering, the micropipettes were held at a vertical position facing directly to the nickel target and rotated at a certain speed during the plasma creation and deposition so that uniform thin film of nickel would be created on the surface of the micropipette. Four batches of sensors were produced by varying the power from 200 to 300 W and deposition time from 30 minutes to 1 hour.

To connect the sensor with a voltmeter, lead wires (Sn alloy and Ni) were constructed at the end (opposite to the tip) of the micropipette by using the same material as the lead wires; this will prevent creation of unwanted signals from additional metal junctions unless otherwise compensated electronically. Therefore the Ni-wire was wound around the surface of the coated Ni and secured with epoxy. Similarly, the same solder wire, with ample length, was soldered to the inner Sn-alloy. To strengthen the connection and minimize the contact resistance between the wound Ni-wire and the thin film Ni on the glass surface, a heat shrink tube was used on top of the junction (Fig. 1e).

Calibration of the fabricated sensors was conducted in a thermally insulated calibration chamber filled with water. The cylindrical body of the chamber was made of aluminum and the covering lid was made of Teflon for thermal insulation.

The temperature of the water-filled chamber was controlled at a constant level with an accuracy of ± 0.01 °C. The temperature of the chamber was varied from room temperature of 21 °C to 40 °C by using an in-house temperature controller integrated with an automated data logging system with LabVIEW. A high-precision digital thermometer and the fabricated sensor were immersed into the water bath. Thus, both of them were in close proximity to indicate nearly the same temperature.

The voltage generated by the sensor was recorded by a voltmeter (Nano Voltmeter, Keithley 2182). During the calibration process, the cold junction (Ni-Cu and Sn-Cu junction) of the sensor was maintained at a constant temperature (e.g. 24.5 °C) which was slightly above the room temperature using another in-house isothermal block made of aluminum so that unwanted additional thermocouple effects due to the cold junction could be removed. A plot of voltage signals provided by the sensor versus temperatures measured by a high-precision digital thermometer was obtained. The standard deviation in the voltage measurement was less than 0.018 μV which is equivalent to temperature rise of 0.002 °C, which is much less than the temperature measurement accuracy of 0.01 °C.

In order to measure the thermal conductivity of the CNT film, a class B laser at 633 nm was irradiated at the center of the CNT film suspended on a polycarbonate substrate and the temperature difference at two radial positions were measured using two different pipette sensors of approximately 3.5 μm sensing tip with seebeck coefficient of 5.67 and 7.44 μV/°C during two separate experiments for the same CNT film under the same experimental conditions.

The optical setup was prepared to guide the laser from the laser source to the film. Combination of anti reflection (AR) coated lenses from THORLABS were used to produce a collimated light which was then diverted at 90° by a beam

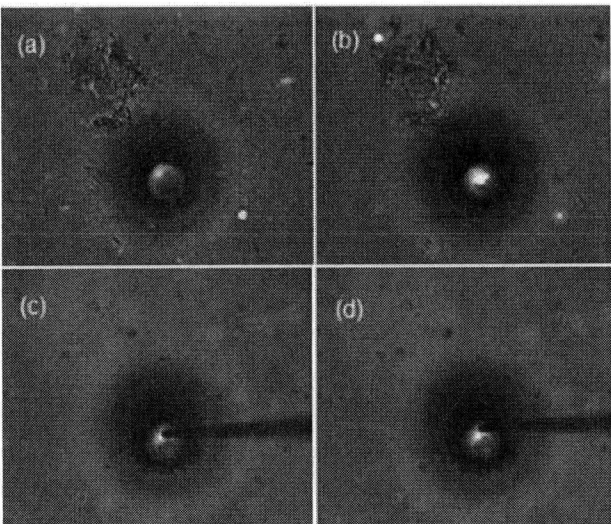

Figure 2 Measurement of temperature with pipette sensor. (a) CNT film at focal plane. (b) laser shined at the center of the film, (c) pipette sensor at position 1 and (d) pipette sensor at position 2.

splitter (THORLABS CM1-BS013) towards CNT film. A 20X objective lens (Mitutoyo MPlan Apo SL) was used to focus the laser into approximate diameter of 3 μm at the focal plane. White light from high intensity illuminator (Edmund optics MI-150) was guided with an optical fiber and delivered through an objective lens to illuminate sample placed at the focal plane. The CNT film on the polycarbonate substrate was positioned horizontally in between two of the objective lens at the focal plane of 20X objective lens with use of XYZ motorized stage. In a similar approach, motorized stage was used to precisely control the movement of pipette sensor to position at a desired location on the surface of the thin film. Special care was taken to approach the sensor on the CNT film as slight extra pressure would break the film.

A variable ND filter was used to control the power of the laser at a desired level. The film was placed at the focal point of the objective lens which was confirmed with the clear image obtained with the CCD camera as shown in Figure 2 (a). Laser was illuminated at the center of the film as shown in Figure 2 (b) and then turned off momentarily to position the pipette sensor. Then the sensor was brought to a position (Figure 2c) such that its radial position r > 6μm in order to avoid the direct laser irradiation at the tip of the sensor. Extensive care was taken while landing the pipette sensor on the film. In order to minimize the contact resistance in between sensor and film, a slight further movement to the sensor was provided by motorized controlled stage. At this position the maximum voltage signal was recorded by the voltmeter which confirms that the sensor and the film were in good contact. For consistency in measurement, the same process was repeated to measure the voltage signal at different locations on the film.

Laser was then irradiated at a fixed power level (200 μW) and at the steady state the voltage generated by the sensor was recorded using nano voltmeter (Keithley 2182). The LabVIEW program controlled the data recording such that the voltmeter would record 3 data points returning their mean

value every 10 seconds for 1 minute at the steady state. The transmitted data to the computer was then recorded in a text file for post data processing. It was observed that voltage signal was significantly higher when the laser was irradiating with sensor touching the film than when sensor was at the same position but not in contact with the film. This is the indication that the sensor is actually measuring the temperature rise due to heat conduction across the film.

The process was repeated to measure the voltage generated at different location. With a known seebeck coefficient of the sensor used, the voltages recorded at the two positions were then converted to temperature difference. It was measured and confirmed with the pipette sensor that the temperature at radius r> 25 μm was approximately close to room temperature.

Power absorbed during the laser irradiation at the center of the film is necessary to determine the thermal conductivity of the film. In order to determine the irradiated power absorbed by the film, conservation of energy was applied. This implies the amount of incident energy is equal to the sum of the absorbed, reflected, and transmitted energy.

With known power before the laser is irradiated on a film, transmitted power reflected power and absorbed power were measured. With a known absorbed power, radial positions and the temperature difference at two locations, thermal conductivity of the CNT film was calculated using simplified Fourier's equation for the radial heat conduction equation below.

$$Q = \frac{2\pi k t (T_1 - T_2)}{\ln\left(\frac{r_2}{r_1}\right)} \qquad (1)$$

## 3. RESULTS

### Calibration of pipette sensor

From four batches of thermal sensors that were produced by varying the power and deposition time during PVD sputtering, we observed that the seebeck coefficient of the sensors produced by the same PVD condition were approximately the same and did not vary much for the different tip size. However, the increase in the deposition time and power i.e, deposition rate of the coating, had significant effect on the seebeck coefficient of the sensors. Higher seebeck coefficient of the sensor resulted from higher deposition rate.

All the sensors were calibrated at room temperature from 21 °C with an increment of 4 °C up to 40 °C. The data obtained from the calibration were fitted with a linear trend line with $R^2$ value of 99% which provide the validation of the linear relationship between voltage and temperature. Therefore the seebeck coefficient of the sensor for this temperature range can be assumed to be constant. Figure 3 to figure 5 are the plots of voltage versus temperature for different sensors produced with the same PVD condition.

The calibration was repeated for some of the sensors to check the repeatability in measurement by the sensor. It showed that variation in seebeck coefficient was within 4%. This is an acceptable value as the error can be generated in the voltage measurement by voltmeter whose error in measurement was designated as ±0.02 µV. Also the instrumental error for the digital thermometer was same i.e, ±0.02 µV. Root Sum Square (RSS) method [6] was used to determine the uncertainty in the value of the seebeck coefficient. This error was calculated to be ±0.05 µV/°C for a unit change in voltage and temperature. During the measurement of voltage, a sensor with seebeck coefficient of 6.55 µV/°C was capable of recording the data with standard deviation of 0.01 which is equivalent to 0.002 °C change in temperature. Therefore the resolution of the sensor was determined as 0.002 °C while the measurement accuracy was limited by the calibration, which is 0.01 °C.

However the seebeck coefficients of the sensors are far below than those of bulk material, the possible cause of which can be related to the thermoelectric behavior of vapor deposited thin metal film. Hill et al. [7] explained the dependency of thermoelectric power of a thin film using a conduction mechanism of free electrons in crystal lattice:

$$S = \frac{\pi^2 K^2 T}{3e\varsigma} \left\{ \frac{d \ln \lambda(E)}{d \ln(E)} \right\}_{E=\varsigma} \quad (2)$$

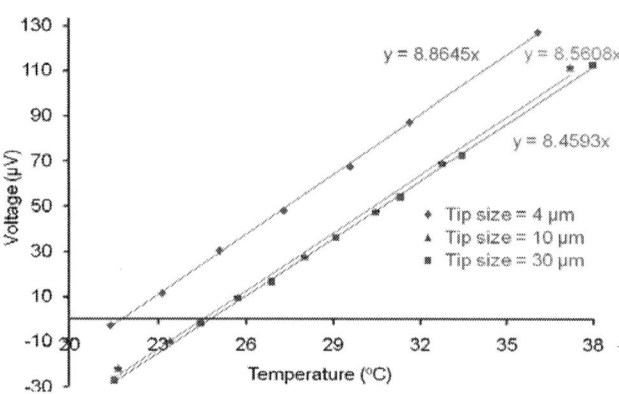

Figure 3 Calibration of sensors varying in tip diameters from 4 to 30 microns that were produced with the same PVD conditions (Power = 300 W, Deposition time = 1 hr).

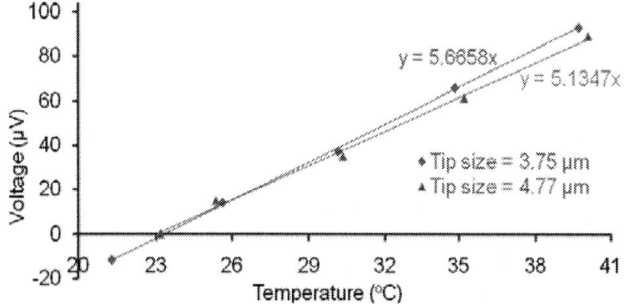

Figure 4 Calibration of sensors varying in tip diameters from 3.75 to 4.77 microns that were produced with the same PVD conditions (Power = 250 W, Deposition time = 45 mins).

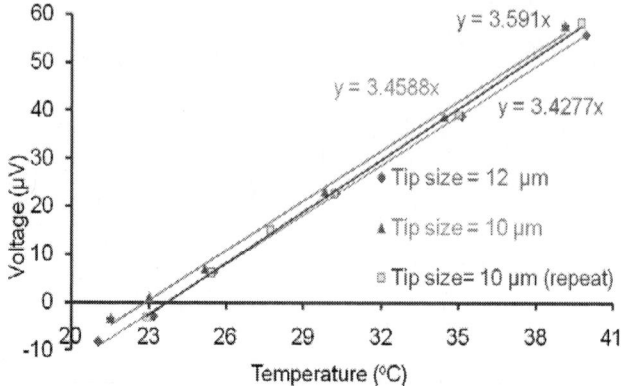

Figure 5 Calibration of sensors varying in tip diameters from 10 to 12 microns that were produced with the same PVD conditions (Power = 200 W, Deposition time = 30 mins).

where S is thermoelectric power, K is boltzman constant, e is electronic charge, T temperature, $\varsigma$ is Fermi energy, $\lambda$ is mean free path, and E is the electron energy.

Here S is a function of Fermi energy and Mean free path (MFP), both of which are dependent on addition of impurities and the thickness of the film. A thickness of 80-90 nm is estimated based on the measurement of film thickness on a quartz substrate using a profile-meter after deposition. At this thickness, there is a possibility of formation of a large number of broad empty channels [8] incorporated with impurities and structural defects. This results in reduction of the mean free path and enhancement of the electrical resistivity, which explains the reduced seebeck coefficient in thin films.

*Thermal Conductivity measurement of CNT film*

As the lateral dimension of the suspended CNT film was 50 µm, there was a necessity to use the smallest tip pipette sensor to measure the temperature precisely within a limited region. An experiment was carried out by irradiating laser at two power levels. Variable ND filter was used to fix the laser power at approximately 200 µW. Voltage at two different locations, approximately 8 µm apart in radial directions, were measured. Laser power at various locations in the optical setup was measured with the power meter in order to determine the power absorbed by the CNT film. Using all the powers measured, the power absorbed by the CNT film was determined at 65.60 µW which was about 33% of the total irradiated power. Then using equation 1 the thermal conductivity was determined as 75.004 W/m°C. The experiment was repeated at the same power level using the same pipette sensor but this time the temperature difference was measured for two new locations. The thermal conductivity of the film was calculated as 76.37 W/m°C. Two more experiments were conducted with the same sensor with slight increment in laser power to approximately 250 µW. Obviously the value of the power absorbed increased to 79.2 µW but the percentage power absorption by the film remain approximately the same. The thermal conductivity of CNT film with this experiment was calculated as 73.48 W/m°C. Similar experiment was completed using pipette sensor with smaller sensitivity. From the measurement the thermal conductivity was determined to be approximately 72 W/m°C. The thermal conductivity of the CNT film determined by two different sensors while differing the power level to irradiate the film were approximately the same value and therefore average value 73.418 W/m°C was accepted as the thermal conductivity of the film. From the measurement the standard deviation in the thermal conductivity values is only 1.94 W/m°C which is 2.64% error in the measurement.

To confirm that the heat transfer in the film was mostly due to the radial heat conduction in the film, the natural convection and radiation heat loss was also evaluated. When the laser was fixed at 250 µW (highest power used during the experiment), the sensor detected that the laser would produce a temperature of 27°C at the center of the film. Since the experiment was conducted at a room temperature of 21°C, the mean film temperature to calculate the convection heat transfer can be at 24°C. At this mean film temperature Rayleigh number was determined at $2.79 \times 10^{-4}$ which confirmed that there will be natural convection. Using Lienhard [9] assumption the convective heat transfer coefficient at mean film temperature of 24°C for air was determined to be 22.85 W/m$^2$ K. Then using Newton's law of cooling the heat loss due to the natural convection was determined to be 0.269 µW which is only 0.36 % of the absorbed power 73 µW (absorbed power assumed as average of the maximum and minimum power recorded during the experiment). Similarly using Stefan Boltzmann law, the heat loss to the surrounding from the film surface was calculated as 0.069 µW which is only 0.09% of the total power absorbed in the film. Therefore the heat transfer by natural convection and radiation was ignored and the temperature difference measured by the sensor was confirmed as the sole effect of the radial heat conduction in the film.

## 4. CONCLUSION

A novel technique for fabricating high-resolution micropipette thermal sensors using inexpensive materials was described. The developed technique can be implemented relatively easily in a laboratory setting. The micropipette sensors were manufactured by filling Sn-based alloy inside a glass micropipette and coating thin films on the outer surface of a glass micropipette by PVD so that a thermocouple junction was created only at the tip of micropipette. With a comparative study of the sensors produced by different PVD coating conditions, it was demonstrated that higher deposition rate would produce sensors with higher sensitivity and the sensitivity of the sensor would not be affected by the tip size. Sensitivity of the sensor was found as low as 3.43 µV/ °C and maximum of 8.86 µV/°C depending on the coating conditions. Measurement performed by two different sensors with different sensitivity produced similar data results for same laser power. The calculated thermal conductivity of the CNT film using the measurement data varied from 71.384 to 76.37 W/m°C. Although the measurements were performed with 2 different sensors with different laser irradiating power, the standard deviation in the calculated value for thermal conductivity was only 1.94 W/m°C which is 2.64 percent of the mean value. Therefore the mean value 73.418 W/m°C was accepted as the thermal conductivity of the CNT film at room temperature.

## REFERENCES

[1] A. Majumdar, 1999. Scanning thermal microscopy, Annu. Rev. Mater. Sci. 29, 505-585.

[2] F. Volklein, F, and D.-P. E. Kessler, 1987. Methods for the measurement of thermal conductivity and thermal diffusivity of very thin films and foils. Measurement Vol 5 No 1 , 38-45.

[3] G. Fish, O. Bouevitch, S. Kokotov, K. Lieberman, D. Palanker, I. Turovets, and A. Lewis, 1995. Ultrafast response micropipette- based submicrometer thermocouple, Rev Sci. Instrum. 66 (5), 3300-3306.

[4] M. S. Watanabe, N. Kakuta, K. Mabuchi and Y. Yamada, 2005. Micro-thermocouple probe for measurement of cellular thermal responses, Proc. 27th Ann. Conf. 2005 IEEE EMB., 4858-4861.

[5] N. Kakuta, T. Suzuki, T. Saito, H. Nishimura and K. Mabuchi, 2001. Measurement of microscale bio-thermal responses by means of micro-thermocouple probe, EMBS International Conference. Istanbul, Turkey: IEEE, 3114-3117.

[6] J. P. Holman, 1994. Experimental Methods for Engineers Sixth Edition. New York: McGraw Hill, Inc.

[7] D. E. Hill, L. Williams, G. Mah and W. L. Bradley, 1997. The effect of physical vapor deposition parameters on the thermoelectric power of thin film Molybdenum-Nickel junctions, Thin solid films 40, 263-270.

[8] M. Adamov, B. Perovic and T. Nenadovic, 1974. Electrical and structural properties of thin gold films obtained by vacuum evaporation and sputtering, Thin Solid Films 24, 89-100.

[9] A. F. Mills, 1995. Heat and Mass Transfer. California: Library of Congress Cataloging-in-Publication Data.

# Application of Thermal Transient Testing for Solar Cell Characterization

A. Vass-Várnai [(1,2)], B. Plesz [(2)], Z. Sarkany [(1)], A. Malek [(2)], M. Rencz [(1,2)]

[(1)] Mentor Graphics Mechanical Analysis Division
H-1117 Gabor Denes st. 2., Budapest, Hungary
E-mail: <andras_vass-varnai; zoltan_sarkany; marta_rencz>@mentor.com

[(2)] Budapest University of Technology and Economics, Department of Electron Devices
H-1117 Magyar tudósok ave. 2., Budapest, Hungary
E-mail: <vassv; plesz; rencz>@eet.bme.hu ; madrian@vipmail.hu

## Abstract

The current paper deals with the application of thermal transient testing as a characterization tool for solar cells and modules. Based on the measurement of a representative samples -including concentrator and non-concentrator solar cells - we prove the applicability of this measurement technique. Metrics such as junction-to-base plate thermal resistance are derived and can serve as a basis of a model for the accurate prediction of the performance of solar modules.

The used technique also enables us to verify the quality of attachment layers in a solar module allowing fair quality control and reliability analysis of these devices.

Solar cell specific measurement problems such as the possibly long initial transient section due to electric switching are investigated; we have found that large area solar cells tend to have electric responses lasting up to several hundred microseconds, covering significant thermal information. As a solution for this problem the thermal information covered by the initial electric transient was regained by fitting simulation results to the measurement data.

## Keywords

Solar cells, Thermal transient testing, Structure functions; Electro-thermal modeling

## 1. Introduction

Solar cells are one of the most prospering fields nowadays with yielding an average growth of over 30 % per year over the last decade. Thanks to this superior growth R&D in this field has increased, and produced a wide variety of different solar cell concepts, e. g. crystalline silicon cells, amorphous silicon cells, semiconductor alloy multi-junction cells for concentrator applications, non-silicon thin film solar cells or organic solar cells. All of these cell types share a common feature, namely that they convert light into electric energy, but due to changes in the ambient conditions (such as temperature and irradiation) their behavior can be extremely varying. As an example, the power characteristics of crystalline silicon and organic solar cells are quite different: while the power generated by a silicon solar cell decreases with temperature [1,2], the temperature dependence of organic cells shows a positive coefficient [3].

Current standard measurements [4,5] do not incorporate the variation of temperature, although the power generated by

photovoltaic devices shows a strong dependence on these parameters. Thus there is a strong need for measurement techniques that can be applied for combined electro-thermal investigations.

Characterization methods where the variation of the temperature is taken into account, but only steady states are investigated, can be summarized as static electro-thermal methods. By extending today's characterization techniques with such testing methods more accurate models for PV structures can be derived.

As for the interlink between electric and thermal behavior of solar cells it is characteristic that the electric response to changing irradiation conditions is by orders of magnitudes faster than the thermal response. Thus for example in case of rapidly changing weather conditions the characteristics measured with static electro-thermal methods can result in false predictions if a feedback on the cell temperature is not available. Since the cells are normally incorporated dust- and waterproof into the module, temperature sensors would have to be integrated during the fabrication process, if feedback on the cell temperature is needed. This would lead to additional costs and changes in manufacturing processes. Prediction models based on static electro-thermal characterization need three input parameters, namely irradiance, ambient temperature and cell temperature. Since the cell temperature is a feedback signal there is no possibility to create explicit black box models.

By using transient electro-thermal measurements and appropriate models, a more sophisticated description of device behavior can be given. Using thermal-transient measurement the heatflow path between the heat generation source (the cell) and the ambient can be measured. In contrary to static electro-thermal characterization methods this way not only the thermal resistance, but also the heat capacitances of the solar module can be measured. If in addition the heat generation in the solar cells is determined from the absorbed irradiation and the ambient temperature is known, a precise prediction can be given at every single moment of operation. With models based on thermal transient measurements no cell temperature feedback is needed, only the irradiance level and the ambient temperature have to be introduced as input parameters to the model.

In addition to precise complex solar module models thermal transient testing can also provide valuable data for the thermal design of concentrator solar cell applications.

Beside the modeling possibilities the methodology described in this paper could also be used for in-line or final product characterization, e.g. for the determination of the quality of die attach layers or the control of proper heat transfer properties in concentrator solar cell modules.

## 2. Experimental

For the accurate thermal modeling of an encapsulated solar module an approach similar to those used for the characterization of power electronics can be applied.

Thermal transient testing is an appropriate tool for mapping the heat conduction path between a semiconductor junction and its ambient. This technique is widely used to derive package parameters such $R_{thJA}$ and $R_{thJC}$ in case of power devices. In these cases the temperature sensor can be a dedicated temperature sensitive diode or a transistor.

Since solar cells are basically PN junctions and thus have a temperature sensitive forward voltage, the thermal transient methodology is suitable for their testing in practically the same way [6].

To prove the applicability of thermal transient testing different samples were used, including single and multijunction solar cells. For the investigation of the die attach layer characterization special samples with simulated errors were prepared.

### 2.1. Specific Issues of solar cell measurements

For the measurement of solar cells some differences to the measurement of power electronics should be considered.

The packaging of solar modules is not primarily designed to enhance the cooling path from the junction, but for transmittance at the upper side and mechanical stability at the backside. For this reason the heat paths in both directions are comparable. Concentrator cells are an exception as due to the high power density of the irradiation additional measures have to be taken to increase the thermal conductivity between the cell and the ambient.

In addition the current of the solar cell varies with illumination, so it is essential to perform the temperature sensitivity calibration under dark conditions. For practical considerations we performed the measurements in a dark environment, too.

### 2.2. Investigated samples

For the actual tests a GaAs concentrator solar cell (CSC) with a 10x10 mm surface area, mounted on an aluminum based MCPCB was used (sample A). The structure shown in Fig. 1 is similar to those used in case of power LED-s, therefore we may assume that the heat-conduction path is mainly one dimensional in this particular case.

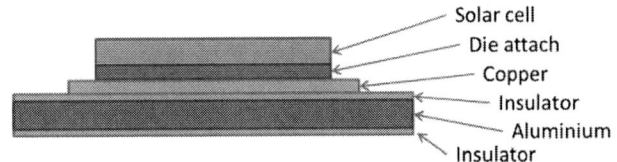

**Fig. 1:** Cross-section of the CSC samples

The samples used to simulate delamination of solar cells were processed from commercially available single junction solar cells. The cells were diced into pieces of 30 x 20 mm² sample size, and were fixed on a copper plate with an ordinary TIM material containing silver balls. The TIM was applied in a thickness of 500 um. To alter the thermal resistance of the die attach spacers cut from 500 μm thick borosilicate glass wafers were used, see Fig. 6. For the investigations three samples were processed. As a reference one cell was attached directly on the copper plate without any TIM. The remaining two samples were processed as described above, with different ratios between the TIM and the glass area. The two samples had a glass to TIM ratio of 0 (grease only) and 1:3 (25 % of the solar cell area) percent respectively. This test represents the delamination of the solar cells in different scales.

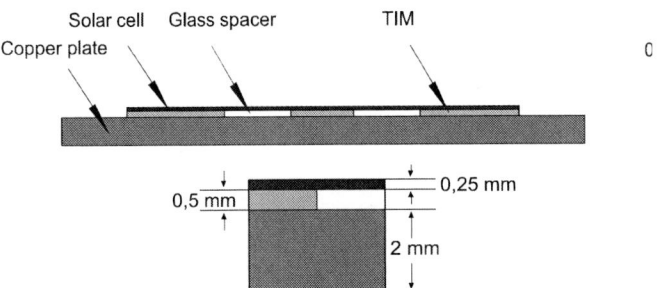

**Fig. 2:** Cross section of the samples used for investigating delamination

For the measurement of the initial electric transient samples prepared from single junctions cells were used. Sample preparation was performed as described above with a sample size of 20 x 20 mm².

### 2.3. Calibration

The temperature sensitive parameter (TSP) value has to be calibrated in case of each device under test to measure the temperature dependence of the forward voltage of the diode. For device calibration we prefer to use a temperature controlled environment to set up different ambient temperature values. A constant sensor current is driven through the diode, while we measure the forward voltage values at each settled temperature point. With trial measurements we selected 10 mA constant sensor current for biasing the diode, to obtain an optimal signal-to-noise ratio.

The temperature of the reference plate was controlled from 25 to 65 °C in 10 °C steps, and the forward voltages were measured. A calibration curve can be viewed in Fig. 3. The constant for the conversion of the measured voltage data to temperature data during thermal transient testing is derived from the slope of the calibration curve.

| Sample identifier | K-factor [mV/K] | Size [mm] | Description |
|---|---|---|---|
| A | 6.2 | 10 x 10 | GaAs multi-junction |
| B | 2.39 | 20 x 30 | 25%glass_6A_5mA |
| C | 2.59 | 20 x 30 | 100% TIM_6A_5mA |
| D | 2.06 | 20 x 30 | Reference_6A_5mA |
| E | 2.35 | 20 x 20 | 100% TIM, modeling |

**Table 1:** k-factor values and sizes corresponding to the solar cell devices shown in this study

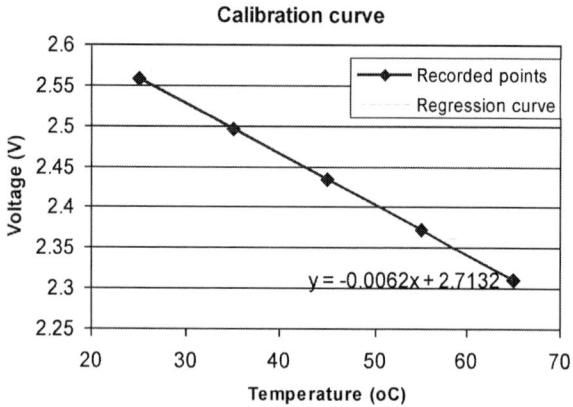

**Fig. 3:** Calibration curve of a triple junction cell

### 2.4. Measurement procedure

In the actual test arrangement we used the modification of the so-called transient dual interface methodology [7] to identify the thermal resistance from the junction to the bottom of the base-plate. In order to follow the standard the measurements were carried out at two different boundary conditions. First good thermal contact was established between the MCPCB and a temperature stabilized cold plate using heat conducting grease. As a second step the grease was removed so the thermal resistance between the MCPCB and the cold plate increased. With the measured two different boundary conditions we can identify the divergence point of the curves, and read the characteristic thermal resistance values.

### 3. Results

Temperature transient curves corresponding to two different boundary conditions are shown in Fig. 4.

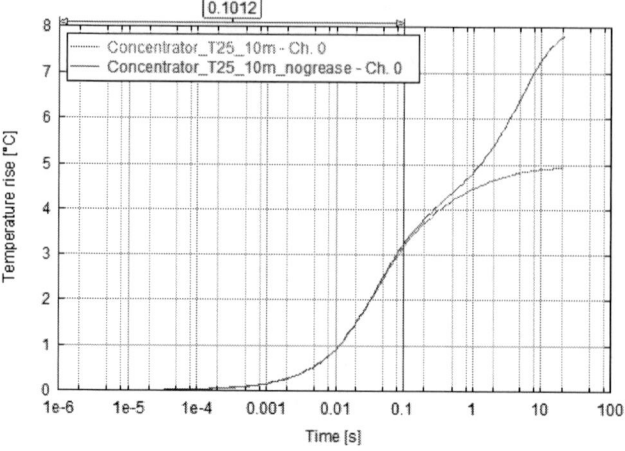

**Fig. 4:** Smoothed temperature transient responses for two different boundary conditions

You can note that until one time point, approximately 0.1 second both functions run together, indicating that the main trajectories of the heat flow spread in the same structure. If a TIM material is used, the junction temperature elevation is 4.9 °C above the cold-plate temperature. If no TIM material is used, the temperature elevation is 2.7 °C higher, i.e. 7.6 °C.

Fig. 5. shows the cumulative structure functions of the curves belonging to the two different boundary conditions. We can identify the point of separation at about 3.14 K/W which is the junction-to-MCPCB thermal resistance of the structure. The total thermal capacitance is about 1.4 Ws/K.

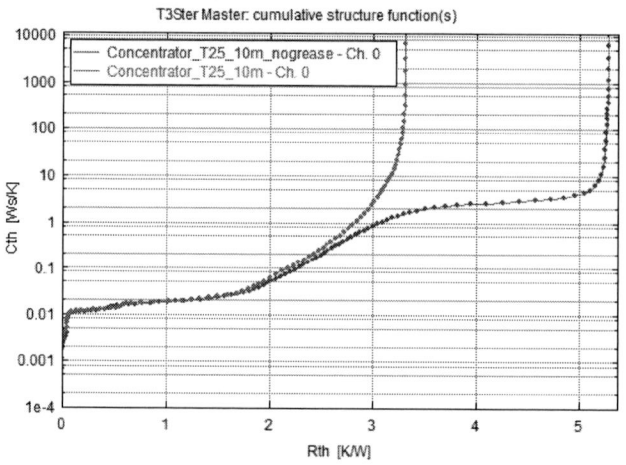

**Fig. 5:** Cumulative structure function for two different boundary conditions

We tried to find the correspondence between the layers of the solar cell structure shown in Fig. 1. and the different parts of the structure functions.

The real dimensions of the chip are (based on our measurements) 5.67x6.88x0.18 mm, giving a total volume of 7.02 mm$^3$.

In comparison the volume of the chip calculated based on the thermal capacitance value identified from the early, steep part of the structure function and the specific heat of GaAs was 7.16mm$^3$ (see point 1 in Fig. 6.).

The approximate Rth of the TIM layer can be measured with the help of the cumulative structure function, too as it can be seen in Fig. 6.

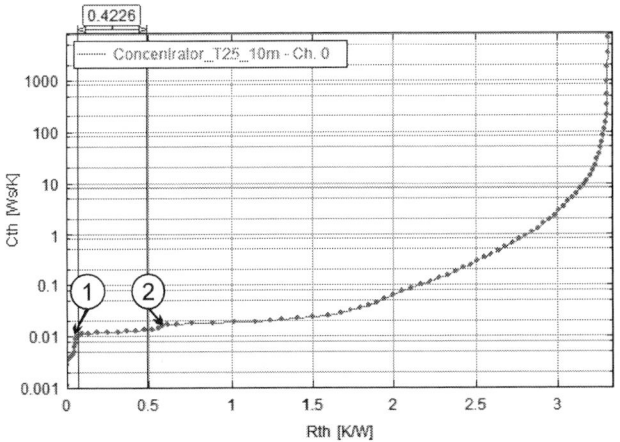

**Fig. 6:** Rth of the TIM layer

The flat section between point 1 and 2 represents the attachment layer as it shows a significant increase in its thermal resistance at a very low capacitance elevation. The next step should represent the copper layer. In order to prove this we determined the actual volume of the copper and compared it to the measurement result. The measured value was 1.47 mm$^3$ while the calculated volume based on the real geometry (0.035 mm thick copper) was 1.4 mm$^3$.

This simple calculation proved that the region between points 1 and 2 is the die attach layer.

### 3.1. Characterization of die attach quality

As we can accurately determine the resistance of the DA layer in a module, we assume that variations and changes of its resistance can be measured too. This results in a method for testing and characterization of the quality of the die attach and other interface materials in the module.

The resulting structure functions can be viewed in Fig. 7. The silicon volume is identical in all three cases, and equals to 150 mm$^3$. As the heat-spreading is ideally one dimensional in this case this value can be precisely read from the structure functions. The reference sample and the sample having 100 % TIM coverage behave similarly, but the reference shows slightly better performance.

The sample with 25 % glass coverage and 75 % TIM coverage shows a definite increase in its die attach resistance. The change in the die attach resistance can be easily observed by comparing thermal resistance values corresponding to a characteristic thermal capacitance in the system. In our example the 10 Joule/K value is characteristic to the copper base plate. The curve corresponding to the 25 % glass coverage reaches this value approximately 0.1 K/W "later" than the sample with total TIM coverage.

The overall thermal resistance of this system is very low, and the thermal conductivity difference between the glass and the TIM is approximately 1:4.

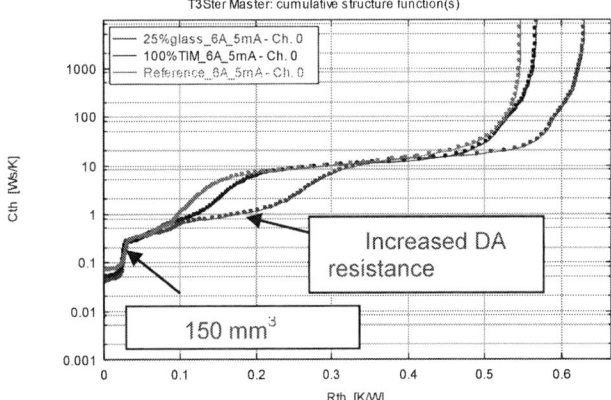

**Fig. 7:** Structure functions corresponding to the solar cell models

The fact that despite all these the difference can be well measured shows the good resolution of the presented method. Thus we have proved the applicability of thermal transient testing for die attach characterization in solar cells in a non-destructive way.

### 3.2. Investigation of electrical transient length

Due to the power step forced on the sample during thermal transient testing, each measured curve starts with an electric transient section. This is characteristic to the measurement process, originating from the discharge of the diffusion capacitance. The typical length of the electrical response is in the range of 10-20 microseconds, however in case of some devices it may last up to a few hundred microseconds. During evaluation the electric transient is substituted by a constant or square-root function. In case the substituted section is too long, we may lose information describing structures close to the heat-source. Figure 8 shows a thermal transient measurement including the initial electric transient.

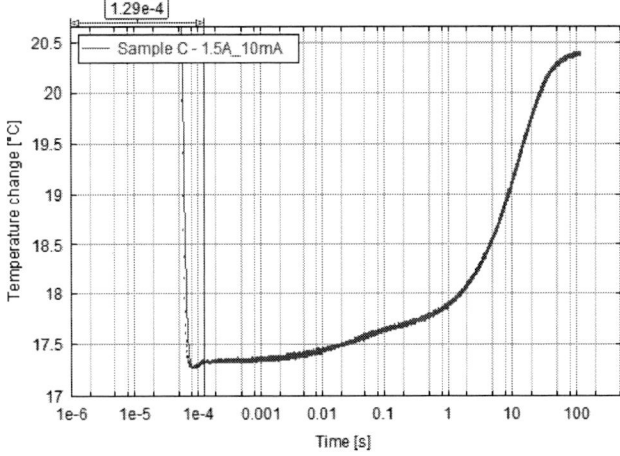

**Fig. 8:** Electric and thermal transient response of a solar cell

During the tests we have observed that the initial electric response may significantly differ among different samples and test conditions. For this reason we have investigated the factors influencing the length of the initial electrical transient.

Comparing the electric transient length measured on samples of different size, we have found that the length of the electric transient increases with the surface area of the sample, see Table 2.

We have also investigated the effect of the measurement and heating current on the transient length. For both single junction cells and multi junction cells, with the increase of the measurement current the electrical transient length decreased. The increase of the heating current however has an opposite effect, higher heating current levels result in increasing electric transient length, see Fig. 9.

Even though the tendencies shown in the figure appear clearly in all cases, from the aspects of the measurements the electric transient length is not influenced significantly.

| Sample size [mm] | Electric transient length [us] |
| --- | --- |
| 10 x 10 | 22 |
| 20 x 20 | 79 |
| 20 x 30 | 129 |

**Table 2:** Relationship between the sample area and the electric transient length

### 3.3. Simulations

As the increase of the size of the solar cell results in longer electric transient responses, in case of large area cells a significant part of the thermal information may be covered by this parasitic effect. We have shown that the proper selection of the measurement and heating current levels may reduce the length of the electric response. In case this methodology cannot lead to sufficient results, the simulation of the early part of the temperature response may be a suitable solution.

The thermal model however has to be calibrated as it is very difficult to prepare the numerical model that way that it behaves exactly as the physical structure. The most challenging structures to simulate are the thermal interface materials as often both their thermal conductivity and bond-line thickness are unknown parameters. In order to make sure that the simulation yields correct results, the model should always be calibrated against the real measurement.

Stable structures, which are easy to model, can be the silicon itself and the underlying metal base. The resistance of the die-attach layer has to be calibrated separately by tuning its material parameters until the best fit with the measured curve is achieved both in the time and the structure function space.

For modeling purposes we have selected a single junction cell with a long electric transient, over 100 microseconds.

The detailed numerical model of the selected sample can be seen in Fig.10.

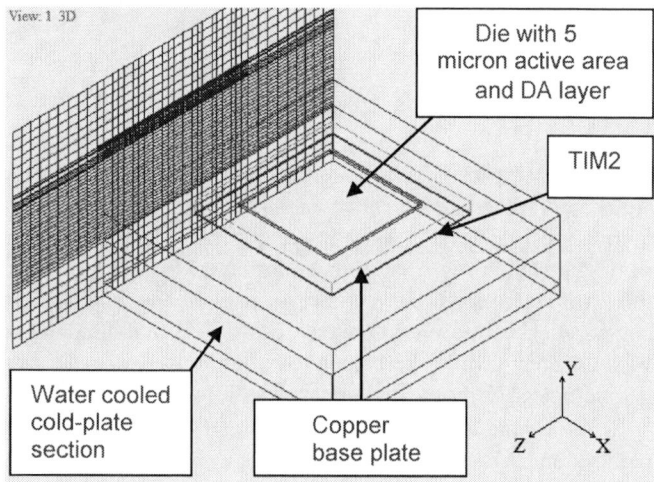

**Fig. 10.:** Detailed numerical model of the solar cell structure shown in Fig.2.

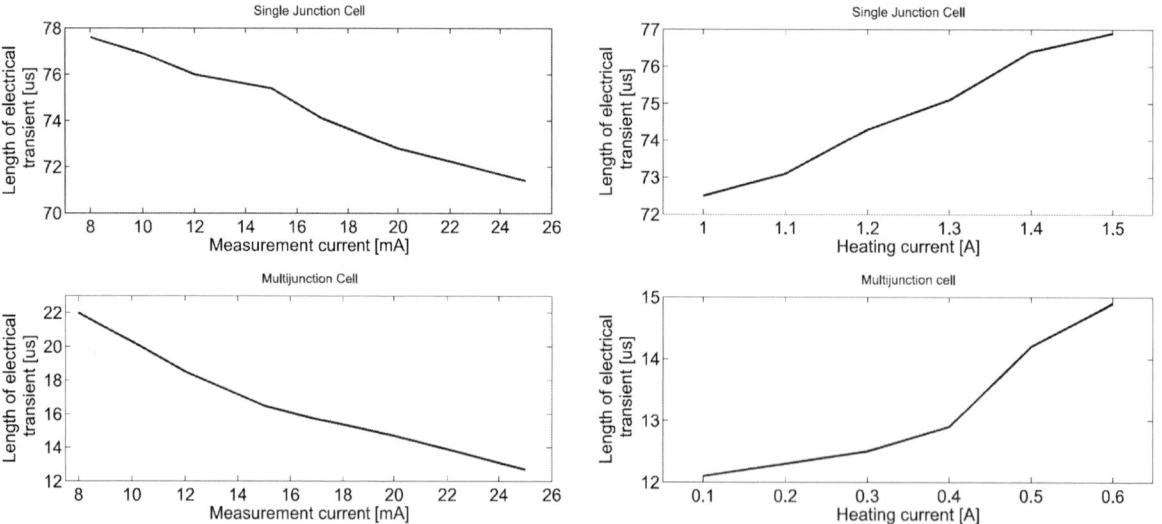

**Fig. 9:** Length of the electric transient as a function of the measurement and heating currents

At the creation of the numerical model the geometry was built up with the highest precision and most of the material parameters were known. Still it took 14 iterations to refine the model that way that the simulated transient response matched the measured one with a satisfying accuracy. Major alignments were done at both TIM layers, the resulting transient responses can be observed in Fig. 11, and the early section of the cumulated structure function describing the cell, the die attach and the copper base-plate regions can be observed in Fig. 12.

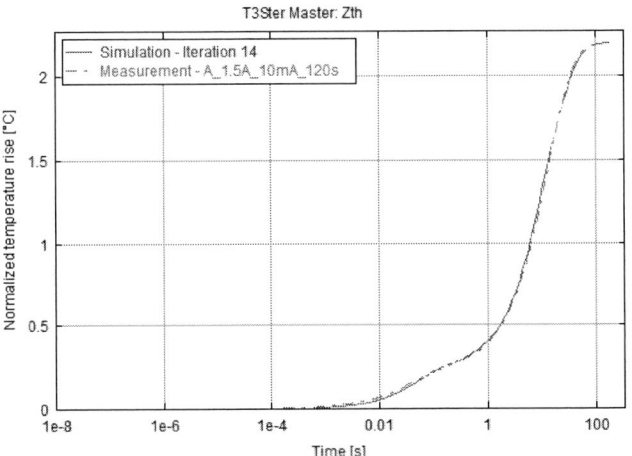

**Fig. 11.:** Measured and simulated transient responses.

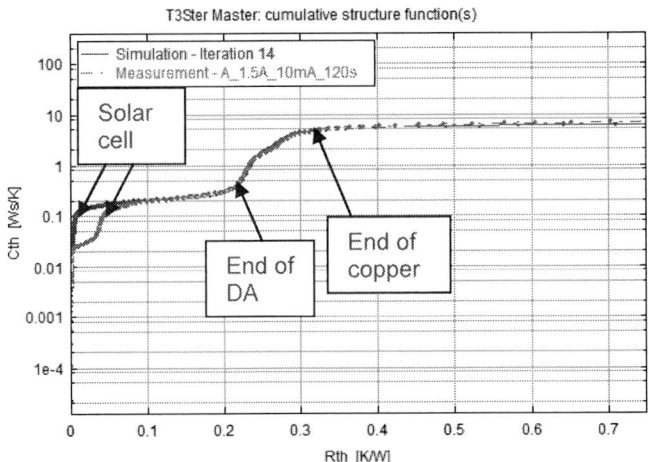

**Fig. 12.:** Structure functions of measured and simulated results

As it can be seen in the figures above, the simulation resembles the original measurement results well, except for the early sections of the curves. The difference in the part of the structure function describing the solar cell is caused by the correction used to substitute the initial electric transient, that was applied in case of the measured response. As all sections except for the one describing the solar cell fit well, we may assume that the simulation has been correctly calibrated for the structural elements of our interest. From this we can conclude that the thermal information covered by the long

electrical transient can be regained using proper simulation techniques. The prerequisite of this is the thorough calibration of the model to well known elements in the assembly, such as the chip and base plate material and geometry.

## 4. Conclusions

We demonstrated that thermal transient testing is applicable to the characterization of solar modules. As it was shown thermal transient measurements performed on solar cells can provide cell-to-base plate thermal resistance metrics. This can be used as a base for sophisticated solar module models providing precise predictions for a broad spectrum of operating conditions.

In addition the inner structural layers can be characterized – as shown in the case of the die attach in the discussed test arrangement. Due to this thermal transient testing can be applied to e.g. quality management of fully assembled modules. Such a test can be essential in cost- and reliability sensitive applications.

We found that the length of the initial parasitic electric transient shows a slight dependence on measurement parameters such as the measurement and heating current. With the increase of the solar cell surface, the initial electric transient also increases, making the measurement of large surface cells difficult. In order to overcome this, we used simulation to regain the thermal information covered by the electric transient.

## Acknowledgments

This work was supported by the SE2A ENIAC Joint Undertaking Project of the EU No. 12009. This work is connected to the scientific program of the "Development of quality-oriented and harmonized R+D+I strategy and functional model at BME" project. This project is supported by the New Széchenyi Plan (Project ID: TÁMOP-4.2.1/B-09/1/KMR-2010-0002). The research was partially funded by the project TECH-08-D/2-2008-0101 of the National Development Agency.

## References

1. Green, M. A. (2003), General temperature dependence of solar cell performance and implications for device modelling. Progress in Photovoltaics: Research and Applications, 11: 333–340.
2. Temperature Dependence of Protocrystalline Silicon/Microcrystalline Silicon Double-Junction Solar Cells, Kobsak Sriprapha, Seung Yeop Myong, Akira Yamada and Makoto Konagai, Jpn. J. Appl. Phys. 47 (2008) 1496
3. Katz, EA; Faiman, D; Tuladhar, SM; Kroon, JM; Wienk, MM; Fromherz, T; Padinger, F; Brabec, CJ; Sariciftci, NS: Temperature dependence for the photovoltaic device parameters of polymer-fullerene solar cells under operating conditions, JOURNAL OF APPLIED PHYSICS, 90 (10): 5343-5350 NOV 15 2001, ISSN: 0021-8979
4. [3] IEC 60904-1:2006, Photovoltaic devices, Part 1: Measuremenet of photovoltaic current-voltage characterstics3. Azar, K, R. S. McLeod, R. E. Caron,

5. IEC 60904-8:1998, Photovoltaic devices, Part 8: Measurement of spectral response of a photovoltaic (PV) devices
6. Siegal, B.; , "Solar Photovoltaic Cell thermal measurement issues," Semiconductor Thermal Measurement and Management Symposium, 2010. SEMI-THERM 2010. 26th Annual IEEE , vol., no., pp.132-135, 21-25 Feb. 2010
7. D. Schweitzer, "The junction-to-case thermal resistance: A boundary condition dependent thermal metric", 26th Annual IEEE SEMI-THERM Symposium, March 2010, San Jose, CA,USA, pp. 151

# Integration of a Phase Change Material for Junction-Level Cooling in GaN Devices

Daniel Piedra[1], Tapan G. Desai[2], Richard Bonner[2], Min Sun[1], Tomás Palacios[1]

[1]Dept. of Electrical Engineering and Computer Science, Massachusetts Institute of Technology
77 Massachusetts Avenue, Rm. 39-623
Cambridge, MA 02139
[2]Advanced Cooling Technologies, Inc.,
1046 New Holland Ave.
Lancaster, PA 17601
dpiedra@mit.edu

## Abstract

Next generation gallium nitride (GaN) RF power transistors offer higher power, higher efficiency and wider bandwidth than competing Si technologies. However, the high power densities available in GaN power transistors create new challenges for heat dissipation. This paper presents a novel micro-scale thermal storage design that involves phase change material (PCM) filled grooves etched in the substrate to remove the heat generated in the active regions of a pulsed-mode GaN transistor. High electron mobility transistors (HEMTs) were fabricated on a GaN-on-Si wafer. Backside patterning and etching were done to thin the Si substrate under the active channel region of the selected transistors. A phase change material (PCM) with a melting temperature of 118°C was deposited in the etched grooves. Electrical measurements were carried out to compare the performance of transistors with and without PCM filled grooves. It was found that the groove etching did not degrade the transistor performance under low power conditions where the junction level heating is not enough to start the PCM melting process. From the current-voltage characteristics at different temperatures ranging from 25°C to 120°C, it was found that at higher temperatures, the current density in the PCM-enabled device was larger than in the reference device, due to the enhanced thermal management. The role of PCM was confirmed when measurements at temperatures well above the melting temperature of the PCM did not show signs of increase in the current density. The maximum current density in the device with PCM material was found to be much more stable under pulsed conditions than in current state-of-the-art devices.

## Keywords

Gallium Nitride, Phase Change Material, Junction Level Cooling

## Nomenclature

$V_g$-Gate voltage (V)
$V_{ds}$-Drain to source voltage (V)
$I_{ds}$-Drain current at a given drain to source voltage (A)
$I_{ds,max}$-Maximum drain current (A)

## Introduction

The need for high power, high frequency transistors is increasing as RF and microwave applications are becoming more important driven by the tremendous demand for wireless telecommunications. AlGaN/GaN High Electron Mobility Transistors (HEMTs) have become the preferred option for solid state amplifiers in the 1-40 GHz frequency range. With an output power density of more than 40 W/mm (output power normalized by gate width) at 4 GHz [1], these devices offer higher power density, higher efficiency levels, lower cooling requirements, and better impedance matching than silicon-based electronics [2]. For a given operating frequency, GaN transistors can run at higher voltages (bias voltages of, for example, 48 V or 65 V in X-band devices) and hence higher output power. Another important application of GaN transistors is high voltage switches in the next generation of power electronics.

In spite of the excellent performance demonstrated by GaN transistors, the extremely high power densities available in GaN devices create new challenges for heat dissipation. The device junction temperature needs to be maintained below 150-175°C to minimize degradation in the transport properties of the semiconductor. Even when operating at the relatively low power densities of 4 W/mm, these devices need to be attached to large heat sinks, which significantly reduce the system scalability.

On a GaN chip, the heat is generated at discrete locations due to the electrical resistance faced by the current as it flows from the source to the drain. In addition, many GaN power amplifiers operate under pulsed mode, i.e., an "ON" or active period followed by an "OFF" or inactive period, which induces rapid changes in channel temperature during each cycle. In this paper we propose a new cooling technology based on a phase changing material (PCM) which enables heat dissipation, operation at much higher power densities and the use of silicon substrates instead of the much more expensive silicon carbide (SiC) substrates typically used in GaN electronics to improve heat dissipation. This approach is based on the increase of the effective heat capacitance of the material layers (via thermal storage material such as PCM) nearest to the transistor junction to reduce the peak temperature and the transient changes.

## Concept

The new cooling approach presented in this paper takes advantage of the heat of fusion in PCMs to the increase the heat capacitance of the transistor structures at the junction level. The melting point of the PCM creates a "stop" for the peak junction temperature provided there is sufficient PCM material to absorb all of the heat generated during the "on" pulse and relatively long inactive time to release the heat and refreeze all of the PCM. An optimal amount of PCM will prevent the chip junction temperature from rising and falling rapidly, making the chip isothermal over the entire duty cycle, which is ideal in terms of increased device reliability. In the proposed chip level cooling application, the transients are fast, so only a small amount of PCM (less than a microgram) is required for this application, which makes the integration of PCM into the device level very feasible.

GaN-based transistors were fabricated on the front side of a GaN-on-silicon wafer and several grooves were etched in the back-side of the silicon substrate using standard microfabrication technology. The groove-type design extends the surface area for heat transfer between the PCM and groove walls and improves the accessibility to hot spots. These grooves were filled with an appropriate amount of PCM, after which the electrical performance of the GaN devices was tested. Figure 1 shows a summary of the mask layout used during the device fabrication. A row of transistors with back Si-etch was followed by a row of transistors without back Si-etch. This design allowed the devices with grooves and without grooves to be tested under the same conditions.

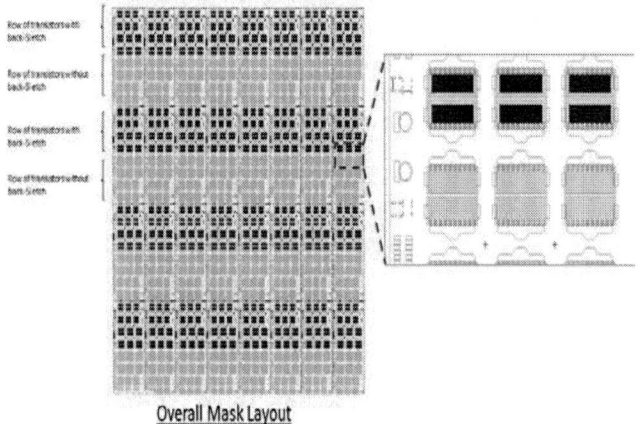

**Figure 1:** Top View of Mask Layout for fabricated GaN transistors with integrated PCM cooling.

## Fabrication

Prototype devices were fabricated to experimentally test the concept of junction level thermal storage in devices operated under pulsed conditions. The epitaxial structure of the wafer consisted of a 25 nm AlGaN barrier layer on a 1.8 μm GaN buffer layer grown by metal organic chemical vapor deposition (MOCVD) on a 550 μm Si substrate with (111) orientation. The process flow of the device fabrication is shown in Figure 2. It starts with the patterning and isolation of individual transistors. The device isolation was performed through mesa etching with a $Cl_2$-based Reactive Ion Etching plasma system. Then, the source and drain ohmic contacts were defined and deposited through electron beam deposition. A Ti/Al/Ni/Au metal stack was used and annealed at 870 °C for 30s to form the ohmic contacts. A Ni/Au/Ni gate electrode was then defined by optical lithography and deposited by electron beam deposition. The devices were passivated with a 25 nm $Al_2O_3$ dielectric deposited by atomic layer deposition, and openings were etched in the dielectric using buffered oxide etch.

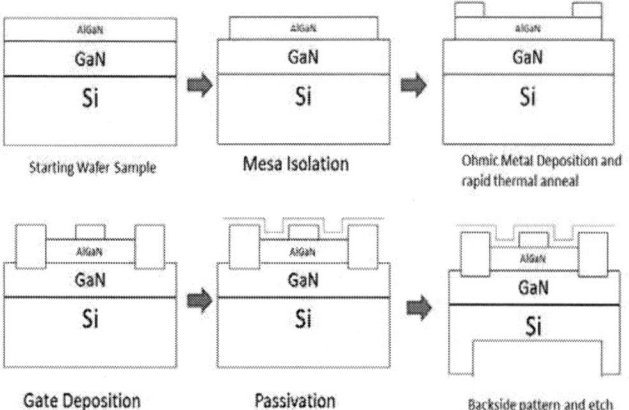

**Figure 2:** Process flow for the GaN HEMT

Then, the PCM thermal storage structure was fabricated by first patterning the region where the PCM material would be deposited. The silicon substrate was etched through a Bosch dry etch process, and the photoresist was removed by standard cleaning with solvents. The device fabrication was finished with the deposition of the PCM into the silicon grooves by melting the PCM material in the grooves. Adequate care was taken to remove the excess PCM lying on the chip outside the groove. The PCM was then encapsulated with a thin indium foil. Figure 3 shows the cross-section of a fully fabricated device, as well a top-view and a bottom-view image of the fabricated chip.

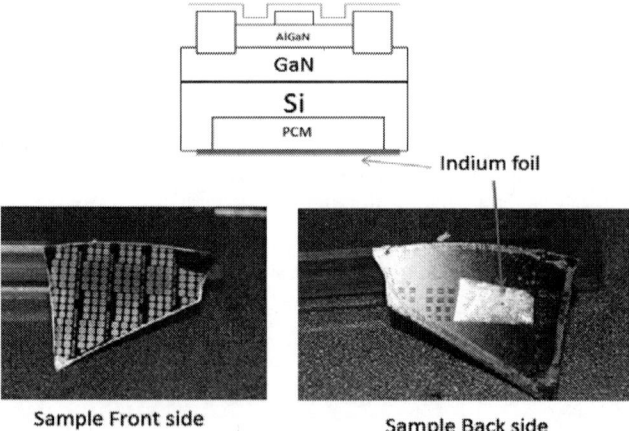

**Figure 3:** Fabricated GaN sample--(Top) Cross-sectional image of fully fabricated AlGaN/GaN HEMT with an integrated PCM cooling system, (bottom-left) Top-view image of a fabricated chip, (bottom-right) Bottom-view image of the fabricated chip showing the indium foil encapsulation

## Breakdown Voltage Measurements

The first test on these fabricated devices was to confirm that under low power conditions the groove fabrication process does not degrade the transistor performance. Under these conditions, the PCM does not play a role as the junction level temperature does not rise above the melting temperature of PCM. The 3-terminal off-state breakdown voltages of devices with PCM and devices without grooves were measured. To keep the device in the "off-state", the gate voltage was kept constant at $V_g$=-8 V and the drain-to-source voltage was swept up until the drain current reached 1mA/mm. The values of breakdown voltage, $V_{bk}$, for the two cases, devices with PCM filled grooves (127.5 V) and devices without grooves (122.2 V) were similar, as shown in Figure 4.

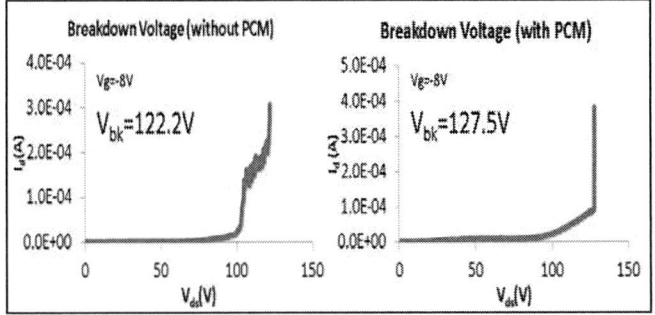

**Figure 4**: Three-terminal off-state breakdown voltage of GaN transistors with and without PCM

## DC $V_{ds}$-$I_{ds}$ Measurements at Different Chuck Temperatures

Negative differential resistance in the saturation region of the $V_{ds}$-$I_{ds}$ curve of GaN transistors is an indication of self-heating [3]. The magnitude of the decrease of drain current in the saturation region correlates to the magnitude of device self-heating. Figure 5 shows the current-voltage (I-V) characteristics of devices with PCM-filled grooves (blue empty circles) compared to control devices without grooves (red lines). At room temperature, the normalized values of drain current ($I_{ds}$/$I_{ds,max}$, the value of the drain current at a given drain-to-source voltage divided by the maximum drain current) for the PCM filled devices are similar to the devices without grooves. With increasing chuck temperature (100°C and 120°C), the difference between the red and blue lines increases. This indicates that the difference in the decrease of relative drain current between PCM and non-PCM devices is larger at higher temperatures. The increase in drain voltages results in increased self-heating and higher junction temperature. The current density in the PCM-enabled devices is higher than the value in the reference devices; this difference is associated with the enhanced thermal management of the PCM device and the subsequent reduction in self-heating. The role of PCM is confirmed when the chuck temperature is increased to 140°C which is well above the melting temperature of PCM. Under these conditions, the $V_{ds}$-$I_{ds}$ curve characteristics are similar to those at room temperature, because the melted PCM no longer acts as a thermal storage.

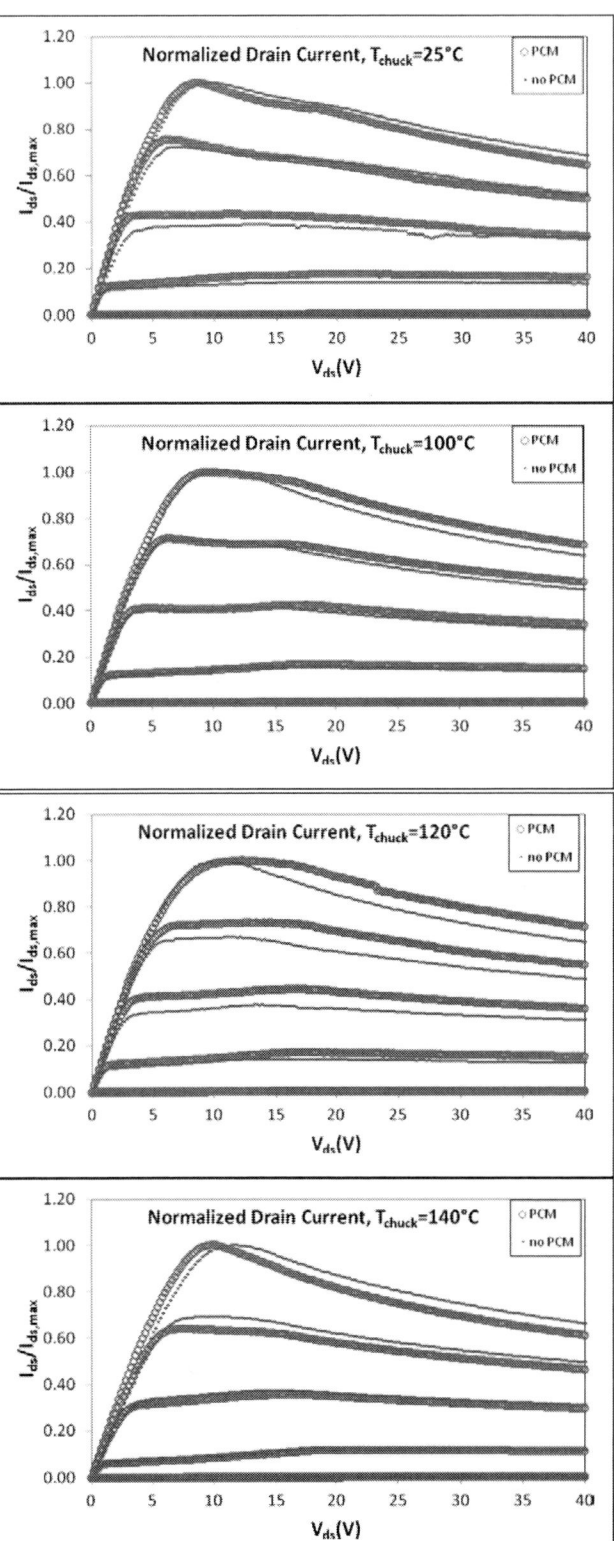

**Figure 5:** Normalized values of drain current ($I_{ds}$/$I_{ds,max}$, the value of the drain current at a given drain-to-source voltage divided by the maximum drain current) to illustrate the self-heating effect.

## Pulsed IV Measurements

Pulsed current-voltage (I-V) measurements were performed on the devices with PCM-filled grooves and devices without grooves. The chuck temperature was set to 95°C to be able to operate the small devices that were fabricated in this project under thermal conditions similar to what scaled devices encounter. At each pulse width, the PCM filled groove measurement was followed by the measurements on devices without grooves, before going to the next pulse width. This procedure allowed for enough time (in the range of minutes) for the re-solidification of the melted PCM. Each data point in Figure 6 represents the maximum drain current at the respective pulse width divided by the overall maximum drain current for a single pulse. As the pulse width decreases, the maximum current through the device increases due to reduced device self-heating. As evident from Figure 6, the maximum current density in the device with PCM material is more stable with the pulse width than in the standard device. The device performance in terms of current density shows 10% improvement due to enhanced thermal management. This improved performance demonstrates the role of PCM as a thermal storage to reduce the temperature increase in the device. It is expected that the performance improvement will be substantially higher after further optimization in terms of groove size and location. A computational model is being developed to perform this optimization.

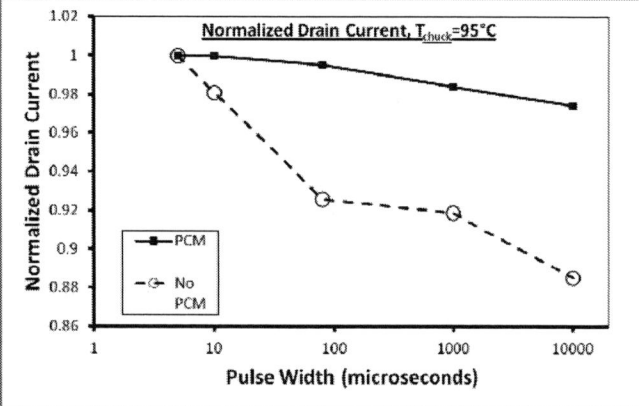

**Figure 6:** Maximum drain current at the respective pulse width divided by the overall maximum drain current from pulsed IV measurements for devices with and without PCM.

## Conclusions

This paper presents a novel micro-scale thermal storage design to remove heat generated in the active regions of a pulsed mode gallium nitride transistor. High electron mobility transistors were fabricated on GaN grown on a Si substrate. Backside patterning and etching were done to thin the Si substrate under the active channel region of selected transistors and a PCM was deposited in the etched grooves. The groove etching did not degrade the transistor performance under low power conditions where the junction level heating is not enough to start the PCM melting process. From the DC current-voltage characteristics at different temperatures, it was seen that the difference in the decrease of relative drain current between PCM and non-PCM devices is larger at higher temperatures (as much as a 7% difference). The role of PCM was confirmed when measurements at temperatures well above the melting temperature of PCM did not show signs of increase in the current density. The maximum current density in the device with PCM material was found to be much more stable under pulsed conditions than in current state-of-the-art devices.

## Acknowledgments

This work was supported by the Small Business Innovation Research (SBIR) grant from the National Science Foundation under the Award No: IIP-1047111. The device fabrication was performed in the cleanroom facilities of the Microsystems Technology Laboratories (MTL) at MIT and at the Center for Nanoscale Systems (CNS) at Harvard Univeristy.

## References

1. U. K. Mishra et.al., *Proc. of the IEEE* 96, 287 (2008)
2. T. Palacios, U.K. Mishra. "GaN-Based Transistors for High-Frequency Applications," in *Comprehensive Semiconductor Science and Technology*, P. Bhattacharya, R. Fornari, and H. Kamimura, Ed. Amsterdam: Elsevier, 2011, pp. 242-298.
3. Y.-F. Wu, B.P. Keller, S. Keller, D. Kapolnek, P. Kozodoy, S.P. Denbaars and U.K. Mishra, "High Power AlGaN/GaN HEMTs for Microwave applications", *Solid-State Electronics*, Volume 41, Issue 10, Proceeding Proceedings of the Topical Workshop on Heterostructure of Microlectronics, October 1997, Pages 1569-1574

# Thermoreflectance CCD Imaging of Self Heating in AlGaN/GaN High Electron Mobility Power Transistors at High Drain Voltage

Kerry Maize
Department of Electrical Engineering
University of California, Santa Cruz
Santa Cruz, CA, USA
kerry@soe.ucsc.edu

Eric Heller and Donald Dorsey
Materials and Manufacturing Directorate
Air Force Research Laboratory
Wright-Patterson Air Force Base OH, USA, 45433-7707

Ali Shakouri
Birck Nanotechnology Center
Purdue University
West Lafayette, IN, USA 47907-2057
shakouri@purdue.edu

## Abstract

Thermoreflectance CCD imaging with sub-micron spatial resolution was used to characterize self heating nonuniformity in two finger AlGaN/GaN high electron mobility power transistors at equivalent power (1.27 W) for different combinations of drain and gate voltage. Thermoreflectance images of device surface temperature revealed formation and redistribution of local hotspots as transistor drain voltage increased from $V_D$=10.7 V to 50 V. For all bias points, heating between the two fingers was not fully uniform. At high drain voltage, heating migrated toward the drain side of the channel and increased thermal nonuniformity was observed along symmetric drain and source fingers. Direct microthermocouple measurements confirmed the spatial temperature nonuniformity that was observed in thermoreflectance images. Results demonstrated the usefulness of fast thermoreflectance imaging to inspect self heating in GaN thin film power devices with high with high spatial resolution.

## Keywords

component; Thermoreflectance imaging; GaN; HEMT; Power Transistors; Self Heating

## I. INTRODUCTION

There is increasing demand for integrated microelectronic power devices that can operate at extremes of frequency, power, and temperature. To extend beyond the limitations of silicon based power electronics, engineers are exploring novel materials such as gallium nitride (GaN) for integrated power transistors. GaN has several material properties that make it advantageous in power electronics. Of primary significance is the ability to engineer modulation doped AlGaN/GaN interfaces with very high electron mobilities. These films can be incorporated in high electron mobility transistors (HEMT). Other significant material advantages of GaN are its large bandgap of 3.4 eV, which permits high breakdown voltage due to a much higher electric field at which impact ionization becomes a limiting factor, and much lower intrinsic carrier concentrations allowing a maximum operating temperature of up to 970 K. [1]

Thermal processes, however, remain a significant factor in performance and reliability of GaN based transistor devices. GaN HEMTs have shown saturation current dependence on temperature [2]. Due to the high power levels reachable in GaN HEMTs and the high electric fields achievable that can force this power to be dissipated in a small volume, temperature gradients can be extreme [3]. Temperature stress affects many degradation phenomena observed in GaN power devices such as current collapse during fast switching [3] and pit formation [4,5]. For these reasons self heating in GaN devices remains an active area of study, and robust and versatile thermal characterization methods are needed. While Raman thermometry is a scanning laser method that has been used to measure both temperature and stress in GaN heterostructures at specific points near the surface and with some limited spatial resolution below the GaN surface [3,6], one method that is particularly well suited to measure heating across the entire surface of microelectronic devices is thermoreflectance CCD imaging [7-9]. Thermoreflectance imaging measures reflected visible wavelength illumination to provide two dimension maps of surface temperature distribution with submicron spatial resolution and 50 mK temperature resolution. The superior spatial resolution allows thermal imaging of GaN device features on a scale much smaller than is possible using infrared imaging. This paper presents results of thermoreflectance CCD images of a GaN HEMT under pulsed operation at various stages of transistor pinch-off for equivalent power. Drain voltage was varied between $V_D$=10.7 V to 50 V with gate voltage adjusted to maintain the same total device power of approximately 1.27 W. In extreme pinch-off with drain voltage above 47.5 V, thermoreflectance images of the active GaN HEMT revealed formation of local hotspots along the ohmic metallization and GaN channel and general temperature redistribution across the transistor. This redistribution at high drain voltage was also confirmed using direct thermocouple measurements of the active area of the GaN transistor.

### A. Device Description

The device was an AlGaN/GaN high electron mobility transistor (HEMT) consisting of two ~5 µm long by 150 µm wide channels (two fingers) designed for power RF applications. The unpackaged sample was grown on a ~500 µm SiC wafer, and contacted with Cascade Microtech ACP40-L coplanar probes to allow low Ohmic loss and fast RF switching, as shown in Fig. 1. Each finger of the transistor (Fig. 1 inset) was contacted at top and bottom by a grounded source metal and biased drain metal with the active channel (SiN passivated AlGaN/GaN stack) in between.

---

This work was supported in part by a grant from Air Force Research Laboratory

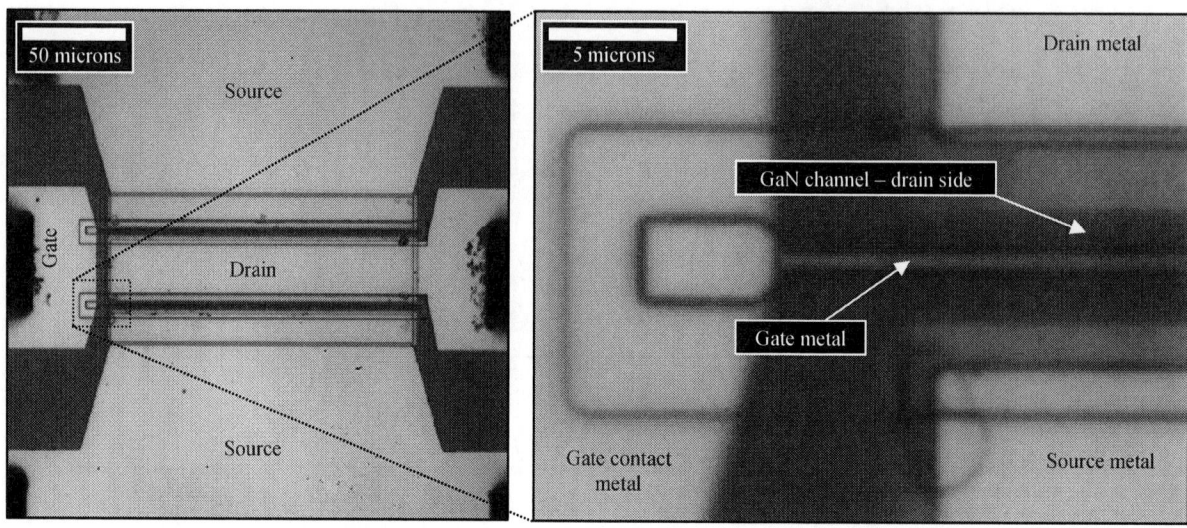

Figure 1. Optical image of two channel AlGaN/GaN high electron mobility power transistor. The gate width is 150 μm.

With no bias applied to the gate (a Schottky diode contact), the AlGaN/GaN interface forms a ~2 nm thin highly conductive layer. This layer is often referred to as a two-dimensional electron gas (2-DEG). Negative gate voltage reverse biases the diode and depletes the conducting layer of carriers by electrostatic field effect and turns the transistor off. As such, this is often the site of a lot of power dissipation, especially when the transistor is only partially turned on. Drain current for the HEMT in this study was negligible for $V_G < -4$ V.

## II. METHOD

### A. Thermoreflectance Imaging

Fig. 2 shows a diagram of the thermoreflectance CCD imaging setup used to acquire thermal images of the active GaN HEMT. The top surface of the HEMT sample is illuminated under a reflectance microscope using a narrow-band LED light source. The light reflected from the sample material surface is recorded on a variable frame rate, 12-bit scientific grade Dalsa CCD camera with 1024x1024 pixel resolution. Camera operation and device excitation timing is controlled by a LabVIEW program and custom designed hardware trigger board. Repeated device excitation pulses are sent to the HEMT and synchronized to the camera acquisition. For this study, square voltage pulses one millisecond in duration were applied to the drain of the GaN HEMT at 1% duty cycle. For each excitation cycle, the electrical pulse causes a temperature rise in the HEMT, which in turn induces a change in the optical reflectance of the HEMT material surfaces. Images of the HEMT surface reflectance during both the excited (on) and unexcited (off) states are recorded in the CCD camera. Comparing the change in reflectance amplitude for these two images produces a two-dimension thermoreflectance map across the HEMT surface. Thermoreflectance image maps are then converted to temperature maps by applying a thermoreflectance coefficient ($C_{TH}$), which describes a material's quantitative change in optical reflectivity in response to a change in temperature. The thermoreflectance change for most materials is very small, on the order of $10^{-4} C^{-1}$. However, by averaging over many device excitation cycles the thermoreflectance amplitude signal to noise ratio can be sufficient to yield temperature resolution down to 50mK. Averaging times for each thermal image depend on the specific material $C_{TH}$ and desired signal to noise. Most semiconductor and metal materials yield good thermoreflectance images in less than one minute for temperature changes of 1C or greater. For the GaN HEMTs in this study, good signal to noise was obtained for the ohmic contact metal regions in about thirty seconds. However, each image was averaged for 30 minutes to improve signal to noise on the GaN, which has a smaller $C_{TH}$ than the metal .

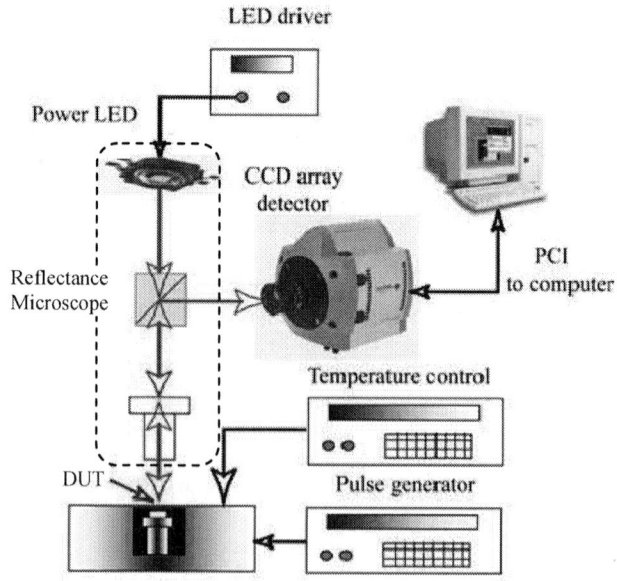

Figure 2. Thermoreflectance CCD imaging and device pulsing configuration.

TABLE I. HEMT BIAS POINTS FOR THERMOREFLECTANCE IMAGES

| $V_G$ (V) | $V_D$ (V) | $I_D$ (mA) | Power (W) |
|---|---|---|---|
| 2 | 10.7 | 119 | 1.27 |
| -1.43 | 18.1 | 70.8 | 1.28 |
| -2.56 | 30 | 41.8 | 1.25 |
| -3.05 | 40.1 | 31 | 1.24 |
| -3.08 | 42.5 | 29.8 | 1.27 |
| -3.12 | 45.2 | 28.5 | 1.29 |
| -3.17 | 47.5 | 26.8 | 1.27 |
| -3.41 | 50 | 25.5 | 1.28 |

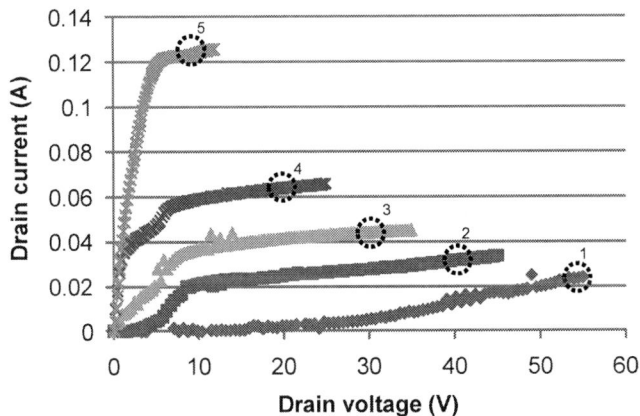

Figure 3. GaN HEMT steady state drain current versus drain voltage for several different gate voltages. The indicated points correspond to equivalent power of 1.27W.

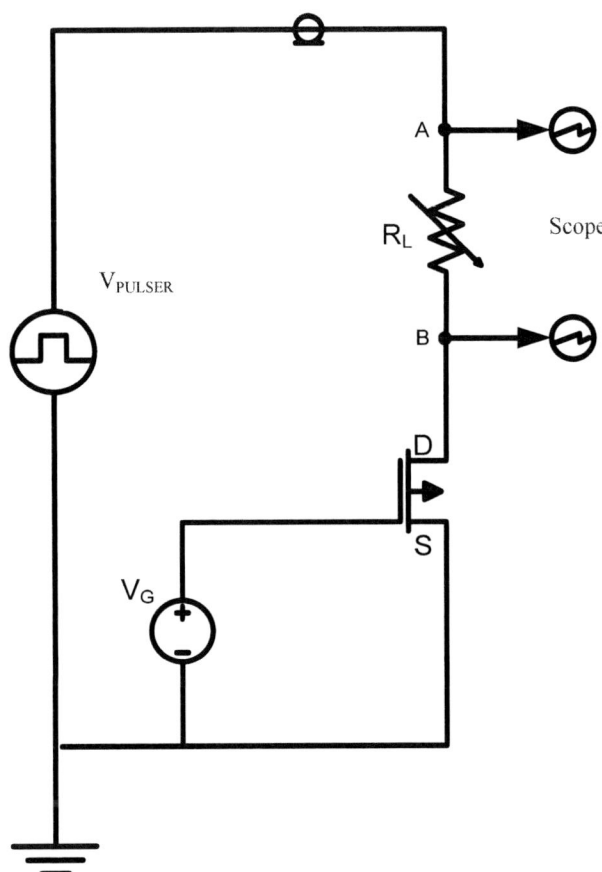

Figure 4. Thermoreflectance pulsing circuit for the GaN HEMT.

Experimental calibration of the GaN HEMT material $C_{TH}$ is performed in a procedure separate from device thermal imaging. Calibration involves heating the entire sample uniformly using an external micro-thermoelectric stage. The temperature induced change in optical reflectance over the entire sample is recorded by the CCD while temperature is measured simultaneously using a type E microthermocouple. $C_{TH}$ for each material region of interest on the sample are obtained by comparing the calibration image and thermocouple measurements. $C_{TH}$ depends on the wavelength of external illumination [9], and a green 530 nm LED was found to be a good choice for the HEMT sample due to its relatively high $C_{TH}$ on the drain and source metal. For a 50X magnification microscope objective with 0.35 numerical aperture, experimental calibration yielded a $C_{TH}$ on the surface metal (Fig. 1 "Drain", "Source", "Gate contact" metals) of $2.6 \times 10^{-4}$ C$^{-1}$. Calibration of $C_{TH}$ for GaN using the LED wavelengths available for this study did not yield results consistent with separate direct thermocouple temperature measurements of the active HEMT device. Consequently, thermoreflectance results for the GaN regions of the HEMT are presented in arbitrary units.

## B. Device Pulsing Parameters

Self heating in the HEMT was studied for different combinations of drain voltage and current at equivalent power. Fig. 3 shows the GaN HEMT steady state drain current versus drain voltage curves for several different gate voltages. The steady state measurements were obtained using an HP 4156B parametric analyzer, with the device probed on wafer. The highlighted bias points show different combinations of drain voltage and drain current all corresponding to the same 1.27 W power dissipation, but some open channel and some mostly closed. Thermoreflectance CCD images were obtained for the bias points in Table 1 to compare HEMT surface spatial temperature distribution at the various stages of pinch-off.

978-1-4673-1110-6/12 $31.00 © 2012 IEEE

Fig. 4 shows the thermoreflectance pulsing circuit. For each bias point during thermoreflectance imaging drain voltage was pulsed while gate voltage was held constant to maintain the transistor at a specific level of pinch-off. Thermal images were averaged for 30 minutes per image. Drain voltage was applied using a Berkeley Nucleonics (BNC) 202H high voltage pulser. The specified rise time for the pulser is 3 ns rising edge and 10ns falling edge. Drain current was measured from the voltage drop across a variable resistor placed in series with the device. The variable resistor was set to 50 ohms while measuring the highest drain currents (71 and 119 mA) and to 500 ohms to improve accuracy while measuring the lower currents in pinch-off (25-41 mA). Drain voltages reported represent the potential drop from drain to source across the transistor, and do not include the voltage drop across the resistor $R_L$. The drain voltage pulse waveform was monitored during thermoreflectance image acquisition to ensure correct bias of the HEMT.

Studies have revealed that GaN based transistors can experience both permanent and temporary performance degradation after extended electrical stress due to formation of trapping states. Because acquisition of a single thermoreflectance image can require averaging over thousands of HEMT excitation cycles, it was important to quantify any device degradation during imaging. Device integrity was measured by comparing the HEMT steady state drain IV before and after each thermoreflectance image was obtained. These checks revealed less than 2% change in HEMT IV performance

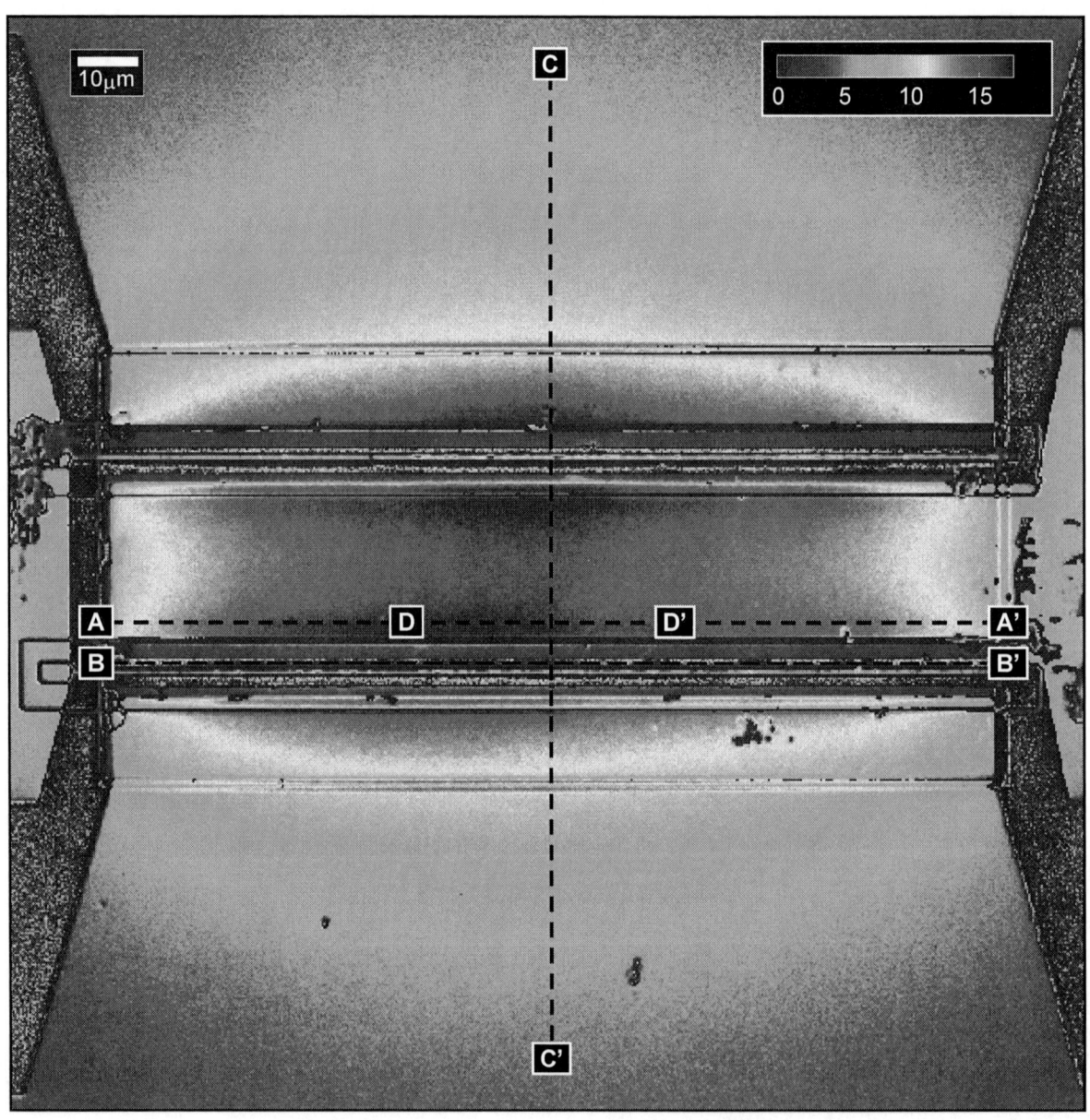

Figure 5. Thermoreflectance image of the GaN HEMT for the bias point $V_D = 50$ V, $I_D = 25$ mA, and $V_G = -3.41$ V. Location of profile lines are indicated.

due to imaging electrical stress, which is within the noise threshold of the DC measurement apparatus.

## III. RESULTS

Analysis of thermoreflectance images of the active GaN HEMTs revealed a redistribution of spatial surface temperature for the cases of extreme transistor pinch-off ($V_D = 47$ V, 50 V.) Experiment results demonstrating hotspot dependence on drain voltage are presented in the following section. Spatial temperature distribution is analyzed using both thermoreflectance data calibrated for temperature on the contact metal, and uncalibrated thermoreflectance amplitude profiles for the GaN material regions. Heating distribution is analyzed along both a horizontal and a vertical dimension of the HEMT, revealing hotspot dependence on drain voltage for profiles both parallel to and perpendicular to the gate width. Results of direct microthermocouple measurements are also presented showing local HEMT temperature increase at high $V_D$ (extreme transistor pinch-off.)

### A. Temperature Profiles Parallel to The Gate Width

Fig. 5 shows the thermoreflectance image of the GaN HEMT for the single bias point of $V_D = 50$ V and $I_D = 25$ mA. Gate voltage is -3.4 V. Fig. 6 plots the calibrated horizontal temperature profiles measured on the drain contact metal near the lower of the two HEMT fingers for the full range of drain voltages $V_D$=10.7-50 V at equivalent power (1.27 W.) The location of the measured profiles is indicated by line A-A' in the thermal image of Fig. 5. The profiles are taken parallel to the gate width. Two temperature trends are noticeable in these profiles. First, the average temperature of the drain contact metal profiles increases with drain voltage. This is expected behavior in field effect transistors, with the strongest electric field, and therefore greatest power density, concentrated on the drain side of the transistor channel. Second, temperature becomes redistributed nonuniformly along the horizontal profile at extreme pinch-off. The profiles corresponding to lower drain voltage ($V_D = 10$ V - 45.2 V) show an overall symmetric temperature distribution, hotter at the center of the

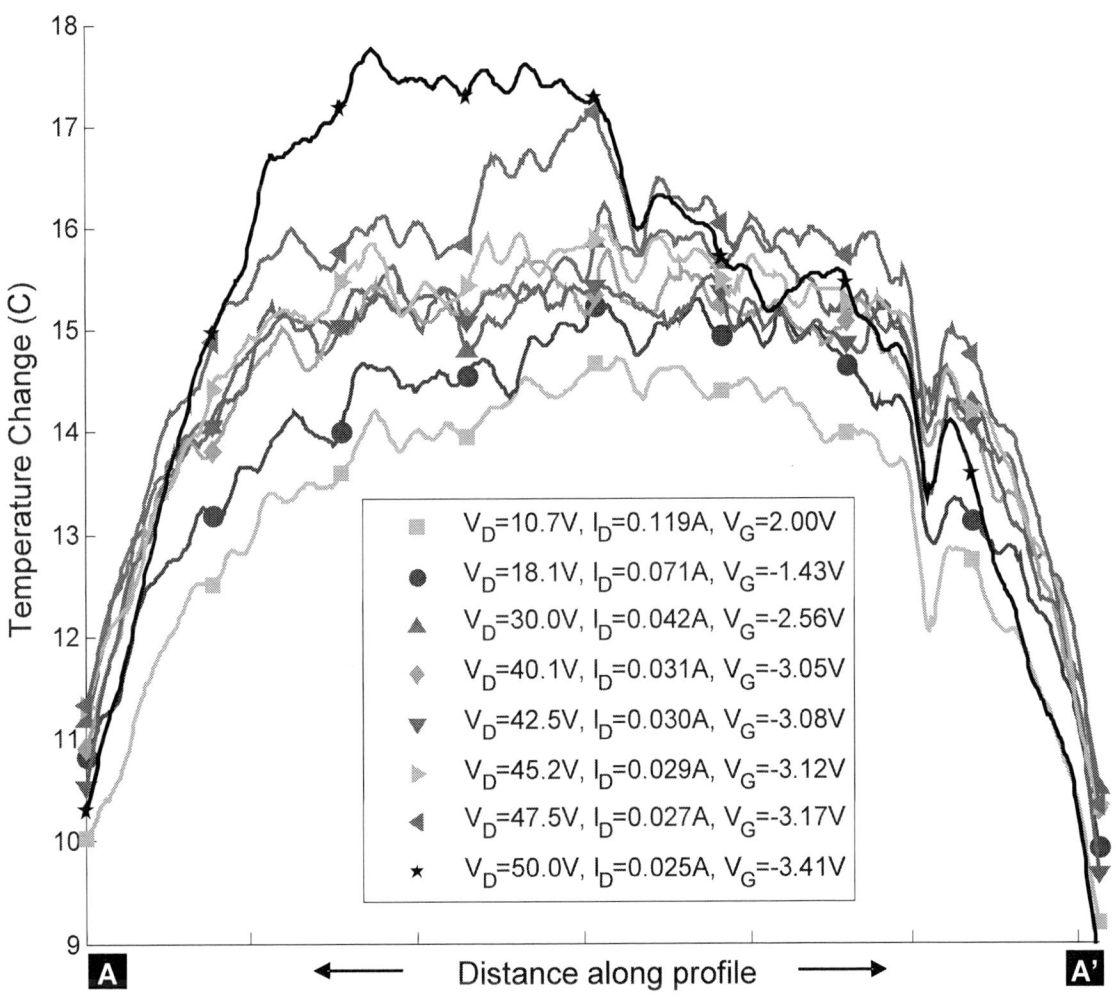

Figure 6. Calibrated horizontal temperature profiles (A-A') measured on the drain metal near the lower of the two channels for the full range of drain voltages $V_D = 10.7$-50 V at equivalent power (1.27 W.)

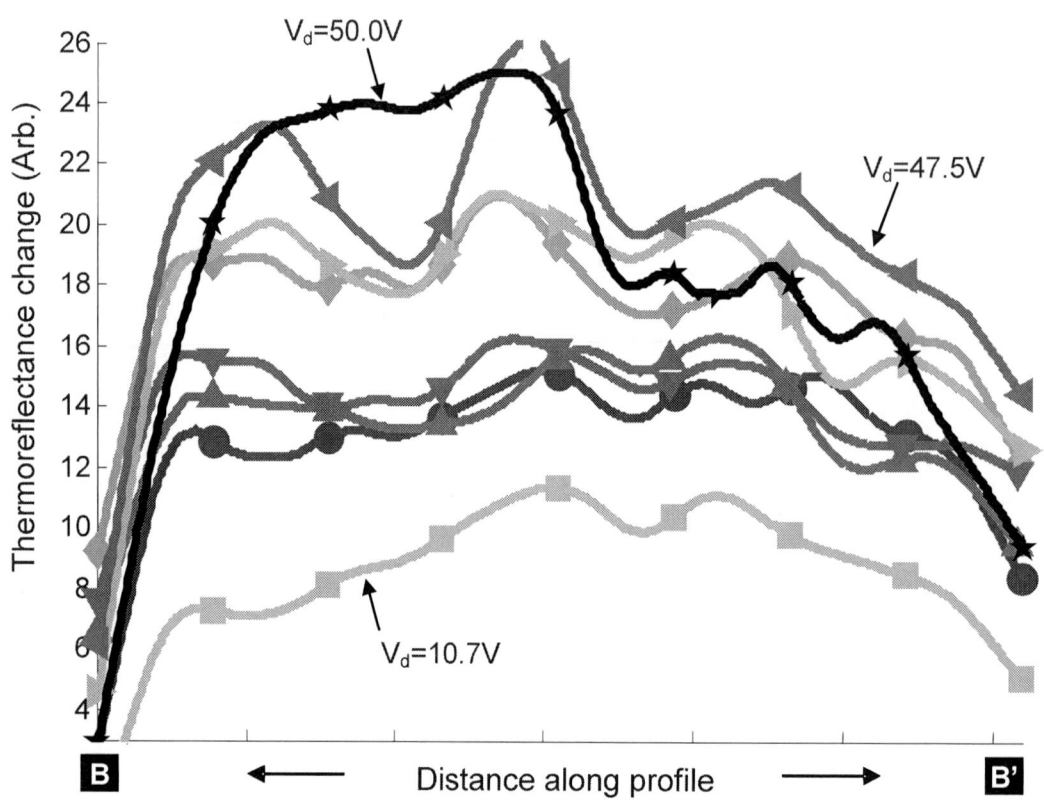

Figure 7. Thermoreflectance change profiles for the gallium nitride material along the channel (profile B-B'). The gallium nitride profiles are not calibrated for temperature.

device and cooler at the edges. No abrupt hotspots are visible. At higher drain voltage, $V_D = 47.5V$, $V_G = -3.17$ V, a local hotspot first appears near the center of the horizontal profile. The hotspot is seven percent (1.3 °C) hotter than the adjoining region of the profile. Temperature nonuniformity becomes more apparent for the next increase in pinch-off bias at $V_D=50V$, $V_G = 3.41$ V. This distribution shows a clear temperature "hump" on the left side of the horizontal profile, as much as 12 percent (2.1 °C) hotter than the similar region for the profile at next lower pinch-off bias point of $V_D = 47.5$ V. Furthermore, for the highest drain voltage ($V_D = 50$ V) the increase in temperature on the left side of the drain finger coincides with a decrease in temperature on the right side of the finger. The average temperature on the right side of the profile at $V_D = 50$ V is four percent lower than the average temperature for the same region at $V_D = 47.5$ V. One possible cause of temperature redistribution at extreme pinch-off may be device processing defects that are noticeable only at high drain voltage. These thermal defects are visible due to the high spatial resolution of the thermoreflectance method. Such detail may have appeared blurred under infrared thermal microscopy.

*1) Heating in GaN Channel*

Horizontal profiles of the GaN surface thermoreflectance amplitude were obtained for the lower finger channel. Profile line B-B' from Fig. 5 was taken on the GaN channel. (See Fig. 1 for enlarged optical images of the nitride channel and gate

regions.) The profiles were taken on the drain side of the gate metal. Both the metal profiles and the GaN profiles were obtained from the same series of thermoreflectance images. Fig. 7 plots the thermoreflectance change profiles for the GaN material along the channel width for the same range of drain voltage as the metal profiles shown in Fig. 6. Because an accurate value of the GaN $C_{TH}$ was not obtained by the time of this writing, the profiles of Fig. 7 are presented in arbitrary units representing the magnitude of the measured thermoreflectance change on the GaN channel. The thermoreflectance signal to noise is weaker on the GaN material than on the contact metal, so the channel profiles were smoothed using a moving average of 20 pixels in the horizontal (channel width) direction.

The thermoreflectance profiles along the channel agree with the trends seen in the temperature profiles on the metal. Drain side heating increases with increasing drain voltage. Also, the formation of hotspots in the channel for extreme pinch-off bias points, $V_D = 47.5$ V and 50 V, match the horizontal location of the hotspots seen in the metal profiles for the same drain voltages.

978-1-4673-1110-6/12 $31.00 © 2012 IEEE

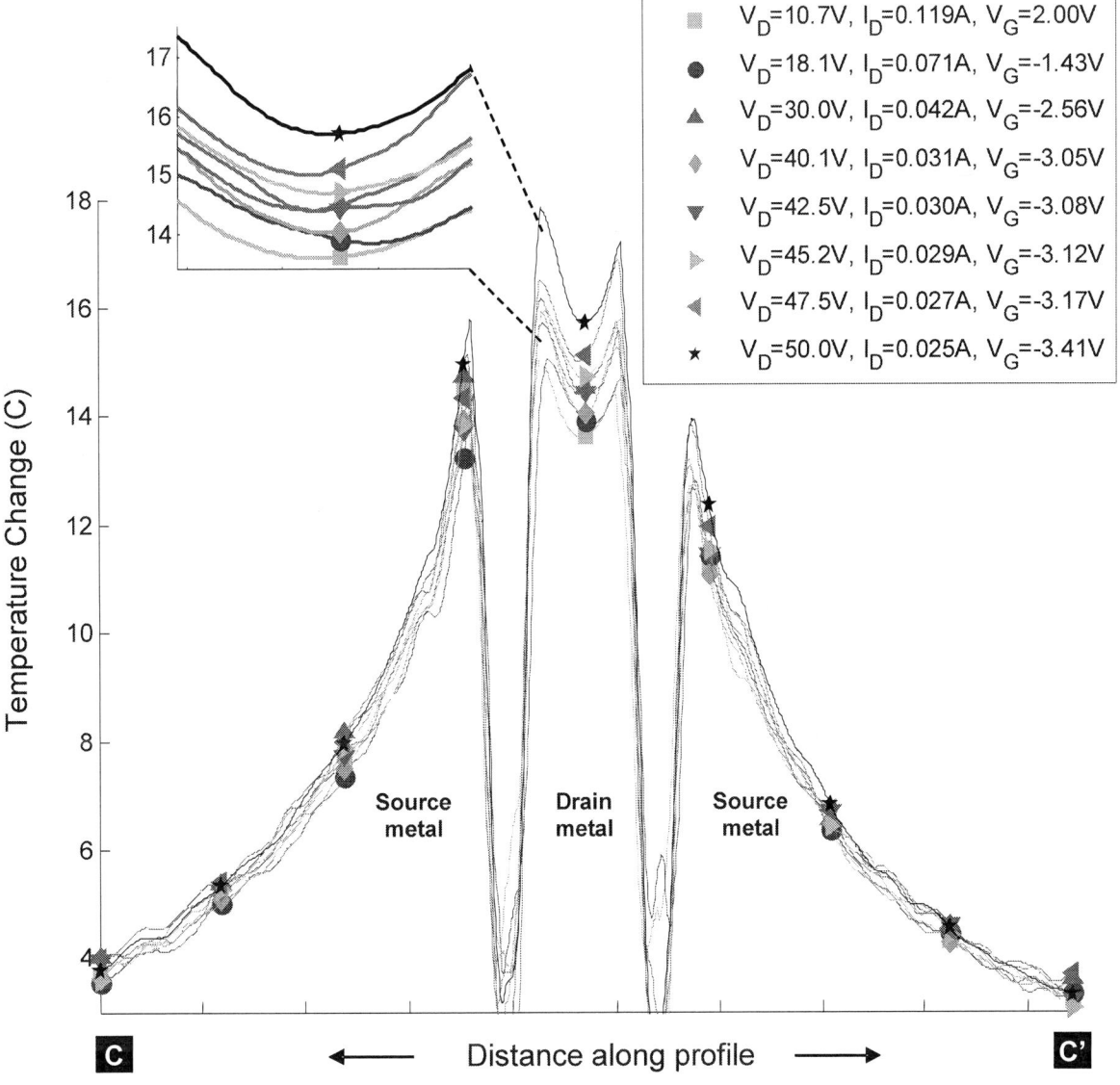

Figure 8. Vertical thermoreflectance profiles (C-C') across the HEMT. Profiles are calibrated only for the metal regions.

*B. Temperature Profiles Perpendicular to Gate Width*

Thermoreflectance profiles were also analyzed perpendicular to the gate width. The vertical profile line, indicated by line C-C' in Fig. 5, crosses in order: upper source contact metal, upper finger channel and gate, drain contact metal, lower finger channel and gate, and finally lower source contact metal. The vertical thermoreflectance profiles across the HEMT are plotted in Fig. 8 for the same range of drain voltage bias points ($V_D = 10.7 - 50$ V) as the horizontal profiles. The vertical profiles are temperature calibrated only for the source and drain contact metal regions. These regions are indicated in Fig. 8. The profiles reveal HEMT heating is asymmetric along the vertical axis as well as the horizontal axis. Along profile C-C' the upper source contact metal is on average 8% (1 °C) hotter than the lower source metal. Similarly the drain contact metal near the upper finger is 6% hotter than

the same contact near the lower finger. Temperature distribution along the vertical profile also varied with drain voltage, and the trend was similar to the one seen for the horizontal profiles. However, this variation is more likely due to the horizontal redistribution of temperature with increasing $V_D$ as shown in the horizontal profiles of Fig. 6. The most noticeable redistribution of temperature occurs for the highest drain voltage. For example, at $V_D=50$V, the maximum temperature change along the vertical profile (occurring at the drain metal near the upper gate) was ~18% greater than the temperature change at the same location for $V_D=10.7$V. Vertical profiles along on the right side of the HEMT (not plotted) revealed a concurrent *decrease* of ~8% in maximum temperature change on the drain metal as $V_D$ increased from 10.7 to 50V. This result is consistent with the hypothesis of power redistribution in the HEMT at high $V_D$.

Microscope photograph showing placement of thermocouple

Optical image of GaN HEMT from thermoreflectance system

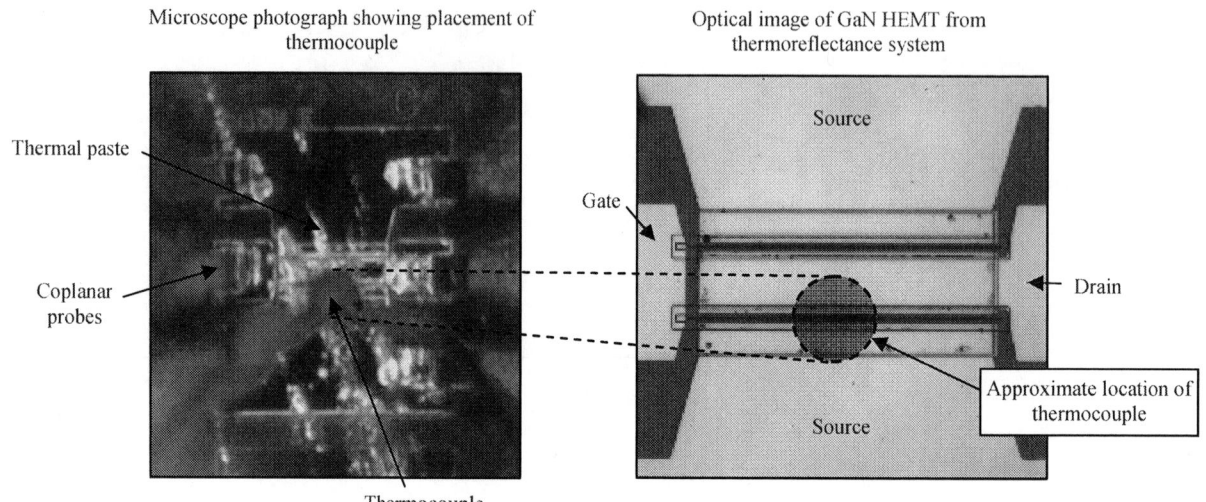

Figure 9. Microscope and CCD optical image showing HEMT and location of thermocouple measurement.

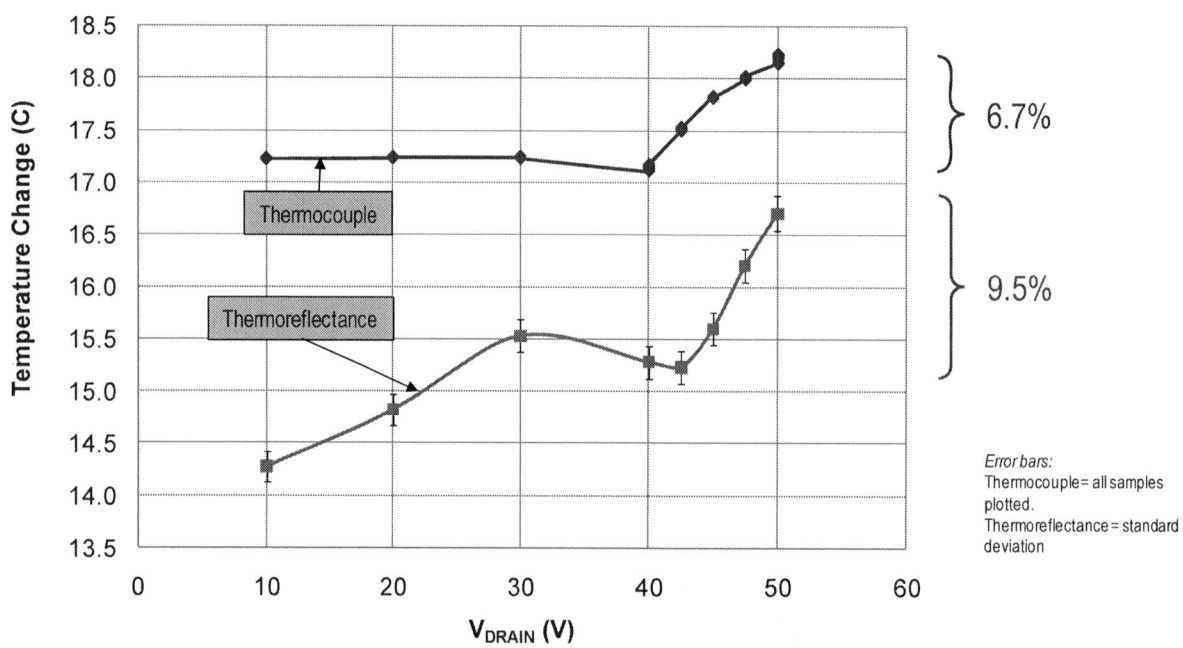

Figure 10. Comparison of single point thermocouple temperatue measurements for location indicated with average thermoreflectance measurement for same area.

## C. Thermocouple Measurements

Direct thermocouple measurement was used to confirm point temperature fluctuations for different HEMT pinch-off bias conditions. Although direct thermocouple measurements do not provide as much spatial temperature information as thermoreflectance images, they can be used for temperature measurement at a single location with spatial resolution approximately equal to the size of the thermocouple junction tip. For this experiment a 25 μm type E Omega thermocouple was used. The thermocouple junction tip was brought into contact with the top surface of the GaN HEMT. Thermal paste was used to improve the thermal interface. The junction tip, approximately 50 μm in diameter, was positioned in the center of the HEMT lower finger as shown in the microscope photograph of Fig. 9. Notably, the placement of the thermocouple did not appear to affect the measurements, as seen by the electrical data. The approximate location and diameter of the thermocouple interface is indicated by the circle in the optical CCD image of the HEMT in Fig. 9. For thermocouple measurement, the HEMT was biased using constant voltage on both the gate and drain. Thermocouple temperature was acquired using a voltmeter and LabVIEW system accurate to 50 mK.

Fig. 10 shows the thermocouple temperature on the HEMT for the same range of pinch-off bias points that were used during thermoreflectance imaging. Also plotted are the average thermoreflectance temperatures measured on the drain metal. The thermoreflectance temperatures are the average of the central third of the horizontal profile, as indicated along the line D-D' in Fig. 5. Both the thermocouple and thermoreflectance measurements show sudden fluctuation in temperature for drain voltage above $V_D = 40$ V. The thermocouple results provide additional evidence of temperature and therefore power redistribution in the HEMT at extreme pinch-off.

## IV. CONCLUSION

Thermoreflectance CCD imaging was used to characterize self heating with sub-micron spatial resolution across the surface of a two finger gallium nitride high electron mobility transistor (GaN HEMT) at various stages of transistor pinch-off for equivalent device power. Two dimension temperature maps calibrated to the transistor drain and source contact metal revealed variation in temperature spatial distribution (implying non-uniform operation of the device) for the cases of greatest pinch-off and highest drain voltage.

## REFERENCES

[1] Moran, J., "The effects of temperature [0-300K] and electron radiation on the electrical properties of AlGaN/GaN heterostructure field effect transistors" Thesis, Air Force Institute of Technology, Mar 2009.

[2] D. Donoval, et al., "High-temperature performance of AlGaN/GaN HFETs and MOSHFETs", Microelectronics Reliability, (2008)

[3] Simms RJT, Pomeroy JW, Uren MJ, Martin T, Kuball M. Current collapse in AlGaN/GaN transistors studied using time-resolved Raman thermography. Applied Physics Letters. 2008;93(20):203510.

[4] Chowdhury U, Jimenez JL, Lee C. TEM observation of crack-and pit-shaped defects in electrically degraded GaN HEMTs. Electron Device. 2008;29(10):1098-1100.

[5] Alamo J a del, Joh J. GaN HEMT reliability. Microelectronics Reliability. 2009;49(9-11):1200-1206.

[6] Beechem T, Christensen A, Graham S, Green D. Micro-Raman thermometry in the presence of complex stresses in GaN devices. Journal of Applied Physics. 2008;103(12):124501.

[7] Ju S, Kading OW, Leung YK, Wong SS, Goodson KE. Short-timescale thermal mapping of semiconductor devices. IEEE Electron Device Letters. 1997;18(5):169-171.

[8] Tessier G, Holé S, Fournier D. Quantitative thermal imaging by synchronous thermoreflectance with optimized illumination wavelengths. Applied Physics Letters. 2001;78(16):2267.

[9] D. Luerssen, J. A. Hudgings, P. M. Mayer, and R. J. Ram, IEEE Semiconductor Thermal Measurement and Management Symposium, 2005. p. 253-258.

# Thermal Factors Influencing the Reliability of GaN HEMTs

Jason A. Carter[1], Jeremy Acord[1], Daniel Hoffmann[1], Andrew Trageser[1], Charles Pagel[2]

[1]Pennsylvania State University – Electro-Optics Center
Freeport, PA, USA

[2]Naval Surface Warfare Center, Crane Division
Crane, IN, USA

jcarter@eoc.psu.edu

## Abstract

Gallium Nitride high-electron mobility transistors (HEMT) devices show great promise in their ability to tolerate the high temperature environments of advanced radar systems. This paper examines how GaN HEMT junction temperature determination can vary, owing to factors such as packaging variability, measurement error, and uncertainty in material property data. To demonstrate the impact of these variables, this paper uses practical examples of infrared thermography, micro-Raman thermography, device transient electro-thermal response analysis on GaN HEMT devices, and finite element analysis (FEA). These variations in temperature are combined into a probability model to estimate how life prediction will change as a function of these various factors.

## Keywords

## 1. Introduction

As a nascent technology compared to GaAs, Si, or non-solid state technology, GaN-on-SiC transistors have not established a history of reliability from which end-users of the technology can establish its long term replacement and refurbishment costs. [1] Nonetheless, GaN provides a number of distinct advantages over older technologies, including improved heat transfer properties, wider bandgap energy, higher operational temperatures, and higher frequency performance. [2]

In lieu of historical reliability information, the consumers of this technology must depend on accelerated lifetime testing (ALT) of parts where a predicted operational lifetime, on the order of millions of hours, is extrapolated from faster failures (hundreds of hours) achieved at highly elevated temperatures. The validity of this extrapolation is dependent on three assumptions: 1) that the physics of failure for the GaN device is analogous to previous technologies, allowing for a log-linear extrapolation (the Arrhenius model) through time-temperature space, 2) that the ALT is exciting the same predominant failure as occurs in fielded devices under standard operating conditions, and 3) that the operational temperature of the device is known. [3]

This paper focuses on this third assumption, using empirical (micro-Raman thermography, transient thermal testing using the T3ster from Mentor Graphics, and midwave infrared thermography) and finite-difference modeling (ANSYS-Fluent) techniques to assess the measure, spatial-uniformity, and statistical variability in temperature measurements on GaN transistor devices.

## 2. Background

### 2.1. The Arrhenius Reliability Model

Historically, the activation of thermally-induced failures has been shown to follow an Arrhenius model whereby the time-to-failure and device operating junction temperature are related by the relationship:

$$t_{failure} = Ae^{\frac{E_a}{RT}}$$ (1)

where $t_{failure}$ is the time-to-failure, Ea is the failure activation energy, R is the Boltzmann constant, and T is the device junction temperature. If the log of both sides of (1) is taken, then the following lognormal relationship is determined:

$$\log t_{failure} = \log A + BT^{-1}$$ (2)

where $B = E_a/R$. [2]

In practice, the operational lifetime of a device is predicted by stressing the device at elevated temperatures well beyond the typical operating temperature. At these elevated temperatures, the device fails faster than it would at operational conditions allowing researchers to complete the tests in time spans of tens to thousands of hours as opposed to the millions to tens of millions of hours one expects the device to last under fielded operating conditions. The rate of those failures is used to determine the values for A and B, which are the y-intercept and slope of equation (2), respectively.

However, the accuracy of the Arrhenius model is dependent on the certainty with which one ascertains the device junction temperature. The extrapolation of the Arrhenius model along a lognormal plot across several decades of time is highly sensitive to the placement of the temperature-failure time data under the accelerated life testing. As shown in this paper, the ability to sample a device set and measure device temperature to a high degree of confidence can lead to a large uncertainty as to the predicted lifetime of the device.

### 2.2. Computational Model

In conjunction with the empirical testing performed on these devices, a computational heat transfer model was built using ANSYS-Fluent. Such models are commonly used in the determination of a packaged device's thermal resistance. [5] [6] [7] The models are useful in providing thermal information throughout the whole volumetric domain of the packaged device, as opposed to the limited scope of empirical measurements. However, such models are limited by how much is known about material properties, the thermo-electrical conditions that affect heat generation in the devices, and knowledge of device details and variability, limiting model accuracy in predicting junction temperature.

Temperature-dependent material properties for the device die were provided by the device manufacturer and geometric detail was provided both by the manufacturer and analysis via optical microscopy and focused ion beam sectioning. Engineering estimates were made as to the die attach thickness and the copper-molybdenum package properties were attained from literature ([8], [9]) and from vendor specifications.

A fully-3D model was constructed, taking advantage of a symmetry plane between the middle two gate fingers in the device. For simplification of these initial models, some of the peripheral metallization to which external electrical contacts were made by wire-bond were ignored. Ohmic metalizations in immediate proximity to the gate fingers were included in detail. The heat flux was added to the model as a volumetric heat source bounded by the volume immediately under the gate metallization at the two-dimensional electron gas (2DEG) layer that defines the device junction. [4] The back side of the device Cu-Mo package was modeled with a convection boundary condition.

Given these assumptions and conditions, a simulation was executed. Figure 1 shows the temperature contours of the top surface of the device. The simulation reveals that a thermal gradient should exist both 1) along an individual gate finger, with the hottest location at the center of the gate and coolest location at the ends and 2) perpendicular to the gate fingers from the hottest center-most finger near the symmetry edge out to the coolest gate finger farthest from the center.

The computational model also allows for a sensitivity analysis which investigates how important uncertainty in geometry and material properties is with respect to junction temperature determination.

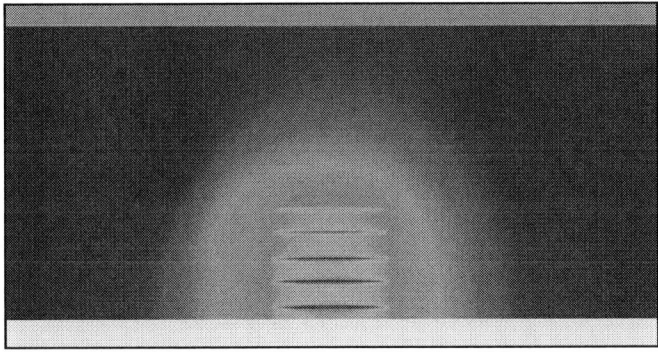

**Figure 1. Simulated temperature contours (using ANSYS-Fluent) on the active surface of the device. The bottom edge of the figure represents the symmetry plane. Note the clear thermal gradient that exists both parallel to the gate fingers and across from finger-to-finger.**

For example, a common design decision in the packaging of devices is the kind of die attach. Epoxies offer the significant advantage of reworkability, allowing end-users to replace failed devices without scrapping an entire module or system. However, they also pose the significant disadvantage of poor thermal conductivity when compared to metal eutectic bonds.

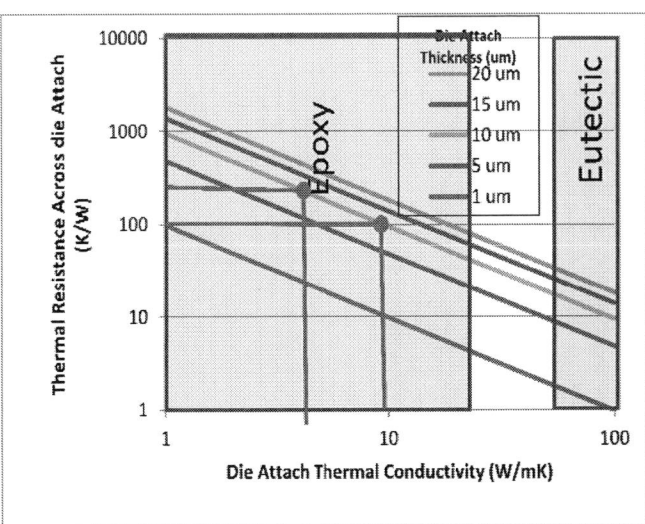

**Figure 2. Die attach thermal resistance as a function of die attach thermal conductivity and thickness.**

Figure 2 shows the sensitivity of die attach thermal resistance (and in turn, device junction temperature) to both 1) the choice of die attach material and 2) uncertainty in die attach material properties. If a 10μm thick epoxy layer has a thermal conductivity of 5 W/mK compared to 10 W/mK, the junction temperature of the device will increase 200°C for every watt dissipated. A similar sensitivity exists if the epoxy thickness is 10μm instead of 5μm. The sensitivity is far less severe if eutectic is used rather than epoxy.

Such examples of typical uncertainty in property and geometric variability limit the confidence in the accuracy of such computational models for each individual packaged device.

## 3. Experimental Description

In order to assess accuracy of the computational model and to determine the uncertainty in empirical temperature measurement techniques, two hundred discrete GaN-on-SiC high electron mobility transistors (HEMTs) were purchased from a commercial source and then packaged by a separate commercial entity. Twenty of those packaged devices were sampled from the population and run through a battery of three empirical tests: 1) mid-wave infrared (IR) thermography, 2) transient thermal testing (TTT), and 3) micro-Raman thermography (μRT).

For all three tests, the devices were operated in a custom made water-cooled fixture was designed and fabricated (Figure 3)

The fixture was designed to accommodate a number of critical aspects of testing. First, both the IR and μRT testing require an optical line-of-sight to the device, which can be seen in the Top view in Figure 3. Second, the devices needed to control the thermal environment of the device, and thus included a channel through which chilled water could be passed. Third, the fixture had to accommodate the exchange of 20 device packages undergoing a battery of three different tests. So the fixture had to allow for easy device handling and consistent electrical contact with the power sources driving the device. As a result of this requirement, no thermal interface material was used between the device package and the water-cooled block on which it contacted. The optimal thermal contact was made by using hinged doors on the opposite side of the device package to press the package down onto the block, the force of which was maximized and held constant by using machine screws to hold the doors down relative to the block.

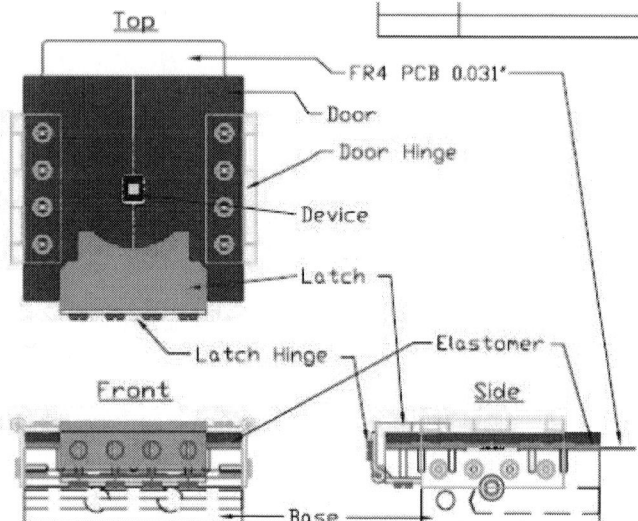

**Figure 3. Schematic of custom test fixture in which all testing described was performed.**

### 3.1. Mid-Wave IR Thermography

IR thermography offers the advantage of rapid temperature measurement and spatial gradient analysis, including the identification of thermal anomalies (e.g., hot spots). However, particularly for this application where gate lengths are submicron in dimension, IR does not provide adequate resolution (~8μm spot size) to determine temperature at or near the device junction.

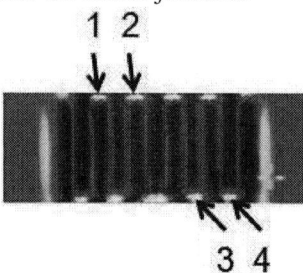

**Figure 4. IR image of device with the location of the direct temperature measurements**

An IRCameras mid-wave camera equipped with a Flir-Janos microscope lens was used to image the 20 sampled devices over a range of operating conditions. Each device was first imaged in a pinched-off condition where no current is passing through the device channel and then under a load of varying drain voltage (up to 15V) and gate voltage (-3 < Vd < -1V) resulting in a drain current not exceeding 450mA and a total dissipated power not exceeding 7W. At each condition, an image was obtained and the temperature determined at four prescribed locations on the exposed semi-conductor material just off the edge of the transistor finger (Figure 4)

The camera was calibrated by recording data from the IR image of the device as the temperature of the device was increased isothermally using a hot plate and the recorded radiosity-temperature relationship was established.

### 3.2. Transient Thermal Tester

Another technique that can be used to determine a spatially-averaged thermal resistance in a device package is transient thermal testing. Summarizing briefly, the technique is based on the assessment of the thermal network as analogous to an RC electrical network (Figure 5).

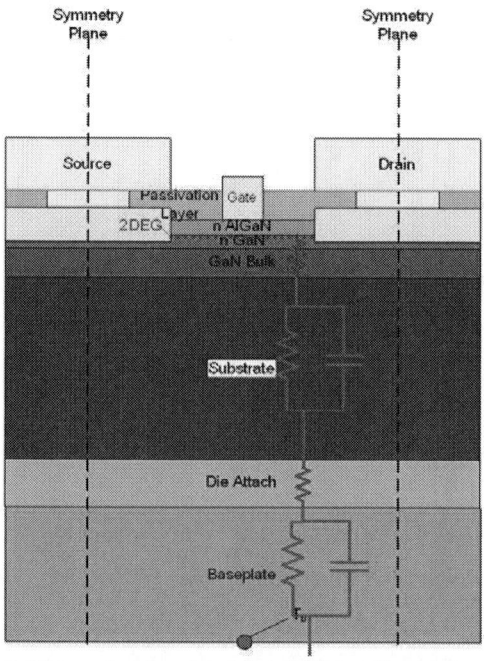

**Figure 5. Thermal network schematic for a discrete transistor.**

As the heat is generated at the device junction, it travels through various materials (thermal resistors and capacitors) and across interfaces (purely resistive). Each of these network elements impacts the response of the thermal network to a transient input, thus the TTT approach infers the thermal network that must exist between junction and heat sink to result in the measured response. A commercial transient thermal tester (T3ster from Mentor Graphics) was used to assess the same 20 sampled devices. For more details on the theoretical and operational details of this technique, see [9].

For this set of tests, the devices were stepped down from a drain current of 850mA to 25mA, with the gate voltage held at 0V, to maximize the signal voltage across the device. The drain voltage was then measured for a duration of 100 seconds after the step down. Each device was tested three times with each test consisting of three tests averaged together; thus each device was stepped and monitored nine times. As with the IR tests, the voltage response of the device was calibrated using a heater block accompanied with fine-gauge thermocouples placed on the backside of the device package measuring temperature.

The transient thermal test approach has an advantage as an approach that does directly measure junction temperature (as opposed to proximate measurements limited by optical line-of-sight and spatial resolution). The approach is limited by

978-1-4673-1110-6/12 $31.00 © 2012 IEEE

the fact that the obtained thermal information is necessarily averaged across all of the gates and does not capture peak localized temperature. Furthermore, this technique requires the careful elimination of non-thermal electrical effects from the purely electro-thermal response of the device to the electrical step.

### 3.3. Micro-Raman Thermography

Micro-Raman techniques are becoming increasingly valuable in electronics thermal characterization because of the superior spatial resolution (~800nm spot size) it offers when compared to IR. It is a technique widely described in the literature. [10] The disadvantage of the technique is that it is a time-intensive point-by-point measurement. It is also highly sensitive to topological variability of the test device, resulting in variable scattering of the signal back to the spectrometer.

The system used to test the 20 devices described here used a 488nm Ar+ laser. The laser was focused on 13 prescribed locations as shown in Figure 6.

**Figure 6. Thirteen prescribed location where μR measurements were taken on HEMT device**

The locations of the measurements were designed to provide an indication of the thermal spatial gradients along device gate fingers and across the device periphery.

The Raman effect, whereby a small fraction of injected photons interact with atoms fixed within their crystallographic lattice, either adding or subtracting energy from those atoms, is fundamentally a stress measurement technique. [11] To eliminate the conflation of crystallographic stress from thermal stress, a calibration was made at each of the 13 prescribed points in Figure 6. Thereby if a defect in the crystallographic lattice existed by chance near one of the 13 points, the effect of that stress on the measurement would not bias the thermal measurement.

Each device was scanned using the μRT technique three times, with the spectra of the three measurements averaged together to provide the final measurement.

### 4. Results

Since each of the empirical techniques measures temperature at different locations on the device surface (Figure 7), the expectation is that each method will capture different temperatures.

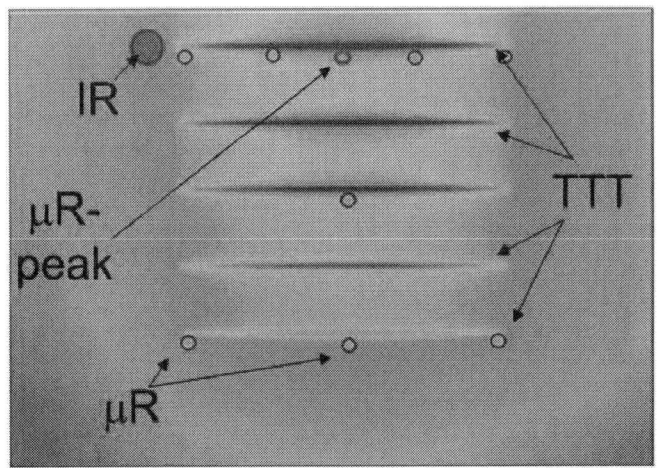

**Figure 7. Location of empirical measurements for each technique – Infrared (IR), micro-Raman (μR), and transient thermal tester (TTT)**

The IR measure is farthest from the device gate where most of the heat is generated, and so should result in the lowest temperatures. Conversely, the μRT measure should come closest to capturing the peak device junction temperature. The TTT and spatially averaged μRT measurements should compare closely, as both sample across the gates and provide device junction temperature from both colder and warmer regions of the device.

### 4.1. Results: Mid-Wave IR Thermography

The most significant advantage of using IR thermography for devices with spatial scales below the physical limits of IR imaging resolution is the quick identification of thermal hot spots, as seen in Figure 8.

**Figure 8. Mid IR image of sampled device showing hot spot on right-most finger.**

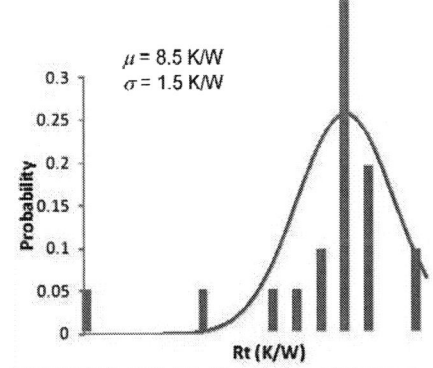

**Figure 9. Probability histogram and Gaussian distribution based on the derived thermal resistance from IR measurements**

The peak measured temperature from the IR tests is plotted versus power dissipated through the device for all 20 devices to obtain Figure 10. Except for two devices, the other 18 devices are grouped about a line that represents the average thermal resistance of 8.5°C/W for the sample set. A probability distribution based on the data in Figure 10 is seen in Figure 9.

The average thermal resistance, given by the following equation

$$\frac{T_{meas} - T_{coolant}}{P_{diss}} \qquad\qquad 3$$

where $T_{meas}$ is the measured temperature, $T_{coolant}$ is the ambient coolant temperature of the heat sink, and $P_{diss}$ is the power dissipated across the device (drain current times drain voltage), for the IR measurements was 8.5 K/W with a standard deviation of 1.5K/W. Correspondingly, if a device were dissipating 4W of energy, the variability represented in Figure 9 results in only 68% confidence that the measured device temperature will be between 28 and 40 °C above the baseplate temperature.

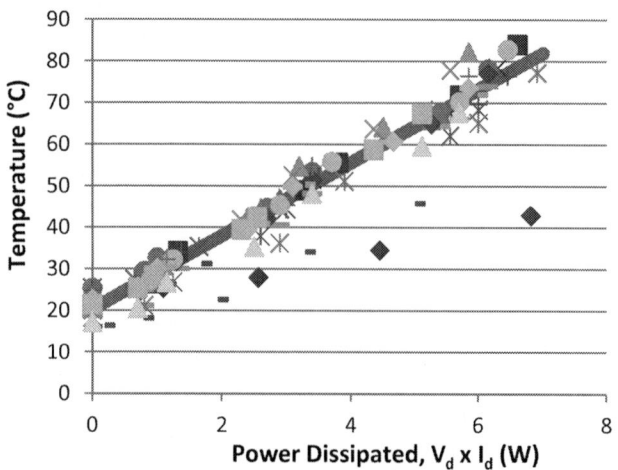

**Figure 10. Temperature of the peak measured temperature versus power dissipated for 20 tested devices**

## 4.2. Results: Transient Thermal Tester

The spatially-averaged temperature was measured for each of the 20 sample devices using the TTT approach. The result of those measurements was compiled into a probability distribution histogram in Figure 11. The mean thermal resistance using this method works out to 11.4 K/W, higher than the thermal resistance measured via IR, as expected. The standard deviation using the TTT approach was 1.1 K/W, meaning that if the device is dissipating 4W, a 68% confidence exists that the average device junction temperature is between 41 and 50°C above baseplate.

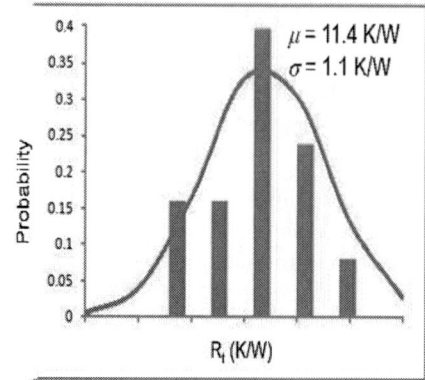

**Figure 11. Probability histogram and Gaussian distribution based on derived thermal resistance from TTT measurements.**

Since the devices are experiencing a step down in drain current, the device temperature should decrease steadily from a hot state to a cool state, corresponding to a steady decline in voltage. For a majority (13 of 20) of the devices, this steady decline was evidenced as seen in Figure 12. Each tested device corresponds to a color and each trace of a particular color represents one of the three tests for that device. Note that the colors are tightly spaced, which indicates good experimental repeatability. The spread of the thermal transient curves after 100 seconds is about 13°C across the 13 tested devices. This corresponds to a range in thermal resistance across the sample of 9-13 K/W.

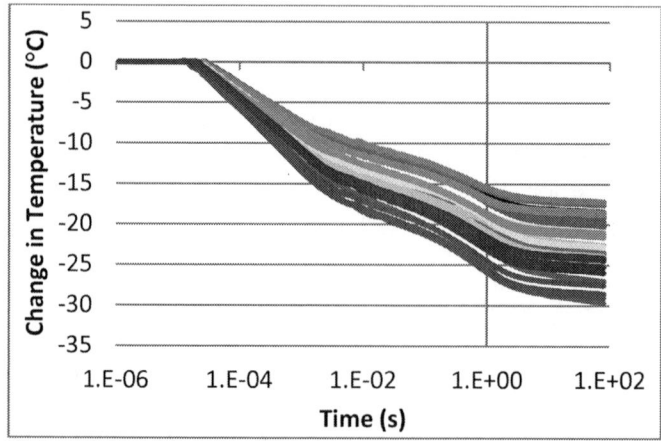

**Figure 12. Thermo-electrical response of devices to the electrical step using the TTT approach.**

However, a subsample of the devices exhibited a non-thermal phenomenon at a time on the order of 10 ms to 1 s after the step down, where the drain voltage increased (Figure 13). Since there is no source of additional heating that could cause the voltage to increase, the phenomenon must be a non-thermal electrical phenomenon. At the point of this writing, the cause of that phenomenon is not clear.

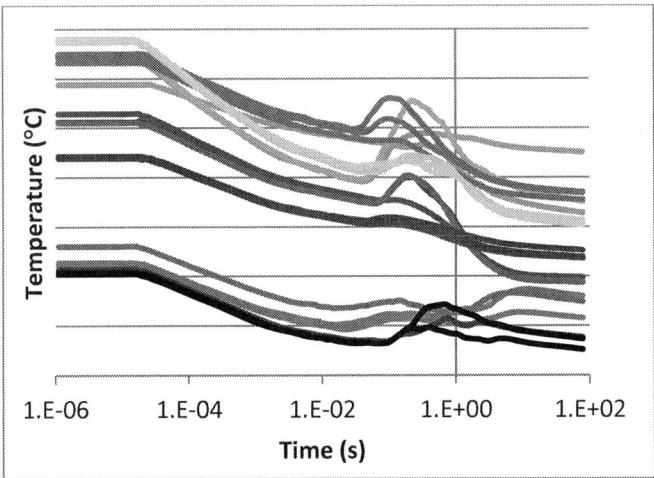

Figure 13. Thermo-electrical response of subset of tested devices showing non-thermal electrical response between 10 ms and 1s

### 4.3. Results: Micro-Raman Thermography

The micro-Raman results were studied in two ways, 1) attaining the peak point temperature measured of the 13 prescribed locations and 2) a spatially-averaged measure across the 13 measured points. The peak temperature should be indicative of the hottest temperature on the active device, which in theory, is thought to drive device failure. Based on an idealized model (Figure 1), the expectation is that the peak temperature should be the center-most location on the center-most gate finger. In practice, the Raman measurements do not necessarily follow that paradigm. Across this sample of 19 devices (one device proved impossible to measure due to topological variability scattering the Raman signal), the peak temperature for the device occurred at 7 of the 13 prescribed points. Most of the peak temperature measurements were made along the center gate finger (12 of 19) and only 1 occurred along the perimeter finger. If the temperature at each of the 13 locations is averaged across the 19 devices, the thermal resistance measurement for each point appears in Table 1.

| Finger Loc: | 1 | 3 | 5 | 8 | 10 |
|---|---|---|---|---|---|
| **Source End** | 10.45 | | 13.55 | | 10.12 |
| **1/4** | | | 13.17 | | |
| **Center** | 11.75 | 12.65 | 13.01 | 12.75 | 11.08 |
| **3/4** | | | 12.47 | | |
| **Drain End** | 9.51 | | 10.26 | | 8.69 |

Table 1. Thermal resistance (K/W) for each prescribed micro-Raman measurement location and averaged across 19 devices

When averaged across the device sample set, the predicted thermal gradient perpendicular to the gate fingers is seen empirically. However, the predicted parallel gradient is not seen along the center finger (5). Along that finger, the hottest location does not appear to be at the finger center, but at the source end of the finger. This deviation from prediction may be due to the model simplification where metalizations off the ends of the gate fingers were not included.

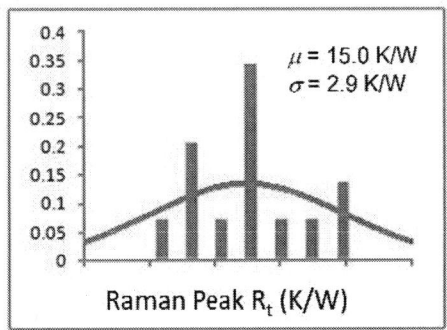

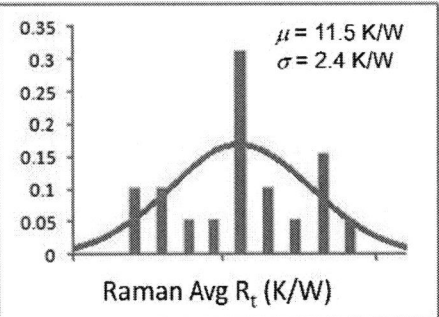

Figure 14. Probability histogram and Gaussian distribution based on derived thermal resistance from μR measurements. The peak Raman measurement is at the top and the 13-point average is at the bottom.

The Raman measurements have the highest standard deviation of the three empirical techniques, due several factors: 1) the high precision of the technique combined with a limited sampling of spatial data points on the device surface (as compared to the TTT technique which is an analog measurement across the entire device) and 2) a high sensitivity to topological variability across the device surface. Given that sensitivity, if the device is dissipating 4W, there is 68% confidence that the peak device junction temperature is between 48 and 72°C above baseplate temperature.

Nonetheless, the peak Raman measurement assesses a thermal resistance within 5% of the device manufacturer's specified thermal resistance for this family of devices. Furthermore, it is worth noting how closely the average Raman measured thermal resistance (11.5 K/W) compares to the TTT approach (11.4 K/W).

### 5. Conclusions

A multi-tool approach to assessing device junction temperature and thermal performance has been shown. Each tool provides different information and uncertainties so that such a broad empirical assessment is important to determine junction temperature and thermal performance.

The thermal variability of packaged devices can be high. As such, an end user must view the assumption that the device junction temperature is known in the assessment of the Arrhenius reliability model with a degree of skepticism. To increase confidence in accounting for such uncertainty one should consider

- Instituting part-specific thermal models
- Increasing statistical sampling – acquisition cost versus refurbishment cost

- Monitoring quality improvements in manufacturing and packaging processes

## Acknowledgments

This material is based upon work supported by Naval Surface Warfare Center, Crane Division through the Naval Sea Systems Command under Contract No. N00164-09-C-GR34

Any opinions, findings and conclusions, or recommendations expressed in this material are those of the author(s) and do not necessarily reflect the views of NSWC Crane or the Naval Sea Systems Command.

## References

[1] Interdepartmental Committee for Meteorological Services and Supporting Research, "Federal Research and Development Needs and Priorities for Phased Array Radar (FCM-R25-2006)," Office of the Federal Coordinator for Meteorological Services and Supporting Research, Washington, DC, 2006.

[2] W. L. Pribble, J. W. Palmour and etal, "Applications of SiC MESFETs and GaN HEMTs in power amplifier design," *Microwave Symposium Digest,* pp. 1819-1822, 2002.

[3] D. S. Green, B. Vembu and etal, "GaN HEMT thermal behavior and implications for reliability testing and analysis," *Physica Status Solidi,* vol. 5, no. 6, pp. 2026-2029, 2008.

[4] M. Ohring, Reliability and Failure of Electronic Materials and Devices, San Diego: Academic Press, 1998.

[5] J. C. Freeman, "Channel Temperature Model for Microwave AlGaN/GaN HEMTs on SiC and Sapphire MMICs in High Power, High Efficiency SSPAs," in *International Microwave Symposium*, Fort Worth, TX, 2004.

[6] A. Prejs, S. Wood and etal, "Thermal analysis and its application to high power GaN HEMT amplifiers," *Microwave Symposium Digest,* pp. 917-920, 2009.

[7] J. L. Jimenez and U. Chowdhury, "X-Band GaN FET Reliability," *Reliability Physics Symposium,* pp. 429-435, 2008.

[8] "Copper Molybdenum Heatsinks," Plansee -- Thermal Management Solutions, [Online]. Available: http://www.plansee-tms.com/Copper%20Molybdenum%20Heatsinks.pdf. [Accessed 1 December 2011].

[9] "Table of Specialty Solders and Alloys," Indium Corporation, [Online]. Available: http://www.indium.com/products/alloy_sorted_by_indalloy_number.pdf. [Accessed 1 December 2011].

[10] E. Heller and A. Crespo, "Electro-thermal modeling of multifinger AlGaN/GaN HEMT device operation including substrate effects," *Microelectronics Reliability,* vol. 48, no. 1, pp. 45-50, 2008.

[11] V. Szekely, A. Ress and etal, "New approaches in the transient thermal measurements," *Microelectronics Journal,* vol. 31, no. 9-10, pp. 727-733, 2000.

[12] M. Kuball, J. M. Hayes and etal, "Measurement of temperature in active high-power AlGaN/GaN HFETs using Raman spectroscopy," *Electron Device Letters,* vol. 23, no. 1, pp. 7-9, 2002.

[13] J. W. Ager III and M. D. Drory, "Quantitative Measurement of residual biaxial stress by Raman spectroscopy in diamond grown on a Ti alloy by chaamical vapor deposition," *Physical Review B,* vol. 48, no. 4, pp. 2601-2607, 1993.

# A Framework Theory for Dynamic Compact Thermal Models

Mohamed-Nabil Sabry[1], Mohamed Dessouky[2]
[1]Mansoura University, [2]Mentor Graphics
[1]Mansoura, [2]Cairo, Egypt
mnabil.sabry@gmail.com

## Abstract

This fundamental work aims at obtaining a new topology of Dynamic Compact Thermal Models (DCTM), which is capable of capturing all problem physics. It solves problems encountered by the widely used thermal impedance model, which is only a special case of the proposed new topology. The latter starts from governing partial differential equations of transient conduction heat transfer. It uses analytical procedures based on the Green's function to obtain the required DCTM topology, which is thus the most general one. This work generalizes an already presented approach for static problems based on the flexible profile technology to obtain the most general topology of static compact models. Some simple examples are treated to show that the widely used DCTM based on the thermal impedance concept fails to capture some aspects of transient problems, while the approach proposed here correctly considers them.

## Keywords

Conduction, Transient, Compact Model, Topology

## 1. Introduction

In order to design modern complex electronic systems, manifesting multiple physical effects that are usually coupled, involving as well a huge number of components, Compact Thermal Models (CTMs) are indispensible tools. As a design tool, a compact model should:

(a) Describe observed thermal behavior of an object viewed as a black box, i.e. without having to reveal its internal structure.

(b) Offer a good compromise between precision and calculation speed.

(c) Be universally usable, i.e. gives reasonable predictions for any set of external conditions (any usage scenario).

The first condition (a) is important to keep IP rights of the supplier of an object, which will be used by another team as a component in a larger system. Object supplier should also supply its CTM. Importing team, who needs to design the overall system, which may contain a huge number of components, expects a "sufficiently good" precision, without heavily affecting CPU time (condition b), due to system complexity.

This paper mainly addresses the last requirement (condition (c)), while respecting the other two. In fact, a model that can only be used in a given set of external conditions is of limited value. It can be replaced by tabulated results of verification experiments. One can always fit any model equation on any set of experimental data, called the generating set, using enough adjustable parameters. However, if this equation did not capture problem physics, its predictions may have large errors when used for a case that is outside the generating set. Of course, a compact model, containing by definition a limited number of degrees of freedom (*DoF*), cannot be as precise as the detailed model. *Hence, it is crucial that this limited number of DoF be selected such as to cover all problem physics, even if each aspect is addressed approximately. That is why having a general topology, emanating from the governing partial differential equation, is a valuable modeling tool.* This has already been done in a previous work for steady problems. The objective of this paper is to generalize this approach for transient problems, which will pave the way for a new generation of more precise models with controlled errors.

Governing partial differential equations are:

$$\partial[\rho(\mathbf{r})c(\mathbf{r})T(\mathbf{r},t)]/\partial t = -\nabla \cdot \mathbf{q}(\mathbf{r},t) + q_v(\mathbf{r},t) \quad (1)$$

$$\mathbf{q}(\mathbf{r},t) = -k(\mathbf{r})\nabla T(\mathbf{r},t) \quad (2)$$

where $T$ is the temperature, $\rho$ is the density, $c$ is the heat capacity, $k$ is the thermal conductivity, $\mathbf{q}$ is the heat flux density, $q_v$ is a volumetric heat source and $\mathbf{r}$ and $t$ are respectively space and time independent variables.

The following section resumes already available results for static 3D models [1], which have to be included in the dynamic model as well. It will be followed by an introductory section for transient 1D models, in order to present the core contribution of this paper in a simple case, which will help better understanding new fundamental issues introduced in this work. Generalization to transient 3D models will then be made in a following section before formulating conclusions.

## 2. Overview of the general form of static 3D models

A series of publications [2, 3] have already addressed CTM fundamentals for 3D static problems. A thermal port is defined as a zone (surface or volume) in the object to be modeled that exchanges heat with external sources. Theoretically speaking, heat flow due to temperature differences is analogous to electric current flow due to electric potential differences. Nevertheless, analogy fails at the practical level. In fact, the ratio of electric conductivity of a conductor over that of an insulator is of the order of $10^8$. The corresponding ratio of thermal conductivity of a thermal conductor over that of a thermal insulator is of the order of 100 only. This means one can safely assume that an electric port is an equipotential surface, i.e. a contact 'point'. However, such an assumption for thermal nodes is only an approximation, which is not always justified. Temperature $T$ and heat flux density $q$ are never uniform over any thermal port, unlike the case of its 'analogous' electric conduction. Early CTMs assumed perfect analogy. In its crudest form, the CTM defines only two thermal ports (a heat source: the so-called "junction" and a heat sink labeled either "ambient" or

978-1-4673-1110-6/12 $31.00 © 2012 IEEE

"case") with uniform $T$ and $q$ on each, resulting in a single thermal resistor model. It has long been recognized [4] that such models fail to give reasonable predictions under different usage scenarios, i.e. different external conditions. They are not BCI (Boundary Conditions independent), which is a term that was forged to qualify universally usable models. Compact models are never fully BCI, but should be close. A network of resistors joining a set of thermal ports, each viewed as a node of uniform $T$, has been suggested [4] as a better model topology. It has produced a significant improvement in precision. Such models can accommodate a non-uniform $T$ over each node, *provided the profile remains fixed.*

Modern applications are characterized by multiple heat sources (MCM, SoC, 3D stacked dies …), which can be triggered independently. This gives rise to a variable $T$ and/or $q$ profile over each thermal port, depending on which source is triggered. Classical CTMs, based on thermal resistive network, would fail in correctly modeling such problems. A rather recent breakthrough, although not yet a standard practice, was achieved by switching to flexible profile models [1]. In this new generation, on each thermal port two finite sets of state variables describe $T$ and $q$ fields over the port. These state variables can be for instance coefficients of polynomials in space variables. Hence, such model allows taking into consideration variable space profiles of both $T$ and $q$. Unlike classical models, which are extracted in such a way that restricts them to be usable for a fixed or almost fixed profile, the new 'Flexible profile' approach incorporates $DoF$ that describe the profile. Hence, it is universally valid, especially for multiple heat sources. The number of state variables needed to describe the profile on each port depends on precision requested and can vary between different ports. It will be assumed though for simplicity that the number was fixed at the value $M$ for each of the $N$ thermal ports constituting the CTM. Hence, the total number of degrees of freedom ($DoF$) is $NM$. The vectors $\mathbf{T}$ and $\mathbf{q}$ (each of size $NM$) will group all state variables over all ports describing $T$ and $q$ fields. The CTM will thus take any of the general forms:

$$\mathbf{T} = \mathbf{Rq} \quad \mathbf{YT} = \mathbf{q} \tag{3}$$

where $\mathbf{R}$ is a resistance matrix and $\mathbf{Y}$ is an admittance matrix. System size is $NM$, i.e. it is more general than the resistive network model. Nevertheless, system (3) size is significantly less than that of a detailed FEM (Finite Element Method) model, for a precision that is closer to that of FEM than that of classical methods. Such a general model has been obtained starting from the governing equations (1, 2) (after omitting time derivative) using a modified Green's theorem. It is universally usable (condition (c) above), satisfying thus BCI requirements. The flexible profile approach [1] is outside the scope of this paper. However, an extremely simple example will be given below to show how profile may have an influence on the 'equivalent' thermal resistance. Assume that the body to be modeled was internally composed of two boxes having different thermal conductivities, $k_1$ and $k_2$, as shown in Figure 1. Both parts have the same cross-sectional area $A$ and length $L$. A total amount of heat $Q$ flows over the whole body from top to bottom surface, while sidewalls are kept insulated. Suppose that the heat flux profile was not uniform, such that heat imposed over the top surface of part 1 was $aQ$, ($a<1$), the remainder $(1-a)Q$ being imposed on part 2. As a result, temperature profile will also be non-uniform. Temperature differences over each part are:

$$\Delta T_1 = aQ L / k_1 A \quad \Delta T_2 = (1-a)Q L / k_2 A \tag{4}$$

The average temperature is:

$$\Delta T_{avg} = \left( a/k_1 + (1-a)/k_2 \right) Q L / 2A \tag{5}$$

The equivalent thermal resistance, i.e. the ratio of average temperature difference $\Delta T_{avg}$ to input heat flux $Q$, depends on $a$, i.e. depends on the profile.

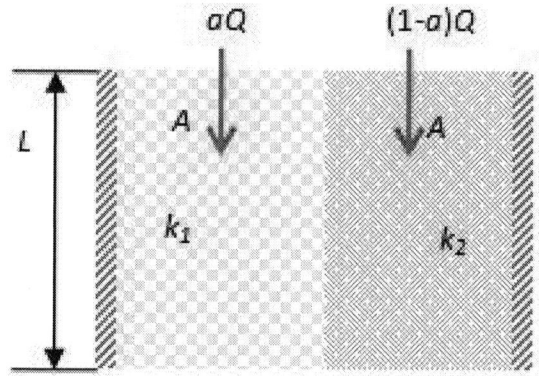

**Figure 1**. Effect of profile on equivalent resistance

## 3. A preliminary analysis of transient 1D problems

### 3.1. Problem formulation

In order to better understand physics of transient conduction, a simple 1D problem (see Figure 2) will be defined as follows. Problem is governed by equations (1, 2), in which it is assumed that all physical properties as well as volumetric heating are piecewise continuous in space. This implies, by integrating (1) over a vanishingly small volume, that the heat flux density $\mathbf{q}$ is continuous in space.

Let us consider the temperature field at a given instant of time $t_0$ and take any arbitrary point $\mathbf{r}_0$ as a reference. Starting from that point it is possible to draw a 'reference' heat flow path, which is a curve that is tangent at each point to $\mathbf{q}$. The arc length $x$ along the reference heat flow path will be taken as a coordinate. A group of heat flow paths lying at the immediate neighborhood of the one passing through $\mathbf{r}_0$ constitutes an infinitesimal heat flux tube. The arbitrarily small cross sectional area $A(x)$ of the infinitesimal heat flux tube normal to the reference heat flow path is in general space dependent. Due to the smallness of this area, $\rho c$, $k$, $\mathbf{q}$ and $q_v$ can be considered as uniform over any cross-section, although still variable with $x$. This also applies to $T$, from (2), since $\nabla T$ is normal to $A$. Heat conduction within the infinitesimal heat flux tube is thus one dimensional. Heat conduction in a whole finite domain can be considered as one dimensional provided a certain number of conditions, enumerated below, were satisfied. Let us first, consider a point lying on the reference heat flow path at a distance $x$ from the origin. Construct a surface, passing by this point, which is perpendicular to **q** at all its points. Such a surface is by construction isothermal. A family of such surfaces can be constructed at different values of $x$. Conditions for one-dimensional heat conduction in the domain are:

- The domain is delimited by surfaces that are either tangent to heat flow paths (hence adiabatic) or normal (hence isothermal). Delimiting isotherms can only be two in number: a single heat inlet surface at $T = T_1$ and a single heat outlet surface at $T = T_2$.
- The following parameters are uniform over isothermal surfaces: $\rho c$, $k$, **q** and $q_v$.
- The initial temperature field was itself uniform over each isothermal surface.

Consider two isothermal surfaces at $x$ and $x+dx$. Initially, they are at temperatures $T$ and $T+dT$ respectively. By virtue of (2), assuming conditions enumerated above hold, it is evident that the distance along any heat flow path, other than the reference one, between both surfaces is also $dx$. Hence, a single space coordinate $x$ can be defined along any heat flow path curve, ranging from 0 at node 1 to $L$ at node 2.

If, at each of the delimiting isotherms (at $T_1$ and $T_2$), boundary conditions on **q**, if any, remain uniform in space but variable with time, then isothermal surfaces will still have the same shape at subsequent instants, but the uniform temperature value on each of them may be time dependent. Hence, if we integrate equation (1) over the infinitesimal volume lying between isothermal surfaces at $x$ and $x+dx$ and delimited by the outer adiabatic surface, we get using (2):

$$\rho(x)c(x)A(x)\partial T(x,t)/\partial t =$$
$$\frac{\partial}{\partial x}\left[k(x)A(x)\partial T(x,t)/\partial x\right]+q_v(x,t)A(x) \qquad (6)$$

At each of the delimiting isotherms, the following total heat fluxes (considered as positive if entering) can be defined:

$$Q_1 = -\left\{k(x)A(x)\partial T(x,t)/\partial x\right\}\big|_{x=0} \qquad (7)$$

$$Q_2 = \left\{k(x)A(x)\partial T(x,t)/\partial x\right\}\big|_{x=L} \qquad (8)$$

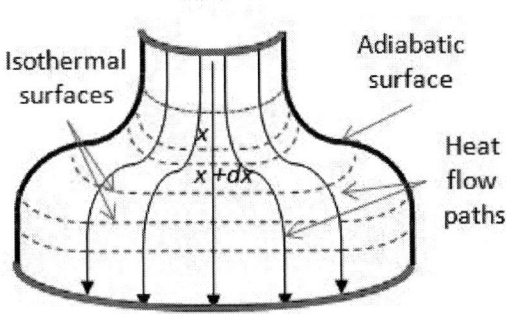

**Figure 2.** The simple 1D problem

### 3.2. Problem solution: obtaining the new model topology

Variable physical properties complicate problem solution. The following change of variables eliminates this difficulty:

$$dx/[k(x)A(x)] = dy/[kA]_{ref} \qquad (9)$$

$$dt/[\rho(x)c(x)A(x)][k(x)A(x)] = du/[\rho cA]_{ref}[kA]_{ref} \qquad (10)$$

Quantities having the subscript *ref* are the average over $x$ from 0 to $L$ of the quantity between square brackets. This change of variables transforms the governing equations into:

$$C_{ref}\,\partial T/\partial u = \left(L^2/R_{ref}\right)\frac{\partial^2 T}{\partial y^2}+Q_b \qquad (11)$$

where:

$$C_{ref} = [\rho cA]_{ref}\,L \qquad R_{ref} = L/[kA]_{ref} \qquad (12)$$

$$Q_b = q_v A(x)L\,\frac{k(x)A(x)}{[kA]_{ref}} \qquad (13)$$

Taking the Laplace transform of (1) gives:

$$d^2 T_s/dy^2 - \beta^2 T_s = -S_s \qquad (14)$$

$$S_s = \left(R_{ref}Q_{bs}+\tau_{ref}T_0(y)\right)/L^2 \qquad (15)$$

In which subscript $s$ means the Laplace transformed variable, $\beta^2 = s\tau_{ref}/L^2$, $\tau_{ref} = R_{ref}C_{ref}$, and $T_0$ is the initial temperature. A General analytical solution of such a problem in terms of $T_{1s}$, $T_{2s}$ (Laplace transforms of temperatures at nodes 1 and 2) and $S_s$, can be readily obtained:

$$T_s = \frac{\left(T_{1s}\,sinh(\beta(L-y))+T_{1s}\,sinh(\beta y)+Q_{cs}(y)\right)}{sinh(\beta L)} \qquad (16)$$

where $Q_{cs}$ is an involved expression, which takes a simple form for uniform volumetric heating and initial temperature:

$$Q_{cs} = \left(sinh(\beta L)-sinh(\beta y)-sinh(\beta(L-y))\right)S_s/\beta^2 \qquad (17)$$

Heats entering nodes 1 and 2 ($Q_1$ and $Q_2$) can be obtained from (16, 17) using (7–10). After some algebra, results can be recast in the form of a "Π" circuit as shown in Figure 3, in which:

$$\psi_1 = sinh(\beta L)/\beta L \qquad \psi_2 = \frac{2(cosh(\beta L)-1)}{\beta L\,sinh(\beta L)} \qquad (18)$$

$$Q_{b0s} = Q_{bs}+C_{ref}T_0 \qquad (19)$$

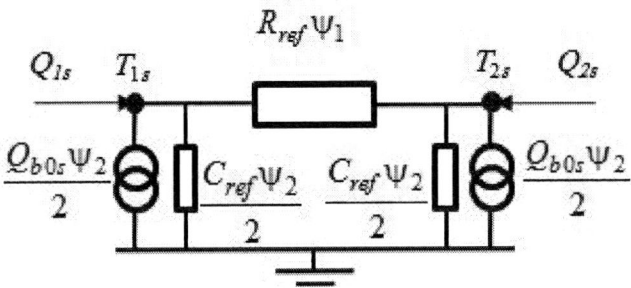

**Figure 3.** The proposed "Π" model for the 1D problem

The "Π" topology has been obtained starting from the governing equation without any assumption other than that of a 1D transient conduction. Note that volumetric heating is not necessarily uniform; it was introduced to simplify resulting expression. This topology is thus the most general form capturing all dynamical effects for a 1D problem. It will be compared in the next section with the widely used model for 1D problem based on the so-called thermal impedance $Z_{th}$.

### 3.3. Comparing with classical model topology

The classical model, which is based on the thermal impedance $Z_{th}$, is the impulse response in either time or Laplace domains for a pulse applied at node 1 while node 2 is grounded. It takes the form of grounded complex impedance (Figure 4). The $Z_{th}$ model fails to capture three physical effects that are correctly captured by the general $\Pi$ model proposed here:

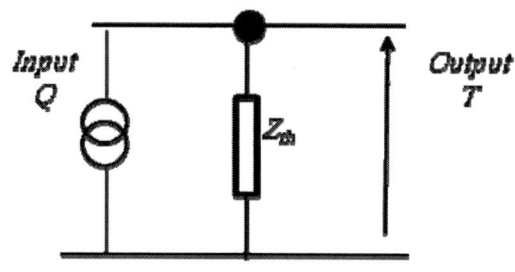

**Figure 3.** The widely used thermal impedance model

(a) The thermal impedance model ignores thermal inertia effects of one of the two nodes. In fact, by grounding node 2 in figure 2, ignoring $T_0$ and $q_v$, we get the same model as that of figure 3. Grounding has eliminated the thermal capacitance attached to node 2. If the system contained only one object to be modeled, both models ($\Pi$ and $Z_{th}$) are equivalent. If, however, the system contained more than one object, which is almost always the case, the general model proposed here will allow joining models for two objects, while correctly capturing dynamics at interface nodes. The widely used thermal impedance model will fail at this level.

(b) The $Z_{th}$ model ignores volumetric heat generation $q_v$. Even if it was replaced by an "equivalent" heat source localized at the heat source node, it will never induce the same temperature temporal variation, due to the body thermal inertia. For a uniform heating distributed over a finite volume, phase delays between $T$ and $q$ can never be the same as those of a point source.

(c) The thermal impedance model ignores initial conditions $T_0(x)$, by setting them always to zero. In the "$\Pi$" model, $T_0$ together with heat generation $q_v$ are included in $Q_{b0s}$.

The following very simple test cases will quantitatively prove the superiority of the proposed $\Pi$ model, as compared to the $Z_{th}$ model.

**Case 1:**

Consider a box of uniform cross-section heated at its top over the whole surface and cooled at its bottom at the whole surface as well. Physical properties are assumed constant. No volumetric heating is applied. Initial conditions are homogeneous. If we consider the whole body, grounding node 2, eliminating volumetric heating and initial conditions, the $\Pi$ model reduces to frequency dependent R-C pair in parallel, having the following impedance:

$$Z_{full-box} = \frac{R_{ref}\,sinh(\beta L)}{\beta L\,cosh(\beta L)} \qquad (20)$$

This value is the same for both approaches $\Pi$ and $Z_{th}$. The difference appears if we decide to break the box into two halves of equal height $L/2$. Model equations will be applied for each part and join them in series. The $\Pi$ model has a topology shown in Figure 4, while that of $Z_{th}$ is shown in Figure 5. After lengthy algebra, the resulting expression for the combined two halves for each modeling topology is:

$$Z_{2halves-\pi\,model} = \frac{R_{ref}\,sinh(\beta L)}{\beta L\,cosh(\beta L)} \qquad (21)$$

$$Z_{2halves-Zth\,model} = \frac{R_{ref}\,sinh(\beta L/2)}{(\beta L/2)\,cosh(\beta L/2)} \qquad (22)$$

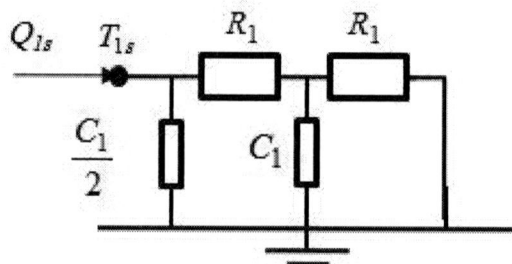

**Figure 4.** The $\Pi$ model for two half boxes

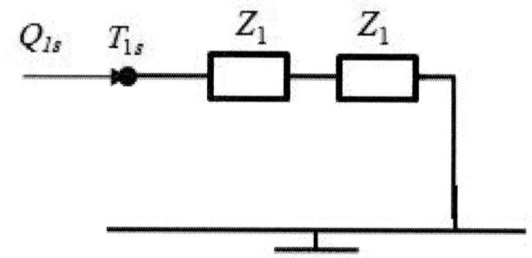

**Figure 5.** The $Z_{th}$ model for two half boxes

It is clear that the $\Pi$ model topology allows joining models of two different objects together, while the $Z_{th}$ model topology cannot. The latter has a time constant that is 4 times smaller than the exact solution. The step response due to a step $Q_0$ applied at node 1 is shown in figure 6 for both cases. The abscissa is the dimensionless time $t\,/\,\tau_{ref}$, the ordinate is the dimensionless temperature $T\,/\,R_{ref}Q_0$.

**Case 2:**

Consider the same full box as that of the previous case, but let heating be uniformly distributed over the volume, instead of being concentrated at the top node. The $\Pi$ model considers this case by applying an equivalent heating at both ends (Figure 3). The $Z_{th}$ model does not consider this case directly. The only way to do that, while respecting the energy conservation principle, is to replace the volumetric step by two steps of equal strength localized at both nodes, each

978-1-4673-1110-6/12 $31.00 © 2012 IEEE

having half the full heating. Of course, thermal dynamics cannot be respected in this case, because heat applied inside the object cannot affect both ends instantaneously. Figure 7 compares results of both models. Note that the Π model is an exact solution of the governing equation containing the volumetric heating.

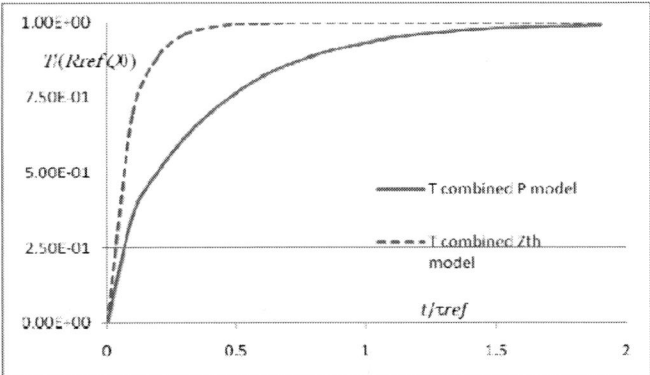

**Figure 6.** Comparing results for case 1

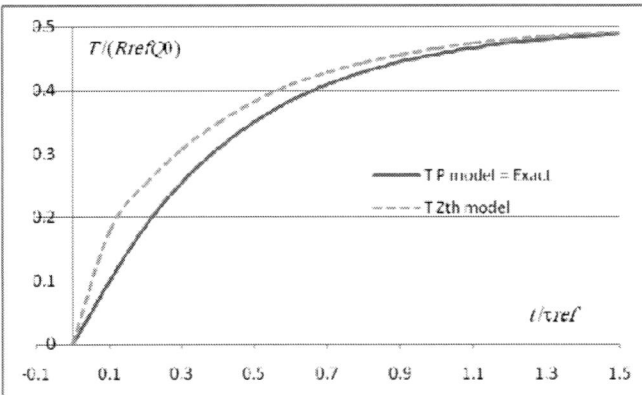

**Figure 7.** Comparing results for case 2

Last, but not least, the fact that the Π model correctly handles non-zero initial conditions, while the $Z_{th}$ model cannot, needs not be proved. In fact, initial conditions are treated the same way as volumetric heating. The reason is simple: an impulse volumetric heating applied at $t=0$ can always produce any initial temperature distribution requested. Hence, correctly treating an initial distribution is a consequence of correctly treating volumetric heating.

## 4. Generalization to transient 3D models

The standard procedure to deal with transient 3D problems is to define a set of nodes and construct a network of thermal impedances joining each pair of nodes. This standard approach suffers from serious deficiencies in both its spatial and its temporal aspects. Temperature gradients in both space and time have a higher effect on reliability than the absolute temperature [5, 6].

In order to show spatial problems, consider the steady state, where the thermal impedance network transforms into a thermal resistance network, relating $T$ and $q$ for different nodes. The main problem with such a representation is that on each node, only one value of $T$ and one value of $q$ can be

defined. Hence, it gives rise to a rather large error if the temperature profile over each thermal port was not fixed. Large variations in temperature profile are expected in multiple heat source problems, especially when each one is triggered independently, as mentioned above.

In addition, it suffers from deficiencies mentioned in section 3 for temporal dependence. Namely, a thermal impedance network cannot incorporate volumetric heating, non-zero initial temperature field as well as inability to join models of two different objects. Transient effects have severe consequences on reliability [7 – 10]. Such phenomena are particularly important in MEMS [11, 12]

In the sequel, the procedure used to construct the general topology will be briefly sketched. The governing partial differential equation is obtained from (1, 2) assuming constant physical properties:

$$k\nabla^2 T(\mathbf{r},t) = \rho c\, \partial T(\mathbf{r},t)/\partial t - q_v(\mathbf{r},t) \qquad (23)$$

Problems with a temperature dependent thermal conductivity $k$ can be transformed, via a change of variables [13] to a problem with constant $k$. The general solution of steady heat transfer by conduction can be obtained using the Green's function $G$ formulation. The later $G$ satisfies:

$$k\nabla^2 G(\mathbf{r},\mathbf{r'}) = -\delta(\mathbf{r} - \mathbf{r'}) \qquad (24)$$

$$k\mathbf{n}.\nabla G = \begin{cases} -1/A_{ref} & \mathbf{r} \in \partial\Omega_{ref} \\ 0 & \mathbf{r} \notin \partial\Omega_{ref} \end{cases} \qquad (25)$$

where $\delta$ is the Dirac distribution, $\partial\Omega_{ref}$ is an arbitrary portion of the boundaries taken as reference and $A_{ref}$ is its area. Multiplying (23) by $G$ and (24) by $T$, subtracting and integrating over the whole domain $\Omega$, we get using (25) as well as Green's theorem:

$$T_e(\mathbf{r'},t) + \rho c\int_\Omega G\frac{\partial T}{\partial t}(\mathbf{r},t)d\mathbf{r} = $$
$$+ \int_\Omega Gq_v(\mathbf{r},t)d\mathbf{r} + \int_{\partial\Omega} Gq_{surf}(\mathbf{r},t)d\mathbf{r} \qquad (26)$$

$$T_e = T - T_{ref}(t); T_{ref}(t) = \int_{\partial\Omega_{ref}} T(\mathbf{r},t)d\mathbf{r} \Big/ \int_{\partial\Omega_{ref}} d\mathbf{r} \qquad (27)$$

where $q_{surf}$ is a surface heat flux density at the boundaries. In the sequel, both $q_v$ and $q_{surf}$ will be denoted by $q$. In order to 'compact' equation (26), assume a complete set of space functions $\phi_{in}(\mathbf{r})$ was assumed on each node $\Omega_i$ and expand over it both $T$ and $q$:

$$T(\mathbf{r},t)\Big|_{\mathbf{r}\in\Omega_i} = \sum_{n=0}^N T_{in}(t)\phi_{in}(\mathbf{r}) \qquad (28)$$

$$q(\mathbf{r},t)\Big|_{\mathbf{r}\in\Omega_i} = \sum_{n=0}^N q_{in}(t)\phi_{in}(\mathbf{r}) \qquad (29)$$

The vector of functions $\mathbf{T}(t)$ contains all elementary time functions $T_{in}(t)$. Same goes for the vector of functions $\mathbf{q}(t)$. Substituting in (26), multiplying by $\phi_{in}$ and integrating gives, after some algebra that goes largely beyond the paper interest, the following set of ordinary differential equations:

$$\mathbf{C}\, d\mathbf{T}(t)/dt + \mathbf{Y}\, \mathbf{T}(t) = \mathbf{q}(t) \qquad (30)$$

Above model is the required general form of DCTM. It can be transformed into an algebraic system, if needed; using results obtained in the previous section for the general temporal 1D model. It is universally usable (condition (c)) for transient problems.

## 5. Conclusion

The most general model topology for Dynamic Compact Thermal Models (DCTM) has been obtained starting from governing partial differential equation of transient heat conduction, using Green's theorem.

*The advantage of such an achievement is to create models that capture all physical effects participating in the phenomenon of transient 3D conduction, which will pave the way for a more precise and a more efficient generation of new models.*

The new topology captures a certain number of effects that were absent in classical topology based on the well-known thermal impedance $Z_{th}$. In particular:

- The new topology allows combining models obtained for different components of a system, in order to simulate multi-component systems, which are the rule in modern electronic systems. The $Z_{th}$ model cannot do the same, due to its failure to capture thermal inertia effects at interface nodes of different components.
- The new topology captures transient changes in temperature and/or heat flux profiles over each heat exchange surface, while classical models are limited to the unrealistic fixed profile during transient operation.
- The new topology takes into consideration distributed heating within body volume, while the $Z_{th}$ model cannot.
- The new topology can handle non-zero initial conditions, while the $Z_{th}$ model cannot.

Of course, compact models will have fewer Degrees of Freedom (DoF) than corresponding detailed FEM models, in order to achieve high performance at the expense of precision. The new topology will allow distributing the few DoFs efficiently among relevant physical effects to obtain the best possible trade off between precision and speed.

## References

[1] Sabry, M.-N., Abdelmeguid, H.S., "Compact thermal models: A global approach", Journal of Electronic Packaging, Transactions of the ASME 130 (4), pp. 0411071-0411076, 2008

[2] Sabry M.N., "Dynamic Compact Thermal Models Used for Electronic Design: A Review of Recent Progress", Proc.Interpack '03, Maui, July 6-11, paper # Interpack2003-35185, 2003

[3] Lasance C., "Ten Years of Boundary-Condition-Independent Compact Thermal Modeling of Electronic Parts: A Review", Heat Transfer Engineering, pp.149-169, 2008

[4] Lasance, C., Den Hertog, D., and Stehouwer, P., "Creation and Evaluation of Compact Models for Thermal Characterisation Using Dedicated Optimisation Software," 15th Annual IEEE Semiconductor Thermal Measurement and Management Symposium, SEMI-THERM, p. 1, 1999.

[5] Lall P., Pecht M., Hakim E., Influence of Temperature on Microelectronics and System Reliability, ISBN 0-8493-9450-3 (CRC Press, NY, 1997)

[6] Parry J., Rantala J., Lasance C., "Enhanced Electronic System Reliability – Challenges for Temperature Prediction", *IEEE CPT* Vol.25, No.4, pp. 533-538, 2002

[7] McShane, E. and Shenai, K.: "Package effects on avalanche rating of power MOSFETs", IEEE International Workshop on Integrated Power Packaging, IWIPP 2000, pp. 93 – 96, 2000

[8] Yong Li Xu; Stout, R. and Billings, D.: "Electronic package thermal response prediction to power surge", The 7th Intersociety Conference on Thermal and Thermomechanical Phenomena in Electronic Systems, ITHERM 2000, Vol. 2, pp. 366 – 371, 2000

[9] Bergogne, D.; Hammoud, A.; Tournier, D.; Buttay, C.; Amieh, Y.; Bevilacqua, P.; Zaoui, A.; Morel, H.; Allard, B.; "Electro-thermal behaviour of a SiC JFET stressed by lightning-induced overvoltages", 13th European Conference on Power Electronics and Applications, 2009. EPE '09, pp 1-8, 2009

[10] Castellazzi, A.; Wachutka, G.; "Low-voltage PowerMOSFETs used as dissipative elements: electrothermal analysis and characterization" 37th IEEE Power Electronics Specialists Conference, 2006. PESC '06, June 2006, No. 9121596, 2006

[11] Alexeenko, A.A.; Fedosov, D.A.; Gimelshein, S.F.; Levin, D.A.; Collins, R.J.; "Transient heat transfer and gas flow in a MEMS-based thrusters", Journal of Microelectromechanical Systems, Vol. 15, No. 1, pp. 181 – 194, 2006

[12] Yun-Ze Li; Yu-Ying Wang; Kok-Meng Lee; "Dynamic Modeling and Transient Performance Analysis of a LHP-MEMS Thermal Management System for Spacecraft Electronics", IEEE Transactions on Components and Packaging Technologies, Vol. 33, No. 3, pp. 597 – 606, 2010

[13] Batty, W.; Christoffersen, C.E.; Panks, A.J.; David, S.; Snowden, C.M. and Steer, M.B.: "Electrothermal CAD of power devices and circuits with fully physical time-dependent compact thermal modeling of complex nonlinear 3-d systems", IEEE Trans. on Components and Packaging Technologies, Vol. 24, no. 4, pp. 566 – 590, 2001

# Heat Sink Design Optimization Using the Thermal ShortCut Concept

Byron Blackmore, John Parry & Robin Bornoff
Mentor Graphics Mechanical Analysis Division
81 Bridge Road, Hampton Court, Surrey UK
Byron_blackmore@mentor.com

## Abstract

Calculation and display of a thermal 'ShortCut' scalar field as an integrated part of a CFD simulation enables a practitioner to visualize and understand the physical mechanisms by which heat is removed from an electronics system and where opportunities exists to introduce new heat transfer paths. By applying the characteristics of this thermal ShortCut scalar to heat sink design aspects, one can identify near optimal solutions with a minimal number of simulations. This work will demonstrate a correlation between the ShortCut scalar and the local Nusselt number, and use this correlation to determine the optimal areas where fin material can be removed with minimal impact on the thermal performance of the heat sink. The results will be compared to that obtained by more traditional Design of Experiments techniques.

## Keywords

Thermal, Shortcut, Nusselt

## Nomenclature

BN – BottleNeck Number
$D_h$ – hydraulic diameter = 2 x fin gap
DoE – Design of (Numerical) Experiment
Re – Reynolds Number
Nu – Nusselt Number
RSM – Response Surface Modeling
SC – ShortCut Number
SO – Sequential Optimization

## 1. Introduction

Electronics thermal management involves the design of an electronics system to facilitate the effective removal of heat from the active surface of an integrated circuit (the heat source) out to a colder ambient surrounding. As the heat travels from the source it passes through various objects and length scales; from the die through the package to the board, into a chassis and out to an operating environment.

How 'easily' the heat passes from the source(s) to the ambient will determine the temperature rise at the source and all points in-between. The often complex 3D heat flow paths carry proportions of the heat with varying degrees of ease. Those paths that carry a lot of heat and which offer large resistances to that heat flow are considered 'thermal bottlenecks'. Identifying and relieving these bottlenecks through a redesign will allow the heat to pass to the ambient more easily, thus reducing temperature rises along the heat flow path, all the way back to the heat source.

The addition of new heat flow paths that allow the heat to bypass these bottlenecks and pass directly to colder areas and on to the ambient more easily will also result in a decrease in temperature rise. Identification and implementation of such thermal 'shortcut opportunities' also allow targeted design changes to be made with maximum effect.

## 2. BottleNeck and ShortCut Numbers and Their Characteristics

### 2.1. BottleNeck Number (BN)

The dimensionalized BN number is the dot product of the heat flux and temperature gradient vectors (Fig. 1).

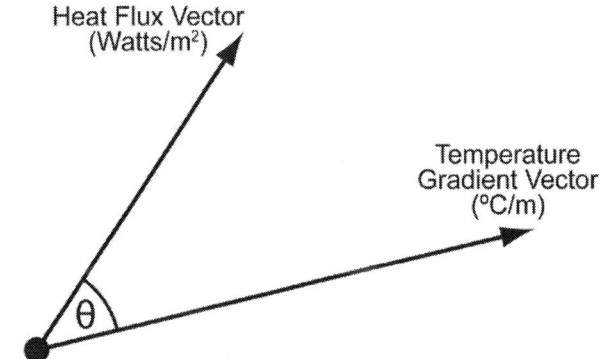

**Figure 1:** Misaligned Heat Flux and Temperature Gradient vectors

In vector notation: BN = Heat Flux • Temperature Gradient. In scalar notation: BN = |Heat Flux x Temperature Gradient x $\cos(\Theta)$ |.

If the angle between the two vectors is zero, i.e. the heat flux is aligned with the temperature gradient as it would be for conductive heat flow in a homogenous thermally isotropic material, then BN is the product of the magnitudes of the vectors, since $\cos(0°) = 1$ .

Large values of this BN scalar, computed as a part of the thermal simulation, pinpoint areas of high heat flow experiencing a large local thermal resistance (characterized by a large, aligned temperature gradient), and thus identify the thermal bottlenecks in a design. Normalizing this scalar by the maximum value in a model will provide an indication of the relative levels of thermal bottleneck in a single simulation model.

### 2.2. ShortCut Number (SC)

The dimensionalized SC number is also calculated from the heat flux and temperature gradient vector fields. The SC scalar value at any point is calculated as the magnitude of the cross product of the two vector quantities. In vector notation: SC = Heat Flux x Temperature Gradient. In scalar notation: SC = | Heat Flux x Temperature Gradient x $\sin(\Theta)$ |.

If the temperature gradient is orthogonal to the heat flux, then SC is simply the product of the vector magnitudes, since $\sin(90°) = 1$.

Large values of the SC field pinpoint areas where large heat flux vectors are misaligned with large temperature gradient vectors (i.e., the heat is not moving directly toward a significantly cooler area), and thus identify locations where the benefit in establishing a new heat transfer path to shortcut the heat to colder areas of the design is highest. Normalizing this scalar by the maximum value in a model will provide an indication of the relative levels of shortcut opportunities in a single simulation model.

## 3. Literature Review

Examples of the application of the BN and SC Numbers to thermal management applications have been previously published [1, 2, 3, 4]. In [1], BN and SC are introduced, and distributions of both are shown for a single TO 263 device on a board to illustrate where the optimal locations in the package to implement thermal design changes are to be found. In [2], BN and SC distributions are used to derive a sequence of thermal design changes for a board in a forced convection situation. BN distributions suggested the shape of the optimal copper pad for two overheating packages, while SC distributions suggested the use of a heat sink and the optimal locations for an array of thermal vias. In [3], both BN and SC parameters were used to identify design changes for an electronics module in the context of 'Define, Identify, Design, Optimize, and Verify' version of Design for Six Sigma. Both design changes were confirmed via simulation to reduce the targeted component temperatures.    In [4], the BN number was used to optimize several aspects of heat sink design, in particular the fin widths of a plate fin heat sink, and the insertion of a copper slug in the heat sink base.

## 4. Correlation of ShortCut number to Nusselt Number

The ShortCut number is a useful tool in heat sink design primarily because it shows a strong correlation with local Nusselt number for convective surfaces. This characteristic of the SC number is fundamental to the heat sink design applications that will be discussed in later sections. This section will demonstrate this qualitative correlation by comparing Nu and SC distributions for the case of flow over a fixed temperature plate, and the case where only a small section of the plate is held at an elevated temperature. Both of these selected cases are representative of typical flow and temperature conditions found within a heat sink and/or a printed circuit board.

A local Nusselt number provides a measure of the effectiveness of heat transfer from a surface to the surrounding fluid. It is notoriously difficult to fully automate the extraction of a 'local ambient' temperature used in the Nusselt number definition for the general case. By showing a correlation of Nusselt number with SC number, and the fact that the SC number is itself much simpler to define and calculate, use of the SC number to identify and appreciate areas of effective convective heat transfer can be proposed. Note that this appears to be in contradiction to the previous statements that the SC number is indicative of where there is an opportunity to insert new heat transfer paths where currently there exists none. However the nature of the SC

number [1], involving a temperature gradient orthogonal to a predominant heat flux path, when examined at solid/fluid interfaces where the air flow is parallel to the surface, is also indicative of the local effectiveness of the heat transfer.

As illustrated in Figure 2, as the air heats up as it flows through the channel the perpendicular temperature gradient will reduce thus the SC number should also reduce. So, although the SC number is NOT a direct analogy to the local Nusselt number, it should be high where there are symptoms that are indicative of high Nusselt number conditions.

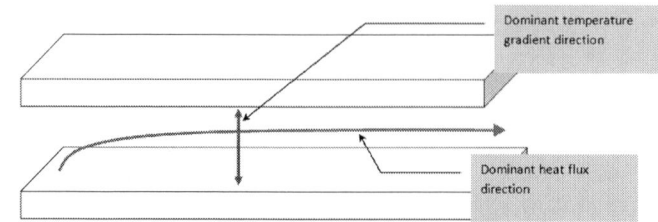

**Figure 2:** Parallel Fixed Temperature Plate Heat Flux and Temperature Gradient Directions

### ShortCut-Nu Correlation - Fixed Temperature Wall Example

The geometry for this example is shown in Figure 3. A solid wall at a fixed temperature of 0 °C is defined with a length of 0.2 m. A constant velocity of air is introduced over the wall at a speed of 1 m/s and at a temperature of -100 °C [chosen for mathematical convenience]. Symmetry boundary conditions are applied on the upper plane of the model at a height of 5 mm above the fixed temperature wall. This equates to a parallel wall channel type flow with a channel gap of 10 mm. A Reynolds number of 1261 (laminar flow) is realized. Symmetry boundary conditions are applied on the two sides of the model representing an infinitely wide channel.

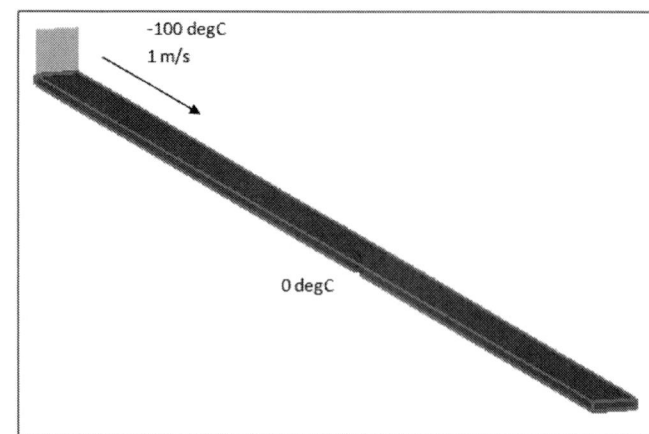

**Figure 3:** Parallel Fixed Temperature Plate Model Geometry

For a fixed wall temperature, the effectiveness of heat transfer will be seen as the ratio of the amount of heat that is transferred to the air and how close that air temperature is to the wall temperature. More heat moved for less temperature difference is the result of effective heat transfer. This effectiveness will be high in the entrance to the channel as the fluid is still relatively cold compared to the wall and the local velocity gradient is high thus the wall friction will be high and thus, due to the Reynolds' analogy, heat transfer will be more effective.

A Nusselt number is defined as:

$$Nu = \frac{q'L}{(T_{wall} - T_{ambient})k}$$

Where:

$L$ = characteristic length (m)

$k$ = thermal conductivity of fluid (W/mK)

$q'$ = heat flux (W/m$^2$)

$T_{wall}$ = local wall temperature

$T_{ambient}$ = local ambient temperature**

**taken as flow weighted average vertically across the channel at a plane adjacent to where the local heat flux is recorded*

Of course, the effectiveness of heat transfer is inherent in its definition, i.e. the ratio of q' and dT.

## Results

A plot of the SC number is shown in Figure 4 (for a few channel heights in the flow direction only for clarity). Figure 5 shows the variation in normalized Nusselt number and SC number plotted against the flow direction.

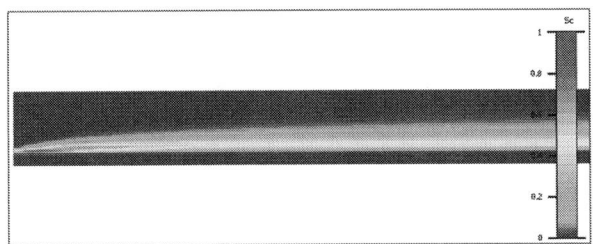

**Figure 4:** Parallel Fixed Temperature Plate: SC Number Distribution

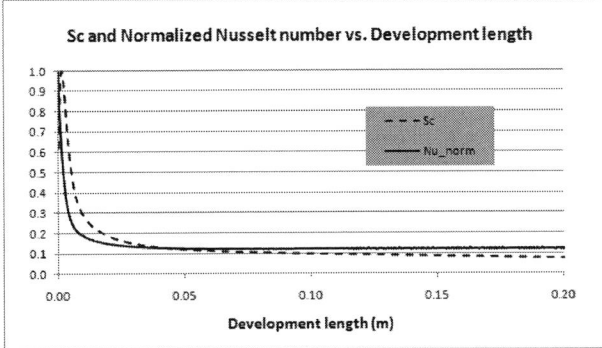

**Figure 5:** SC and Normalized Nu Number vs. Development Length

The local Nusselt number variation on the surface of the fixed temperature wall is calculated for each grid cell face at that solid-fluid interface in the model and plotted against the flow direction. Similarly the SC number is plotted for the row of air cells immediately adjacent to the wall surface, again in the flow direction. As we are looking for a qualitative correlation between the two parameters, the Nusselt number is normalized along that profile. The SC number is itself already normalized.

The main difference is seen in the local spike in SC very near the entrance to the channel. This is due to the detachment of the maximum SC area from the wall surface as seen in Figure 4 where the line of maximum SC number is not parallel to the surface whereas the Nusselt number line is by its definition is taken along the solid/fluid interface. Apart from this entrance effect difference there is a strong enough correlation between the two parameters, enough to consider SC variation near walls as a parameter that will indicate where the local solid to fluid heat transfer is effective.

The next example is almost identical to the first one, apart from only a small section of the wall is being held at a fixed temperature. The wall is conductive allowing heat spreading within it. This configuration is more indicative of heat transfer from a locally heated PCB surface.

## SC-Nu Correlation – Local Fixed Temperature Wall Example

The geometry for this example is shown in Figure 6 and a plot of the SC number distribution is shown in Figure 7.

The normalized Nusselt number variation at the wall surface and the SC variation in the first row of grid cells in the air adjacent to the wall are plotted in Figure 8. The extent of the fixed temperature region is shown by the two vertical lines.

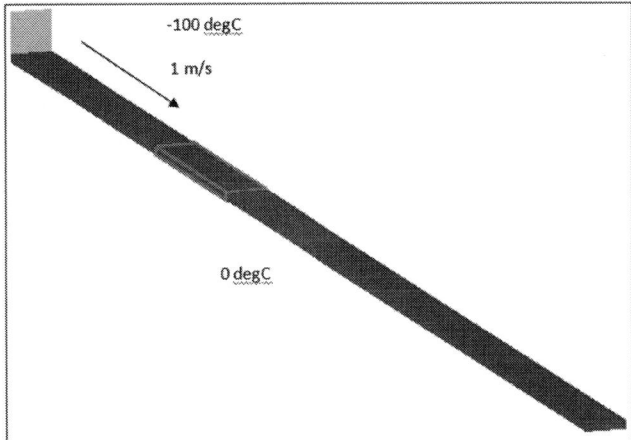

**Figure 6:** Parallel Fixed Temperature Region Model Geometry

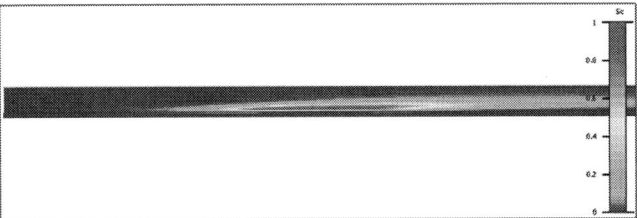

**Figure 7:** Parallel Fixed Temperature Region: SC Number Distribution

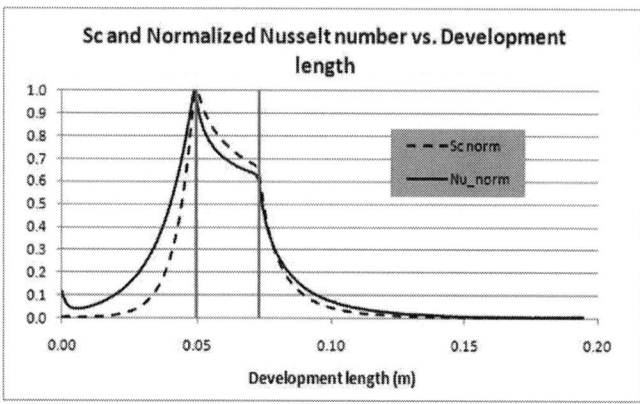

**Figure 8:** SC and Normalized Nu Number vs. Development Length

Overall an excellent correlation is seen, more than enough to give confidence in using the SC variation in a model to determine the areas that are being most effective in transferring heat from a solid surface to the air. The following sections will apply this correlation to heat sink design, in particular to identify the surfaces of the heat sink that are ineffective as candidates for removal in the interest of reducing heat sink mass.

## 5. Optimization vs. Insight for Heat Sink Design

The BN and SC Numbers can be used individually, or together with any electronics thermal design. In this paper we will focus on the application of the SC Number to heat sink design, again referring the reader to [4] for the discussion of the BN field in this area. Heat sink design is a generic challenge that now forms part of the thermal design of many electronics systems, from consumer electronics to military avionics. The traditional approach to heat sink design optimization is automated parametric design optimization, using numerical Design-of-Experiment (DoE) techniques coupled with Response Surface Modeling (RSM) or Sequential Optimization (SO) in conjunction with CFD [5]. While these approaches have been shown to be very effective, they can also consume prohibitive amounts of engineering time and computing resource. The basic methodology is as follows. A base model is defined in the analysis software following standard modeling practices for the tool in use. Then, the parameters to be varied as part of the DoE have to be selected and their ranges defined. Next a sufficiently large set of design points needs to be generated within the design space, usually requiring 5-10 design points per design parameter. Each of these designs then has to be solved to a converged state. Once all the designs have been solved a Response Surface can be created by fitting the data points for a given objective (or cost) function, comprised of quantities that should be optimized, such as weight and temperature rise. The minimum value of the Response Surface can then be found within the design space, and this design created and solved to check the performance of the design. An alternative to using RSM is use SO, which entails starting from the best design found from the DoE set of simulations, and stepping sequentially towards the optimum using analysis of previously solved models to guide the step size and direction within the design space.

In contrast to the traditional approach, using the SC Number field and the physical insights it offers allows designers to create near-optimal design configurations directly from a single simulation.

The following section will discuss the implementation of both techniques for a selected subset of heat sink design, and draw conclusions about the results and calculation time.

## 6. Heat Sink Profile Optimization

The first heat sink design task considered is the optimization of individual fin heights and lengths in the flow direction for an extruded plate fin heat sink. The test case for this design task is described in Table 1 and Figure 9.

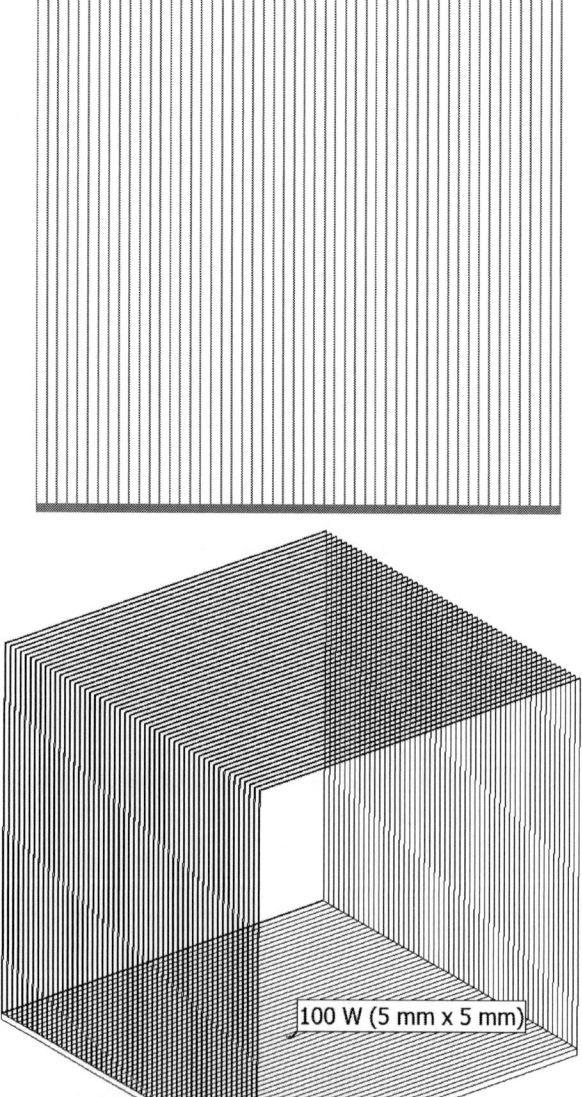

**Figure 9:** Over-sized Heat Sink and 100 Watt Source

| Material | Aluminum, k=137 W/mK |
|---|---|
| Base Thickness | 5 mm |
| Base Length | 300 mm |
| Base Width | 300 mm |
| Fin Thickness | 1 mm |
| Fin Height | 295 mm |
| Number of Fins | 52 |
| Device Air Speed | 1 m/s |
| Heat Source Size | 5 x 5 mm |
| Heat Source | 100 W |

**Table 1:** Base Case Heat Sink Parameters

This is a simplified case which consists of a deliberately oversized heat sink with a small heat source at the center of a thick base. We have also fixed the device velocity, i.e. the velocity of the flow entering the finned region is constant and uniform across the fin gap. Note that the deliberate over sizing of this heat sink was necessary to effectively isolate the effect of the fin height and length changes on heat sink performance, i.e., the conclusions of the study would be obfuscated by base spreading effects, non-uniform flow rate distributions in the channels were a more practical initial heat sink design selected for analysis. This approach is the same as that taken in [4].

### 6.1. DoE Description

We first adopt a classical DoE approach, where the parameters that describe the shape of the volume to be removed from the heat sink are defined as independent variables. The shape of the heat sink to be designed is constrained to be prismatic as shown in Figure 9 to reduce the number of simulations required (i.e. it is prohibitively expensive to parametrically describe a heat sink with a fully general geometric profile). This is in itself a limitation of the DoE approach, though insight into the 'general' dimensions of the optimal heat sink may look like can still be ascertained by this approach to some degree. The parameters used are described in Figure 10 and Table 2.

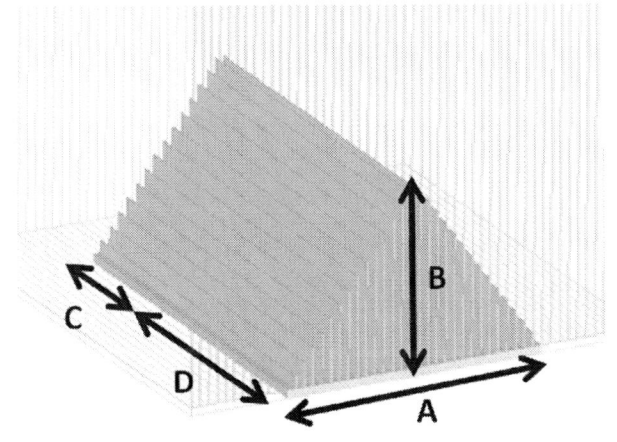

**Figure 10:** Parameters Used to Describe a Prismatic Heat Sink Profile in Relation to Base Case Heat Sink

| Parameter | Range |
|---|---|
| A – Heat Sink Base Width | 30 – 300 mm |
| B – Heat Sink Maximum Height | 15 – 300 mm |
| C – Distance from Heat Source Center to Upstream Edge | 15 – 150 mm |
| D – Distance from Heat Source Center to Downstream Edge | 15 – 150 mm |

**Table 2:** Prismatic Heat Sink Parameter and DoE Ranges.

A DoE of 500 designs (representing approximately 125 designs per design parameter) was constructed so that the range of each parameter was varied independently over the range shown in Table 2. This is a larger number of simulations per parameter than described in Section 5 and was selected to more completely fill the design space and avoid comparing results later with data interpolated from a surface response. For each scenario, the temperature at the heat source and the volume of the heat sink were recorded. Instead of forming cost functions with various degrees of emphasis on source temperature and heat sink mass, the data was organized to sort the simulation results by thermal performance (in the form of heat sink thermal resistance increase from the base heat sink), and then by heat sink mass reduction (again compared to the base case). For any given 'acceptable thermal performance' level, we can then quickly determine the smallest heat sink that is able to deliver that performance based on the DoE results.

The results of the DoE are plotted in Figure 11, showing only data points with an increase in thermal resistance of less than 100% for clarity.

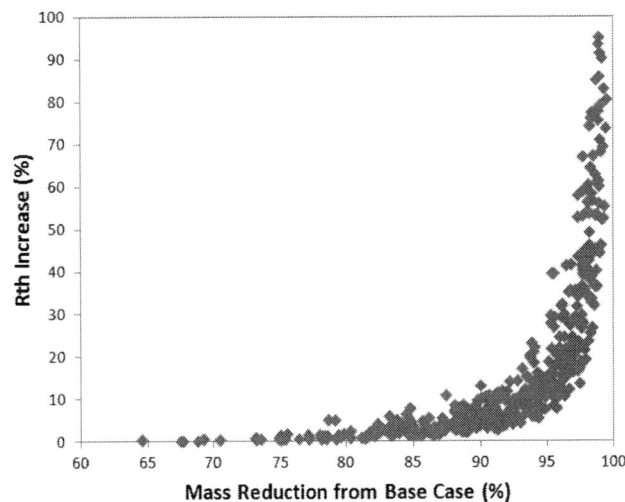

**Figure 11:** DoE Results. All percentage changes are in reference to the base heat sink results.

### 6.2. SC-based Optimization

For a general thermal design application, use of the SC Number for heat sink design optimization proceeds as follows. The correlation with local Nusselt numbers means that on convective surfaces, regions with larger values of SC are deemed to be operating more efficiently than regions with smaller values of SC. These relatively inefficient convective surfaces can then be identified as candidates for removal in the interest of mass and cost reduction, as well as increasing overall volume flow and decreasing pressure drop.

The identified design change is implemented in the computational model, and then solved to re-calculate the SC distribution, which is then used to identify other potential design changes. This process is repeated, until a thermally satisfactory design is arrived at. In this section, we will work though one such design change derivation, and in the process validate that the SC distribution can replicate the most efficient designs found in the DoE.

The SC field for the base case is plotted in Figure 12.

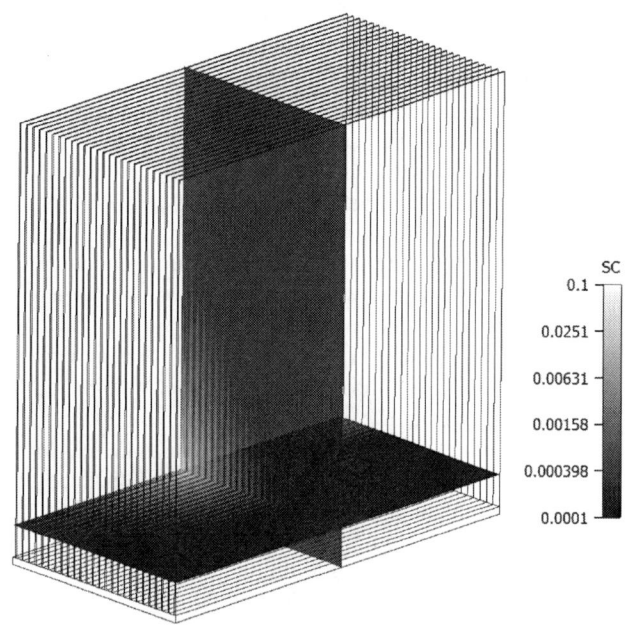

**Figure 12**: SC Distribution for Base Case

For this type of analysis it is useful to define a threshold, under which any value of SC on a convective surface is deemed to be ineffective and should be removed. Any point on a surface with SC greater than the selected threshold should be retained as a part of the optimal heat sink design. This is most conveniently displayed with an isosurface plot. Figure 13 is an example of such a plot, where all points in the model with normalized SC = 0.001 are connected and displayed for the base case model. The color of the plot is unimportant and chosen to add texture to the plot and improve visibility.

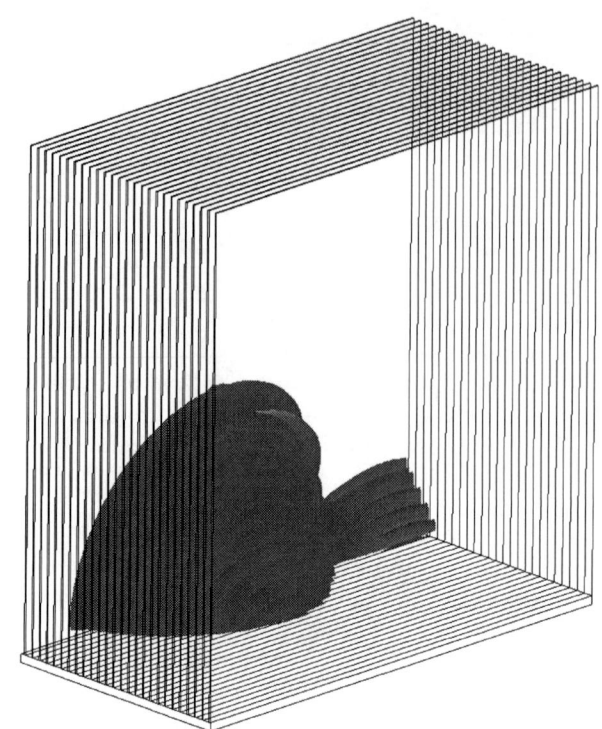

**Figure 13**: Normalized SC = 0.001 Isosurface Plot

Using the data observed in Section 6.1, we'll select the best heat sink design for two selected thermal performance levels, and determine the most equivalent SC threshold value for that thermal performance level. The expectation is that there will be an SC isosurface that compares well with each DoE optimal result. This of course assumes that constraining the DoE heat sink methodology to use prismatic heat sinks was an acceptable course of action as is suggested by the results presented in [6].

_1% Increase in Rth_heatsink_

The first value selected is an acceptable increase of 1% for heat sink thermal resistance. The DoE results yield the optimal prismatic heat sink as shown in Figure 14 and Table 3.

| Parameter | Range |
|---|---|
| A – Heat Sink Base Width | 226 mm |
| B – Heat Sink Maximum Height | 202.5 mm |
| C – Distance from Heat Source Center to Upstream Edge | 93.3 mm |
| D – Distance from Heat Source Center to Downstream Edge | 126.8 mm |
| Heat Sink Thermal Resistance Increase | 0.99% |
| Heat Sink Mass Reduction | 81.35% |

**Table 3:** DoE Optimal Heat Sink for Rth Increase < 1%

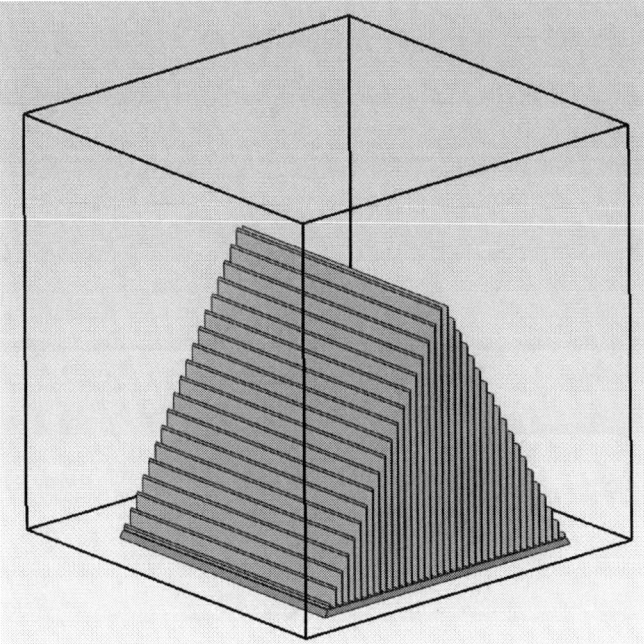

**Figure 14**: DoE Optimal Heat Sink for Rth increase < 1%. Wireframe box shows the extent of the base case heat sink.

By adjusting the SC threshold used to create the SC isosurface for the base case, a value of SC = 0.00008 was judged to be the best qualitative comparison with the ideal prismatic heat sink as shown in Figures 15-16. This is a good result, confirming expectations that an SC isosurface does predict a heat sink geometry topology that is very similar to a DoE optimized results, albeit with some discrepancies due to the prismatic assumption built into the DoE setup.

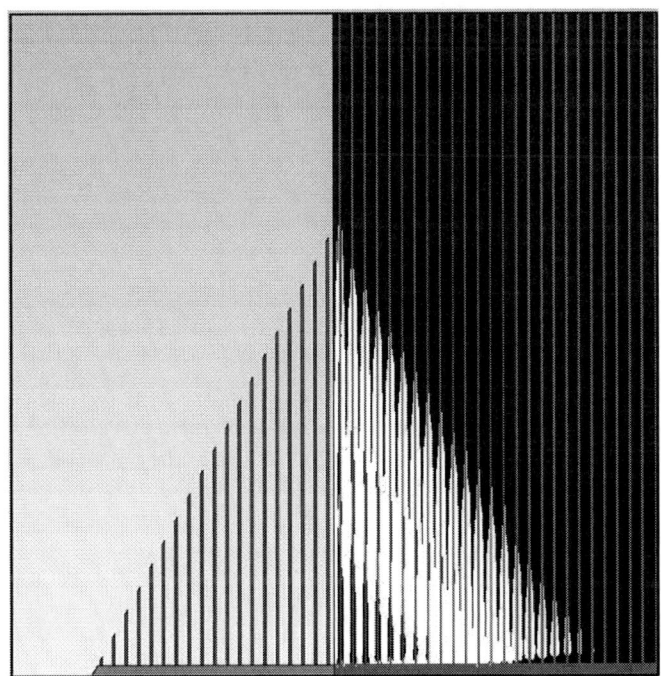

**Figure 15**: Left half shows the geometry for the DoE best result for Rth increase < 1%. Right half shows SC = 0.00008 isosurface for base model. View is from the front.

**Figure 16**: Top half shows the geometry of the DoE best result for Rth increase < 1%. Bottom half shows SC = 0.00008 isosurface for base model. View is from above.

<u>12% Increase in Rth_heatsink</u>

The second value selected is an acceptable increase of 12% for heat sink thermal resistance. The DoE results yield the optimal prismatic heat sink as shown in Figure 17 and Table 4.

| Parameter | Range |
|---|---|
| A – Heat Sink Base Width | 129.9 mm |
| B – Heat Sink Maximum Height | 121 mm |
| C – Distance from Heat Source Center to Upstream Edge | 61.2 mm |
| D – Distance from Heat Source Center to Downstream Edge | 63.6 mm |
| Heat Sink Thermal Resistance Increase | 11.38% |
| Heat Sink Mass Reduction | 96.37% |

**Table 4:** DoE Optimal Heat Sink for Rth Increase < 12%

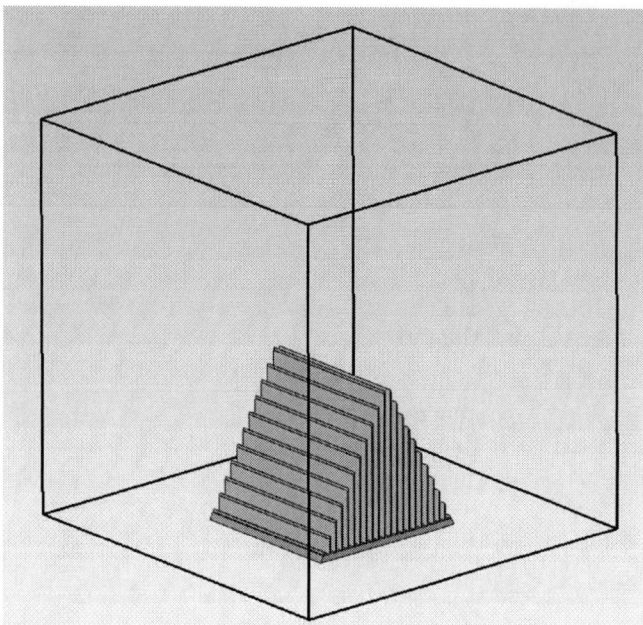

**Figure 17**: DoE Best Heat Sink for Rth increase < 12%. Wireframe box shows the extent of the base case heat sink.

By adjusting the SC threshold used to create the SC isosurface for the base case, a value of SC = 0.0007 was judged to be the best qualitative comparison with the ideal prismatic heat sink as shown in Figures 18-19. This is another good result, confirming expectations that the two means to determine the ideal heat sink for a certain minimum thermal performance criteria yield similar results.

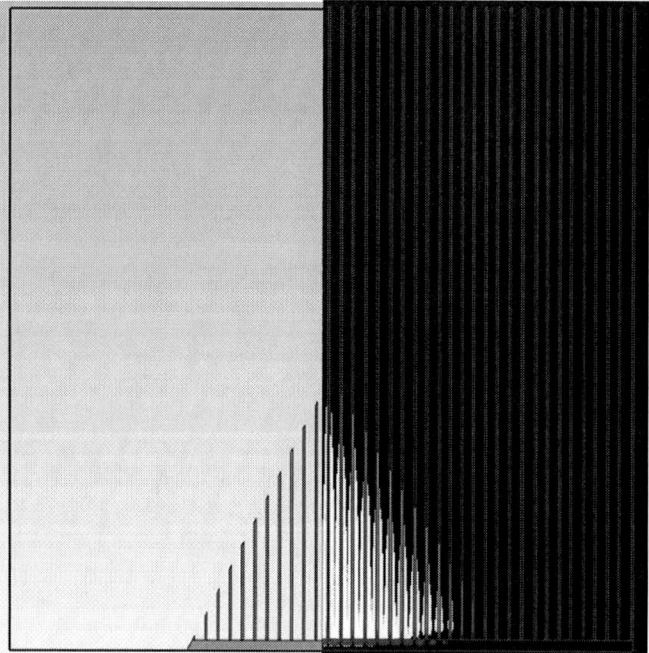

**Figure 18**: Left half shows the geometry of the DoE best result for Rth increase < 12%. Right half shows SC = 0.0007 isosurface for base model. View from the front.

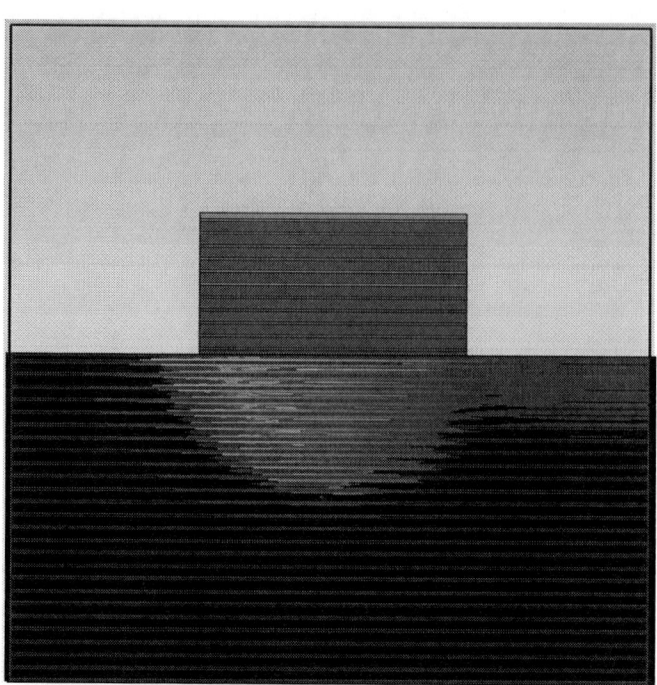

**Figure 19**: Top half shows the geometry of the DoE best result for Rth increase < 12%. Bottom half shows SC = 0.0007 isosurface for base model. View from the top.

It's important to note that while the results are similar, there are differences in time to calculate and profile geometric flexibility. The DoE approach in general can require hundreds of simulations to produce reliable answers, and that is required even when restricting the shapes to be investigated. The SC approach requires a single simulation to produce similar results, and as demonstrated by the isosurfaces in Figures 13, 15, 16, 18, and 19, does not require any restrictions or assumptions about allowable heat sink topology to begin. Rather, the SC threshold isosurface can form any shape, and leaves the designer to determine which level of optimal shape approximation is best from an analysis and cost of manufacturing perspective. Note that were the shape described by the SC isosurface manufacturable, it would present less of an obstruction to the approaching flow, reducing both pressure drop and the wake downstream of the heat sink.

## 7. Heat Sink Application 2

This SC threshold isosurface approach will next be considered for a more detailed and realistic electronics cooling application, a case involving many heat sources and components. The geometry for this application is shown in Figure 20. It consists of a sealed aluminum, ribbed enclosure with 19 fins on each side. Inside the enclosure are a printed circuit board with components on both sides (30 W total dissipation) and a 20 W power supply. The cooling strategy is to include conduction posts to the enclosure for the largest dissipation components and force air over the external heat sink fins.

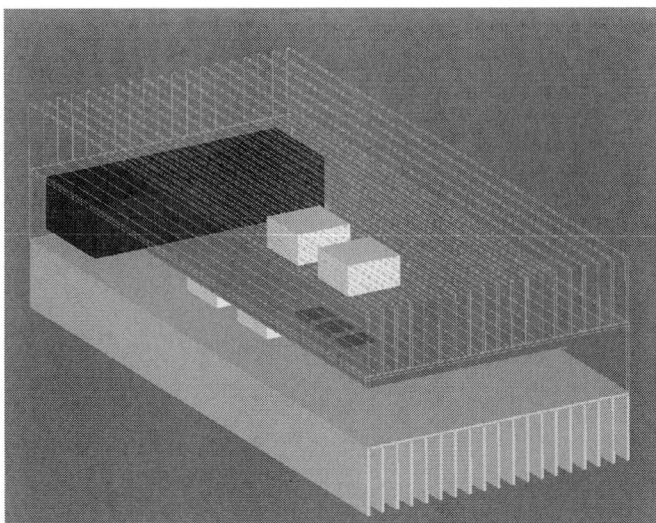

**Figure 20**: Geometry of Sealed Ribbed Enclosure and Internal Contents.

A SC = 0.0015 isosurface around the fins on the top heat sink is shown in Figures 21-23 from several viewing perspectives. The SC threshold value chosen was determined by varying the value of SC used to create the isosurface plot from 0.0001 to 0.1 and selecting the SC value that produced a plot that partially covered most of the fins on the top of the chassis to some degree. By doing so, we can use the relative coverage of the fins by the isosurface plot to draw conclusions about the relative effectiveness of each fin. In the previous section, the created isosurface was used to create a new heat sink profile. In this case we'll use the isosurface shape to determine which fins should be removed completely.

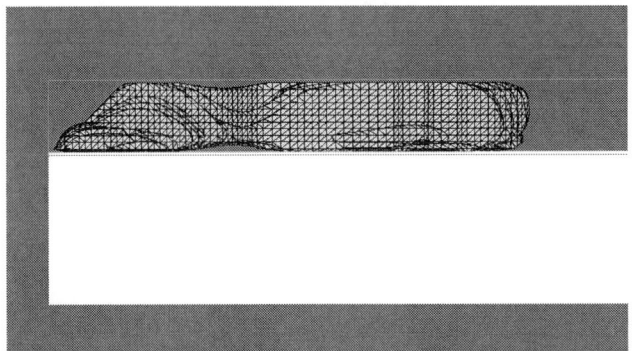

**Figure 21**: SC = 0.0015 isosurface viewed from side.

**Figure 22**: SC = 0.0015 isosurface. Isometric View.

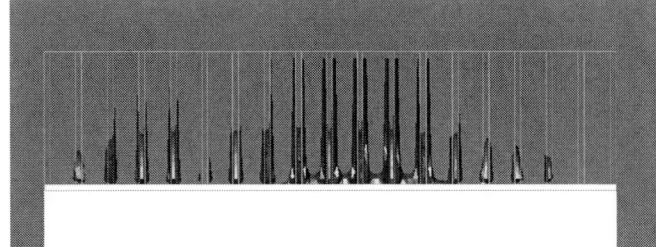

**Figure 23**: SC = 0.0015 isosurface. Front View.

Considering the isosurface geometry, and moving from left to right in Figure 23, fins 9-13 have the largest proportion of their surfaces operating efficiently, with a large fraction of these fins covered by the isosurface. Fins 1-8 and Fins 14-19 do not have as much surface area covered by the SC isosurface, and the priority of removing each of these fins is determined by observing the relative amount of each fin that is covered by the SC isosurface. The end fins, in particular Fins 1, 18, 19, are not covered at all and we would expect to see little thermal performance degradation if they were to be removed.

To confirm these observations, each one of the heat sink fins on the top surface of the enclosure was numerically removed in sequence (but only one removed at a time). The temperature of several components on the top side of the PCB was recorded for each scenario as a measure of the deleterious effects of removing that heat sink fin. The results are shown in Figure 24.

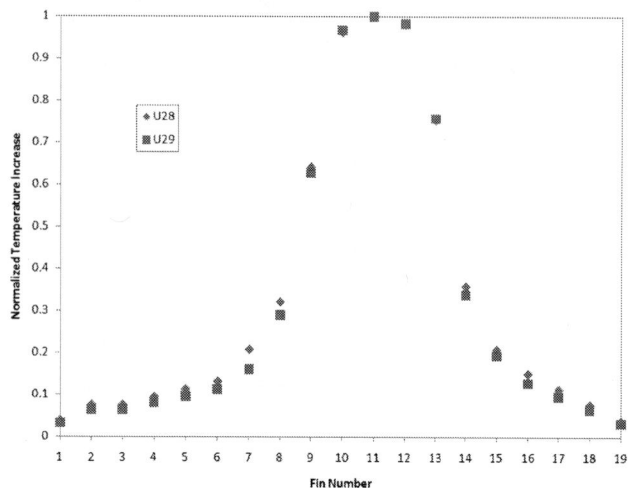

**Figure 24**: Normalized Component Temperature Increase for the successive removal of each individual fin in turn.

If we consider Figure 23 and Figure 24 together, we can observe that the fins that exhibited small values of SC in the original calculation could be removed with the least impact on these component temperatures. This again agrees with the concept of using SC as a means to evaluate which fins are the least effective.

## 8. Conclusions

This work has demonstrated how the SC number correlates with the local Nusselt number, and therefore how the distribution of the SC scalar field around convective surfaces provides insight into the relative effectiveness of these surfaces. The visualization of the SC field was demonstrated to have utility in the design of heat sinks, in particular by identifying those fins and surfaces that could be removed with minimal impact on thermal performance. The results compared well with traditional methods of approaching these design tasks, such as DoE or simply calculating all possible design alternatives, and did so with a single simulation.

This design methodology for heat sinks complements the design insights provided by the BN field [4]. The usage of BN and SC distributions is not limited to heat sink design of course, with the principles discussed here being easily applied to other thermal management applications.

## Acknowledgments

CFD simulations and BN and SC post-processing were carried out with Mentor Graphics' FloTHERM V9.2 software.

## References

1. John Parry, Robin Bornoff, Byron Blackmore, "Thermal BottleNecks and ShortCut opportunities; innovations in electronics thermal design simulation" Electronics Cooling Magazine, Vol. 16, No. 3, Fall 2010, pp. 24-25.
2. Byron Blackmore, John Parry and Robin Bornoff, "New 3D thermal quantities help designers address thermal problems as they arise" Cover Story in Printed Circuit Board Design & Fabrication Magazine, Vol. 29, No. 11, pp. 30-32.
3. Wendy Luiten, "Thermal Design in the Design for Six Sigma – DIDOV Framework", Proceedings of 27th IEEE SEMI-THERM Symposium, San Jose, CA, March 2011, pp. 272-279.
4. Bornoff, Robin., Blackmore, Byron., Parry, John., "Heat Sink Design Optimization Using the Thermal Bottleneck Concept", Proceedings of 27th IEEE SEMI-THERM Symposium, San Jose, CA, March 2011, pp. 76-80.
5. J. Parry, Robin Bornoff, P. Stehouwer, Lonneke Driessen and Erwin Stinstra, "Simulation-Based Design Optimisation Methodologies Applied to CFD", Proceedings of 19th SEMI-THERM Symposium, San Jose CA, March 2003, pp. 8-13.
6. Kakaç, Sadik, Ramesh K. Shah & Win Aung, Handbook of Single-Phase Convective Heat Transfer, John Wiley & Sons (New York 1987) Ch. 3 pp. 35.
7. Byron Blackmore, Robin Bornoff, "Thermal Bottlenecks and ShortCut Opportunities; Innovations in Electronics Thermal Design by Simulation", Mentor Graphics SupportNet, May 20, 2011.

# A Method to Adapt Zth-Junction-to-Ambient Curves to Varying Ambient Conditions

Dirk Schweitzer

Infineon Technologies AG

Am Campeon 1-12, 85579 Neubiberg, Germany

dirk.schweitzer@infineon.com

## Abstract

The transient temperature response to a unit power step, commonly referred to as Zth-curve, is an important thermal characteristic for semiconductor devices. Zth-junction-to-ambient curves can be found in most component data sheets and are often used to calculate the time-dependent junction temperature response to a given power pulse or profile. The influence of ambient temperature and power dissipation on Zth-JA however is often neglected in these calculations since it is not possible to provide a Zth-JA curve for each ambient temperature and power dissipation. This paper presents an algorithm based on clipping a part of the structure function which allows transforming a given Zth-JA curve such that it can be adapted to arbitrary ambient temperature and power dissipation levels.

## Keywords

Transient thermal resistance, Rth-JA, Zth-JA, radiation and convection heat transfer, structure function analysis.

## 1. Introduction

For a semiconductor device which is heated with constant power $P_H$ starting at time $t = 0$ the transient junction-to-ambient thermal resistance $Z_{th-JA}(t)$ is defined as

$$Z_{th-JA}(t) = \frac{T_J(t) - T_{amb}}{P_H}, \qquad (1)$$

where $T_J(t)$ is the time dependent junction temperature and $T_{amb}$ the (constant) ambient temperature. At time $t = 0$ the whole device is at ambient temperature, i.e. $T_J(t = 0) = T_{amb}$. For $t \to \infty$ the Zth-junction-to-ambient converges to the steady-state thermal resistance $R_{th-JA}$:

$$R_{th-JA} = Z_{th-JA}(t \to \infty) = \frac{T_J(t \to \infty) - T_{amb}}{P_H}. \qquad (2)$$

Since $R_{th-JA}$ and $Z_{th-JA}(t)$ describe the whole heat-flow path from junction to ambience, they are determined to a large part by the environment of the semiconductor, e.g. the printed circuit board (PCB) the device is soldered to. Therefore JEDEC standard JESD51 [1] defines standardized environments for the measurement of these thermal characteristics with respect to test-board layout and boundary conditions. The standardized thermal resistances $R_{th-JA}$ and $Z_{th-JA}(t)$ according to JESD51 are included in the datasheets of most semiconductor devices. They provide valuable information on the thermal performance of a device and they are often (mis-) used to calculate first estimates of the expected junction temperature for given operating conditions.[*]

If the transient thermal resistance $Z_{th-JA}(t)$ is known, the junction temperature response $T_J(t)$ can be calculated for arbitrary power profiles $P(t)$ by a convolution integral [2, 3]:

$$T_J(t) = T_{amb} + \int_0^t P(\tau) \cdot \dot{Z}_{th-JA}(t - \tau) \, d\tau \qquad (3)$$

Of course, the junction temperature calculated by (3) is valid only for the same environment the Zth-JA has been determined for. Nevertheless, equation (3) is quite useful, since it allows computing the thermal response for any power dissipation scenario much faster and easier than could be obtained by direct simulation or measurement with the given power profile $P(t)$. A fast numerical solver for the convolution equation (3) has been implemented in Infineon's transient thermal multisource simulator (TTM) [4].

It is well known – and yet often neglected – that not only the "hard" environment (e.g. PCB, heat sink) has an influence on the Rth-JA and Zth-JA values, but also "soft" boundary conditions such as ambient temperature $T_{amb}$ and heating power $P_H$. Since these latter influences are second order effects, resulting mainly from the temperature dependence of radiative heat transfer, they are often underestimated. Figure 1 shows three Zth-JA curves which have been computed for the same system, a MOSFET device in TO263 package on a JEDEC 2s0p PCB, but for different ambient temperatures: 25°C, 75°C, and 125°C (see section 2 for details of the simulation). Changing the ambient temperature from 25°C to 125°C reduces the steady-state Rth-JA by 26%, i.e. in this example the influence of the ambient temperature is significant.

While the steady state thermal resistance Rth-JA for arbitrary temperatures in between 25°C and 125°C could be easily found by interpolation of the known Rth-JA values at 25°C, 75°C, and 125°C, it remains unclear how to interpolate between the three Zth-curves shown in figure 1. This paper presents a simple algorithm based on clipping of the structure function which allows transforming a given Zth-JA curve such that it can be adapted to varying ambient temperature and power dissipation levels.

After a brief recapitulation of the basics of heat transfer from a PCB to the environment the structure functions describing this heat transfer and the changes they undergo as the boundary conditions are varied are investigated. It is shown how this can be used to transform a Zth-JA curve to match varying ambient conditions. Finally several examples demonstrate the applicability and accuracy of this method.

---

[*] Strictly speaking Rth-JA and Zth-JA are valid only for exactly the environment and boundary conditions they have been determined for. They should not be applied to compute the device temperature for very different conditions.

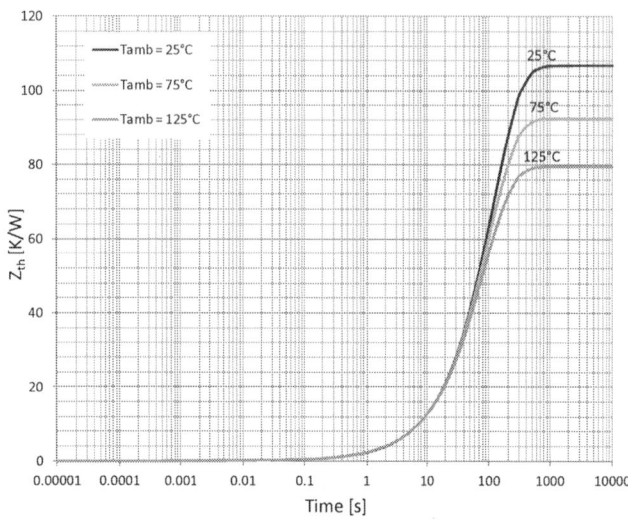

**Figure 1**: Zth-JA curves of a MOSFET in TO263 package on a JEDEC 2s0p PCB for different ambient temperatures.

## 2. Convective and radiative heat transfer

The basic laws of heat transfer from a PCB in JEDEC JESD51 standard environment to the surroundings shall be briefly recapitulated since they explain the influence of the boundary conditions on the Zth-JA curve. For a more comprehensive introduction to the theory of heat transfer the reader is referred e.g. to [5]. Two different physical phenomena are responsible for the heat transfer, convection and radiation.

The convection heat flow $q_{conv}$ from the surface of a solid body to the surrounding gas or fluid (i.e. air in most cases) is proportional to the surface area $A$ and to the difference between surface temperature $T_S$ and ambient air temperature $T_{amb}$,

$$q_{conv} = h_{conv} A (T_S - T_{amb}). \qquad (4)$$

The convection heat transfer coefficient $h_{conv}$ is the factor of proportionality. In the following it is assumed that the surface temperature of the solid is higher than the surrounding air temperature, $T_S > T_{amb}$, i.e. heat is transferred from the surface to the environment. $T_S$ and $T_{amb}$ denote the Kelvin temperature.

The convection heat transfer coefficient is not constant; it depends on the temperature difference $T_S - T_{amb}$ and on the physical properties of the surrounding gas (which depend again on $T_{amb}$). For forced convection in moving air the heat transfer coefficient is also a function of the air velocity. In the literature a number of semi-empirical models can be found which describe these correlations. Table 1 contains some formulas which are often used in thermal analysis to compute the heat transfer coefficient for natural convection [5].

In addition to the heat transfer by convection heat energy is dissipated from a hot surface to the environment also by radiation. The net radiation heat loss $q_{rad}$ from a body with surface temperature $T_S$ to a much larger "envelope" at ambient temperature $T_{amb}$ (e.g. the walls and ceiling of the surrounding room) is described by the Stefan-Boltzmann law

$$q_{rad} = \varepsilon \sigma A (T_S^4 - T_{amb}^4) \qquad (5)$$

with the Stefan-Boltzmann constant $\sigma = 5.67 \cdot 10^{-8}$ W/(m²K⁴) and the emissivity $\varepsilon$ of the surface. Equation (5) can be further transformed

$$q_{rad} = \varepsilon \sigma A (T_S - T_{amb})(T_S + T_{amb})(T_S^2 + T_{amb}^2), \qquad (6)$$

i.e. the radiation heat transfer can formally be described in the same way as the convective heat transfer by a radiation heat transfer coefficient $h_{rad}$:

$$q_{rad} = h_{rad} A (T_S - T_{amb}), \qquad (7)$$

with

$$h_{rad} = \sigma \varepsilon (T_S + T_\infty)(T_S^2 + T_\infty^2). \qquad (8)$$

Figure 2 shows an example for the temperature dependence of convection and radiation heat transfer coefficients. The curves in this graph have been calculated using equation (8) and the laws in table 1 (which apply strictly speaking only to an isothermal plate in still air).

We use these (and similar) formulas in Finite Element simulations of standard Rth- and Zth-values successfully to calculate local heat transfer coefficients for the surface of a JEDEC-PCB with non-uniform temperature distribution. Simulation results with an error < 5% for Rth-JA compared to measurements justify this approach.

The total heat transfer coefficient $h = h_{conv} + h_{rad}$ increases very quickly with increasing ambient and surface temperature, mainly due to the strong temperature dependence of the radiation part given by equation (8). Thus, if the ambient temperature is increased, the heat transfer from the PCB to the environment is enhanced, which results in a lower Rth-JA value (figure 2).

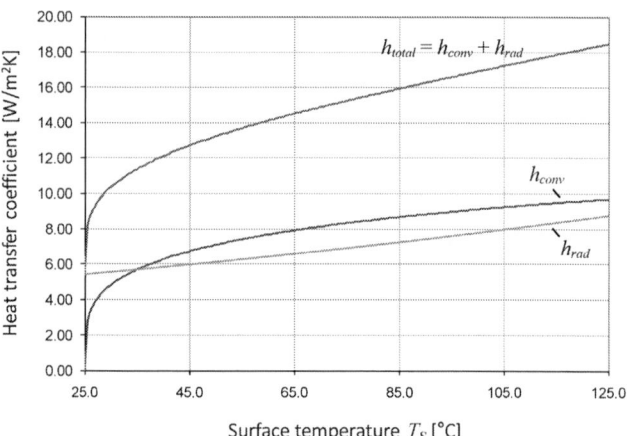

**Figure 2**: Calculated convection, radiation, and total heat transfer coefficient as function of the surface temperature $T_S$ for an ambient temperature $T_{amb} = 298$K (25°C). The convection heat transfer coefficient has been computed according to the formulas in table 1.

Similarly, the heat transfer is enhanced if the power dissipation in the component is increased, because this also increases the surface temperature of the component and the surrounding PCB.

| | |
|---|---|
| $h_{conv} = \dfrac{k}{L} \cdot Nu$ | Convection heat transfer coeff. |
| $k$ : | thermal conductivity of air $\approx 2.6 \times 10^{-2}$ W/(mK) at 25°C |
| $L$ : | typical length of the system (e.g. edge of PCB) |
| $Nu$: | Nusselt number |

| | |
|---|---|
| $Nu = 0.53 \cdot (Pr \cdot Gr)^{0.25}$ | Nusselt number for laminar flow |
| $Pr$: Prandtl number | ($\approx 0.7$ for most gases) |
| Gr: Grashoff number | |

| | |
|---|---|
| $Pr = \dfrac{\nu}{\alpha}$ | Prandtl number ($\approx 0.7$) |
| $\nu$ : | kinematic viscosity of air $\approx 1.55 \times 10^{-5}$ m²/s at 25°C |
| $\alpha$ : | temperature diffusity of air $\approx 2.2 \times 10^{-5}$ m²/s at 25°C |

| | |
|---|---|
| $Gr = L^3 g \beta (T_S - T_{air})/\nu^2$ | Grashoff number |
| $g$: | acceleration of gravity $= 9.81$ m/s² |
| $\beta$: | volumetric thermal expansion coefficient of air $\approx 1/T_{air} = 3.4 \times 10^{-3}$ K⁻¹ for $T_{air} = 298$K (25°C) |

**Table 1**: Some semi-empirical laws commonly used in thermal analysis to calculate the (natural) convection heat transfer coefficient. For forced convection there exist similar formulas. See [5] for further details.

## 3. Structure function analysis

Since the early publication [6] by V. Szekely et al. the structure function analysis has become a very useful and often employed means to analyse Zth-curves. For details on the transformation of a Zth-curve into the corresponding structure function and the interpretation of the latter the reader is referred e.g. to [6-9].

Figure 3 briefly outlines the necessary steps to transform a Zth-curve into its corresponding structure function. First, the time-constant spectrum $R(z)$ is computed by numerical deconvolution. If $z$ denotes logarithmic time $z = \ln(t)$ and $a(z)$ the unit step response as a function of $z$ (i.e. a(ln $t$) = $Z_{th}(t)$) the following equation holds true [8]:

$$\frac{da}{dz} = R(z) \otimes w(z)$$

with  (9)

$$w(z) = \exp(z - \exp(z)),$$

i.e. the time-constant spectrum $R(z)$ can be computed by deconvolution of the time derivative $da(z)/dz$ of the Zth-curve with the function $w(z)$. By discretization of the time-constant spectrum one obtains a thermally equivalent Foster RC-network which in the next step is transformed into a Cauer RC-network. The cumulative thermal capacitance of the Cauer ladder plotted vs. its cumulative thermal resistance is the cumulative structure function (or, more accurately, its discrete approximation).

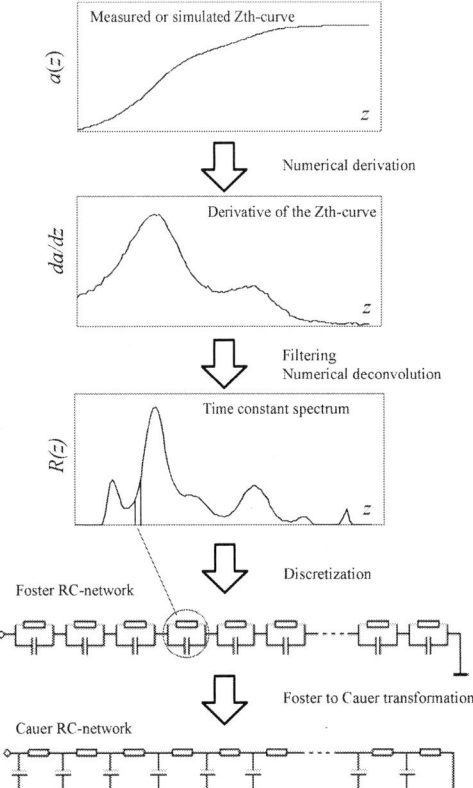

**Figure 3**: Schematic view of the transformation of the Zth curve into a Cauer-model [9].

A step-by-step guideline how to compute the structure function can be found e.g. in [10]. It should be mentioned that, because of the large number of RC-stages involved, the Foster to Cauer transformation must be implemented with extended floating point precision as offered e.g. by the GNU Multiple Precision Arithmetic Library GMP [11].

Figure 4 shows the cumulative structure functions for the Zth-JA curves from figure 1. The structure functions perfectly

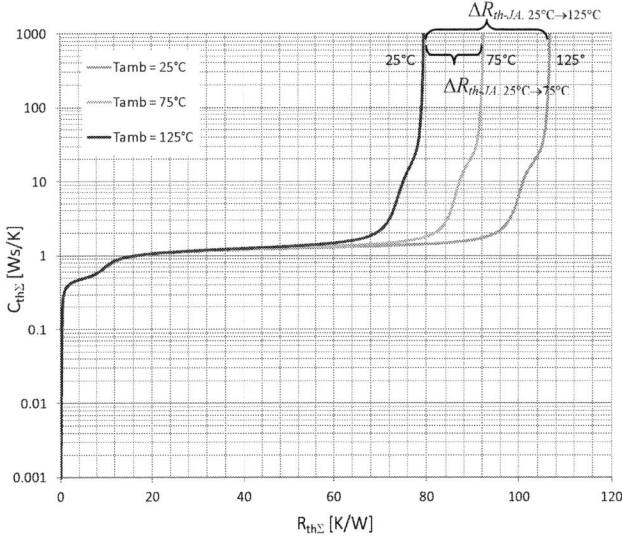

**Figure 4**: Cumulative structure functions of the MOSFET on a JEDEC 2s0p PCB for different ambient temperatures; obtained by transformation of the Zth-curves in figure 1.

match each other at the beginning, indicating that the heat conduction inside the MOSFET and into the PCB is not influenced by the change in ambient temperature.[†] The long, almost horizontal plateaus, before the structure functions finally diverge to infinite heat capacity (i.e. ambience), correspond to the heat flow inside the thermally low conducting PCB and the heat transition from the PCB surface into the surrounding space. At higher ambient temperatures this heat transition is enhanced and the length of the plateau in the structure function becomes smaller. The thermal resistance $\Delta R_{th\Sigma}$, by which the length of plateau is reduced, equals the change $\Delta R_{th\text{-}JA}$ of the junction-to-ambient thermal resistance Rth-JA.

## 4. Structure function clipping algorithm

This observation gives rise to the following idea: If the structure function for a given ambient temperature is known, it should be possible to obtain the structure function for higher ambient temperatures by simply cutting a part of length $\Delta R_{th\text{-}JA}$ from the plateau. Of course the steady-state thermal resistance Rth-JA at target ambient temperature must be known. Figure 5 schematically presents the proposed algorithm. Starting at some thermal resistance value $R_{th,cut}$ an interval of length $\Delta R_{th\text{-}JA}$ is removed from the plateau (or region of constant slope / low curvature) preceding the upper end of the structure function. The upper remaining part of the structure function is then shifted by the vector $\mathbf{v} = (-\Delta R_{th\text{-}JA}, -\Delta C_{th\Sigma})$ which is defined by the structure function clipping points at the begin and end of the interval. By transforming the shortened structure function back into a Zth-curve one obtains the Zth-JA for the desired ambient temperature.

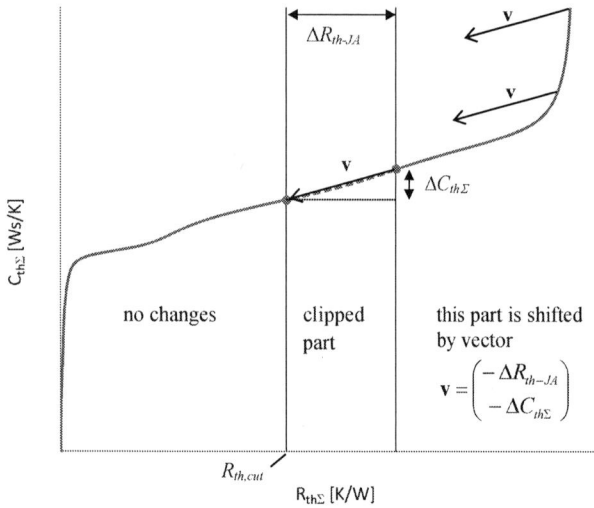

**Figure 5**: Schematic presentation of the structure function clipping algorithm.

---

[†] Strictly speaking the ambient temperature has also some influence on the heat flow path inside the semiconductor device since the thermal properties of some materials (especially silicon) are temperature dependent. However the corresponding change in thermal resistance is so small compared to Rth-JA that this effect is neglected here.

The question remains how to choose the position $R_{th,cut}$ of the clipping interval. The physics requires that the structure function should be clipped near its upper end since that is the region which is subject to changes when the heat transition from the PCB to the surrounding space changes. However clipping the structure function in a region of strong curvature would not correctly transform the structure functions for different ambient temperatures as can be seen e.g. in figure 3. Therefore the clipping interval should be choosen such that it falls in the region of minimum curvature closest to the upper end of the structure function.

For the back-transform of the clipped structure function into a Zth-curve the associated Cauer network (the thermal resistances and capacitances of which are obtained simply as the differences of the cumulative resistances and capacitances of subsequent structure function points) must be transformed back into a Foster network. An algorithm for this purpose is presented e.g. in [12]. Again it should be noted that also the back-transform from Cauer to Foster RC-network requires extended floating point precision. If $R_{thF,i}$ and $C_{thF,i}$, $i = 1 \ldots N$, denote the thermal resistances and capacitances of the thus obtained Foster-network with $N$ RC-stages, the Zth-curve can be easily computed using

$$Z_{th}(t) = \sum_{i=1}^{N} R_{thF,i} \left( 1 - \exp(-\frac{t}{\tau_i}) \right) \qquad (10)$$

with time-constants

$$\tau_i = R_{thF,i} \cdot C_{thF,i}.$$

## 5. Application of the algorithm - some examples

The applicability of this algorithm shall now be demonstrated by means of some examples.

### 5.1 JEDEC 2s0p board, varying ambient temperature

Figure 6 shows again the structure functions of the MOSFET device on a JEDEC 2s0p PCB for 25°C, 75°C, and 125°C ambient temperature (continuous lines). Additionally the transformed structure functions obtained by clipping and shifting the 25°C curve, are shown (dashed lines). As can be seen, transformed and original structure functions match very well, both for 75°C and 125°C.

The final goal is of course to obtain the Zth-JA curve for varying ambient temperatures. Figure 7 compares the true Zth-JA curves for 75°C and 125°C (continuous lines) to the back-transforms of the shortened 25°C structure function into their corresponding Zth-curves (dashed lines). They match almost perfectly; there is hardly any difference visible. Figure 8 shows the relative error in a time range from 1s to steady-state. For shorter times Zth-JA is almost independent of ambient temperature (compare e.g. figure 1); i.e. in that time range the values of the Zth-curve for 25°C can be used with negligible error for all other ambient temperatures as well. For further discussion of this issue see also section "6. Some remarks" below. The maximum deviation between true and transformed Zth-JA curves in this example is 0.6 K/W with a maximum relative error of less than 0.8% in the time range from 1s to 10 000s.

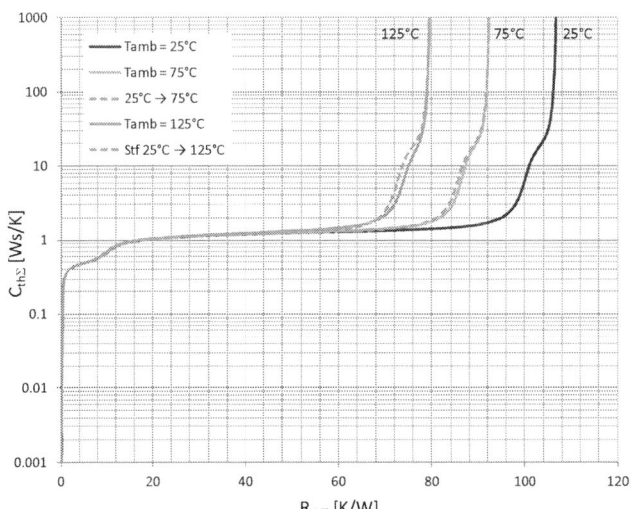

**Figure 6**: Structure functions of the MOSFET device on a JEDEC 2s0p board. The dashed curves have been obtained by clipping and shifting the 25°C structure function.

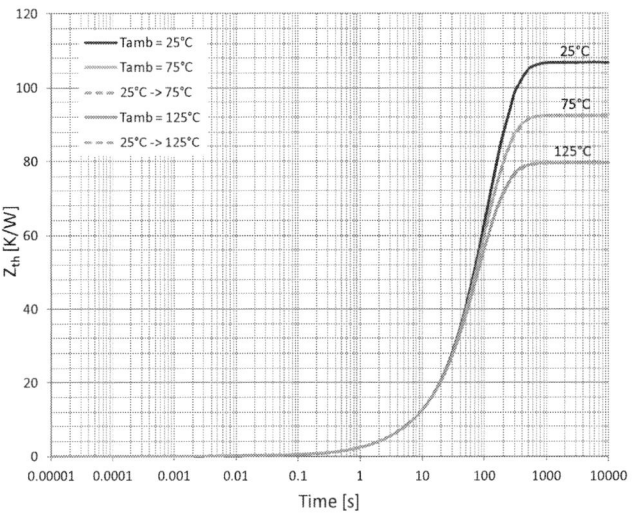

**Figure 7**: The back-transforms of the clipped 25°C structure functions (dashed lines in figure 6) almost perfectly match the true Zth-JA curves.

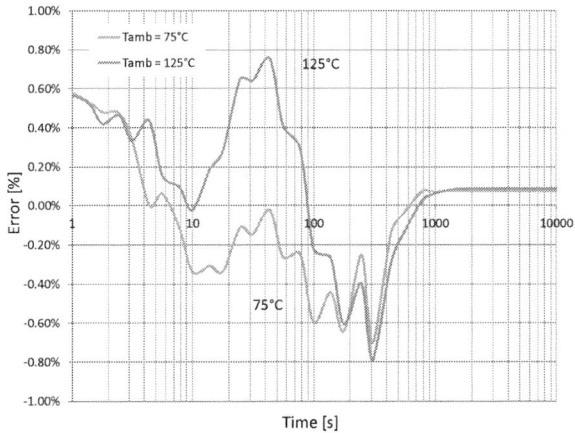

**Figure 8**: Relative error of the Zth-JA curves obtained by structure-function clipping (compared to the true Zth-curves).

**5.2 JEDEC 2s2p board, varying ambient temperature**

In the next example the same MOSFET device has been placed on a JEDEC 2s2p board, i.e. compared to the previous example the PCB contains now two additional copper layers which improve the thermal conductivity of the board. Thermal vias connect the device to the first inner copper plane. Figure 9 shows the simulated Zth-JA curves for this set-up for 25°C, 75°C, and 125°C ambient temperature. The steady-state Rth-JA values are now considerably lower, thanks to the additional copper layers in the PCB. The application of the structure function clipping algorithm yields the Zth-curves shown in figure 11. Again the match between transformed (from the 25°C curve) and true Zth-curves is excellent, with a maximum absolute error of 0.4 K/W and a maximum relative error of less than 3% (figure 12).

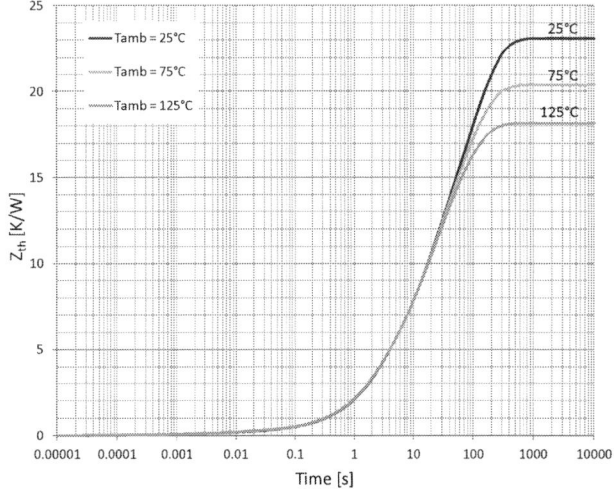

**Figure 9**: Zth-JA curves of the MOSFET device package on a JEDEC 2s2p PCB for different ambient temperatures.

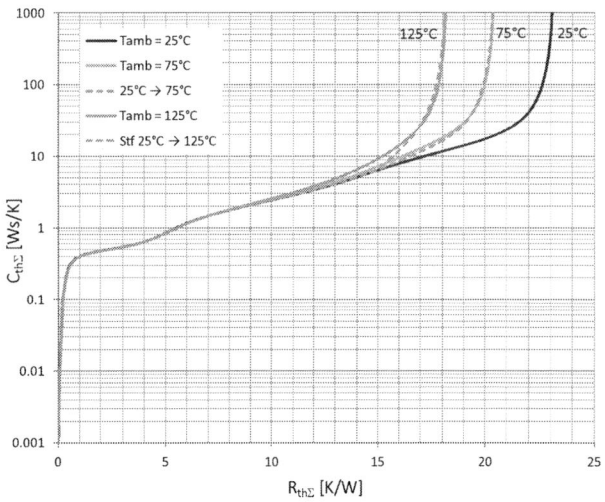

**Figure 10**: Structure functions of the MOSFET device on a JEDEC 2s2p board. The dashed curves have been obtained by clipping and shifting the 25°C structure function.

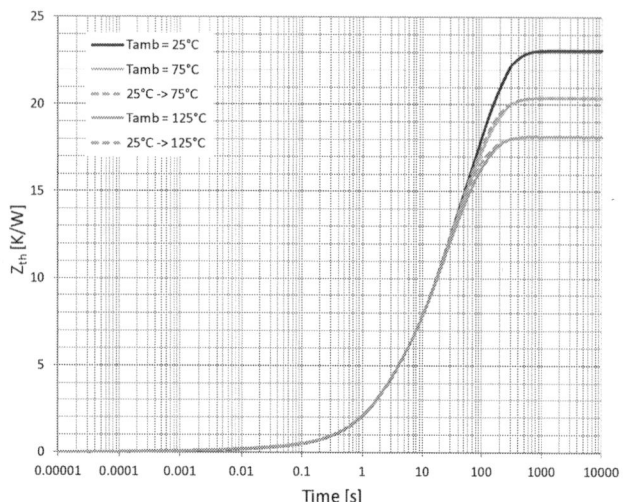

**Figure 11**: Example of the MOSFET device on a JEDEC 2s2p board. Again the back-transforms of the clipped structure functions almost perfectly match the true Zth-JA curves.

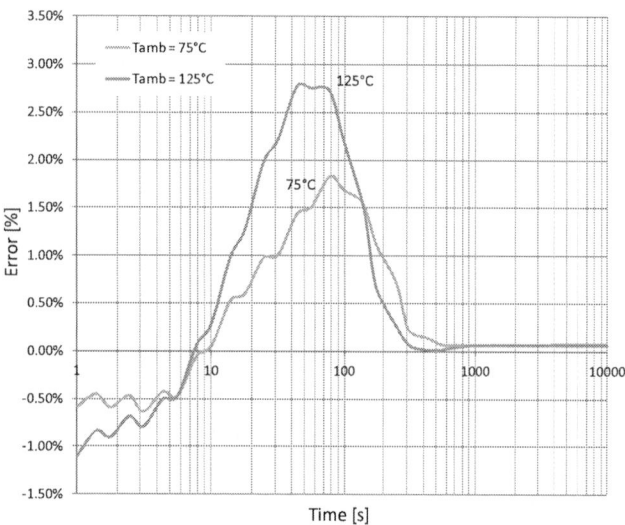

**Figure 12**: Relative error of the Zth-JA curves obtained by structure-function clipping for the JEDEC 2s2p board.

### 5.3 JEDEC 2s0p board, varying power dissipation

In the last example the MOSFET device shall be soldered to a JEDEC 2s0p board again. However, now the power dissipation is changed whereas the ambient temperature remains constant. Figure 1 3 shows the simulated Zth-JA curves for 1W and 2W power dissipation at an ambient temperature of 25°C. The deviation between clipped 1W-structure function and 2W-structure function is larger than in the previous examples (figure 14). The reason is probably the very inhomogeneous change of the surface temperature caused by the increasing power dissipation. On the component itself the temperature increase is much higher than on the surrounding PCB. Consequently also the heat transfer coefficient increases mainly locally at the component site which changes not only the strength of the heat flow but also its spatial distribution. As a result not only the length of the plateau of the structure function changes but also its characteristic shape (figure 14).

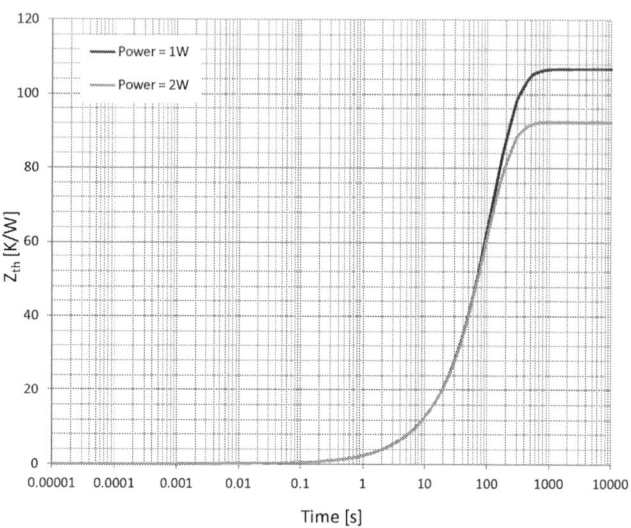

**Figure 13**: Zth-JA curves of the MOSFET device on a JEDEC 2s0p PCB for 1W and 2W power dissipation.

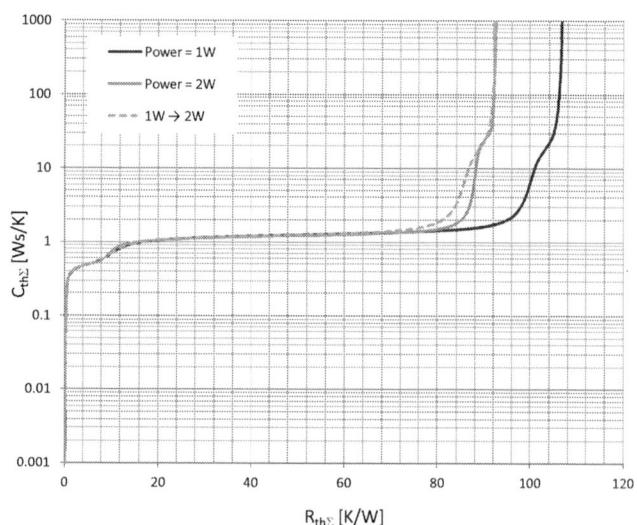

**Figure 14**: Structure functions of the MOSFET device for 1W and 2W. The dashed curve has been obtained by clipping and shifting the 1W structure function.

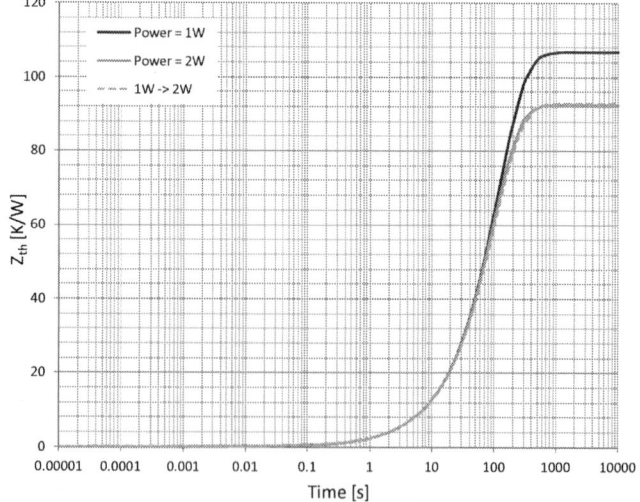

**Figure 15**: True vs. transformed (dashed) Zth-curve for 1W.

978-1-4673-1110-6/12 $31.00 © 2012 IEEE

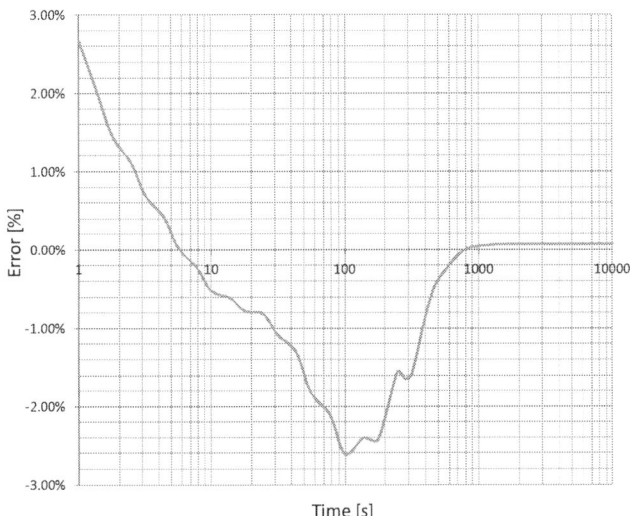

**Figure 16**: Relative error between true Zth-JA and the transformed Zth-curve for 1W.

However the match between the associated Zth-JA curves with a maximum relative error < 3% (figure 16) is still quite well.

## 6. Some remarks

In order to apply the structure function clipping algorithm the Rth-JA of the target ambient conditon must be known. As mentioned above this can be achieved for a wide range of ambient temperatures and/or power dissipation levels by interpolation of a few measured or simulated Rth-JA values.

In principle the back-transform of the clipped structure function into a Zth-curve should correctly describe the new Zth-JA over the whole time-range. However, due to numerical reasons the accuracy of the thus obtained Zth-curve is often quite poor in the short time range. Since the Zth-values in the short time range are (almost) independent of ambient temperature and power dissipation (see also footnote [†]), the values of the original Zth-curve should be used instead for short times $t < t_x$, where $t_x$ is determined by the device and PCB. In many cases $t_x = 1s$ will be a good choice; until this time the heat flow is mainly governed by heat conduction inside device and PCB and the heat transition from component and PCB to the surroundings has not yet an influence.

## 7. Conclusions

As has been demonstrated the proposed algorithm can be used to adapt a given Zth-JA junction-to-ambient curve to changes in ambient temperature and power dissipation. In the given examples the transformed Zth-curves describe thermal behavior for the new ambient conditions very accurately.

In principle the same method could be applied in any situation where the steady-state junction-to-ambient thermal resistance Rth-JA of a semi-conductor device changes due to varying ambient conditions while the heat flow inside the device (and therefore its Zth-curve in the short time range) remains more or less unaffected. Further possible applications, though not tested yet, could be the adaption of Zth-curves to different air flow conditions (air velocity) in forced convection environments or to changes in PCB size or layout of copper areas on PCB connected to the device (size of convective area).

Not suitable however is the algorithm if the varying ambient condition also changes the heat flow distribution inside the or very close to the device (e.g. if the same device is used with and without top-mounted heat-sink). In that case the structure function changes in a more complex way which cannot be described by simply clipping a part of it.

Instead of generating a whole set of Zth-curves for a range of possible boundary conditions it is often sufficient to compute or measure just one Zth-curve (together with a small set of Rth-JA values) which can then be transformed accordingly. Apart from time savings the main advantage of the presented method lies in its greater flexibility and the possibility of its implementation in thermal calculators.

## References

1. Electronic Industries Association, "Methodology for the Thermal Measurement of Component Packages (Single Semiconductor Device)", EIA/JEDEC Standard, JESD51, 1995 [www.jedec.org].
2. Y.C. Gerstenmair and G. Wachutka, "A New Procedure for the Calculation of the Temperature Development in Electronic Systems", EPE'99 conference, Lausanne, Switzerland, 1999.
3. Y.C. Gerstenmair and G. Wachutka, "Calculation of the Temperature Development in Electronic Systems by Convolution Integrals", Proc.16th SEMITHERM, San Jose, pp. 50-59, 2000.
4. D. Schweitzer, "A Fast Algorithm for Thermal Transient Multisource Simulation Using Interpolated Zth-Functions", IEEE Trans. Comp. Packag. Technol., vol. 32, no. 2, pp. 478-483, 2009.
5. H. Pape, "Treatment of Convection and Radiation without CFD in Thermal Resistance Calculations", Proc. 7th THERMINIC, Paris, pp. 43-49, 2001.
6. V.Szekely, Tran Van Bien, "Fine structure of heat flow path in semiconductor devices: a measurement and identification method", Solid State Elec., Vol. 31, pp. 1363-1368, 1988.
7. V. Szekely, M. Rencz, A. Poppe, B. Courtois, "New Way for Thermal Transient Testing", Proc.15th SEMITHERM, San Diego, pp. 182-188, 1998.
8. V.Szekely, "Identification of RC networks by deconvolution: Chances and Limits", IEEE Trans. On Circuits and Systems – I: Fundamental Theory and Applications, Vol. 45, No. 3, pp. 244-258, 1998.
9. D. Schweitzer, H. Pape, and L. Chen, "Transient Measurement of the Junction-to-Case Thermal Resistance Using Structure Functions: Chances and Limits", *Proc. 24th SEMITHERM*, San Jose, CA, pp. 193-199, 2008.
10. Electronic Industries Association, "Transient Dual Interface Test Method for the Measurement of the Thermal Resistance Junction-to-Case of Semiconductor Devices with Heat Flow Trough a Single Path", EIA / JEDEC Standard, JESD51-14, 2010 [www.jedec.org].
11. GNU Multiple Precision Arithmetic Library (GMP) [www.gmplib.org].
12. Y.C. Gerstenmair and G. Wachutka, "Combination of Thermal Subsystems Modelled by Rapid Circuit Transformation", Proc.13th THERMINIC, Budapest, pp. 115-120, 2007.

# Server Liquid Cooling with Chiller-less Data Center Design to Enable Significant Energy Savings

Madhusudan Iyengar[1*], Milnes David[1], Pritish Parida[2], Vinod Kamath[1], Bejoy Kochuparambil[1], David Graybill[1], Mark Schultz[2], Michael Gaynes[2], Robert Simons[1], Roger Schmidt[1] and Timothy Chainer[2]

IBM System & Technology[1] and Research[2] Divisions

*mki@us.ibm.com

## Abstract

This paper summarizes the concept design and hardware build efforts as part of a US Department of Energy cost shared grant, two year project (2010-2012) that was undertaken to develop highly energy efficient, warm liquid cooled servers for use in chiller-less data centers. Significant savings are expected in data center energy, refrigerant and make up water use. The technologies being developed include liquid cooling hardware for high volume servers, advanced thermal interface materials, and dry air heat exchanger (chiller-less with all year "econom izer") based facility level cooling systems that reject the Information Technology (IT) equipment heat load directly to the outside ambient air. Substantial effort has also been devoted towards exploring the use of high volume manufacturable components and cost optimized cooling designs that address high volume market design points. Demonstration hardware for server liquid cooling and data center economizer based cooling has been built and is operational for a 15 kW rack fully populated with liquid cooled servers. This design allows the use of up to 45 °C liquid coolant to the rack. Data collection has commenced to document the system thermal performance and energy usage using sophisticated instrumentation and data collection software methodologies. The anticipated benefits of such energy-centric configurations are significant energy savings at the data center level of as much as 30% and energy-proportional cooling in real time based on IT load and ambient air temperatures. The objective of this project is to reduce the cooling energy to 5% or less of a comparable typical air cooled chiller based total data center energy. Additional energy savings can be realized by reducing the IT power itself through reduced server fan power and potentially less leakage power due to lower device temperatures on average for most locations.

This paper focuses on the server liquid cooling, the rack enclosure with heat exchanger cooling and liquid distribution, and the data center level cooling infrastructure. A sample of recently collected energy-efficiency data is also presented to provide experimental validation of the concept demonstrating cooling energy use to be less than 3.5% of the IT power for a hot summer day in New York.

## Keywords

Data center, liquid cooling, energy efficiency.

## 1. Background

Information Technology (IT) data centers consume a large amount of electricity in the US and world-wide. Cooling uses a significant portion of this energy. Figure 1(a) shows a facility level schematic for the cooling system that is used to transfer heat from the server exhaust air, to the ambient outdoor air, which is the ultimate heat sink. The transfer of heat from the IT equipment to the room level coolant flow is depicted in Figure 1(a) via the sketch labeled as the data center building. As shown in Figure 1(a), racks of IT equipment are arranged in rows to form several aisles in which two rows of racks face each other at their inlets. These racks are usually air cooled inside the servers. Thus, the racks require a continuous and reliable supply of cool air for their operation which is supplied by the Computer Room Air Conditioning (CRACs). The chilled air enters the room via perforated floor tiles, passes through the racks getting heated in the process, then finds it way to the intake of the room CRACs which cool the hot air and blow it into the under floor plenum. The chilled water from the chiller is usually pumped through a network of under floor pipes which supply and remove water to and from the CRACs. The air supplied to such equipment is typically in the 15-32°C range for allowable temperatures with 18-27°C being the nominal long term recommended band [1].

As further seen in Figure 1(b), the IT equipment usually consumes about 50% of the total energy that is required to operate a data center. The cooling energy consumption is roughly 25-30% of this total energy use [2-4]. Thus, using these typical values, the cooling energy use is about 50% of the IT energy use. This is an important base line metric regarding the well documented inefficiencies of air-cooled data centers [5] and subsequent results presented herein will relate to this metric to document the energy savings realized through the innovative cooling designs demonstrated. The various cooling infrastructure components that are made up of three elements: the refrigeration chiller plant (including the cooling tower fans and condenser water pumps, in the case of water-cooled condensers), the building chilled water pumps, and the data center floor air-conditioning units (CRACs). are depicted in Figure 1 (b) as the cooling pie segment and in Figure 1 (c) as the HVAC cooling and the HVAC fans and blowers pie segments. About half the cooling energy used is consumed at the refrigeration chiller compressor and about a third is used by the room level air-conditioning units for air movement, making them the two primary contributors to the data center cooling energy use. This paper primarily focuses on an innovative server and data center cooling design that eliminates the energy usage of room air conditioning devices and the chiller compressor by using liquid cooling at the server and by operating at coolant temperatures that are above the ambient for the entire year.

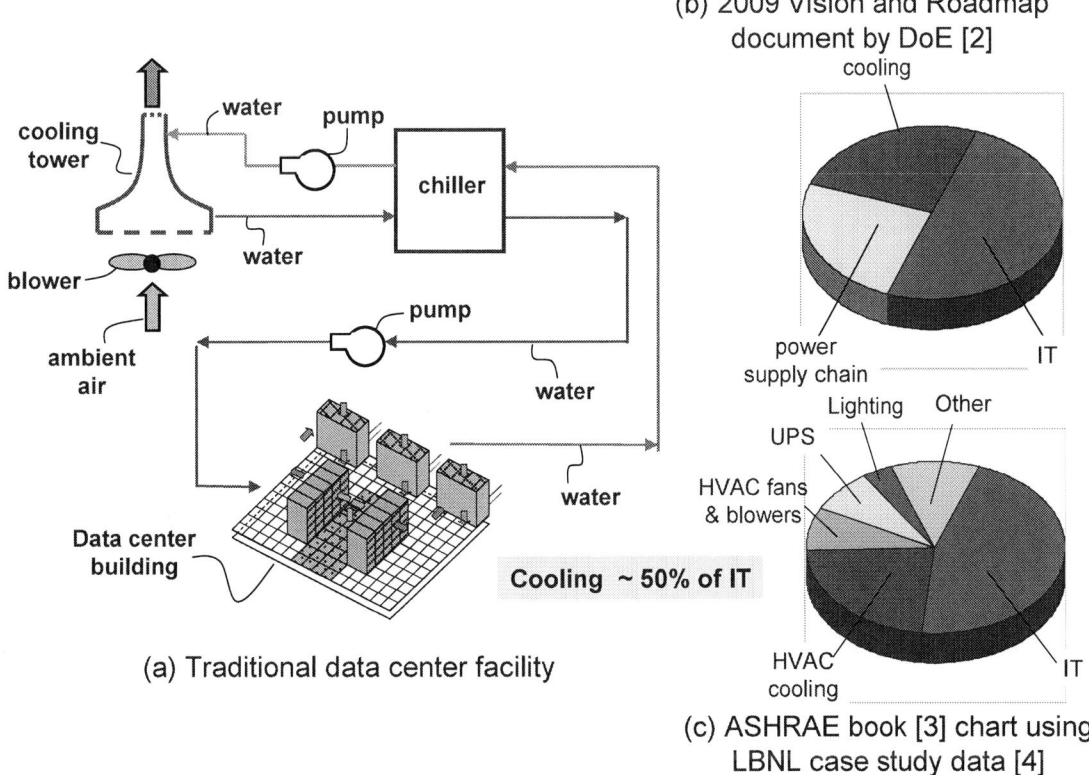

(b) 2009 Vision and Roadmap document by DoE [2]

(a) Traditional data center facility

Cooling ~ 50% of IT

(c) ASHRAE book [3] chart using LBNL case study data [4]

Figure 1: Traditional Chiller Plant Based Data Center Cooling Loop
(a) Schematic of data center cooling, (b) Typical data center energy breakdown per DoE [2]
(c) Data center energy breakdown per ASHRAE book [3] using and LBNL case studies [4]

In addition to significant energy usage, traditional data center cooling also results in refrigerant and m ake up water consumption that are addressed i n the new cooling designs discussed.

## 2. Innovative Data Center Liquid Cooling Design

A schematic of the novel highly energy-efficient data center cooling concept that ha s been named as the Dual-Enclosure-Liquid Cooling (DELC) unit is shown in Figure 2. On the left of Figure 2, the external dry cooler unit is depicted with a heat exchanger coil and a fan to reject the data center cooling load into the ambient air flowing through the air-to-liquid heat exchanger coi l. An ext ernal pump is also shown which circulates the coolant through the dry cooler unit and buffer coolant distribution unit. The reci rculation valve allows part of t he hot coolant to bypass the dry cooler in winter months to ensure t hat the liquid entering the lab is above the dew point temperature of the lab so as t o prevent any condensation in the vicinity of the servers. The right side of Figure 2 shows the data center cooling design inside the lab which includes a buffer coolant distribution unit and the 100% liquid cooled rack. Thus, in contrast to the traditional cooling design illustrated via Figure 1(a), there is no refrigeration unit and onl y three powered cool ing devices: the external dry cooler fans , the external pum p, and the internal pump. The rack l iquid cooling includes both air and liquid cooling devices and will b e described in subsequent sections.

Figure 3 shows phot ographs of t he data center cooling loop infrastructure. Fi gure 3(a) shows the external dry cooler loop with five fans m ounted above a l arge air to liquid heat exchanger coil. The air flow is drawn into the heat exchanger from below and t hrough the sides, and the hot air expelled upwards from the fans. To t he right in Figure 3(a), an auxiliary enclosure is seen which houses the external pump and recirculation valve that was described above via Figure 2. This enclosure also includes some instrumentation including a pressure meter and t emperature sensors t o measure coolant temperature leaving the dry cooler as well as before and after the valve "tee". Thus, using valve operation (winter), it would be possi ble to determine the coolant temperatures before and after mixing of hot coolant that bypasses the dry cooler. Figure 3(b) shows a phot ograph of the piping layout inside the lab. The piping coming into the lab can be seen from the outside when looking at the top right of Figure 3(a). Inside the lab there is a by pass loop to allow coolant bypass and device servicing for the 50 micron filter. Also seen in Figure 3(b) are the temperature sensors, the flow m eters for the internal and ext ernal loops, pressure sensors and the buffer coolant distribution unit. Thi s buffer unit allows separation of the internal and external coolant loops.

978-1-4673-1110-6/12 $31.00 © 2012 IEEE       213

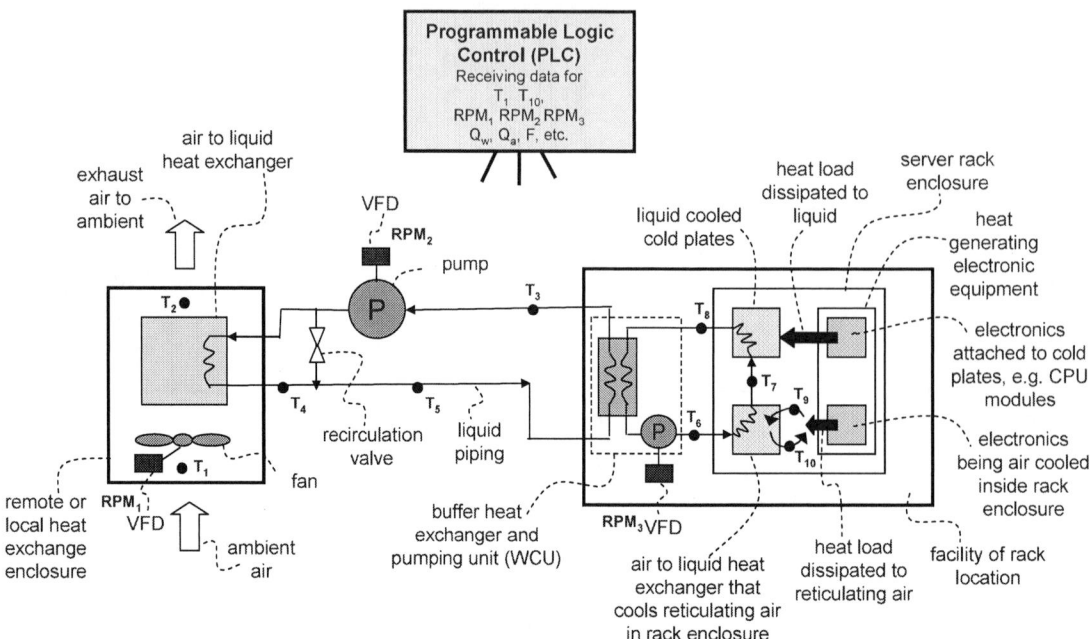

Figure 2: Schematic of new data center cooling design – Dual Enclosure Liquid Cooling (DELC)

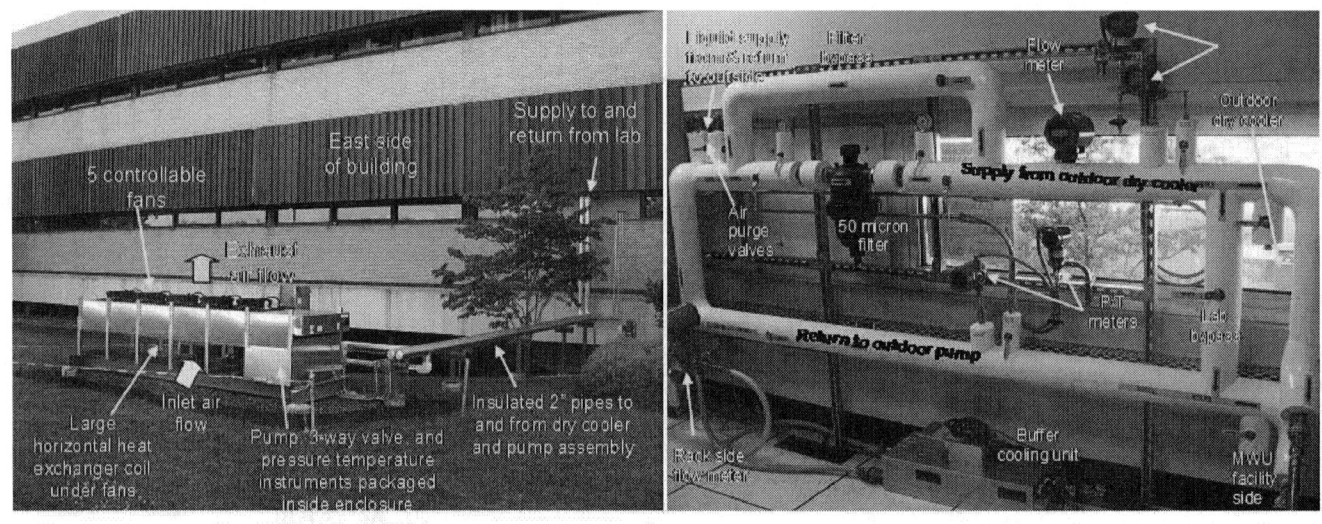

(a) External dry cooler unit with fans, heat exchanger coil, pump, recirculation valve, instrumentation and controls

(b) Internal (lab) piping with instrumentation and buffer coolant distribution unit

Figure 3: Photographs of data center cooling hardware
(a) External dry cooler unit, (b) Internal (lab) piping layout and instrumentation

Such separation allows the use of water in side the lab even in winter months when it may be desirable to fill the external loop with a wat er-glycol mixture to guard against freezing during cold New York wi nters. The buffer unit also allows for the use of specially    treated water on the system side, i.e. the water flowing through the rack cool ing devices, which results in a greater tolerance at the data center level in allowing the use of less clean water in the external loop which is often the case. The buffer unit seen at the bottom of Figure 3(b) is comprised of a pum p and a pl ate type liquid to liquid heat exchanger and i f desired could be packaged i nto the bottom of a server rack [6].

## 3. Server Rack Liquid Cooling Design

The liquid cooled server rack i s shown in Figure 4. The rack includes front and rear covers that duct the air flow to and from the servers t hrough a side air-to-liquid heat exchanger coil (Side Car). This Side Car unit is comprised of an air to liquid heat exchange r that cools the hot server exhaust air to a satisfactory temperature for intake into the servers. In Figure 4(a) the front of the rack can be seen with the Side Car unit on the right. Figure 4(b) shows a schematic section plan view of the rack and depicts the server node as well as the side car unit. The water flow from the buffer unit first enters the Side Car heat exchanger to cool the re-circulating server air flow. After flowing through the Side Car air to liquid heat exchange r, the partially heated water then enters the rack inlet manifold which distributes the water to each of the liquid cooling s ub-assemblies inside the nodes. A flexible hose i s attached to the inlet of t he node l iquid cooling assembly with a one-eared hose clamp. The other end of the hose which has a quick disconnect coupler supplies the water to the node from the rack m anifold. A si milar hose returns the warm water from the node t o the rack l evel exit manifold. Thus, there are two rack manifolds, one for l iquid supply and the other for return. Each m anifold has 42 port s with quick disconnect couplers for connect ion to the node liquid cooling devices using the flexible hoses. Fi gure 4(c) illustrates the plumbing from the rack manifold to the node as well as the numerous ports on the manifold. The rack cooling

design described above accom modates both air and liquid cooled devices at the server level while transferring the entire rack heat load into the water at th e rack level. This is an important attribute that provides flexibility in the equipment to be housed inside the rack. However, it should be noted that while there are both air and wate r cooled devices in the rack, both of these sets of components must accept warm air or warm water to allow the data center design discussed in this paper to work. The Side Car unit has been appl ied to fully air cooled racks prior to the application [7] discussed herein.

A key feature of the Server Rack Design is the elimination of any heat load to the data center. An other important feature is that the economizer based cooling isolates the rack from the outside air environment, thus minimizing the risk of com ponent contamination through particulate, chemical or gaseous matter that may be present in the outdoor ambient from time to time.

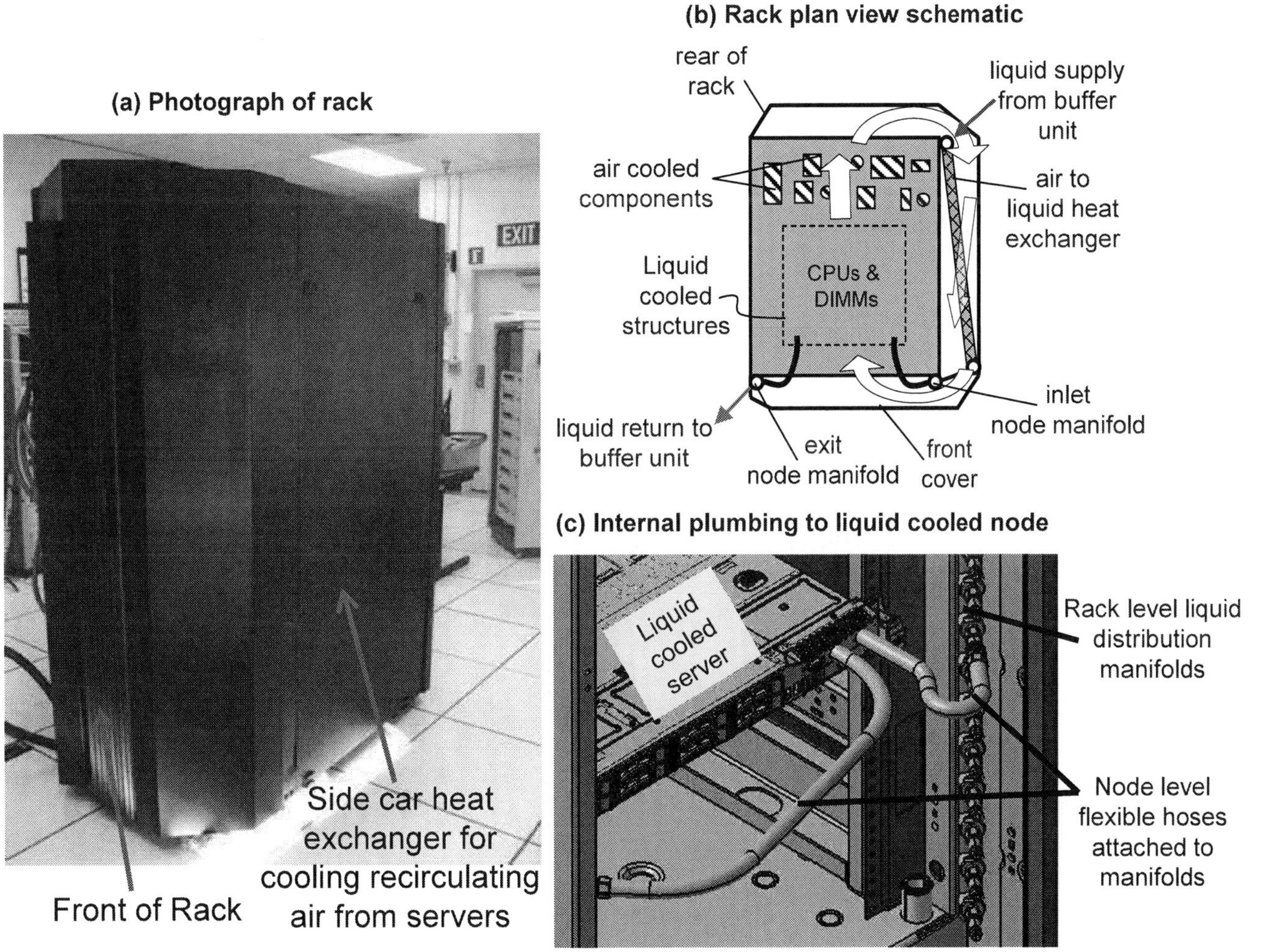

Figure 4: Rack liquid cooling design
(a) Photograph of the front of the rack, (b) Plan view schematic of rack internals [8],
(c) Plumbing of server node liquid cooling to rack manifolds.

## 4. Server Liquid Cooling Design

A perspective view of the hybrid cooled server comprised of air cooled and water cooled devices is shown in Figure 5. The server is an IBM x3550 M3 server which is 1U tall (1.75 inches or 44.45 mm) and fits in a standard 19 inch (483 mm) rack that is actually about 24 inches wide (609.5 mm). The microprocessor modules are cooled using cold plate structures. The Dual-In-Line-Memory Module (DIMM) cards are cooled by attaching them to a pair of conduction spreaders which are then bolted to a cold rail that has water flowing through it. The microprocessors and DIMMs have a typical maximum power of 130 W and 6 W each, respectively, and the server node power is about 400 W for its maximum. All the other components in the server shown in Figure 5 are air cooled. These devices include the storage disk drives, the power supply and various surface mounted components on the Printed Circuit Board (PCB). While the air cooled version of this server had six fan packs (two fans per pack), in the water cooled server three of these six fan packs were removed. The fan control algorithm was also modified to allow the server to operate up to 50 °C inlet air temperatures compared to existing commercial servers which typically enforce power down procedures if the inlet air temperature rises above 40°C.

## 5. Server Liquid Cooling Components

The water cooling structures that are used to construct the hybrid air/water cooled node described in the preceding section are shown in Figure 6. The cold plate used to extract heat from the microprocessor modules is shown in Figure 6(a). It is made up of two parts: the copper core with the water flow channels that make up a parallel plate fin array, and the aluminum frame that is soldered to the copper core and provides the structural support during clamping of the cold plate to the module. The two part design allows the use of the higher thermal conductivity copper material for heat transfer while using the less expensive aluminum for structural function.

Figure 6(b) displays the sub-assembly made of two copper spreader plates that are mechanically attached to a DIMM card using spring clips with a thermal interface material between the spreader and the DIMM. The spreader-DIMM sub-assembly is inserted into an electrical socket on the PCB and then the spreaders are bolted on both sides to liquid cooled cold rails which extract the heat from the DIMMs through the spreaders. There is another thermal interface material between the spreader and the cold rails. Figures 6(c)-6(e) show three different cold rails designs that were required to meet the "keep out" dimensional constraints of the server. These constraints included capacitors mounted on the PCB, cards mounted on sockets, and fans located at the front of the server. The cold rails have tapped screw holes in their solid portions. The distinct cold rail shapes shown in Figures. 6(c)-6(e) result from the unique geometry constraints. The cold rails shown terminate in hose barb connections and represent the individual prototypes made for testing and not the design used in the full node cooling assembly.

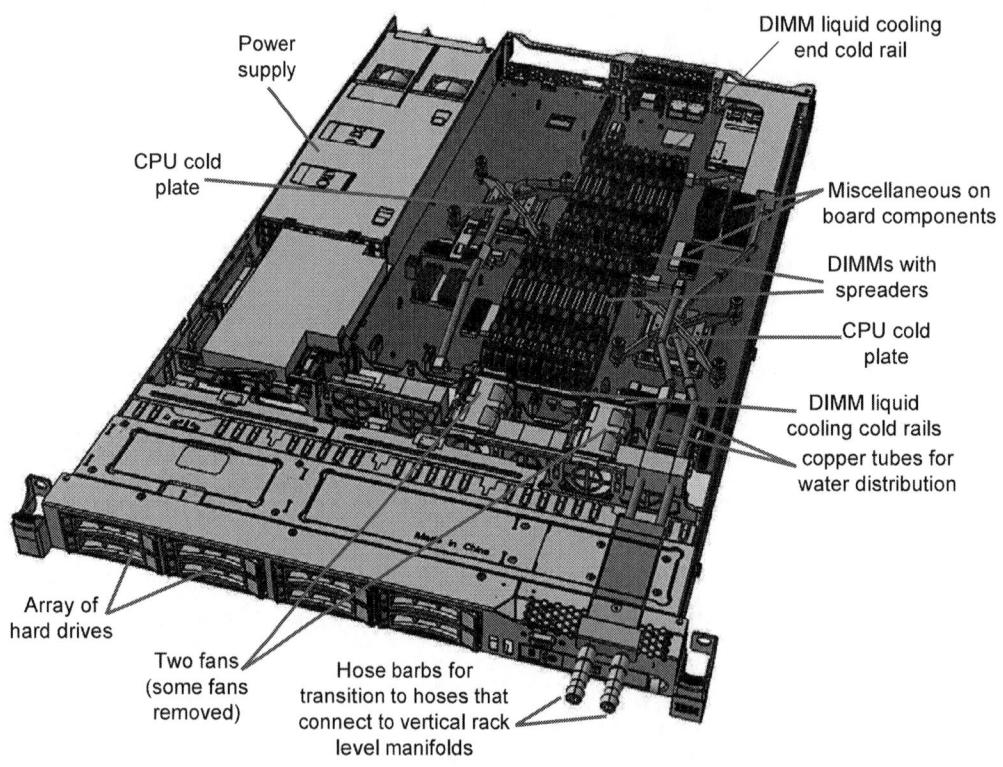

Figure 5: Hybrid air-water cooled 1U server designed for intake of 45°C water and 50°C air.

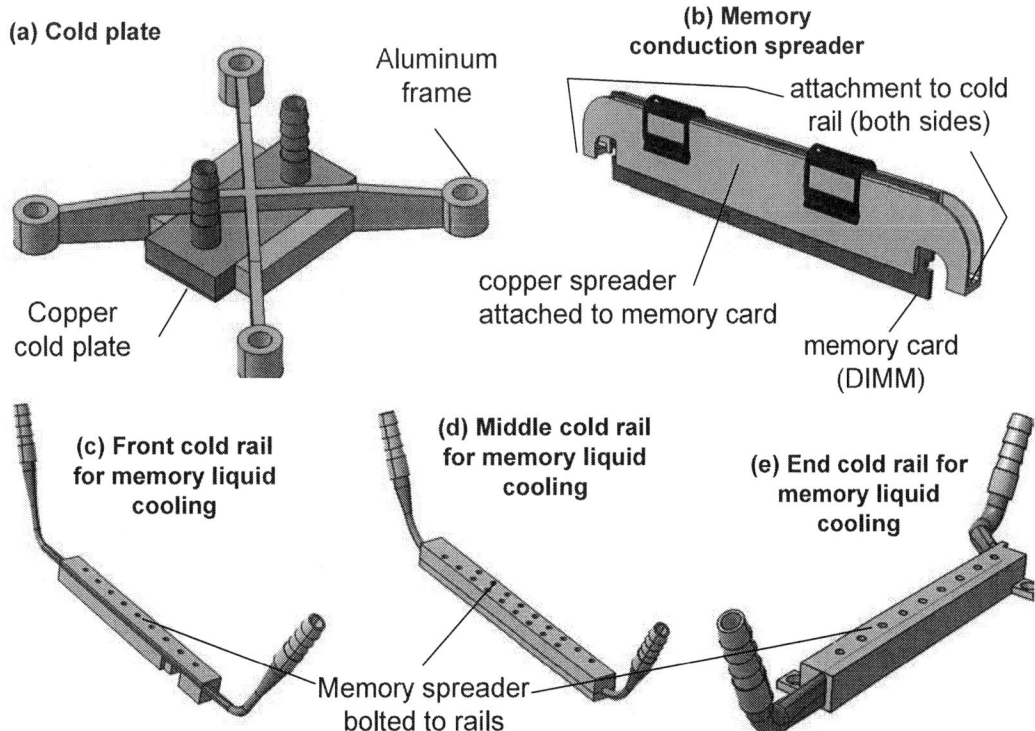

Figure 6: Liquid cooled components inside the server; (a) cold plate for the microprocessor module, (b) DIMM with conduction spreader, (c) front cold rail for memory liquid cooling, (d) middle cold rail for memory liquid cooling, (e) end cold for memory liquid cooling.

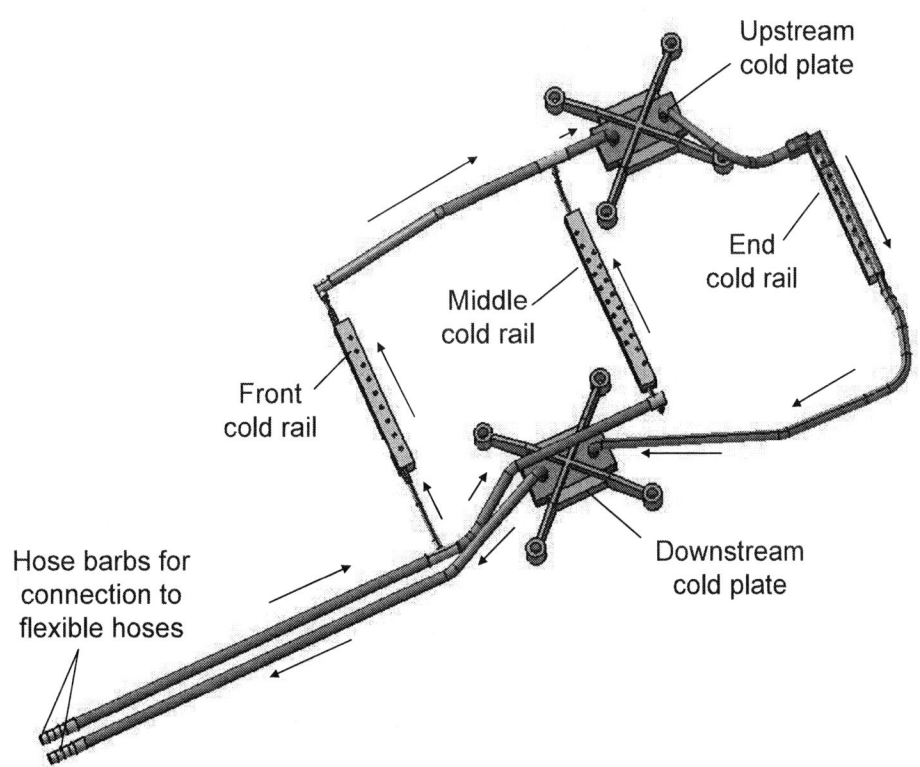

Figure 7: Node cooling sub-assembly for partially liquid cooled server

978-1-4673-1110-6/12 $31.00 © 2012 IEEE          217

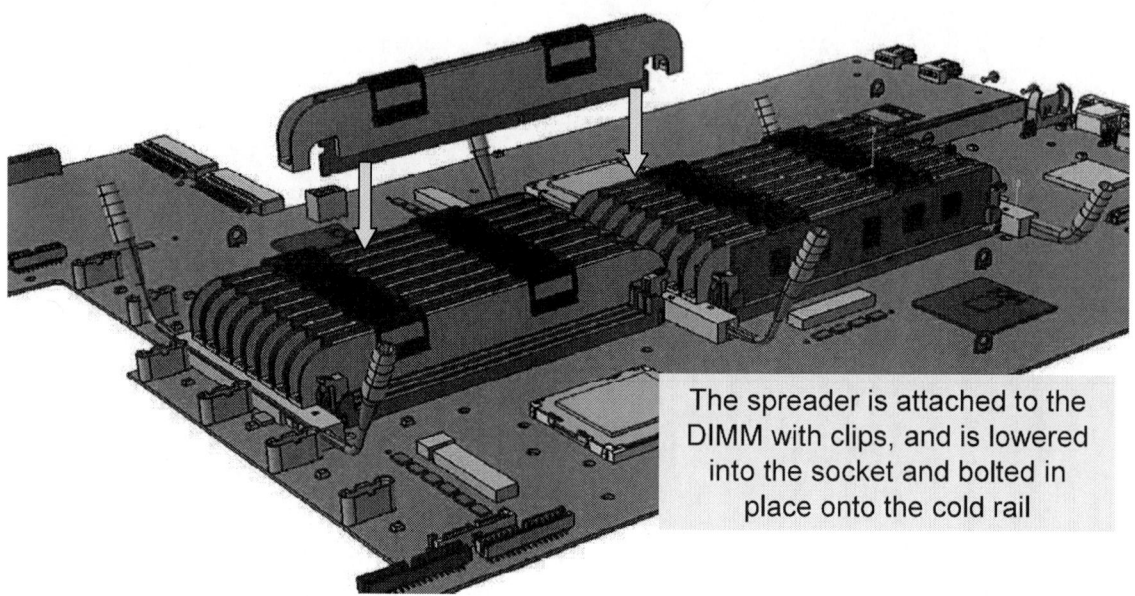

The spreader is attached to the DIMM with clips, and is lowered into the socket and bolted in place onto the cold rail

Figure 8: Illustration of DIMM-spreader assembly into liquid cooled server node

Figure 7 shows the assembled node cooling sub-assembly that is made up of two cold plates, three cold rails, and copper tubing. The structure shown in Figure 7 is lowered onto a server board and is attached at the cold plate frames (shown in Figure 6(a)) to the server PCB. The water flow path is also depicted in Figure 7 with the inlet being at the copper tube shown at the bottom left of the image (left of the pair of ports). The water flow splits at the first junction resulting in parallel flow through the front and middle cold rails with the flow then joining in front of the first cold plate (upstream). After flowing through the first cold plate the water passes through the end cold rail and then the second cold plate (downstream) after which it exits the node.

Figure 8 illustrates the assembly of the DIMM with the spreader attached into the server node. Similar to the drawings seen in Figs. 6(c) - 6(e), the cold rails in Figure 8 are special prototypes manufactured for testing purposes. Only the cold rails of the cooling loop appear in Figure 8. However, viewing the complete loop in Figure 7 in conjunction with Figure 8, provides a better understanding of the assembly sequence that results in the liquid cooled server displayed in Figure 5. After the loop is assembled, the DIMM and spreader sub-assembly is inserted into the electrical sockets to make good electrical contact. Then the two ends of the spreader are bolted to the cold rail using screws. As mentioned earlier, thermal interface materials are used between the spreader and the DIMM chips as well as between the spreader ends and the cold rails.

## 6. Thermal Chamber Test Data For Air And Water Cooled Servers

Tables 1 and 2 below provide thermal data collected in a controlled test chamber environment to characterize both the base line air-cooled and new partially water cooled server designs. The goals of the characterization were to collect both thermal and power data as well as to validate the partially water cooled server for the high air and water coolant temperatures for which it had been designed.

Table 1 provides data for two air-cooled node configurations, namely for a typical 25°C air inlet as may commonly occur in a data center and a 35 °C air inlet temperature as might occur in a hot spot area of a data center or in a facility that sought to operate servers at a significantly high ambient environment with the intention of realizing cooling energy savings at the facility level. Table 2 provides comparable data for two partially water cooled server configurations, namely, for a "cool" 20.1°C water inlet temperature and a "hot" 45 °C water inlet temperature. The former 20°C water temperature would be representative of a typical data center and the latter 45°C water temperature might be a worst case condition for many parts of the globe. It should be noted that the statements mentioned above assume the data center design previously described in detail in this paper where the air and water temperatures entering the server are closely coupled to the outdoor air temperature. For the water cooled server tests, the air temperature was maintained at 5 °C above the water inlet temperature in anticipation of such a difference as a typical condition in the DELC cooling system where the water to the rack will first cool the re-circulating air via the Side Car heat exchanger and then enter the node to cool the water cooled components as shown in Figure 1(a). Figure 9 provides more detailed data for the server powers, the fan powers, the CPU lid temperatures, and the DIMM temperatures for a number of inlet coolant conditions. The two exerciser settings used for the data provided in Figure 9 are for the CPU and the DIMM memory. In reality, it is unlikely that a real workload would exercise only the CPU or only the memory, but will be some combination of both.

978-1-4673-1110-6/12 $31.00 © 2012 IEEE         218

Table 1: Thermal test chamber data for air cooled servers

| CASE A | CASE B |
|---|---|
| **"Cool" air cooled node** | **"Hot" air cooled node** |
| ▪ 25.3°C inlet air temperature | ▪ 35.4°C inlet air temperature |
| ▪ Exerciser setting at 90% | ▪ Exerciser setting at 90% |
| ▪ 12 fans running at 7242 rpm (avg.) | ▪ 12 fans running at 11978 rpm (avg.) |
| ▪ System power = 395 W | ▪ System power = 423 W |
| ▪ Fan power = 19.1 W | ▪ Fan power = 56.8 W |
| ▪ CPU lid temps. = 65.3 °C, 74 °C | ▪ CPU lid temps. = 68.9 °C, 71.9 °C |
| ▪ DIMM temperatures = 35-46 °C | ▪ DIMM temperatures = 35-46 °C |
| ▪ 12 x 8 GB DIMMS | ▪ 12 x 8 GB DIMMS |

Table 2: Thermal test chamber data for partially water cooled servers.

| CASE C | CASE D |
|---|---|
| **"Hot" water cooled node** | **"Cool" water cooled node** |
| ▪ 49.9 °C inlet air temperature | ▪ 24.9 °C inlet air temperature |
| ▪ 45.2 °C inlet water temperature | ▪ 20.1 °C inlet water temperature |
| ▪ Exerciser setting at 90% | ▪ Exerciser setting at 90% |
| ▪ 3 fans running at 12612 rpm (avg) | ▪ 3 fans running at 5838 rpm (avg) |
| ▪ System power = 411 W | ▪ System power = 354 W |
| ▪ Fan power = 30.9 W | ▪ Fan power = 8.3 W |
| ▪ CPU lid temps. → 62.8°C, 61.9°C | ▪ CPU lid temps. → 36.8°C, 35.9°C |
| ▪ DIMM temperatures → 53-56 °C | ▪ DIMM temperatures → 28-33 °C |
| ▪ 12 x 8 GB DIMMs | ▪ 12 x 8 GB DIMMs |

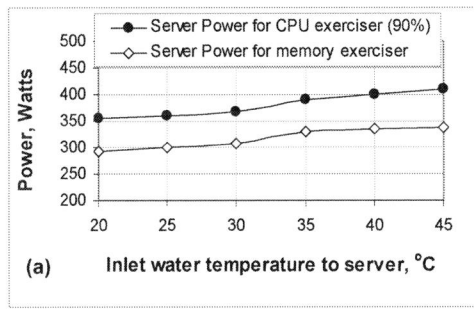
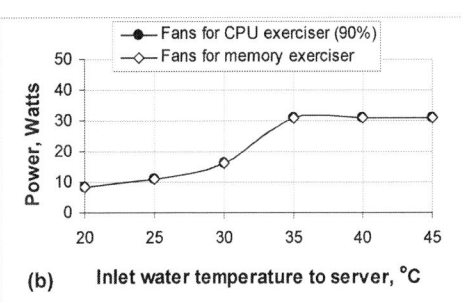
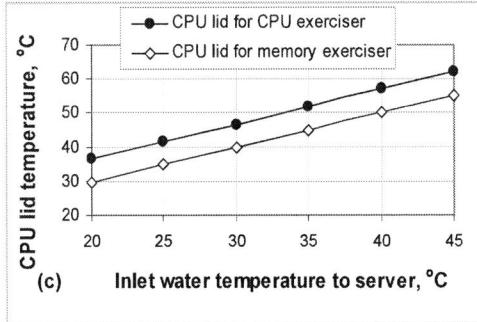
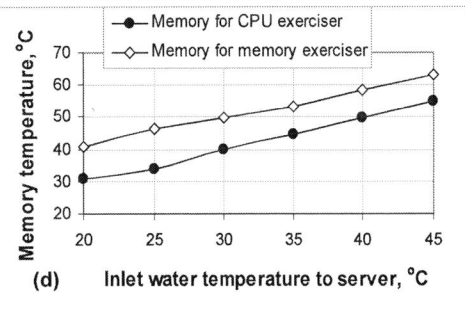
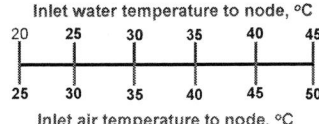

Figure 9: Power and device temperature data for partially water cooled node,
(a) Server power, (b) Fan power, (c) CPU lid temperature, (d) DIMM temperature

It should be noted that in Figure 9(b) both lines for the CPU and memory exerciser settings are nearly identical, thus appearing to be a single line.

It is interesting to compare the performance of Case A for a typical "cool" air cooled node configuration representative of most air cooled data centers to Case C and Case D for the newly proposed partially liquid cooled nodes representing cool and warm water, respectively.

Comparing Cases A and C, the typical "cool" air cooled node consumes 395 W of power while the "hot" water cooled node uses 411 W which is a difference of 16 W (4.1%). That may be explained due to a difference of 11.8 W in fan power since the CPU lid temperatures (and thus CPU power) are comparable. The increase in fan power is due to the fact that the fan speeds for both nodes are programmed to ramp fan rpm with inlet air temperature. Even though the partially liquid cooled node has three less fans than the air cooled node they are running at higher rpm due to the higher air ambient temperature. With a change to the fan speed algorithm, the water cooled node could use less fan power even at elevated air ambient temperatures compared to an air-cooled node at typical air ambient temperatures.

Comparing Cases A and D, the "cool" water cooled node consumes 10.4% less power (41 W) which is partly because it has three fans running at low speed compared to six fans running at a slightly higher speed for Case A. Of this 41 W, 10.8 W is attributed to fan power and the remaining is likely due to reduced leakage power from the 30-40°C much cooler temperatures at the CPU.

While an actual test in a specific location would be needed for confirmation, it may be speculated that for many locations only a small percentage of the time would be spent at the "hot" coolant temperatures represented by Case C and that the server will likely experience the Case D conditions for most of the year. If this were the case, then in addition to the significant data center level energy savings, there can be substantial IT power related energy savings of the order of 10% due the combination of leakage power and fan power reductions. In contrast to these gains, operating the air-cooled server at elevated ambient temperatures of 35°C can result in a 7.1% (28W) increase in IT power while there will be some data center level cooling energy reductions.

## 7. Temperature Excess Between Outdoor Air And Server Node Coolant

The preceding section highlighted the effect of coolant temperatures (air or water) on the server power, the fan power, and the CPU and DIMM device temperatures. Therefore it is important to estimate what these coolant temperatures will be for a given location and time of the year. To do this, it is necessary to characterize the temperature difference between the entering cold fluid and the exiting hot fluid (approach temperature difference) on either side of the various heat exchanger devices in the loop including the dry cooler air-to-liquid heat exchanger, the buffer unit liquid-to-liquid heat exchanger, and the Side Car air-to-liquid heat exchanger. Figure 10 provides data based calculations for the approach temperature difference, $\Delta T_{Approach}$ for the three heat exchanger devices that are in the loop which was described in the schematic shown in Figure 2.

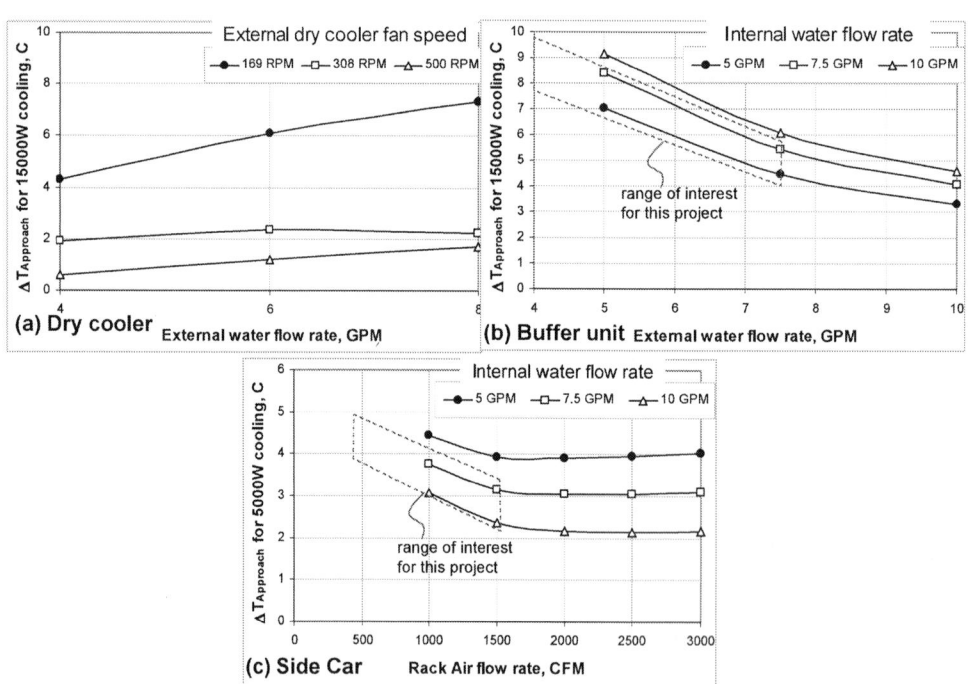

Figure 10: Approach temperature differences for the three heat exchanger devices in the loop, (a) Dry cooler external air-to-liquid heat exchanger coil, (b) Buffer unit liquid to liquid heat exchanger, (c) Rack Side Car air to liquid heat exchanger

Figure 10(a) shows the approach tem perature difference for the dry cooler which in this case will b e the difference between the outdoor air temperature entering the coil and the liquid tem perature leaving the coil. Assum ing a sample dry cooler external fan speed of 500 RPM and an external water flow rate of 8 GPM , the $\Delta T_{Approach}$ for the dry cooler would be 1.7 °C for a heat load of 15 kW . For a heat load of 30 kW , the $\Delta T_{Approach}$ would be 3.4 °C which is twice that for the 15 kW case.

Figure 10(b) shows the approach t emperature difference for the buffer unit. For an external water flow rate of 7.5 GPM and an i nternal (rack si de) water flow rate of 5 GPM, the $\Delta T_{Approach}$ for the buffer heat exchanger unit will be 4.5 °C for a heat load of 15 kW.

Figure 10(c) shows the approach tem perature difference for the Side Car. For an internal flow rate of 5 GPM, a rack ai r flow rate of 1500 cfm and an ai r side heat load of 5 kW the $\Delta T_{Approach}$ is 3.9°C The air heat load at the Side Car heat exchanger will b e less th an the total heat load because only part of the rack heat load is rejected to the air flow circulating inside the rack.

Thus, as a sam ple calculation, addition of all these approach temperature differences at the three heat exchange devices results in a total $\Delta T_{Approach}$ of 1.7 + 4.5 + 3.9 = 10.1 °C for the air temperature entering the server node. Si nce the water supplied to the rack will absorb 5 kW in heat load at the Side Car air-to-liquid heat exchanger, its temperature for the 5 GPM flow rate will rise by 3.8 °C, and thus the total $\Delta T_{Approach}$ for the water entering the node can be cal culated by 1.7 + 4.5 + 3.8 = 10 °C.

The calculations presented in Figure 10 are for a total heat load of 15 kW out of which the rack air heat load is 5 kW. In the actual system, the total load as well as the air load can vary based on t he actual computational workload as wel l as the temperature differential between the rack coolant temperatures and the lab ambient which will drive some heat loss into the room. There are ot her factors that can influence these heat loads including the heat loss in all the piping in the entire loop even though the piping in this case is insulated. In addition to these factors, the se rver fan speed is a function of server device temperatures and the inlet air tem perature as measured by the server air inle t temperature sensor. Thus t he rack air flow rate can vary from case to case. W hen the rack air flow rate varies, the $\Delta T_{Approach}$ for the Side Car heat exchanger will also change as sh own in Figure 10 (c). In winter months due to freezing weather outside and extrem ely low outdoor air temperatures, the coolant in the external loop will need to be a water-glycol mixture. The use of an anti-freeze in the external loop will increase the $\Delta T_{Approach}$ of both the dry cooler coil and t he buffer uni t liquid-to-liquid heat exchanger for t he same conditions reported in Figs. 10(a)-10(b).

## 8. Weather Data Analyses For Poughkeepsie, New York

To further the discussion in the previous two sections, weather data from the National Renewable Energy Lab [9] has been pl otted in Figure 11 for Poughkeepsi e, New York, where the data center test facility h as been constructed. Figure 11(a) shows how t he dry bulb temperature varies with the time of t he year where hour 1 i s 12 AM – 1 AM on January 1st. As m ay be seen from the histogram shown in Figure 11(b), for a t ypical year in Poughkeepsie, New York, less than 2% of t he year experiences outdoor dry bulb air temperatures greater than 30°C. Even with the dry cooler air-to-liquid heat exchanger, the buffer uni t liquid to liquid heat exchanger, and the side car air-to-liquid heat exchanger, t he coolant temperature excess from the ambient can be estimated for typical conditions to be less than 10°C for the inlet water to the server and to be less than 15°C for the inlet server air. Thus, using this estimate, 30-35°C outside air will resu lt in 40-45°C water to the node and 45-50°C air to the server; which will occur on average less than 2% of the year.

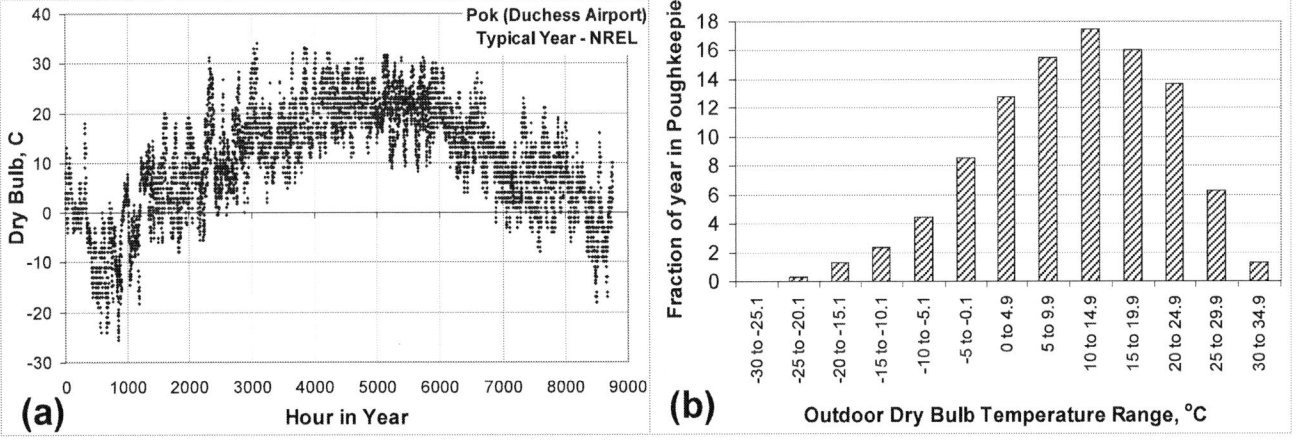

Figure 11: Typical weather data for outdoor dry bulb temperature for Poughkeepsie, New York
(a) Dry bulb temperature vs. time of year [4], (b) Histogram (%) of outdoor dry bulb temperatures.

In contrast, the outdoor dry bulb per the data from Figure 11(b) is less than 10°C for 45 % of the year. Using the same estimated temperature excesses di scussed previously in this section, an outdoor dry bulb air temperature of 5-10°C or less will result in inlet temperature to the node of less than 15-20°C for t he node i nlet water and l ess than 20-25°C for the server inlet air, which will on average occur for 45 % of the year. These resul ts support the argument to use C ase D in Table 2 as opposed t o Case C for t he typical water cooled design point for a fair comparison to Case A in Table 1 as the typical air cooled scenario fo r locations whose weather are similar to Poughkeepsi e, New York. Such a coupled analyses between the node thermal data and the local weather data would of course be require d for each location. It should also be noted that while the NREL [9] typical year is based on data collected for over 10 year s, the maximum temperatures for a specific year could vary fro m year to year. Thus, while longer term energy savings estimates can be based on a typical year, the therm al design itself needs to accom modate variations from the typical weather patterns for a specific location.

## 9. Experimental Thermal/Energy Data - Data Center Loop Performance

Sample results of a 22 hour run t hat was performed on August 4th and 5th, 2011 wi th the server rack popul ated with 38 liquid cooled servers and one net work switch node are shown in Table 3. Thi s run represents a hot summer day in Poughkeepsie, New York. The servers were exercised with a CPU exerciser at a 90% setting. In addition, the server DIMMs were si multaneously exercised using a memory centric script. The data reported below is the hourly average of parameters based on the values from the previous one hour. For this test, the buffer unit pump and the external pump were both set to a fixed speed. Thus, t he water flow rate in both the internal and external loops was a constant value.

As seen in Table 3, the cooling energy use for the 22 hour experiment for t he novel data center loop described in this paper is less than 3.5% of the IT power which is in severe contrast to the traditional data center loop described via Figure 1 in which the cooling power was about 50% of the IT power. W hile best-of-breed traditional data centers may display configurations where the cooling power can be as low as 30% of t he IT power, t he data center and server cooling design described in this paper still d emonstrates significant energy savings opportunities th at can be enabled through liquid cooling of servers and econom izer based design of data centers.

In summary, this 22 hour experi ment provides validation of bot h the system concept and the thermal and energy design of t he servers, the rack, and the data center cooling loop. The data represents a si gnificant milestone in this project and dem onstrates extremely high cooling energy efficiency for an econom izer based dat a center in which the above ambient coolant is still isolated from the outside environment. This particular point is important given the industry trend in cloud computing data centers to use outside air directly for cooling of servers which does pose a reliability risk in certain geographies or under cert ain natural disaster events.

Table 3: Results of a 24 hour run documenting thermal and power data for new data center loop
*last row of values are an average for 47 minutes and not 1 hour*

| Date Stamp | Time Stamp | Ave. outside air temp | Ave. IT Power | Ave. Cooling Power | Cooling power as % of IT | Ave. rack inlet air temp | Ave. rack inlet water temp |
|---|---|---|---|---|---|---|---|
| | | C | kW | kW | | C | C |
| 8/4/2011 | 4:00:30 PM | 28.1 | 13.43 | 0.440 | 3.28 | 38.1 | 36.3 |
| 8/4/2011 | 4:59:56 PM | 28.1 | 13.47 | 0.442 | 3.28 | 38.5 | 36.4 |
| 8/4/2011 | 6:00:22 PM | 27.9 | 13.50 | 0.441 | 3.27 | 38.6 | 36.4 |
| 8/4/2011 | 7:00:54 PM | 27.3 | 13.48 | 0.440 | 3.26 | 38.5 | 36.2 |
| 8/4/2011 | 8:01:25 PM | 26.3 | 13.43 | 0.433 | 3.23 | 38.2 | 35.8 |
| 8/4/2011 | 9:01:54 PM | 24.9 | 13.36 | 0.433 | 3.24 | 37.7 | 35.2 |
| 8/4/2011 | 10:02:17 PM | 23.2 | 13.22 | 0.438 | 3.31 | 36.9 | 34.2 |
| 8/4/2011 | 11:02:41 PM | 22.4 | 13.09 | 0.437 | 3.33 | 36.2 | 33.3 |
| 8/5/2011 | 12:03:04 AM | 21.7 | 12.99 | 0.439 | 3.38 | 35.5 | 32.6 |
| 8/5/2011 | 1:03:27 AM | 20.9 | 12.96 | 0.440 | 3.39 | 35.1 | 32.0 |
| 8/5/2011 | 2:03:49 AM | 20.2 | 12.88 | 0.443 | 3.44 | 34.6 | 31.2 |
| 8/5/2011 | 3:04:12 AM | 19.8 | 12.81 | 0.441 | 3.44 | 34.3 | 30.8 |
| 8/5/2011 | 4:04:34 AM | 19.4 | 12.77 | 0.442 | 3.46 | 34.0 | 30.4 |
| 8/5/2011 | 5:04:58 AM | 19.1 | 12.76 | 0.443 | 3.47 | 33.9 | 30.3 |
| 8/5/2011 | 6:05:24 AM | 19.0 | 12.76 | 0.448 | 3.51 | 33.8 | 30.4 |
| 8/5/2011 | 7:05:46 AM | 19.0 | 12.77 | 0.440 | 3.45 | 33.9 | 30.4 |
| 8/5/2011 | 8:06:10 AM | 20.8 | 12.79 | 0.437 | 3.41 | 34.3 | 31.1 |
| 8/5/2011 | 9:06:41 AM | 23.3 | 13.01 | 0.432 | 3.32 | 35.8 | 33.3 |
| 8/5/2011 | 10:07:10 AM | 25.0 | 13.21 | 0.432 | 3.27 | 37.2 | 35.0 |
| 8/5/2011 | 11:07:40 AM | 27.4 | 13.37 | 0.437 | 3.27 | 38.0 | 35.9 |
| 8/5/2011 | 12:08:12 PM | 29.1 | 13.48 | 0.449 | 3.33 | 38.6 | 36.6 |
| 8/5/2011 | 1:08:44 PM | 30.5 | 13.58 | 0.465 | 3.43 | 39.3 | 37.3 |
| 8/5/2011 | 1:55:08 PM | 31.0 | 13.58 | 0.477 | 3.51 | 39.5 | 37.6 |

978-1-4673-1110-6/12 $31.00 © 2012 IEEE

## 10. Summary

This paper documents the concept design and hardware build of a novel economizer based data center test cell with a liquid cooled server rack. The liquid cooled rack includes both air and water cooled server components. The water cooled components include the microprocessor modules and the memory cards. The air-cooled devices include the storage disk drives, the power supplies, and the surface mounted components on the printed circuit board. The design was validated using test chamber experiments under highly controlled conditions. The water cooled components are cooled using cold plates on the microprocessor modules and using conduction spreaders attached to the memory cards which bolt to cold rails that have water flowing through them. The air cooled components are cooled using fans inside the server. The exhaust air is cooled using a Side Car air-to-liquid heat exchanger that is integrated into the rack to cool the re-circulating server air flow. The partially water cooled server node was designed to accept as high as 45 °C water and 50 °C air into the node. The data center design is comprised of the 100% water cooled rack, a buffer coolant conditioning unit with a liquid-to-liquid heat exchanger to separate the internal and external liquid loops, and an outdoor dry cooler unit to reject the rack heat load into the outside environment. A 22 hour experiment was conducted with a full rack of servers which yielded an extremely energy-efficient data center loop with only 3.5% of the rack power being used for cooling at the data center level on a relatively hot New York summer day.

The anticipated benefits of such energy-centric configurations are significant energy savings at the data center level of approximately 25%, which represents greater than 90% reduction in the cooling energy usage compared to conventional refrigeration based systems. For a typical 1 Megawatt data center this would represent a savings of roughly $90-$240k per year at an energy cost of $0.04 - $0.11 per kWh. The prototype DELC cooling technology being characterized in this program will be evaluated to see how these developments may be incorporated into a portfolio of leading edge energy efficient technologies.

## Acknowledgments

This project was supported in part by the U.S. Department of Energy's Industrial Technologies Program under the American Recovery and Reinvestment Act of 2009, award no. DE-EE0002894. Technical contribution and insight from George Manelski and Pat Coico of the IBM Systems & Technology Group is very much appreciated with respect to their work in the data center and rack cooling hardware build. We also thank the IBM Research project managers, James Whatley and Brenda Horton and DOE Project Officer Debo Aichbhaumik, DOE Project Monitors Darin Toronjo and Chap Sapp and DOE HQ Contact Gideon Varga for their support throughout the project. The authors acknowledge technical help from Robert Rainey, Michael Ellsworth, Corey Vandeventer, James Steffes, Mark Steinke and Gerry Weber of the IBM Systems and Technology Group and last but not the least, our 2010 summer co-op, Dan Simco.

## References

1. ASHRAE, "Thermal Guidelines for Data Processing Equipment – second edition", 2009, available from http://tc99.ashraetcs.org/
2. Vision and Roadmap – Routing Telecom and Data Centers Toward Efficient Energy Use, May 13, 2009, www1.eere.energy.gov/industry/datacenters/.../vision_and_roadmap.
3. ASHRAE book, "Best Practices for Datacom Facility Energy Efficiency", Second Edition.
4. Lawrence Berkeley National Labs, 2007, Benchmarking Data Centers – Charts", http://hightech.lbl.gov/benchmarking-dc-charts.html.
5. M. Iyengar, R. Schmidt, and J. Caricari, 2010, "Reducing Energy Usage in Data Centers Through Control of Room Air Conditioning Units", Proceedings of the IEEE ITherm Conference in Las Vegas, USA, May.
6. M. Ellsworth, L. Campbell, R. Simons, M. Iyengar, R. Chu, and R. Schmidt, 2008, "The Evolution of Water Cooling for IBM Large Server Systems: Back to the Future", Proceedings of the IEEE ITherm Conference in Orlando, USA, May.
7. R. Schmidt, M. Iyengar, D. Porter, G. Weber, D. Graybill, and J. Steffes, 2010, "Open Side Car Heat Exchanger that Removes Entire Server Heat Load Without any Added Fan Power", Proceedings of the IEEE ITherm Conference, Las Vegas, June.
8. R. Chu, M. Iyengar, V. Kamath, and R. Schmidt, 2010, "Energy Efficient Apparatus and Method for Cooling an Electronics Rack", US Patent 7791882 B2.
9. Typical year hour by hour weather data available on website of the US National Renewable Energy Lab (NREL).

# Experimental Investigation of Water Cooled Server Microprocessors and Memory Devices in an Energy Efficient Chiller-less Data Center

*Pritish R. Parida[1], Milnes David[2], Madhusudan Iyengar[2], Mark Schultz[1], Michael Gaynes[1], Vinod Kamath[3], Bejoy Kochuparambil[3] and Timothy Chainer[1]

[1] IBM T. J. Watson Research Center, Yorktown Heights, NY
[2] IBM Systems & Technology Group, Poughkeepsie, NY
[3] IBM Systems & Technology Group, Raleigh, NC
*prparida@us.ibm.com

## Abstract

Understanding and improving the thermal management and energy efficiency of data center cooling systems is of growing importance from a cost and sustainability perspective. Toward this goal, warm liquid cooled servers were developed to enable highly energy efficient chiller-less data centers that utilize only "free" ambient environment cooling. This approach greatly reduces cooling energy use, and could reduce data center refrigerant and make up water usage. In one exemplary experiment, a rack having such liquid cooled servers was tested on a hot summer day (~32 ℃) with CPU exercisers and memory exercisers running on every server to provide steady heat dissipation from the processors and from the DIMMs, respectively. Compared to a typical air cooled rack, significantly lower DIMM temperatures and CPU thermal values were observed.

## Keywords

Data-center, Chiller-less data centers, energy efficient, liquid cooled servers.

## Nomenclature

BMC    -    Baseboard Management Controller
CPU    -    Central Processing Unit
DIMM   -    Dual Inline Memory Module
DTS    -    Digital Thermal/Temperature Sensor
GPM    -    Gallons per minute
HVAC   -    Heating Ventilation and Air Conditioning
IPMI   -    Intelligent Platform Management Interface
LPM    -    Liter per minute
MWU    -    Modular Water Unit
PECI   -    Platform Environment Control Interface
RPM    -    Revolutions per minute

## 1. Introduction

Exponential growth in the demand for data processing and storage continues to stimulate rapid growth in the U.S. data center industry. In 2005, server driven power usage amounted to 1.2% of total US energy consumption [1]. Over the past six years, energy use by these centers and their supporting infrastructure is estimated to have increased by nearly 100 percent [2]. In the face of growing global energy demand, uncertain energy supplies, and volatile energy prices, innovative solutions are needed to radically advance the energy efficiency of these data center systems. Information Technology (IT) equipment usually consumes about 45-55% of the total electricity in a data center, and total cooling energy consumption is roughly 25-30% of the total data center energy use [2, 3]. In addition to significant energy usage, traditional data center cooling also results in refrigerant and make up water consumption. Thus, understanding and improving the thermal management and energy efficiency of data center cooling systems is of growing importance from a cost and sustainability perspective.

Only recently, efforts have been made to identify the sources of energy inefficiencies and to figure out best practices to improve the efficiencies of these capital and energy intensive data centers [3-8]. Efficiency improvements naturally center on these four highly inter-dependent areas – (i) IT equipment and software, (ii) power supply chain and/or back up power, (iii) cooling and, (iv) overall data center design including allowable operational temperature and humidity ranges for IT equipment [2-5]. Energy efficiency gains that may be achieved in the IT equipment and software will decrease the demand for power, and therefore, the demand for cooling. Also, a larger temperature and humidity operating range would enable longer hours of water/air side economizer mode operation with ambient "free cooling" resulting in reduced cooling power consumption. In terms of cooling design improvement, using liquid at the rack or server level is a far more efficient method of transferring concentrated heat loads than using air, due to much higher volumetric specific heats and higher heat transfer coefficients.

For the present study, warm-liquid cooling infrastructure for the server components was developed along with a dual enclosure air/liquid cooling system to allow "free" cooling from the outdoor ambient environment. A rack having 38 such liquid cooled servers was tested on a hot summer day (~32 ℃) with CPU exercisers and memory exercisers running on each server to provide steady heat dissipation from the processors and from the DIMMs, respectively. This paper discusses the server thermal data for that run. Moreover, a one to one comparison with a typical air cooled server was performed to quantify the benefit in terms of improvement in server electronics temperatures and server power consumption.

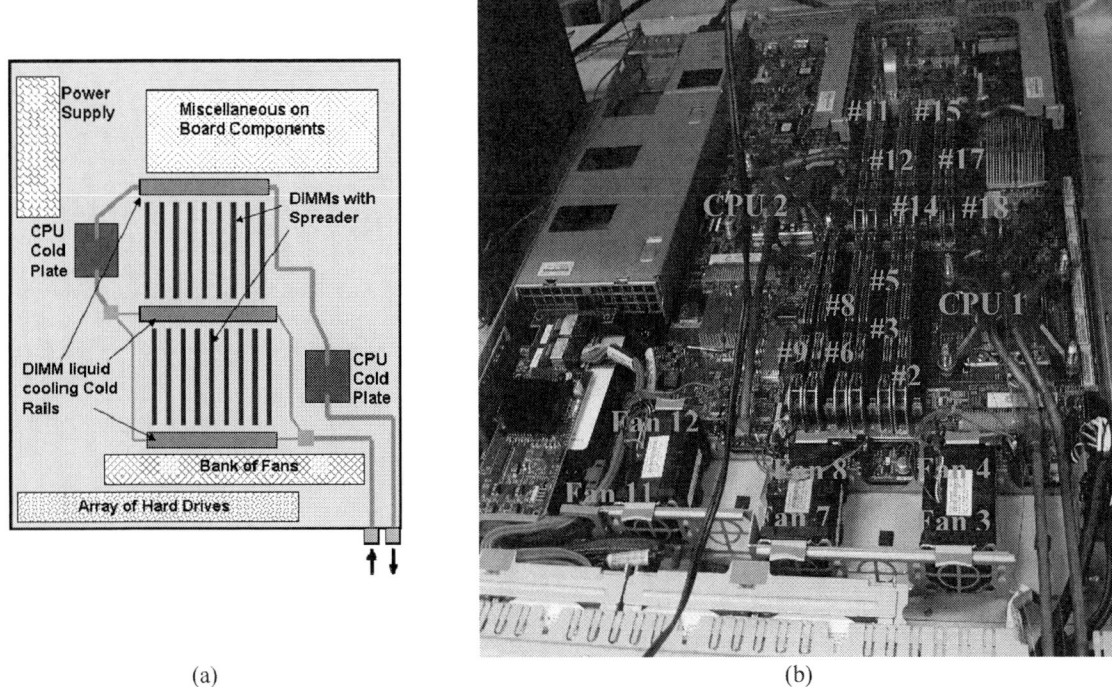

(a)                                      (b)

Figure 1. (a) Schematic of the volume server with node liquid cooling loop and other server components. (b) Node liquid cooling loop, having liquid cooling components for both the processors (CPU 1 and CPU 2) and the 12 DIMMs (numbered 2 through 18), installed in an IBM System X volume server.

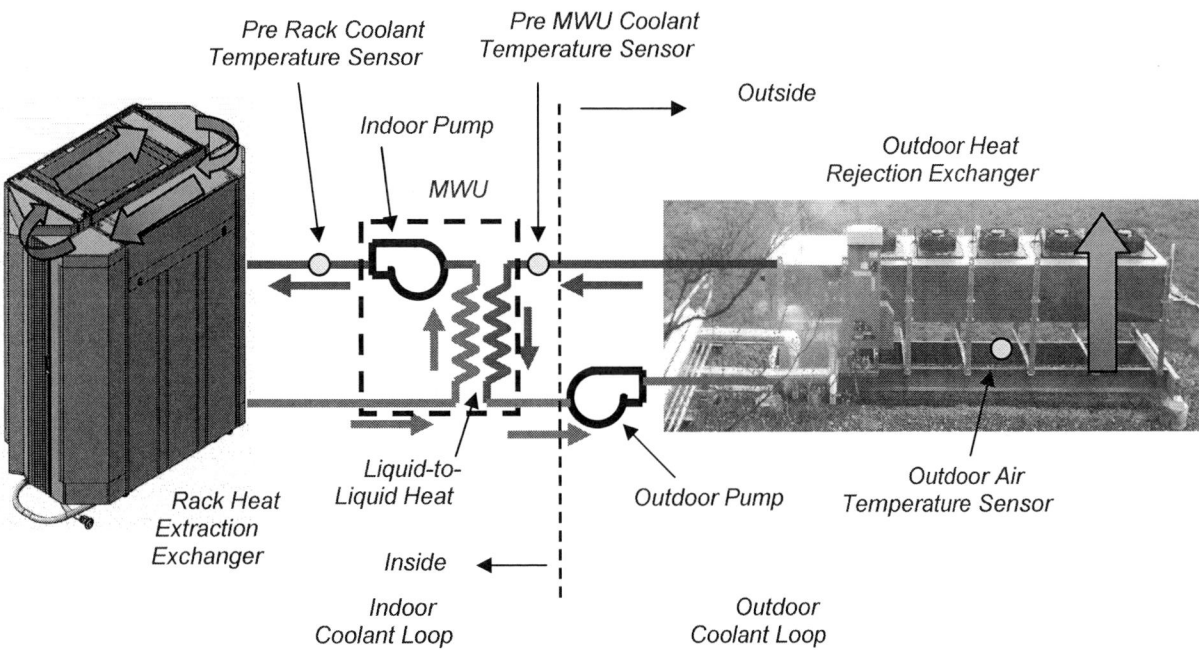

Figure 2. Schematic of the chiller-less data center liquid cooling design.

## 2. Data Center Liquid Cooling Design

Warm liquid cooled servers as shown in Figure 1 were developed to enable highly energy efficient chiller-less data centers that utilize only "free" ambient environment cooling. This approach greatly reduces cooling energy use, and could reduce data center refrigerant and makeup water use.

978-1-4673-1110-6/12 $31.00 © 2012 IEEE          225

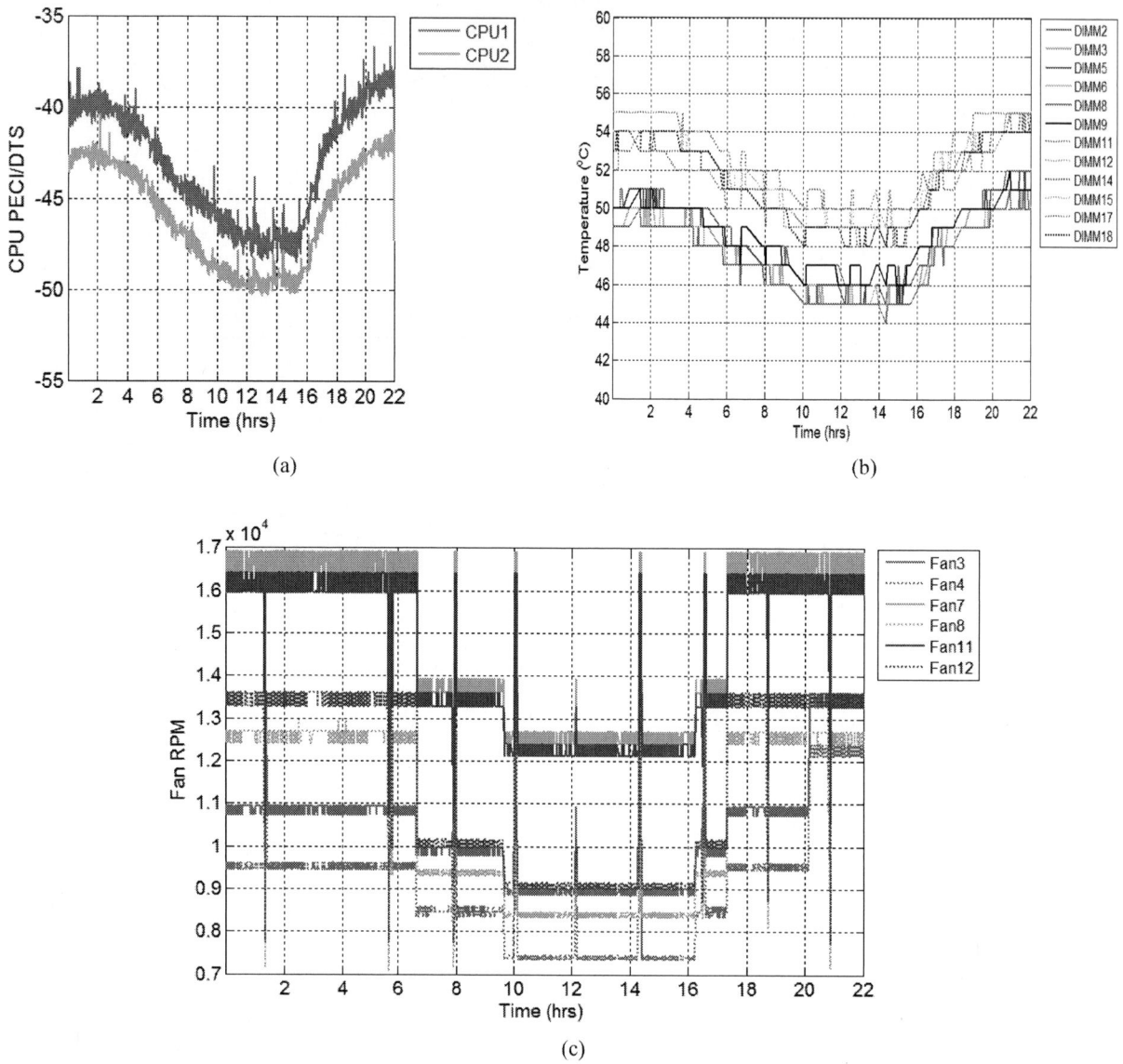

Figure 3. Server component data showing (a) hottest core PECI/DTS numbers for CPU 1 and CPU 2, (b) DIMMs temperature for each of the 12 DIMMs and (c) system fans rpm for one of the server from a sample 22 hours run.

Figure 1 (b) shows an IBM System X 3550 M3 1U volume server with a node liquid cooling loop having liquid cooling components for both the micro-processors and the Dual In-line Memory Modules (DIMMs). The microprocessor modules are cooled using cold plate structures while the DIMMs are cooled by attaching them to a pair of conduction spreaders which are then bolted to a cold rail that has water flowing through it. The loops were designed, modeled and characterized using computational fluid dynamics modeling tools. Every server consisted of two, six core 3.3 GHz micro-processors and 12 8GB DDR3 DIMMs from two memory suppliers which are herein referred to as Supplier1 and Supplier2. DIMMs were installed in slot numbers 2, 3, 5, 6, 8, 9, 11, 12, 14, 15, 17 and 18. The microprocessors and DIMMs have a typical maximum power of 130 W and 6 W, respectively. The modifications to the server included removing three of the

six server fan packs resulting in less IT power at elevated temperature operation. Each server fan pack consists of two coaxial fans – one counter clockwise-rotating and the other clockwise-rotating. Other server components such as the power supply, hard-disk drives and other miscellaneous components were air-cooled. These partially water cooled servers were designed to accept as high as 45 °C water and 50 °C air into the node. The modified servers were installed in a Rack Heat Extraction Exchanger (server rack with liquid cooling manifolds and Side Car air-to-liquid heat exchanger) to completely remove heat at the rack level either by direct thermal conduction or indirect air to liquid heat exchange. The air flow inside the rack enclosure was provided by the server fans. The liquid coolant was circulated between the Rack Heat Extraction Exchanger and an Outdoor Heat Rejection Exchanger to move the heat from the System X servers to the outdoor ambient air environment. A liquid-to-liquid heat exchanger was used to transfer the heat from the indoor Rack Heat Extraction Exchanger loop to the Outdoor Heat Rejection Exchanger loop. Further details on this concept data center design and associated hardware build can be found in the reference articles [9-10].

Figure 2 shows the schematic of the chiller-less data center liquid cooling design that was developed as a part of thisstudy. The Rack Heat Extraction heat exchanger along with the Modular Water Unit or MWU (liquid-to-liquid heat exchanger and one of the pumps) was installed inside the building. The outdoor heat rejection unit and the other pump were installed outside the building. Since the refrigeration based components are completely eliminated in the present approach, the air and the water temperatures entering the server are closely related to the outdoor ambient conditions.

Different workloads such as CPU exerciser, Memory exerciser and Linpack were executed on the servers to provide continuous and steady heat dissipation from the processors and DIMMs and to characterize the system performance. Component information such as processor PECI/DTS value (Platform Environment Control Interface / Digital Thermal Sensor value – which indicates the difference between the current processor core temperature and maximum junction temperature) [11], DIMM temperatures, system fan rpm and other such information

was collected using the IPMI (Intelligent Platform Management Interface) and BMC (Baseboard Management Controller) tools.

## 3. Day Long Operation of Data Center Test Facility

The data center test facility was continuously run for a day (~ 22 hours) with varying outdoor heat rejection exchanger fan speeds and internal and external loop coolant flow rates set to 7.2 GPM (27.2 LPM) and 7.1 GPM (27 LPM) respectively. The outdoor heat exchanger fans were programmed to linearly vary in speed from 169 RPM to 500 RPM as the pre-MWU temperature varied from 30°C to 35°C. For pre-MWU temperatures below 30°C the fans run at a constant speed of 169 RPM.

Figure 3 shows (a) the hottest core PECI/DTS values for each CPU, (b) maximum DIMM temperature for each of the 12 DIMMs and (c) the rpm of the system fans for one of the servers during the sample 22 hours run that began and ended in the afternoons of successive days. Observations such as variation of 5-6 °C in the DIMM temperatures and CPU 2 running relatively cooler than CPU 1, were as expected based on the computational fluid dynamic simulations of the cooling loop. The rpm of the server fans changes predominantly based on the server inlet air temperature. The more normal rpm changes due to load driven processor temperature rise were eliminated as even under full power the processors were running below the temperatures which would normally cause processor driven fan rpm increases.

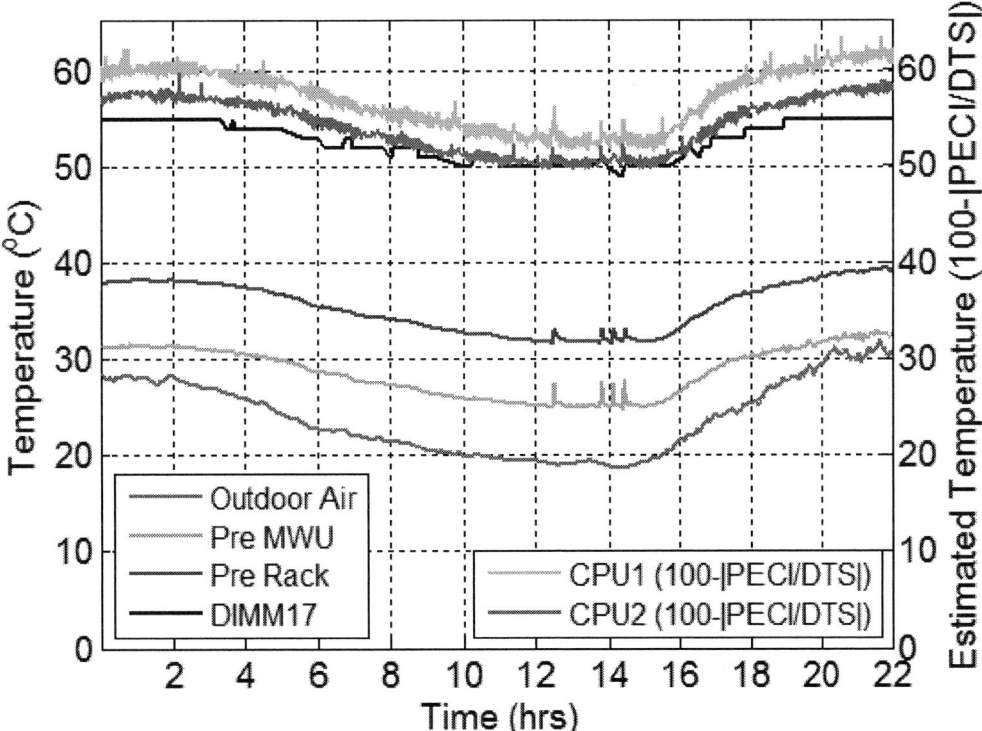

Figure 4. Variation of temperature from the outdoor air to the server components.

978-1-4673-1110-6/12 $31.00 © 2012 IEEE

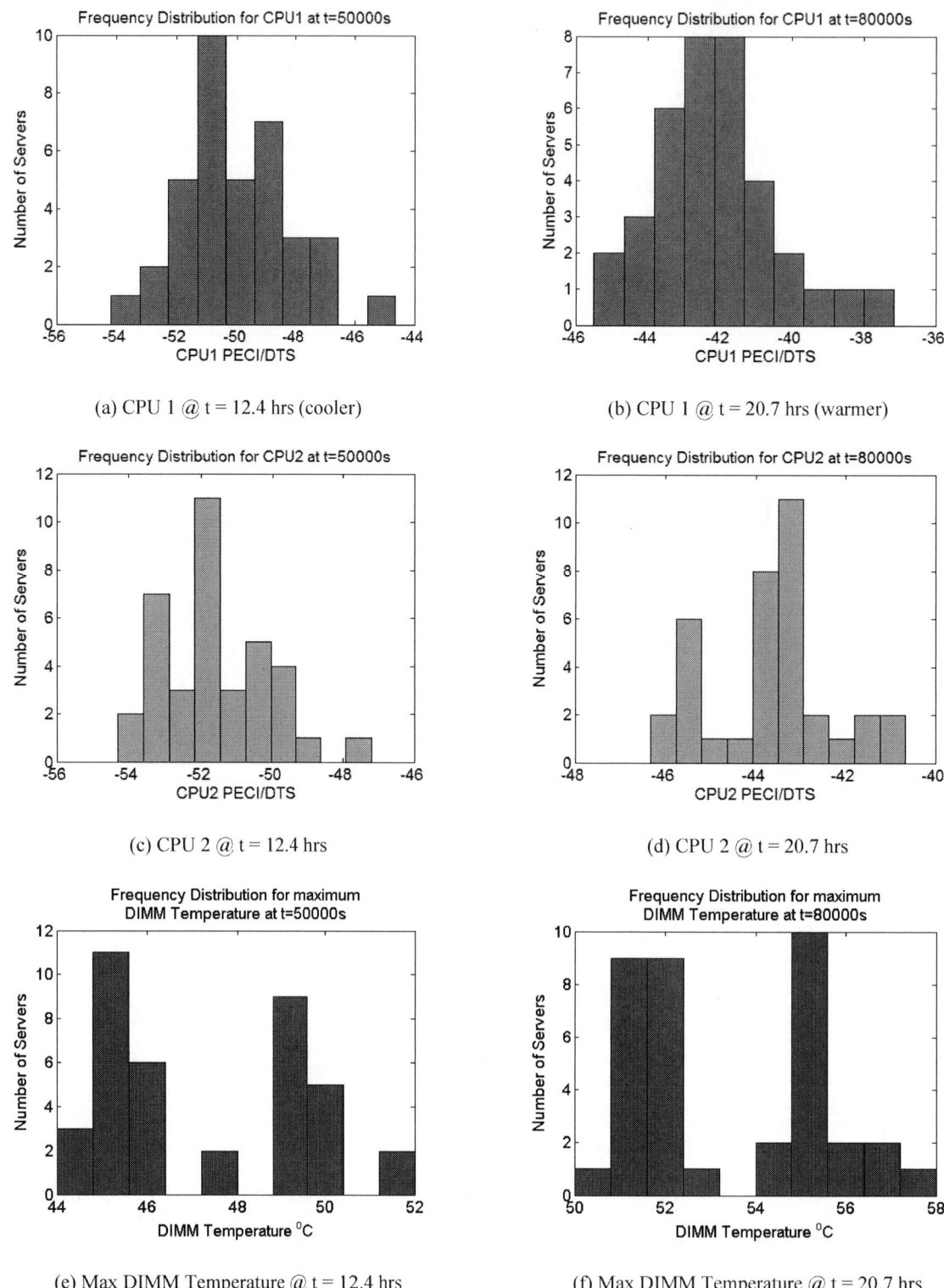

Figure 5. Frequency distribution of CPU 1 and CPU 2 PECI/DTS numbers and maximum DIMM temperatures at t = 12.4 hrs (cooler) and t = 20.7 hrs (warmer) from the 22 hours test run.

Figure 4 shows the outdoor air temperature, the pre-MWU and pre-Rack coolant temperatures for the same 22-hour run. It is to be noted here that because the internal and external loop coolant flow rates are kept constant through the sample run, the temperature delta between the pre-

MWU and pre-Rack temperature remains constant. Also, when the pre-MWU temperature is less than 30 °C, the outdoor heat exchanger fans run at constant rpm causing the temperature delta between the outdoor ambient temperature and the pre-MWU temperature to remain constant. However, when the pre-MWU exceeds 30 °C the outdoor heat exchanger fans starts to ramp up causing a drop in the temperature delta between the outdoor air temperature and the pre-MWU temperature. Hence, over the duration where the pre-MWU temperature is less than 30 °C, temperatures at all the locations of the cooling system and of the cooled electronics (that is, the pre-MWU, the pre-Rack, microprocessors junction temperature, DIMMs temperature, etc.) follow the outdoor ambient temperature profile at an essentially fixed offset.

Figure 4 also shows the hottest DIMM temperature (DIMM 17 for this server) and the hottest core estimated temperature for each CPU. In the absence of a direct calibration between PECI/DTS values and absolute temperature, we choose to approximate the hottest CPU core temperature as 100 minus the absolute value of the PECI/DTS number. There were 38 servers in the rack with CPU exercisers and memory exercisers running on every server to provide steady heat dissipation from the processors and from the DIMMs. Average PECI/DTS for the hottest core in CPU 1 was -43.5 with the max/min values of -36.7/-50.5. Average hottest DIMM (#17 for this server) temperature was 53 °C with the max/min values of 55 °C/50 °C. All the other servers in the rack showed similar temperatures, PECI/DTS values and fan rpm profiles.

From Figure 4, it can also be seen that the minimum temperature occurs around 12.4 hours and the maximum temperature occurs around 20.7 hours. Frequency distributions of the CPU PECI/DTS numbers and maximum DIMM temperatures at these time instances were evaluated and are presented in Figure 5. The mean maximum CPU1 core PECI/DTS number at time = 12.4 hrs was -50 and at 20.7 hrs was -42.1 with a standard deviation of 1.92 and 1.74 respectively. The mean maximum CPU2 core PECI/DTS number at 12.4 hrs was -51.6 and at 20.7 hrs was -43.7 with a standard deviation of 1.53 and 1.36 respectively. The variability in the PECI/DTS numbers can be attributed to the general variability in the performance of each core in a micro-processor. PECI/DTS numbers of each core of each processor were also recorded and evaluated to characterize this core-to-core and processor-to-processor variability. The mean maximum DIMM temperature at 12.4 hrs was 47.2 °C and at 20.7 hrs was 53.4 °C. Note that the variability in the DIMM temperatures is mainly due to the different types of DIMMs. All the servers that reported relatively cooler DIMMs had 8GB DDR3 DIMMs from Supplier 1 while all the servers that reported relatively warmer DIMMs had 8GB DDR3 DIMMs from Supplier 2. This is consistent with the observation that Supplier 1 DIMMs dissipate less heat than Supplier 2 DIMMs for similar performance.

Overall, the system was operated for 22 hours at an average outdoor temperature of 23.8 °C and max/min temperatures of 32 °C and 19 °C. The average IT power was 13.14 kW and average cooling power was 0.44 kW or roughly 3.5% of the IT power. In comparison, HVAC air cooled data centers can have cooling powers of 60% of IT power [3]. Typical outdoor air dry bulb temperature distribution for a number of US cities such as Poughkeepsie, NY can be found at the NREL website [12]. According to this distribution, the outdoor air temperature in Poughkeepsie, NY stays below 25 °C for more than 93% of the year. Upon extrapolating the current result from the 22 hour run where the average outdoor temperature was 23.8 °C, a data center in Poughkeepsie, NY based on the present approach could potentially be cooled using less than 3.5% of the IT Power.

## 4. Comparison with Typical Air Cooled Server

Although the above results validate the well known fact that liquid cooling at the server level is a far more efficient method of transferring concentrated heat loads than air, a one to one comparison with a typical air cooled server is required to quantify the benefit in terms of improvement in temperatures and server power consumption. For that purpose, the thermal data of one of the liquid cooled nodes was compared against the thermal data of its air cooled version. The water flow rate through the liquid cooled servers was maintained at 0.7 lpm.

Figure 6 shows the comparison of the estimated junction temperature of the two processors for a liquid cooled server with an air cooled server cooled by air at 22 °C. For the liquid cooled server, two cases were considered – one with 45 °C server inlet water temperature (and 50 °C server inlet air temperature) and the other with 25 °C server inlet water temperature (and 30 °C server inlet air temperature). Figure 6(a) summarizes the estimated junction temperature comparison when each processor was exercised at 90% while Figure 6(b) summarizes the estimated junction temperature comparison when the memory exerciser was executed. In both the cases, liquid cooled microprocessors showed much lower junction temperatures even with warm liquid coolant. Note that the 45 °C water temperature might be an extreme condition for many parts of the world and even for that condition the microprocessors were at least five PECI/DTS units cooler than the typical air cooled servers. For the memory exerciser case, although the heat dissipation from the processors is not much, the difference in the estimated junction temperature is higher. This is because the server fans are running at a lower rpm consuming lower power (see Figure 8(b)).

Figure 7 shows the comparison of the DIMM temperatures for the liquid cooled server with an air cooled server cooled by air at 22 °C. Here again, 25 °C and 45 °C server inlet water temperature cases were considered. Figure 7(a) summarizes the DIMM temperature comparison when each processor was exercised at 90% while Figure 7(b) summarizes the DIMM temperature comparison when the memory exerciser was executed. Note that the DIMMs in slots 2, 3, 5, 6, 8 and 9 are closer to the fans and are cooled by relatively cooler air while the DIMMs in slots 11, 12, 14, 15, 17 and 18 are away from the fans and are cooled by relatively warmer air due to preheat from DIMMs in the front bank. When only the processors are exercised (Figure 7(a)), the heat dissipation from the DIMMs is very small and thus the DIMM temperatures are closer to the server inlet air temperature (for air cooled server) or server inlet water temperature (for liquid cooled servers). In such cases, the benefit of going to liquid cooling for the DIMMs is negligible. However, when the memory modules are exercised, the benefit of going to liquid cooling becomes prominent. In some cases, DIMMs of a warm liquid cooled server might show lower temperatures than those shown by the DIMMs of a typical air cooled server.

Figure 8 shows the comparison of the server power and server fan power consumption for the liquid cooled server with an air cooled server cooled by air at 22 °C. Here again, 25 °C and 45 °C server inlet water temperature cases were considered. Figure 8(a) summarizes the server power

and server fan power comparison when each processor was exercised at 90% while Figure 8(b) summarizes the server power and server fan power consumption comparison when the memory exerciser was executed. Figure 8(a) shows that the total server power goes up when the server is cooled with a 45 °C warm water and 50 °C server inlet air. Most of this increase in the power is due to the increased power consumption by the server fans as the server sees a 50 °C inlet air temperature. If we

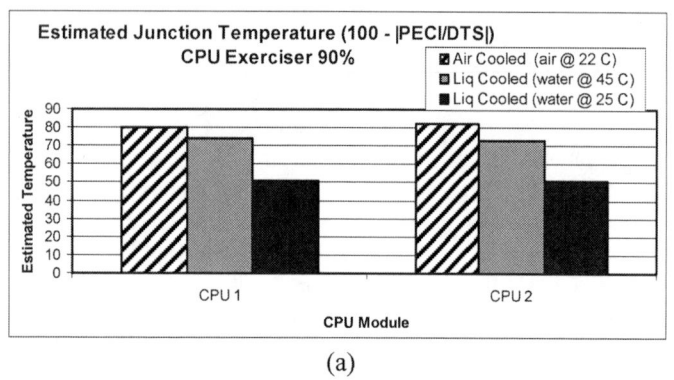

(a)

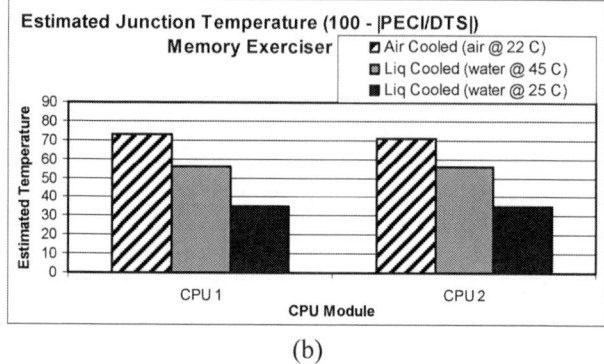

(b)

Figure 6. Comparison of estimated junction temperature for a liquid cooled server with a typical air cooled server (a) when the CPUs are exercised at 90% and (b) when the memory modules are exercised.

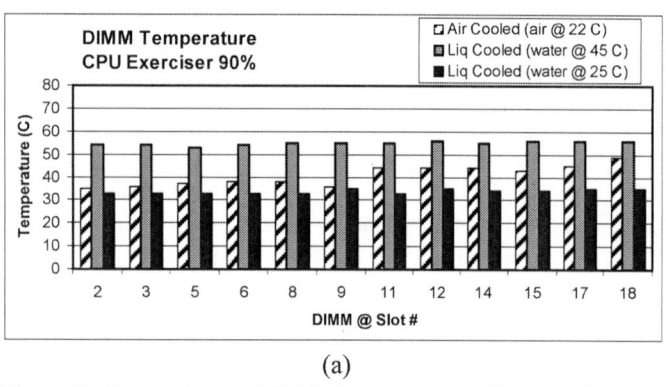

(a)

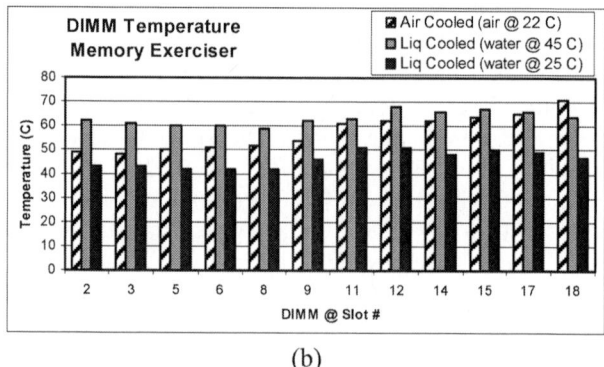

(b)

Figure 7. Comparison of DIMM temperatures for a liquid cooled server with a typical air cooled server (a) when the CPUs are exercised at 90% and (b) when the memory modules are exercised.

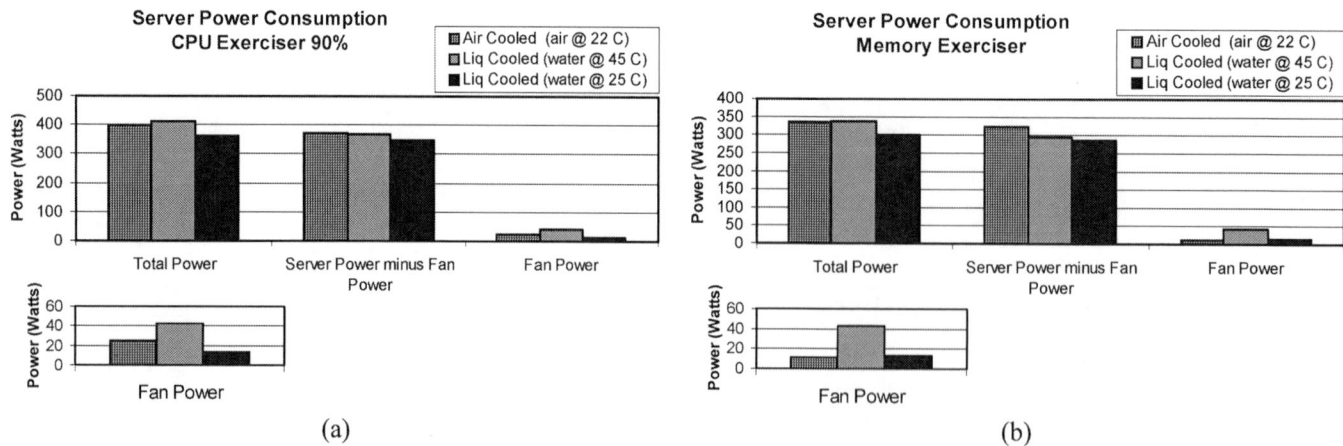

(a)

(b)

Figure 8. Comparison of server power and fan power consumption for a liquid cooled server with a typical air cooled server (a) when the CPUs are exercised at 90% and (b) when the memory modules are exercised.

978-1-4673-1110-6/12 $31.00 © 2012 IEEE

subtract that fan power from the total power, we see that the power consumed by the server electronics is lower than that consumed by the server electronics of a typical air cooled server. This reduction in the server electronics power consumption becomes more prominent for the 25 °C water cooled server where a more than 6% reduction in power consumption was observed. This reduction in power could possibly be due to the reduction in leakage power as the liquid cooled electronics were running at much lower temperatures.

For the memory exerciser case, this reduction in power consumption was observed to be greater than 11% where the improvement in estimated junction temperature was ~40 units. Between the 25 °C and 45 °C inlet water temperature cases, this reduction was greater than 5.5% and 2.5% for the CPU exerciser and memory exerciser cases respectively.

In summary, liquid cooling the servers does provide a significant benefit in terms of lower server electronics temperatures as well as in terms of lower server electronics power consumption. Thus, by going to liquid cooling IT power can be reduced along with the significant reduction in cooling power.

## 5. Conclusions

Warm liquid cooled servers were developed to enable highly energy efficient chiller-less data centers that utilize only "free" ambient environment cooling. This approach greatly reduces cooling energy use, and could reduce data center refrigerant and make up water usage. A rack having 38 such liquid cooled servers was tested on a hot summer day (~32 °C) with CPU exercisers and memory exercisers running on every server to provide steady heat dissipation from the processors and from the DIMMs. Significantly lower processor PECI/DTS values ~ -40 to -55 (that is, 40 to 55 units away from the maximum junction temperature) and DIMM temperatures ~ 45-60 °C were observed. Additionally, a one to one comparison with a typical air cooled server showed that by liquid cooling the servers IT power can be reduced along with the significant reduction in cooling power.

The anticipated benefits of such energy-centric configurations are significant (~25%) energy savings at the data center level which represents a greater than 90% reduction in the cooling energy usage compared to conventional refrigeration based systems. For a typical one megawatt data center this would represent a savings of roughly $90-$240k/year at an energy cost of $0.04 - $0.11 per kWh. The prototype dual-enclosure liquid-cooling (DELC) technology being characterized in this program will be evaluated to see how to incorporate these developments into a portfolio of leading edge energy efficient technologies.

## Acknowledgments

This project was supported in part by the U.S. Department of Energy's Industrial Technologies Program under the American Recovery and Reinvestment Act of 2009, award number DE-EE0002894. Authors of this article would like to thank Robert Rainey from IBM STG Raleigh, NC for his help in extracting the thermal data from the servers. The authors would also like to thank the IBM Poughkeepsie Site and Facilities team for their engineering help in the construction of the data center facility. Technical contribution and insight from David Graybill, Daniel Simco, Robert Simons and George Manelski of IBM STG Poughkeepsie, NY is very much appreciated with respect to their work in the data center and node and rack cooling hardware build. We also thank the IBM Research project managers, James Whatley and Brenda Horton and DOE Project Officer Debo Aichbhaumik, DOE Project Monitors Darin Toronjo and Chap Sapp and DOE HQ Contact Gideon Varga for their support throughout the project.

## References

1. Brown, et. al., "Report to Congress on Server and Data Center Energy Efficiency", Public Law 109-431, U.S. Environmental Protection Agency, ENERGY STAR Program, Aug 2, 2007.

2. "Vision and Roadmap: Routing Telecom and Data Centers toward Efficient Energy Use", US Dept. of Energy, May 13, 2009.

3. Greenberg, et. al., " Best Practices for Data Centers: Lessons Learned from Benchmarking 22 Data Centers", ACEEE Summer Study on Energy Efficiency in Buildings, 2006.

4. ASHRAE, "Thermal Guidelines for Data Processing Environments – Expanded Data Center Classes and Usage Guidance", TC 9.9 Mission Critical Facilities, 2011.

5. ASHRAE Datacom Series 6, "Best Practices for DataCom Facility Energy Efficiency," 2$^{nd}$ Edition, Altanta, 2009.

6. LBNL, "High-Performance Buildings for High-Tech Industries, Data Centers". Berkeley, Calif.: Lawrence Berkeley National Laboratory, 2006, http://hightech.lbl.gov/datacenters.html.

7. Moss et. al., "Chiller-less Facilities: They May Be Closer Than You Think", Dell technical white paper, 2011, http://content.dell.com/us/en/enterprise/d/business~solutions~wh itepapers~en/Documents~chillerless-facilities-white-paper.pdf.aspx.

8. W. Tschudi, "Best Practices Identified Through Benchmarking Data Centers," Presentation at the ASHRAE Summer Conference, Quebec City, Canada, June, 2006.

9. M. Iyengar, M. David, P. Parida, V. Kamath, B. Kochuparambil, D. Graybill, M. Schultz, M. Gaynes, R. Simons, R. Schmidt and T. Chainer, "Server Liquid Cooling with Chiller-less Data Center Design to Enable Significant Energy Savings", IEEE SEMITherm Conference, 2012.

10. M. David, M. Iyengar, P. Parida, R. Simons, M. Schultz, M. Gaynes, R. Schmidt, T. Chainer, 2012, "Experimental Characterization of an Energy Efficient Chiller-less Data Center Test Facility with Warm Water Cooled Servers", IEEE SEMITherm Conference, 2012.

11. Michael Berktold and Tian Tian, "CPU Monitoring with DTS/PECI", White paper, http://download.intel.com/design/intarch/papers/322683.pdf

12. "National Solar Radiation Database: Typical Meteorological Year", http://rredc.nrel.gov/solar/old_data/nsrdb/1991-2005/tmy3/.

# Experimental Characterization of an Energy Efficient Chiller-less Data Center Test Facility with Warm Water Cooled Servers

*[1]Milnes P. David, [1]Madhusudan Iyengar, [2]Pritish Parida, [1]Robert Simons,
[2]Mark Schultz, [2]Michael Gaynes, [1]Roger Schmidt, [2]Timothy Chainer
[1]IBM Systems & Technology Group, Poughkeepsie, NY, USA
[2]IBM Watson Research Center, Yorktown Heights, NY, USA
*milnespd@us.ibm.com

## Abstract

Typical data centers utilize approximately 50% of the total IT energy in cooling of the server racks. We present a chiller-less data center where server-level cooling is achieved through a combination of warm water cooling hardware and re-circulated air; eventual heat rejection to ambient air is achieved using a closed secondary liquid loop to ambient-air heat exchanger (dry-cooler). Several experiments were carried out to characterize the individual pieces of equipment and data center thermal performance and energy consumption. A 22+ hour experimental run was also carried out with results indicating an average cooling energy use of 3.5% of the total IT energy use, with average ambient air temperatures of 23.8°C and average IT power use of 13.14 kW.

## Keywords

Liquid cooled servers, data center, energy efficiency

## 1. Introduction

Data center energy use and energy efficiency are key issues in today's economic and environmental backdrop and are predicted to play a growing role over the coming decades as information technology (IT) energy use begins to rival commercial, residential and industrial energy consumption [1,2]. Typical data centers utilize 25%-30% of the total data center energy (or ~50% of the IT energy) in cooling of the IT server racks with much of this energy being consumed by the computer room air conditioning (CRAC) units and the chiller plant to provide suitably cooled and conditioned air to the IT racks [1,3-5]. Heat rejection is ultimately achieved at the wet cooling tower where the heated facility water is evaporatively cooled and extra water added to replace that lost via evaporation.

A key approach to reducing the total cooling energy use is to eliminate or reduce the need for both CRAC units and chiller plants by: i) using liquid cooling at the server and/or rack level and ii) using water/air side economizers with ambient 'free-air' cooling [6]. The use of liquid cooling allows the heat generated at the servers to be more efficiently transported from the heat source to the sink as compared to air thus reducing energy waste due to the need to provide chilled and conditioned air to an entire room. The use of water or air-side economization enables the use of ambient external conditions to directly provide the necessary cooling without intermediate refrigeration steps. This is particularly attractive in regions where the weather is cool. Use of chiller-less data center cooling in hotter regions or during the hottest parts of the year would require additional evaporative cooling techniques.

Both liquid cooling and economizer based data centers have been independently investigated and presented in the literature. Ellsworth et al. recently documented the water cooling of an IBM Power 775 (P7-IH) Supercomputing system where in excess of 96% of the 180 kW of rack heat load is taken up by the water [7]. Previous work by Ellsworth and Iyengar into the water cooled IBM Power 575 (P6-IH) also determined that the power required to transfer the dissipated heat to the ambient was 45% less for a liquid cooled system [8]. Wei documents the performance of a hybrid liquid and air cooled Fujitsu high end server GS8900 where 60% of the heat dissipated is absorbed by the liquid and the remaining exhausted into the room [9]. Apart from the reduced heat dissipation into the computer room, the use of liquid cooling reduced the junction temperatures and increased performance by 10%. This cooling hardware is now implemented in the Fujitsu K-computer complex which, combined with other system improvements, results in a highly efficient 830 MFlops/W.

At the data center level, both Yahoo! and Facebook have unveiled air-side economizer based data center designs with air-cooled servers. S. Noteboom, Yahoo's VP of Global Data Centers, presented their work on a new air-side economized data center with evaporative cooling in Buffalo, NY. Results indicate a power usage effectiveness (PUE) of 1.08 and an estimated saving of 36 million gallons of water per year compared to their previous chiller based data center design [10]. Facebook's data center in Prineville, Oregon, features direct air-side economization with an evaporative cooling system. These features combined with improvements in the electrical distribution have led to a PUE ranging between 1.06 and 1.1 [11]. However the use of air cooling at the rack and servers requires the handling, filtering and conditioning of a significant quantity of air, and potentially hot/cold aisle containment.

In this study we propose a chiller-less, ambient air cooled data center where cooling at the rack is achieved through a combination of warm liquid and enclosed re-circulated air cooled servers [12]. Heat is ultimately rejected from the warm facility water to the ambient air through the use of a closed loop liquid-to-air heat exchanger. The proposed design is expected to reduce the cooling energy to less than 5% of the IT energy and eliminates the need for make-up water. Additionally, when the outlet water from the server racks exceeds 40°C there is the potential to recover some of this

978-1-4673-1110-6/12 $31.00 © 2012 IEEE

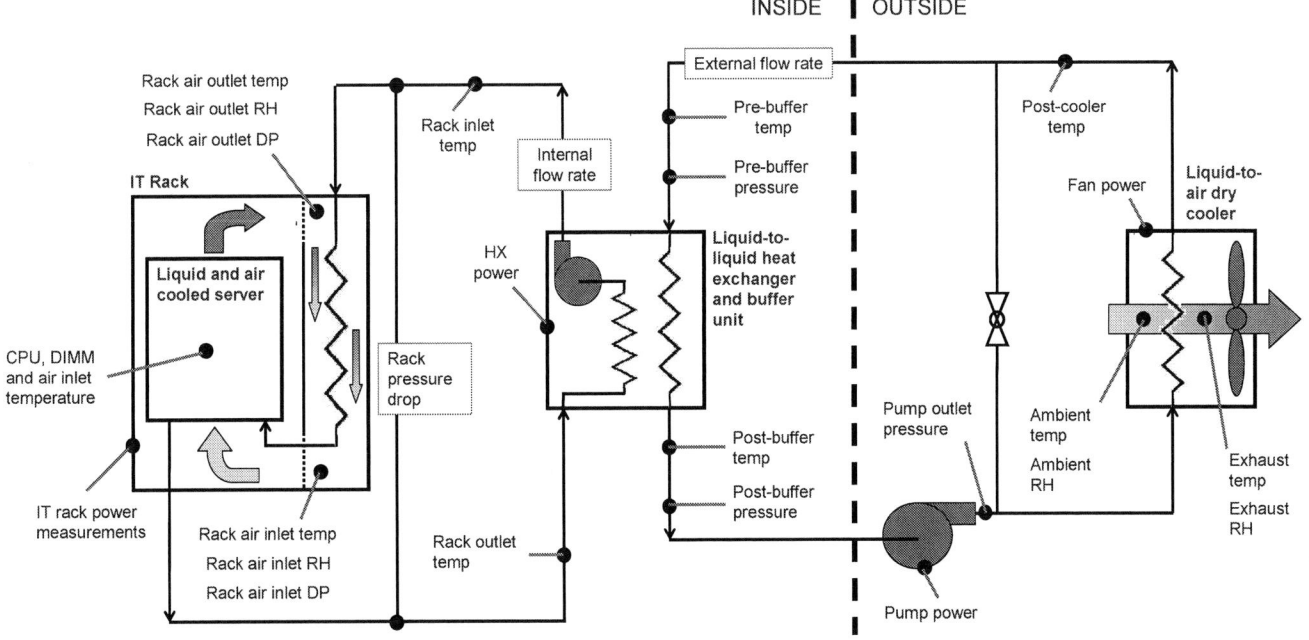

Fig. 1. Schematic representation of the energy efficient data center test facility. IT servers are warm water cooled with the heat ultimately rejected to ambient air via a liquid-to-air heat exchanger.

energy to heat water for use in low grade commercial and residential heating applications. This paper discusses the experimental characterization carried out on the designed energy efficient data center test facility and presents the hydraulic, thermal and power data for the data center at various operational conditions as well as the results obtained for a run carried out over a day with ambient temperatures ranging from 18.6°C to 32°C.

## 2. Description of the Dual Loop Data Center Test Facility

A dual loop data center test facility, shown above in Fig. 1, was constructed to cool a rack of warm water cooled IBM System X volume servers. The liquid cooled servers, shown in Fig. 2, contain cold plates, cold rails and heat spreaders such that the processors and memory modules are primarily water cooled and the remaining server components cooled by re-circulated air. The air is circulated within the rack and is cooled by the incoming water using an air-to-liquid heat exchanger mounted within the rack enclosure. This server and rack arrangement significantly reduces the need for specialized computer room air conditioning units since a large fraction of the heat generated at the servers is absorbed by the liquid. The liquid used in the rack is circulated in an internal loop with heat exchange to a secondary, external loop coolant such as water or water-glycol mixture. The liquid-to-liquid heat exchanger (buffer unit) allows the year-round use of thermally superior water in the internal loop as well as varying the degrees of coolant quality and chemical treatment in the internal and external loops. The external coolant is circulated through a liquid-to-ambient-air heat exchanger (dry-cooler). The closed external loop requires no additional make-up water, as would be the case in a wet cooling tower approach. The use of two coolant loops and physical isolation

of the rack (or computer room) from the external air makes the local rack environment easier to control and maintain and helps ensure cleanliness and reliability of the rack and servers.

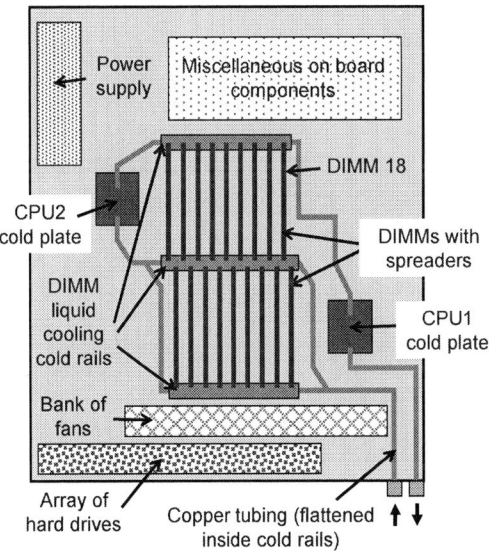

Fig. 2. Schematic representation of the liquid cooled IBM M3 X-3550 servers showing the liquid loop, two mini-channel cold plates for CPU1 and CPU2, and DIMM cold rails. Copper spreaders are attached to the DIMMs and conduct heat to the cold rails.

The described test facility is instrumented at various locations, both within the servers and in the two cooling loops. Temperature, pressure, flow-rate, humidity and power measurements are collected via a programmable logic controller (PLC) box, shown in Fig. 3, using a custom-built program running in Labview.

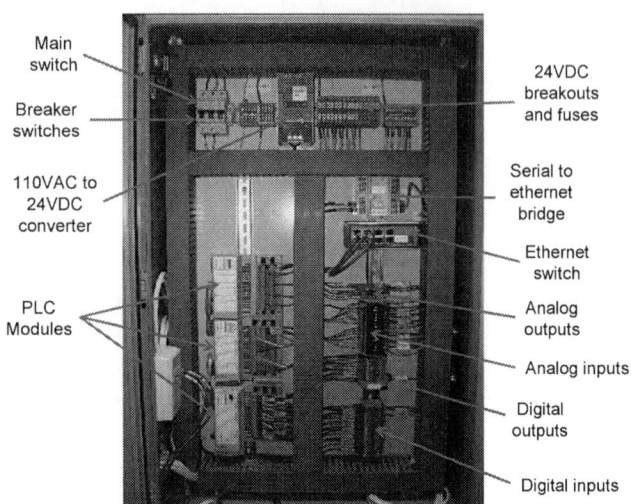

Fig. 3. Photo of PLC box used to monitor the data center test facility. The PLC is connected to a remote computer with the Labview based programs.

## 3. Data Center Test Facility Characterization

A key requirement to gain a more detailed understanding of the test facility behavior is characterization of the major components therein. Nine characterization tests were carried out, details of which are described in Table 1, where the internal and external pump RPM and dry cooler fan RPM were varied to study the thermal behavior of the system. The IT power during these runs varied from 13.4 to 14.5 kW.

Table 1. Test conditions used to characterize the major components of the data center test facility.

| Test # | Int Pump Flow GPM (LPM) | Ext Pump Flow GPM (LPM) | Cooler RPM RPM |
|--------|-------------------------|-------------------------|----------------|
| 1 | 4.1 (15.5) | 3.98 (15.1) | 170 |
| 2 | 4.1 (15.5) | 6.03 (22.8) | 170 |
| 3 | 4.1 (15.5) | 8.08 (30.6) | 170 |
| 4 | 6.1 (23.2) | 3.95 (15.0) | 310 |
| 5 | 6.1 (23.0) | 5.99 (22.7) | 310 |
| 6 | 6.1 (23.1) | 8.01 (30.3) | 310 |
| 7 | 8.1 (30.7) | 3.96 (15.0) | 500 |
| 8 | 8.1 (30.5) | 6.01 (22.7) | 500 |
| 9 | 8.1 (30.7) | 8.03 (30.4) | 500 |

The key thermal data from the cooling loop obtained during the nine characterization runs is shown in Fig. 4(a). The pre-buffer liquid temperature is found to reduce significantly as the cooler fan speeds are increased from 170 to 310 RPM and less so when increasing from 310 to 500 RPM. The pre-rack liquid and pre-rack air temperatures are found to be almost 10°C and 13.5°C higher than ambient at the lowest internal flow rate of 4.1 GPM (15.5 LPM) and found to improve as the external flow rate is increased from 4 GPM (15 LPM) to 8 GPM (30 LPM) at intermediate, 6.1 GPM (23.1 LPM) and higher, 8.1 GPM (30.6 LPM), internal flow rates. At the lowest internal flow rate and dry-cooler fan speed, the lower heat capacity rate of the air flow across the dry-cooler with respect to the external liquid flow and mostly flat dry-cooler effectiveness causes the temperature of the cooling water entering the buffer unit to rise with increasing external liquid flow. This results in the temperature of water in the internal loop leaving the buffer heat exchanger to be relatively unchanged despite the increasing buffer unit effectiveness with external liquid flow rate. At the higher internal flows and fan speeds (tests 4-9), increasing the external liquid flow only results in a slight increase in the liquid temperature leaving the dry-cooler as the external liquid now has a lower heat capacity rate with respect to the air flow. This results in a more pronounced lowering of the internal loop liquid (and subsequently air) temperatures as the external liquid flow is increased.

Figure 4(b) highlights the server component temperatures with respect to the approaching liquid temperature over the nine characterization runs. Increase in the internal loop flow rate finds a significant improvement in the component temperatures when moving from 4 GPM to 6 GPM and a smaller improvement when moving to 8 GPM. CPU2 temperatures are lower than CPU1 as CPU1 sees a larger liquid preheat in the server as seen in Fig. 2.

a)

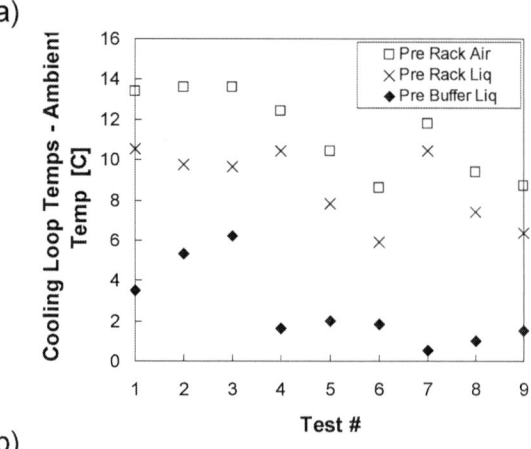

b)

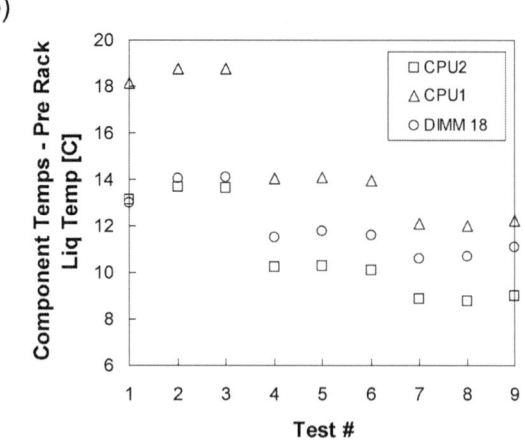

Fig. 4. Data center and server temperatures for the nine loop characterization tests: (a) Cooling loop temperatures (relative to outside ambient) (b) Server component temperatures (relative to pre-rack liquid temperatures) obtained from an instrumented server.

Using the obtained data, heat transfer effectiveness times heat capacity rate ($\varepsilon$-$C_{min}$) surfaces can be constructed (Fig. 5) as a function of the internal pump (within the buffer unit), external pump and dry cooler fan operation rates. These surfaces have value in the development of semi-empirical models of the data center and in understanding the manner in which the system reacts to changes in pumped flow rates and cooler fan speeds. Figure 5(a) shows that the buffer unit heat exchanger's $\varepsilon$-$C_{min}$ increases more strongly with the increase in the internal loop pump RPM at higher external flows than at lower external flows because at higher external pump flows the internal flow is the minimum fluid. Similarly, the rise in $\varepsilon$-$C_{min}$ is larger with increase in external flow rate when the internal pump flow is at the highest flow-rate. Figure 5(b) shows the $\varepsilon$-$C_{min}$ surface obtained for the dry-cooler heat exchanger. In general, we found that increasing dry cooler fan RPM is less effective as compared to increasing the external pump flow rate in improving the $\varepsilon$-$C_{min}$ for the heat exchanger and consequently in reducing loop temperatures (as can be seen in Fig. 4(a)). This is particularly true at the two higher cooler fan speeds of 310 and 500 RPM.

a)

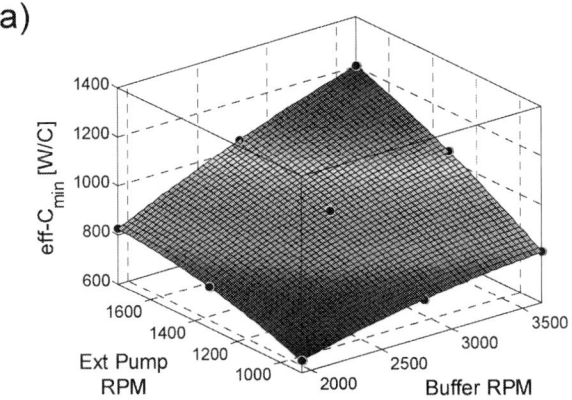

b)

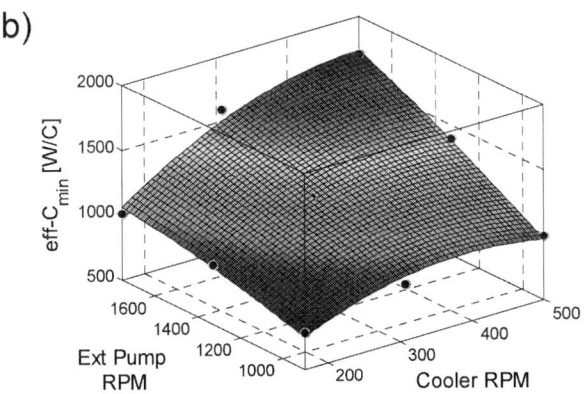

Fig. 5. $\varepsilon$-$C_{min}$ surfaces obtained for the (a) buffer unit and for the (b) dry-cooler as a function of the buffer unit and external pump speed and the cooler fan and external pump speed respectively.

Additional characterization tests were carried out to obtain detailed information regarding the hydraulic and power consumption characteristics of the rack, buffer unit and dry-cooler unit. The pressure drop across the buffer unit and rack

display a quadratic behavior with respect to pump speed and flow-rate. The flow-rates for the two pumps expectedly obey a linear relationship with the input RPM. Power consumption was found to follow a cubic relationship with respect to input RPM for the two pumps and the cooler fan as seen in Fig. 6. It is evident from the figure that the fans can quickly become the leading consumer of cooling energy in the data center at high operation speeds (>700 RPM). The relatively smaller improvement in loop temperatures at fan speeds higher than ~300 RPM results in a high cooling power cost for a given rack inlet liquid temperature improvement as compared to the external loop pump, emphasizing the need to limit cooler fan use. Though Fig. 6 indicates that the internal loop pump consumes less power for a given RPM, the power consumed for a given flow-rate is higher in the internal pump as compared to the external pump due to a lower flow-rate at a given RPM.

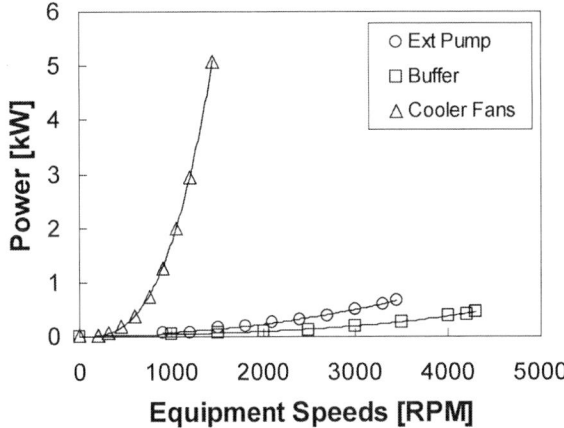

Fig. 6. Power consumption in the three pieces of cooling equipment used in the data center test facility. The cooler fans are clearly the leading power consumer at speeds greater than ~700 RPM thus motivating limited use of the fans when possible.

## 4. Day Long Operation of the Data Center Test Facility

The data center test facility was continuously run for a day (~ 22 hours) with varying cooler fan speeds and internal and external flow rates set to 7.2 GPM (27.2 LPM) and 7.1 GPM (27 LPM) respectively. The dry cooler fans were programmed to linearly vary in speed from 170 RPM to 500 RPM as the pre-buffer temperature varied from 30°C to 35°C. At pre-buffer temperatures below 30°C the fans run at a constant speed of 170 RPM. Figure 7 shows the pre-buffer, pre-rack liquid, pre-rack air and ambient air temperature during the course of the run. The ambient air temperature varied from 18.6°C to 32°C with an average of 23.8°C. The average pre-rack liquid and air temperature were measured to be 33.7°C and 36.4°C. The diurnal cycle can be clearly observed with temperatures dropping after sunset and reaching a low just before sunrise, after which the temperatures begin to rise rapidly. The pre-buffer, pre-rack liquid and pre-rack air temperatures generally follow the ambient temperature but the temperature difference to the ambient markedly reduces as the pre-buffer temperature rises over 30°C and the fans begin to ramp up and reduce the loop-to-ambient temperature delta.

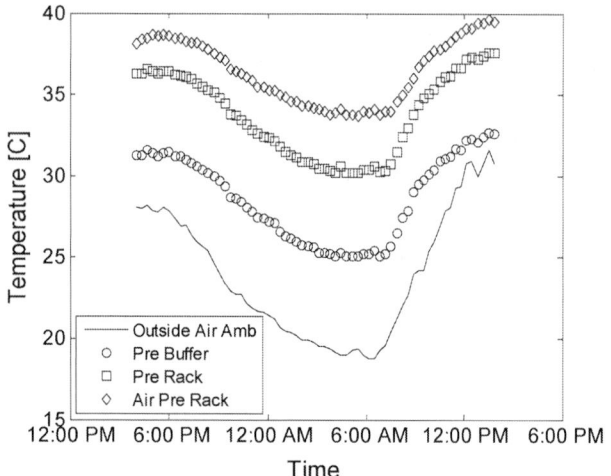

Fig. 7. Temperature data for the data center over the course of 22 consecutive hours. Cooling loop temperatures show that the loop temperatures generally follow the ambient temperature with a reduction in the delta with the ambient as the pre-buffer temperatures exceed 30°C and the dry-cooler fans are engaged.

The power consumption by the three pieces of cooling equipment is shown in Fig. 8(a). The power consumption by the internal and external pump is constant throughout the run but rises for the fans starting at around 10am. This causes the cooling power to rise during the late morning and into the afternoon. The IT power also varies during the day despite the use of a steady CPU and memory exerciser. The IT power consumption exhibits a linear relationship with the pre-rack air (and pre-rack liquid) temperature as shown in Fig. 8(b). This is due to a combination of increased leakage at higher temperatures as well as higher server fan power use. The server fan's RPMs are set by the measured server air inlet temperature. This implies that though cooling energy could be conserved by using warm water, the IT energy use may actually rise and potentially reduce the overall energy benefit. This also suggests that running IT servers as hot as allowable may not necessarily result in the best energy savings. Other factors such as the potential for energy recovery from hot water also need to be considered before the appropriate thermal operating point can be correctly identified.

Figure 9 shows a plot of the Cooling PUE, where Cooling PUE = (heat absorbed by liquid + cooling power) / (heat absorbed by liquid). The industry standard PUE number cannot be determined for our test facility due to the use of single rack situated in a mixed use lab space that is not optimized for use as a computer room. However, the Cooling PUE is a useful metric to determine the energy efficiency of the cooling solution used to cool the IT rack with values approaching unity indicating a more energy efficient cooling solution. The average Cooling PUE for this 22 hour run was calculated to be 1.035 (with an equivalent Coefficient of Performance, COP, of 29). The average IT power dissipated

during the course of the run was measured to be 13.14 kW with the cooling power varying from 430 W to 490 W. The heat dissipated from the IT rack that was lost to the lab environment, and not taken up by the cooling water, was found to be a maximum of 18% of IT power with an average of just 4%. The Cooling PUE rises during the late morning and into the afternoon due to the increased power use at the external cooling fans. The power consumed by the monitoring and control enclosure is not accounted for in this calculation as it is treated as a fixed energy cost whose impact on a data center is minimal for multi-rack configurations. The power consumed by this enclosure is determined to be approximately 82 W.

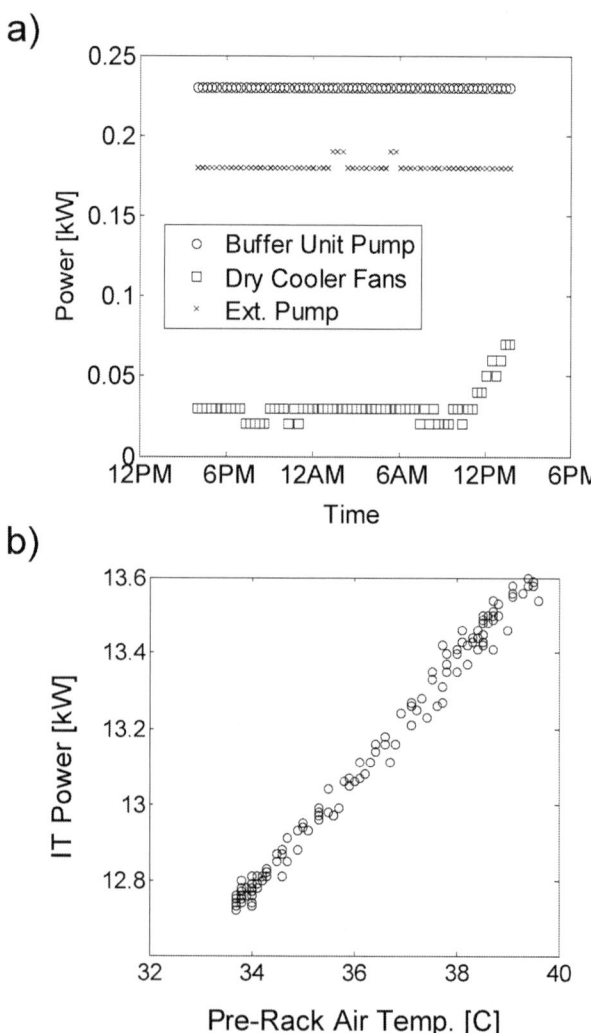

Fig. 8. (a) Cooling equipment power use over the course of the 22 hour run. Fan power increases as fans ramp up in response to increasing pre-buffer liquid temperature. (b) IT power draw as a function of rack air temperature showing increasing power use as temperature rises due to leakage and server fan speed up.

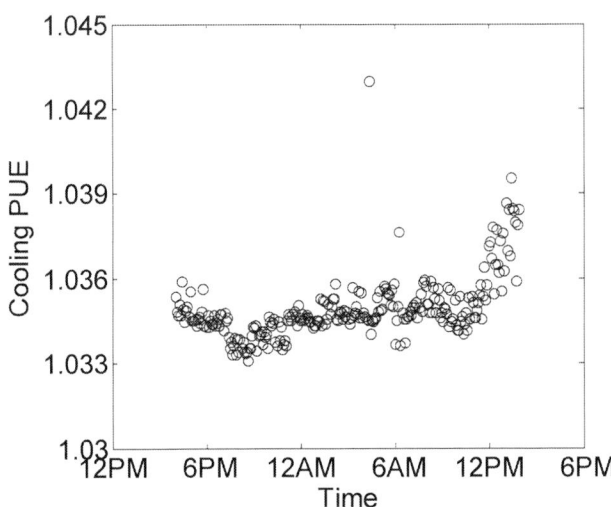

Fig. 9. Cooling PUE for the data center over the course of 22 consecutive hours. Cooling PUE can be maintained below 1.05 for most of the day with an average of 1.035 (i.e. cooling power = 3.5% of IT power absorbed by fluid). Cooling PUE rises starting late morning due to increased power use by the dry cooler fans.

## 5. Conclusions

An ambient air cooled, energy efficient, chiller-less data center test facility with warm water and re-circulated air cooled volume servers was designed and constructed to reduce cooling energy use below 5% of the IT energy use. This test facility has been characterized to determine the thermal, hydraulic and power consumption characteristics of the system and to determine the air-to-liquid rack heat exchanger, liquid-to-liquid buffer heat exchanger and liquid-to-ambient air dry cooler performance. This characterization is necessary for accurate modeling of the data center test facility. A one day test run was also carried out to characterize the energy efficiency of the test facility. Results from the one-day run found the cooling energy use to be on average 3.5% of the total IT energy use, with ambient air temperatures an average of 23.8°C and IT power an average of 13.14 kW. The anticipated benefits of such energy-centric configurations are significant energy savings at the data center level of approximately 25% which represents greater than 90% reduction in the cooling energy usage compared to conventional refrigeration based systems. For a typical 1 Megawatt Data Center this would represent a savings of roughly $90-$240k/year at an energy cost of $0.04 - $0.11 per kWh. The prototype dual-enclosure liquid-cooling (DELC) technology being characterized in this program will be evaluated to see how these developments may be incorporated into a portfolio of leading edge energy efficient technologies.

## Acknowledgments

This project was supported in part by the U.S. Department of Energy's Industrial Technologies Program under the American Recovery and Reinvestment Act of 2009, award no. DE-EE0002894. The authors would like to thank George Manelski of IBM Systems & Technology Group (STG) for his help in building the data center, David Graybill (STG) for the mechanical design and assembly of the servers and rack, Sal Rosato and Yun Lau of Grubbs and Ellis for their invaluable help in building the data center monitoring platform, and Bejoy Kochuparambil, Robert Rainey and Vinod Kamath (STG - RTP, NC) for their help, technical contribution and experience with the SystemX servers used in this work. The authors also acknowledge the technical advice and assistance of Michael Ellsworth, Levi Campbell, Corey Vandeventer and James Steffes (all of IBM STG, NY). We also thank the IBM Research project managers, James Whatley and Brenda Horton, DOE Project Officer Debo Aichbhaumik, DOE Project Monitors Darin Toronjo and Chap Sapp, and DOE HQ Contact Gideon Varga for their support throughout the project.

## References

1. M. Iyengar, R. Schmidt and J. Caricari, "Reducing Energy Usage in Data Centers Through Control of Room Air Conditioning Units", Proceedings of the IEEE ITherm Conference in Las Vegas, USA, May, 2010.
2. Brown et al., "Report to Congress on Server and Data Center Energy Efficiency: Public Law 109-431," Lawrence Berkely National Laboratory, August 2007.
3. ASHRAE Datacom Series 6, "Best Practices for DataCom Facility Energy Efficiency," 2nd Edition, Altanta, 2009.
4. Vision and Roadmap – Routing Telecom and Data Centers Toward Efficient Energy Use, May 13, 2009, www1.eere.energy.gov/industry/dataceters/pdfs/vision_and_roadmap.pdf.
5. ASHRAE, "Thermal Guidelines for Data Processing Equipment" 2nd Ed, 2009, http://tc99.ashraetcs.org/.
6. S. Greenberg et al., "Best Practices for Data Centers: Lessons Learned from Benchmarking 22 Data Centers," ACEEE Summer Study on Energy Efficiency in Buildings, 2006.
7. M. J. Ellsworth et al., "An Overview of the IBM Power 775 Supercomputer Water Cooling System," Proceedings of ASME InterPACK 2011, Porland, Oregon, July, 2011, Paper# IPACK2011-52130.
8. M. J. Ellsworth and M. Iyengar, "Energy Efficiency Analyses and Comparison of Air and Water Cooled High Performance Servers," Proceedings of ASME InterPACK 2009, San Francisco, California, July 19-23, 2009, Paper# IPACK2009-89248.
9. J. Wei., "Hybrid Cooling Technology for Large Scale Computing Systems – From Back to the Future," Proceedings of ASME InterPACK 2011, Porland, Oregon, July, 2011, Paper# IPACK2011-52045.
10. S. Noteboom, "Lower Cost. Higher Performance. Faster to Build: Creating an Efficient Data Factory" (Yahoo), The 7x24 Conference, Boca Raton, Florida, June, 2010.
11. V. Mulay, "Open Compute Project – Server and Data Center Design" (Facebook), Panel presentation at ASME InterPack 2011, Portland, Oregon, 2011.
12. R. Chu, M. Iyengar, V. Kamath, and R. Schmidt, "Energy Efficient Apparatus and Method for Cooling an Electronics Rack", US Patent 7791882 B2, 2010.

# A Multiple Vibrating-Fan System Using Interactive Magnetic Force and Piezoelectric Force

H. K. Ma*, W. F. Luo, H. C. Su

Department of Mechanical Engineering, National Taiwan University
No. 1, Sec. 4, Roosevelt Road
Taipei, Taiwan
skma@ntu.edu.tw

## Abstract

A vibrating fan system is being pursued as a means to create a heat dissipation system. To improve the efficiency of a heat dissipation system, a novel multiple vibrating-fan system that is actuated by both piezoelectric effect and magnetic force has been developed. The performance of the system is affected by the geometry of the fans and the distance between the fans and the heat sink. The surface temperature of a 20W heat sink can be reduced from $67°C$ to $50°C$ using this multiple vibrating-fan cooling system; The power consumption of this system is only 0.027 W, which is lower than a single piezoelectric (PZT) fan. Thus, the novel design of a cooling system with multiple fans shows effective thermal dissipation as well as low power consumption.

## Keywords

piezoelectric fan, magnetic force, cooling, heat dissipation

## 1. Introduction

There has been considerable interest in heat dissipation systems based on straight fin structures because of their low cost, high reliability, and feasibility. However, as the power consumption of electronic components increases, the amount of heat that must be removed from microelectronic devices also increases. Rotary fans are commonly attached to the fin structure to improve thermal performance. While these rotary fans can improve the heat convection by forced convection, they require extra high-power input and produce noise.

A number of studies on piezoelectric (PZT) actuators used for cooling applications have been presented, including liquid cooling [1, 2] and air cooling [3, 4]. Toda and Osaka [5, 6] first proposed the concept of piezoelectric fans for a cooling system because of their small size, low power consumption, and long lifetime. Rather than combining a vibrating fan with a static heat sink, Ma et al. [7] proposed the concept of a simple vibrating-fan system. Through the vibration of the fans, heat can be removed by force convection, and thus an additional rotary fan for cooling is not necessary. Though the system provides a remarkable heat dissipation rate and has low power consumption, it is not widely applicable because it is difficult to manufacture.

On the contrary, a piezoelectric fan is manufactured easily by attaching a cantilever beam to a piezoelectric plate. Thus, the piezoelectric fan can operate at a high ratio of fan tip deflection to power consumption [8], inducing oscillating flow efficiently. Açıkalın et al. [9] developed a vibrating-fan system to induce force convection and replace additional rotary fans. This system shows that enhancement in the heat transfer coefficient of fins can reach 102%. Furthermore, according to the latter experiment [10], the capacity of heat dissipation by the oscillating fan is five times that of one without fans.

The theoretical vibration model of a piezoelectric bimorph with a thin elastic plate was derived from a lumped-mass system [11]. This indicates that the damping effect on the mechanical wobbling of the attached elastic plate cannot be ignored. Kimber et al. [12] presented experimental methods to measure the pressure and flow rate of the piezoelectric fans. The Reynolds number was also defined to evaluate the performance of the fans. Yoo et al. [13] researched the influences of the dimension and the material on a vibrating fan, finding that phosphor bronze and aluminum are also good choices for a vibrating fan.

In this study, a novel multiple vibrating-fan system actuated by both piezoelectric and magnetic forces was developed. The novel design is composed of four passive fans, actuated by magnetic force, and one piezoelectric fan. Because the magnetic force acts on each fan, the fans can vibrate synchronously and produce an oscillating flow with low power consumption. The multiple vibrating-fan system is placed in a heat sink to improve the thermal performance of heat sink without changing the original heat dissipation system.

The effects of vibrating frequency, actuated voltage, geometry of fans, position between fans and heat sink, and power consumption are analyzed to assess the optimal performance of the multiple vibrating-fan system. Overall, this innovative system offers improvement in the application of a heat sink given its advantages of simple structure, high heat dissipation, and low power consumption.

## 2. Design and experimental set-up

Four magnetic fans and one piezoelectric (PZT) fan are used in this novel multiple vibrating-fan system, as shown in Fig. 1. The magnetic fans are made of phosphor copper. A magnet is attached on the tip of both fans so the repulsive magnetic forces can be applied, causing the fans to vibrate simultaneously. The geometry of the magnetic fan in each case is shown in Table 1.

Figs. 1 and 2 show the schematic view of the system. Each fan is fixed on a clamp that can be moved along a track. Thus, the fans can be adjusted to a suitable position.

The PZT fan, assigned as No. 1, vibrates by applying an alternating current. Then, the magnetic fan can be actuated by the interactive magnetic force. Moreover, the multiple vibrating-fan system is integrated with a heat sink to be a cooling system, as shown in Fig. 3 (a). Five thermocouples are pasted on the center of the fin base and $\lambda$ is defined as the distance between the fin base and the edge of fan, as shown in Fig. 3 (b).

978-1-4673-1110-6/12 $31.00 © 2012 IEEE

In order to prevent the perturbation of external convective currents, the system is placed in a large plastic box. A 20 W heat source is attached to the back side of the heat sink and a steady-state surface temperature of 68°C is reached, while the ambient temperature is 20°C. An alternating voltage from 40 V to 100 V is applied to the PZT plate. The fan amplitude is observed with a high speed camera (JVC-HM550) and the temperature is detected by the thermocouples attached on the surface of the heat sink.

|  | Fan geometry (mm) | | | Magnet geometry(mm) | |
|---|---|---|---|---|---|
| Name | L | H | W | D | t |
| Case1 | 75 | 0.3 | 20 | 8 | 2 |
| Case2 | 75 | 0.2 | 20 | 8 | 2 |
| Case3 | 60 | 0.2 | 20 | 8 | 2 |

**Table 1**: Specification of each design

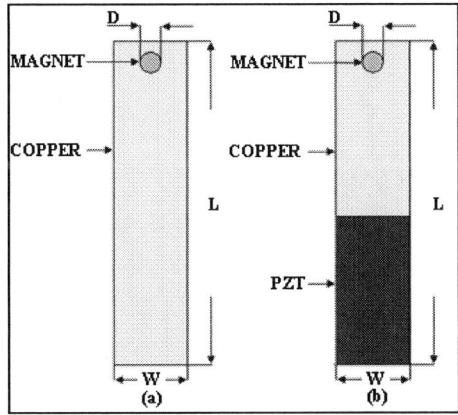

**Fig. 1:** (a) The magnetic fan (b) The PZT fan.

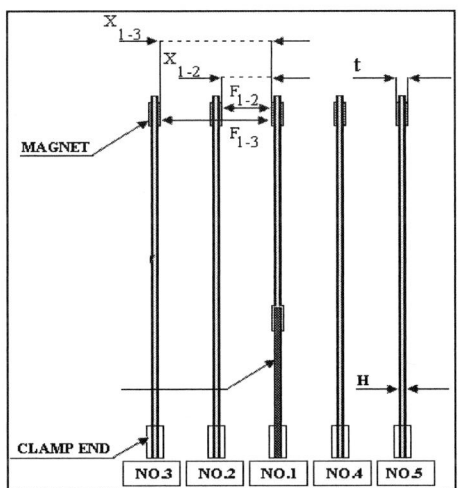

**Fig. 2:** Schematic view of the multiple vibrating-fan system.

## 3. Theory background

The vibrating amplitude of the magnetic fans depends on the length, thickness, and width of the magnetic fan and the vibrating frequency. The magnetic fan can be represented as an end-loaded cantilever beam. In order to find the maximum

vibrating amplitude, the resonance frequency is considered carefully. The magnetic fan can be represented as a spring with a stiffness $K_{beam}$, as shown in Equation (1),

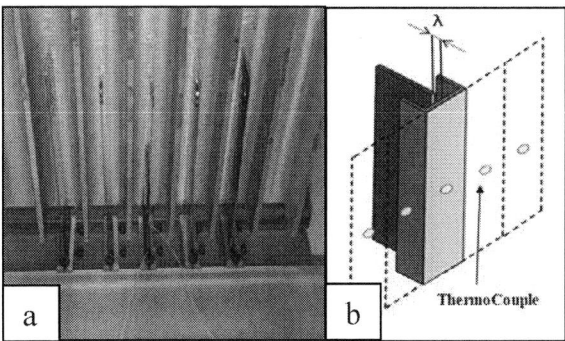

**Fig. 3:** The multiple vibrating-fan system: (a) actual view and (b) schematic view.

$$K_{beam} = \frac{EwH^3}{4L^3} \quad (1)$$

where E is Young's modulus, W is the width, H is the thickness, and L is the length. The effective mass of the magnetic fan can be approximated as $m_{eff}$. For the magnetic fans, the effective mass is $m_{mag} + 0.24m_b$, where $m_b$ is the effective mass of the beam and $m_{mag}$ is the effective mass of the magnet. When an end mass, the magnet, is added on the tip of the fan, the resonance frequency of the magnetic fan can be expressed as Equation (2):

$$\omega_{beam} = \sqrt{\frac{K_{beam}}{m_{mag}+0.24m_b}} \quad (2)$$

However, the magnetic force is also an additional stiffness that is introduced within the system [14]. Thus, this effect on the resonance frequency of the magnetic fan should be considered.

The cylindrical magnets are used to apply the desired magnetic force on the magnetic fan. The magnetic force between any two cylindrical magnets is given as [15]

$$F_{mag}(x) = \left[\frac{B_0^2A^2(4t^2+D^2)}{4\pi\mu_0H^2}\right]\left[\frac{1}{x^2} + \frac{1}{(x+2H)^2} - \frac{2}{(x+H)^2}\right] \quad (3)$$

where $B_0$ is the magnetic flux density near each pole, A is the area of the magnets, t is the thickness of the magnet, D is the diameter of the magnet, x is the distance between the magnets, and $\mu_0$ is the permeability of the intervening medium.

According to Equation (3), the magnetic force depends on the characteristics of the magnets as well as the distance between the magnets. Fig. 2 show that $F_{1-2}$ and $F_{1-3}$ are the interactive force between two specific magnets. If each spacing distance between the magnetic fans is the same, $X_{1-3}$ will be twice as large as $X_{1-2}$. When $X_{1-2}$ is 10 mm, $F_{1-2}/F_{1-3}$ can be calculated to be 11.67. Therefore, the interactive force $F_{1-3}$ can be eliminated to simplify the analysis; the effective stiffness between No. 1 fan and No. 3 fan can be eliminated. The effective stiffness between No. 1 fan and No. 5 fan can also be eliminated. We only need to consider the magnetic force between adjacent fans.

978-1-4673-1110-6/12 $31.00 © 2012 IEEE       239

As the fans start to vibrate, the repulsive magnetic force can be represented as a spring in compression, which induces an additional stiffness that influences the resonance frequency of each fan. The magnitude of the magnetic stiffness $K_{mag}$ can be written in terms of the change in magnetic force as a function of distance as Equation (4):

$$K_{mag} = \left| \frac{dF_{mag}}{dx} \right| \qquad (4)$$

$K_{mag}$ can be expressed as Equation (5) by inserting Equation (3) into Equation (4),

$$K_{mag}(x) = \left[ \frac{B_0{}^2 A^2 (H^2 + R^2)}{\pi \mu_0 H^2} \right] \left[ -\frac{2}{x^3} - \frac{2}{(x+2H)^3} + \frac{4}{(x+H)^2} \right] \qquad (5)$$

where the sign of $K_{mag}$ is dependent on the direction of total magnetic force. With the effects of variable stiffness resulting from the magnetic force, the total stiffness is expressed as Equation (6):

$$K_{eff} = K_{beam} + \Delta K_{mag} \qquad (6)$$

Thus, the resonance frequency is a function of both the beam stiffness and the stiffness associated with the magnetic force. The effective resonance frequency can be derived, as shown below:

$$\omega_{eff} = \sqrt{\frac{k_{eff}}{m_{mag} + 0.24 m_b}} \qquad (7)$$

## 5. Results and discussion

The heat dissipation of the multiple vibrating-fan system is affected by the vibrating amplitude. In order to find an optimal operation condition, magnetic fans with different geometries have been developed and the temperature and power consumption have been measured to study the performance of each design.

### 5.1 Effect of magnetic fan thickness on vibrating amplitude

The thickness (H) of magnetic fans has a big influence on the maximum vibrating amplitude as well as the resonance frequency. As shown in Figs. 4 and 5, the resonance frequency of the thicker magnetic fan (Case 1)–is 36.4 Hz, which is higher than that of Case 2. At a fixed actuating force, the maximum vibrating amplitude of a thinner magnetic fan is bigger than a thicker one. However, the thinner magnetic fans may have less durability.

### 5.2 Effect of magnetic fan length on vibrating amplitude

According to Equation (2), the resonance frequency will decrease by increasing the length of the fan. The influence of magnetic fan length can be found by comparing the resonance frequencies of Case 2 and Case 3, as shown in Figs. 5 and 6. The resonance frequency of magnetic fans in Case 2 is 21.8 Hz, while it is 30 Hz in Case 3. The experiment results of changing the length of the magnetic fans show a good agreement between the experiment and theorem.

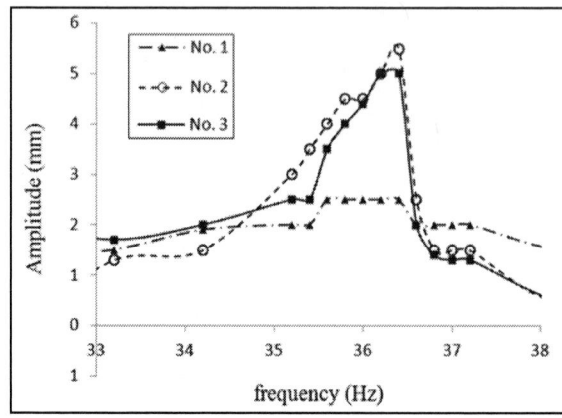

**Fig. 4:** The amplitude of the magnetic fans (Case 1) under different frequencies.

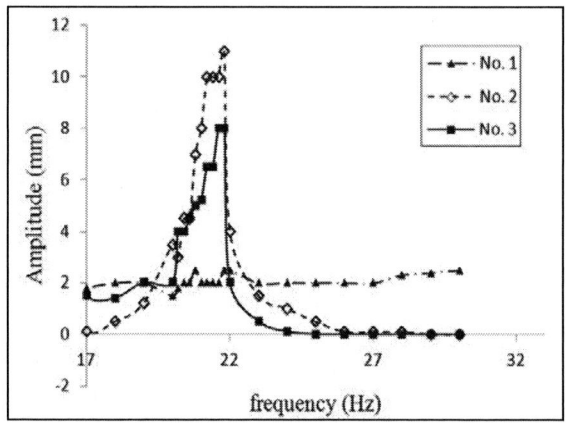

**Fig. 5:** The amplitude of the magnetic fans (Case 2) under different frequencies.

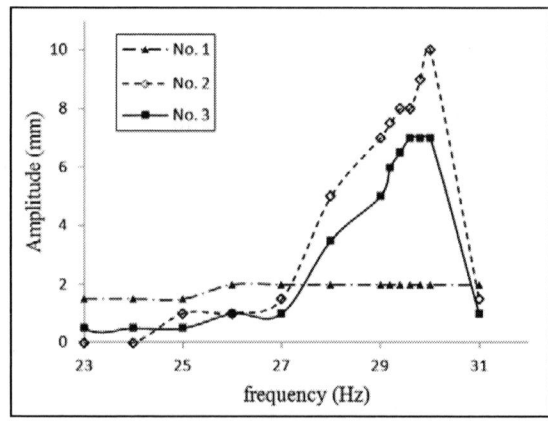

**Fig. 6:** The amplitude of the magnetic fans (Case 3) under different frequencies.

### 5.3 Effect of spacing distance between the magnetic fan and fin base (λ) on the temperature drop at the fin surface

The thermal performance of a multiple vibrating-fan system can be found by comparing the temperature drop on the fin base. To compare the influence of different fan and fin positions, multiple vibrating-fan systems with different spacing distance between magnetic fan and fin base (λ) have been designed and tested. Both systems are attached to the same fin combined with a 20W heat source. As shown in Fig.

7, the temperature drops in both systems are measured and the temperature drop is about 15℃ by using smaller λ=2 mm system, resulting in a better performance.

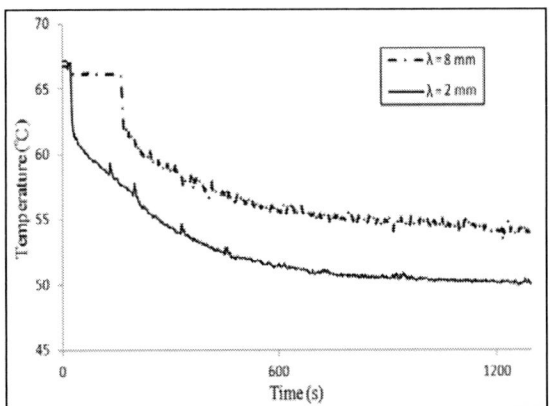

**Fig. 7:** The average surface temperature drop at the fin base under different λ.

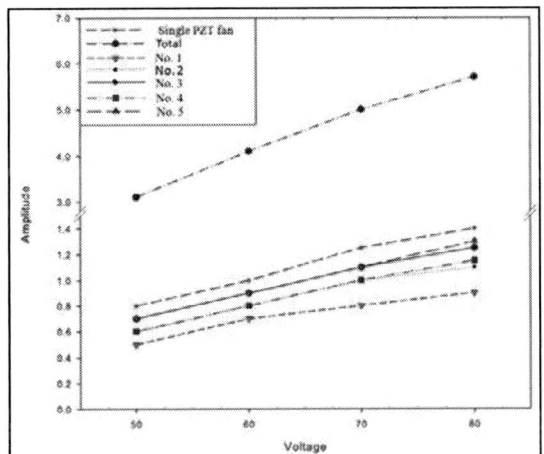

**Fig. 8:** Comparison of the maximum frequency of a single PZT fan and the multiple fan (Case 1) under different voltages

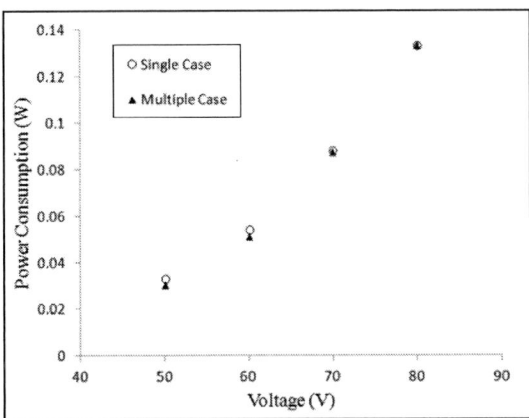

**Fig. 9:** Comparison of power consumption of a single PZT fan and the multiple fan system under different voltages (both at resonance frequency)

**5.4 Performance of the multiple vibrating-fan system**

A comparison of the maximum vibrating amplitude of the multiple vibrating-fan system and a single PZT fan under different voltages is shown in Fig. 8. To find the better performance, the fans are operated at a resonance frequency of 36.4 Hz for the multiple fan system and 40 Hz for a single PZT fan. Although the amplitude of each fan is smaller than the single PZT fan, the total amplitude of the five fans is much higher than that of the single PZT fan. The total amplitude of the multiple vibrating-fan system is about 400% more than the single PZT fan. Furthermore, the lower resonance frequency leads to the lower power consumption, as shown in Fig. 9. With a 50 V operating voltage, the power consumption of the multiple fan system is only 0.028 W. In brief, the multiple vibrating-fan system can induce larger vibrating amplitude with lower power consumption.

**6. Conclusions**

A multiple vibrating-fan system with low power consumption actuated by both piezoelectric effect and magnetic force has been developed to improve the efficiency of the systems. The major conclusions from this study are summarized below:

1. When the piezoelectric fan is combined with passive magnetic fans, the total amplitude of the multiple vibrating-fan system is about 400% more than that of a single PZT fan under the same power consumption.
2. Compared to a single PZT fan, the multiple vibrating-fan system has lower resonance frequency, resulting in low power consumption. With a 50 V actuated voltage, the power consumption of the multiple vibrating-fan system is only 0.028 W.
3. Increasing the length or reducing the thickness of the magnetic fans leads to a lower resonance frequency and larger vibrating amplitude.
4. For a 20 W heat source, the surface temperature of the heat sink with straight fins is reduced about 15°C by the multiple vibrating-fan system.

**Acknowledgment**

This research is funded by the National Science Council of Taiwan. (NSC 99-2622-E-002-029-cc3)

**References**

1. H. K. Ma, B. R. Hou, C. Y. Lin, J. J. Gao, The improved performance of one-side actuating diaphragm micropump for a liquid cooling system, International Communications in Heat and Mass Transfer 35 (2008) 957-966.
2. H. K. Ma, B. R. Chen, C. Y. Lin, J. J. Gao, Development of an OAPCP-micropump liquid cooling system in a laptop, International Communications in Heat and Mass Transfer 36 (2009) 225-232.
3. B. G. Loh, S. Hyun, P. I. Ro, C. Kleinstreuer, Acoustic streaming induced by ultrasonic flexural vibrations and associated enhancement of convective heat transfer, Acoustical Society of America 111 (2) (2002) 875-883.
4. T. Acikalin, S.V. Garimella, J. Petroski, A. Raman, Optimal design of miniature piezoelectric fans for cooling light emitting diodes, in: The Ninth Intersociety Conference on Thermal and Thermomechanical

Phenomena in Electronic Systems, Las Vegas (2004) 663-671.

5. M. Toda, Theory of air flow generation by a resonant type PVF2 bimorph cantilever vibrator, Ferroelectrics 22 (1979) 911-918.

6. M. Toda, Voltage-induced large amplitude bending device-PVF2 bimorph—its properties and applications, Ferroelectrics 32 (1981) 127-133.

7. H. K. Ma, B. R. Chen, H. W. Lan, K.T. Lin, and C.Y. Chao, Study of an LED Device with Vibrating Piezoelectric Fins, 25th Semiconductor Thermal Measurement, Modeling, and Management Symposium (SEMI-THERM-25), San Jose, California, USA, (2009).

8. I. Sauciuc et al., Key challenges for the piezo technology with applications to low form factor thermal solutions, Proceedings of ITHERM (2006) 781-785.

9. T. Acikailn, S. V. Garimella, A. Raman, J. Petroski, Characterization and optimization of the thermal performance of miniature piezoelectric fans, International Journal of Heat and Fluid Flow 28 (2008) 806–820.

10. J. Petroski, M. Arik, M. Gursoy, Optimization of piezoelectric oscillating fan-cooled heat sinks for electronics cooling, IEEE Transactions on Components and Packaging Technology 33 (1) (2010) 25-31.

11. K. Yao, K. Uchino, Analysis on a composite cantilever beam coupling a piezoelectric bimorph to an elastic blade, Sensors and Actuators 89 (2001) 215-221.

12. M. Kimber, K. Suzuki, N. Kitsunai, K. Seki, S. V. Garimella, Quantification of piezoelectric fan flow rate performance and experimental identification of installation effects, Proceedings of ITHERM (2008) 471-479.

13. J. H. Yoo, J. I. Hong, W. Cao, Piezoelectric ceramic bimorph coupled to thin metal plate as cooling fan for electronic devices, Sensors and Actuators 79 (2000), 8-12.

14. V. R Challa, M. G. Prasad, Y. Shi, F. T. Fisher, A vibration energy harvesting device with bidirectional resonance frequency tunability, Smart Materials and Structures 17 (1) (2008)

15. V. R Challa, M. G. Prasad, F. T. Fisher, A coupled piezoelectric–electromagnetic energy harvesting technique for achieving increased power output through damping matching, Smart Materials and Structures 18 (9) (2009)

# Study of a Cooling System with a Piezoelectric Fan

H. K. Ma*, C. L. Liu, H. C. Su, W. H. Ho

Department of Mechanical Engineering, National Taiwan University, No. 1, Sec. 4, Roosevelt Road, Taipei, Taiwan

*Corresponding author tel: +886-23629976; fax: +886-2-23632644; e-mail address:skma@ntu.edu.tw

## Abstract

Previous studies show that the performance of a piezoelectric (PZT) fan is strongly affected by length, vibrating frequency, and fan amplitude. This study examines a cooling system, which is composed of a heat sink made of aluminum and a piezoelectric fan. An oscillating airflow can be generated and induced by the fan deformation. The piezoelectric fan between two fins may break the thermal boundary layer and enhance the heat dissipation rate of the cooling system with forced convection. In order to estimate the optimum design of the cooling system, the effects of operating frequency, fan amplitude, fan arrangement, Ri $(Gr/Re_{PZT}^2)$, and power consumption are analyzed. Moreover, the relationship between the dimensionless PZT-convection number $(M_p)$ and Ri is investigated to analyze the performance of the cooling system. A three-dimensional, transitional model has been successfully built to account for the flow field of the cooling system. The optimum cooling system shows a dimensionless PZT-convection number of 2.3.

Keywords: Piezoelectric fan, cooling, heat sink, force convection

## 1. Introduction

A number of studies on the piezoelectric actuators used for cooling applications are presented recently, including liquid cooling [1] and air cooling [2]. Toda and Osaka [3] proposed the concept of piezoelectric fans for cooling systems because of their small size, lower power consumption, and long life. A piezoelectric fan is manufactured by attaching a cantilever beam to a piezoelectric plate. An alternating current is applied at the resonant frequency of the piezoelectric fan. Thus, the piezoelectric fan can operate at a high ratio of fan tip deflection to power consumption [4]. The vibrating cantilever beam can operate with enough air-moving capability to cool electronic equipment as small fans. The theoretical vibration model of a piezoelectric bimorph with a thin elastic plate was derived from a lumped-mass system [5], which indicated that the damping effect could not be ignored on the mechanical wobbling of the attached elastic plate.

Acikalin et al. [6] presented a piezoelectric fan mounted to a constant heat flux surface. Several experimental parameters, including configuration, fan amplitude, fan length, and frequency, as well as the distance between the fan and the heat source, were explored to assess the cooling ability of the piezoelectric

fan. The results showed that the fan frequency offset from resonance and fan amplitude were the crucial parameters of the design. Before 2010, almost all studies were about piezoelectric fans used to enhance the average heat transfer coefficient on a heated surface. In 2010, Petroski et al. [7] proposed a cooling system, which combined two piezoelectric fans with a heat sink. The heat dissipation rate of the cooling system was 560% higher than that of a cooling system with natural convection.

Ma examined an innovative design of a vibrating fins cooling system. A three-dimensional simulated model for the cooling system was built to investigate the effects of the vibrating fins [8]. Simulated results showed that the performance of the vibrating fins cooling system was strongly affected by dimensions, operating frequency, pitch, and amplitude of the fins. In this study, the piezoelectric fan is positioned inside the heat sink, thus allowing it to cool the inner surfaces of the heat sink. The effects of operating frequency, fan amplitude, fan arrangement, the importance of natural convection relative to the forced convection Ri $(Gr/Re_{PZT}^2)$, and power consumption of the fan are analyzed to assess the performance of the cooling system. Moreover, the relationship between the dimensionless PZT-convection number $(M_p)$ and the importance of natural convection relative to the forced convection (Ri) are also investigated. A three-dimensional, transitional model has been successfully built to account for the flow field of the cooling system. This study provides a comprehensive technique to analyze the piezoelectric fan cooling system. By taking advantage of this technique, the performance of a cooling system can be assessed conveniently.

## 2. Design and experimental set-up

The schematic view of the experimental set-up is shown in Fig. 1. The dimensions of the fin base were 100 x 30 x 1 (mm) and the dimensions of the extruded fin were 100 x 20 x 1 (mm). Fifteen thermocouples were pasted on the inner surfaces of the heat sink. A heat source device, which could dissipate 1W~3W power was attached to the outer surface of the heat sink. The outer surfaces of the heat sink were covered by insulation block. The piezoelectric fan was driven in the range between 60V ~ 100V by an AC power supply. The system was placed in an acrylic box to prevent perturbation from external convective currents. A thermal meter was used to record the temperature data. The fan amplitude could be observed by a high-speed camera (JVC-HM550).Thus, the power consumption, the operating frequency, and the

978-1-4673-1110-6/12 $31.00 © 2012 IEEE      243

Fig. 1 The experimental set-up.

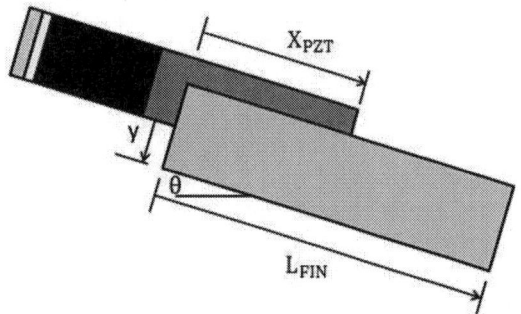

Fig. 2 Parameters of the piezoelectric fan and the heat sink.

fan amplitude could be recorded to analyze the performance of the cooling system.

The piezoelectric fan, composed of a rectangular piezoelectric plate and a polyvinyl chloride (PVC) sheet, was placed inside the heat sink. As shown in Fig. 2, the parameters included the vertical orientation y, the horizontal orientation $X_{PZT}/L_{FIN}$, the inclined angle of the cooling system $\theta$, the vibrating frequency $\omega$, and the fan amplitude $\alpha$.

## 3. Data Analysis

In order to calculate the overall convection heat transfer coefficient, the fin surface was divided into fifteen equal sections. The thermocouples were arranged in each section as shown in Fig. 3. Eq. (1) shows the energy balance of the cooling system where $Q_H$ is the input heat from the heat source, $Q_{FIN,Total}$ is the total dissipated heat from the fin surfaces by convection, and $Q_{FIN,i}$ is the dissipated heat from each section.

$$Q_H = Q_{FIN,Total} = \sum_{i=1}^{15} Q_{FIN,i} \tag{1}$$

Eq. (1) can be taken to be the following Eq. (2) by substituting the notations.

$$Q_{FIN} = \bar{h} \times A_{FIN} \times (\bar{T}_S - T_\infty) \tag{2}$$

It is denoted that $A_{FIN}$ is the total area of the inner fin surface, $\bar{T}_s$ is the average temperature of the inner surface, $T_\infty$ is the ambient air temperature, and $\bar{h}$ is the overall convective heat transfer coefficient. Thus, the overall convective heat transfer coefficient can be computed by Eq. (3).

$$\bar{h} = \frac{Q_{FIN}}{A_{FIN}(\bar{T}_S - T_\infty)} \tag{3}$$

By dividing the bottom and the side fin surface equally into five sections, as shown in Fig. 3, the central temperature of a section can be assumed the average temperature of the section. Thus, the average temperature of the inner surface can be derived by using the weighting method, as shown in Eq. (4),

$$\bar{T}_s = \frac{\sum_{i=1}^{5} T_{base,i} * A_{base,i} + 2 * \sum_{i=1}^{5} T_{side,i} * A_{side,i}}{A_{FIN}} \tag{4}$$

where $T_{base,i}$ is the temperature measured by the thermocouples of each section at the bottom fin surface, and $T_{side,i}$ is the temperature measured by the thermocouples of each section at the side fin surface; $A_{base,i}$ is the area of each section at the bottom fin surface, and $A_{side,i}$ is the area of each section at the side fin surface. According to Eq. (4), the overall convective heat transfer coefficient of the heat sink with the vibrating piezoelectric fan can be calculated as Eq. (5) and the overall heat convection coefficient of the heat sink with natural convection can be calculated as Eq. (6).

$$\bar{h}_{PZT} = \frac{Q_{FIN}}{A_{FIN}(\bar{T}_{S,PZT} - T_\infty)} \tag{5}$$

$$\bar{h}_0 = \frac{Q_{FIN}}{A_{FIN}(\bar{T}_{S,0} - T_\infty)} \tag{6}$$

In order to demonstrate the convection ability improved by the piezoelectric fan, the dimensionless PZT-convection number ($M_p$) is defined as shown in Eq. (7) to assess the convection ability.

$$M_p = \frac{\bar{h}_{PZT}}{\bar{h}_0} = \frac{\text{Forced convection coefficient with the PZT fan}}{\text{Natrual convection coefficient}} \tag{7}$$

The Reynolds number and the Nusselt number are also calculated to estimate the convective ability of the cooling system. In this study, the fan amplitude is used as the length scale. The velocity terms of the Reynolds number and the Nusselt number are based on the maximum fan tip velocity. Thus, the Reynolds number and the Nusselt number can be expressed as Eq. (8) and Eq. (9) [9],

$$Re_{PZT} = \frac{\omega \alpha L_{pzt}}{\nu} \tag{8}$$

$$Nu_{PZT} = \frac{\bar{h}_{PZT}L_{pzt}}{k} \tag{9}$$

where $\nu$ is the dynamic viscosity, $\alpha$ is the fan tip amplitude, $\omega$ is the vibrating frequency, k is the conductivity, and $L_{pzt}$ is the characteristic length. The characteristic length is chosen by employing the hydraulic diameter of the vibrating fan envelope as shown in Eq. (10) [10],

$$L_{pzt} = \frac{4\alpha w}{2(\alpha + w)} \tag{10}$$

where w is the width of the piezoelectric fan. In addition, the cooling effect of this cooling system is the combination of natural convection and forced convection. The importance of natural convection relative to the forced convection (Ri) is defined as Eq. (11) by the Grashof number (Gr) and the Reynolds number (Re) [11].

$$Ri = \frac{Gr}{Re_{PZT}^2} \tag{11}$$

By replacing the notations of Gr and $Re_{L,PZT}$, Eq. (11) can be expressed as Eq. (12).

$$Ri = \frac{g\beta(\bar{T}_{S,PZT} - T_\infty)L_c{}^3}{\nu^2 L_{pzt}{}^2} \tag{12}$$

In Eq. (12), the characteristic length ($L_c$) is defined as the height of the fin (H) for buoyancy effect and $\beta$ is the expansion coefficient. For an ideal gas, the thermal expansion coefficient $\beta$ is expressed as 1/T, where T is assumed as 273 K. Therefore, Ri can be rewritten as,

$$Ri = \frac{g\beta H^3(\bar{T}_{S,PZT} - T_\infty)}{\omega^2 \alpha^2 L_{pzt}{}^2 T}. \tag{13}$$

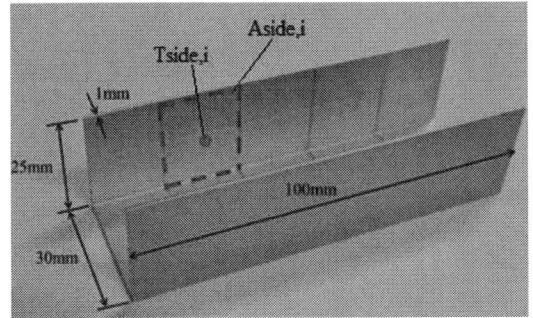

Fig. 3 Thermocouples are mounted on center of the enclosed section.

## 4. Simulation model

Using CFD-GEOM and CFD-ACE$^+$, a three-dimensional, transitional model was built to account for the flow field of the cooling system. In order to simplify the model, the vibrating motion of the piezoelectric fan was defined as an equation of sine wave, and expressed as Eq. (14),

$$y = \frac{\alpha}{2} * \left(\frac{x}{L}\right)^2 \sin(2\pi\omega t), \tag{14}$$

where $\omega$ is the vibrating frequency and L is the length of the piezoelectric fan. The vibration period is divided into ten equal time steps. The time step is set as Eq. (15).

$$\Delta t = \frac{1}{10\omega} \tag{15}$$

Other major assumptions of the model are as follows:
(1) The stable temperature of the cooling system without a piezoelectric fan is set as the initial condition of simulation.
(2) The isothermal temperature of 340 K is set as the boundary condition of the heat source.
(3) The fluid and solid properties are set as air and copper, as shown in Table 1.
(4) The density and viscosity of fluid are assumed constants.
(5) There is no slip condition on the wall.
(6) The effect of gravity is considered.
(7) The inlet and outlet pressure of control surfaces are fixed pressures.

## 5. Results and discussion

In order to choose the best arrangement of the piezoelectric fan, different orientations for the piezoelectric fan were investigated. The simulation model was also used to examine the correlation between the flow field induced by the piezoelectric fan and the temperature drop on the inner fin surface.

As mentioned in the introduction, low power consumption is one of the characteristics of the piezoelectric fan. Thus, the piezoelectric fan's power consumption was investigated in this study. Moreover, other dimensionless parameters were chosen to analyze the experimental data. By following these methods, the orientation and operating conditions of the piezoelectric fan cooling system could be established more efficiently.

### 5.1 The arrangement of the piezoelectric fan

The arrangement of the piezoelectric fan may influence the generated flow field and the performance of the cooling system. In order to find the best arrangement of the piezoelectric fan, the vertical orientation y, the horizontal orientation $X_{PZT}/L_{FIN}$, and the inclined angle of the cooling system $\theta$ were considered when investigating the cooling system.

According to previous studies, the gap between the fan and the heated surface strongly affected the performance of the cooling system. In this study, three different vertical orientations, y (5 mm, 10 mm, and 15 mm), were tested under the condition of $X_{PZT}/L_{FIN}$=0.5 and θ=0°. As shown in Fig. 4, the results indicated that the temperature drop on both the bottom and the side fin surfaces obviously increased when the piezoelectric fan approached the bottom fin surface. Thus, y=5 mm was chosen as the vertical orientation.

Tests of the inclined angle θ (at 0°, 15°, 30°, 45°,60°, 75°, and 90°) were completed under the conditions of $X_{PZT}/L_{FIN}$=0.5, y=5 mm. The performance of the cooling system with θ=0° was superior to those with other inclined angles.

A three-dimensional model was built to investigate the flow field of the cooling system. Fig. 5 shows that the maximum velocity near the bottom surface and the side surface both occur at $X_{PZT}/L_{FIN}$= 0.3~0.5. Theoretically, the high velocity leads to a larger temperature drop. Fig. 4 shows good agreement with this anticipated phenomenon. The maximum temperature drop occurs at the position where the maximum velocity occurs. The three-dimensional simulation model may help to decide the position of the heat source and the arrangement of the piezoelectric fan.

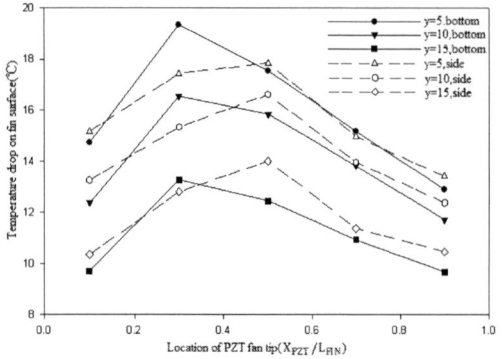

Fig. 4 The temperature drop on the bottom and side fin surfaces ( $X_{PZT}/L_{FIN}$=0.5, θ=0°, frequency=30 Hz, Amplitude=18 mm)

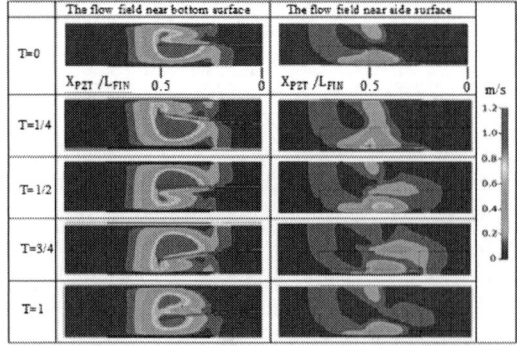

Fig. 5 The flow field near the bottom and the side surface ($X_{PZT}/L_{FIN}$=0.5, y= 5 mm, θ=0°)

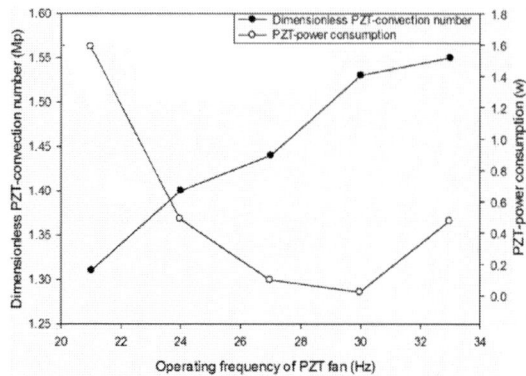

Fig. 6 The power consumption of the piezoelectric fan ($X_{PZT}/L_{FIN}$=0.5, y= 5 mm, θ=0°, amplitude=10 mm)

## 5.2 The power consumption of the piezoelectric fan

In the fan cooling system, the power consumption of the PZT fan is also an important issue. Fig. 6 shows the power consumption and the $M_p$ when the piezoelectric fan vibrates at different frequencies. The power consumption was 0.511 W and the $M_p$ was 1.4 under the operating frequency of 24 Hz. However, the power consumption decreased to 0.022 W and the $M_p$ changed to 1.53 at the operating frequency of 30 Hz because 30 Hz is the resonant frequency that allows the PZT fan to vibrate at its maximum amplitude. According to this result, power consumption should be considered, rather than $M_p$ when selecting an operating frequency. An inappropriate operating frequency will increase the power consumption by twenty times.

## 5.3 The correlation between Re number and Nu number

According to Eq. (8), the frequency and the fan amplitude are directly proportional to $Re_{PZT}$. The two parameters can be easily observed. They also play important roles in enhancing $Nu_{PZT}$. However, the $Nu_{PZT}$ cannot be calculated directly from the frequency and the amplitude, so the correlation between $Re_{PZT}$ and $Nu_{PZT}$ may help us to estimate the $Nu_{PZT}$ conveniently. Fig. 7 shows the $Nu_{PZT}$ and the $Re_{PZT}$ of the experimental data at different amplitudes and frequencies. When the $Re_{PZT}$ is below 700, the $Nu_{PZT}$ number increases almost linearly with the increasing $Re_{PZT}$ number.

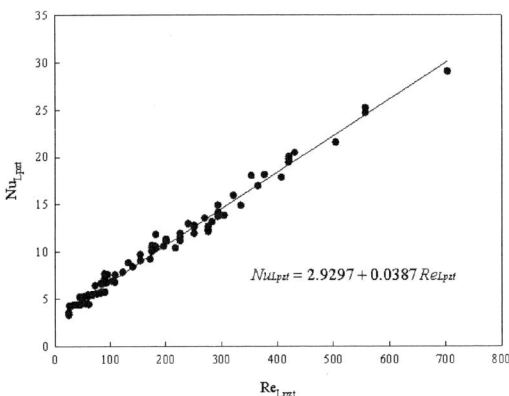

Fig. 7 The correlation between Re number and Nu number of the cooling system.

### 5.4 The relationship between $M_p$ and Ri

The dimensionless parameter is defined as Ri ($Gr/Re_{PZT}^2$), representing the importance of natural convection relative to the forced convection. When Ri<1, natural convection is negligible. However, when Ri>10, forced convection is negligible [18]. Fig. 8 shows the dimensionless analysis of the experimental data at different frequencies and amplitudes. In Fig. 8, two fitting curves are drawn according to the experimental data. The $M_p$ can be correlated with the Ri and then generalized as the following equations. The equations may be used to analyze and assess the performance of the piezoelectric fan cooling system more conveniently.

$$M_p = 1.268 - 0.057 \ln Ri \qquad Ri > 10$$

$$M_p = 1.389 - 0.276 \ln Ri \qquad Ri < 1$$

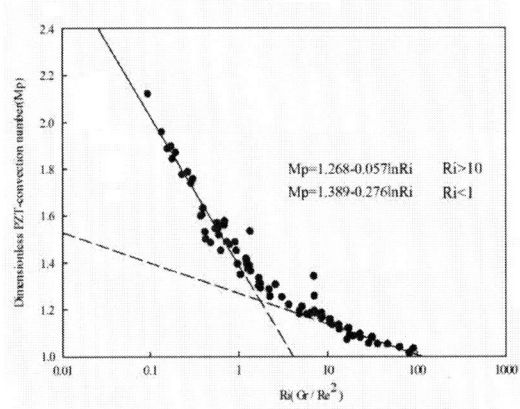

Fig. 8 The relationship between $M_p$ and Ri ($X_{PZT}/L_{FIN}=0.5$, y= 5 mm, $\theta = 0°$)

## 6. Conclusions

This study evaluated the performance of a piezoelectric fan cooling system using experiments, simulation, and non-dimensional analysis. The major findings are as follows:

1. Using a three-dimensional, transitional model to account for the flow field induced by the piezoelectric fan, maximum velocity usually occurs at the position 20 mm before the fan tip; the results show good agreement between the temperature drop and the flow field.

2. The orientations of the piezoelectric fan strongly affect the $M_p$. The $M_p$ of the system which operates with a frequency of 30 Hz and an amplitude of 18 mm reaches 2.3 ( $X_{PZT}/L_{FIN}=0.5$, y=5 mm, $\theta=0°$)

3. In the case ( $X_{PZT}/L_{FIN}=0.5$, y=5 mm, $\theta=0°$), the $Nu_{pzt}$ almost linearly increases with an increasing $Re_{pzt}$. According to this correlation, the $Nu_{pzt}$ can be derived from the amplitude and the operating frequency that determines $Re_{pzt}$.

$$Nu_{pzt} = 2.9297 + 0.0387 Re_{pzt} \text{ when } Re_{pzt} < 700$$

4. The $M_p$ can be correlated with the dimensionless ratio of natural convection to forced convection (Ri) and generalized as the following equations:

$$M_p = 1.268 - 0.057 \ln Ri \qquad Ri > 10$$

$$M_p = 1.389 - 0.276 \ln Ri \qquad Ri < 1$$

The equations can be used to analyze the performance of the piezoelectric fan cooling system conveniently.

## Acknowledgment

This research is funded by the National Science Council of Taiwan. (NSC 99-2622-E-002-029-cc3)

## References

1. H. K. Ma, B. R. Chen, C. Y. Lin, J. J. Gao, Development of an OAPCP-micropump liquid cooling system in a laptop, International Communications in Heat and Mass Transfer 36 (2009) 225-232.

2. T. Acikalin, S.V. Garimella, J. Petroski, A. Raman, Optimal design of miniature piezoelectric fans for cooling light emitting diodes, in: The Ninth Intersociety Conference on Thermal and Thermomechanical Phenomena in Electronic Systems, Las Vegas (2004) 663-671.

3. M. Toda, Voltage-induced large amplitude bending device-PVF2 bimorph — its properties and applications, Ferroelectrics 32 (1981) 127-133.

4. I. Sauciuc et al., Key challenges for the piezo technology with applications to low form factor thermal solutions, Proceedings of ITHERM (2006) 781-785.

5. K. Yao, K. Uchino, Analysis on a composite cantilever beam coupling a piezoelectric bimorph to an elastic blade, Sensors and Actuators 89 (2001) 215-221.

6. T. Acikalin, S. M. Wait, S. V.Garimella, A. Raman, Experimental investigation of thermal performance of piezoelectric fans, Heat Transfer Engineering 25(1)(2004)4-14.

7. James Petroski, Mehmet Arik, and Mustafa Gursoy, Optimization of piezoelectric oscillating fan-cooled eat sinks for electronics cooling, IEEE Transactions on Components and Packaging Technology 33 (1) (2010)

8. H. K. Ma, B. R. Chen, H. W. Lan, K. T. Lin, C.Y. Chao, Study of an LED device with vibrating piezoelectric fins, IEEE Semi-Therm (2008)

9. Mark Kimber, Suresh V. Garimella, Measurement and prediction of the cooling characteristics of a generalized vibrating piezoelectric fan, International Journal of Heat and Mass Transfer 52 (2009) 4470-4478

10. M. Kimber, S. V. Garimella, A. Raman, Local heat transfer coefficients induced by piezoelectrically actuated vibrating cantilevers, Journal of Heat Transfer 123 (2007) 1168-1176

11. Y. A. Cengel, Heat transfer, New York (2004) 460-466

## Nomenclature

| | |
|---|---|
| $Ri$ | The importance of natural convection relative to the forced convection. |
| $Gr$ | Grashof number of the PZT fan cooling system. |
| $Re_{pzt}$ | Reynolds number of the PZT fan cooling system. |
| $Nu_{pzt}$ | Nusselt number of the PZT fan cooling system. |
| $L_{pzt}$ | The characteristic length of the PZT fan cooling system |
| $y$ | Vertical orientation. |
| $X_{PZT}$ | The length of the fan inside the heat sink. |
| $L_{FIN}$ | The length of the heat sink. |
| $X_{PZT}/L_{FIN}$ | Horizontal orientation. |
| $\omega$ | The vibrating frequency. |
| $\alpha$ | Fan tip amplitude. |

| | |
|---|---|
| $Q_H$ | Input heat from the heat source. |
| $Q_{FIN,To}$ | Total dissipated heat from the fin surface by convection. |
| $Q_{FIN,i}$ | The dissipated heat from each section. |
| $A_{FIN}$ | Total area of the inner fin surface. |
| $\overline{T_s}$ | The average temperature of the inner surface. |
| $T_\infty$ | The ambient air temperature. |
| $\overline{h}_{PZT}$ | Forced convection coefficient with the PZT fan |
| $\overline{h}_0$ | Natural convection coefficient |
| $T_{base,i}$ | Temperature of each section at the bottom fin surface. |
| $T_{side,i}$ | Temperature of each section at the side fin surface. |
| $A_{base,i}$ | The area of each section at the side bottom surface. |
| $A_{side,i}$ | The area of each section at the side fin surface. |
| $k$ | The conductivity of air. |
| $\nu$ | The dynamic viscosity |
| $w$ | The width of the piezoelectric fan. |
| $M_p$ | Dimensionless PZT-convection number |

# Modeling of Fan Failures in Networking Enclosures

Susheela Narasimhan
Cisco Systems Inc
250 W. Tasman Drive
San Jose, CA 95134
snn@cisco.com

Gokul Shankaran and Shankar Basak
Ansys Inc
Austin, TX
Sankar.basak@ansys.com
Gokul.Shankaran@ansys.com

## 1.0 Abstract

Modeling of fan failures in networking chassis is a challenging task. There is not enough data or literature available to accurately model fan failures. This paper embarks on a study consisting of both modeling and experimental cases to investigate how to accurately model fan failures. The study will include CFD simulations in different ways to model fan failures and also real life experimental measurements to verify the simulation concepts. Recommendations will then be made about the exact and accurate ways of modeling fan failures. The study also involves cases of fan failures for both front to back airflow (Pull Systems) and back to front airflow (Push Systems)

Normally the fans have been modeled as two dimensional entities. The fan curve measured by the vendor is used in the fan during modeling. The problem that arises with this kind of a fan modeling especially during fan failures is that the three dimensional effect of the rotor and stator blades of the fan is not taken into account. In reality, the fan blades provide a big obstruction to the flow reversal that happens due to pressure imbalance during fan failures.

In this paper, we start with modeling a single fan in an AMCA wind tunnel. The complete rotor and stator geometry of the fan is modeled. We run a MRF (Multiple Reference Frame) model to generate the fan curve for the fan and compare it with the experimental fan curve.

After we validate the fan curve in an AMCA model for a single fan, the paper discusses three different sets of temperature and flow data:

i. Temperature and flow data in a real system with four fans modeled with two dimensional fans

ii. Temperature and flow data in a real system with four fans modeled with MRF fans (full 3 dimensional rotor and stator blade geometry)

iii. Experimental comparisons with the simulated data

Conclusions will be drawn based on this modeling and experimental data about accurate ways of modeling fans during fan failures in real systems.

## 2.0 Introduction

The power dissipation of electronic components consisting of CPUs (Processors) and also Application Specific Integrated Circuits (ASIC) have burgeoned from being 20-30 W in the year 2000 to about 80-100 W today. The prime air movers (fans) have increased their speeds from being 9000 RPM to about 20000 RPM for small fans and from 4000 RPM to about 10000 RPM for bigger fans. All these higher speeds and increased power dissipations lead to

fan blades being exposed to higher temperatures and the failure rates of fans are increased.

It has become important to look at fan failures during thermal design early in the design cycle. The main question that arises is how to model fan failures. Most of the software tools available model fan failures using a 2-dimensional lumped fan object. There is no clear answer as to how to model a fan failure. The air flow reversal that occurs due to the fan failure is subjected to the resistance of the free-wheeling of blades due to the flow reversal. It is hard to characterize the impedance during reversed airflow and free-wheeling of the blades. The Multiple Reference Frame (MRF) fan modeling technique [1,2], which is based on the actual rotor geometry, is likely to better capture this physics. This study goes through a methodology to model the fan failures in networking enclosures. The ability of the MRF fan model to capture the effects of fan failure is also investigated.

## 3.0 Methodology

The study focuses on modeling fan failures using two dimensional fans and fans that have real blade geometries to capture the flow obstructions accurately. The results from the simulation will also be compared against some experimental data for two hot ASICs in a networking top of the rack switch. The study involves the following steps:

1) Obtaining fan blade geometry from the fan vendor

2) Characterizing and compare the measured fan curve for a single fan in a numerical wind tunnel with a Multiple Reference Frame (MRF) fan [1]

3) CFD modeling of a top of rack networking switch with 4 of these fans modeled as 2-dimensional lumped fans. The model will include cases of (i) all fans running, and (ii) fan failure of one of these fans at a time, ie; a total of 5 cases are solved.

4) CFD modeling of the same networking switch with the 4 fans modeled using the MRF approach for all 5 cases.

5) Comparison of temperature and airflow results from the 2-dimensional lumped fan simulations (3 above), MRF simulations (4 above), and experimental results, for all cases. Experiments are performed in an environmental chamber.

## 4.0 Fan Blade Geometry and Fan Curve Characterization

The axial fan geometry of the 40mm x 40mm x 28mm, was obtained from the fan vendor. The fan blade geometry consists of the rotor blades as shown in Figure 1 and the stator blades as shown in Figure 2.

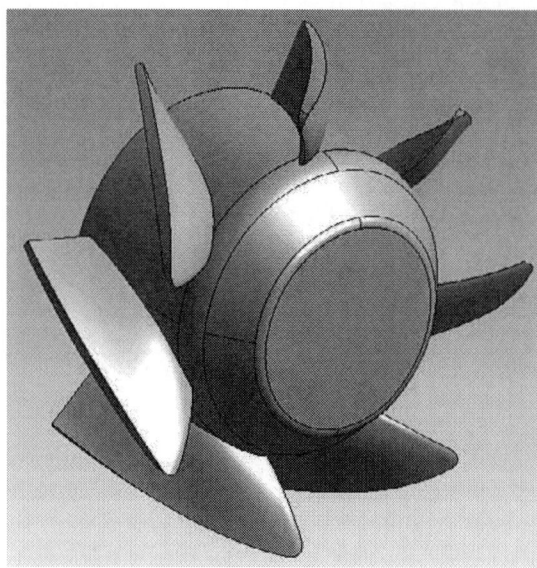

**Figure 1. Rotor Blade Geometry of the fan**

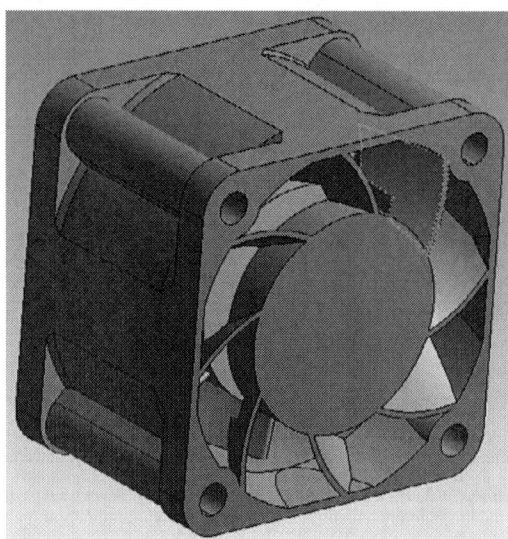

**Figure 2. Stator Blades of the Fan Geometry**

The rotor blades have an airfoil shaped cross-section, which varies from the root to the tip of the blade. It is important to use this exact blade geometry for good accuracy. In this case, the vane blades also have a complex shape which must be incorporated in the model. These are best obtained from the fan vendor.

The fan blade geometry was then put into a numerical wind tunnel, similar to the exhaust duct wind tunnel AMCA standard [1, 4]. This wind tunnel was then used to characterize the fan under different speeds to obtain the fan curve. Figure 3 below shows the comparison between the vendor's experimentally measured fan curve and the MRF generated fan curve

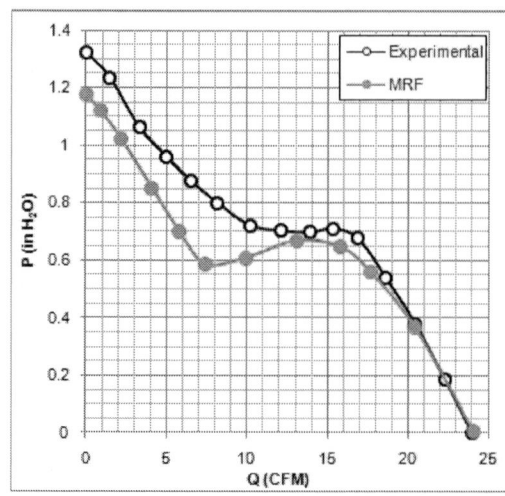

**Figure 3. Comparison Between Vendor Curve and MRF Generated Fan Curve**

The curves compare well at higher flow (lower pressure) regions and fairly well at the lower flow (higher pressure) regions. There is a discrepancy that is displayed near the knee region.

## 5.0 Modeling of a Networking Switch with 2D Fan and MRF Fan

The system considered for this study consisted of a top of the rack switch in which the airflow traverses from front to back (cold isle to hot isle in a data center) and also back to front ( hot isle to cold isle in a data center). The system has a very narrow inlet (0.13") that is formed by a system of two plates. The system inlet is shown below in Figure 4. The system of perforated plates is shown in Figure 5.

**Figure 4. Picture showing the thin slit opening for the top of the rack switch**

978-1-4673-1110-6/12 $31.00 © 2012 IEEE     250

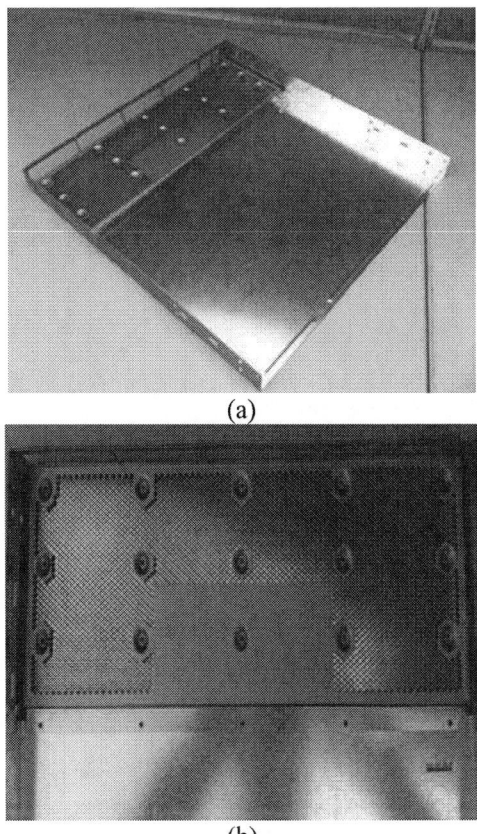

(a)

(b)

**Figure 5. The two plate system that forms the channel for airflow path (a) oblique view (b) top view**

**PULL/Front to Back System** :- Air enters the switch through the thin slit and traverses the narrow channel. It is then pulled through the switch by a system of four 40 mm x 40 mm x 28 mm fans and also two power supplies which have their own 40 mm fans. The system fans are numbered Fan1-Fan4 from left to right. It is important to mention that the system fans had a max RPM of 14700 whereas the power supply fans had a max RPM of 9500. In this study, only the left power supply slot was filled. The right power supply (PS) slot was filled with a blank. Five different sets of thermal simulations were carried out.

The first simulation included all 4 fans running together. Subsequent simulations were conducted with each of the fans failing sequentially, with the remaining 3 fans running. The case temperatures of two important switching ASICs were compared between the 2D fan case, MRF fan case and also experimental values. The two switching ASICs are shown as Forwarding Engine (FE) and Packet Processor (PP) as shown Figure 6. The forwarding engine FE is located very close to the fan tray whereas the Packet Processor PP resides right behind PP. FE dissipates about 28 W while the PP dissipates about 25 W. The system fans were run at a RPM of 11,100 RPM whereas the power supply fan was run at a RPM of 6500.

**PUSH / Back to Front System** :- In case of the Push / Back to Front airflow system, the air is pushed by the four fans into the system and exhausts thro the narrow slit. The power supply fans also push the airflow from the back to the front through the narrow slit. For the purpose of this study, the system fans were run at a RPM of 12700 and the power supply fans were run at a RPM of 11100. The left bay of the power supply was filled with a dead power supply and the right bay was filled with a regular power supply.

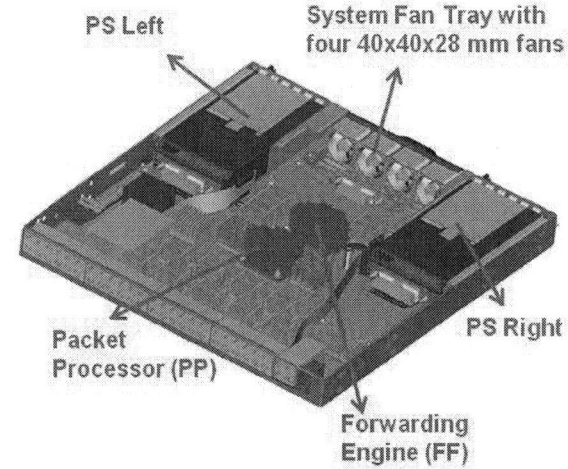

**Figure 6. Diagram showing the top of the rack networking switch with details of components and fan tray**

## 6.0 Results and Discussion

Simulations were run with both the 2D fan, MRF fan for the cases of all fans running and also each of the four fans failed sequentially. Experiments were conducted in a similar manner inside an environmental chamber. The ambient temperature used for the simulations and also the experiments was 40 °C.

The results will be discussed under two perspectives for both the front to back system and the back to front system.
   a) Temperature
   b) Airflow

### PULL SYSTEM – FRONT TO BACK SYSTEM

#### A) TEMPERATURE:

The component temperatures of PP and FE were compared for all fans running and also each of the fan failure cases. The comparison is as shown below in Table 1 and all temperatures are shown in C. The % error is calculated based on rise over ambient (40 °C)

Two different conclusions are very clear from the above result. For the case of all fans running, both 2D fan model and the MRF model seem to give close result when compared with the experiment. For fan failure cases, the MRF model gives more accurate prediction for the Forwarding Engine (FE) which is the one closest to the fan blades. The Packet Processor (PP) temperatures are better predicted by the 2D fan. In general, the increase in temperature due to fan failure seems to be underpredicted by the MRF fan and over predicted by the 2D fan.

978-1-4673-1110-6/12 $31.00 © 2012 IEEE

## Table 1. Case temperatures of PP and FE components of Figure 6

| | All fans On | | | | |
|---|---|---|---|---|---|
| Component | Expt | 2d Fan | %err | MRF | %err |
| PP | 70.91 | 69.50 | -4.55 | 66.65 | -13.76 |
| FE | 80.79 | 82.20 | 3.46 | 77.88 | -7.12 |

| | Fan 1 Failed | | | | |
|---|---|---|---|---|---|
| Component | Expt | 2d Fan | %err | MRF | %err |
| PP | 77.51 | 77.90 | 1.05 | 68.31 | -24.52 |
| FE | 87.57 | 97.00 | 19.82 | 81.32 | -13.14 |

| | Fan 2 Failed | | | | |
|---|---|---|---|---|---|
| Component | Expt | 2d Fan | %err | MRF | %err |
| PP | 77.64 | 77.90 | 0.68 | 68.09 | -25.38 |
| FE | 87.52 | 97.80 | 21.63 | 80.98 | -13.77 |

| | Fan 3 Failed | | | | |
|---|---|---|---|---|---|
| Component | Expt | 2d Fan | %err | MRF | %err |
| PP | 77.64 | 78.20 | 1.48 | 68.27 | -24.91 |
| FE | 86.81 | 97.50 | 22.83 | 80.79 | -12.88 |

| | Fan 4 Failed | | | | |
|---|---|---|---|---|---|
| Component | Expt | 2d Fan | %err | MRF | %err |
| PP | 78.20 | 78.00 | -0.53 | 69.37 | -23.11 |
| FE | 86.38 | 93.80 | 15.99 | 80.39 | -12.93 |

## B) AIRFLOW:

The airflow for both the MRF fan and the 2D fan for the cases f all fans running, and Fan1 failed are shown below, in Table 2.

## Table 2. Flow rate (in CFM) comparisons between MRF and 2D fan models

| | All fans working MRF Fan | All Fans Working 2D Fan | Fan 1 failed, MRF Fan | Fan 1 Failed, 2 D fan |
|---|---|---|---|---|
| Fan 1 | 12.2 | 7.23 | -7 | -4.95 |
| Fan 2 | 6.4 | 7.25 | 12.4 | 8.36 |
| Fan 3 | 5.3 | 7.207 | 13.2 | 8.54 |
| Fan 4 | 12.8 | 7.29 | 13.7 | 8.59 |

- For the case of all fans working, the 2D fan predicts uniform airflow for all fans, whereas the MRF fan predicts non-uniform airflow between fans. The total airflow of 36.7 CFM for MRF model is closer to the experimentally measured value of about 36 CFM. Figure 7 shows the streamlines in the MRF simulation.
- When the Fan1 was failed, a flow reversal of about 5-7 CFM was predicted by the 2D fan and MRF fan models.
- Figure 8 shows airflow reversal from the MRF model when Fan4 was failed. The airflow reversal through the failed fan can be seen clearly.

## PUSH SYSTEM – BACK TO FRONT SYSTEM

### A) TEMPERATURE

The temperatures of PP and FE were compared for all fans running and also the case of individual fan failed. The summary is shown in Table 3 below

| | All Fans Working | | | | |
|---|---|---|---|---|---|
| Component | Expt | 2d Fan | % Error | MRF Fan | % Error |
| PP | 67.9 | 69 | 3.9 | 68.98 | 3.9 |
| FE | 59.7 | 59.3 | 2.0 | 57.1 | 13.2 |

| | Fan 2 Fail | | | | |
|---|---|---|---|---|---|
| Component | Expt | 2d Fan | % Error | MRF Fan | % Error |
| PP | 81.5 | 78.8 | 6.5 | 82.5 | 2.4 |
| FE | 68.4 | 64 | 15.5 | 73.3 | 17.3 |

| | Fan 3 Fail | | | | |
|---|---|---|---|---|---|
| Component | Expt | 2d Fan | % Error | MRF Fan | % Error |
| PP | 74.6 | 76.5 | 5.5 | 76 | 4.0 |
| FE | 61.9 | 65.2 | 15.1 | 60.6 | 5.9 |

| | Fan 4 Fail | | | | |
|---|---|---|---|---|---|
| Component | Expt | 2d Fan | % Error | MRF Fan | % Error |
| PP | 73.1 | 70.8 | 6.9 | 76.4 | 10.0 |
| FE | 61.6 | 60.2 | 6.5 | 62.5 | 4.2 |

### Table 3 Temperature Rise over ambient Comparison between 2D fan and MRF fan for Back to Front System

As can be seen from the table above, for the component far away from the fans (PP), the error is within 10% of the experimental values for both the 2D and the MRF models. For the component close to the fans (FE), the error is slightly more of the order of 15-17% for both 2D and MRF models.

### B) AIRFLOW

Table 4 below shows the aiflow comparison between 2D fan and MRF fan for the PUSH ( back to front airflow system)

Table 4 Airflow Comparison between 2D fan and MRF fan for Back to Front Airflow System

| | All Fans Working MRF Fan | All Fans Working 2D Fan | Fan 2 Failed MRF Fan | Fan 2 Failed 2D Fan |
|---|---|---|---|---|
| Fan 1 | 15 | 7 | 15.1 | 9 |
| Fan 2 | 4.5 | 7 | -7.7 | -6 |
| Fan 3 | 4.2 | 7 | 5 | 9 |
| Fan 4 | 4.3 | 7 | 15.5 | 9 |

The following conclusions can be drawn from the airflow compilation

a) 2D fans predict uniform flow for each of the fans whereas MRF fans predict non-uniform flow for the fans.

b) A flow reversal of 7 CFM is seen through the failed fan in both 2D and MRF fan models.

Figure 9(a) and 9(b) show the airflow streamlines when all fans are working for the MRF model.

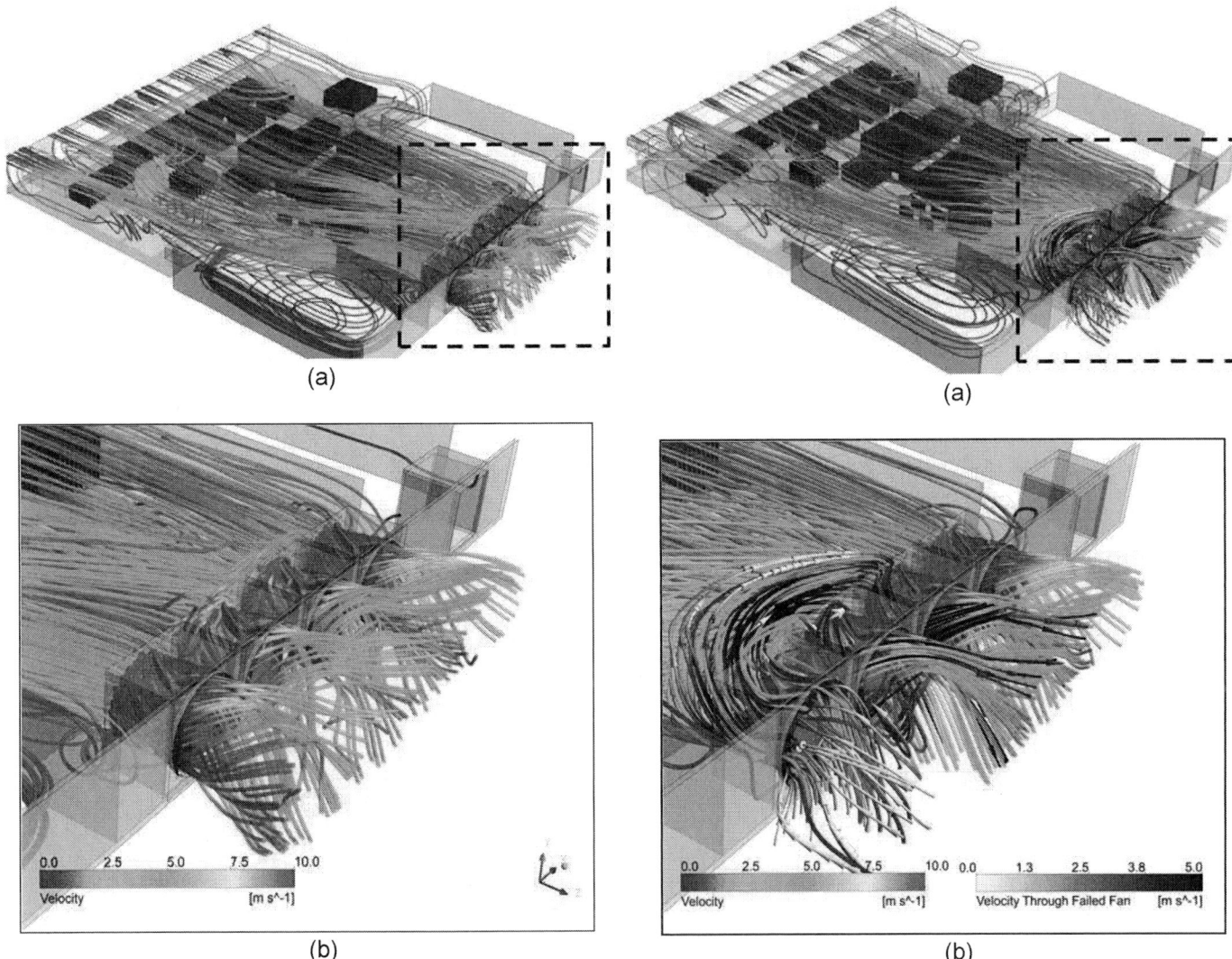

**Figure 7. Streamlines with all fans on (a) Full model, (b) zoom-in of boxed region**

Figures 10(a) and 10(b) show the streamlines for the MRF model when fan4 was failed.

**Conclusions**

A detailed study consisting of modeling a 2D fan, MRF fan and experimental studies was undertaken to predict temperatures and airflow due to fan failures. Following conclusions were drawn:

**PULL System**

- Predictions with both fan models were acceptable for the case of all fans running.
- While the MRF model predicted better correlation to the chip temperature closer to the fan blades, the 2D fan better predicted the temperature rise in the PP package due to fan failures.

**Figure 8. Streamlines with Fan 4 failed, all other fans are on (a) Full model, (b) zoom-in of boxed region**

- The MRF fan model predicted the airflow better than the 2D fan when compared to the experiment.

**PUSH System**

- Both MRF and 2D fan models gave temperatures within engineering approximation when compared to the experimental values under the conditions of all fans working and also fan failure scenarios.
- The 2D fan predicted uniform airflow for the fans whereas the MRF fan predicted a non-uniform airflow for the fans.
- The airflow in the environmental chamber, and exact component heat dissipation values, are some factors that could have affected the chip temperatures during experimental testing.

Modeling adequate environment around the network switch may affect the CFD model results. More detailed investigations are needed to accurately measure, model and predict fan failures. It is also recommended to use a wind tunnel measured airflow versus pressure drop curve for a failed fan if using a 2D model for fan failures.

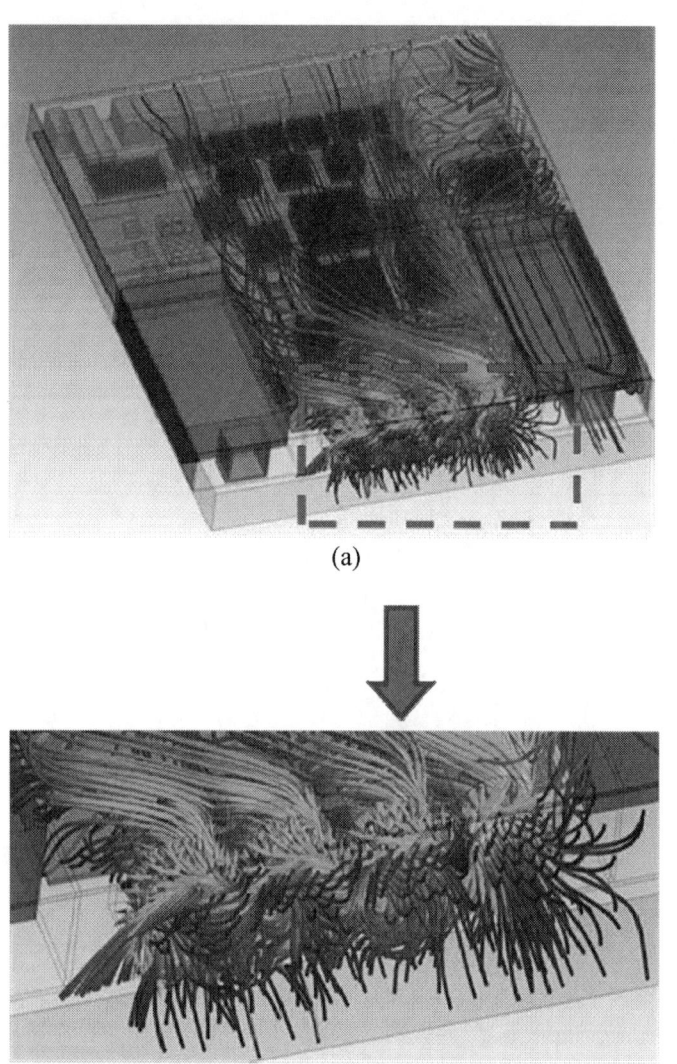

(a)

(b)

**Figure 9 Airflow Streamlines for the MRF model with all fans working a) Full Model b) Zoomed in Box Region**

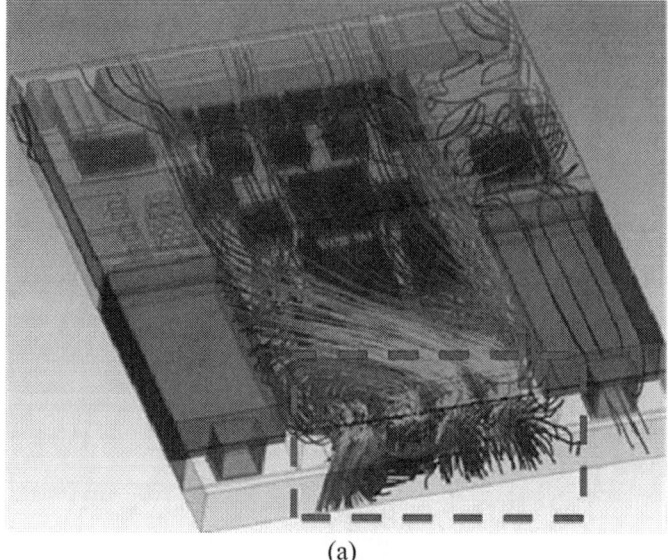

(a)

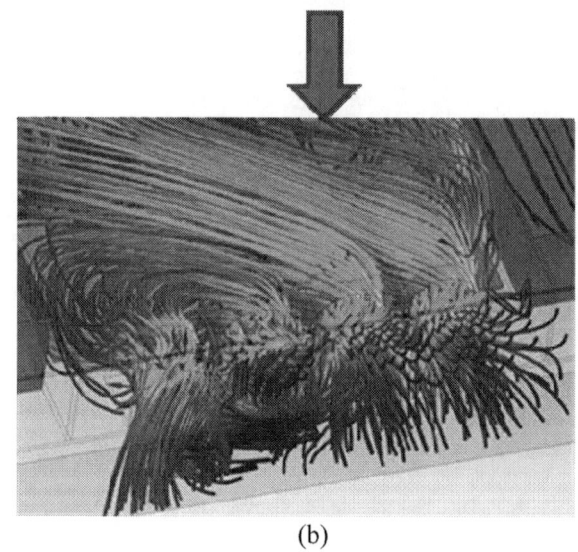

(b)

**Figure 10 Airflow Streamlines for the MRF model with fan4 failed a) Full Model b) Zoomed in Box Region**

## References

[1] G. V. Shankaran, and M. B. Dogruoz, "Validation of an Advanced Fan Model with Multiple Reference Frame Approach," Proc. ITHERM - 12th IEEE Intersociety Conference on Thermal and Thermomechanical Phenomena in Electronic Systems, June 2010, Las Vegas, NV, USA

[2] T. Q. Dang, and P. R. Bushnell, "Aerodynamics of Cross-Flow Fans and Their Applications to Aircraft Propulsion and Flow Control," Progress in Aerospace Sciences, Chapter 45, Elsevier, 2009.

[3] G. K. Batchelor, An Introduction to Fluid Dynamics, Cambridge University Press, Cambridge, UK, Reprinted 1992.

[4] ANSI/AMCA Standard 210-07 | ANSI/ASHRAE 51-07, "Laboratory Methods of Testing Fans for Certified Aerodynamic Performance Rating", AMCA International Inc.

[5] ANSYS Icepak 13 Documentation, ANSYS Inc., 2010.

# Synjet Augmented Cooling of a 1U Security Chassis

Raghav Mahalingam and Brandon Noska
Nuventix Inc.
Ph: 512-736-9523
Email: raghav@nuventix.com

Susheela Narasimhan
Cisco Systems Inc
snn@cisco.com

## Abstract

In this paper, we present a case study for an approach that uses a combination of system fans and localized Synjets to augment the cooling performance in a 1U security chassis. We show that the improvements in thermal performance achieved by using a Synjet augmented fan approach can be translated into significant improvements in cooling system power consumption and acoustics. We also show that overall system reliability can be increased by using an appropriate selection of Synjets and fan speeds, despite the fact that additional active components are introduced into the system. Finally, some remarks are made on the system cost implications of localized Synjet cooling implementation.

## Keywords

Synjet, fan augmentation, power, acoustics, reliability

## 1. Introduction

With a rapid increase in cloud computing over the last few years, cooling needs of data centers and servers have grown rapidly. The use of large banks of fans is quite common in such systems and cooling system power is a significant fraction of the data center operational costs. Additionally, noise level from fans is a strong function of fan speed and this limits the use of higher speed fans. As computing needs increase it is getting harder for system architects to stay with air cooling, though it is commonly preferred due to its simplicity. A methodology to extend the use of air cooling has been developed at Nuventix, where extremely high reliability local synthetic jets (also called Synjets) are used in conjunction with system fans to improve the overall system performance, i.e., lower acoustics, power consumption and higher reliability.

Synthetic jets are formed by periodic suction and ejection of fluid out of an orifice bounding a cavity by the time periodic motion of a diaphragm that is built into one of the walls of the cavity (Figure 1). During the ejection phase (the first three frames in Figure 1), a coherent vortex, accompanied by a jet, is created and convected downstream from the jet exit. Once the vortex flow has propagated well downstream, ambient fluid from the vicinity of the orifice is entrained (the last two frames in Figure 1). The bulk of the high speed air has moved away from the orifice, avoiding re-entrainment, while quiescent air from around the orifice is sucked into the orifice. Thus, a synthetic jet is a "zero-mass-flux" jet

comprised entirely of the ambient fluid and can be conveniently integrated with the surfaces that require cooling without the need for complex plumbing. The evolution of a two-dimensional synthetic jet has been studied in detail by Smith and Glezer[1]. The far field characteristics (e.g. rate of lateral spreading and streamwise decay of centerline velocity) are similar to conventional turbulent jets. Electronic cooling with Synjets is dealt with in detail by Mahalingam et al. [2,3]

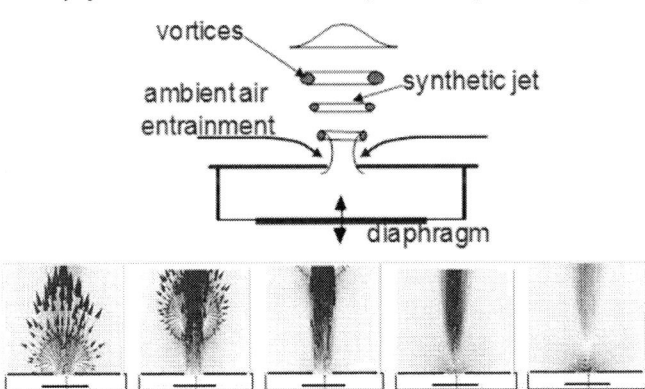

**Figure 1.** Schematic of synthetic jet formation and particle Image Velocimetry data of the formation process

Synthetic jets have been shown to be very effective in control of mean flows in aerodynamic flow control applications [4]. A similar principle can be applied to flow bypass control as shown in Figure 2, which describes the use of synthetic jets for controlling a fan induced flow passing over a heat sink. The dotted lines in the figure show the streamlines observed using smoke visualization. In the top figure, an un-ducted heat sink is subject to a fan flow that is pulling air through the heat sink. As can be seen from the streamlines a significant portion of the incoming flow tends to flow over the heat sink. Adding Synjets upstream of the inlet to the heat sink enables controlled re-entrainment of the inlet flow to reduce flow bypass while simultaneously breaking up the local boundary layers on the walls of the heat sink.

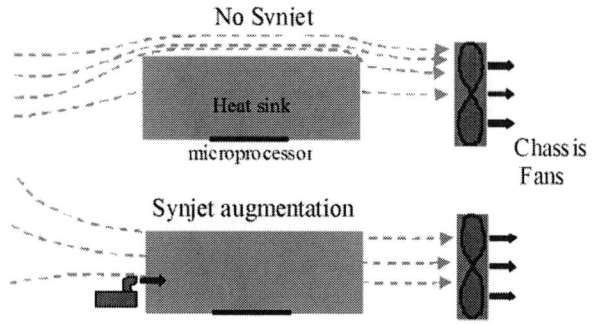

**Figure 2.** Schematic of fan augmentation with Synjets

Mahalingam et al. [5] showed the efficacy of Synjet augmentation in Newisys 4300 quad-socket, 3U, AMD Opteron rack-mounted model server. Figure 3 shows the case to ambient thermal resistance of one of the CPU's with and without the jet augmentation at different fan speeds. At the idling speed of 5500 RPM the thermal resistance drops from about 0.43 C/W to about 0.35 C/W, while at the full speed of 9000 RPM, the performance goes from 0.33 to 0.3 C/W.

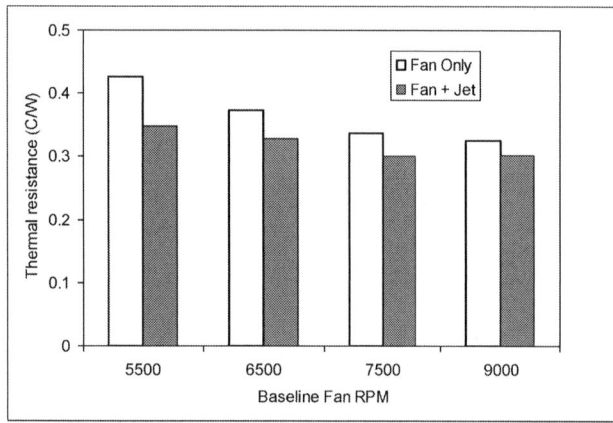

**Figure 3.** Decrease in thermal resistance due to the Synjet augmentation

Figure 4 shows the improvement in the noise levels to the jet augmentation. The data shows the sound pressure level for the fan only and the jet augmentation case at the same heat dissipation level. Thus, by operating the fans at lower speed and using the jets to augment the performance of the fans, the SPL of the system was effectively dropped by about 9dBA.

| Configuration | System power consumed | System Acoustics |
|---|---|---|
| 9000 RPM Fans Only | 108W | 75dBA |
| 6500 RPM Fans Plus Synjets | 62W | 65dBA |

**Figure 4.** Improvements in System sound pressure level and cooling power consumption for equivalent thermal performance.

In addition to improving the acoustic emissions, the addition of Synjets has the potential to improve the system reliability (lifetime) by enabling reduced fan speeds. Reducing the system fan speed will greatly affect the system lifetime by reducing wear of mechanical fan components, by reducing the rate at which airborne contaminants foul the system and the fan, and by reducing vibrations on the system imposed by the fan. Synjets are created using electromagnetic actuators that have no moving parts in friction. This is the primary reason for their higher reliability. Synjets tested in the Nuventix reliability facility have shown lifetimes of several hundred thousand hours in accelerated life tests[6], as seen in Figure 5 and thus the combined lifetime of a SynJet and a lower speed fan has the potential to be better than the same fans running at higher speed.

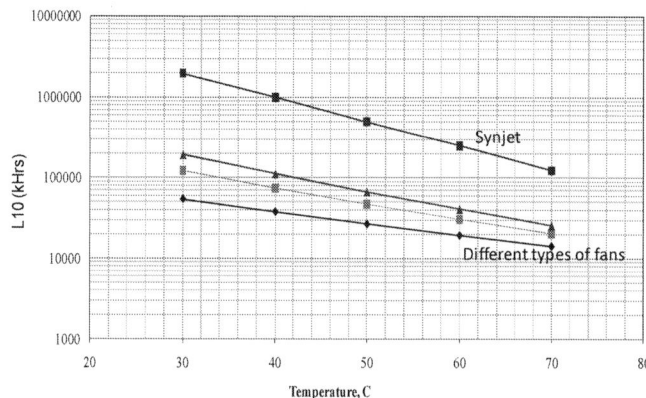

**Figure 5.** L10 lifetime of Synjet compared to fans

In Section 2, we will demonstrate the benefits of using localized augmentation from an analytical standpoint, using known relations for effect of fan RPM on fan acoustics, power and reliability. Following that, in Section 3, we will show the experiments performed on a Cisco 1U security chassis where localized Synjets are used to augment the muffin fans used in the system. Section 4 will present conclusions from this work.

## 2. Augmentation of system fans with localized Synjets

The principle of augmentation is practical only when the addition of the localized cooling element does not adversely affect system parameters such as acoustics, reliability and power consumption. In this section, we will show that under certain operating conditions, it is indeed beneficial to use a localized cooling solution.

The fundamental idea behind localized cooling is that if there are a few hot spots in the system that drive up the system fan speeds, those hot spots can be cooled locally, and the system fan speeds can be lowered. Lowering the fan speeds, or the fan RPM, has the benefit of lowering fan power, fan acoustics and increasing fan reliability. There are established relationships between fan RPM and power and acoustics[7]. It was assumed that the number of failures in time (and therefore, L10) is proportional to the RPM of the fan. These relationships are shown in Table 1 below.

978-1-4673-1110-6/12 $31.00 © 2012 IEEE

| Power(P) | $P_2 = P_1 (RPM_2 / RPM_1)^3$ |
|---|---|
| Noise (N) | $N_2 = N_1 + 50 \log_{10}(RPM_2 / RPM_1)$ |
| Failure Rate (FIT) | $FIT_2 = FIT_1 (RPM_2 / RPM_1)$ |

**Table 1.** Relationships between fan performance characteristics and RPM

Figure 5 shows the effect of adding localized cooling on system acoustics. The amount of reduction that can be achieved in fan RPM is represented as the ratio of augmented to un-augmented RPM on the x-axis and is a critical parameter to achieve acoustic mitigation with localized cooling. The dBA level difference between the Synjets and the fans is a strong factor in determining the reduction achieved in acoustics.

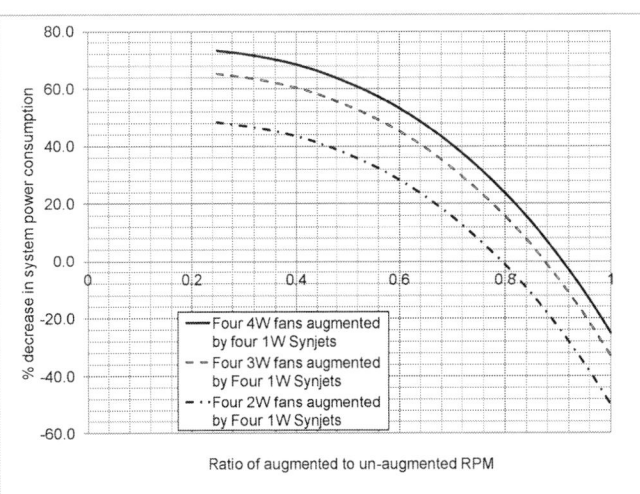

**Figure 7.** Effect of addition of localized Synjets to system power.

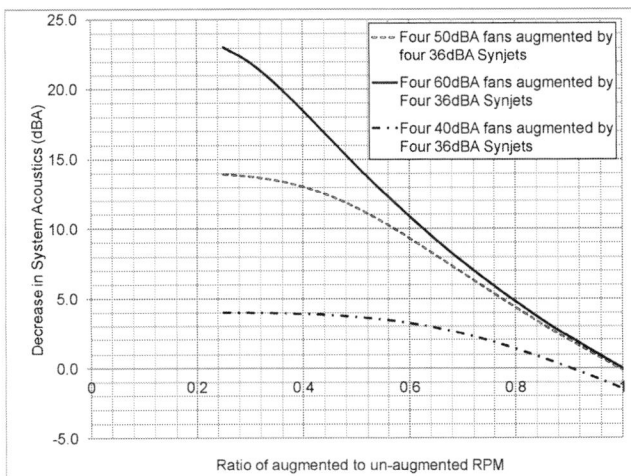

**Figure 6.** Effect of addition of localized Synjets to system acoustics.

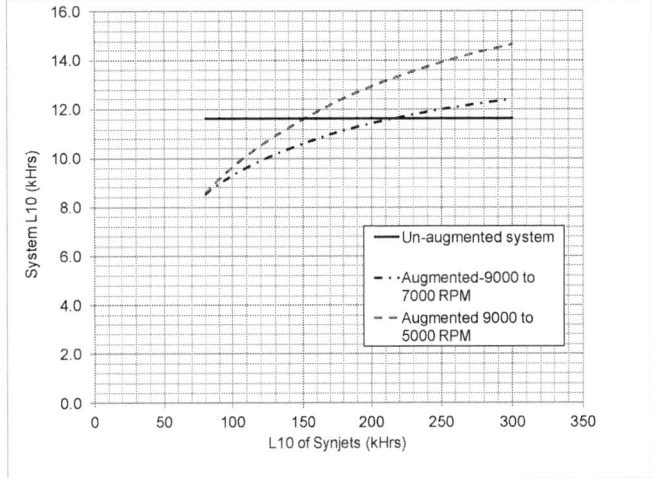

**Figure 8.** Effect of addition of localized Synjets to system reliability.

Figure 7 shows the effect of localized cooling on system power. Again, as was the case with system acoustics, the amount of reduction that can be achieved in the fan RPM is critical. Since most Synjets consume 1W or less power, they provide significant benefits in power consumption.

Traditionally, it has been assumed that localized active cooling is impractical since the addition of any extra moving parts in the system will decrease system life. However, this is only true if the life of the added cooling solution is comparable to the life of existing fans. As can be seen from Figure 8, the lifetime of the system depends strongly on the lifetime of the added localized cooling system. For example, even if the fan RPM can be reduced from 9000 to 5000, the localized cooling solution has to have an L10 of at least 150kHrs for each added component (for 4 fans at 40kHrs each and 4 added localized Synjets). Fans typically do not have such large L10 lifetimes. However, Synjets have been shown to have L10 lifetimes in excess of 300kHrs and thus make localized augmentation feasible.

From the charts it can be inferred that under the right conditions, it is possible to achieve one or more system benefits from localized cooling.

## 3. Augmentation demonstration on a Cisco 1U security chassis

A 1U security chassis with two Intel 3 GHz Prescott CPU at 90 W was cooled using Synjet augmentation. The heat sinks have embedded heat pipes and three fans provided 150 to 200 LFM upstream of the heat sinks. Commercially available Synjets were retrofitted into this setup, and the jets were not specifically designed for this application. The setup is shown in Figure 9.

**Figure 9.** 1U Security Chassis cooled by Synjet augmentation.

### 3.1. Results

Figure 10 shows the improvement in thermal performance as a function of incoming mean velocity created by the fans. The curves show a strong dependence on incoming velocity, but a significant improvement in theta for all incoming velocities. The more significant way to look at the data is to compare, for a given theta, what the mean flow velocity would be with and without Synjets. For example, at the 0.4 C/W theta point, the incoming fan velocity can be reduced from 0.95m/s down to about 0.75m/s. This suggests that the fans flow has to be about 30% higher if Synjets are not used in the system. Similarly, in Figure 11, the fans can be operated at 7000RPM instead of 9000RPM at the 0.4C/W point. These reductions in fan RPM translate into significant improvements in system acoustics, power and reliability.

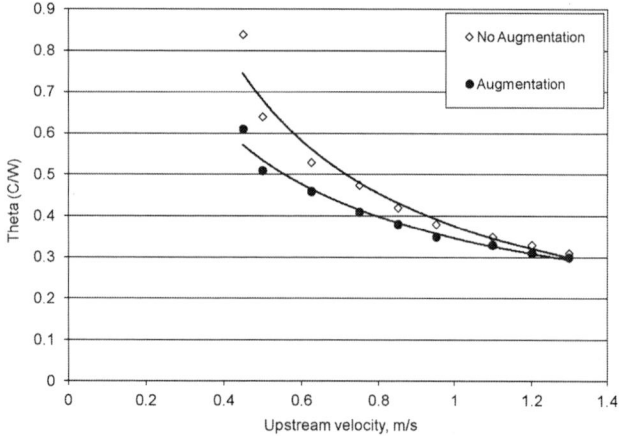

**Figure 10.** Improvements in system cooling performance as a function of system flow velocity

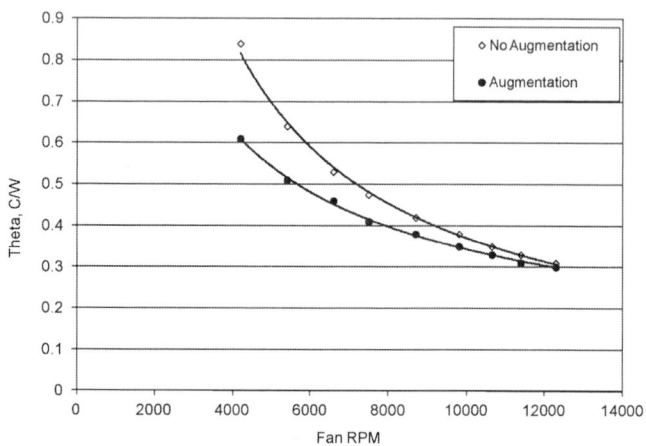

**Figure 11.** Improvements in system cooling performance as a function of system fan RPM

### 3.2. Effect on system acoustics, power and reliability

The improvements in thermal performance can be translated into system performance benefits, such as acoustics, power and reliability. Figure 12 shows the effect of lowering the fan speed on the system acoustics. These data were measured in a calibrated acoustic chamber with a noise floor of 15dbA. For example, a system cooling of 0.4 C/W can be achieved at about 50dBA with localized augmentation, as opposed to 55dBA when the only the system fans are used.

Figure 13 shows the effect on system power. At low thetas around 0.3-0.4 C/W, the system power consumption is reduced between 15-20% when augmented with localized cooling. For example, at 0.35C/W, the system power drops from 6W to 5W. While this power reduction may seem small for this particular system, but when it is realized in a large number of systems the overall energy savings become substantial. It is also important to note that the Synjets used here were not design specifically with this system in mind, but were retrofitted using existing products.

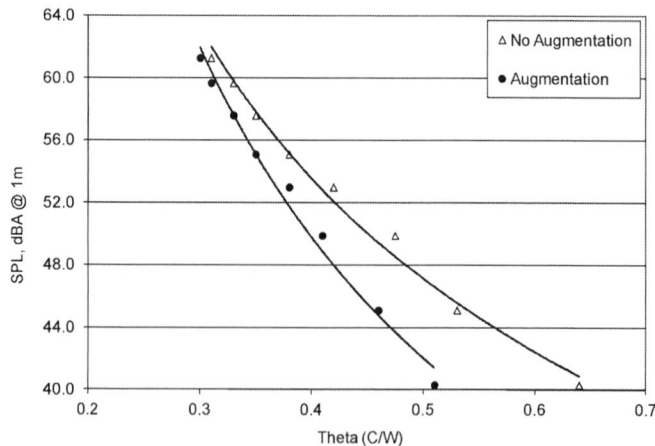

**Figure 12.** Improvements in system acoustics due to localized Synjet augmentation

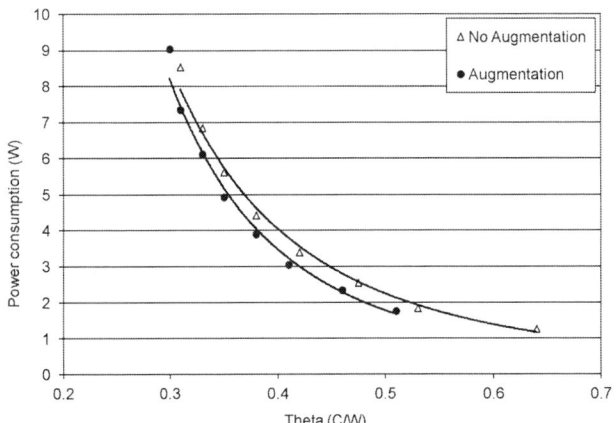

**Figure 13.** Improvements in system power due to localized Synjet augmentation

Figure 14 shows the system L10 as a function of the L10 of the added localized cooling. For the sake of matching the experimental data, the results use an un-augmented system fan RPM of 9000 and an augmented system fan RPM of 7000 RPM. These correspond to a system theta of about 0.4 C/W. It is also assumed that each individual fan has an L10 of ~40kHrs. Also, the number of fans and Synjets are accounted for in this calculation. As seen in the graph, the system reliability increases as the localized cooling reliability increases. The crossover point for when the system reliability is higher with localized augmentation occurs when the localized cooling element has a reliability of at least 210kHrs. As was seen in Figure 5, the L10 life of Synjets is significantly higher than 210kHrs, so in this system adding Synjets would increase the lifetime of the system. An assumption of 300kHrs for the Synjet L10, increases the system L10 from ~12kHrs to ~13kHrs.

### 3.3. Remarks on system cost

From a system standpoint, it is important to not ignore the additional cost of the localized Synjet. Synjets are comparable in cost to fans and therefore the increased cost of Synjets has to be recovered from lowering fan power consumption. If there are specific system components which are hot spots that drive up the fan speeds, local Synjets can be used to reduce the fan RPM. This reduction in fan RPM decreases fan power usage, which recovers the cost of the additional Synjet components over time. This has to be calculated as an ROI for each specific system.

### 4. Conclusions

A methodology is presented that uses a combination of system fans and localized Synjets, where the Synjets are used to augment the cooling performance of the fans in a 1U security chassis. Relations for effect of fan RPM on acoustics, power and reliability show that it is possible, under certain conditions, to improve one or more system performance parameters at a targeted thermal performance level. In the specific case study described here, improvements in thermal performance achieved by using a combined Fan+Synjet approach are translated into a ~5dBA reduction in acoustics, ~15% reduction in cooling power and 1kHrs increase in overall system L10. Depending on the level of reduction in power for a specific system, the overall system cost benefits can be calculated as an return on investment over time.

### References

1. Smith, B. L., and Glezer, A., "The Formation and Evolution of synthetic jets", Phys. Fluids, Vol. 10, No. 9, Sept. 1998, pp. 2281-2297.
2. Mahalingam, R., Rumigny, N., and Glezer, A., "Thermal Management using Synjet Ejectors", IEEE-CPMT, 27, 3, Sept. 2004.
3. Mahalingam, R., and Glezer, A., " Design and Thermal Characteristics of a Synjet Ejector Heat Sink", Journal of Electronic Packaging, v127, n1, June 2005.
4. Glezer, A., and Amitay, M., "Synthetic Jets", Ann. Rev. Fluid Mech., 34, pp. 503, 2002.
5. Mahalingam, R., Heffington, S., Jones, L., and Schwickert, M., "Newisys Server Processor Cooling Augmentation Using SynJet Ejectors", ITherm 2006.
6. Schwickert, M., "SynJet Thermal Management Technology Increases LED Lighting System Reliability", IEEE Reliability Society Annual Activities Report, http://www.ieee.org/reliabilitysociety, 2010.
7. http://www.comairrotron.com/airflow_note.shtml

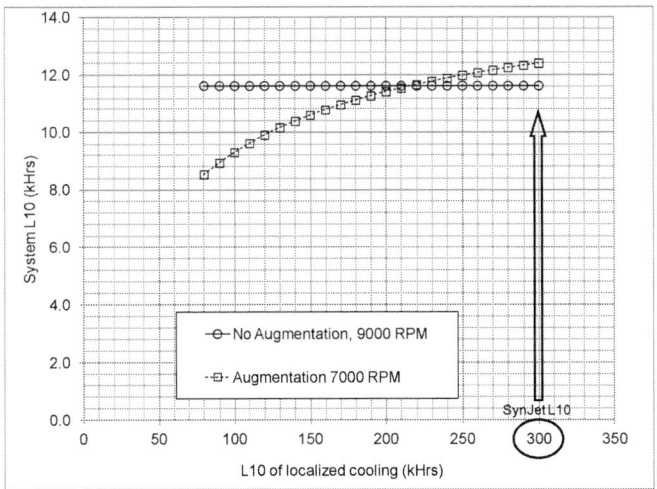

**Figure 14.** Improvement in system reliability due to localized Synjet Augmentation

# Volume Averaging Theory (VAT) Based Modeling and Closure Evaluation of Scale-Roughened Plane Fin Heat Sink

Feng Zhou, David A. Vasquez, George W. DeMoulin, David J. Geb and Ivan Catton

Department of Mechanical and Aerospace Engineering, University of California, Los Angeles, USA

48-121 Engineering IV, 420 Westwood Plaza, Los Angeles, CA 90095

zhoufeng@ucla.edu, catton@ucla.edu

## Abstract

The present paper describes an effort to model a plane fin heat sink (PFHS) with scale-roughened surfaces based on Volume Averaging Theory (VAT) and evaluate the closure terms of the model using CFD code. Modeling a PFHS as porous media based on VAT, specific geometry can be accounted for in such a way that the details of the original structure can be replaced by their averaged counterparts and the VAT based governing equations can be solved for a wide range of heat sink designs. To complete the VAT based model, proper closure is needed, which is related to a local friction factor and a heat transfer coefficient of a Representative Elementary Volume (REV). The terms in the closure expressions are complex and sometimes relating experimental data to the closure terms is difficult. In this work we use CFD code to obtain detailed solutions of flow and heat transfer through an element of the scale-roughened heat sink and use these results to evaluate the closure terms needed for a fast running VAT based code, which can then be used to solve the heat transfer characteristics of a higher level heat sink. The objective is to show how heat sinks can be modeled as porous media based on Volume Averaging Theory and how CFD can be used in place of a detailed, often formidable, experimental effort to close the VAT based model.

**Keywords** Scale roughened surface, heat sink, Volume Averaging Theory, closure

## Nomenclature

| | |
|---|---|
| $A_w$ | Wetted surface, m$^2$ |
| $A_{wp}$ | The cross flow projected area, m$^2$ |
| $c_p$ | Specific heat, J/(kg·K) |
| $D$ | Diameter of the scale, m |
| $D_h$ | Hydraulic diameter based on VAT, m |
| $d_h$ | Hydraulic diameter defined by Chang, m |
| $d_p$ | Diameter of the spherical particles, m |
| $e$ | Scale height, m |
| $F_1, F_2$ | Blending function |
| $F_p$ | Fin pitch, m |
| $f$ | Friction factor |
| $H$ | Channel height, m |
| $h$ | Heat transfer coefficient, W/(m$^2$K) |
| $k$ | Turbulence kinetic energy per unit mass, m$^2$/s$^2$ |
| $k_f, k_s$ | Thermal conductivity of fluid and solid, W/(m·K) |
| $k_t$ | Turbulent heat conductivity, W/(m·K) |
| $L$ | Length of the channel, m |
| $\langle m \rangle$ | Porosity |

| | |
|---|---|
| $Nu$ | Nusselt number |
| $P_k$ | Shear production of turbulence |
| $Pr$ | Prandtl number |
| $P_{r_t}$ | Turbulent Prandtl number |
| $P$ | Scale pitch, m |
| $\Delta p$ | Pressure drop, Pa |
| $q$ | Heat flux, W/m$^2$ |
| $Re$ | Reynolds number |
| $S$ | An invariant measure of the strain rate |
| $S_w$ | Specific Surface, 1/m |
| $S_{wp}$ | The cross flow projected area per volume, 1/m |
| $T_s$ | Channel wall temperature, K |
| $T_f$ | Fluid temperature, K |
| $u_m$ | Average velocity through the channel, m/s |
| $W$ | Width of the channel, m |

**Greek**

| | |
|---|---|
| $\alpha$ | Turbulence model constant or scale attack angle |
| $\beta, \beta^*$ | Turbulence model constant |
| $\lambda_f$ | Thermal conductivity of the fluid, W/(m·K) |
| $\mu$ | Viscosity, Pa·s |
| $\mu_t$ | Turbulent eddy viscosity, Pa·s |
| $\nu$ | Kinematic viscosity, m$^2$/s |
| $\nu_t$ | Turbulent kinematic viscosity, m$^2$/s |
| $\rho$ | Density, kg/m$^3$ |
| $\sigma_\varepsilon$ | $k$-$\varepsilon$ turbulence model constant |
| $\sigma_k$ | Turbulence model constant for the k equation |
| $\sigma_\omega$ | $k$-$\omega$ turbulence model constant |
| $\phi_1$ | Constant in the original k-ω model $\sigma_{k1}$,L |
| $\phi_2$ | Constant in the transformed k-ε model $\sigma_{k2}$,L |
| $\phi$ | Represent the constant in the SST model $\sigma_k$,L |
| $\tau_{wL}$ | Laminar shear stress, [Pa] |
| $\tau_{wT}$ | Turbulent shear stress, [Pa] |
| $\Delta\Omega$ | The volume of the REV, m$^3$ |
| $\omega$ | Specific turbulence dissipation rate |

***Subscripts and Superscripts***

| | |
|---|---|
| $\sim$ | A value averaged over the representative volume |
| $-$ | An average of turbulent values |
| $\wedge$ | Fluctuation of a value |
| $\langle f \rangle_f$ | Means the superficial average of the function |
| $f$ | Fluid phase |
| $t$ | Turbulent |
| $s$ | Solid phase |

978-1-4673-1110-6/12 $31.00 © 2012 IEEE

## 1. Introduction

With the trend towards increasing levels of integration density, thermal management has become increasingly challenging to meet the elevated power dissipation duty. A variety of techniques [1] for heat transfer enhancement have been developed, including ribs [2-6], pin fins [7-9], dimpled surfaces [10-12], surfaces with arrays of protrusions [13], and surface roughness. The design objective of these techniques is to significantly enhance convective heat transfer coefficients without substantial increases in streamwise pressure drop penalties.

A new heat transfer enhancement techniques was developed by Chang et al. [14-18] using scale roughened surfaces. The geometrical details of the proposed scale roughened surface are shown in Figure 1. The scales are arranged in the staggered manner and the two opposite scale-roughened walls were in line arrangement. The authors then compared $\overline{Nu}/Nu_\infty$ and $\overline{f}/f_\infty$ of the scale roughened surface with the flow and heat transfer results reported by different research groups for rib-roughened channels [2-6] and dimpled surfaces [10]. The heat transfer enhancement of the scale-roughened surface is surprisingly good compared with rib-roughened and dimpled surfaces.

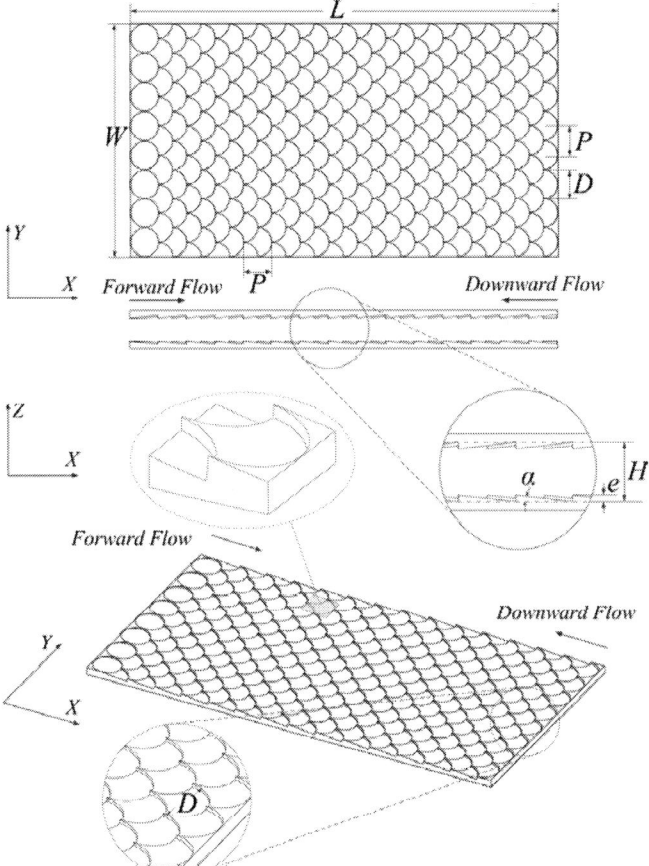

**Figure 1:** Geometrical details of scale-roughened surface (redrawing according to Fig. 1 in [14]).

Considering the scale-roughened surface is a promising heat transfer enhancement techniques, it could be applied to plane fin heat sinks to augment their heat dissipation capability, see Figure 2, but a further optimization is required. If one wants to optimize such a heat transfer device, simple equations are the only answer but they need to be made more rigorous. It is proposed that Volume Averaging Theory (VAT) [19-26] be used to develop the simple equations allowing clear rigorous statements to be made that define how the friction factor and heat transfer coefficient are to be determined. By modeling heat sinks as porous media, specific geometry can be accounted for in such a way that the details of the original structure can be replaced by their averaged counterparts and the governing VAT equations can be solved for a wide range of heat sink designs. This 'porous media' model, which is a function only of porous media morphology, represented by porosity and specific surface area, and its closure, can easily be adapted to many different structures.

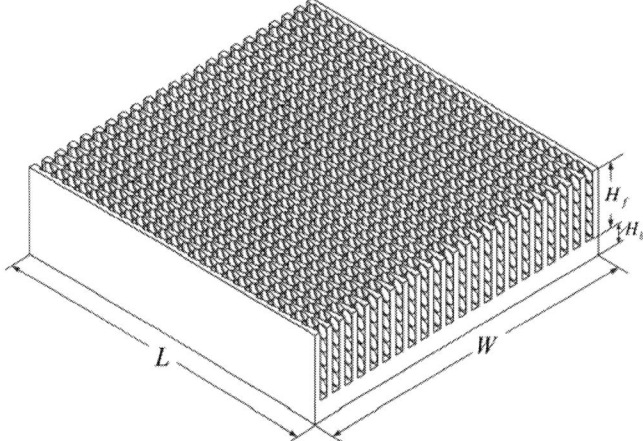

**Figure 2:** A plane fin heat sink with scale roughened surfaces.

Closure theories for transport equations in heterogeneous media have been the primary measure of advancement and for measuring success in research on transport in porous media. Obtaining closure for the VAT based governing equation set is the trickiest part while using VAT to model and optimize the heat transfer device. The porosity and specific surface area are geometrically defined terms. The closure terms, which are related to a local friction factor and a heat transfer coefficient, can be obtained in two ways. The first is to rescale the available experimental data reported for fully developed flow, using the 'porous media' length scale suggested by VAT [22-24, 26]. However, Chang et al. [14] only tested one specific scale roughened surface, which is not enough to obtain the closure. At this time, Computational Fluid Dynamics (CFD) is an alternative approach to evaluate these closure terms [27-31]. It should be noted that if CFD is used to obtain the closure, the friction factor and heat transfer will be calculated more rigorously by integrating the complete closure formula over the REV.

In the following presentation, a plane fin heat sink with scale roughened surfaces is first modelled based on Volume Averaging Theory. After that, 3-D numerical calculations are run to simulate the heat transfer and fluid flow across the channels which consist of 15 REVs. In the end, the rigorously derived closure terms are evaluated over one of the selected REVs, and two correlations for friction factor and Nusselt

978-1-4673-1110-6/12 $31.00 © 2012 IEEE

number are proposed which are the closures we are looking for.

## 2. VAT Based Modeling

A schematic diagram of a plate fin heat sink with scale roughened surface is shown in Figure 2. Generally, the air is forced to flow through the channels between the fins by a fan. The scales act to increase secondary flows and turbulence levels to enhance mixing, and to form coherent fluid motions in the form of streamwise oriented vortices, and also provide some heat transfer augmentation by increasing surface areas for convective heat transfer [1]. This is a problem of conjugate heat transfer within a heterogeneous hierarchical structure. It is not easy to optimize this kind of problem since many parameters are required to describe the geometry. Simple equations are the only answer if one wants to find the optimum configuration for these kinds of conjugate heat transfer devices.

### 2.1. VAT based governing equations

Based on rigorous averaging techniques developed by Whitaker [32] who focused on solving linear diffusion problems and by Travkin and Catton [24, 26] who focused on solving nonlinear turbulent diffusion problems, the thermal physics and fluid mechanics governing equations in heterogeneous porous media were developed from the Navier-Stokes equation and the thermal energy equations. This is the starting point for studying flow and heat transfer in porous media and also the basis of the present work.

In this section, a model based on Volume Averaging Theory is developed to describe transport phenomena in plane fin heat sinks. The air flow is considered as 'porous flow', in which the term 'porous' is used in a broad sense.

The momentum equation is

$$0 = -\frac{1}{\rho_f}\frac{d\langle \bar{p}\rangle_f}{dx} + f^* S_w \frac{\tilde{\bar{u}}^2}{2} + \frac{\partial}{\partial z}\left(\langle m(z)\rangle(v+v_t)\frac{\partial \tilde{\bar{u}}}{\partial z}\right) \quad (1)$$

The energy equation for fluid phase is

$$\rho_f c_{pf}\langle m\rangle \tilde{\bar{u}}\frac{\partial \tilde{\bar{T}}_f}{\partial x} = \frac{\partial}{\partial z}\left(\langle m\rangle(k_f+\tilde{k}_t)\frac{\partial \tilde{\bar{T}}_f}{\partial z}\right) + h^* S_w\left(\tilde{T}_s - \tilde{T}_f\right) \quad (2)$$

The energy equation for solid phase is

$$\frac{\partial}{\partial x}\left[(1-\langle m\rangle)k_s\frac{\partial \tilde{T}_s}{\partial x}\right] + \frac{\partial}{\partial z}\left[(1-\langle m\rangle)k_s\frac{\partial \tilde{T}_s}{\partial z}\right] = h^* S_w\left(\tilde{T}_s - \tilde{T}_f\right) \quad (3)$$

### 2.2. Closure Terms of the VAT Equations

To complete the VAT based model, four closure terms need to be closed. It is believed that the only way to achieve substantial gains is to maintain the connection between porous media morphology and the rigorous formulation of mathematical equations for transport.

Two of the closure terms, the averaged porosity and the specific surface area are geometrically defined and it is quite easy to define them if one selects the REV correctly. The selection for a PFHS with scale roughened surface, see Figure 3 is seen to repeat in both the cross-stream and flow directions. Therefore, the porosity is

$$\langle m\rangle = 1 - \frac{\Delta\Omega_s}{\Delta\Omega} = 1 - \frac{P_t P_l\left(F_p - H\right) + 2\kappa e P_t P_l}{P_t P_l F_p} \quad (4)$$

in which, $\Delta\Omega$ is the volume of the REV defined as

$$\Delta\Omega = P_t P_l F_p \quad (5)$$

$\Delta\Omega_s$ is the volume of the solid part of the REV defined by

$$\Delta\Omega_s = P_t P_l\left(F_p - H\right) + 2\kappa e P_t P_l \quad (6)$$

in which $\kappa$ is the ratio of the solid volume of the scale to the total volume of the $e \times P \times P$ slab, see the shaded part in Figure 3. It can easily be shown that $\kappa$ is a constant and $\kappa = \frac{3}{4} - \frac{\pi}{8}$.

The specific surface area is defined as

$$S_w = \frac{A_w}{\Delta\Omega} = \frac{2P_t P_l + \pi D e}{\Delta\Omega} = \frac{2P_t P_l + \pi D e}{P_t P_l F_p} \quad (7)$$

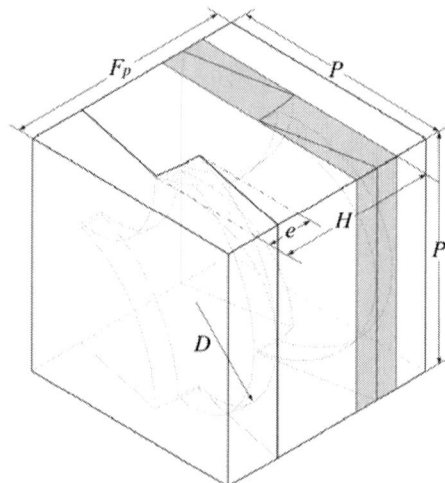

**Figure 3:** Representative Elementary Volume (REV) for a PFHS with scale roughened surfaces.

At this point, the VAT based model of scale roughened PFHS is still not fully closed. The other two closure terms, the local friction factor, $f^*$, in the momentum equation and the local heat transfer coefficient, $h^*$, in the VAT energy equations still need to be evaluated. To evaluate the closure terms, a commercial Finite Volume Method (FVM)-based code, CFX [33], is used to analyze the convective heat transfer in three-dimensional channels with two opposite scale-roughened walls.

## 3. Numerical Method and Procedures

### 3.1. Computational Domain and Boundary Conditions

Due to the reason that the local values or values for fully developed flow and heat transfer are the only two kinds of values that have a physical meaning when describing transport phenomena with VAT based equations, attention should be paid to the selection of physical model. The computational domain should be long enough, so that the closure could be evaluated over the selected REV that is not affected by the entrance or re-circulation at the outlet [34]. A computational domain with fifteen REVs was selected as the computational domain, see Figure 4.

Because of the thickness of the fin, the air velocity profile at the entrance is not uniform. The computational domain is then extended upstream a distance of five stream-wise REV length so that a uniform velocity distribution can be ensured at the domain inlet. The computational domain is extended downstream 15 times the stream-wise REV length, so that at the outer flow boundary no flow recirculation exits and the local one-way method can be used for the numerical treatment of the outer flow boundary condition [35]. The boundary conditions applied on the computational domain are tabulated in Table 1.

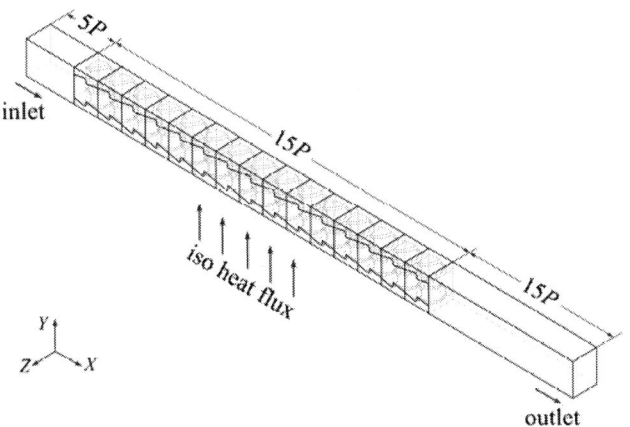

**Figure 4:** Computational domain. The length of the extended region was not drawn in scale.

### 3.2. Grid System

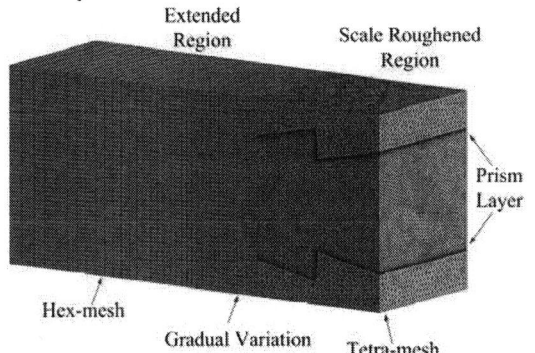

**Figure 5:** Example of the grid system. Only part of the whole model is shown.

The grid systems for all the scale-roughened channel models are built by Ansys Meshing [33]. Due to the roughness of the wall, unstructured tetra-mesh is created for the scale-roughened test channel, with prism layers being inserted in the near wall region. In the extended regions, a coarser and structured hex-mesh is adopted to conserve computational resources. A grid system with a gradual variation in and after the scale-roughened test channel is used to avoid the undesirable effect of an abrupt grid width change in the computing region. The grid system for one of the model is shown in Figure 5.

**Table 1:** Boundary conditions

| | |
|---|---|
| Inlet | $u$=const, $v$=$w$=0, $T$=const |
| Outlet | $p = p_{atm}$ |
| Surfaces of extended regions | Slip and adiabatic wall |
| Interface between air and solid | No-slip |
| Bottoms of the channel wall | Iso heat flux |
| The other surfaces | Symmetric |

Grid independence tests were made carefully by recursive refinement and comparison between the numerical simulation results. The above process was repeated until the variation of Nusselt number and friction factor was less than 0.5%, so that the numerical predictions can be regarded as grid-independent. With the turbulence predictions employed, the meshes near the fluid solid interface are fine enough to resolve the flow behavior close to the no-slip wall. For all the simulation cases, $y^+$ values in the near-wall region are less than 1.

### 3.3. Mathematical Model

The air flow is assumed to be three-dimensional, incompressible, steady state and turbulent. Buoyancy and radiation heat transfer effects are not taken into consideration. The three-dimensional governing equations for continuity, momentum and energy are as follows:

(1) Continuity equation

$$\frac{\partial \rho u_i}{\partial x_i} = 0 \tag{8}$$

(2) Momentum equation

$$\rho u_j \frac{\partial u_i}{\partial x_j} = \frac{\partial}{\partial x_j}\left[(\mu + \mu_t)\frac{\partial u_i}{\partial x_j}\right] - \frac{\partial p}{\partial x_i} \tag{9}$$

(3) Energy equation

$$\rho u_j \frac{\partial T}{\partial x_j} = \frac{\partial}{\partial x_j}\left[\left(\frac{\mu}{\text{Pr}} + \frac{\mu_t}{\text{Pr}_t}\right)\frac{\partial T}{\partial x_j}\right] \tag{10}$$

The $k-\omega$ based Shear-Stress-Transport (SST) model with automatic wall function treatment [36] is used to predict the turbulent flow and heat transfer along the heat sink channel. The model blends the robust and accurate formulation of the $k-\omega$ model in the near-wall region with the free-stream independence of the $k-\varepsilon$ model in the far field. The SST model gives a highly accurate prediction of the onset and the amount of flow separation under adverse pressure gradients by the inclusion of transport effects into the formulation of the eddy-viscosity [37]. This results in a major improvement in terms of flow separation predictions. The superior performance of the SST model has been demonstrated for high accuracy boundary layer simulations in a large number of validation studies [37].

Menter [38, 39] proposed the equations for the SST model as

$$\frac{D(\rho k)}{Dt} = \tilde{P}_k - \beta^* \rho k \omega + \frac{\partial}{\partial x_j}\left[(\mu + \sigma_k \mu_t)\frac{\partial k}{\partial x_i}\right] \tag{11}$$

$$\frac{D(\rho\omega)}{Dt} = \alpha\rho S^2 - \beta\rho\omega^2 + \frac{\partial}{\partial x_i}\left[\left(\mu + \sigma_\omega\mu_t\right)\frac{\partial\omega}{\partial x_i}\right]$$
$$+2\left(1-F_1\right)\rho\sigma_{\omega_2}\frac{1}{\omega}\frac{\partial k}{\partial x_i}\frac{\partial\omega}{\partial x_i} \tag{12}$$

where the blending function $F_1$ is defined by:

$$F_1 = \tanh\left\{\left\{\min\left[\max\left(\frac{\sqrt{k}}{\beta^*\omega y},\frac{500\nu}{y^2\omega}\right),\frac{4\rho\sigma_{\omega_2}k}{CD_{k\omega}y^2}\right]\right\}^4\right\} \tag{13}$$

in which

$$CD_{k\omega} = \max\left(2\rho\sigma_{\omega_2}\frac{1}{\omega}\frac{\partial k}{\partial x_j}\frac{\partial\omega}{\partial x_j},10^{-10}\right) \tag{14}$$

The turbulent eddy viscosity is computed from:

$$\nu_t = \frac{a_1 k}{\max\left(a_1\omega, SF_2\right)} \tag{15}$$

where $S$ is the invariant measure of the strain rate and $F_2$ is a second blending function defined by

$$F_2 = \tanh\left\{\left[\max\left(2\frac{\sqrt{k}}{\beta^*\omega y},\frac{500\nu}{y^2\omega}\right)\right]^2\right\} \tag{16}$$

To prevent the build-up of turbulence in stagnation regions, a production limiter is used in the SST model:

$$P_k = \mu_t\frac{\partial u_i}{\partial x_j}\left(\frac{\partial u_i}{\partial x_j}+\frac{\partial u_j}{\partial x_i}\right) \rightarrow \tilde{P}_k = \min\left(P_k,10\cdot\beta^*\rho k\omega\right) \tag{17}$$

Each of the constants is a blend of the corresponding constants of the $k$-$\varepsilon$ and the $k$-$\omega$ model:

$$\phi = F_1\phi_1 + \left(1-F_1\right)\phi_2 \tag{18}$$

The constants for this model take the values

$$\beta^* = 0.09,$$
$$\alpha_1 = 5/9, \beta_1 = 3/40, \sigma_{k1} = 0.85, \sigma_{\omega1} = 0.5, \tag{19}$$
$$\alpha_2 = 0.44, \beta_2 = 0.0828, \sigma_{k2} = 1, \sigma_{\omega2} = 0.856.$$

A commercial Finite Volume Method (FVM)-based code, CFX [33], is used to analyze the turbulent convective heat transfer in the scale-roughened channels. This code solves the Reynolds-averaged Navier–Stokes equations with a high resolution scheme for the advection terms as well as turbulence numerics. The fully coupled momentum and energy equations are solved simultaneously. The RMS type residual for solution convergence criteria is set to be $10^{-5}$ for the momentum balance and $10^{-6}$ for the energy equation.

## 4. Closure Evaluation

The closure evaluation in this section consists of three parts. First, two different length scales used to evaluate the flow and heat transfer characteristics of the scale-roughened channels are defined. After that, the computational model and the method adopted in current numerical simulations are verified and validated by comparing the CFD results with the experimental data. Finally, two correlations which serve as the closures of the present VAT based model is proposed based on the simulation results.

### 4.1. Length Scales

Before evaluating the closure terms, it is interesting to note that using a particular length scale leads to a parameter that is very beneficial when evaluating the heat transfer coefficient and friction factor. It was shown by Travkin and Catton [26] that globular media morphologies can be described in terms of $S_w$, $\langle m\rangle$ and $d_p$, and can generally be considered to be spherical particles with

$$S_w = \frac{6\left(1-\langle m\rangle\right)}{d_p} \tag{20}$$

$$D_h = \frac{2}{3}\frac{\langle m\rangle}{\left(1-\langle m\rangle\right)}d_p \tag{21}$$

This expression has the same dependency on equivalent pore diameter as found for a one diameter capillary morphology leading naturally to

$$S_w = \frac{6\left(1-\langle m\rangle\right)}{d_p} = \frac{6\left(1-\langle m\rangle\right)}{\dfrac{3}{2}\dfrac{\left(1-\langle m\rangle\right)}{\langle m\rangle}D_h} = \frac{4\langle m\rangle}{D_h} \tag{22}$$

This observation leads to defining a simple "universal" porous media length scale

$$D_h = \frac{4\langle m\rangle}{S_w} \tag{23}$$

that meets the needs of both morphologies: capillary and globular. This was also recognized by Whitaker [22] when he used a very similar (differing by a constant) length scale to correlate heat transfer for a wide variety of morphologies. Zhou et al. [40] also showed that using the 'porous media' length scale is very beneficial in collapsing complex data yielding simple heat transfer and friction factor correlations. Therefore, the Reynolds number defined by $D_h$ is

$$Re_{D_h} = \frac{\rho u_m D_h}{\mu} \tag{24}$$

To validate the simulation result, the Reynolds number is defined the same as Chang et al.[14]

$$Re = \frac{\rho u_m d_h}{\mu} \tag{25}$$

in which $d_h$ is the hydraulic diameter defined by

$$d_h = 2\cdot H\cdot W /\left(H+W\right) \tag{26}$$

The averaged Nusselt number in the developed region for validation is defined as

$$Nu = \frac{qd_h}{\lambda_f\left(\overline{T_s}-\overline{T_f}\right)} \tag{27}$$

The friction factor for validation is defined as

$$f = \frac{2\Delta p}{\rho u_m^2}\frac{d_h}{4L} \tag{28}$$

## 4.2. Validation and Verification

To verify the computational model and the method adopted in numerical simulation, preliminary computations were first conducted for a scale roughened channel which had the same dimensions as the one experimentally tested by Chang et al. [14], see Figure 6. From the figure we can see that the maximum deviation of the Nusselt number and the friction factor from experiment are 6.3% and 12.1% with the average deviation being around 3.5% and 5.8% respectively. Our predicted results and the experimental data agree very well, thereby showing the reliability of the physical model and the adopted numerical method.

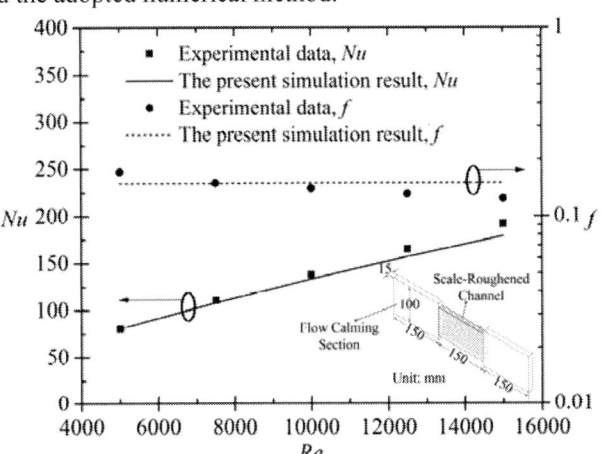

**Figure 6:** Validation of the present CFD simulation by comparing with experimental data by Chang et al. [14]

## 4.3. Closure

Travkin and Catton [26] rigorously derive the closure terms for VAT based model from lower scale governing equations. The closure term in the VAT momentum equation, $f^*$, has the form as

$$f^* = 2\frac{\int_{\partial S_w} \overline{p} \cdot d\overline{s}}{\rho_f \widetilde{\overline{u}}^2 A_{wp}} \frac{S_{wp}}{S_w} + 2\frac{\int_{\partial S_w} \tau_{wL} \cdot d\overline{s}}{\rho_f \widetilde{\overline{u}}^2 A_w}$$

$$+ 2\frac{\int_{\partial S_w} \tau_{wT} \cdot d\overline{s}}{\rho_f \widetilde{\overline{u}}^2 A_w} - \frac{\frac{\partial}{\partial x_j}\left\langle \hat{\overline{u}}_i \hat{\overline{u}}_j \right\rangle_f}{\frac{1}{2}\rho\widetilde{\overline{u}}^2} + \frac{\frac{\partial}{\partial x_j}\left(\left\langle \tilde{v}_T \frac{\partial \widetilde{\overline{u}}_i}{\partial x_j}\right\rangle_f\right)}{\frac{1}{2}\rho\widetilde{\overline{u}}^2} \quad (29)$$

The first three terms are form drag, laminar and turbulent contributions to skin friction, respectively. The fourth term represents the spatial flow oscillations, which are a function of porous media morphology and tells one how flow deviates from some mean value over the REV. The fifth term represents flow oscillations that are due to Reynolds stresses and are a function of porous media morphology and its time averaged flow oscillations.

The closure term in the VAT energy equation, $h^*$, can be defined in various ways and in general will depend on how many of the integrals appearing in the VAT equation one uses and lumps into a single transport coefficient, see Travkin and

Catton [26]. The nature of the equation shows that the energy transferred from the surface is integrated over an area and then divided by the chosen REV volume; therefore, the heat transfer coefficient is defined in terms of porous media morphology, usually described by specific surface and porosity. The complete form of the closure term $h^*$ is

$$h^* = \frac{\frac{1}{\Delta\Omega}\int_{\partial S_w}(k_f + k_t)\nabla T_f \cdot dS}{S_w(\tilde{T}_s - \tilde{T}_f)}$$

$$- \frac{\rho_f c_{pf}\nabla \cdot \left(\langle m \rangle \widehat{\tilde{u}_f \tilde{T}_f}\right)}{S_w(\tilde{T}_s - \tilde{T}_f)} + \frac{\nabla \cdot \left(\frac{k_f}{\Delta\Omega}\int_{\partial S_w} T_f dS\right)}{S_w(\tilde{T}_s - \tilde{T}_f)} \quad (30)$$

In most engineered devices, the geometry is regular and a well-chosen REV will lead to only the first term being needed. However, when in doubt, one should use the complete form given by Eq.(30).

After solving the three-dimensional governing equations (8), (9) and (10) with appropriate boundary conditions, the closure for the VAT based momentum equation and energy equation is obtained by integrating Eqs. (29) and (30) over the selected REV to compute the friction factor and heat transfer coefficient.

**Table 2:** Dimensions of the numerically tested models.

| Model | e/D | H/D | Model | e/D | H/D |
|-------|------|------|-------|------|------|
| 1 | 0.05 | 0.75 | 11 | 0.15 | 1.75 |
| 2 | 0.05 | 1.25 | 12 | 0.15 | 2.25 |
| 3 | 0.05 | 1.75 | 13 | 0.2 | 0.75 |
| 4 | 0.05 | 2.25 | 14 | 0.2 | 1.25 |
| 5 | 0.1 | 0.75 | 15 | 0.2 | 1.75 |
| 6 | 0.1 | 1.25 | 16 | 0.2 | 2.25 |
| 7 | 0.1 | 1.75 | 17 | 0.25 | 0.75 |
| 8 | 0.1 | 2.25 | 18 | 0.25 | 1.25 |
| 9 | 0.15 | 0.75 | 19 | 0.25 | 1.75 |
| 10 | 0.15 | 1.25 | 20 | 0.25 | 2.25 |

To make the correlations applicable to relatively wide range of dimensions of scale roughened channels, 20 different models, see Table 2, were simulated at different Reynolds number, ranging from 500 to $3\times10^4$. Attempts are made to correlate the present simulation results using a multiple regression technique. The correlations of Nusselt number and friction factor are proposed as follows.

The Nusselt number is correlated as

$$Nu^* = \frac{h^* D_h}{\lambda_f}$$

$$= 0.144\,\mathrm{Re}_{D_h}^{0.765}\left[\left(\frac{e}{D_h}\right)^{0.695} + 0.457\right]\left(\frac{H}{D_h}\right)^{1.018} + 8.235 \quad (31)$$

The friction factor is correlated in the form of [23]

$$f^* = \frac{A}{\mathrm{Re}_{D_h}} + B \quad (32)$$

in which

978-1-4673-1110-6/12 $31.00 © 2012 IEEE

$$A = 94.53$$

$$B = 0.0019 \, \mathrm{Re}_{D_h}^{0.217} + 3.544 \left(\frac{e}{D_h}\right)^{1.465} \left(\frac{H}{D_h}\right)^{0.0232} \quad (33)$$

Figure 7 and Figure 8 show the comparison between the numerical simulation results and the results predicted by the proposed correlations. The proposed heat transfer correlation, Eq.(31), can describe all the simulation results within a deviation of 15% and 98.2% of them within 10%. The correlation of friction factor Eqs.(32) and (33) can predict 90.5% of data within a deviation of 10% and all of them within a deviation of 20%. The correlations of Nusselt number and friction factor have an average deviation of 3.2% and 5.4% respectively.

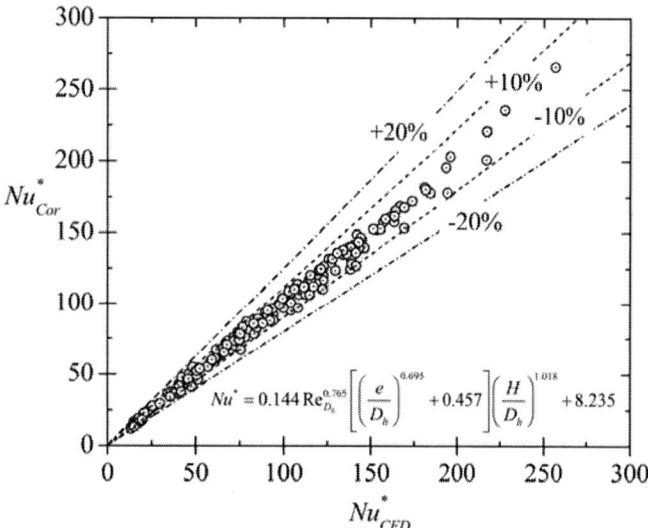

**Figure 7:** Deviation of the proposed Nusselt number correlation.

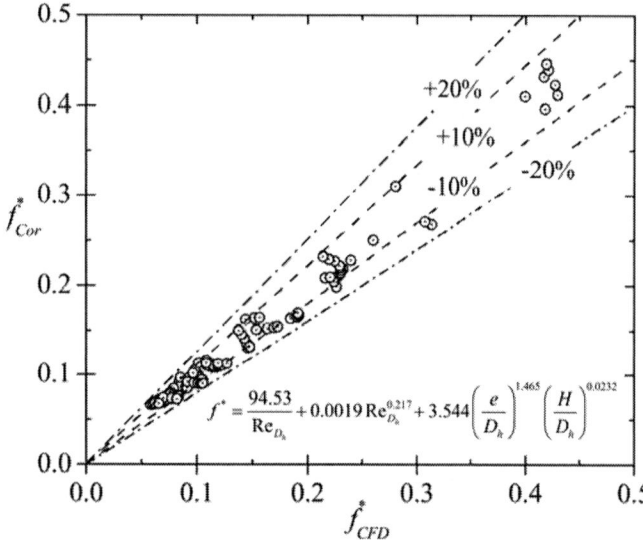

**Figure 8:** Deviation of the proposed friction factor correlation.

At this point, the VAT based model of PFHS with scale roughened surfaces is fully closed. With the closure correlations, the governing equation set is relatively simple and could be solved discretely in a minute. With the help of a statistical tool for Design of Experiments (DOE) or Genetic

Algorithm (GA), a scale roughened PFHS could be designed and optimized in an hour, instead of days of CFD or experimental work. How to design and optimize a scale roughened PFHS based on VAT will be presented in another paper.

## 5. Concluding Remarks

Volume Averaging Theory is little more than a judicious application of Green's and Stokes' theorems to carry out the integration needed to average the point-wise conservation equations in a rigorous way. Many everyday engineered devices are hierarchical and heterogeneous and can be effectively treated by application of VAT. It is an approach that can be applied to many different types of transport phenomena, see Travkin and Catton [26].

The present paper describes an effort to develop a VAT based hierarchical model for a plane fin heat sink with scale roughened surfaces and obtain the closure for the model by CFD code. A length of 15 REVs was selected to be the computational domain. The rigorously derived closure terms of heat transfer coefficient and friction factor were evaluated over the carefully selected REV. Two correlations of friction factor and Nusselt number were established based on the simulation results.

With closure of the friction factor and the heat transfer coefficient, the problem is closed and the porous media governing equations derived from VAT are

$$\tilde{M}\left(\langle m \rangle, S_w, f^*\right) \quad (34)$$

$$\tilde{T}_s\left(\langle m \rangle, S_w, h^*\right) \quad (35)$$

$$\tilde{T}_f\left(\langle m \rangle, S_w, h^*\right) \quad (36)$$

where $\tilde{M}$ stands for averaged momentum equation variables, $\tilde{T}_s$ and $\tilde{T}_f$ stand for averaged energy equation variables for solid and fluid phase.

From the statements above, the macro scale equations are functions only of porous media morphology, represented by porosity and specific surface area, and its closure. Furthermore it was shown that by proper scaling, closure is a function of the "porous media" as well, which further generalizes macro scale porous media equations.

**Acknowledgments**

The support of a DARPA grant within the MACE program is gratefully acknowledged. The views, opinions, and/or findings contained in this article are those of the author and should not be interpreted as representing the official views or policies, either expressed or implied, of the Defense Advanced Research Projects Agency or the Department of Defense.

**References**

1 Ligrani, P.M., Oliveira, M.M., and Blaskovich, T.: 'Comparison of heat transfer augmentation techniques', AIAA Journal, 2003, 41, (3), pp. 337-362

2 Taslim, M.E., Li, T., and Kercher, D.M.: 'Experimental heat transfer and friction in channels roughened with

angled, V-shaped, and discrete ribs on two opposite walls', J. Turbomach.-Trans. ASME, 1996, 118, (1), pp. 20-28

3   Han, J.C., Zhang, Y.M., and Lee, C.P.: 'Augmented Heat Transfer in Square Channels With Parallel, Crossed, and V-Shaped Angled Ribs', Journal of Heat Transfer, 1991, 113, (3), pp. 590-596

4   Gao, X., and Sunden, B.: 'Heat transfer and pressure drop measurements in rib-roughened rectangular ducts', Experimental Thermal and Fluid Science, 2001, 24, (1-2), pp. 25-34

5   Park, J.S., Han, J.C., Huang, Y., Ou, S., and Boyle, R.J.: 'Heat transfer performance comparisons of five different rectangular channels with parallel angled ribs', International Journal of Heat and Mass Transfer, 1992, 35, (11), pp. 2891-2903

6   Cho, H.H., Wu, S.J., and Kwon, H.J.: 'Local Heat/Mass Transfer Measurements in a Rectangular Duct With Discrete Ribs', Journal of Turbomachinery, 2000, 122, (3), pp. 579-586

7   Park, K., Choi, D.-H., and Lee, K.-S.: 'Optimum design of plate heat exchanger with staggered pin arrays', Numerical Heat Transfer, Part A: Applications: An International Journal of Computation and Methodology, 2004, 45, (4), pp. 347 - 361

8   Khan, W.A., Culham, J.R., and Yovanovich, M.M.: 'The Role of Fin Geometry in Heat Sink Performance', Journal of Electronic Packaging, 2006, 128, (4), pp. 324-330

9   Zhou, F., and Catton, I.: 'Numerical evaluation of flow and heat transfer in plate-pin fin heat sinks with various pin cross-sections', Numerical Heat Transfer, Part A: Applications, 2011, 60, (2), pp. 107-128

10  Mahmood, G.I., Hill, M.L., Nelson, D.L., Ligrani, P.M., Moon, H.K., and Glezer, B.: 'Local Heat Transfer and Flow Structure on and Above a Dimpled Surface in a Channel', Journal of Turbomachinery, 2001, 123, (1), pp. 115-123

11  Mahmood, G.I., and Ligrani, P.M.: 'Heat transfer in a dimpled channel: combined influences of aspect ratio, temperature ratio, Reynolds number, and flow structure', International Journal of Heat and Mass Transfer, 2002, 45, (10), pp. 2011-2020

12  Burgess, N.K., and Ligrani, P.M.: 'Effects Of Dimple Depth on Channel Nusselt Numbers and Friction Factors', Journal of Heat Transfer, 2005, 127, (8), pp. 839-847

13  Mahmood, G.I., Sabbagh, M.Z., and Ligrani, P.M.: 'Heat Transfer in a Channel with Dimples and Protrusions on Opposite Walls', Journal of Thermophysics and Heat Transfer, 2001, 15, (3), pp. 275-283

14  Chang, S.W., Liou, T.-M., and Lu, M.H.: 'Heat transfer of rectangular narrow channel with two opposite scale-roughened walls', International Journal of Heat and Mass Transfer, 2005, 48, (19-20), pp. 3921-3931

15  Chang, S.W., Liou, T.M., Chiang, K.F., and Hong, G.F.: 'Heat transfer and pressure drop in rectangular channel with compound roughness of V-shaped ribs and

deepened scales', International Journal of Heat and Mass Transfer, 2008, 51, (3-4), pp. 457-468

16  Chang, S.W., Yang, T.L., Liou, T.-M., and Fang, H.G.: 'Heat transfer in rotating scale-roughened trapezoidal duct at high rotation numbers', Applied Thermal Engineering, 2009, 29, (8-9), pp. 1682-1693

17  Chang, S.W., Yang, T.L., Liou, T.-M., and Hong, G.F.: 'Heat transfer of rotating rectangular duct with compound scaled roughness and V-ribs at high rotation numbers', International Journal of Thermal Sciences, 2009, 48, (1), pp. 174-187

18  Chang, S.W., and Lees, A.W.: 'Endwall heat transfer and pressure drop in scale-roughened pin-fin channels', International Journal of Thermal Sciences, 2010, 49, (4), pp. 702-713

19  Anderson, T.B., and Jackson, R.: 'Fluid Mechanical Description of Fluidized Beds. Equations of Motion', Industrial & Engineering Chemistry Fundamentals, 1967, 6, (4), pp. 527-539

20  Slattery, J.C.: 'Flow of viscoelastic fluids through porous media', AIChE Journal, 1967, 13, (6), pp. 1066-1071

21  Whitaker, S.: 'Diffusion and dispersion in porous media', AIChE Journal, 1967, 13, (3), pp. 420-427

22  Whitaker, S.: 'Forced convection heat transfer correlations for flow in pipes, past flat plates, single cylinders, single spheres, and for flow in packed beds and tube bundles', AIChE Journal, 1972, 18, (2), pp. 361-371

23  Travkin, V., and Catton, I.: 'A two-temperature model for turbulent flow and heat transfer in a porous layer', Journal of Fluids Engineering, 1995, 117, (1), pp. 181-188

24  Travkin, V.S., and Catton, I.: 'Porous media transport descriptions -- non-local, linear and non-linear against effective thermal/fluid properties', Advances in Colloid and Interface Science, 1998, 76-77, pp. 389-443

25  Travkin, V.S., and Catton, I.: 'Turbulent Flow and Heat Transfer Modeling in a Flat Channel with Regular Highly Rough Walls', International Journal of Fluid Mechanics Research, 1999, 26, (2), pp. 159-199

26  Travkin, V.S., and Catton, I.: 'Transport phenomena in heterogeneous media based on volume averaging theory', Advances in Heat Transfer, 2001, 34, pp. 1-144

27  Horvat, A., and Catton, I.: 'Numerical technique for modeling conjugate heat transfer in an electronic device heat sink', International Journal of Heat and Mass Transfer, 2003, 46, (12), pp. 2155-2168

28  Horvat, A., and Catton, I.: 'Application of Galerkin Method to Conjugate Heat Transfer Calculation', Numerical Heat Transfer, Part B: Fundamentals: An International Journal of Computation and Methodology, 2003, 44, (6), pp. 509 - 531

29  Horvat, A., and Mavko, B.: 'Hierarchic modeling of heat transfer processes in heat exchangers', International Journal of Heat and Mass Transfer, 2005, 48, (2), pp. 361-371

30  Horvat, A., and Mavko, B.: 'Calculation of conjugate heat transfer problem with volumetric heat generation

using the Galerkin method', Applied Mathematical Modelling, 2005, 29, (5), pp. 477-495

31 Vadnjal, A.: 'Modeling of a Heat Sink and High Heat Flux Vapor Chamber, PhD Thesis', University of California Los Angeles, 2009

32 Whitaker, S.: 'The method of volume averaging' (Kluwer Academic Publishers, 1999. 1999)

33 ANSYS, I., Canonsburg, Pennsylvania, 2009

34 Zhou, F., Hansen, N.E., Geb, D.J., and Catton, I.: 'Determination of the Number of Tube Rows to Obtain Closure for Volume Averaging Theory Based Model of Fin-and-Tube Heat Exchangers', Journal of Heat Transfer, 2011, 133, (12), pp. 121801

35 Patankar, S.V.: 'Numerical heat transfer and fluid flow' (Hemisphere Publishing Corp., 1980, 1st edn. 1980)

36 Menter, F.R., and Esch, T.: 'Elements of industrial heat transfer predictions'. Proc. 16th Bazilian Congress of Mechanical Engineering (COBEM), Uberlandia, Brazil2001 pp. Pages

37 Bardina, J.E., Huang, P.G., and Coakley, T.J.: 'Turbulence modeling validation, testing, and development', in Editor (Ed.)^(Eds.): 'Book Turbulence modeling validation, testing, and development' (NASA Technical Memorandum, 1997, edn.), pp.

38 Menter, F.R.: 'Two-equation eddy-viscosity turbulence models for engineering applications', AIAA Journal, 1994, 32, (8), pp. 1598-1605

39 Menter, F.R., Kuntz, M., and Langtry, R.: 'Ten years of industrial experience with the SST turbulence model', Turbulence, Heat and mass transfer, 2003, 4, pp. 625-632

40 Zhou, F., Hansen, N.E., Geb, D.J., and Catton, I.: 'Obtaining Closure for Fin-and-Tube Heat Exchanger Modeling Based on Volume Averaging Theory (VAT)', Journal of Heat Transfer, 2011, 133, (11), pp. 111802

# Two-layer Heat Spreading Approximations Revisited

Clemens J.M. Lasance
Philips Research Laboratories, Emeritus
Consultant @ SomelikeitCool
Nuenen, the Netherlands
lasance@onsnet.nu

## Abstract

When confronted with system level thermal analysis, the designer needs at least a first guess about the thermal behavior of the whole system, even if she[1] is only responsible for a single part, such as the PCB or the heat sink. A useful strategy is to check globally whether some claims of vendors make sense or not. For example, some vendors of metal-core PCBs state that their enhanced (and hence more expensive) thermal-conductive dielectrics are required for reliable performance. A spreadsheet-based Calculator is discussed that allows the designer to assess the dominant thermal resistances in a series network comprising the thermal path from junction to ambient for a two-layer case such as an LED on a metal core PCB (MCPCB).

An earlier paper has shown that it is possible to adapt the approximate one-layer heat spreading equations to accommodate two layers. However, while building the Calculator, it was found that these equations did not allow for a correct estimation of the spreading resistances only, which is required to get insight in the importance of this element in the total chain. *The main purpose of this paper is to share this acquired knowledge.*

## Keywords

Heat spreading, MCPCB, LED

## 1. Introduction

Parts of this paper have been published earlier [1,2] but writing a 'white paper' for this year's APEC/IPC conference has forced the author to revisit the topic of how to deal with multi-layer heat spreading, more specifically two-layer heat spreading. This kind of heat spreading is quite common in practice, especially for LEDs. One of the applications, an LED mounted on a submount fixed to a heat sink, was covered in earlier papers [1,2]. This paper deals with the analysis of an LED on a MCPCB where we again meet two layers: the dielectric, and the metal core. Earlier papers showed an extension to two layers of an analytical solution based on the pioneering work by Song, Lee and Au [3,4] and Lee [5] who derived an analytical expression for one-layer heat spreading, An interesting alternative is offered by the University of Waterloo (UoW) through a user-friendly web-based tool that allows users to calculate heat spreading effects for multilayered substrates [6]. Applying the same method to the MCPCB two-layer combination resulted in the realization that we face a problem, due to the fact that the standard description of averaging the air-side layer temperatures cannot be used for the MCPCB case due to the dielectric layer. A way of dealing with this problem is proposed.

The paper starts with repeating some topics discussed in earlier papers, such as the fact that heat spreading is by no means a trivial issue. When multiple layers and sources are part of the problem, there is no other way than to rely on dedicated tools. For simple LED applications the role of heat spreading will be extensively discussed.

The paper shows several examples of how to use simple formulae and excel-based sheets (the Calculator [7]) for design purposes[2]. One of the conclusions is that in the majority of the cases, especially when dealing with natural convection-driven applications, the thermal properties of the PCB are not the major bottleneck, and hence heat spreading is not a big issue. It is often easier and cheaper to improve other elements in the thermal resistance chain. *From a thermal point of view only*, in many cases of practical interest it will turn out that the thermal performance of the PCB is relevant only for high heat flux cases (e.g. liquid cooling) and for top-of-the-bill LEDs.

The new element for which the paper was written is the following. As pointed out in ref. [2] the approach for extending the one-layer equations to two layers is in calculating an effective heat transfer coefficient and area for the second layer to be used as a boundary condition for the first layer. The problem with both the S&L equations and the UoW tool is that they result in the correct answer for the total thermal resistance, but not for the individual contributions of the spreading thermal resistance and the convective thermal resistance. The reason is that the convective resistance is based upon the total area. This is the right approach for a metal spreader, but not for a thin dielectric layer, as can easily be seen when considering a small heat source on a thin layer with low thermal conductivity. This causes a hot spot of about the same size as the heat source on the other side, which is the area available for heat transfer. Obviously, this thermal resistance is much higher than when taking the whole area. For our Calculator we need this split, because we want to see all elements of the chain in order to make the right design decisions.

---

[1] In case you wonder why I use 'she' everywhere, you may wonder why you wouldn't wonder when I would use 'he' everywhere.

---

[2] At this moment in time the web-based Calculator does not offer the heat spreading calculation.

## 2. Heat spreading: *not* a trivial issue

As argued before, designers should know upfront if heat spreading is an issue or not. Unfortunately, no simple rules exist in order to make an early decision. Except for the simplest of cases, the equations describing heat spreading physics do not have an explicit mathematical solution. Hence, we have to rely on clever approximations or suitable computer codes. The following section discusses the basics of heat spreading physics. For a more in-depth discussion the reader is pointed to Lasance [2]. A distinction is made between single source and multiple sources heat spreading.

### *Single source*

Heat spreading is essentially area enlarging: the larger the area, the more heat can be removed at the same temperature difference (subject to certain size limits). Contrary to what is believed by many designers: heat spreading is *not* a trivial issue. Consider the simple configuration shown in Figure 4, left. A square source with zero thickness of size $A_s$ centrally located on a square plate of size A, thickness $d$ and thermal conductivity $k$ dissipates $q$ W. The top and sides of the plate are adiabatic (insulated), the bottom 'sees' a uniform heat transfer coefficient $h$ (W/m$^2$K).

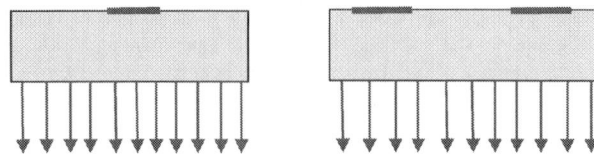

**Figure 1** Heat spreading from single source (left) and two sources (right)

The remarkable thing is that even for this simple configuration no explicit solution is known for the description of heat spreading. Observing the exact *implicit* solution of the governing differential equations reveals the source of the complexity of heat spreading: it is *not* possible to separate the convection and conduction parts. Readers who want to dig into these matters may consult the following additional papers: Punch and Davies [8 9], Ellison [10]. In other words, changing the heat transfer coefficient changes also the value of the spreading resistance. Consequently, it is not possible to write the problem in terms of one conduction resistance describing the heat spreading inside the solid and one convection resistance describing the boundary condition because the two are dependent. While equation 1, describing the temperature rise of the source due to dissipation q via a total thermal resistance $R_{total}$ from junction to ambient is physically correct, splitting up this resistance according to equations 2 and 3 is *not* correct.

$$\Delta T = q \cdot R_{total} \tag{1}$$

with

$$R_{total} = R_{conduction} + R_{convection} \tag{2}$$

and

$$R_{conduction} = g \cdot \frac{d}{k \cdot A} \qquad R_{convection} = \frac{1}{h \cdot A} \tag{3}$$

with $g = f(A_s, A)$ some geometrical factor representing the spreading (often based on the 45° angle rule).

There is one exception: the analysis becomes much more straightforward when the temperature gradients over the area that is in contact with the environment can be neglected. In other words, when a uniform temperature may be assumed. Such is often the case with relatively small heat sinks and spreaders, when the thermal conductivity is relatively high, and when the heat transfer coefficient is relatively low.

A final complexity stems from the fact that decreasing the thickness of the plate does *not* automatically result in a decrease in temperature, caused by the fact that a smaller thickness also implies a decrease in spreading capability. Hence, for a certain combination of thickness and thermal conductivity given the boundary conditions and the dimensions a minimum in the total thermal resistance may be found.

### *Multiple sources*

*Multiple* sources (Figure 2 right) add another layer of complexity because the coupling between the sources is not only dependent on the dimensions and physical properties but also on the boundary conditions and, worst of all, on the dissipation of the sources themselves. Even the definition of thermal resistance is lost when more than one source is present. The reason is that the second condition for a correct definition is violated; the fact that the same flux has to enter and leave the resistance. The flux of one source will enter the source node, the flux of two or more sources the ambient node. The essential point to understand is that when dealing with multiple sources the concept of thermal resistance becomes meaningless, except in the situation where the multiple sources/spreader assembly is subdivided into many resistances for each of which the definition holds. This is mathematically equivalent to a finite volume discretization.

### *How to address heat spreading*

In growing order of complexity, we may distinguish the following four approaches to calculate heat spreading:

a)  1D series resistance network with or without a geometrical correction factor
b)  Analytical solution-based approximate equations
c)  Software based on analytical solutions
d)  Conduction-only finite volume/element based codes

All four approaches are extensively discussed in ref. [2]. For the purpose of this paper the following summary is presented:

Approach a) is illustrated in equations 1) and 2), and it has been demonstrated that it is only valid in limited cases. For situations where one is dealing with a single source, predominantly one-sided heat transfer, and one heat spreading layer, the analytical solution-based approximate equations (easy to embed in a spreadsheet) demonstrate an order of magnitude higher accuracy over the 1D series resistance network approach. This is approach b).

For situations where double-sided heat transfer plays a role, or multiple sources, or multiple layers, the problem becomes intractable from an approximate analytical point of view and we have to rely on computer codes.(approach c). Implicit solutions are known for multi-layer cases with multiple sources and uniform boundary conditions, even when time is a parameter. User-friendly software exists that is based on these solutions [11] with the big advantage that also people with little background in heat transfer can get insight in the physics underlying heat spreading by simply changing a few parameters. An additional advantage is that no mesh generation is required. Another already mentioned source of information can be found on the website of the University of Waterloo [6]. A number of easy to use calculators can be addressed of which one in particular is very useful for this case: the two-layer metal-core PCB. Also of interest: one of their papers (Culham and Yovanovich [12]) contains a couple of graphics showing clearly the errors a designer may encounter by using a simple series resistance approach. However, the user should be aware of the limitations of both software codes. For more practical cases for which layers consist of more than one material or for which the boundary conditions cannot be considered uniform, more advanced conduction-only codes should be used. Figure 2 shows some cases that can and cannot be handled by the analytical software discussed in this section.

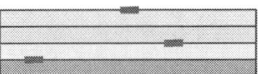 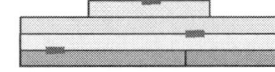

*htop, hbottom: all values*
*hside: only 0 and ∞*

*htop, hbottom, hside: all*
*values*

**Figure 2** Left: cases that can be solved analytically. Right: cases that cannot be solved analytically

When one is dealing with problems that resemble the case shown in Figure 2, right, one has to rely on more sophisticated codes (approach d). In principle, all Finite Volume/Finite Element/etc. codes can be used that solve the heat diffusion equation. In practice, only those user-friendly codes are useable that enable a designer to get results in an hour or so. Some popular CFD codes used in conduction-only mode are examples of such a code. It is recommended to validate the model in an early stage by comparing the results with those obtained analytically using the software described in this section.

In summary, the author is of the opinion that using analytical solutions, including the 1D series resistance network, in one way or another has its main merits in getting insight, hence is second to none from an educational point of view. However, when accuracy is at stake in the final design

stages the recommended approach for solving real-life problems is in using a 3D conduction solver. To support the designer in acquiring the insight mentioned, a spreadsheet Calculator has been developed, discussed in more detail in the following section.

## 3. The Calculator

The Calculator [7] is a spreadsheet based tool with the objective to get quick insight in the main parameters that govern the temperature rise of the LED. To realize this, the following input is required:

- Number of LEDs
- Dimensions of LED source and PCB area
- Power dissipation
- LED thermal data from datasheets
- PCB dielectric and board thermal conductivities, area and thicknesses
- TIM thermal data between PCB and heat sink
- Area enlargement factor for heat sink
- Heat transfer coefficient to ambient air
- Maximum allowed LED and ambient temperatures

Dependent on the question, the user has a choice of options. Often, the starting point is a given PCB area with a number of LEDs. First calculation is the area per LED (however, the user should realize that this is only valid if the dissipation of all LEDs is approximately the same. If not, you have to consult an expert). Two extreme cases can be considered:

Best Case

One-dimensional heat transfer is assumed, hence no heat spreading. In other words, effectively the LED area is equal to the PCB area.

Worst case

We assume no heat spreading in the dielectric layer direct under the LED, and ideal heat spreading in the metal part. In other words: the LED area is used for calculating the thermal resistance of the dielectric layer, the PCB area is used to calculate the thermal resistance of the metal layer.

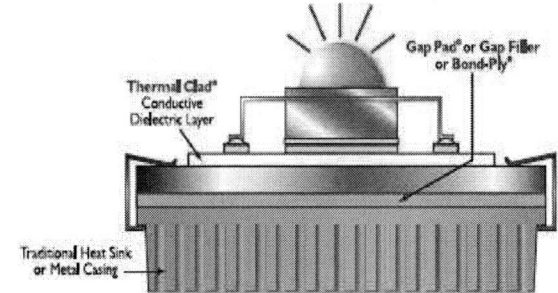

**Figure 3** Layout for examples showing e.g. the dielectric layer and the TIM (in between the MCPCB and the heat sink)

978-1-4673-1110-6/12 $31.00 © 2012 IEEE      271

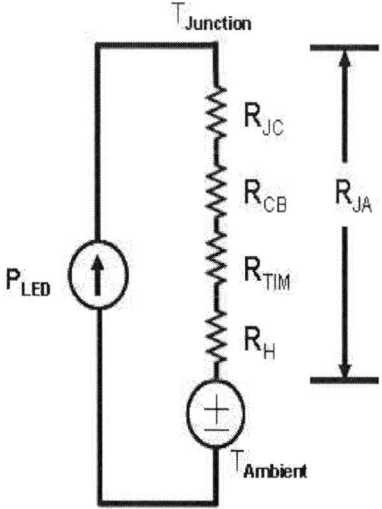

**Figure 4** Thermal network for layout shown in Figure 3.

$R_{JC}$: junction to case

$R_{CB}$: MCPCB, encompassing the dielectric layer and the metal core

$R_{TIM}$: interface material

$R_H$: heat sink to ambient

It turns out that for many practical cases such as the layout shown in Figure 3 the best and worst cases are quite close to each other. The reason is that the LED thermal resistance ($R_{JC}$ in Figure 4) and the airside resistance ($R_H$) are often an order of magnitude larger than the MCPCB and TIM thermal resistances ($R_{CB}$ and $R_{TIM}$).

While the Calculator is very easy to use and the principles can be understood by many designers, these conclusions are based on the assumption of 1D heat transfer, and hence the question needs to be addressed: do the conclusions change when we take heat spreading into account? If so, the Calculator would become much more complex as will be evident from the following section.

### 4. The role of heat spreading in simple LED applications

Let us make a distinction between two cases that often occur in practice when dealing with LED applications: a single heat spreader, and a heat spreader with a thin dielectric layer on top.

#### 4.1. Single layer

As is demonstrated in Ref. [2], for this kind of heat spreading cases we may apply the following approximate equation:

$$R_{thj-a} = \frac{1}{h_{eff} A_2} + \frac{\ln(\frac{A_2}{A_1}) - 2\gamma}{4\pi kd} \quad (4)$$

where $\gamma$ is Euler's constant, 0.58, $h_{eff}$ the effective h caused by the increased area of the heat sink, d the thickness of the PCB, $A_2$ the PCB area and $A_1$ the LED area. With some

care the first term of the right-hand side could be interpreted as a convective term and the second as a conductive term.[3] The heat spreading becomes manifest through the logarithmic term and the fact that the thickness appears in the denominator. However, a warning should be issued here: don't use this equation before you have read the quoted paper, there are limitations to its use, related to limitations in the ratios of $A_1/A_2$ and k/dh. The beauty of using this simple equation is that it is easily seen what the effect is of heat spreading, namely the relative magnitude of the two terms, plus the difference in the second term as compared to the simple 1D heat transfer underlying the series resistance approach:

$$R_{thj-a} = \frac{1}{h_{eff} A_2} + \frac{d}{kA_2} \quad (5)$$

Let us insert some 'standard' input data. The PCB area allocated to the LED is 1cm², the enhanced area by the heat sink is 20 times this area and the thickness of the PCB is 1.6 mm. Let us also assume for the sake of simplicity, there is no TIM and the LED area is one tenth of the PCB area. For the convective term we get: 500/h. For the spreading term we find: 60/k, for the 1D term: 16/k. What does this mean in practice? Assume natural convection (h=10 W/m²K) and k=160W/mK for the heat spreader. The convective resistance becomes 50 K/W, the spreading resistance 0.5 K/W, and for the 1D resistance we find 0.1 W/mK. While it is clear that heat spreading causes the conductive part to rise significantly (from 0.1 to 0.5 K/W), compared to the convective part it is peanuts. When we switch to forced convection (h=50 W/m²K), the conclusion is the same. We should go to much higher heat transfer coefficients or much lower thermal conductivities before heat spreading becomes an issue.

The much more accurate SLA equations are not reproduced here, but lead to the same conclusions, see ref. [2].

#### 4.2. Two layers

For the problem at hand, the MCPCB case, we need to include two layers, see Figure 6. The question is: can we adapt the equations formulated for a single layer for two layers? The answer is: yes we can [13]. The idea is to replace the convective part *1/hA* of the submount single-layer equation by the equation for the spreader, for which the source area equals the submount area $A_2$.

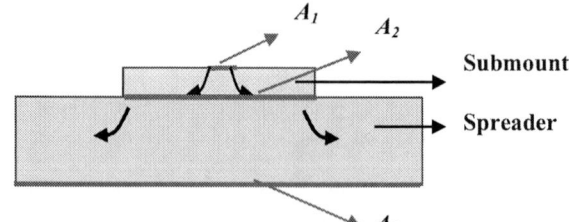

**Figure 6** Two-layer case with a submount, can be replaced by a dielectric layer

---

[3] Due to the negative sign this equation does *not* represent a simple series resistance network.

Applying the idea to equation 4 leads to the following set of equations:

$$R_{thj-a} = \frac{1}{hA_2} + \frac{\ln(\frac{A_2}{A_1})}{4\pi k_1 d_1} - \frac{\gamma}{2\pi k_1 d_1} \quad (6)$$

applied to the first layer (submount or dielectric layer), in which the convective term is replaced by the spreading resistance of the second layer:

$$\frac{1}{hA_2} = \frac{1}{hA_3} + \frac{\ln(\frac{A_3}{A_2})}{4\pi k_2 d_2} - \frac{\gamma}{2\pi k_2 d_2} \quad (7)$$

with the submount base area $A_2$ as the new source area.

The implicit assumption is that the heat flux profile leaving layer 1 may be approximated by a uniform heat flux source applied to layer 2. In most cases, the associated errors are of second-order importance.

While equations 6 and 7 show the principle of two-layer approximate heat spreading, their use is not recommended because the limiting conditions mentioned above are frequently reached. Instead, the SLA equations should be used that can be adapted in the same way.[4]

In summary, the approach for extending the one-layer equations to two layers is in calculating an effective heat transfer coefficient and area for the second layer to be used as a boundary condition for the first layer. However, we face a problem here. When we want to get insight into the relative contributions of the spreading resistances of the two layers compared to the convective resistance the problem with both the SLA equations and the UoW tool is that they cannot give this answer. However, they do result in the right answer for the total thermal resistance. The reason is that the convective resistance is based upon the total area. As already has been pointed out in the introduction, this is the correct approach for the metal core but not for the thin dielectric layer.

Hence, the problem boils down to the following. While the SLA equations and the UoW tool can be used for two-layer heat spreading, we cannot get a feeling for the magnitude of the individual contributions. The question remains: what are the errors for the two-layer case for both the simple series resistance approach and the adapted SLA-equations?

## 5. Validating the Calculator

The method chosen is the following: for the thin dielectric we use a spreading resistance based on some spreading angle rule, from which the source area follows that is subsequently used to calculate the heat spreading in the metal using the series resistance approximation and the SLA equations. The

results were compared with two analytical-based codes (THERMAN and the UoW tool) and a CFD code in conduction-only mode checked for grid-independence. It was found by trial and error that the best angle rule that matched the detailed results was about half the (in)famous 45° spreading angle rule (see e.g. Guenin [14]) for a whole range of practical boundary conditions and material properties to within 5%. While the UoW tool provides only the average results, not the maximum values, this is a reasonable assumption for our case. The LED is represented by a flat heat source, but in reality, the LED itself acts as a kind of heat spreader, hence reducing the temperature gradients.[5] The most important conclusion that can be drawn from using the more sophisticated results of the two-layer heat spreading approach is that the worst-case approach discussed above is sufficiently accurate to be useful for first-order analysis. The reason is the often large source dimension/thickness-of-dielectric ratio. All examples given deal with a Luxeon Rebel of footprint size 3*4,5 mm.

The foregoing exercises lead to the following conclusion:

*In the majority of the cases, especially when dealing with natural convection-driven applications, the thermal properties of the PCB are not the major bottleneck. It is often easier and cheaper to improve other elements in the thermal resistance chain.*

Hence, there is often no reason to buy an MCPCB because of its better thermal performance. Of course, there may be other reasons such as CTE mismatch or breakdown voltage requirements to choose a more sophisticated PCB.

In summary, from a thermal point of view only, in many cases of practical interest it will turn out that the thermal performance of the PCB is relevant only for high heat flux cases (e.g. liquid cooling) and for top-of-the-bill LEDs.

Caveat: Be aware that the conclusions may change considerably when the size of the LED is reduced to say 2*2 mm, introducing a spreading resistance of the dielectric that cannot be neglected. In such cases it does pay when increasing the thermal conductivity of the dielectric.

## 6. Conclusions

Regarding thermal management of LED applications, it is argued that the designer needs at least a first guess about the thermal behavior of the whole system, even if she is only responsible for a single part, such as the PCB or the heat sink. Various approaches are discussed to treat heat-spreading effects in practice, to be used in early and final design stages. A spreadsheet-based Calculator is demonstrated that allows

---

[4] As pointed out in ref. [2], SLA has also its limitations but these are rarely encountered in practice.

---

[5] The maximum junction temperatures found using the Calculator were compared with THERMAN and CFD results. The author found it remarkable that THERMAN that is based on the same set of equations as the UoW tool and additionally can handle maximum temperatures, more layers, arbitrary sources and transients, generated average results that differed from the UoW results by more than 10%. Must be a software error, because a check using the CFD code returned the UoW results to within 1%.

the designer to assess the dominant thermal resistances in a series network comprising the thermal path from junction to ambient for a two-layer case, in particular the case of an LED on a metal core printed circuit board consisting of a thin dielectric layer on top of a metal layer. While building the Calculator it was found that the earlier-published approximate two layers equations did not allow for a correct estimation of the individual spreading resistances, required to assess their importance in the chain. A simple correction factor is proposed. Using these simplified equations, it can easily be shown that in many practical cases the thermal conductivity of the dielectric of a metal core PCB does not play a significant role.

Additionally, it is argued in the Appendix that the use of the word 'thermal impedance' to characterize thermal interface materials should be forbidden by law.

## References

[1] Lasance C., Heat spreading: not a trivial problem, ElectronicsCooling, Vol. 14, May issue, pp. 24-30 (2008)

[2] Lasance C., How to Estimate Heat Spreading Effects in Practice, J. Electron. Packag., **132**, 031004, (2010)

[3] Song S., Lee S. and Au V., Closed-Form Equations for Thermal Constriction/Spreading Resistances with Variable Resistance Boundary Condition, IEPS Conference, pp. 111-121 (1994)

[4] Lee S., Song S., Au V., Moran K., Constriction/Spreading Resistance Model for Electronics Packaging, ASME/JSME Thermal Engineering Conf., Vol.4, 199-206 (1995)

[5] Lee S., Calculating spreading resistance in heat sinks, ElectronicsCooling, January issue (1998)

[6] MHTL Simulation Tools. 2012. MHTL Simulation Tools. [ONLINE] Available at:
http://mhtlab.uwaterloo.ca/RScalculators.html.

[7] Calculator. 2012. Calculator. [ONLINE] Available at:
http://www.saturnelectronics.com/calculator/

[8] Punch, J. and Davies M., "Three-Dimensional Heat Conduction through Rectangular Multiple Layer Plane Substrates," ASME HTD-Vol. 249, Heat Transfer Measurements and Analysis, pp. 89-100 (1993)

[9] Punch, J. and Davies M.,. "A Parametric Study of Heat Spreading Through Multiple Layer Substrates," Thermal Management of Electronic Systems: Proc. of EUROTHERM Seminar 29, Editors, Hoogendoorn, C. J., Henkes, R. A. W. M., and Lasance, C. J. M., Kluwer Academic Publishers, pp. 156-169 (1994)

[10] Ellison G., Maximum Thermal Spreading Resistance for Rectangular Sources and Plates With Nonunity Aspect Ratios, IEEE Transactions on Components and Packaging Technologies, vol. 26, no. 2, p.439 (2003)

[11] THERMAN, by Micred, Hungary

[12] Culham J. and Yovanovich M., Factors affecting the calculation of effective conductivity in printed circuit boards, Proc. ITHERM '98, Seattle, pp.46-467 (1998)

[13] Courtesy of B. Obama

[14] Guenin B., The 45°Heat Spreading Angle – An Urban Legend?, ElectronicsCooling Nov issue, 2003

## Appendix: Thermal impedance revisited

The author would use this opportunity again to point at the inappropriate use of the term 'thermal impedance'. It is important that all people involved use the same terminology to define the performance of TIMs. The problem is that part of the people (mostly US vendors) uses the word 'thermal impedance' as shorthand for 'unit area thermal resistance'. This violates the electrothermal analogy commonly in use because of two reasons. First, in the electrical world 'electrical resistance' and 'electrical impedance' have the same unit, namely Ohm. Consequently, 'thermal impedance' should have the dimension K/W, not K m$^2$/W. Second, 'electrical impedance' is a time-dependent quantity. In limiting cases, for frequency zero or large enough times approaching steady state, the impedance becomes equal to the resistance. Sticking to the current definition of 'thermal impedance' will cause a lot of confusion in the future, because the use of dynamic test methods is the obvious choice for application-specific tests, one output of which is thermal impedance. When quoting the performance of a TIM per area, we propose to use 'R-value' (universally accepted in the building field), 'unit area thermal resistance' or simply 'unit thermal resistance'. It is hoped for that the JEDEC JC15 committee takes its responsibility in this respect.

*Note added in proof*: Fortunately, there seems to be light at the end of this tunnel. An increasing number of papers use the phrase 'unit thermal resistance', so the point seems to be taken.

# Heat Spreading from a Small Source on a Thin Plate

Ir. G. A. (Wendy) Luiten

Philips Research

High Tech Campus 37, 5656 AE Eindhoven, The Netherlands

Wendy.Luiten@Philips.com

## Abstract

In view of the trend towards higher power densities in ever smaller package envelopes, heat spreading of a small source on a thin plate is gaining in importance. In this paper, an alternative description is presented to the popular Song-Lee-Au (SLA) approach. A length scale is derived, depending on geometry, heat transfer coefficient, and material parameters, and the relevance of this length scale is investigated. It is shown that the length scale has physical significance both for the distance that heat spreads out over the plate and for the total amount of heat cooled away. The work is the extension of this author's earlier work on Characteristic Length and Cooling Circle to the domain of finite thin plates and small sources. Thermal resistances were calculated with the Cooling Circle and with SLA, and the results were compared to numerical simulations. In the numerical comparison the difference between the two approaches emerged as different assumptions as to the thermal conductivity of the heat source. Cooling Circle assumes a thin plate, and a uniform temperature, thus very large thermal conductivity, in the source area. SLA incorporates the effect of spreading in the plate thickness, and for the thin plates comparison this boils down to the use of baseplate conductivity in the source area. Both methods match well with the numerical results. For larger plates, the Cooling Circle approach continues to match well with the numerical values. SLA values for large thin plates resulted in higher resistances, contrary to both numerical results and physics based expectations. This was attributed to the SLA correlations being used outside the range for which they were validated. The Cooling Circle approach has great engineering significance as it enables quick engineering estimations and provides a helpful mental image.

## Keywords

*Heat spreading, thin plate, SLA, Cooling Circle, characteristic length, constriction resistance, spreading resistance.*

## Nomenclature

| $A_p$ | plate area | $m^2$ |
|---|---|---|
| $A_s$ | source area | $m^2$ |
| $f(\eta)$ | factor according to eq.4 | (-) |
| $h$ | heat transfer coefficient | $W/(m^2 K)$ |
| $k$ | thermal conductivity | $W/(mK)$ |
| $L$ | length | $m$ |
| $L_c$ | Characteristic Length | $m$ |
| $q$ | transferred heat | $W$ |

| $Q$ | scaled transferred heat | (-) |
|---|---|---|
| $r$ | radius | $m$ |
| $R$ | thermal resistance | $K/W$ |
| $R_0$ | SLA average plate resistance | $K/W$ |
| $R_c$ | SLA constriction resistance | $K/W$ |
| $r_i$ | source radius | $m$ |
| $r_o$ | plate radius | $m$ |
| $R_t$ | SLA total resistance | $K/W$ |
| $t$ | plate thickness | $m$ |
| $w$ | plate width | $m$ |
| $\eta$ | scaled radius | (-) |
| $\lambda$ | parameter in SLA | (-) |
| $\theta$ | scaled temperature | (-) |
| $\theta_o$ | $\theta$ at the rim of the plate | (-) |
| $\omega$ | scaled radial spreading distance | (-) |
| $i$ | (subscript) Inner | |
| $o$ | (subscript) Outer | |

## 1. Introduction

Miniaturization and increasing performance continue as trends in the development of new generations of electronic components. In combination, the trend towards more performance in smaller component envelopes leads to smaller and smaller heat sources of increasing power densities. At the module and system level, these components need to be cooled to realistic temperatures and power densities; heat needs to be spread out from the small source. This is commonly achieved in a flat thin form factor like a printed circuit board or a plate heat sink.

This paper discusses a novel approach of determining the cooling characteristics of a heat source of finite size, placed on a thin plate, with a constant heat transfer coefficient as a boundary condition.

Heat spreading from a source on a plate has been the subject of extensive studies. A multitude of approaches exist, ranging from simple to complex, with varying accuracy [1-4].

The simplest approach is the one taught in elementary heat transfer courses: Convective heat transfer is proportional to the plate area [1-2]. Heat spreading considerations are simply ignored, and it is assumed that the entire plate will be at the same uniform temperature. Depending on the specific conditions, this approach can result in reasonable answers or spectacular failures. This is especially unfortunate as defining the conditions where this approach is valid is usually not a part of the elementary heat transfer course.

978-1-4673-1110-6/12 $31.00 © 2012 IEEE

The popular "Song-Lee-Au" (SLA) geometry consists of a heat source on top of a baseplate [4]. Cooling is on the bottom of the base plate. The SLA approach splits the case in two parts, with each part assigned a thermal resistance [4-6]. The first resistance is based on the surface area of the plate. This is the average temperature rise of the plate divided by the source heat load. The second resistance is the so-called constriction resistance. The average constriction resistance is the average temperature rise of the source with respect to the average plate temperature, divided by the source heat load. Likewise, the maximum constriction resistance is defined as the maximum temperature rise of the source with respect to the average plate temperature, divided by the source heat load. The constriction resistance combines the effect of conduction over the plate thickness and spreading in the in-plane direction. By putting the constriction resistance and the average resistance in series, the corresponding temperature of the heat source is calculated. SLA is a correlation based method. The validity was tested for source/sink length ratios between 2 and 20, and thickness/sink length ratios between 0.01 and 10 [5].

Variations on the 'series' approach exist, e.g. as mentioned in [3]. In the author's opinion, the problem with the series approach is that it invokes the wrong mental picture, as the constriction resistance is often mentally associated with the source and the plate directly below it. Consider the case of a small heat source on a large plate: If the plate is enlarged further, one would wrongly assume that the constriction resistance (wrongly attributed to the source and its direct surroundings) would remain the same, and the average resistance (attributed to the plate) would change, resulting in a different source temperature due to the series connection. In reality, this only happens for small plates and thus the series approach is not only not validated but also invalid for large plates.

In the authors opinion it is strange to describe heat transfer from a small source on a plate in terms of length ratios scaled to the size of the plate. An IR thermograph picture of a heat source on a badly conducting plate shows that only a limited area around the source is heated, and as only the heated area is above ambient temperature, only the heated area can contribute to cooling. Enlarging the plate beyond this size will only add material that is not heated by the source, thus remains at ambient temperature, and cannot contribute to cooling. As long as the plate is sufficiently larger than the hotspot, the size of the plate is irrelevant. Clearly, it is not the size of the plate, but the size of the hotspot, or specifically how far the heat spreads out from the rim of the source onto the plate, which is the relevant length scale.

This concept was first explored in [7] using a 1D radial coordinate system. The rectangular geometries of the source and the sink are converted to the cylindrical coordinate system in the customary manner, based upon the corresponding areas:

$$r_i = \sqrt{A_s/\pi}$$

(1a)

$$r_o = \sqrt{A_p/\pi}$$

(1b)

Temperature differences over the thickness of the thin plate are assumed to be negligible. The geometry is shown in Figure 1. In the 1D radial coordinate system, the amount of heat lost from the (circular) plate around the (circular) source can be attributed to a ring shaped area around the source. Combined with the surface area of the source itself, this means that the source is cooled with a so-called Cooling Circle of radius $r_i+L$, where for large sources L equals the so-called Characteristic Length defined in the same paper [7]. Since the Characteristic Length was derived for the Cartesian case, it was found that $L=L_c$ was only valid for large sources. For small sources on infinite thin plates, it was necessary to apply a correction factor.

In the Cooling Circle approach [7], the effect of enlarging a plate depends on the size relative to the Cooling Circle. If the plate is smaller than the Cooling Circle, enlarging the plate will increase the cooling area of the source, and thus the temperature of the source will drop. If the plate is much larger than the Cooling Circle, the cooling is limited to the area of the Cooling Circle, enlarging the plate will add material that is not heated, and thus will not influence the temperature of the source. The Cooling Circle concept is sound from a physics point of view and also visually aligns with thermal images from experimental or computational sources.

In [7] the Cooling Circle approach was limited to the case of large heat sources on infinite large thin plates. For small sources a correction factor applied. In the present paper, the subject of Characteristic Length and Cooling Circle is revisited for the case of small sources and finite plates.

## 2. Derivation of the Length scale

In [7] the Characteristic Length for the Cartesian heat spreading case was derived directly from the differential equation. An alternative approach is suggested in the work of Bejan [8-9]: In Constructal theory it is shown that relevant length scales often appear when competing effects are of the same magnitude. In heat spreading in a thin plate, the competing mechanisms are conductive heat spreading in the in-plane direction and the convective heat transfer from the thin plate to the ambient.

Assume a longitudinal thin fin of thickness t, width w, conductivity k, and heat transfer coefficient h as boundary condition. The plate is sufficiently thin that the temperature is uniform over the plate thickness. Assume heat spreading in the longitudinal direction over distance L. Equating the resistance for longitudinal in-plane conduction to the resistance for convection leads to the emergence of the Characteristic Length.

$$R_{conduction} = L/(twk)$$

(2a)

$$R_{convection} = 1/(Lwh)$$

(2b)

$$L = \sqrt{tk/h}.$$

(2c)

Equation (2c) is the same expression as was derived from the differential equation in [7], so this suggests that likewise equating the conduction and convection resistances for the 1D radial heat spreading in the cylindrical case will result in a length scale which is valid for the circular geometry.

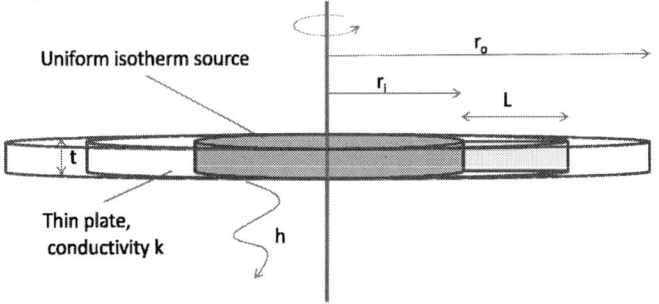

**Figure 1: Circular geometry**

The circular geometry is shown in Figure 1. A circular source with radius $r_i$ (m), is embedded in a thin plate of thickness $t$ (m) and thermal conductivity $k$ (W/mK). The applied boundary condition for the plate is cooling with a heat transfer coefficient $h$ W/m$^2$K. Heat spreads radially from the source over distance $L$ (m). The plate is sufficiently thin that the temperature is uniform over the thickness.

In this case, the thermal resistances are given by:

$$R_{conduction} = \frac{\ln\left(\frac{r_i + L}{r_i}\right)}{2\pi kt}$$

(3a)

$$R_{convection} = \frac{1}{\pi((r_i + L)^2 - r_i^2)h}$$

(3b)

Equating the resistances and defining $\eta_i = {r_i}/{_L}$ leads, after some re-arranging, to

$$L^2 = \frac{tk}{h} \cdot \frac{1}{\left(\eta_i + {1}/{2}\right)(ln(\eta_i + 1) - ln(\eta_i))}$$

(3c)

This expression is similar to equation *(2c)*, apart from a factor,

$$f(\eta) = \left(\eta + {1}/{2}\right)(ln(\eta + 1) - ln(\eta))$$

(4)

**Figure 2** shows the factor for 0.001<$\eta$<2. It is clear that the factor converges to 1. For $\eta$>0.5 the factor =1 within 10%, and L≈$L_c$. This confirms the results for use of the correction factor in [7].

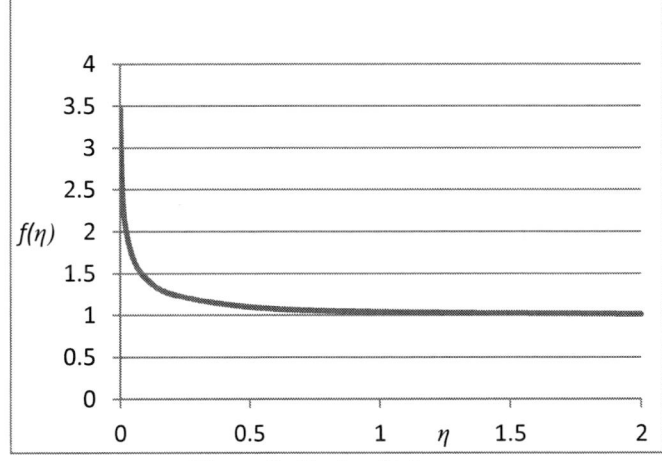

**Figure 2: Factor $f(\eta)$ vs. $\eta$**

## 3. What does the length scale mean?

The physical meaning of the length scale will be investigated. For this purpose it is convenient to use scaled quantities:

$$\eta = {r}/{L_c}$$

(5)

$$\theta = \frac{T - T_{ambient}}{T_{source} - T_{ambient}}$$

(6)

$$L_c = \sqrt{{kt}/{h}}$$

(7)

In the cylindrical case, the total heat transferred is given by

$$q = 2\pi h \int_0^{r_o} (T - T_{ambient})\, r dr$$

(8a)

In scaled variables:

$$q = 2\pi h \int_0^{\eta_o} \theta(\eta)\,(T_{source} - T_{ambient})\, L_c\eta L_c d\eta$$

(8)

So it is convenient to scale $q$ as

$$Q = \frac{q}{hL_c^2(T_{source} - T_{ambient})} = 2\pi \int_0^{\eta_o} \theta(\eta)\ \eta d\eta$$

(9)

*Equation (9) shows that Q equals the rotational volume resulting from rotating the $\theta(\eta)$ curve around the axis $\eta=0$.* The case of the circular isothermal source of radius $\eta_i$ embedded in a thin circular plate of radius $\eta_o$ shown in Figure 1 is a combination of the solution for the source itself and the solution for a thin annular fin with inner radius $\eta_i$, and outer radius $\eta_o$.

Assuming an isothermal source

$$0 \le \eta \le \eta_i$$
$$\theta = 1$$
$$Q_{source} = \pi\eta_i^2$$

For the thin annular fin, [2], and [7] give the exact solutions for the 1D radial case in terms of Bessel functions:

$$\eta_i \le \eta \le \eta_o$$

$$\theta = \frac{K_1(\eta_o)I_o(\eta) + I_1(\eta_o)K_0(\eta)}{K_1(\eta_o)I_0(\eta_i) + I_1(\eta_o)K_0(\eta_i)}$$

(10)

$$Q_{fin} = \int_{\eta_i}^{\eta_o} \theta(\eta)2\pi\eta d\eta = 2\pi\eta_i \cdot \frac{I_1(\eta_o)K_1(\eta_i) - K_1(\eta_o)I_1(\eta_i)}{I_1(\eta_o)K_0(\eta_i) + K_1(\eta_o)I_0(\eta_i)}$$

(11)

For the total heat transfer, the contributions of the annular fin and source itself have to be added

$$Q = \pi\eta_i^2 + Q_{fin}$$

It was shown earlier that Q equals the rotational volume resulting from rotating the $\theta(\eta)$ curve around the axis $\eta=0$. This volume is the sum of the cylindrical volume with radius $\eta_i$ and height $\theta=1$, contributed by the source, and a ring shaped volume contributed by the annular fin. This ring volume extends from $\eta=\eta_i$ to $\eta=\eta_o$, and is bounded at the top by the rotated $\theta$ curve, dropping from $\theta=1$ to $\theta=\theta_o$.

### 3.1 Infinite case

In the case of an infinitely large thin plate, $K_1(\eta_o)$ and $K_0(\eta_o)$ approach $0$, so $\theta$ and $Q$ become:

$$\theta = 1 \; for \; \eta < \eta_i, else \; \theta = \frac{K_0(\eta)}{K_0(\eta_i)}$$

$$(12)$$

$$Q = \pi\eta_i^2 + 2\pi\eta_i \frac{K_1(\eta_i)}{K_0(\eta_i)}$$

$$(13)$$

It was noted in [7] that for the large circular source on the infinite thin plate, cooling was as if the source had the Characteristic Length added to the radius.

In scaled form: $Q_{circle} = \pi(\eta_i + 1)^2$

This suggests that the length scale derived in equation (3c), could be used in the same manner.

$$Q_{circle} = \pi(\eta_i + \frac{1}{\sqrt{f(\eta_i)}})^2$$

$$(14)$$

With $f(\eta)$ as defined in equation (4).

An analytical spreadsheet calculation confirms that indeed this is a good approximation over a large range of $\eta_i$. In the range $\eta_i$ $10^{-4} \leq \eta_i \leq 2$, the value by equation (14) approaches the value as given by equation (13) to within 6%. For $\eta_i > 2$, it is correct to within 1%.

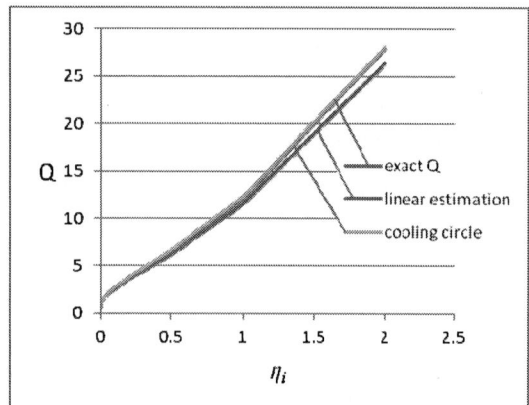

**Figure 3a: Comparison of calculated Q for sources of varying size on an infinite plate**

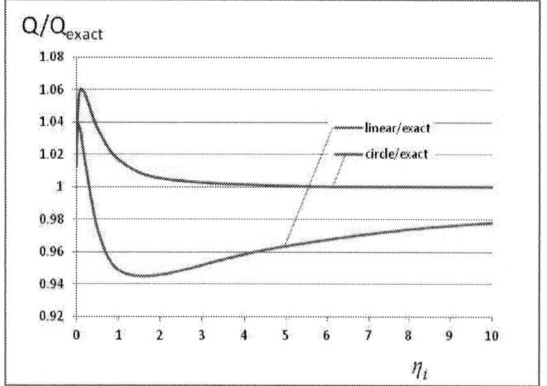

**Figure 3b: Relative accuracy of the linear and the cooling circle Q with respect to the exact Bessel solution**

Figure 3a shows the exact 1D radial Q from equation (13) compared to the value from equation (14) and a linear approach, detailed below. Figure 3b shows both approximate values relative to the exact 1D radial Q. The graphs show that both the Cooling Circle approach and the linear approach are good approximations to the Bessel function based value of Q.

*Linearization*

The shape of the temperature field is a plateau with $\theta=1$ at the location of the source, followed by a drop down to a plateau at $\theta=0$. This suggests that the temperature profile can also be approximated by the two plateaus, and a linear drop over a certain unknown radial distance $\omega$. Q is the volume resulting from rotating this temperature profile around the $\eta=0$ axis. The resulting rotational volume representing Q is a cone with base radius $\eta_i + \omega$ and the tip cut off horizontally at $\theta=1$. The central cylindrical surface of the cut off cone is the cylinder representing the contribution of the source with base $\eta=\eta_i$ and height $\theta=1$.

In the limit to vanishingly small sources, the central cylinder vanishes and the truncated cone becomes a full cone with height $\theta=1$, and base radius $\omega$. Consequently Q in the limit for vanishingly small sources becomes the volume of this full cone. It was shown earlier in equation (14) and Figure 3 that Q can also be approximated by a Cooling Circle of radius $\eta_i + \frac{1}{\sqrt{f(\eta_i)}}$ at source temperature $\theta=1$, that is by a cylindrical volume of this radius and height $\theta=1$. Equating the volumes of the cone and the cylinder will result in a solution for $\omega$ that ensures that for small sources the correct cooling Q is represented.

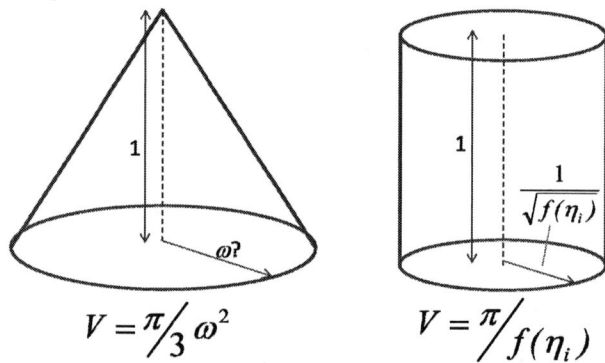

**Figure 4: Equating the volume of the cone to the volume of the cylinder to determine $\omega$**

This is depicted in Figure 4. Equating the volume of the cone [10] to the volume of the cylinder leads to

$$^1/_3 \; \omega^2 = {}^\pi/_{f(\eta_i)}$$

$$\omega = \frac{\sqrt{3}}{\sqrt{f(\eta_i)}} = \frac{\sqrt{3}}{\sqrt{(\eta_i + {}^1/_2)(ln(\eta_i + 1) - ln(\eta_i))}}$$

$$(15)$$

This is the linear estimation used earlier and shown in the Q graphs in Figure 3.

The corresponding temperature distribution is given by:

$\eta \leq \eta_i \quad \theta(\eta) = 1,$

$\eta_i < \eta \leq \eta_i + \omega, \; \theta(\eta) = 1 - (\eta - \eta_i)\omega$

$$\eta > \eta_i + \omega, \quad \theta(\eta) = 0$$

(16)

In other words, $\theta=1$ in the source area, and decreases from there with a slope $1/\omega$.

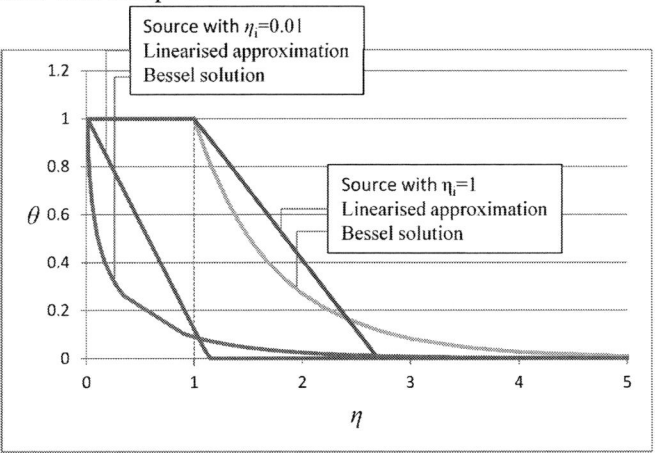

**Figure 5 Comparison of the linearised and the Bessel (scaled) temperature distribution for a small and a large source on an infinite plate**

The exact and the linear approximation of the temperature distribution are shown in Figure 5**Error! Reference source not found.** for a small source with $\eta_i=0.01$, and a larger source with $\eta_i = 1$.

Q for the linearised case is given by the volume of the cone with base radius $\eta_i + \omega$, slope $1/\omega$, and truncated at $\theta=1$.

$$Q = \frac{\pi}{3\omega}((\eta_i + \omega)^3 - \eta_i^3)$$

(17)

It can be seen from the longitudinal fin expressions in [7] that for large sources, when the radius of curvature becomes so large that the annular case equals the longitudinal case, the limit value for $\omega = 2$, while taking the limit of equation (15) results in $\omega = \sqrt{3} \approx 1.73$, an underestimation by 14%. However, this does not result in significant errors in Q: the 14% error occurs in the ring shaped volume around $Q_{source}$ representing $Q_{fin}$. At large sources, the contribution of the source itself is much more important in the total Q, and thus the 14% error in $Q_{fin}$ does not invalidate the total, as is confirmed Figure 3.

### Small cylindrical sources on Finite plates

Since the linearised solution proved so successful in case of the infinite plate, its extension to the case of the finite plate is investigated.

*Temperature solution.*

The linear approximation is $\theta=1$ in the source area, and decreases from there with a slope $1/\omega$ until $\theta=0$. For $\eta_o < \eta_i + \omega$, $\theta_o$ is

$$\theta_o = (\frac{\eta_i + \omega - \eta_o}{\omega})$$

Figure 6 shows the linear approximation and the Bessel solution according to equation (10) for a source with $\eta_i = 0.1$ on a plate of $\eta_o = 0.8$. Figure 7 shows composite curves for three sizes of circular sources with $\eta_i = 0.1$, $\eta_i = 1$, and $\eta_i = 2$,

for different sizes of finite circular cooling plates. The results show that in all cases, the linear temperature approximation deviates significantly from the actual temperature values, but that it acts as a good first estimate.

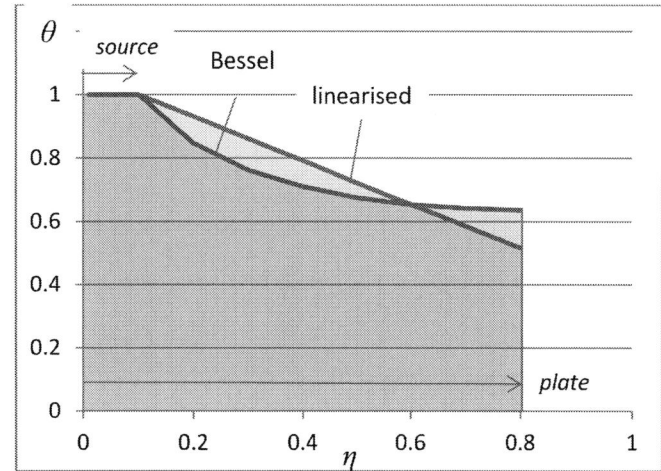

**Figure 6: Comparison between the linear approximation and the Bessel temperature distribution a source with $\eta_i = 0.1$ on a finite plate with $\eta_o = 0.8$**

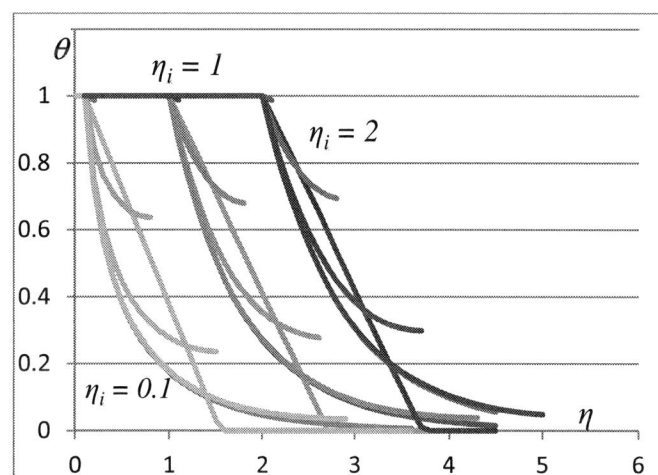

**Figure 7: Composite temperature curves for circular sources with $\eta_i = 0.1$, $\eta_i = 1$, and $\eta_i = 2$, on different sizes of finite circular cooling plates.**

*Heat transfer:*

For the finite plate $Q = Q_{fin} + \pi\eta_i^2$, with $Q_{fin}$ given in equation (11).

Q for the linearised case is again given by the volume of revolution for the $\theta$-curve around the central $\eta = 0$ axis: Again, this is a cone with slope $1/\omega$, truncated at $\theta=1$, and base $\eta_o$ in case $\eta_o < \eta_i + \omega$. In this case, $\theta_o \neq 0$, so the cone sits on top of a cylinder with base $\eta_o$ and height $\theta_o$. The case $\eta_o \geq \eta_i + \omega$ reverts to the infinite plate case.

After some mathematics:

$$Q = \frac{\pi}{3\omega}(\eta_o^3 - \eta_i^3) + \pi\eta_o^2 \left(\frac{\eta_i + \omega - \eta_o}{\omega}\right) for \ \eta_o < \eta_i + \omega$$

$$Q = \frac{\pi}{3\omega}((\eta_i + \omega)^3 - \eta_i^3) \ for \ \eta_o \geq \eta_i + \omega$$

*(18)*

Figure 8 shows the exact 1D radial Bessel function based Q values and the linearised Q values for three sizes of circular sources with $\eta_i = 0.1$, $\eta_i = 1$, and $\eta_i = 2$, and for sizes of finite circular cooling plates i.e different values of $\eta_o$. The results show that the linearised approach matches well with the Bessel function results.

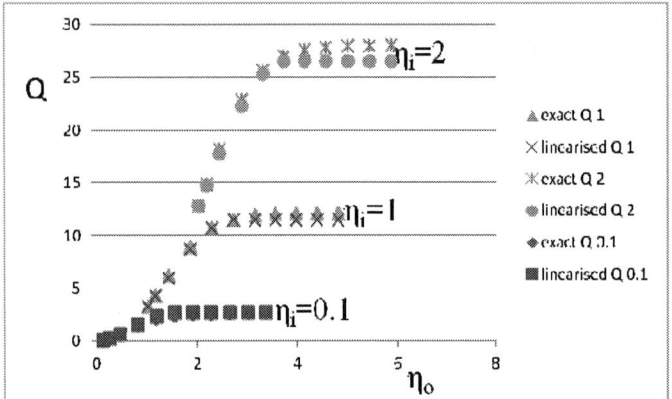

**Figure 8: Linearised vs Bessel based Q**

Table 1 contains the % errors for the linearised values of Q over a wide range of source sizes, and plate sizes extending up to 10ω from the source. The results show that the infinite results are valid within ~5% from distances larger than 2ω. This is in keeping with the results for the longitudinal case. Furthermore the results show that for all source sizes, the linearised Q values are accurate to within 5% if the outer radius is less than ½ ω away from the source. For values in between, the linearised Q values are accurate within 20% for small sources[1], and within 10% for sources over $\eta_i$=0.25.

| $\eta_0 = \eta_i + c\,\omega$ | % error (Q-exact - Q-linearised)/Q-exact | | | | | | | |
|---|---|---|---|---|---|---|---|---|
| c | 0.002 | 0.01 | 0.1 | 0.25 | 0.5 | 1 | 2 | 5 |
| 0.01 | -0.6% | -0.4% | -0.1% | -0.1% | 0.0% | 0.0% | 0.0% | 0.0% |
| 0.1 | -5.1% | -5.1% | -3.9% | -2.7% | -1.8% | -1.1% | -0.6% | -0.3% |
| 0.25 | -6.3% | -7.0% | -7.1% | -6.1% | -4.9% | -3.5% | -2.2% | -1.0% |
| 0.5 | 5.0% | 3.2% | 0.0% | -1.6% | -2.7% | -3.0% | -2.6% | -1.5% |
| 0.75 | 19.8% | 17.7% | 12.4% | 8.0% | 3.8% | 0.5% | -1.0% | -1.1% |
| 1 | 20.1% | 18.6% | 14.2% | 9.2% | 4.3% | 0.2% | -1.7% | -1.7% |
| 2 | 3.6% | 4.0% | 3.8% | 0.9% | -2.5% | -5.0% | -5.3% | -3.6% |
| 10 | 2.0% | 3.0% | 3.3% | 0.6% | -2.7% | -5.2% | -5.4% | -3.6% |
| infinite | 2.0% | 3.0% | 3.3% | 0.6% | -2.7% | -5.2% | -5.4% | -3.6% |

**Table 1: % error linearised Q vs exact Bessel based Q**

## 4. Case study and Comparison to SLA

In order to assess the validity of the Cooling Circle approach, a case study is calculated using both the Cooling Circle approach and the SLA method. The results are validated against numerical (Flotherm [11]) simulations.

*Case*: The case is the one posed by Seri Lee in [4]:"Consider an aluminium heatsink (k=200 W/mK) with base plate dimensions 100x100x1.3 mm thick. According to the heatsink catalog, the thermal resistance of this base plate

---

[1] 0.002 is the smallest source size that the author can realistically think off. It is the equivalent of a 3mmx3mm source on a flat heatpipe, k=10000 W/mK, with 1 sided h=10 W/m²K

is 1.0 C/W. Find the maximum resistance of the heat sink if used to cool a 25 x 25 mm device".

*Summarized SLA approach [4]*:

$$\lambda = \frac{\pi^{3/2}}{\sqrt{A_p}} + \frac{1}{\sqrt{A_s}} = 95.68$$

$$R_c = \frac{\left(\sqrt{A_p} - \sqrt{A_s}\right)\left(\lambda k A_p R_0 + \tanh(\lambda t)\right)}{k\sqrt{\pi A_p A_s}\left(1 + \lambda k A_p R_0 \tanh(\lambda t)\right)} = 0.66$$

$$R_{max} = R_c + R_0 = 1.66 \; C/W$$

*Cooling Circle*:

It follows from equation *(9)* that

$$R = \frac{1}{QhL_c^2}$$

*(19)*

The effective h=100 W/m²K from $R_0$=1=1/($A_p h$).

From equation *(2c)*: $L_c = \sqrt{kt/h} = 0.051 \; m$

$r_i = \sqrt{A_s/\pi} = 0.014 \; m$, so $\eta_i = 0.28$

$r_o = \sqrt{A_p/\pi} = 0.056 \; m$, so $\eta_o = 1.11$

From equation *(15 )*: ω=1.59,
By equation *(18)*: Q= 2.70
From equation *(19)*: R= 1.42 C/W

Linear drop is over distance $\omega L_c$=1.59 x 51 = 81 mm

For a central source, the distance $(r_o-r_i)$=42 mm. This is about halfway the linear drop, so the sink is not isothermal. Further enlarging the sink up to $r_o$=$r_i$+$\omega L_c$=14 + 81 = 95 mm will lower R. This is equivalent to a sink of area $A_p = \pi r_o^2$ =168x168 mm. Larger sinks will not result in significantly lower R values.

*Numerical model*:

The numerical model consists of an aluminum plate of the specified dimensions, thickness 1.3 mm, k=200 W/(mK). The source, 25 x 25 mm is located central on the plate. On the top and on the bottom of the plate, the system boundary coincides with the plate surfaces. Cooling boundary condition is a heat transfer coefficient on the system boundary. The reference temperature is 0 and the power of the source is 1 W, therefore the calculated temperature in the centre of the source corresponds one-on-one to the R values (K/W). Two cases were calculated. In the first case, the thermal conductivity of the source was taken as k=10000 W/(mK), resulting in an isothermal source with a thermal resistance of 1.40 K/W. This is the case calculated by the Cooling Circle, where the source area is assumed to be isothermal. In the other case, the thermal conductivity of the source is equal to the thermal conductivity of the plate heatsink, i.e. in case of a component mounted on top of the heatsink, it is assumed that the component itself is not contributing to the thermal conduction. This is the case calculated by the SLA method. The case geometry and the resulting temperature distribution over the central axis are shown in Figure 9.

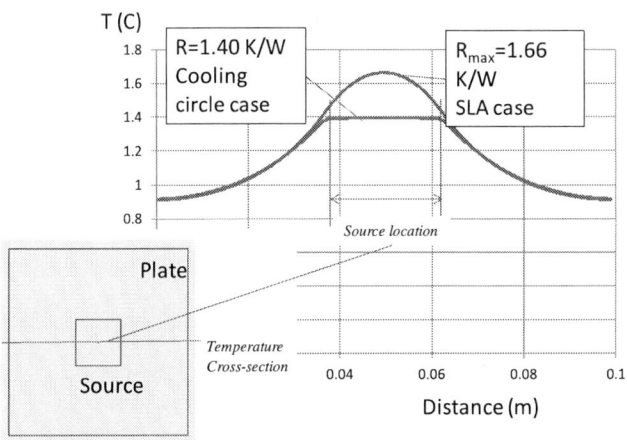

**Figure 9: Numerical results**

A comparison of the results is shown in Table 2. Both Cooling Circle and SLA agree well with the numerical values.

| source | source conductivity W/(mK) | calculated R$_{max}$ K/W | | | % difference |
|---|---|---|---|---|---|
| | | numerical | cooling circle | SLA | |
| isothermal | 10000 | 1.40 | 1.42 | | -1% |
| base material | 200 | 1.66 | | 1.66 | 0% |

. **Table 2: Comparison of calculated R$_{max}$ (K/W)**

*Extension to large plates*

Results for successively larger plates are presented in Table 3. The results for the Cooling Circle approach confirm the prediction that up to a base plate of ~170x170 mm, improvements in R can be expected, but that beyond that size larger base plates will not contribute to further cooling. Cooling Circle and Flotherm show an improvement if the plate is extended from 100 x 100 to 200 x 200, but no significant improvement for larger plates. Furthermore, Cooling Circle results match numerical results within 1%.

In contrast, the results for larger plates with the SLA approach show progressively worse alignment with the numerical results. At larger plate sizes the thermal resistance starts increasing, which is not possible from a physics point of view. Upon examination of the original paper [5] it was discovered that the larger plates quickly fall outside the parameter range for which SLA was validated: at 200 x 200 x 1.3 mm and 400 x 400 x 1.3 mm, the plate is thinner than 0.01 x plate size, and the plates over 600 x 600 x 1.3 also are larger than 20 x the source size. Apart from that, examination of the original comparison of the SLA correlation with numerical results [5] shows a 25% over prediction for the most similar case.

| plate dimensions | source | source conductivity W/(mK) | calculated R$_{max}$ K/W | | | % difference |
|---|---|---|---|---|---|---|
| | | | flotherm | cooling circle | SLA | |
| 100 x 100 mm | isothermal | 10000 | 1.396 | 1.416 | | 1% |
| | base material | 200 | 1.666 | | 1.657 | -1% |
| 200 x 200 mm | isothermal | 10000 | 0.930 | 0.902 | | -3% |
| | base material | 200 | 1.213 | | 1.286 | NA |
| 400 x400 mm | isothermal | 10000 | 0.893 | 0.902 | | 1% |
| | base material | 200 | 1.178 | | 1.398 | NA |
| 600 x 600 mm | isothermal | 10000 | 0.893 | 0.902 | | 1% |
| | base material | 200 | 1.177 | | 1.488 | NA |
| infinite | isothermal | 10000 | 0.893 | 0.902 | | 1% |
| | base material | 200 | 1.177 | | 1.750 | NA |

**Table 3: Comparison of calculated R$_{max}$ (K/W) for large plates**

## 5. Summary and conclusion

The Cooling Circle approach from [7] was extended to small cylindrical sources both in infinite and finite thin plates. The derivation of the relevant length scale for small cylindrical sources was inspired by Constructal theory.

It was shown that for large and small circular sources the cooling in case of an infinite thin plate can be approximated by the cooling of a Cooling Circle with radius $(r_i+L)$, with

$$L = \frac{L_c}{\sqrt{(1/2 + \eta_i)(\ln(\eta_i + 1) - \ln(\eta_i))}}$$

(20)

$$\eta_i = r_i/L_c, \text{ and } L_c = \sqrt{kt/h}$$

For $\eta_i>0.5$ $L=L_c$ within 10%.

For a finite thin plate the cooling can also be approximated as resulting from a linear temperature drop from the rim of the source outwards. The slope of the temperature drop is such that ambient temperature is reached over a radial distance of

$$\omega = L\sqrt{3} = \frac{L_c\sqrt{3}}{\sqrt{(1/2 + \eta_i)(\ln(\eta_i + 1) - \ln(\eta_i))}}$$

(21)

The thermal resistance of the source on a finite thin plate with outer radius $r_o$ is given by:

$$\eta_o = r_o/L_c$$

$$Q = \frac{\pi}{3\omega}(\eta_o^3 - \eta_i^3) + \pi\eta_o^2\left(\frac{\eta_i + \omega - \eta_o}{\omega}\right) for \ \eta_o < \eta_i + \omega$$

$$Q = \frac{\pi}{3\omega}((\eta_i + \omega)^3 - \eta_i^3) \ for \ \eta_o \geq \eta_i + \omega$$

$$R = \frac{1}{QhL_c^2}$$

(22)

The Cooling Circle approach has considerable practical value as it enables quick engineering estimations and provides a visual aid as to how far the heat goes.

The Cooling Circle approach was compared to the SLA approach and to numerical calculations. For small base plate dimensions, both the Cooling Circle and SLA matched well with the numerical results. In the numerical comparison the difference between the two approaches emerged as different assumptions as to the thermal conductivity of the heat source. Cooling Circle assumes a thin plate, and a uniform temperature in the source area. SLA incorporates the effect of spreading in the plate thickness, and for the thin plates comparison this boils down to using the baseplate conductivity in the source area. For larger plates, the Cooling Circle aligned well with numerical results and with physics based expectations. The calculated SLA values resulted in higher resistances for larger plates, contrary to both numerical results and physics based expectations. This was attributed to the SLA correlations being used outside the range for which they were validated.

## Acknowledgments

The support of Philips Electronics, especially of the Consumer lifestyle High impact innovation site Eindhoven and the Bruges development site are gratefully acknowledged.

## References

1. Lasance, C and Luiten, W , Course manual for "Cooling of electronics from an industrial point of view", Semitherm 25 short course, San Jose 2009

2. H. Y. Wong, "heat transfer for engineers", Longman London, 1977

3. Lasance, C, "heat spreading, not a trivial problem" electronics cooling magazine, may 2008

4. Lee, S., "Calculating Spreading Resistance in Heat Sinks," Electronics Cooling, Vol. 4, No. 1, January, 1998

5. Song, S., Lee, S., Au, V., "Closed Form Equation for Thermal Constriction/Spreading Resistances with Variable Resistance Boundary Condition", Proceedings of the 1994 IEPS Conference, 1994, pp. 111-121.

6. Lee, S., Song, S., Au, V., and Moran, K.P., "Constriction/Spreading resistance model for electronics packaging" ASME/JSME Thermal Engineering Conference 1995, volume 4, pp 199-205.

7. Luiten, W, "Characteristic Length and Cooling Circle", Semitherm 26, san jose 2010

8. Bejan, A. "Shape and structure, from engineering to nature", Cambridge University press, 2000

9. Bejan, A. and Lorente, S."Design with Constructal theory" John Wiley & sons, 2008

10. Bronshtein & Semendyayev, "a guide book to mathematics", Verlag harri deutsch, zurich 1973

11. http://www.mentor.com/products/mechanical/products/flotherm

# Design Considerations for Heat Spreader in High Heat Flux Systems

K. Alam, X. Shen and R. Taposh
Ohio University
Athens, OH 45701
alam@ohio.edu

## Abstract

Over the last several years, there has been a strong emphasis on developing heat sinks for high heat flux systems. Consequently new materials are being developed and novel thermal transport mechanisms are being investigated. Some of the novel materials are composites with thermal conductivity exceeding 500 W/mK. In this paper we examine the effect of very high thermal conductivity on the heat spreader performance with reference to high convective heat fluxes. It is shown that the thermal characteristics at high heat fluxes increases the importance of optimization of the heat spreader resistance in the design of the heat sink.

## Keywords

Heat sink, high heat flux, optimization

## 1. Introduction

One of the important components of the heat sink is the heat spreader which improves the heat dissipation from the chip into the environment. The heat spreader tends to be the most significant mass in the heat sink. In view of the increase in heat flux requirements for power electronics, the design of the heat spreader requires detailed analysis of the heat sink components. To reduce the heat sink conductive thermal resistance, diamond-copper composite heat spreader material have been produced having thermal conductivity greater than 500 W/mK [1]. Studies have been conducted to improve the convective heat transfer from the heat spreader by using foams, both metallic and non-metallic [2-3], and effective heat transfer coefficients of more than 20 kW/m²K at the heat spreader have been demonstrated [4]. These developments enable the design of heat sinks for high heat fluxes.

It is well understood that the heat spreader enhances the heat sink performance by spreading of the heat flux over a wider area; but this is partially counteracted by the spreading resistance, i.e., resistance due to the longer path taken by the heat flux. Several studies have been carried out to determine and optimize the spreading resistance [5-9]. Song et al. [7] and Lee et. al. [8], have developed a set of simple formula based on a axi-symmetrical cylindrical geometry which can give reasonably accurate results for non-circular geometries. However, for purposes of looking at the optimum values, it is advantageous to start with an accurate analytical solution, such as the solution developed by Feng et al. [9] based on three dimensional rectangular geometry using the method of Fourier expansion. Lasance [10] has provided guidelines for selecting heat spreaders with a comparative analysis of approximate and analytical solutions. His results for typical heat source of 5 mm square show that a high thermal conductivity spreader will approach optimum at or above 2 mm thickness. In the present study, we examine the optimum for very high heat fluxes.

## 2. Analytical solution

In this study, the approach of Feng et al. [9] is used as the basis for the optimization of a heat spreader in terms of the non-dimensional geometric parameters of the heat spreader. Since this is an accurate analytical solution, the optimum can be determined accurately by numerical differentiation. The geometry shown in Figure 1, will be analyzed with typical heat source and heat spreader dimensions.

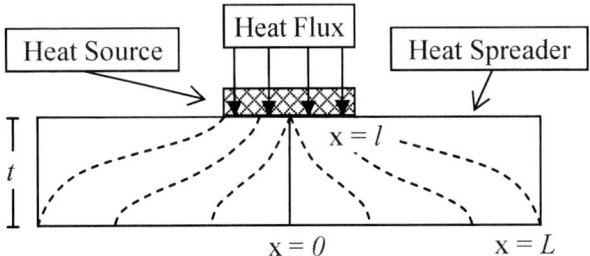

**Figure 1.** Schematic of the heat spreader system

The solution for the temperature field obtained by Feng et al. [9] is derived in terms of non-dimensional variable $T^*$ by solving the following equation:

$$\frac{\partial^2 T^*}{\partial X^2} + \frac{1}{\beta^2}\frac{\partial^2 T^*}{\partial Y^2} + \frac{1}{\tau^2}\frac{\partial^2 T^*}{\partial Z^2} = 0 \qquad (1)$$

where, $X$, $Y$, and $Z$ are the $x$, $y$ and $z$ coordinates non-dimensionlized with respect to the heat spreader dimensions.

The boundary conditions are

$$\left.\frac{\partial T^*}{\partial X}\right|_{X=0} = \left.\frac{\partial T^*}{\partial X}\right|_{X=1} = \left.\frac{\partial T^*}{\partial Y}\right|_{Y=0} = \left.\frac{\partial T^*}{\partial Y}\right|_{Y=1} = 0 \quad (2)$$

$$\left.\frac{\partial T^*}{\partial Z}\right|_{Z=0} = B_i T^* \qquad (3)$$

$$\left.\frac{\partial T^*}{\partial Z}\right|_{Z=1} = \begin{cases} 1 \,...\, (0 < X \le \varepsilon, 0 \le Y \le \gamma) \\ 0 \,...\, (\varepsilon < X \le 1, \gamma \le Y \le 1) \end{cases} \quad (4)$$

In the above boundary equations, the parameters are also non-dimensionalized by the heat spreader dimensions, as defined below:

$2L$ = Length of heat spreader
$2l$ = length of heat source in contact with heat spreader
$t$ = thickness of heat spreader
$k$ = thermal conductivity of the heat spreader

$h$ = convective heat transfer coefficient at the heat spreader

$\varepsilon$ = Non dimensional heating contact length (length of heat source/length of heat spreader)

$\tau$ = Non dimensional thickness of the heat spreader ($t/L$)

$\gamma$ = Non dimensional contact width

$\beta$ = Ratio of width to length of the heat spreader

$Bi = h\,t/\,k$ (Biot number)

In the above definition of the Biot number, $h$ is the effective heat transfer coefficient $h$ at the bottom of the heat spreader. The effective value $h$ incorporates the total heat transfer effect due to presence of fins or a foam in the fluid flow. With air flowing through a graphitic foam, the value of $h$ has been shown to exceed 4000 W/m$^2$K [4]; when water was used, the value exceeded 20,000 W/m$^2$K. The non dimensional temperature is obtained by combining the temperature, Biot number, the heat transfer coefficient and the source heat flux; and the solution can be written as,

$$T^* = T^*(X, Y, Z, \varepsilon, \gamma, \tau, \beta, B_i) \qquad (5)$$

The goal is to optimize the heat spreader dimensions so as to reduce the total thermal resistance. In this paper, the total thermal resistance will be studied for the simplified case of a square heat spreader centered under a square heated area and a uniform heat flux. The total thermal resistance $R_t$, defined on the basis of the maximum temperature is given as:

$$R_t = R_m + R_f + R_s = \frac{t}{kA} + \frac{1}{hA} + R_s \qquad (6)$$

where A is the area of the heat spreader base, $R_m$ is the conductive resistance of the heat spreader material, $R_f$ is the convective resistance from the heat spreader, and $R_s$ is the spreading resistance arising from the spreading out of the heat flux from the source into the heat spreader.

Feng et al. [9] has derived an expression for the spreading resistance; which, for a square heat spreader, can be expressed in the following non-dimensional form:

$$\psi_s(\varepsilon, \gamma, \tau, \beta, B_i) = \frac{R_s}{2kl} \qquad (7)$$

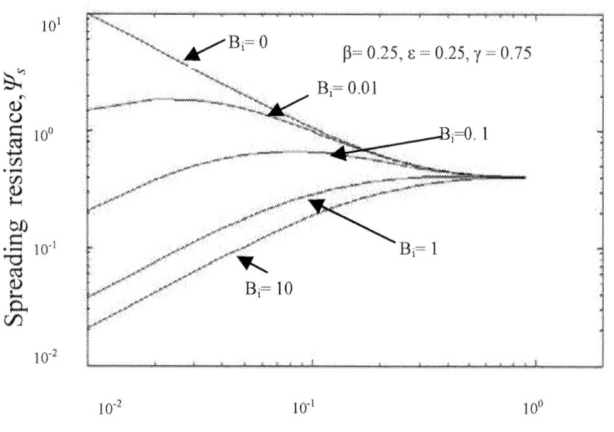

**Figure 2.** Non-dimensional spreading resistance as a function of the non-dimensional thickness of heat spreader and the Biot number.

A typical plot of the non-dimensional spreading resistance is shown below in Figure 2. The variation of the spreading resistance indicates that an optimal value of the total resistance can be found for different Biot numbers.

Figure 2 has been plotted using a corrected version of the solution; and so it differs from the plot originally shown in Feng et al. [9]. It shows the dependence of the spreading resistance on the geometric parameters. However, this figure is difficult to use for selecting the thickness, since the Biot number changes with the thickness parameter on the x-axis. This issue is addressed in the subsequent figures.

### 3. Optimum thickness

To examine the optimum thickness of the heat spreader, it is convenient to study the total thermal resistance for the simplified case of $\beta=1$, i.e. a square shape, with the heat source centered on the heat spreader. For this geometry, the characteristic dimension of the heat source and the heat spreader will be their length values ($2l$, $2L$).

The results of calculation using solution [7] of Equation (1) are shown in Figures 3(a) and 3(b), where the total resistance is plotted as function of heat spreader thickness in non-dimensionalized the heat source length ($2l$). Figure 3 is plotted with the heat source area $A_s=12\times12$ mm$^2$, and $h/k$ taken to be (a)10 m$^{-1}$, or (b) 20 m$^{-1}$. These parameters represent the middle range of convection cooling, and are selected to demonstrate the minimum in the solution curve. The plots are invariant if the ratio of h/k is the same. Therefore, h/k is an important dimensional parameter for the design of the system.

From the two plots in Figure 3, it can be observed that the total resistance decreases rapidly with thickness, and reaches a minimum, but the minimum is not well defined, especially for the typical values of the ratio $l/L$ (expected to be below 0.25). For these situations depicted in Figure 3, the total resistance approaches the minimum value as the thickness value increases to about $(t/2L)\approx0.1$. Therefore, it can be concluded that for this typical heat source dimension and the medium range of h/k, the optimum thickness of the heat spreader is approximately one-tenth of the heat spreader length. Increasing the thickness beyond this value will not help the heat transfer; and may actually reduce the heat transfer. An optimum greater than 10 mm is generally not practical.

In practice, the thickness of heat spreaders tends to be different than optimum because of considerations of weight, space, and cost-effectiveness. Lasance [10] has discussed the difference between two cases of h=250 W/m$^2$K and h=10,000 W/m$^2$K, He concludes that the thickness of the heat spreader is not critical at high thermal conductivity values, and thicker is better for heat dissipation.

However, in demanding applications where very high heat fluxes must be established through the heat sink, it is advisable to examine the optimum thickness of the heat spreader. In such cases, the h/k ratio can approach a value of 100 m$^{-1}$ or more.

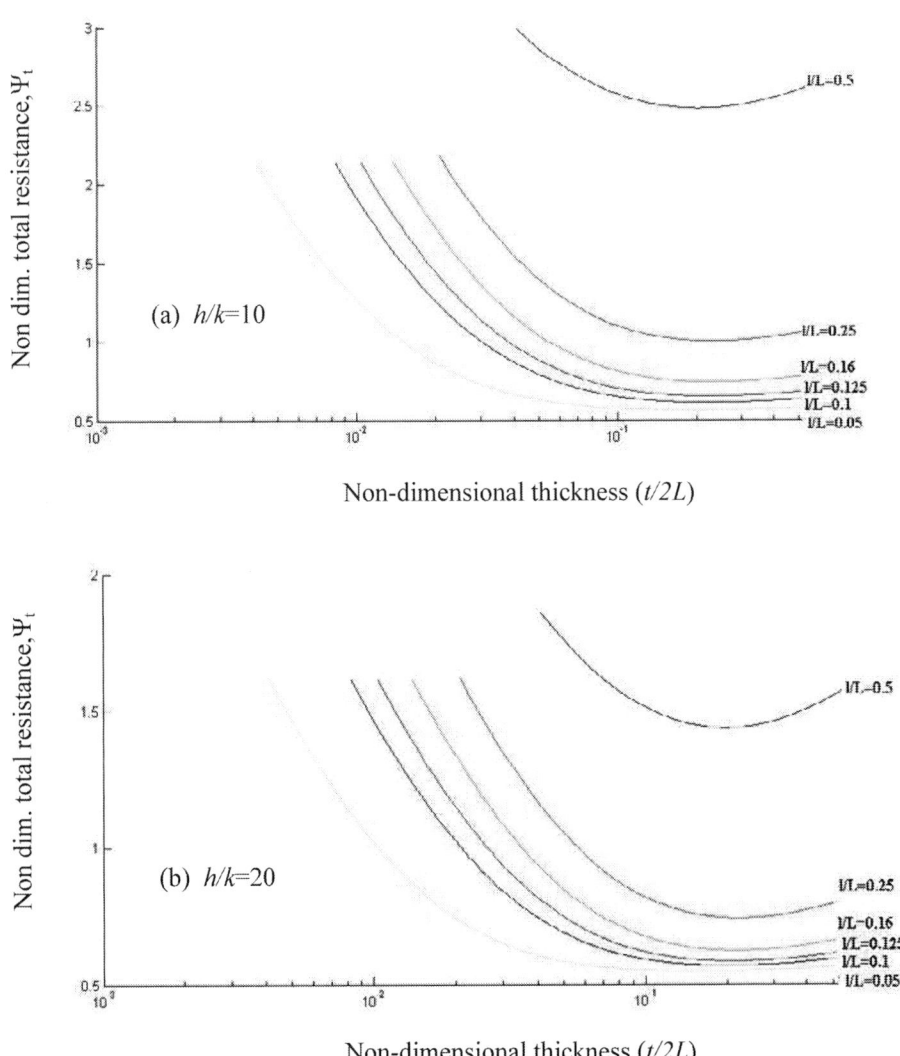

**Figure 3.** Variations of total resistance as a function of the spreader thickness in non dimensional form; and the ratio of the source length to the length of the heat spreader($l$/L). Two plots show the behavior for two different ratios of $h/k$.

Using the analytical solution of Equation (1), two plots has been drawn in Figure 4 to illustrate the particular example discussed above by Lasance [10], which has a heat source of 5mmX5mm and a heat spreader of 30mmX30mm. In Figure 4(a), the plots for the total resistance indeed shows that for high conductivity (300 W/mK or higher) materials, the total resistance values are very closely grouped together, and not much is gained by either changing the conductivity, or increasing the thickness above 2 mm. In this case, the Bi numbers are small, so convective resistance dominates.

When the convective heat transfer coefficient is increased to 10,000 W/m$^2$K, which can be achieved by liquid cooling [4], the results change. The total resistance values are much lower, and the curves show higher percentage differences; and it appears that some improvement may be possible with thicknesses higher than 2 mm.

It is also interesting to note that the optimum value is more well defined at the high convection rate, as seen is Figure 4(b). Therefore, for extreme heat fluxes, it is advantageous not to exceed the optimum thickness calculated from the analytical solution.

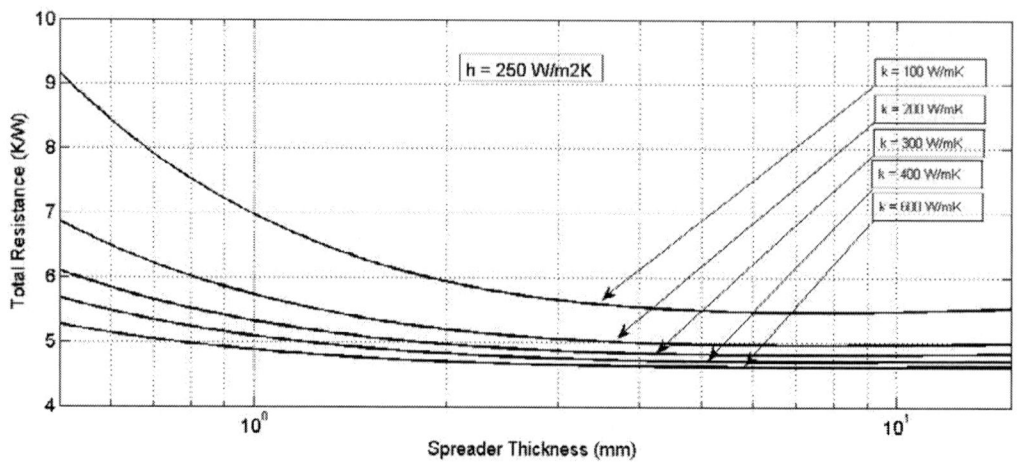

(a) Convective heat transfer coefficient = 250 W/m²K

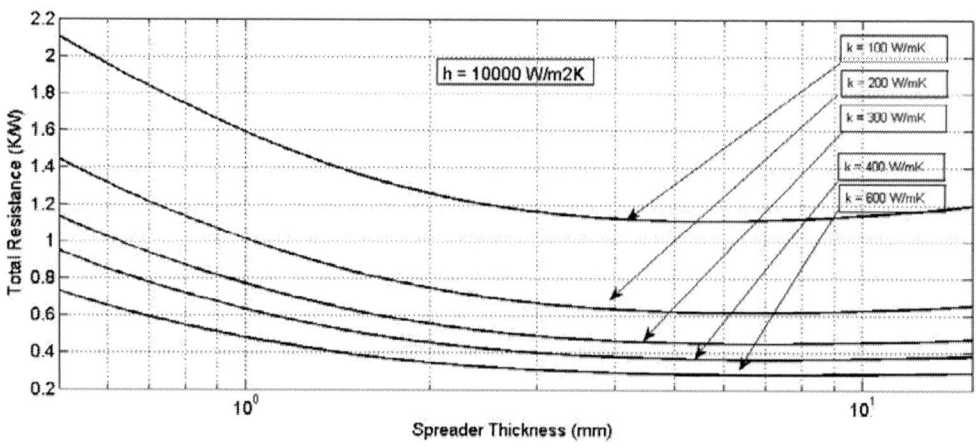

(b) Convective heat transfer coefficient = 10,000 W/m²K

**Figure 4.** Total resistance of a heat spreader of different thermal conductivities as a function of its thickness. Two plots have been drawn with different convective heat transfer coefficients.

Table 1 shows the optimum thickness for heat spreaders at different convection rates on two high conductivity heat spreaders. k=350 WmK and 500 W/mK. This range was selected since heat spreaders with this conductivity have been developed [1] with a reduced CTE to match the device or package material.

From Table 1, it can be concluded that the optimum is approximately 7 mm for the most of the range of convective heat transfer coefficient, although there is some divergence at very high heat fluxes.

**TABLE 1**: Optimum thickness of heat spreader

| Convective Heat Transfer Coefficient (h) W/m²K | Optimum Thickness of Heat Spreader (mm) | |
| --- | --- | --- |
| | K=350 W/mK | K=500 W/mK |
| 250 | 7.38 | 7.38 |
| 1000 | 7.33 | 7.08 |
| 10000 | 6.80 | 7.03 |
| 40000 | 5.39 | 5.97 |

In Figure 5, the total thermal resistances for the optimum thickness has been compared with 1mm thick heat spreader. .

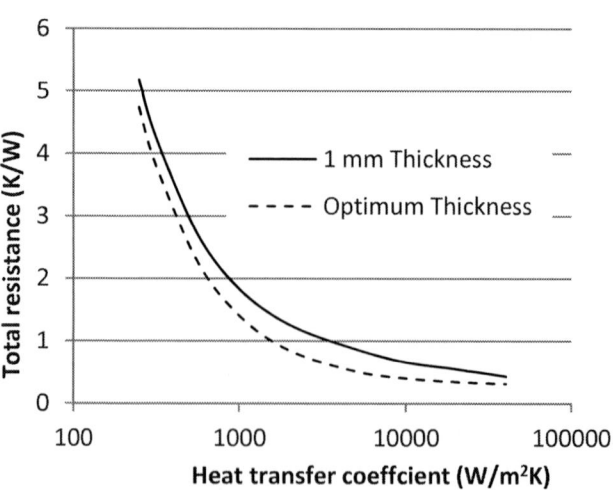

**Figure 5.** Total resistance for 350 W/mK conductivity material plotted as a function of the heat transfer coefficient.

The thermal conductivity value of the heat spreader in Figure 5 was chosen to be 350 W/mK; which would give an optimum thickness in the range of 5.4 to 7.4 mm. The reduction in the total resistance is about 25% with optimum thickness at h=10,000 W/m$^2$K. It can be seen that at high values of $h$, the total resistance will decrease to below 1 K/W, which will make it comparable to the resistance produced by thermal interface materials.

The difference between the total resistances for two different thermal conductivities (selected in Table 1) is illustrated in Figure 6. Each curve is plotted for 1 mm thickness of the heat spreader. It can be seen that there is very small difference in the absolute numbers between the heat spreaders, although the percentage difference can be 20% at high heat fluxes. With the combination of high conductivity materials, and high convective heat flux, the total resistance can become comparable or less than the resistance due to a TIM layer between the heat source and the heat spreader [11].

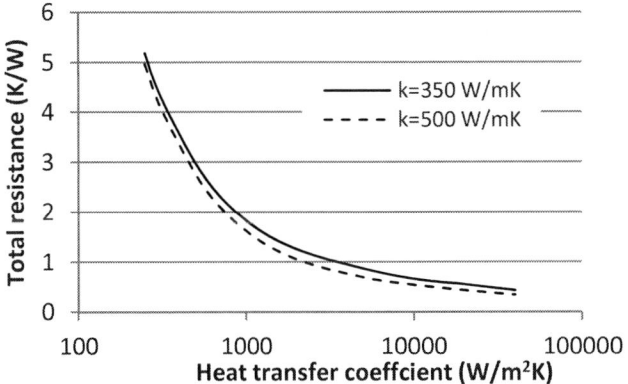

**Figure 6**. Total resistance for two different conductivities of heat spreader as a function of the heat transfer coefficient.

For a 5mmX5mm heat source, the resistance due to a TIM layer is typically about 2 K/W [11]. Therefore, to achieve the high heat flux under these conditions, the traditional TIM should be replaced by soldering for direct attachment of the heat spreader to the device or heat source. This is best done with a high conductivity heat spreader that has low CTE; and solder bond can be a good alternative if the heat spreader is made from a metal matrix composite material. With a solder bond, the interface resistance could drop below 0.2 W/K [11].

## 4. Conclusions

An analytical solution was used to determine the optimum values for the thickness of a high conductivity heat spreader under conditions of high heat flux. It was observed that the performance of the heat sink is not expected to improve greatly when thermal conductivity exceeds 350 W/mK; and the optimum value of the thickness does not change significantly when the heat transfer coefficient is in the range of 1,000 to 20,000 W/mK. At lower values of the convection heat transfer coefficient (below 10,000 W/mK), the high conductivity spreaders show a flat profile and the optimum thickness is not well defined. Because the resistance curve is rather flat at high thermal conductivity and high convective

heat transfer coefficient, the heat spreader with the optimum thickness at the highest expected convective heat flux condition can be expected to provide good performance at lower heat fluxes. At the highest convection rate, the optimum of the total resistance is most well defined; therefore it advisable to determine this optimum value, and use this as the upper limit of the spreader thickness.

A combination of high conductivity and high convection heat transfer coefficient can produce a low resistance that is comparable or less than the typical TIM resistance. In such cases, the heat spreader should be directly attached to the device or heat source with a solder so that the overall resistance is reduced. Resistances of other thermal interfaces must also be considered in the design of the heat sink.

## Acknowledgments

The authors would like to acknowledge the help by Mr. K. Drummond of Ohio University in drafting the paper, and the reviewers' helpful comments.

## References:

1. T. Schubert, B. Trindade, T. Weibgarber, B. Kieback, "Interfacial design of Cu-based composites prepared by powder metallurgy for heat sink applications", Materials Sci. Engg. A, Vol. 475, pp. 39-44, 2004.
2. K. Lim, H. Roh, "Thermal Characteristics of Grpahite Foam Thermosyphon for Electronics Cooling", J. Mechanical Sci. Tech., Vol. 9(10), pp. 1932-1938, 2005.
3. S. Mancin, C. Zilio, L. Rossetto, A. Cavallini, "Foam height effects on heat transfer performance of 20 ppi aluminum foams", App. Thermal Engg., in press, 2011.
4. N. C. Gallego, J. W. Klett, "Carbon Foams for Thermal Management", Carbon, Vol 41, pp. 1461-1466, 2003.
5. G. Maranzana, I. Perry, D. Maillet, S. Rael, "Design optimization of a spreader heat sink for power electronics", Int. J. Thermal Sciences, Vol. 43, pp. 21-29, 2004.
6. S. Yu, K. Lee, S. Yook, "Optimum design of a radial heat sink under natural convection", Int. J. Heat and Mass Transfer, Vol. 54, pp. 2499-2505, 2011.
7. S. Song, S. Lee, and V. Au, 1994, "Closed-Form Equations for Thermal Constriction/Spreading Resistances With Variable Resistance Boundary Condition," Proc. IEPS Conference, pp. 111–121.
8. S. Lee, S. Song, V. Au, and K. P. Moran, 1994, "Constriction/Spreading Resistance Model for Electronics Packaging," Proc. ASME/JSME Thermal Engineering Conference, pp. 111–121.
9. T.Q. Feng, J.L. Xu, "An analytical solution of thermal resistance of cubic heat spreaders for electronic cooling", App. Thermal Engg., Vol. 24, pp. 323-337, 2005.
10. C.J.M. Lasance, "How to Estimate Heat Spreading Effects in Practice", J. of Electronic Packaging, Vol. 132, 031004-1.
11. D. Blazej., "Thermal Interface Materials," Electronics Cooling Magazine, November, 2003.

# Achieving Energy Efficient Data Centers Using Cooling Path Management Coupled with ASHRAE Standards

Matthew Green[*], Saket Karajgikar[*], Philip Vozza[L] Nick Gmitter[L] and Dan Dyer[L]
[*]Future Facilities, Inc., 2055 Gateway Place, Suite 110, San Jose, CA 95110
[L]DLB Associates, 265 Industrial Way West, Eatontown, NJ 07724
E-mail: matt.green@futurefacilities.com

## Abstract

Power trends for data center facilities continue to grow at an alarming rate. In response to this, data center operators are now implementing various practices such as adding blanking plates, hot aisle/cold aisle containment, etc. Also, in response to ASHRAE's expanded thermal guidelines [1], many data center operators are now raising the supply air temperature in an effort to further conserve energy. However, airflow in a high-density thermal environment is a complex phenomenon and requires an appropriate engineering analysis to understand all of the factors. Improperly implementing such practices, which are now known as "Industry Standard Practices", may result in overheating of IT equipment and thus negatively affect data center operations. In the following scenarios, Cooling Path Management (CPM) is used to understand the airflow / thermal environment of the facility. ASHRAE thermal guidelines are then followed to increase the average temperature of the room. In later scenarios it is shown that with proper airflow management, ASHRAE recommendations can be implemented using a full engineering analysis. Results from the final scenario show that the CRAH unit set point can be increased by 23.8°F from the baseline scenario and 5.5 kW of stranded capacity can be regained. Results also show that "Industry Standard Practices" only enable the CRAH unit set point to be increased by 13.4°F. Following the CPM method and ASHRAE thermal guidelines will enable operators to make tangible changes to their data center using objective data.

## Introduction:

In 2004, ASHRAE Technical Committee 9.9 (Mission Critical Facilities, Technology Spaces, and Electronic Equipment) created a task group comprised of IT equipment manufacturers with the goal of defining a set of temperature and humidity envelopes for the reliable operation of data centers. The initial guidelines developed as a result of this effort aimed to strike a balance between IT equipment performance, reliability and energy efficiency. However, in past years data center energy consumption has increased very sharply. According to the 2007 US EPA report [4] presented to congress on server and data center energy efficiency (also according to an LBNL study [3]), about 1.5% of the total energy in the US is consumed by data centers. This report helped to create awareness for improving energy efficient operations of data centers. Many techniques such as the use of blanking plates, hot aisle/cold aisle containment etc. were proposed and have now become industry standard practices. These techniques attempt to eliminate problems such as hot spots, hot air recirculation etc.

## 2. ASHRAE Guidelines and CPM:

In the latest whitepaper (2011) released by ASHRAE TC 9.9 two new data center classes were added, each with expanded limits on the thermal environment. These new data center classes are shown in Table 1. The additional classes were created in an effort to remove the focus of strictly operating data centers within the recommended range, and promote the use of airside or waterside economizers to reduce compressorized-cooling solutions and ultimately save energy. The TC 9.9 whitepaper also provides supplemental information on the various factors that affect IT equipment performance, reliability and energy efficiency in order to educate operators on the costs and benefits of operating at higher supply air temperatures. This data places more emphasis on selecting the operating envelope that matches the individual business values of the data center owner in order to achieve the best Total Cost of Ownership (TCO).

In following the ASHRAE guidelines, owner / operators often neglect the impact of raising supply temperature and industry standard practices on the server fans. Following these practices blindly may overheat the IT equipment and results in increased downtime and decreased performance. The fundamental airflow practices referenced in the guidelines were intended to help data center operators attain better energy efficiency but were not meant to be interpreted as a "one size fits all" solution. Airflow in a high-density thermal environment is a complex phenomenon and a full engineering analysis is needed to understand the impact of many variables. Thus we couple an airflow management technique, Cooling Path Management (CPM), with the ASHRAE guidelines in order to fully understand the cause and effect of any changes made to the data center environment.

CPM is the process of stepping through the full route taken by the cooling air and systematically minimizing or eliminating cooling breakdowns and inefficiencies. The ultimate goal of CPM is to meet the air intake requirement for each unit of IT equipment. CPM is broken down into three areas: grille flows, supply air bypass, and recirculation; as shown in Figure 1. In order to properly implement CPM, each of these areas needs to be analyzed and then corrected based on the corresponding room layout and IT equipment. The payback for using this in-depth engineering approach can save not only operation expenses but may save significant capital costs as well.

| Class | Dry Bulb (°F) | Humidity Range |
|---|---|---|
| Recommended | | |
| A1 to A4 | 64.4 to 80.6 | 41.9°F DP to 60% RH & 59°F DP |
| Allowable | | |
| A1 | 59 to 89.6 | 20% to 80% RH |

**Table 1**: ASHRAE TC 9.9 2011 guidelines data center thermal operation [1].

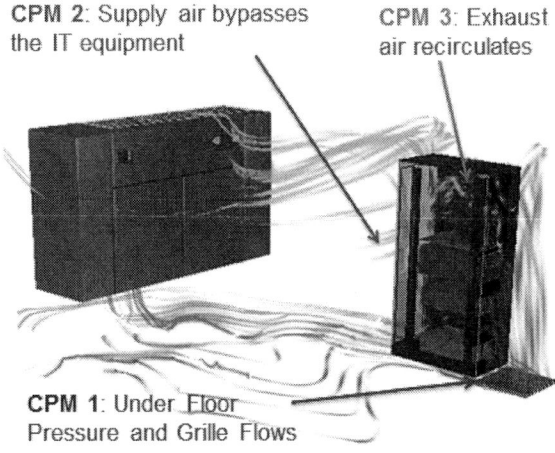

CPM 2: Supply air bypasses the IT equipment

CPM 3: Exhaust air recirculates

CPM 1: Under Floor Pressure and Grille Flows

**Figure 1**: Cooling Path Management breakdown.

Seven different scenarios have been modeled utilizing Computational Fluid Dynamics (CFD). CPM has been applied to each scenario conforming to the ASHRAE guidelines:

- Scenario 1 is the initial model that will be used as the baseline.
- Scenario 2.1 & 2.2 shows the implementation of blanking and containment as one would see in standard practices.
- Scenarios 3 through 5 use CPM to detect inefficiencies in the room. Various methods are then deployed using CPM to increase the supply temperatures.
- Scenario 6 shows using CPM can not only increase the supply temperature but also regain lost capacity by installing new IT equipment.
- Scenario 7 presents a failure scenario and uses CPM to show that the inlet temperatures remain unchanged.
-

**Scenario 1 – Base Data Center Model:**

In order to demonstrate the concept of cooling path management, a 560 ft² virtual data center with a 25" raised floor is considered. The room consists of 12 cabinets with installed IT equipment. Power and temperature dependent fan curves are defined for each piece of IT equipment. A single CRAH of 70kW maximum cooling sensible capacity delivering 16,000 CFM is considered. The minimum supply temperature is set at 55°F. Figures 2 and 3 show the room layout and IT equipment fan curve respectively.

From the CFD results, it is observed that none of the cabinets exceed ASHRAE allowable temperatures (89.6°F). However, one of the cabinets violates the ASHRAE recommended limit. Refer to Figure 4. With the supply temperature set at 55°F the results are very convincing for an owner operator to increase the average room temperature. In order to increase the room temperature, CPM is needed to make changes to the airflow / thermal environment.

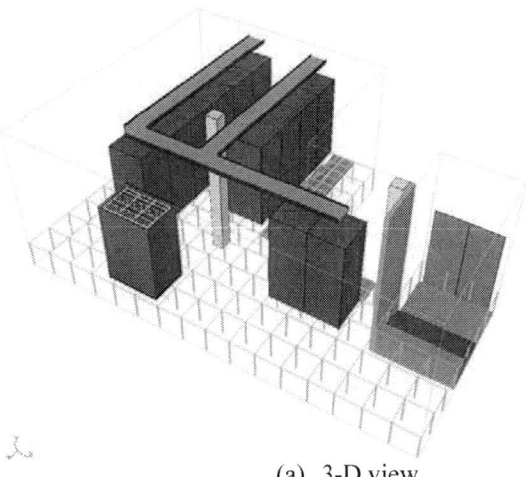

(a) 3-D view

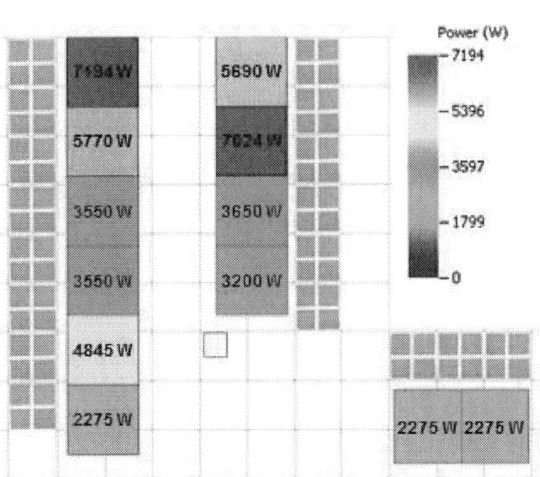

(b) Heat Load

**Figure 2**: Room layout with cabinet heat load

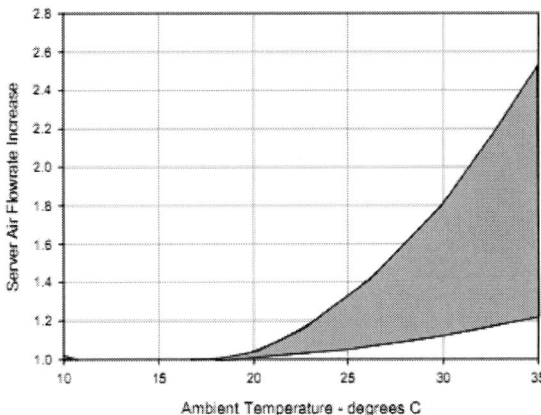

**Figure 3**: Server airflow rate dependence on ambient temperature increase given in ASHRAE TC 9.9 guidelines [1]

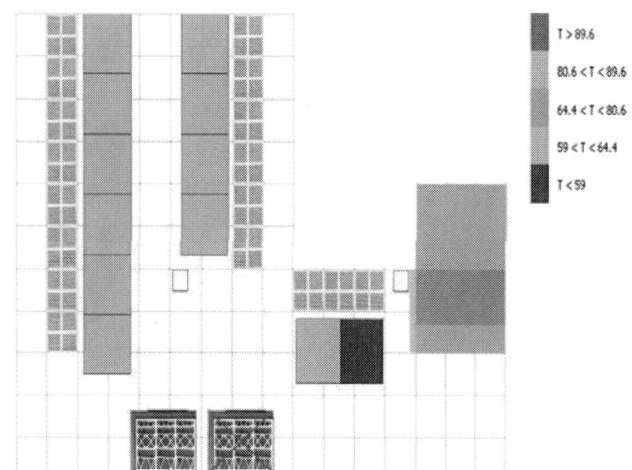

Figure 4: Scenario 1 ASHRAE Temperature Compliance

## Scenarios 2.1 & 2.2 - Implementation of Blanking and Containment:

In this scenario blanking plates and containment have been added. Refer to Figure 5. With the introduction of blanking and containment, all of the cabinets are now seeing constant inlet temperatures below 59°F. Refer to Figure 7 (a). Based on ASHRAE thermal guidelines this represents a significant waste of energy. Observing these results, the operator decides that increasing the supply set point to 75°F will offer energy savings utilizing warmer chilled water temperatures and more available economizer hours, with minimal increased risk to equipment reliability (Scenario 2.1). Accounting for miscellaneous heat gains, the resulting set point increase will change the inlet temperatures to a value just below 79°F.

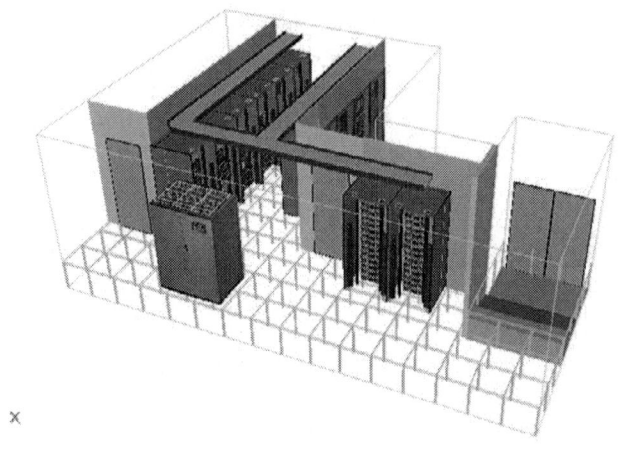

Figure 5: Room layout with containment

However, the operator neglects to fully understand the airflow management. In this case, temperature dependent server fans ramp up, see Figure 6, resulting in inadequate supply airflow and ultimately inducing recirculation.

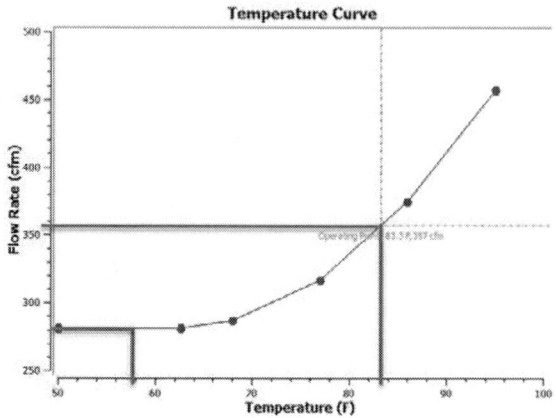

Figure 6: Server Fan Flow Curve

CFD analysis indicates that this causes one cabinet to exceed the ASHRAE allowable temperature limit. Refer to Figure 7 (b). To properly raise the air temperature the operator would use ASHRAE measurement guidelines to set up a series of temperature sensors at the IT inlets (Scenario 2.2). These sensors are used to calibrate the CRAH supply air temperature in order to arrive at an acceptable maximum IT inlet temperature of 80.7 °F. Refer to Figure 7 (c). However, the supply temperature could not be set above 68.4°F to achieve this, which still leaves room for potential energy savings by an increase of 12.2°F to the CRAH set point.

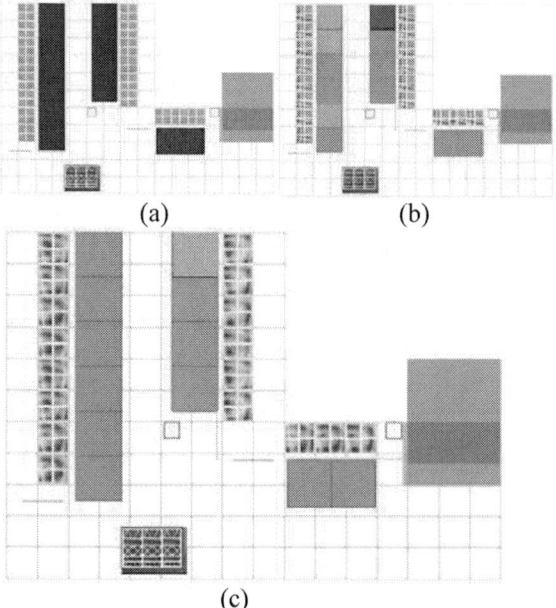

Figure 7: Scenario 2 - ASHRAE temperature compliance

## Scenario 3 – CPM Using Specialized Blanking and Matched Grille Airflow to IT Equipment Airflow Needs:

Using CPM, high volume grilles were added to address the lack of airflow in high load cabinets. Refer to Figure 8. Problem areas in blanking were addressed; accounting in particular for side breathing equipment and excess flow under the cabinets. With these improvements the supply air temperature was increased to 77.8°F, again based on a series of calibration sensors, with all resulting equipment inlet temperatures falling within the ASHRAE recommended range. Refer to Figure 9.

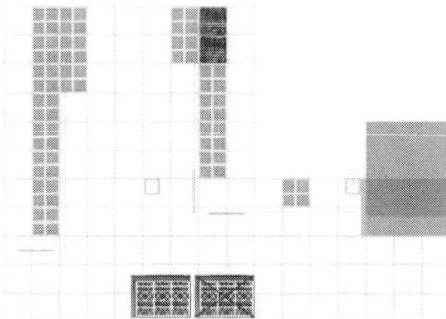

**Figure 8**: Scenario 3 grille modifications

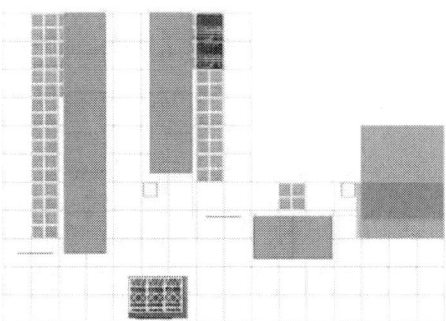

**Figure 9**: Scenario 3 ASHRAE temperature compliance and grille modifications

### Scenario 4 – CPM Used to Target Grille Airflow for Exact IT Equipment Airflow Needs:

In this scenario, blanking and containment have been removed in an attempt to address the cooling path solely by manipulating grille flows. Slotted grilles have been strategically placed based on the specific IT loading and inlet locations. Refer to Figure 10. These slotted grills have angled dampers to direct the airflow at specific problem areas such as side inlets and regions exhibiting recirculation from equipment outlets. Using temperature sensors to adjust the CRAH set point, the resulting supply temperature has been set to 78.6°F and all equipment max inlet temperatures are within the ASHRAE recommended envelope. Refer to Figure 11.

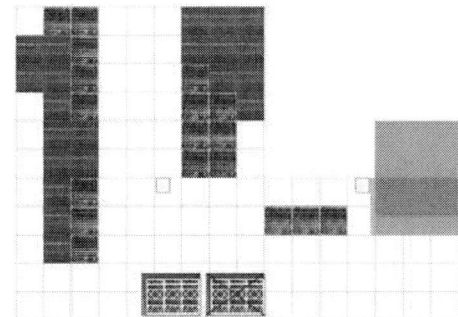

**Figure 10**: Scenario 4 grille modifications

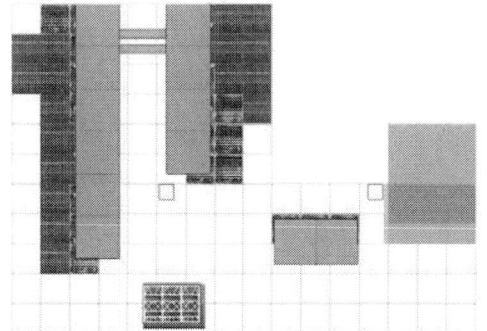

**Figure 11**: Scenario 4 ASHRAE temperature compliance and grille modifications

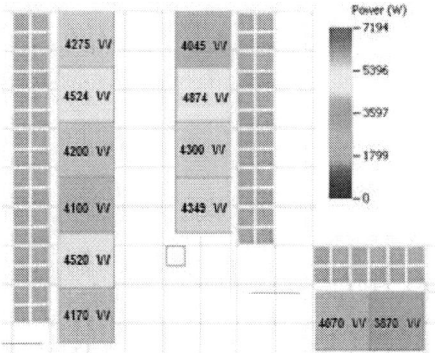

**Figure 12**: Scenario 5 New heat load

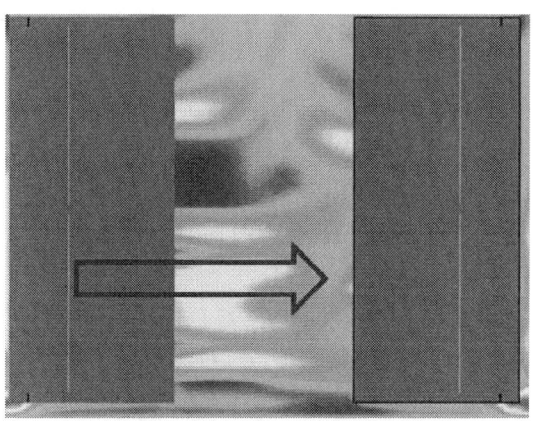

(a)   Scenario 3 Rack Outlets

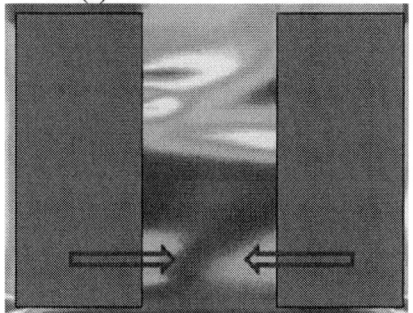

(b)   Scenario 5 Rack Outlets

**Figure 13**: Exhaust Outlet Velocity Comparison

### Scenario 5 – Redistribute IT Equipment using CPM:

The original equipment distribution, shown in **Figure** 3 (b), is currently heavily concentrated in one area of the data center. The goal in this scenario is to distribute the IT load as evenly as possible by moving equipment away from the lower

U slots, and matching exhaust jets to reduce cross cabinet recirculation. See Figures 12 and 13 (a) and (b). The original grille layout, blanking and containment have been preserved, leaving the IT equipment locations as the only variable. The resulting supply set point is 77.7°F to ensure all equipment inlet temperatures fall within the recommended ASHRAE temperature envelope.

### Scenario 6 – Deploying New IT and the Resulting Effects:

The solution in Scenario 3 has been implemented. Power, cooling and space are available and the operator wants to deploy new equipment. Five new pieces of IT equipment are proposed for deployment totaling an additional 5.5 kW of load. The flow demands of the new equipment have caused changes to the airflow pattern, inducing recirculation and starving old equipment of their required airflow. Refer to Figure 14. Once again cooling path management needs to be addressed. A mixture of techniques discussed in Scenarios 3, 4, and 5 (grille location, grille airflow, grille targeted cooling, changing IT equipment U slot location, and specialized blanking / containment) are used to return the max inlet temperatures to within the ASHRAE recommended envelope with the supply temperature set at 78.8°F. Refer to Figure 15.

### Scenario 7 – Failure of Primary CRAH and Secondary CRAH Activates:

This scenario tests to see if the changes made to the data center will produce the same results when the CRAH fails and the backup unit is turned on. In cases where the under floor has insufficient pressure, switching CRAH unit supply locations (Failure Mode) can result in different grille airflows. Results show that all cabinets still remain within the ASHRAE recommended range. Refer to Figure 16.

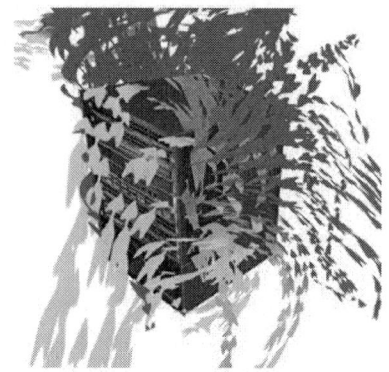

**Figure 14**: Scenario 6 IT equipment changes would cause localized recirculation

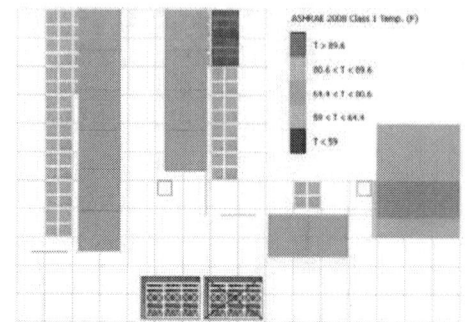

**Figure 15**: Scenario 6 ASHRAE temperature compliance

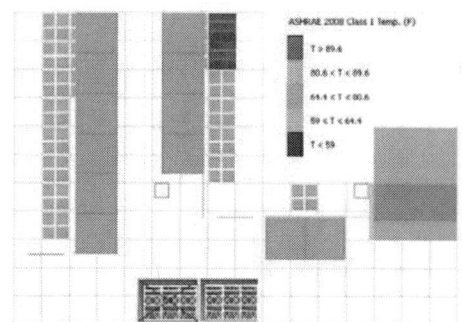

**Figure 16**: Scenario 7 ASHRAE temperature compliance

### Conclusion:

Following the ASHRAE TC 9.9 guidelines coupled with Cooling Path Management allows the owner / operators to make scientific and tangible changes to their data center resulting in significant energy savings. As shown in the preceding scenarios this may be achieved through a multitude of CPM techniques. Ultimately, the supply air was increased from 55°F to 78.8°F, signifying a large opportunity to increase economization hours, see Table 2.

While ASHRAE Thermal Guidelines provide a target, there is no one fix solution for every data center. CPM provides direction for operators to achieve their desired operating condition and can analyze the impact of major changes made to the data center e.g., deploying new IT. Attempting to implement energy saving measures without fully understanding the thermal environment can have detrimental effects on the reliability and performance of IT equipment. Following these principles owner / operators will be able to achieve significant savings due to the increased efficiency of their data center.

| Scenario | Max. Inlet Temperature (°F) | Under-floor Pressure (in. of H$_2$O) | Supply Bypass (%) | Recirculation (%) | Supply Set Point (°F) | Potential Temperature Increase (°F) |
|---|---|---|---|---|---|---|
| 1 | 82.7 | 0.232 | 62.2 | 39.6 | 55 | 25.6 |
| 2 | 58.4 | 0.236 | 43.9 | 11 | 55 | 25.6 |
| 2.1 | 91.5 | 0.238 | 40.5 | 13.9 | 75 | 5.6 |
| 2.2 | 80.7 | 0.237 | 42.8 | 12.1 | 68.4 | 12.2 |
| 3 | 80.5 | 0.154 | 37.1 | 4.7 | 77.8 | 2.8 |
| 4 | 80.4 | 0.001 | 45 | 7.2 | 76.1 | 4.5 |
| 5 | 80.5 | 0.247 | 35.9 | 5.8 | 77.7 | 2.9 |
| 6 | 80.2 | 0.139 | 31.1 | 4.6 | 78.8 | 1.8 |
| 7 | 80.4 | 0.141 | 32.0 | 5.7 | 78.8 | 1.8 |

**Table 2:** Summary of results for all the scenarios

**References:**

[1] ASHRAE TC 9.9, "2011 Thermal Guidelines for Data Processing Environments – Expanded Data Center Classes and Usage Guidance", ASHRAE, 2011

[2] Docca A., Ikemoto S., "Cooling Path Design; a simulation-based methodology designed to maximize IT equipment resilience and cooling energy efficiency", 2008

[3] Kooney J.G., "Estimating Total Power Consumption By Servers In The US and The World", A report by the Lawrence Berkeley National Laboratory, February 2007.

[4] U.S. Environmental Protection Agency ENERGY STAR Program, "Report to Congress on Server and Data Center Energy Efficiency Public Law 109-431," 2007.

# Use of Ducting to Improve Inlet Air Conditions for Side-to-Side Airflow Switches in Data Centers

Jim Fleming
Panduit
412 Rockwell Court
Burr Ridge, IL 60527
jnf@panduit.com

Susheela Narasimhan
Cisco Systems
250 W. Tasman Drive
San Jose, CA 95134
snn@cisco.com

## Abstract

Switches deployed in data centers often utilize side-to-side airflow cooling. This airflow pattern can cause heated exhaust air to flow to the inlet of adjacent equipment in open rack installations or re-circulate to the switch inlet in cabinet applications. Through the use of CFD and testing, ducting solutions have been developed that improve inlet air conditions for both open rack and cabinet applications. Additionally, the ducting solutions make the side-to-side airflow switches compatible with hot aisle/cold aisle layouts as well as air containment solutions.

## Keywords

Data center, Switches, Side-to-side airflow, Ducts

## Nomenclature

| | |
|---|---|
| CFD | Computational Fluid Dynamics |
| CRAH | Computer Room Air Handler |
| HVAC | Heating, Ventilation, Air-conditioning |
| I/O | Input / Output |
| $P_o$ | Fan pressure at stagnation |
| RU | Rack Unit; equal to 1.75 inches |
| $V_o$ | Fan open volume flow rate |

## 1. Thermal Challenges Deploying Side-to-Side Airflow Switches in Open Racks and Cabinets

Network switches in data centers are often cooled using side-to-side airflow. Unlike the more common front-to-back airflow seen in servers, air enters one side of the chassis, passes laterally through the main body of the switch, and exhausts on the opposite side. This architecture allows the front surface of chassis to be dedicated to I/O, maximizing port density and helping to minimize overall switch height. Front-to-back airflow switches are available but have traditionally required larger chassis for equivalent port densities. It should be noted that the switch power supply modules may utilize a separate front-to-back airflow path. A typical side-to-side airflow is shown in Figure 1.

Data center equipment is installed into either open racks or cabinets with perforated doors. The racks or cabinets are arranged in rows with open aisles to their front and rear. Supply air from the HVAC system is delivered to the cold aisle on the front side of the equipment row where it is drawn into the active equipment. Heated exhaust vents to the hot aisle behind the row. This arrangement is repeated throughout the room creating the hot aisle/cold aisle arrangement prevalent in data centers [1].

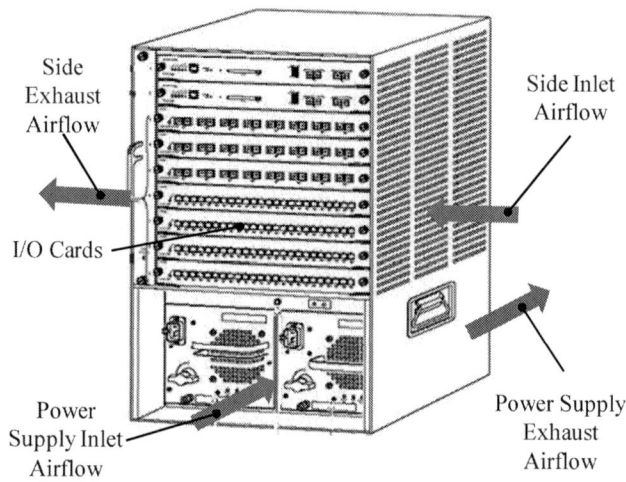

**Figure 1**: Common side-to-side airflow switch.

Switches are designed to cool themselves over a prescribed range of operating conditions, commonly from 0 to 40°C [2-4]. However, the nature of side-to-side airflow in typical hot aisle/cold aisle installations can have adverse effects on the conditions at the switch inlet. In general, the side inlet is not immediately adjacent to the cold aisle. A mix of air from both the cold aisle and the hot aisle is drawn to the inlet, raising average inlet temperatures.

In open racks, switches are frequently installed adjacent to one another. Heated air from the first switch exhausts laterally, in parallel with the row. The heated exhaust travels directly towards the adjacent switch where it is drawn in, resulting in raised inlet temperatures. The process may be repeated if additional switches are installed in the row, continually increasing inlet air temperature. Figure 2 shows an application of switches installed in an open rack environment. Figure 3 illustrates the effects of heated air exhausting from one switch directly into another.

In enclosed cabinets, exhaust leaves the side of the switch and quickly impacts the side of the cabinet. The heated air, no longer travelling in a coherent stream, fills the cabinet reaching all the way back to the inlet on the opposite side. Large bundles of cables from the switch ports may fill the open areas between the front equipment rails and the sides of the cabinet, impeding air from the cold aisle and increasing the proportion of air drawn from the hot aisle. The heated exhaust air may flow forward towards the cold aisle. Hence, a large fraction of the inlet air may actually be comprised of heated exhaust that has re-circulated, driving up inlet temperatures as illustrated in Figure 3.

978-1-4673-1110-6/12 $31.00 © 2012 IEEE

**Figure 2:** Rear (hot aisle) view of switches installed in an open rack deployment.

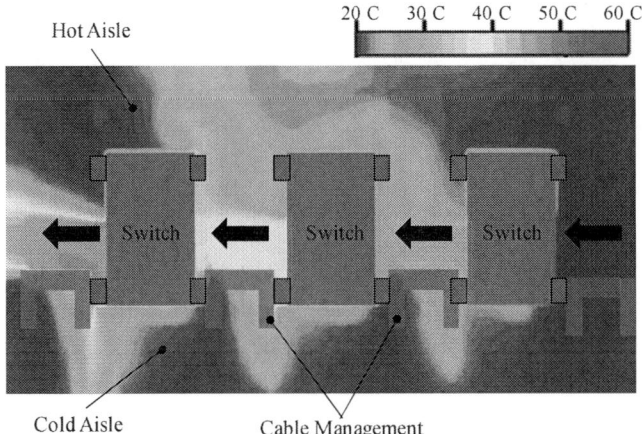

**Figure 3:** Top view illustrating inlet and exhaust air temperatures for a side-to-side airflow switch in an open rack application – top view.

Due to the tendency of side-to-side airflow switches to draw in preheated air, their inlet temperatures are often significantly greater than the cold aisle temperature. Data centers have historically been operated with relatively low supply air temperatures, as low as 15 to 20°C. Assuming a temperature increase of 15°C between the cold aisle and switch inlet, the switch will be ingesting air between 30 and 35 °C. Despite the relatively low temperature air supplied to the data center, the switch could be running within 5 °C of its operating limit.

The data center HVAC systems used to remove the heat generated by servers, switches, and storage equipment can consume between 30% and 50% of the total facility power [5, 6]. Decreasing the amount of power needed to cool the facility has obvious economic benefits and has become a major trend for data centers [7].

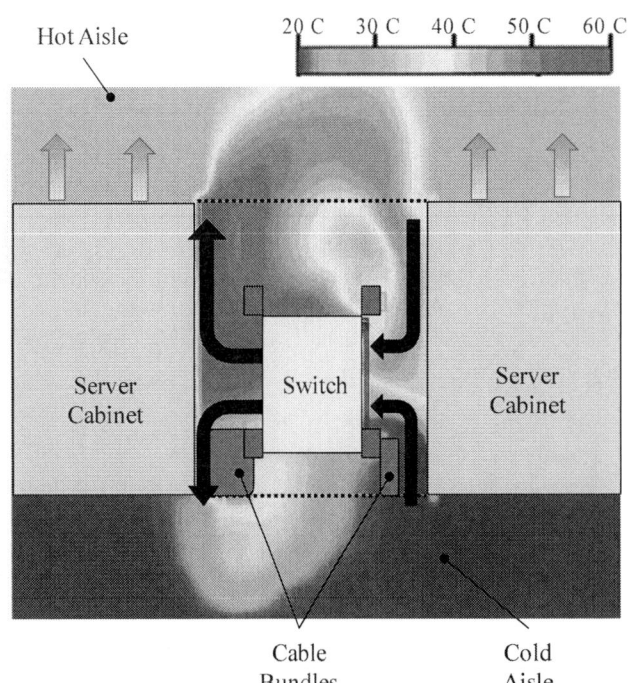

**Figure 4**: Illustration of inlet and exhaust air temperatures for a side-to-side airflow switch in a cabinet application – top view with switch cabinet panels hidden.

One method of reducing HVAC system energy consumption is to raise the temperature of the air supplied to the data center. Chiller-based HVAC systems may reduce energy usage by 1 to 4% for every 1°C the temperature is increased [8, 9]. A second major trend is the use of economizers to cool the data center. When operational, economizers utilize outside air to provide cooling, eliminating the need to run the HVAC system compressor. The energy savings are dependent on how many hours per year the outside air is cool enough to allow the use of the economizer. Raising supply air temperature allows economizers to be used for warmer outside air conditions leading to less compressor use and greater energy savings [10]. These new cooling trends in data centers will create additional challenges for side-to-side airflow switches.

Now, assume the cold aisle temperature is increased to 25°C to take advantage of HVAC system energy savings. Using the same 15°C rise in air temperature from the cold aisle to switch inlet, the switch will be drawing in air at 40°C, its maximum operating temperature. Hence, the need to eliminate or at least minimize the temperature rise from the cold aisle to the switch inlet is apparent.

A "best-practice" in data centers using cabinets is to physically separate the inlet and exhaust airflow [1]. This can be done both inside and outside of the cabinet. Inside cabinets, barriers are formed from blanking panels and other structures typically placed in line with the front equipment rails. With side-to-side airflow switches, this separation will block access to cool air. All inlet air will be drawn from the rear of the cabinet and the hot aisle.

Exterior to the cabinet, physical structures may be constructed to fully contain either the hot or cold aisle,

978-1-4673-1110-6/12 $31.00 © 2012 IEEE

preventing exterior exhaust re-circulation or cold air bypass. Alternatively, each cabinet may use solid rear doors to trap and direct exhaust air through a chimney to the HVAC air return system. These exterior containment structures require the implementation of the previously described internal separation features (blanking panels and air dams) to be effective. Otherwise, exhaust re-circulation and cold air bypass will simply occur within the cabinet. Again, this separation will prevent conditioned air from reaching the switch inlet. It can therefore be concluded that side-to-side airflow switches are subject to thermal challenges if implemented with containment.

## 2. Development of Ducting Solutions

Ducting offers a potential solution for delivering cool air to the switch inlet or directing exhaust away from the inlet of the active equipment. The duct must physically fit to the switch while avoiding interferences with rack/cabinet structures, cabling, and other physical infrastructure. The airflow impedance of the duct must be minimized so it does not affect system fan performance and forced convection cooling of components. Similarly, the duct must not fundamentally alter airflow patterns within the switch.

Initial concepts were evaluated using a commercially available computational fluid dynamics (CFD) software package. The models focused on the balance between system airflow and duct size. Each model included row level physical infrastructure (e.g. rack/cabinet, cabling, cable management structures, perforated floor tiles, etc.) to ascertain their effects on the switch.

The analysis focused on a 14RU side-to-side airflow switch shown in Figure 5. The switch chassis was modeled to reflect expected worst-case heat dissipations as follows:

- 2x Supervisor Cards: 500 W each
- 7x I/O Cards: 820 W each
- 5x Fabric Cards: at 150W each
- 2x Power Supplies: 400 W each

Two sets of fans drive side-to-side airflow:

- Main Fans: bottom 3 rows of 4x 120mm fans
  - Max speed: $V_0$ = 270 CFM; $P_0$ = 1.39 in $H_2O$
  - Min speed: $V_0$ = 117 CFM; $P_0$ = 0.26 in $H_2O$
- Supervisor Fans: Top row of 6x 80 mm fans
  - Max speed: $V_0$ = 90 CFM; $P_0$ = .92 in $H_2O$
  - Min speed: $V_0$ = 45 CFM; $P_0$ = 0.23 in $H_2O$

Airflow impedance for the cards was simulated with simple planar resistances placed in the middle of each slot, orientated perpendicular to flow. Resistance values were calibrated to match measured volumetric flows through the switch. The completed switch model was then placed into a detailed rack or cabinet with cabling and other physical infrastructure. Cabling was modeled using simple cuboids with depth and width equal to the diameter of the cable bundle. Finally, a potential duct concept was added to complete the CFD model.

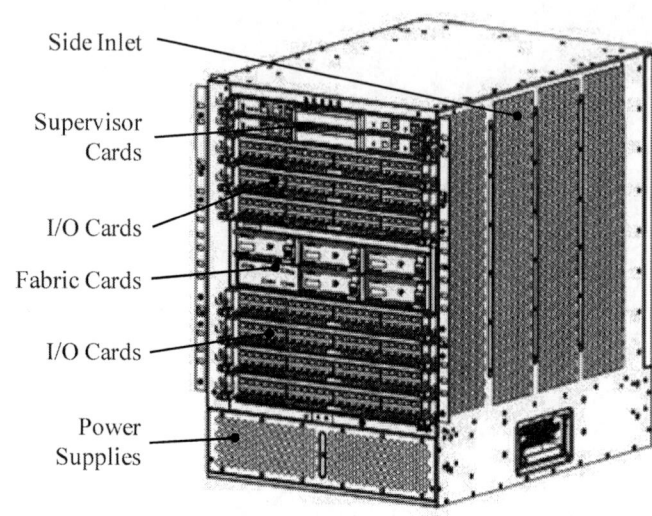

**Figure 5**: 14RU side-to-side airflow switch with cards and power supplies.

Once various concepts were evaluated, the most promising designs were prototyped for testing. Testing occurred in a dedicated lab constructed to simulate typical data center environments. A computer room air handler (CRAH) unit delivered conditioned air to a raised floor plenum. Air was supplied to the racks/cabinets via perforated tiles in the raised floor. Exhaust air exited the room via return vents to the drop ceiling plenum before returning to the CRAH.

Prior to testing the various design concepts, the switch was installed in an open rack to establish baseline performance under ideal conditions. The ideal test set-up was configured to eliminate any airflow resistances near the switch's inlet or exhaust. Additionally, recirculation issues were eliminated by supplying excess conditioned air to the switch via a perforated floor tile to the right (inlet side) of the rack. No other equipment was installed in the lab during this testing.

For all tests, the switch was configured to operate at maximum heat dissipation. Switch fan speed and volumetric airflow was controlled manually based on inlet air temperature. The manual setting was based on the fan control algorithm. Testing was performed for cold aisle conditions of 32 to 52°C. The high end of the test range exceeded the operational specifications of the switch. However, the switch was designed to survive temperature excursions beyond its specified limits. The high temperature testing was performed to quantify performance in over-temperature conditions.

## 3. Open Rack Deployment

An initial concept using an inlet duct placed to the right of the switch and generally residing between the racks was considered. However, as can be seen in Figures 2 and 6, cable management structures typically used between racks would interfere with this concept. The focus then moved to using an exhaust duct to direct air towards the hot aisle. Based on the physical constraints of the switch and rack, a 167mm wide by 914mm deep by 622mm tall duct was developed. The duct resides on the left side of the switch fully covering the perforated exhaust opening. Air passes through the switch from right to left and is vented into the exhaust duct. The duct then directs the air to the rear where it is vented into the hot aisle.

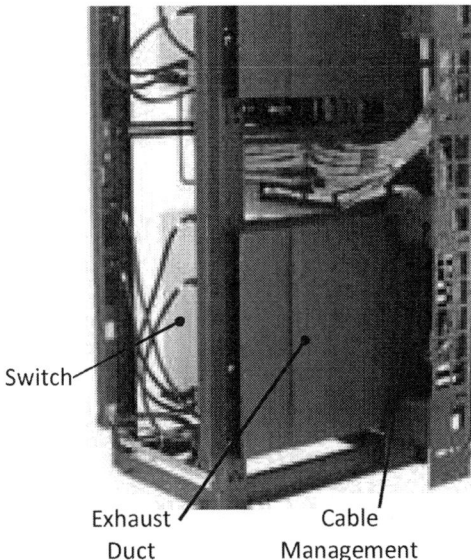

**Figure 6:** Rear perspective of exhaust duct on open rack including physical infrastructure.

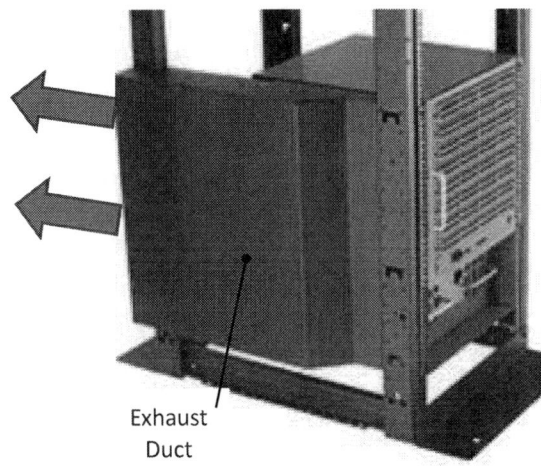

**Figure 7:** Side view exhaust duct on open rack shown without cable management.

CFD analysis of the initial design found high pressure in the exhaust duct, decreasing the operating point of the fans. The effect was most significant on the top row of smaller 80mm Supervisor Fans which are less capable of overcoming pressure than the 120 mm Main Fans. Physical constraints prevented increasing the cross-sectional area of the duct to reduce impedance and pressure. Therefore, a horizontal plate was placed into the duct, 90 mm below the top of the switch, to separate the airflow of the smaller fans from that of the larger fans. The goal was to prevent the higher volumetric flow main fans, and the associated high pressure airflow in the duct, from affecting the performance of the smaller supervisor fans. The lower flow of the Supervisor Fans, once

Fleming, Use of Ducting to Improve Inlet Air Conditions...

isolated from the Main Fan airflow, created less pressure in its portion of the exhaust duct. Figure 8 shows the pressure distribution in the ducts for a horizontal plane at the height of the Supervisor Fans. The addition of the horizontal plate lowered the pressure in the duct downstream of the supervisor fans by as much as 28%.

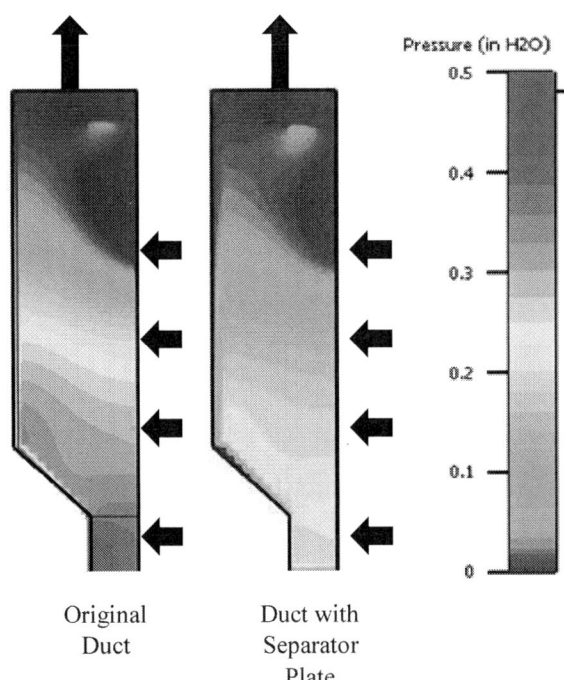

**Figure 8**: Pressure distribution in the exhaust duct. Top view of horizontal cut plane at the height of Supervisor Fans, above separator plate.

CFD analysis predicted the addition of the duct (with and without the separator plate) would reduce volumetric flow through the switch as summarized in Table 1. The loss of flow, while not insignificant, was deemed to be acceptable. It was decided to move forward and test the exhaust duct with the separator plate.

|  | Original Duct | Duct with Separator Plate |
|---|---|---|
| Line Cards | 11.9% | 12.2% |
| Supervisor Cards | 18.3% | 14.4% |

**Table 1**: Predicted decrease in switch volumetric airflow with exhaust duct at maximum fan speed.

Testing was performed with two switches. A single switch was installed in each of two adjacent racks. Tests were run first without the ducts and then with the ducts installed. Data collection focused on the leftmost switch positioned directly downstream from the rightmost switch's exhaust vent.

As expected, the duct did slow airflow resulting in slightly increased component temperatures compared to the previously tested ideal scenario, as shown in Table 2.

The duct did demonstrate a benefit for the downstream switch in the two rack scenario. Table 3 shows the inlet air and component temperature improvement gained from use of the exhaust duct compared to operation without a duct.

| Cold Aisle Temperature | 32°C | 42°C | 52°C |
|---|---|---|---|
| Increase in Component Temp. vs. Ideal Case | 4.2C | 3.9°C | 2.3°C |

**Table 2**: Increase in component temperatures from ideal case to ducted case.

| Cold Aisle Temperature | 32°C | 42°C | 52°C |
|---|---|---|---|
| Decrease in Inlet Air Temp. | 7.5°C | 6.6°C | 4.5°C |
| Decrease in Component Temp. | 5.5°C | 4.5°C | 3.8°C |

**Table 3**: Average decrease in temperature obtained from the installation of the exhaust duct in the two rack scenario.

Figure 9 compares component temperature for an I/O card for the three scenarios. The significant difference between the ideal and the no ducting scenario is clearly evident. Although the installation of the duct does not allow the switch to perform as well as in the ideal case, it clearly provides a benefit.

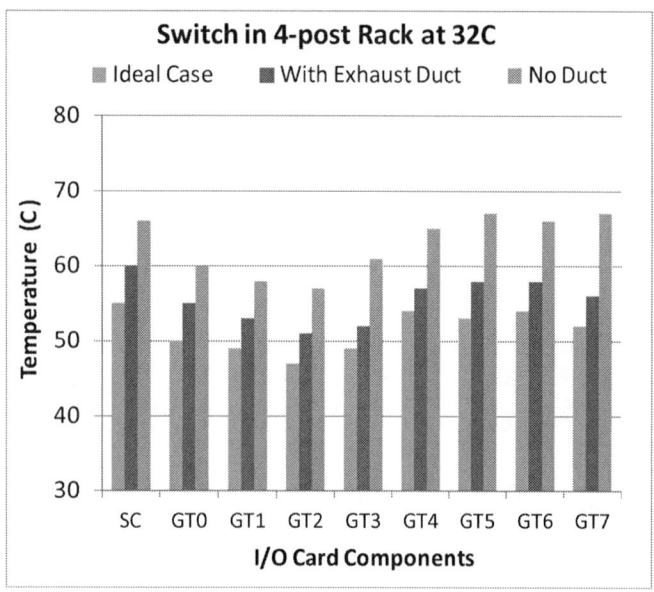

**Figure 9:** Specific components temperatures for the ideal vs. the no duct vs. the ducted case.

Testing was only performed on a two rack line-up. The thermal benefits for a row of racks, such as shown previously in Figure 2, would likely be cumulative. Hence the decrease in inlet and component temperatures for a switch at the end of a row may be significantly greater than the results seen in testing.

## 4. 800mm Wide Cabinet Deployment

The duct design for cabinet applications must deliver improved inlet air conditions while accommodating the switch's power and connectivity requirements. Typically, Fleming, Use of Ducting to Improve Inlet Air Conditions...

cables are routed from the switch to side of the cabinet outside the equipment mounting space as shown in Figure 10. The cables then run vertically, in front of the vertical air dams, to patch panels or out of the cabinet. In the worst-case cabling configuration, using 336 Cat-6A Ethernet cables per switch, the cable bundle has a cross-sectional area of 257 cm$^2$. The cable bundle essentially occupies the entire side area of the cabinet in front of the vertical air dam. Initial duct design concepts drew air from the front of the cabinet, through a hole in the right vertical air dam to the right side of the switch. It was determined that cable management requirements, especially for a two switch deployment with twice the number of cables, made this ducting concept impractical. The cable mass could completely block the duct inlet.

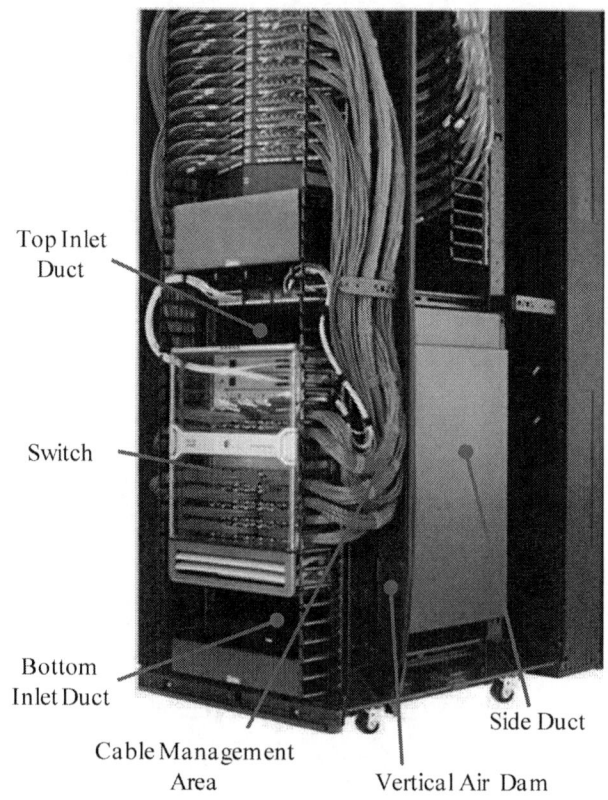

**Figure 10**: Cable management for a single switch deployed in an 800mm cabinet with ducting solution. Front door and cabinet side panel removed for clarity.

Instead, a duct system with inlets between the equipment rails was developed. Duct boxes reside above and below the switch, each 2 RU in height. Each box is open to the front and the right side. A third duct box resides to the right of the switch. Air is drawn in from the front of the top and bottom ducts. The air turns to the right and enters the side duct. The switch then draws air in from the side duct. Exhaust air exits the switch on the left side. Blanking panels and vertical air dams force the exhaust air out the rear of the cabinet. The physical design can be seen in Figure 10 and the airflow path is illustrated in Figure 11.

**Figure 11:** Airflow path for cabinet duct design. Cabinet doors, side panels, and blanking panels are hidden for clarity.

CFD analysis investigated the effect of the duct design on both airflow patterns and volumetric flow. The entire volumetric flow for the switch must pass through the two relatively small inlet ducts and make several sharp turns. Table 4 shows the predicted loss of airflow for the line and supervisor cards.

|  | Inlet Duct |
|---|---|
| Line Cards | 14.6% |
| Supervisor Cards | 16.5% |

**Table 4:** Predicted loss of volumetric flow for switch inlet duct at maximum fan speed.

The duct design significantly altered the flow path of air approaching the switch. In the top duct, the air travels from left to right then rapidly turns 180 degrees through the side duct and into the top most cards in the switch, similarly, air in the bottom duct travels from left to right before turning upwards and then turning again to the left to enter the bottommost cards of the switch. Figure 12 shows the specific areas of concern.

Although the distribution of air to the top and bottom cards was not ideal, the duct design was deemed acceptable for creation of a prototype and testing. Specific data collection was performed to verify thermal performance for cards at the top and bottom of the switch that may be affected by problematic airflow.

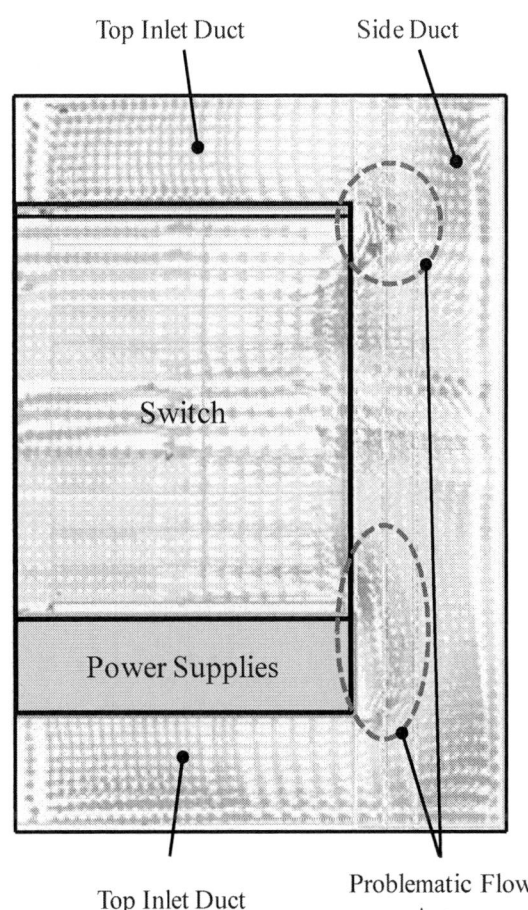

**Figure 12:** Airflow vectors through inlet ducts to switch. Front view of vertical cut plane through middle of switch.

Testing was performed in an 800mm cabinet with two switches installed. Blanking panels were installed in all open rack units within the cabinet. Vertical air dams were placed outside the equipment rails to complete the physical separation of inlet and exhaust air. Additional obstructions were placed outside of the equipment rails in front of the air dams to simulate data cabling.

The first test investigated thermal performance of the switch in a cabinet without the inlet ducts installed. During the initial test at a cold aisle temperature of 32°C, air temperatures at the switch inlet were reported as high as 62°C. The extremely high re-circulation temperatures prevented additional tests at higher cold aisle temperatures.

Testing then proceeded with the inlet ducts installed on both switches in a chimney cabinet. A chimney cabinet uses solid rears doors to contain the switches' hot exhaust air. The contained exhaust vents through the chimney in the roof of the cabinet where it is directed to the room's drop ceiling hot air return plenum. This type of deployment prevents exhaust from re-circulating outside of the cabinet back to the equipment inlets. However, the reduced area for airflow can increase pressure on the exhaust side of the switch, degrading fan performance. Utilizing the chimney cabinet allowed the opportunity to test the inlet ducting in the worst-case containment deployment.

When compared to the previously tested ideal case, the inlet ducts did degrade volumetric airflow leading to higher component temperatures as shown in Table 5. The loss of airflow was greater than observed with the open rack exhaust duct.

| Cold Aisle Temperature | 32°C | 42°C | 52°C |
|---|---|---|---|
| Increase in Component Temp. vs. Ideal Case | 10.8° C | 13.1° C | 11.6° C |

**Table 5**: Average increase in component temperatures from ideal case to the inlet duct case in a chimney cabinet.

However, when comparing the non-ideal cabinet deployment scenarios, the duct did provide a significant thermal benefit. The improvement in inlet air temperatures more than offset the decrease in airflow, as shown in Table 6. Figure 13 shows component level data for an I/O card.

| Cold Aisle Temperature | 32 °C |
|---|---|
| Decrease in Inlet Air Temp. | 14.5°C |
| Decrease in Component Temp. | 11.3°C |

**Table 6**: Average decrease in temperature gained from the installation of the inlet duct.

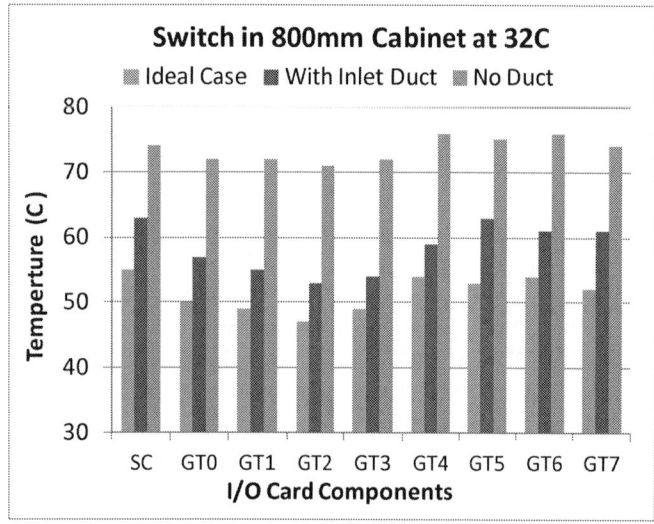

**Figure 13**: Specific components temperatures for ideal vs. the no duct vs. the ducted case.

It should be noted that the lower temperatures for the ducted scenario were achieved despite a significant difference in fan speed. The very high inlet temperatures for the no duct case required the fans to operate at maximum speed. Inlet temperatures for the 32 °C ducted scenario were low enough that the fans were run at minimum speed. The lower fan speed resulted in a 300W decrease in switch power consumption.

To evaluate the changes to the airflow patterns entering the switch, temperature data was analyzed for the I/O cards at the bottom, middle, and near the top of the chassis. As can be seen in Figure 14, the components in the topmost card ran 2 to

8°C warmer than the middle and bottom card. However, these temperatures were still lower than in the "no duct" case.

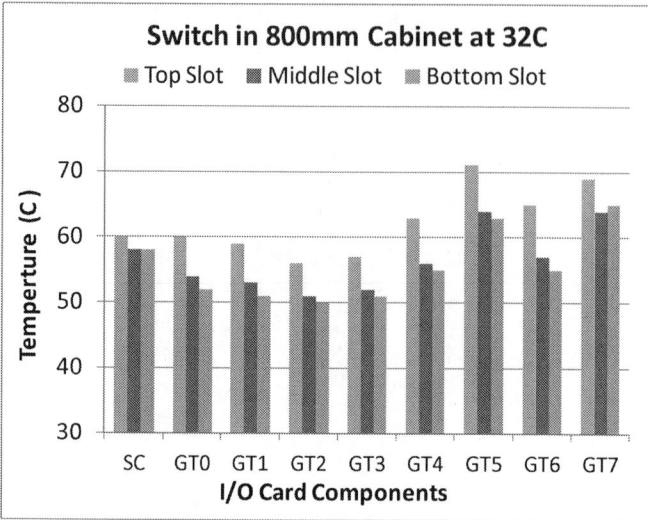

**Figure 14:** Components temperatures for the I/O when installed in the top, middle and bottom of the switch.

As demonstrated by the test data, the switch's thermal performance generally improved, while operating at lower fan speed, with the installation of the inlet duct in chimney cabinets. Hence, it can be seen that the inlet duct design enables side-to-side airflow switches to be used in containment deployments.

## 5. 1000mm Wide Cabinet Deployment

The 800mm cabinet with inlet ducts was capable of housing two of the 14 RU switches and associated cabling. However, an additional goal was to support the deployment of three switches in a 42 RU cabinet. Since three switches would occupy all 42 RU of rack space in a cabinet, there was no room for inlet ducts utilizing space between the equipment rails. Inlet ducting would have to draw in air from outside the equipment rails. Additionally, three switches can support as many as 1008 Ethernet cables, requiring up to 771cm² of open horizontal area for cable routing. The 800mm cabinet simply did not have enough available space for cable routing and side inlet airflow. Therefore a 1000mm wide cabinet was pursued. The space between each equipment rail and the outside of the cabinet was increased by 100mm, creating significant additional area for cabling and airflow.

The newly added space to the outside of the right side equipment rail was used for inlet airflow. The vertical air dam, normally used to seal the area between the right equipment rail and the right side of the cabinet, was removed. Additional sheet metal panels were added to the right of the switches to form a box, open only to the front and the left. This box formed a side duct that allowed the switches to draw cool air in from the front of the cabinet, as shown in Figure 15.

Routing all of the inlet air through the space between the rightmost equipment rail and the side of the cabinet created a restriction, causing the air to accelerate through the constriction. The increased momentum of the air tended to

978-1-4673-1110-6/12 $31.00 © 2012 IEEE 299

carry it further to the rear of the cabinet before being drawn into the switch. The result was decreased flow through the front portions of the switch as shown in Figure 16. Overall volumetric flow also was predicted to decrease by a moderate amount as shown in Table 7.

|  | Inlet Duct |
|---|---|
| Line Cards | 11.9% |
| Supervisor Cards | 13.2% |

**Table 7**: Predicted decrease in switch volumetric flow with inlet duct compared to ideal case.

The lower airflow in the front of the switch could result in higher component temperatures. Temperatures distributions from the front to the back of the switch were examined using CFD to evaluate the magnitude of this concern. It was found that the elevated temperatures did occur near the very front face of the I/O cards as shown in Figure 17. The localized hot spots saw air temperature increases of 8 to 12°C at the lowest airflow setting. Although this was a concern, it was expected that the benefits of the duct would outweigh the airflow issues. As was done for all testing, numerous component temperatures were monitored to determine if the changes in airflow created unacceptable performance.

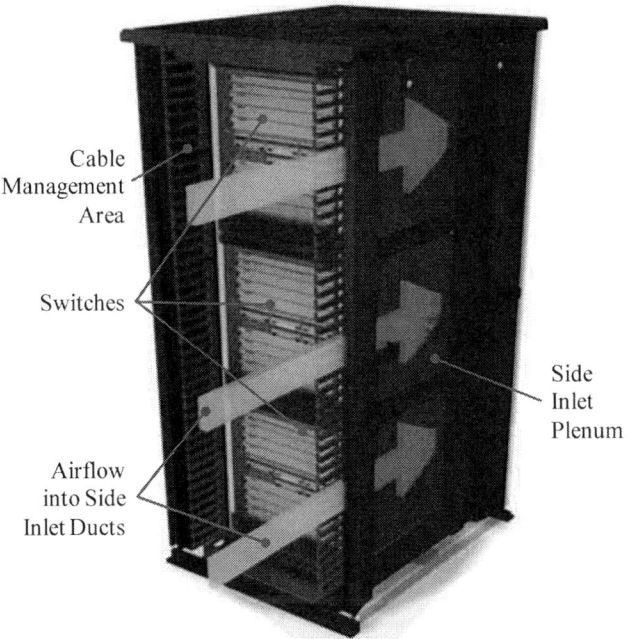

**Figure 15**: Airflow path for 1000mm wide cabinet with three switches. Cabinet doors and side panels are hidden for clarity.

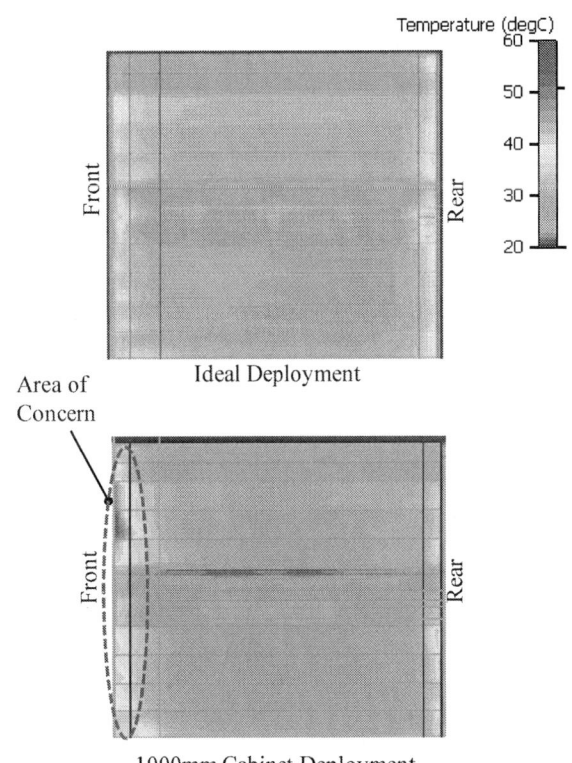

**Figure 17**: Left-side view of the switch showing the temperature distribution at a vertical cut-plane in the center of the chassis.

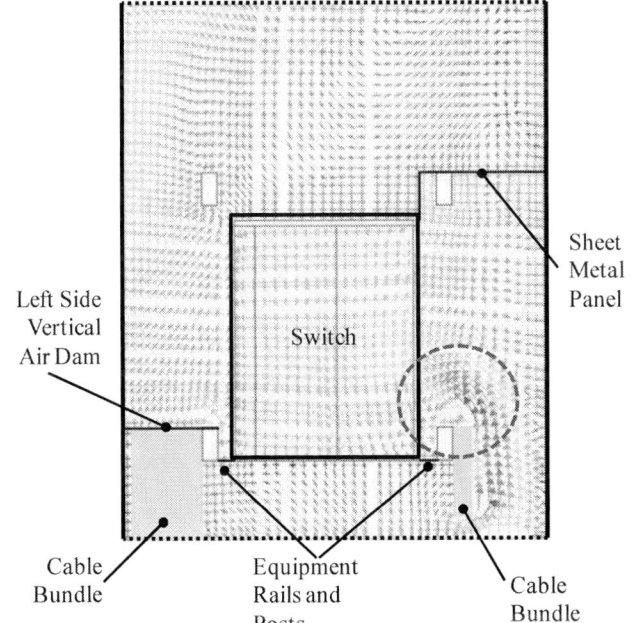

**Figure 16**: Airflow vectors through inlet duct to switch in 1000mm cabinet. Top view showing horizontal cut plane through middle of topmost switch.

It may be noted that this issue further justified the need for a 1000mm cabinet. In an 800mm cabinet, the side inlet would be smaller, creating higher velocities as the air passed through the restriction. The higher velocities would increase the tendency of the airflow to bypass the front of the switch.

Again, testing was performed to quantify performance of the ducting and switches. Three switches were installed in the 1000mm cabinet. Blockages simulating cabling were placed

978-1-4673-1110-6/12 $31.00 © 2012 IEEE          300

to the left and right of the equipment rails. The left blockage was 530 cm$^2$ and the right was 240 cm$^2$.

Logistical and scheduling issues prevented testing the "no duct" case in the 1000mm wide cabinet. However, without the ducting, re-circulation similar to that observed in the 800mm cabinet testing was expected. Testing with the ducting installed was performed in both standard and chimney cabinets.

The increase in component temperatures from the ideal case to the ducted case was small. Very little difference in thermal performance was observed between the standard and chimney cabinet deployments.

| Cold Aisle Temperature | 32°C | 42°C | 52°C |
|---|---|---|---|
| Increase in Component Temp. vs. Ideal Case for Standard Cab. | 2.9°C | 2.3°C | 3.1°C |
| Increase in Component Temp. vs. Ideal Case for Chimney Cab. | 3.6°C | 2.1°C | 2.2°C |

**Table 8**: Average increase in component temperatures from ideal case to inlet duct cases.

Additionally, of the 29 components monitored per I/O card, the maximum increase in temperature observed was 7°C. There was no evidence that the increased temperatures observed in the CFD analysis were problematic in the prototype deployment.

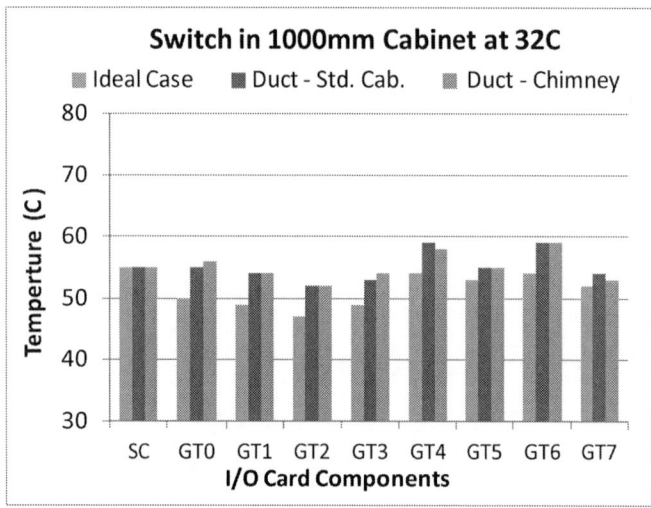

**Figure 18**: Specific components temperatures for the top switch with inlet duct in standard and chimney cabinets.

## 6. Conclusions

Increased inlet air temperatures caused by recirculation of exhaust air in side-to-side airflow switches can be improved through the use of ducting. In open rack installations, exhaust ducts direct the heated air towards the hot aisle and away from the inlet of adjacent switches. For cabinet deployments, inlet ducting provides a path for air to travel from the cold aisle to the switch. The ducting did increase airflow resistance, decreasing total volumetric airflow through the switch. However, this negative was outweighed by the decrease in component temperatures made possible by the lower inlet air temperatures.

Inlet ducts allow switches in cabinets to draw air directly from the cold aisle while providing physical separation of the inlet and exhaust sides of the cabinet. This separation within the cabinet allows the cabinet to be deployed in containment scenarios.

The improved thermal performance enabled by the ducting allows switches to operate at higher cold aisle temperatures. The increase in supply air temperature can improve the efficiency of traditional chiller-based HVAC systems or increase the hours of operation for economizer systems, decreasing the amount of energy required to cool the data center.

## References

1. ASHRAE TC9.9, Best Practices for Datacom Facility Energy Efficiency, American Socity for Heating, Refridgeration, and Air-Conditioning Engineers (Atlanta, GA, 2008), pp. 12, 187.
2. Cisco Systems, "Cisco Nexus 7000 Series Hardware Installation and Reference Guide", August 2011.
3. Cisco Systems, "Cisco Catalyst Catalyst 6500 Series Switches Installation Guide", September 2010
4. Juniper Networks, "EX8208 Ethernet Switch Data Sheet", 2011.
5. Muroya, K., Kinoshita, T., Tanaka, H.; Youro, M., "Power Reduction Effect of Higher Room Temperature Operation in Data Centers," Proc. of IEEE/IFIP Network Operations and Management Symposium, pp. 661-673, 2010.
6. The Green Grid, "The Green Grid Power Efficiency Metrics: PUE an DCiE", www.thegreengrid.com, 2007
7. ASHRAE TC9.9, "2011 Thermal Guidelines for Data Processing Environments – Expanded Data Center Classes and Usage Guidance" www.tc99.ashraetcs.org.
8. Wulfinghoff, D., Energy Efficiency Manual, Energy Institute Press (Maryland, 2010), pp. 264-266
9. Dubin, F., and Long, C., Energy Conservation Standards for Building Design Construction and Operation, McGraw Hill (New York, 1978).
10. Anubhav, K.; Yogendra, J., "Use of Airside Economizer for Data Center Thermal Management", Proc. Of ThETA2, Thermal Issues in Emerging Technologies, pp. 115-124, 2008.

# Enabling Power Density and Thermal-Aware Floorplanning

Ehsan K. Ardestani, Amirkoushyar Ziabari, Ali Shakouri, and Jose Renau
School of Engineering
University of California Santa Cruz
1156 High Street
Santa Cruz, CA, 95064
{eka, aziabari, ali, renau}@soe.ucsc.edu

*Abstract*—With temperature being one of the main limiting factors in design of high performance processors, early evaluation of thermal effects in design stages is becoming a necessity. Floorplanning is an imperative step in the design process where thermal effects can be taken into account. This work studies a thermal-aware floorplanning scheme, with the goal of increasing both reliability and performance measures of the design. We show that a majority of thermal emergencies can be averted by a) leveraging the lateral heat transfer effects (as has been shown previously), and b) by reducing the power density of thermally critical blocks. The former becomes possible through moving, and modifying the aspect-ratio of the blocks in the floorplanning process. The latter, one of the key contributions of this work, is carried out through resizing of functional blocks in a controlled way. We also propose a selective power map generation method for the floorplanning process. In this method the time windows in which thermal emergencies occur guide the power map generation. As a result, we observed an 8.8% performance improvement, and a 40% reliability increase with the area overhead of just 3%.

*Index Terms*—Floorplanning, Power Blurring, thermal simulation, architectural level thermal simulator.

## I. INTRODUCTION

Floorplanning is an essential component of any successful integrated circuit design, particularly in the design of high performance processors. Today, we are experiencing a shift in the problem formation from focusing primarily on area utilization and timing, to one where design objectives, such as power, temperature and reliability are major concerns. This is because power and temperature are first order design constraints. The thermal characteristics affect multiple aspects of processor design including frequency, leakage, performance throttling and cooling cost. In this work, we focus on the thermal impact of floorplanning on performance and reliability measures in an integrated circuit.

Thermal effects on an integrated circuit (IC) impact both reliability and performance. Processors with higher thermal cycles have shorter mean time to failures (MTTF) [1]. Also, high thermal gradients across the chip could adversely impact wire delay [2], and narrow the frequency margin. In addition, all modern processors are equipped with dynamic thermal management (DTM) to keep temperature in the safe range and prevent catastrophic break down of the integrated circuit. At high temperatures, DTM triggers global or local actions to preemptively lower the power consumption. This typically comes at the cost of performance. Hence, a thermal-aware floorplan could potentially improve both reliability and performance of a processor.

With the shrinking trend in very large scale integration (VLSI) circuits, the ratio of leakage to total power keeps increasing. As a result leakage power is gaining more attention as a first order design parameter. Leakage power is temperature dependent, and reducing the temperature across the chip results in less leakage.

The lateral heat spreading and interaction among adjacent functional blocks impacts the hotspot formation on the chip. To either balance the thermal distribution on the chip, or to reduce the hotspots and peak temperature, thermal aware floorplanning methods have been proposed [3]–[5]. The proposed methods run through iterations of floorplan assignments, and thermal evaluation of the processor, with the goal of finding an optimum floorplan in terms of area, timing and thermal profile.

Not all hotspots can be reduced to safe levels by leveraging the lateral heat spreading effect. Thermal behavior of blocks is highly correlated with power density, and in thermal-aware floorplanning, power density is a simple metric that can be used to guide the exploration of design space. Our technique allows for the resizing of individual blocks in the floorplan to address issues related to high power density blocks. This is an issue that cannot be rectified by simply re-placing individual blocks.

In this work, we aim at reducing the number of thermal emergencies in the processor through floorplanning. Through a simulated annealing based method, we study the floorplanning impact on performance and reliability of the chip. Thermal throttling is a DTM method that we implement. We show how a resulting floorplan improves processor performance by lowering the amount of time it spends in thermal throttling. We also consider a set of temperature dependent reliability metrics [1] to relatively characterize the quality of each floorplan.

We use an integrated performance, power and temperature toolchain to study the impact of thermal-aware floorplanning on reliability and performance. Previous works do quantify the reliability of the chip in terms of mean time to failure, nor show the performance impact of the improved floorplan due to thermal effects. This paper has the following contributions:

- It proposes a selective method of generating representative power value for floorplanning,
- Allows for the resizing of functional blocks to add power density exploration capability,
- It quantifies the reliability improvement achieved by the floorplanning scheme, and show how a floorplan could

978-1-4673-1110-6/12 $31.00 © 2012 IEEE

improve the performance of a processor by reducing thermal emergencies.

## II. RELATED WORKS

Thermal-aware floorplanning has been studied in several works. Hung *et al.* [5] use a genetic algorithm to explore different floorplans. They aim at reducing the hotspots and distributing the temperature evenly across the chip, while maintaining the area of the chip. Their study is more focused on circuit level floorplanning, and therefor they use a selected circuit-level benchmarks. The power map for the blocks is also randomly assigned. The temperature of blocks is estimated using HotSpot [6].

Han *et al.* [3] and Sankaranarayanan *et al.* [4] use a simulated annealing based approach to explore the floorplan design space. Their focus is architecturally based. They use a processor floorplan and common CPU benchmarks for their experiments. In addition to the basic block moves, they also explore different aspect ratios for each block in a controlled way. However, they maintain the original area of the block. To estimate the temperature, Han *et al.* use heat diffusion between adjacent blocks. Given the distance of any two blocks and their thermal resistance and power map, the heat diffusion measure is computed and used as temperature estimation. Sankaranarayanan *et al.* run Hotspot at each iteration to compute the steady state temperature of each block.

In our work, we use the same simulated annealing approach as Sankaranarayanan *et al.* [4]. We use a processor floorplan. In addition to the *move* and *aspect ratio*, we also allow for a controlled *resizing* of the block area. While these methods use random power values or compute average power across the benchmarks, we choose to compute the representative power map in a different manner. Our temperature estimation is carried out in grid mode by implementing a Power Blurring thermal simulation method.

## III. METHODOLOGY

### A. Thermal Throttling

Designing a package for the worse case thermal behavior of the chip could inflate the cost. A package could be designed for the worst typical application [7]. Applications that dissipate more heat than what the designed package can tolerate should trigger a runtime thermal management technique to keep the processor's temperature in the safe range. Among these techniques are power gating, throttling the clock or issue logic, or changing the power state through DVFS[1].

We implement the throttling policy by gating the clock. For that, a configurable trigger threshold is defined. We choose $100°C$ as it is close to the junction temperature. When a block in the processor reaches the trigger threshold, the processor stops processing. This in turn reduces the power consumption of the chip down to the leakage power. The processor stays throttled until the critical temperature goes below the trigger threshold. The time a processor spends in throttling could have been spent executing instructions. Hence, the throttling adversely impacts the performance of the processor.

[1]Dynamic Voltage and Frequency Scaling

### B. Power Map Selection

The power map used to obtain temperature estimation for the chip in the floorplannning process plays a critical role in the quality of resulting floorplan. Previous works mainly use an average of selected benchmarks. A set of power maps are generated by running a set of benchmark applications (i.e SPEC CPU benchmark suite). Each power map contains average power consumptions for the functional blocks, computed during the execution of the benchmark. Then the resulted average power for all the benchmarks are averaged together one more time to generate a single power map that contains a power value for each functional blocks. We call this method the *average-based* power selection.

Given the fact that each benchmark utilizes the resources in the processor in a different way, averaging the power traces of different benchmarks could hide a great deal of information about thermal distribution across the chip. The accuracy of these methods could be improved by carefully studying the impact of each benchmark and using the detailed temperature distribution results. However, the rather heavy thermal computation, which in turn results in slow simulation, has been preventing the designers from considering each benchmark, leaving them no choice but averaging all power maps together. Our methodology proposes a more efficient way of selecting power values.

Our power selection method differs in two ways. First, for each benchmark, we only compute the average power around the time during which the processor triggers DTM responses to high temperature. In our case, this means we average power values during the time in which thermal throttling happens. This is to identify the power distributions that becomes critical for reliability or performance.

Second, for each functional block, the maximum power among the benchmarks is selected to form the power map for the chip. The reason for selecting the maximum block power is that different benchmarks (*e.g.*, integer vs. floating point benchmarks) exercise different blocks and as a result have different hotspot distributions. Averaging power maps across the benchmarks could result in high power consumption of a block in one benchmark cancels out by low power consumption of the same block in another benchmark.

We call this method the *selective average-max* power generation (SAM). To show the impact of power selection methods, we generate a power map with each of these methods. For each method, we run the simulation for 4 billion (B) instructions, skipping first 1B and simulating the rest to gather the power numbers. Then we run the floorplanner for both power maps. The results show around 20% improvement by having less thermal throttling in the floorplan resulted from SAM compared to the average-based method..

### C. Floorplanning

Our floorplanner is based on simulated annealing approach. We use a modified version of HotFloorplan [4], with the same simulated annealing parameters. The cost function is a linear combination of the total area, maximum temperature, and the estimated wire delay. During each iteration, the tool

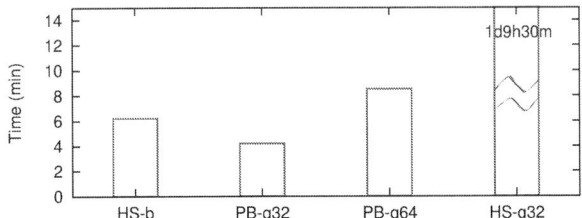

Fig. 1. Execution time for different thermal solver methods.

can change the floorplan through three primitive actions; it can either randomly move, change the aspect ratio or resize a block. The resizing is designed to resolve the blocks with high power density that end up with critical temperature. Those blocks could lead to reliability threatening hotspots which cannot be rectified by re-placing. The current implementation only includes increasing the size up to 50% of the original block size. There is also a constraint of 10% maximum increase in the total area of the chip.

### D. Wire Delay

The floorplanning scheme uses the first order wire delay model. In this model, the delay is a linear function of the wirelength. This is simple enough to be used at the architecture level. To consider the impact of the wire delay in performance simulation, we make sure there is no connected blocks with unacceptable wire delay. Should two blocks have high wire delays, we consider the effect in latency of the block in the performance simulator.

### E. Power Blurring Thermal Simulation

All the previous works are based on block model temperature estimation. However, the block model approximates the temperature of an entire functional block with a single node, and it could potentially lead to inaccuracy in the estimated temperature when the modeled blocks have very high aspect ratios. To show that, we try a floorplan with two blocks among the entire floorplan having a high aspect ratio. We estimate the steady state temperature with a sample power map. We configure HotSpot once for block model and once for grid model to get the temperature estimations. The results for the two high aspect ratio blocks differ up to 10°C. The error increases at higher temperatures which is the range we are most interested in for thermal-aware floorplanning.

Power Blurring methods has been shown to be fast and accurate for steady state temperature modeling [8]–[13]. This method intrinsically solves the thermal equations in grid mode. With implementation of Power Blurring based thermal modeling, we are able to avoid the aspect ratio sensitive problems that raises by using block model.

Figure 1 shows the time it takes to run the floorplanner using different temperature estimation methods. The experiments are run on an AMD *Opteron*[tm] Processor 6172. The first part of the labels for each method refers to the solver (HS or PB) and the second part indicates the model type (*b* for block and *gxx* for grid with size *xx × xx*). The PB solver is as fast as the block model solver used in HotSpot. However, using HotSpot

with grid model takes two orders of magnetude more time to solve the same equations, which makes it impossible to use for this purpose.

### F. Performance Simulator

To study the impact of a floorplan on performance, thermal profile, reliability, power consumption, and performance we configured the following toolchain. For performance simulation we used a modified version of SESC [14] that uses QEMU [15] as the functional emulator executing arm instructions. We configured SESC to pass activity counters to McPAT [16] (every 100K instructions max) which we used for calculating power. We modified McPAT to save the state that it calculates during initialization so that we could call it many times from our simulator. The power numbers from McPAT were used with a modified version of SESCTherm [17] to scale leakage power consumption according to temperature and device properties, and to generate the thermal metrics.

### G. Metrics

We use RAMP [1] as a quantitative basis for reliability. This work describes 5 Mean Time To Failure (MTTF) wear out failure models: Electro Migration (EM), Stress Migration (SM), Time-Dependent Dielectric Breakdown (TDDB), Thermal Cycling (TC) and Negative Bias Temperature Instability (NBTI).

Since MTTF is not additive, the average Failures in Time (FIT) per block is estimated as the application executes. The FIT is proportional to the area. At the end of the execution, we add the area-weighted FITs to report the overall FIT value for the entire processors. Like the RAMP authors, we assume that all the different failure mechanisms have the same contribution to the overall FIT value, which is adjusted to a preset value. In our case, we adjust the FIT value for all the SPEC applications to 10,000. This is approximately equivalent to a MTTF of 11 years which is a short but reasonable lifetime for a processor. Table II shows the selected parameters.

**Electro migration:** occurs when atoms migrate from one end of the interconnect to the other, eventually leading to increased resistance and shorts. The model used in this work for EM is defined as follows:

$$MTTF_{EM} \propto (J)^{-n} \times e^{\frac{E_{a_{EM}}}{kT}}. \qquad (1)$$

**Stress migration:** Materials differ in their thermal expansion rate, and this difference causes thermo mechanical stress, referred to as Stress Migration. We use the following SM model:

$$MTTF_{SM} \propto |T_0 - T|^{-n} \times e^{\frac{E_{a_{SM}}}{kT}}. \qquad (2)$$

**Time-dependent dielectric breakdown:** It is the result of the gate dielectric gradual wear out, which leads to transistor failure. Ramp uses TDDB model

$$MTTF_{TDDB} \propto \left(\frac{1}{V}\right)^{(a-bT)} \times e^{\left(\frac{X + \frac{Y}{T} + ZT}{kT}\right)}. \qquad (3)$$

**Thermal cycling:** Thermal Cycling is another reliability factor since the temporal thermal gradients, *e.g.*, power on and off and high frequency changes in power due to changes in workload behavior, affect the lifetime of the processor. There is no validated model for high frequency thermal cycles, but the effects of low frequency cycling can be modeled via:

$$MTTF_{TC} \propto (\frac{1}{T - T_{amb}})^q. \qquad (4)$$

**Negative bias temperature instability:** NBTI leads to upward shifts in the transistors' threshold voltage that leads to timing violations. Ramp uses NBTI model

$$MTTF_{NBTI} \propto ((ln(\frac{M}{1+2e^{\frac{N}{kT}}}) - ln(\frac{M}{1+2e^{\frac{N}{kT}}} - H)) \times \frac{T}{e^{\frac{-I}{kT}}})^{\frac{1}{\beta}}. \qquad (5)$$

We report one augmented reliability metrics. For that, we combine all the reliability metrics. We compute the Fault In Time (FIT) measure for each individual reliability metrics, and then use them to compute MTTF for the chip as follows:

$$MTTF = 1/(FIT_{EM} + FIT_{SM} + FIT_{TC} + FIT_{NBTI} + FIT_{TDDB}). \qquad (6)$$

In addition to the reliability metrics, we report the gradient temperature across the chip. This is because of thermal gradient impact on interconnect delay [2]. We also report maximum temperature.

## IV. SIMULATION SETUP

While both [3], [4] evaluate the impact of the floorplan on processor performance through wire delay, [4] goes further and studies this impact on the processor performance in terms of IPC[2]. IPC is commonly accepted metric for processor performance in a given clock frequency, and is the main reported metric in all the architectural level evaluations.

Nonetheless, these evaluations are merely based on the wire delay, and are carried out to ensure quality of the placement of blocks in the floorplan. None of these methods show how the improved thermal profile of the chip benefits processor's performance.

To evaluate the impact of the thermally-aware improved floorplan on the performance and reliability of the processor, we start with a manually generated floorplan for a processor as our base configuration. Then our automated floorplanner improves the floorplan. We use 8 SPEC2000 CPU workloads in our experiments, namely *applu, crafty, gzip, mesa, mcf, mgrid, swim,* and *twolf. applu, mesa, mgrid,* and *swim* belong to the floating point workload category. Both the base and improved floorplans are run through an integrated performance, power and temperature estimation simulator. The simulator supports thermal throttling. The simulation estimates the IPC for each of the processor configurations that only differ in their floorplan.

For the evaluation, we study two methods and we compare three floorplans. The original floorplan makes the base configuration and hence is called *base*. The two methods that we evaluate are *MAR-A* and *MARS-SAM*. The first set of abbreviations in the naming stands for the supported primitives

in the floorplan scheme. The second set of abbreviations refers to the power selection method. *MAR-A* implements 2 basic primitives in the floorplanning scheme, Move, and Aspect-Ratio, and it uses the Average-based power selection method. *MARS-SAM* has one additional primitive, reSizing of a block area, in addition to the primitives supported by *MAR-A*. It also uses the Selective-Average-Max power generation method. Figure 3(a) shows the floorplan manually generated as the base configuration. Area of each block is also indicated in the figure. The unit for the numbers is $MM^2$. L2 cache is not shown to save space. It is placed on top of the core with area of $3.90 \times 10^{-6}\ mm^2$.

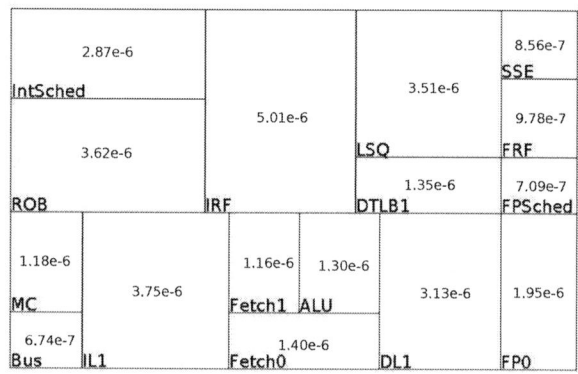

Fig. 2. The processor's core floorplan used as the base configuration. L2 cache is not shown. Total area is $72.36 \times 10^{-4}\ mm^2$

The processor configuration parameters are shown in Table I. Table II shows the reliability metric parameters.

| Parameter | Value |
| --- | --- |
| Frequency | 2.0 GHz |
| ICache | 32KB 2-way (2 cycle hit) |
| DCache | 32KB 8-way (3 cycle hit) |
| L2Cache | 1MB 16-way (12 cycle hit) |
| Mem. Lat. | 180 cycles |
| Branch Pred. | Hybrid 76Kb total memory |
| Issue width | 4 |
| ROB | 192 |
| Inst. Win. | 48 |
| Phy. Reg. (I-F) | 128-128 |

TABLE I
ARCHITECTURAL PARAMETERS.

| Metric | Parameters |
| --- | --- |
| EM | $a_{EM} = 0.9$, $k = 8.617343 \times 10^{-5}$ |
| SM | $n = 2.5, a_{SM} = 0.9$ |
| TDDB | $V = 1.1$, $a = 78$, $b = -0.081$, |
| | $X = 0.759$, $Y = -66.8$, $Z = -8.37$ |
| TC | $T_{amb} = 293$, $q = 2.35$ |
| NBTI | $M = 1.6328$, $N = 0.07377$ |
| | $I = -0.06852$, $\beta = 0.3$, $H = 0.01$ |

TABLE II
THERMAL METRICS CONSTANTS.

[2]Instruction Per Cycle

## V. EVALUATION

Table III presents the results in terms of the impact of the resulted floorplan on percentage of time the processor spends in thermal throttling, and also the impact on it's performance. The original floorplan results in a processor that on average spends 5.96% of the execution time of the benchmarks being throttled because of high temperature. *MAR-A* improves the results by reducing the time spent in throttling by around 6 fold to 1.1%. However, *MAR-A* cannot completely rectify the thermal throttling issue. *MARS-SAM* on the other hand, is able to completely rectify all the throttling issues. As a result, *MARS-SAM* achieves better performance results. While *MAR-A* improves the performance by 6.2% on average across the selected benchmarks, *MARS-SAM* improves it by 8.8%.

| Benchmark | | % in Throttling | | % IPC Improvement | |
|---|---|---|---|---|---|
| | base | *MAR-A* | *MARS-SAM* | *MAR-A* | *MARS-SAM* |
| applu | 1.7 | 0.2 | 0.0 | 1.9 | 2.3 |
| crafty | 11.2 | 0.5 | 0.0 | 13.3 | 14.0 |
| gzip | 0.1 | 0 | 0.0 | 2.2 | 2.2 |
| mcf | 0 | 0 | 0.0 | 0 | 0 |
| mesa | 14.8 | 2.5 | 0.0 | 13.8 | 22.2 |
| mgrid | 0 | 0 | 0.0 | 0 | 0 |
| swim | 3.5 | 0.5 | 0.0 | 4.1 | 6.3 |
| twolf | 16.4 | 5.1 | 0.0 | 14.3 | 23.4 |
| Avg. | 5.96 | 1.1 | 0.0 | 6.2 | 8.8 |

TABLE III

IMPACT OF EACH FLOORPLAN ON THE PERFORMANCE. TT STANDS FOR THERMAL THROTTLING.

| Method | Area ($mm^2$) | overhead |
|---|---|---|
| *base* | $72.36 \times 10^{-4}$ | - |
| *MAR-A* | $72.36 \times 10^{-4}$ | - |
| *MARS-SAM* | $74.51 \times 10^{-4}$ | Total:3% LSQ:50% Bus:50% |

TABLE IV

AREA OF EACH FLOORPLAN AND THE OVERHEAD COMPARED TO THE *base*.

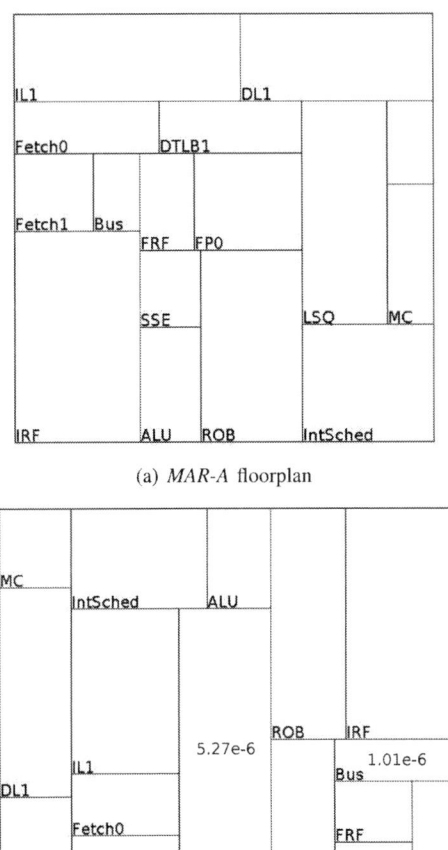

(a) *MAR-A* floorplan

(b) *MARS-SAM* floorplan

Fig. 3. Floorplan resulted from different methods. L2 is not shown to save space. It is placed on top of the core for *MAR-A*, and on the left for *MARS-SAM*. Blocks that have different area than the base floorplan are indicated by their area inside the block (unit is $mm^2$).

Figure 3 shows the floorplan resulted from each of the methods. L2 is not shown to save space. For *MAR-A*, it is placed on top of the core, and it is located on the left for *MARS-SAM*. Blocks that have different area than the base floorplan are indicated by their area inside the block. *MARS-SAM* could resolve the high power density problem of a block by increasing it's area. As a result the floorplan takes more area. While *MAR-A* increases the total area by 0.06%, *MARS-SAM* has to expand two blocks (LSQ and Bus) to resolve the

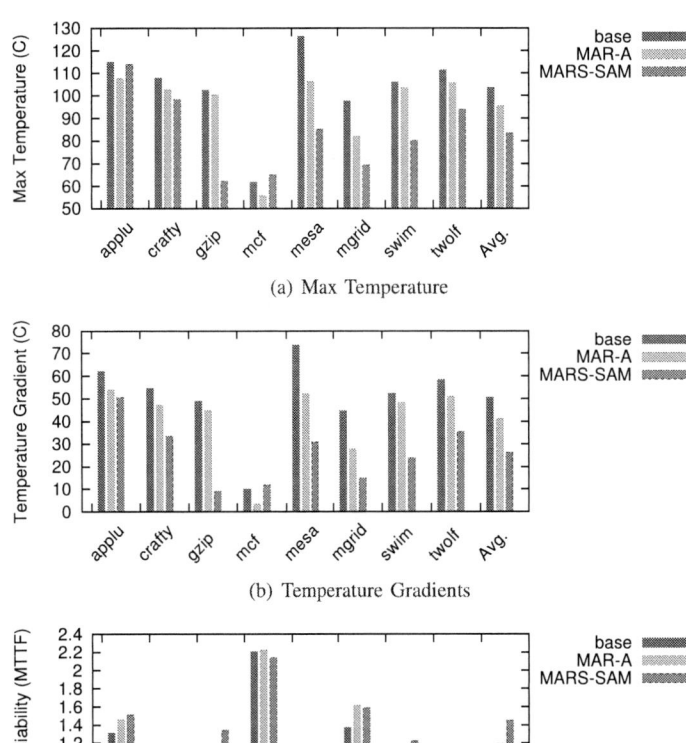

(a) Max Temperature

(b) Temperature Gradients

(c) Reliability

Fig. 4. Thermal metric results.

high density problem to the point that there is no thermal throttling. This results in 3.0% increase in the total area. However, the silicon real state is less critical factor in the design of microprocessors compared to parameters like clock frequency. Hence trading off performance and reliability for a small increase in the area is acceptable. Table IV summarizes the area measures for all the evaluated methods.

Figure 4(a) shows around 20°C reduction of the maximum temperature on average across the benchmarks. Gradient temperature across the chip could influence both reliability and performance. High thermal gradient across the chip could cause timing failure. Figure 4(b) shows the same range of reduction on the thermal gradients across the chip on average.

Estimation of exact mean to time failure of a chip in a simulation environment is not simple. But normalized results can be used to relatively compare two methods. Figure 4(c) shows the normalized reliability metrics. *MARS-SAM* results in a floorplan that on average has more than 40% longer mean time to failure compared to *base*. Note that the average reliability of the chip is computed as the inverse of average FIT measure for each benchmark. One might argue that the reliability of the chip is determined by the worse case benchmark. But since the collection of these benchmarks represent the typical usage of the chip, averaging the FIT measure of all of them shows the overall effect.

Not all the benchmark workloads show improvement. The reason for that is two fold. First, thermal throttling limits the maximum temperature, and caps the reliability degradation. Second, lowering maximum temperature in hotspots might come at the cost of increase in the average temperature in the region. To show the reliability of the chip without capping it through throttling, we disable the throttling and rerun the experiments. The average normalized reliability measure for the chip degrades down to 0.3 compared to the *base* configuration.

Leakage is the temperature dependent component of power consumption. The *Leakage/Total power* ratio increases as the technology size shrinks. In our simulation setup, leakage accounts for around 30% of the total power consumption. Our results show that *MARS-SAM* reduces the leakage by 7% compared to *base*.

## VI. CONCLUSION

This study shows the importance of thermal-aware floorplanning in improving both reliability and performance of a processor. The improvement in the reliability comes from less hotspots across the chip. Lower temperature leads to less thermal emergencies on the chip that would have caused the processor performance to suffer. We use a Power Blurring thermal model to estimate temperature in the floorplanning process. The Power Blurring method fundamentally works in grid model, and it is as fast as the block model solver used in HotSpot. We implement a simulated annealing based floorplanning scheme, empowered with three primitive actions: move, aspect-ratio, and resizing. While the first three primitives leverage lateral heat transfer, the last one helps averting critical hotspots by expanding area of the block in a controlled way. We introduce a more efficient method of selecting a power map for the floorplanning process. Our experiments show around 8.8% increase in performance as a result of averting all thermal emergencies, at the cost of 3% increase in the chip total area. The reliability of chip is improved by 40% as a result of lower hotspots. Power consumption is another design parameter that has been improved by 7% reduction in the leakage component.

## VII. ACKNOWLEDGMENT

We would like to Thank Dr. Xi Wang and Rigo Dicochea for their valuable comments and feedback toward the improvement of the paper.

## REFERENCES

1. J. Srinivasan, A. Sarita, P. Bose, and R. Jude. *Lifetime reliability: toward an architectural solution*, In IEEE MICRO, volume 25, pages 70-80, 2005.
2. A. H. Ajami, K. Banerjee and M. Pedram, *Modeling and Analysis of Non-uniform Substrate Temperature Effects on Global ULSI Interconnects*, IEEE Trans. Comput.-Aided Design Integr. Circuits Syst., vol. 24, no. 6, pp. 849-861, 2005.
3. Y. Han, I. Koren, and C. A. Moritz, *Temperature aware floorplanning*, in Second Workshop on Temperature-Aware Computer Systems(TACS-2), held in conjunction with ISCA-32, June 2005.
4. K. Sankaranarayanan, S. Velusamy, M. Stan, and K. Skadron. *A case for thermal- aware floorplanning at the microarchitectural level*, The Journal of Instruction-Level Parallelism, 7, 2005.
5. W. Hung, Y. Xie, N. Vijaykrishnan, C. Addo-Quaye, T.Theocharides, and M. J. Irwin, *Thermal-aware floorplanning using genetic algorithms*, in Sixth International Symposium on Quality of Electronic Design (ISQED?05), March 2005.
6. K. Skadron and M. R. Stan and W. Huang and S. Velusamy and K. Sankaranarayanan and D. Tarjan,*Temperature-Aware Microarchitecture*, in Proceedings of the 30th Annual International Symposium on Computer Architecture, (ISCA), 2003, pp. 2-13,
7. D. Brooks and M. Martonosi, *Dynamic Thermal Management for High-Performance Microprocessors*, Proc. 7th International Symposium on High Performance Computer Architecture (HPCA01), IEEE CS Press, 2001, pp. 171-182.
8. T. Kemper, Y. Zhang, Z. Bian, and A. Shakouri, *Ultrafast Temperature Profile Calculation in IC chips*, Proc. of 12th International Workshop on Thermal investigations of ICs (THERMINIC), Nice, France, pp. 133-137, 2006.
9. J. H. Park, X. Wang, A. Shakouri, and S.-M. Kang, *Fast Computation of Temperature Profiles of VLSI ICs with High Spatial Resolution*, Proc. 24th Semi-Therm, San Jose, CA, pp. 50-54, 2008.
10. J. H. Park, A. Shakouri, and S.-M. Kang, *Fast Evaluation Method for Transient Hot Spots in VLSI ICs in Packages*, International Symposium on Quality Electronic Design, San Jose, CA, pp. 600-603, 2008.
11. J. H. Park, S. Shin, J. Christofferson, A. Shakouri, and S.-M. Kang, *Experimental Validation of the Power Blurring Method*, Proc. 26th Semi-Therm, Santa Clara, CA, pp. 240-244, February 21-25, 2010.
12. A. Ziabari, Z. Bian, and A. Shakouri, *Adaptive Power Blurring Techniques to Calculate IC Temperature Profile under Large Temperature Variations*, International Microelectronic and Packaging Society (IMAPS) ATW on Thermal Management, Sept 28-30, 2010.
13. A. Ziabari, E. K. Ardestani, J. Renau, and A. Shakouri, *Fast Thermal Simulators for Architecture Level Integrated Circuit Design*, SEMI-THERM 2011.
14. J. Renau, F. Basilio, J. Tuck, W. Liu, M. Prvulovic, L. Ceze, S. Sarangi, P. Sack, K. Strauss, and P. Montesinos, *SESC simulator*, January 2005, http://sesc.sourceforge.net.
15. F. Bellard, *QEMU, a fast and portable dynamic translator*, in Proceedings of the USENIX Annual Technical Conference, ATEC, 2005, Anaheim, CA, pp 41-41
16. Sheng Li, Jung Ho Ahn, Richard D. Strong, Jay B. Brockman, Dean M. Tullsen, and Norman P. Jouppi, *Mcpat: an integrated power, area, and timing modeling framework for multicore and manycore architectures*, in Proceedings of the 42nd Annual IEEE/ACM International Symposium on Microarchitecture, 2009, MICRO 42, pp. 469-480.
17. J. Nayfach and J. Renau, *SOI, Interconnect, Package, and Mainboard Thermal Characterization*, International Symposium on Low Electronics and Design (ISLPED), pp. 327-330, Aug. 2009.

# Cache Leakage Power Estimation Using Architectural Model for 32 nm and 16 nm Technology Nodes

Piotr ZAJAC, Marcin JANICKI, Michal SZERMER, Cezary MAJ, Piotr PIETRZAK, Andrzej NAPIERALSKI

Department of Microelectronics and Computer Science, Technical University of Lodz

Wolczanska 221/223, 90-924 Lodz, Poland

E-mail: pzajac@dmcs.pl; tel: +48 42 6312653; fax +48 42 6360327

## Abstract

The constant increase of subthreshold current of nanometer transistors due to technology scaling may hinder the evolution of high-performance chips in the near future. This evokes the need of accurate leakage power modeling for new nanometer technologies. In this paper, we present an improved subthreshold current model, which was integrated it into an architectural-level power simulator. Using this simulator, we estimated the leakage power in a 2 MB cache memory for 32 nm and 16 nm technology nodes. Our results show that the cache leakage power dissipation for 2 MB 2-way cache at 100 °C fabricated in the 32 nm technology is around 1 W. For the 16 nm technology, we demonstrate the importance of maintaining high threshold voltage to keep leakage power density at the acceptable level.

## Keywords

Leakage power, subthreshold current, architectural modeling, cache, nanometer technologies

## 1. Introduction

Current processors are built using 32 nm technology and this trend is expected to continue down to even sub-10 nm technologies. With shrinking transistor channel length, the leakage power due to subthreshold and gate leakage current is becoming a serious problem for high-performance chips. The semiconductor industry faced this problem several years ago, when leakage power consumption was supposed to inhibit the development of chips already in 45 nm or 32 nm technologies. However, owing to the fact that researchers signaled this problem very early, it was possible to find a solution – the use of high-k dielectric. It allowed reducing gate current by the factor of 1000 for PMOS and 25 for NMOS transistors [1]. However, the miniaturization continues and the problem related to leakage power will soon reappear, mostly due to the subthreshold current, which increases with each technology node. Hence the need of accurate architectural modeling which will allow us to predict when leakage power may become a serious obstacle for the development of high-performance chips. Consequently, we decided to estimate the leakage power dissipation for SRAM L2 cache for processors fabricated in the 32 nm and 16 nm technologies. We used the well-known simulator HotLeakage [2] for calculating leakage power. Since HotLeakage model does not include parameters for 32 nm and newer technologies, we extracted these parameters using SPICE simulation of Predictive Technology Models (PTM) [3] and incorporated them into the simulator. We also improved HotLeakage subthreshold current model. This paper is organized as follows: Section 2 presents the related work, in Section 3 we describe our subthreshold current model and

Section 4 shows that the model is valid for a wide range of parameter values. In Section 5 simulation results are presented and Section 6 summarises the paper.

## 2. Related work

The current modeling in HotLeakage is based on the relatively simple Butts-Sohi model [4] which can be summarised by a single equation:

$$I_{leak} = V_{dd} N k_{design} I_{leak}^i \qquad (1)$$

where $V_{dd}$ is the supply voltage, $N$ is the number of transistors in the circuit, $I_{leak}^i$ is the leakage per transistor and $k_{design}$ is an architectural constant. This model was improved in HotLeakage to take into account the changes of $I_{leak}^i$ and $k_{design}$ when modifying parameters such as temperature, threshold voltage, transistor aspect ratio and supply voltage. In its original version, this tool allows the estimation of leakage for 180 nm, 130 nm, 90 nm and 70 nm technologies. Each technology is represented by a set of experimentally derived parameters. The values of these parameters are calculated by correlating the model with the results obtained from SPICE simulator.

There are many published works based on simulations with HotLeakage tool. Nevertheless, their authors perform simulations for built-in technologies, i.e., for the 180 nm down to the 70 nm technology node. For example, in [5] authors explore the methods of leakage power reduction in cache memories for these built-in technologies. Similarly, in [6] and [7] authors analyze and compare various leakage-control techniques in 70 nm technology. However, to our knowledge, no HotLeakage models for newer technologies have been presented in literature. In [8] authors used the method based on CACTI [9] simulator to estimate the static and dynamic power in several technologies, from 90 nm to 32 nm. However, their work was from the year 2006 and therefore based on models which did not take into account the introduction of the high-k dielectric.

## 3. Subthreshold current model

In this paper we focused our research on the subthreshold current modelling since it is by far the most important part of leakage current in modern technologies. Thus, in our work we do not model gate leakage. As mentioned previously, HotLeakage tool was used in our research.

The subthreshold leakage model in HotLeakage is based on the equation (2):

$$I_{leak}^{i} = I_{sub} = I_0 \exp(b(V_{dd} - V_{dd0})) * (1 - \exp(\frac{-V_{dd}}{v_t})) *$$

$$* \exp(\frac{-|V_{th} + c(T - T_0)| - V_{off}}{n \cdot v_t})$$

$$I_0 = \mu_o C_{ox} \frac{W}{L} v_t^2 \qquad (2)$$

where $\mu_0$ is the zero-bias mobility, $C_{ox}$ is the gate capacitance per unit area, $W/L$ is the gate aspect ratio, $v_t$ is the thermal voltage, $T$ is the temperature, $V_{th}$ is the zero-bias threshold voltage at room temperature $T_0$=300 K, $V_{dd}$ is the supply voltage and $V_{dd0}$ is the default supply voltage for the technology. $b$, $c$, $n$ and $V_{off}$ denote experimentally derived parameters. The procedure of calculating these parameters is thoroughly described in [2]. In short, authors trace the leakage current in function of $W/L$, $T$, $V_{th}$ and $V_{dd}$ separately. Then, iterative curve fitting method is applied to fit the functions $I_{sub}(W/L)$, $I_{sub}(T)$, $I_{sub}(V_{th})$ and $I_{sub}(V_{dd})$ to circuit level simulations obtained by SPICE.

Our model was initially very similar; we introduced however some changes:

1) For our research we assume constant power supply voltage so we eliminated the corresponding term from the model. The model, which takes into account changes of $V_{dd}$ will be developed in the future.

2) Instead of $C_{ox}$ we used the constant $C_{ox} \exp(-1.8)$ which changes nothing in the nature of the model but better corresponds to the BSIM4 model [10] from SPICE. The reason is that in the BSIM4 model, instead of $C_{ox}$, other parameter (see Eq.3) is used in the formula for subthreshold current:

$$\sqrt{\frac{q \varepsilon_{si} NDEP}{2 \phi_S}} \text{, where } \phi_S = 0.4 + \frac{k_B T}{q} \cdot \ln\left(\frac{NDEP}{n_i}\right) \qquad (3)$$

In the above equation $n_i$ is the intrinsic doping concentration and $k_B$ is the Boltzmann constant. Note that the surface potential $\phi_S$ is not a constant as it depends on temperature $T$. However, its influence on this parameter is very weak and can be safely neglected. Concluding, the parameter is roughly equal to $C_{ox} \exp(-1.8)$.

Thus, our subthreshold current model is represented by the following formula:

$$I_{sub}' = \mu_o C_{ox} \exp(-1.8) \frac{W}{L} v_t^2 (1 - \exp(\frac{-V_{dd}}{v_t})) *$$

$$* \exp(\frac{-|V_{th} + c(T - T_0)| - V_{off}}{n \cdot v_t}) \qquad (4)$$

To find the values of parameters $c$, $n$ and $V_{off}$ at first we decided to use the same method as the authors of HotLeakage but for the new 32 nm and 16 nm technologies. Therefore, we performed the SPICE simulations of leakage current for both NMOS and PMOS transistor using Predictive Technology Models (PTM) for high-performance 32 nm and 16 nm technology nodes. We then traced the subthreshold current in function of three variables $W/L$, $T$ and $V_{th}$. Basically, with this

procedure we followed the steps of [2] for two different technologies (32 nm and 16 nm) and with a slightly modified model. But here we discovered the main disadvantage of such an approach. Although the model was correct whenever only one parameter ($W/L$, $T$ or $V_{th}$) was changed, it produced considerable errors (even up to 70% compared with SPICE results) when two or all three parameters were modified at the same time. Indeed, the method based on independent fitting of curves for three different parameters is in theory incorrect. Take for example the function $f(x, y)$ which we want to fit to the results represented by the function $g(x, y)$. Assuming $x=x_0$ and fitting the function for every $y$ and then assuming $y=y_0$ and fitting the function for every $x$ does not necessarily mean that the function $f$ will be fitted to the function $g$ for every pair $(x, y)$.

Hence, the model had to be further modified. First, we noticed that an additional term $\Delta I_{sub}$ is needed. Second, to account for changes of $\Delta I_{sub}$ in function of temperature, we suggested a simple linear function. Third, we had to find the influence of $V_{th}$ on $\Delta I_{sub}$, which is strongly non-linear and had to be approximated with a quadratic function, whose coefficients are linear functions of temperature. Thus, our final subthreshold current model formula reads:

$$I_{sub} = I_{sub}' + \Delta I_{sub}$$

$$\Delta I_{sub} = m_1 (\frac{W}{L} - 1) * \left( m_2 \left( \frac{T}{T_0} - 1 \right) + 1 \right) *$$

$$* \left( f_1(T) * \left( \frac{V_{th0}}{V_{th}} - 1 \right)^2 + f_2(T) * \left( \frac{V_{th0}}{V_{th}} - 1 \right) + 1 \right) \qquad (5)$$

where: $f_1(T) = m_3 - m_4(\frac{T}{T_0} - 1)$, $f_2(T) = m_5 - m_6(\frac{T}{T_0} - 1)$ (6)

Note that the new terms in the equation are formed in such a way that they have no influence on the value of the entire equation when $W/L$=1, $T=T_0$ or $V_{th}=V_{th0}$. Consequently, by adding these extra terms we do not perturb the fitted functions that we had obtained previously. Parameters $m_i$ were found experimentally for both NMOS and PMOS transistors by gradually fitting the functions $I_{sub}(W/L)$, $I_{sub}(T)$ and $I_{sub}(V_{th})$ and trying to minimize the relative error between the model and SPICE-based simulations. This process was iteratively repeated until we obtained satisfactory correlation, i.e., when the error could not be reduced further.

Hence, our model for a given technology is described by a set of nine parameters. However, only four parameters ($c$, $n$, $V_{off}$ and $m_1$) are technology-dependent, the others ($m_2$ to $m_6$) are the same for both 32 nm and 16 nm technology. Consequently, one may see that for each new technology, only four parameters need to be tuned to obtain an accurate model.

978-1-4673-1110-6/12 $31.00 © 2012 IEEE

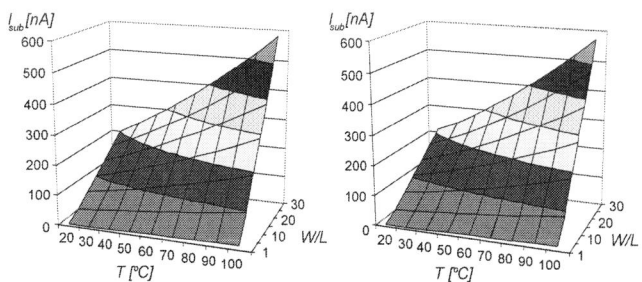

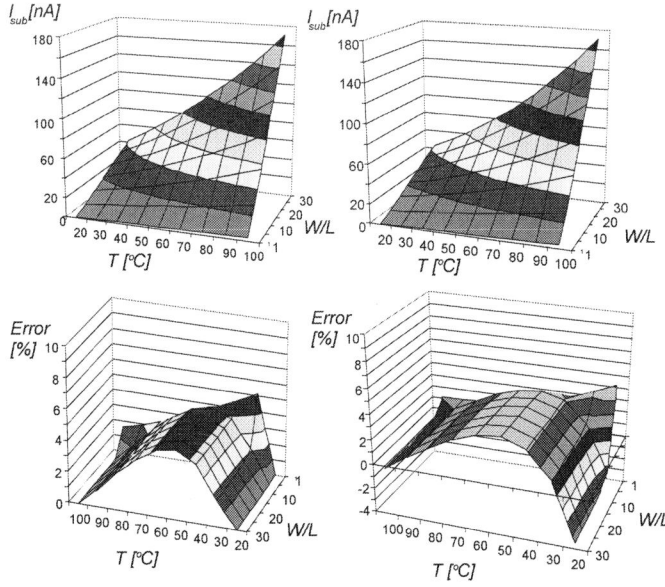

**Figure 2** Subthreshold current of NMOS transistor in the 16 nm technology for $V_{th}$=0.25 V as a function of aspect ratio $W/L$ and temperature $T$ calculated by our model (left) and SPICE (right).

For the 16 nm technology node, fitting the model was much more difficult than for the 32 nm technology. Thus, both maximal and relative error presented in Tables 3 and 4 are higher than those for 32 nm technology. Nevertheless, for the entire analyzed range of parameter values the maximal relative error does not exceed 16%, what again may be considered fairly acceptable.

## 5. Simulation results

The proposed model described in the previous sections was incorporated into the HotLeakage simulator. Moreover, two new technology nodes 32 nm and 16nm were created inside the simulator. For each technology, we set all the parameters (such as supply voltage $V_{dd0}$, $C_{ox}$, $\mu_0$, etc.) according to the data given in PTM model. However, one particular parameter proved to be difficult to estimate, namely the threshold voltage $V_{th}$ for various logic gates. Some works used linear scaling of threshold voltage, but this method may be inaccurate for latest technologies. Especially for the 16 nm node, when new phenomena appear and affect $V_{th}$, it becomes extremely difficult to estimate the value of $V_{th}$ [11]. Consequently, for the 32 nm technology we decided to use the values calculated in SPICE for the PTM NMOS and PMOS model. For each transistor type, we assumed the same threshold voltage regardless of the logic gate type.

**Figure 1** Top: subthreshold current of NMOS transistor in the 32 nm technology for $V_{th}$=0.3 V as a function of aspect ratio $W/L$ and temperature $T$ calculated by our model (left) and SPICE (right). Bottom: relative error of the 32 nm technology model for NMOS (left) and PMOS (right) for $V_{th}$=0.3 V

## 4. Model validation

To prove the validity of our model we compared it with SPICE-based circuit-level simulations in the area of interest, i.e., for $T$=20÷100 °C and $W/L$=1÷30 for both NMOS and PMOS. For the 32 nm model, we assumed that the threshold voltage was in the range of $V_{th}$=0.25÷0.35 V, whereas for the 16 nm model in the range of $V_{th}$=0.2÷0.3 V. Figure 1 shows sample results, i.e., leakage current for NMOS transistor in 32 nm technology for $V_{th}$=0.3 V.

Considering that the subthreshold current in our model depends on four variables and that it changes by three orders of magnitude (from hundreds of picoamperes to hundreds of nanoamperes) within the area of interest, the maximal error of 13% may be considered acceptable.

|  | $V_{th}$=0.25 V | $V_{th}$=0.3 V | $V_{th}$=0.35 V |
|---|---|---|---|
| NMOS | -9.16 % | 5.16 % | 13,21 % |
| PMOS | -7,98 % | 4.15 % | 10,23 % |

**Table 1** Maximal relative error for various $V_{th}$ values for the NMOS and PMOS 32 nm transistor model.

|  | $V_{th}$=0.25 V | $V_{th}$=0.3 V | $V_{th}$=0.35 V |
|---|---|---|---|
| NMOS | 4.25% | 3.21% | 8,3% |
| PMOS | 4.48% | 2.3% | 3.17% |

**Table 2** Average relative error for various $V_{th}$ values for the NMOS and PMOS 32 nm transistor model.

|  | $V_{th}$=0.2 V | $V_{th}$=0.25 V | $V_{th}$=0.3 V |
|---|---|---|---|
| NMOS | 14.4 % | 10.3 % | 15.6% |
| PMOS | 13.9 % | 10.8 % | 15.0 % |

**Table 3** Maximal relative error for various $V_{th}$ values for the NMOS and PMOS 16 nm transistor model.

|  | $V_{th}$=0.2 V | $V_{th}$=0.25 V | $V_{th}$=0.3 V |
|---|---|---|---|
| NMOS | 7.3 % | 3.37 % | 6.9 % |
| PMOS | 7.2 % | 3.0 % | 5.1 % |

**Table 4** Average relative error for various $V_{th}$ values for the NMOS and PMOS 16 nm transistor model.

978-1-4673-1110-6/12 $31.00 © 2012 IEEE

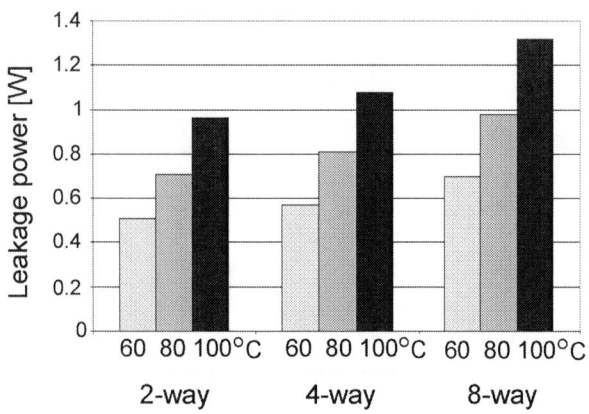

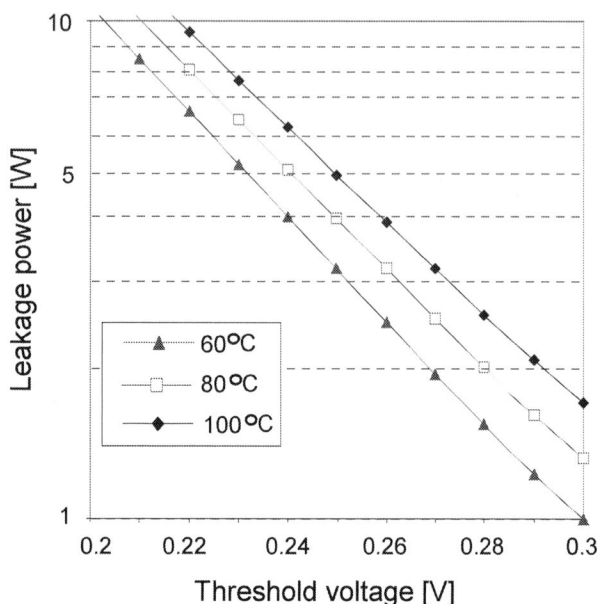

**Figure 3** L2 cache leakage power in function of associativity for various temperatures for the 32 nm technology

However, the value given by SPICE is based on the analytical equation given in [10] and might not always be adequate for new technologies. Other methods, like linear and quadratic extrapolation assume ideal long-channel behaviour which is not true for short channel devices. Moreover, the subthreshold current is very sensitive to the change of threshold voltage and even a small error in the estimation of $V_{th}$ could result in large error when calculating the subthreshold current. Therefore, for the 16 nm technology instead of trying to estimate $V_{th}$ we decided to use it as a parameter and plot the cache leakage power as a function of $V_{th}$. We assumed the same $V_{th}$ for NMOS and PMOS and the range of $V_{th}$ from 0.2 V to 0.3 V.

We configured HotLeakage to simulate an architecture resembling the well-known Alpha 21364 processor [12] with 2 MB L2 cache. We did not implement any leakage reduction techniques. Finally, simulations were run to calculate the leakage power dissipated in the L2 cache as a function of temperature and cache associativity (see Fig. 3) and as a function of temperature and threshold voltage (see Fig. 4).

The increase of power with associativity is perfectly understandable, as the tag array grows in size and therefore dissipates more leakage power. It is also visible that rising the temperature from 60 °C to 100 °C caused the increase of about 90% in the leakage power, regardless of associativity. Overall, our simulations estimate that the cache leakage power stays slightly below 1W for 2-way cache at 100 °C.

In 2009, Intel presented a new SRAM design in the 32 nm technology with integrated power management [13]. They reported 5 mW of leakage power for 1 V power supply at 110°C for 128 kb array, assuming 58% reduction of leakage current due to power-saving techniques. Therefore, one can calculate that without leakage reduction features enabled, the array dissipates 11.9 mW of leakage power. Then, scaling this result with the array size, one can estimate that 2 MB SRAM should dissipate about 1.52 W. In our simulations, for the same conditions but with the supply voltage equal to 0.9 V, the leakage power dissipated in 2 MB cache was 1.11 W from which about 1 W was dissipated in the data array and the rest in decoders, the tag array and other cache elements. The lower power dissipation in our simulation can be however

**Figure 4** Leakage power of 2-way cache in function of threshold voltage for various temperatures for the 16 nm technology.

partially explained by the difference in the supply voltage (0.9 V vs. Intel's 1 V). Note that the supply voltage not only influences power due to $P_{leak}=V_{dd}I_{leak}$ dependence, but also has an important impact on leakage current itself (see Eq.2). Given many assumptions that had to be made during our research, we believe that the obtained results constitute a good estimate of leakage power for architectural simulations.

The results obtained for the 16 nm technology are presented in Fig. 4 which shows that the leakage power rises exponentially when $V_{th}$ decreases. Let us first consider the case when $V_{th}=0.25$ V which seems to be a reasonable value for 16 nm technology. At 100 °C, the leakage power increased by the factor of about 5 in comparison with the 32 nm technology. This result is quite troubling, but the problem becomes even more alarming if we consider the increase of leakage power per entire chip. Note that the leakage power will be additionally increased by the simple fact that in newer technologies the total size of cache will grow. Let us analyze an example for clarity. Current Intel's Sandy Bridge quad-core processors manufactured in the 32 nm technology have 256 kB of L2 cache per core, 1 MB in total. One may suppose that in the 16 nm technology, it will be possible to put 16 such cores on a single die, consequently the amount of L2 cache will rise to 4 MB in total. The conclusion is that not only does the leakage power increase due to higher subthreshold current, but it also doubles with each technology nodes simply because of the area scaling. Consequently, the increase in leakage power in the entire chip when migrating from 32 to 16 nm technology node will be even more dramatic than shown in our results. Take for example the previously analyzed case of $V_{th}=0.25$ V. While power grows by the factor of 5, the power density increases 20-fold which is clearly unacceptable.

There are two solutions to this issue and most likely both will have to be implemented in future technologies to ensure that the power density stays at acceptable level. First, it seems that the threshold voltage cannot continue decreasing, or at least it cannot decrease linearly, with supply voltage. Our results show that the leakage power density may be considered acceptable until $V_{th}$ becomes lower than 0.27 V. This in turn implicates that the power supply will not scale down in future technologies and will stay at the level of 0.7÷0.8 V. Second solution relies on implementing efficient leakage reduction techniques [14] like dual threshold CMOS design, sleep transistors, dynamic $V_{th}$ scaling, forward/reverse body biasing etc.

It is also worth mentioning that recently Intel announced the development of experimental near-threshold voltage processor [15] which indicates that near-threshold computing might be an interesting alternative for future technologies.

## 6. Conclusions

The rising impact of leakage power dissipation on total power consumption evokes the need for accurate leakage modelling. The early discovery that the leakage power dissipation becomes prohibitive may lead to new, more efficient designs and technological improvements. In our paper, we estimated the leakage power for SRAM cache memory fabricated in the 32 nm and 16 nm technology nodes. We used the existing architectural simulator HotLeakage but we modified it to simulate two new technologies. We also changed the subthreshold current model in the simulator and proved it to be accurate by comparing the results with SPICE-based simulations. Our final results show that the cache leakage power dissipation for 2 MB 2-way cache at 100 ºC is around 1 W for the 32 nm technology. For the 16 nm node, we show that the leakage power grows exponentially with threshold voltage and that to maintain leakage power at reasonable level, in future technologies the threshold voltage cannot be scaled down linearly with supply voltage.

Our future research will focus on using our model for developing scheduling techniques for many-core systems which will take into consideration the minimization of leakage power consumption.

## Acknowledgments

The research was supported by the grant of the Polish National Centre of Science No. N515 509140.

## References

1. Intel Technology Journal "45nm High-k+Metal Gate Strain-Enhanced Transistors" Volume 12 Issue 02, June 17, 2008
2. Y. Zhang, D. Parikh, K. Sankaranarayanan, K. Skadron, and M. R. Stan. "HotLeakage: An Architectural, Temperature-Aware Model of Subthreshold and Gate Leakage" University of Virginia Dept. of Computer Science Tech. Report CS-2003-05, Mar. 2003
3. W. Zhao, Y. Cao, "New generation of Predictive Technology Model for sub-45nm early design exploration" IEEE Transactions on Electron Devices, vol. 53, no. 11, pp. 2816-2823, November 2006.

4. J. A. Butts and G. S. Sohi "A static power model for architects" In Proceedings of the 33rd Annual IEEE/ACM International Symposium on Microarchitecture, pages 191–201, Dec. 2000.
5. Y. Meng , T. Sherwood , R. Kastner "Exploring the limits of leakage power reduction in caches" ACM Transactions on Architecture and Code Optimization (TACO), v.2 n.3, p.221-246, September 2005
6. Y. Li , D. Parikh , Y. Zhang , K. Sankaranarayanan , M.Stan, K. Skadron "State-Preserving vs. Non-State-Preserving Leakage Control in Caches" Proceedings of the conference on Design, automation and test in Europe, p.10022, February 16-20, 2004
7. S. Kaxiras , P. Xekalakis , G. Keramidas "A simple mechanism to adapt leakage-control policies to temperature" Proceedings of the 2005 international symposium on Low power electronics and design, August 08-10, 2005, San Diego, CA, USA
8. S. Rodriguez , B. Jacob "Energy/power breakdown of pipelined nanometer caches (90nm/65nm/45nm/32nm)" Proceedings of the 2006 International symposium on low lpower electronics and design, October 04-06, 2006, Tegernsee, Bavaria, Germany
9. P.Shivakumar, N.Jouppi "CACTI 3.0: An integrated cache time, power and area model," WRL Research Report 2001/2, Aug.2001.
10. http://www.idea2ic.com/PlayWithPerl/Bsim_Ref/BSIM4 _manual.pdf
11. A. L. S. Loke, Z.-Y. Wu, R. Moallemi, C. D. Cabler, C. O. Lackey, T. T. Wee, and B. A. Doyle, "Constant-Current Threshold Voltage Extraction in HSPICE for Nanoscake CMOS Analog Design," in Synopsys Users Group (SNUG) 2010 Conference (San Jose, CA), Mar. 2010.
12. R.E. Kessler, E.J. McLellan, D.A. Webb "The Alpha 21264 microprocessor architecture," Proceedings of the International Conference on Computer Design: VLSI in Computers and Processors, 1998. ICCD '98, pp.90-95, 5-7 Oct 1998
13. Y. Wang, U. Bhattacharya, F. Hamzaoglu, P. Kolar, Y. Ng, L. Wei, Y. Zhang, K. Zhang, K.; M. Bohr, "A 4.0 GHz 291Mb voltage-scalable SRAM design in 32nm high-κ metal-gate CMOS with integrated power management" IEEE International Solid-State Circuits Conference - Digest of Technical Papers, 2009. ISSCC 2009., pp.456-457,457a, 8-12 Feb. 2009
14. A. Agarwal, S. Mukhopadhyay, A. Raychowdhury, K. Roy, C.H.Kim, "Leakage Power Analysis and Reduction for Nanoscale Circuits," IEEE Micro, vol.26, no.2, pp.68-80, March-April 2006
15. http://blogs.intel.com/research/2011/09/15/ntvp/

978-1-4673-1110-6/12 $31.00 © 2012 IEEE

# New Simulation Approaches Supporting Temperature-aware Design of Digital ICs

Gergely Nagy[1], András Timár[1], Albin Szalai[1], Márta Rencz[1], András Poppe[1,2]

[1]Budapest University of Technology and Economics, Department of Electron Devices
Magyar tudósok körútja 2, Bldg. Q, wing B, Budapest, H-1117 Hungary

[2]Mentor Graphics Mechanical Analysis MicReD Division
Gábor Dénes utca 2. Infopark Bldg. D, Budapest, H-1117 Hungary

[1]< nagyg|timar|szalai|rencz|poppe>@eet.bme.hu, [2] andras_poppe@mentor.com

## Abstract

Regarding thermal issues in digital IC design a major concern is how timing integrity is affected by the elevated junction temperature and temperature gradients on the chip surface. To predict this in a thermal aware design process one needs a dedicated simulation tool in which the logic simulation of the circuit is coupled to the thermal simulation of the chip and its environment. This paper presents two approaches to this so called logi-thermal simulation. In one of our approaches we rely completely on industry standard EDA tools, standard EDA file formats and interfaces. In the other solution which provides us total freedom in the abstraction level of circuit description and simulation accuracy we use our own logic simulation engine. In both cases the logic simulation engine is connected to our own thermal simulation engines which also use compact thermal models of the IC package during simulation. This paper describes certain implementation aspects and features of our logi-thermal simulation solutions, with emphasizes on modeling the thermal properties of the IC packaging.

## Keywords

Logi-thermal simulation, timing integrity, thermal effect of chip packaging

## 1. Introduction and prior work

Temperature affects circuit operation on component level such as single PN junctions or transistors as well as on the level of more complex functional blocks of e.g. digital IC-s such as different logic gates, flip-flops or even at higher functional levels. As a result of self-heating the local junction temperature increases, inducing changes in the operating characteristics of the devices. In case of digital circuits the operation is treated on higher level of abstraction where the most critical operational characteristics are for example the different timing parameters. Self-heating in digital circuits is connected to the transitions of the logic states, thus dissipation density is directly proportional to the event density of a digital IC.

The importance and necessity of power estimation in modern digital design has been an evidence for the past decade. Papers and studies have been emerging from the mid-1990's on the subject and several methodologies have been proposed. An overview paper [1] written as early as 1996 enumerates the strategies proposed to tackle the problems resulting from the elevated levels of power dissipation and high element densities in modern digital circuits. According to

it, one way to perform power estimation at gate-level is by using the *gate equivalent method*. In cases when the gate-level description is either not available for the elements constituting the circuit or the level of abstraction is willingly kept at the level of boolean descriptor functions, the *activity-based method* can be used [2], where the power of logic elements (combinational and sequential circuits) is approximated using entropy. Power estimation done at gate-level can yield very accurate results with simulation times much shorter than those of a transistor electro-thermal simulation. However designers should be able to acquire information on circuits' dissipation and the possible issues earlier in the design phase when the system is not yet available at gate-level. Furthermore, Crisu et al. [3] state that considerations made at system level can result in a power consumption reduced by 10–100 times, while gate-level strategies lead to a reduction of 2 times merely. Great efforts have been made to find methodologies that enable power estimation at high abstraction levels. Several solutions exist for FPGAs [4] and DPSs [5] at architectural and software levels and nowadays even the test vectors of large scale integrated systems are designed with thermal issues in mind [6]. The thermal behavior of multicore processors are simulated at a very high level, using virtual platforms which enable designers to develop, simulate and evaluate power and temperature management strategies [7], [8]. With an accurate power estimation possible overheated areas (hot-spots) can be detected and thus circuit failure due to thermal problems can be avoided. However, this is not sufficient to avoid all thermal-related issues. Changes in the temperature of logic elements affect their behavior. Gate delays, for example, are dependent on temperature. If the ambient temperature of a circuit is elevated, its operation slows down. The problem with self-heating is that it results in an uneven temperature density as the gate activity in complex systems is inhomogeneous. Accordingly, the delays of the gates can differ substantially over the surface of a chip. Thus a circuit that is dependent on timing may operate impeccably at room temperature when it is switched on, and it might produce errors after a certain time when some parts have warmed up while others not [9].

To treat the operation of digital circuits correctly in a self-consistent manner, the temperature dependence has to be considered together with the switching events, therefore our team suggested the principle of logi-thermal simulation as early as 1997 [10]. In this approach logic simulation of the circuit schematic and the thermal simulation of the physical IC structure and its thermal environment is performed simultaneously, in a self-consistent way. With our first experiments we showed both with simulation and physical measurements that the temperature pattern of the surface of a digital IC is determined by the actual mode of operation. In the logic simulation temperature dependent gate delays were already considered and later some energy models to describe gate dissipation related to a switching event were suggested [11]. As a next step, we were dealing with the resolution issues of the simulation, both in terms of physical space and time and we also studied, how details of packaging such as wire bonds affect the results of electro-thermal and logi-thermal simulation on the surface of the IC chip [12].

## 2. Recent developments towards a multi-abtraction level logi-thermal simulation environment

### 2.1. Detecting thermaly induced timing integrity problems

Most recently we developed our approach into two directions. On one hand, we created a simulation package aimed at standard cell designs in which the principle of logi-thermal simulation is realized by industry standard, commercially available tools and programming interfaces, such as Verilog with its Programming Language Interface (PLI). In this implementation (called CellTherm [13] – see Figure 1) a high speed thermal simulator based on the Fourier-method is used which is completed with a compact model generator as described in [12].

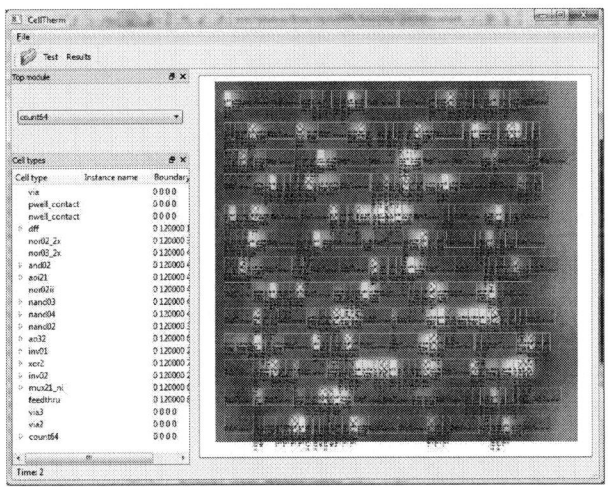

**Figure 1**: Temperature pattern of a standard cell design calculated with the CellTherm tool [13].

The authors of [14] propose a methodology that constitutes the basic idea of the approach used in CellTherm. It deals with temperature map generation of ICs from digital simulations. The work presented in [14] lacks the possibility of back-annotating temperature-dependent delays into the running simulation. Dealing with temperature-dependent delays becomes more and more important as feature size is shrinking and power density is growing. Failure to take the temperature-dependent delays into account can cause timing (e.g. setup- and hold time) violations that can foil correct logic operation. Our methodology also improves the work of Torki et al by coupling the logic and thermal simulator engines with a custom controller and visualization application that can

evaluate logical and thermal calculations and prepare delay back-annotation on-the-fly, in the middle of the simulation.

Classical electro-thermal simulators approach the electro-thermal simulation problem either by FEM simulation, relaxation method or simultaneous iteration. An example of electro-thermal simulation using simulator coupling is presented in [15]. The technique is based on the coupling of a FEM program with a circuit simulator. A pre-RTL temperature-aware design methodology is presented in [16], where a fast, yet accurate architectural thermal model that is able to explore large regions of the design space is proposed. [17] attempts to show that there is a significant peak temperature reduction potential in managing lateral heat spreading through floorplanning. As a demonstration, it uses a wire delay model and floorplanning algorithm based on simulated annealing to present a profile-driven, thermal-aware floorplanning scheme that significantly reduces peak temperature with minimal performance impact that is quite competitive with Dynamic Thermal Management (DTM). The floorplanning tool HotFloorplan is part of the HotSpot software [18] that is developed at DCS, University of Virginia. In addition, simulating the self-heating of the circuit in the early phase of the design before manufacture would make cooling issues less problematic. Self-heating simulations may also eliminate the need for design back-annotation after manufacture. An example of temperature-aware ASIC design flow can be found in [19].

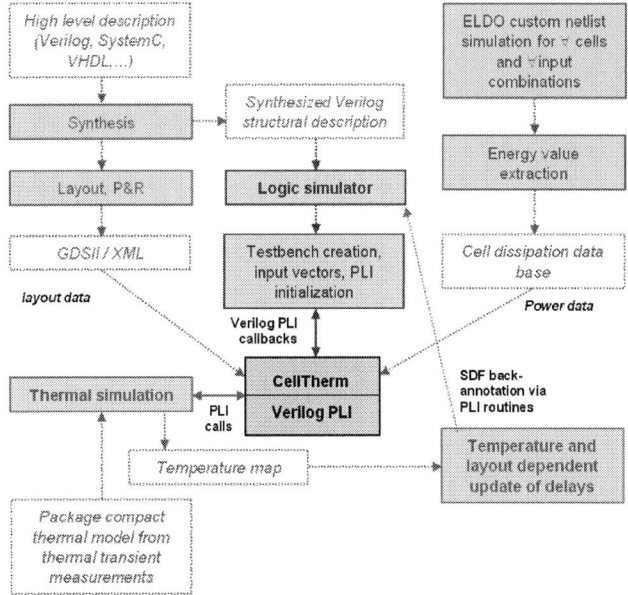

**Figure 2**: Structure of CellTherm and it's connection to a standard cell IC design flow in an EDA environment.

As shown in Fig. 2, our own tools are „glued" to usual EDA tools through standard interfaces such as Programming Language Interface (PLI) of Verilog, Standard Delay Format (SDF) files or XML layout descriptions. We used tools from Mentor Graphics's IC design flow.

In the early phase of the design, when the behavioral description gets synthesized, pre-layout timing data can be approximated and the synthesizer software can extract preliminary delay values from the design. The synthesizer

978-1-4673-1110-6/12 $31.00 © 2012 IEEE

software outputs the predicted post-synthesis pre-layout delay data into a Standard Delay Format (SDF) file, which can later be included in a logic simulation. The logic simulator can take these delay data as startup timing values and it is able to check against basic timing integrity issues. The SDF file contains not just delays of the individual cells of the design but setup and hold timing checks also. This way the simulator can send off an alert if timing requirements are not met. The preliminary delay and timing values are predicted values and do not take physical layout into account. Later on, after floorplanning and Place&Route, real timing data can be extracted from the design in SDF format.

Our recent efforts regarding the CellTherm tool addressed a precise modeling of dissipation and temperature dependence of timing parameters of standard cell library elements. Computer experiments using Mentor Grahics' ELDO program were used to derive such models for elements of the TSMC 0.35µm standard cell library.

The logic simulation is performed by a Verilog simulator. The specific routines constituting the heart of CellTherm are activiated as Verilog PLI callback routines. PLI is also used to activate our thermal simulator when the simulation clock in Verilog is advanced.

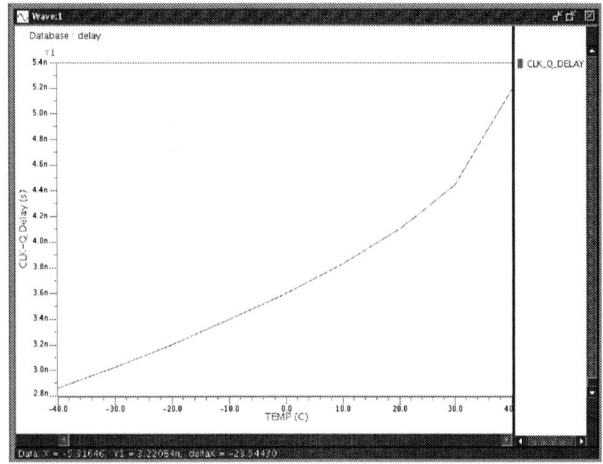

**Figure 3**: Tempareture – delay function of a D flip-flop cell.

The thermal simulator used during the cell level logi-thermal simulation describes the chip and its close environment (such as die attach, part of the leadframe) with a geometry and material properties based model while further parts of the package and its thermal environment are represented by a dynamic compact thermal model, co-simulated with the chip [20]. The power map for the thermal simulator is created from a power database (obtained in the pre-characterization phase of the standard cell library – and as such can be considered as an extension to that library) based on the actual input combinations of the cells. Using power maps corresponding to the actual event distribution of the IC a surface temeparature map is calculated. Based on the local temperatures gate delays are updated using pre-calculated delay vs. temperature characteristics of the cells such as shown in Fig. 3. Updated delays are back-annotated and the logic simulation uses these updated timing parameters in the subsequent simulatuion time instances. Considering

temperature dependent gate delays timing violations can be detected – see Figures 4 and 5.

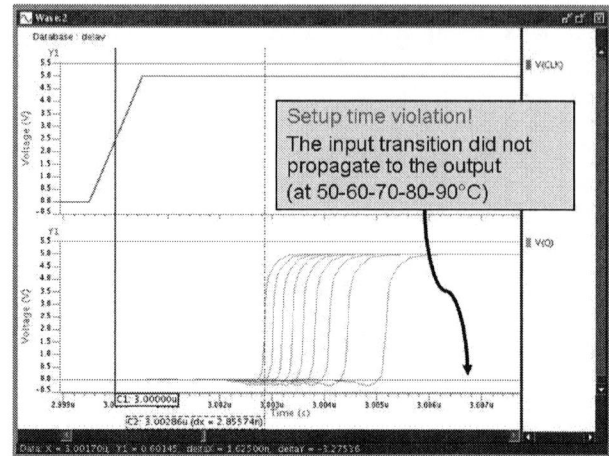

**Figure 4**: Setup time violation is confirmed by transistor level analog simulation above 50 °C local chip temperature.

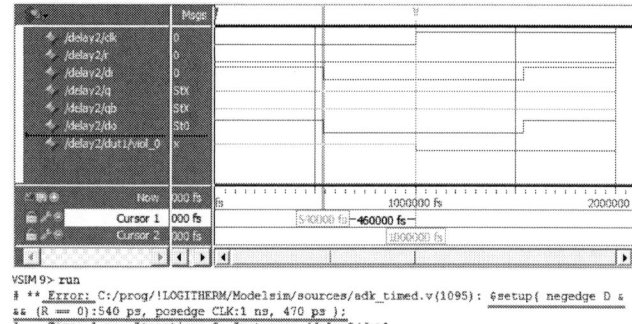

**Figure 5**: Timing violation due to elevated local chip temperature found by log-thermal simulation within the CellTherm simulation package.

### 2.2. Allowing multiple levels of abstraction

Standard cell description in certain cases might be too detailed, sometimes it might be even too rough. Therefore a multi-level approach would be required. In this case, where temperature gradients are flat, coarse grain thermal simulation seems to be appropriate and description of the certain functional blocks of the digital IC being simulated may be sufficient at a higher level of abstraction. Therefore we should allow multi-abstraction level description of the circuit. Our second approach for temperature-aware logic simulation of digital ICs supports this need. For this we developed a multi-level simulation approach where any kind of logic simulation engine and any kind of thermal simulation engine can be used in a logi-thermal simulation framework (Fig. 6). For example so called proxy interfaces allow matching different time resolution of the transient thermal simulation and the time resolution of the logic engine [21]. To support this multi-level abstraction option we developed our own logic simulator engine, forming part of our second logi-thermal simulation tool.

Here we have two proprietary simulator engines with a very strict and thin interface between them. The input of the logic engine is the description of logic elements and the circuit schematic, while that of the thermal engine is the description of the physical realization (layout, package + thermal environment). This can be seen in Fig. 6 along with a rough description of the interface.

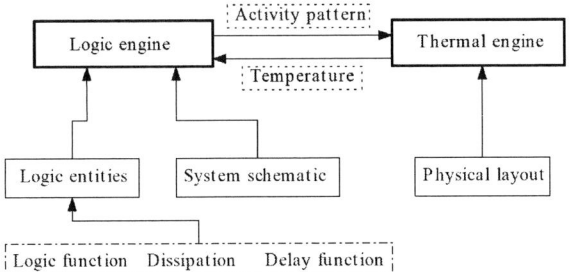

**Figure 6**: The architecture of the logi-thermal simulator which supports different levels of abstraction.

The importance of a thin interface is that it ensures a high degree of flexibility. The coupling of the two engines is realized through the interface and as long as an engine complies with it, the other side doesn't recognize a change. This ensures that a new engine can be inserted in place of the existing ones with ease. The entire idea of the logi-thermal simulation is about making a balance between precision and simulation time. An electro-thermal circuit simulation would yield an absolutely precise result, but the speed at which such a calculation can be performed is not acceptable when working with a large design. So the electric side is replaced with a logic simulator and the thermal side is simplified.

The degree of simplification at each side enhances simulation speed but decreases the precision. In the present architecture this trade-off is not made by the supplier of the software – the designers are able to choose between several thermal models, always picking the one that is in accordance with the logic description's level of detail and yields the desired resolution.

### A. The logic engine

The engines are implemented in C++. Every logic entity is a descendant of one class that represents a generic entity and defines what is common in every logic element. They all have an arbitrary number of digital ports (inputs, outputs or inouts) consisting of an arbitrary number of bits. Every element has a temperature value that is updated by the thermal simulator and two functions defining the temperature-delay and the dissipation function of the element. The first one yields the delay of a certain port at the current local temperature, the second one gives the dissipation value of the element in the current simulation step. Both functions are members of the logic entity class, so they can access every data of the element. This provides the means for the implementation of complex and highly accurate descriptions of the element's thermal behavior. For example, the entire past of the gate can be considered at each step.

The generic entity class has a third virtual member function that describes its logic operation. This function is called every time an input changes and its task is to determine the new value of the output ports for the new input combination. Whatever goes inside this function defines the operation and the level of abstraction of the logic entity.

Combinational logic can be described in very simple functions using the logical and bitwise operators of the C language. Sequential logic circuits can also be realized with ease by using data members for storing the state of the element. Complex digital blocks can be handled similarly: they are described by the algorithm they realize at a very high level. The entire toolset of C++ can be exploited and thus a system defined at software level can already be simulated. All that is needed of a logic element is a digital interface (input and output ports) with which it can communicate with the rest of the elements in the system.

### B. The thermal engine

As mentioned earlier, the logic and the thermal engines are connected through a very thin interface permitting an easy replacement of the engines on both sides. This is particularly advantageous for the thermal calculations as the simulation of a physical structure is more time-consuming than that of the logic operation. This way end-users have the freedom of making tradeoffs between speed and thermal modeling complexity by choosing different thermal simulation codes with different levels of accuracy.

In the present implementation we use a very simple *finite difference* model with adjustable resolution appropriate for most abstraction levels. The physical layout of the system is taken as input and the volume of the chip is divided into cuboids, for each of which an electric equivalent circuit is constructed. The circuit consists of resistors representing the thermal conductivity and capacitors representing the thermal capacitance. A logic element is a dissipator in such a circuit and is represented by one or several current sources. Such a finite difference model can be solved very effectively even for transient cases. We use the most recent implementation [22] of the SUNRED algorithm [23].

### C. Simulation step timing

The two engines are coupled in a way that when one simulation step is evaluated in the logic engine, the activity vector of the elements is produced and the thermal simulator is stepped to determine the effect of the latest dissipations on the thermal map of the circuit. This information is fed back to the logic engine and is converted to the updated values of delays. Then the next logic simulation step is performed and so forth.

The problem with this solution is that the time constants of the two systems might differ in orders of magnitude. A time step that is equivalent to the highest clock frequency in the system is perfectly suitable for the logic engine, but may be a needlessly high resolution for the thermal engine. This results in an excessive amount of calculation taking an impermissibly long time.

As discussed in [12], the time-scales to be covered by the thermal simulator scatter over a wide range, depending on the spatial resolution of the simulation mesh (coarse or fine grain simulation) and the size of the actual chip area that we physically cover by logi-thermal simulation. Another problem is that these time-scales are by orders of magnitude longer then the period of the clock signals in modern digital

978-1-4673-1110-6/12 $31.00 © 2012 IEEE        316

electronics. If the short term, transient behavior of a system is of interest then it is sufficient to use the thermal model of the silicon chip only, in which case the simulation time step of the thermal engine mush closer to the logic engine's time step.

On the other hand, it is very important to find out how a complex system operates in a thermal steady state (in normal operation at an average load). For such simulations, the package and the wider thermal environment of the circuit has to be modeled as well. Consequently, the time constants of the logic and the thermal simulator engines differ in several orders of magnitude. This difficulty can be overcome by placing a so called proxy element between the two engines [9]. The proxy implements the interfaces of both engines thus it appears as a thermal simulator to the logic engine and a digital simulator to the thermal engine. Every query for activity vectors and temperature values arrives to it, and it decides whether to ask for them from the engine on its other side or to simply give a value without performing any calculations. In other words: the proxy resolves the problem of the time-constant mismatch of the logic and thermal side.

With such a proxy it is possible to perform calculations more effectively: the logic engine works at the highest clock frequency and yields the activity vectors to the proxy at the end of each step. The thermal engine is stepped at a much smaller frequency and is given the summed dissipation vector. This way the dissipations are integrated over time and the resolution of the temperature change becomes lower.

## 3. Thermal modeling – including the effect of the package and its environment

As mentioned in the previous section, we are convinced the best approach in thermal simulation is to build a thermal model of the actual die. Considering the high level of abstraction of logic simulation and the corresponding approximations similar trade-offs in thermal modeling are appropriate. Our general simplification is that we neglect the physical depth of the circuit elements on the top of the active IC dice – all the logic gates are considered as 2D shapes on the surface of the 3D die – with a uniform dissipation density within the boundaries of the standard cells. A structure consisting of uniform material layers corresponding to the die, the die attach and part of the lead-frame as shown in [12] seems to be sufficient.

In the circuit design phase when logi-thermal simulation is to be performed, the ultimate chip+packaging structure is not yet available, therefore the thermal simulation model of the chip + packaging can be validated only by using thermal test vehicles (TTVs). Such a simulation model validation example was reported by Skadron et al [24] in which matching surface temperature profiles were used to asses model quality. Most recently, Vass-Várnai et al described a detailed model validation and fine tuning procedure [25] in which structure functions obtained from thermal transient measurements and thermal simulations were used for comparison. In their 2008 paper [26] Perlaky and Farkas presented thermal transient measurement results for a microprocessor and for a RAM chip assembled into a RAM module. In both cases structure functions clearly represent the long term thermal effect of the entire environment of these digital circuits.

In [12] we already showed examples for considering the thermal effects of certain structures of packaging "further away" from the chip: we described the effect of the wire bonding by means of compact thermal models. The problem with this approach is that for circuit designers it is hard to obtain thermal models of packages foreseen for their circuits being designed, even if foreseen packaging solutions are already tested by means of TTVs. The idea of test based compact modeling described by Vass-Várnai et al also in their recent conference paper [25] helps to overcome this difficulty if we reverse it: instead of modeling the internal parts of the package we create the compact thermal model of the package environment from thermal transient measurements. Such modeling was suggested already in 2003 [27] but the element values of the compact thermal model were identified from the measured structure functions in an ad hoc manner. In [25] now a systematic approach is presented. The separation point between the package and the environment (often called the 'case') can be located by the 'dual interface method' used in the JESD51-14 standard, as Vass-Várnai et al suggest in their paper [25]. Our suggestion is the following:

On one hand one describes the chip and its close environment within the thermal simulator being used in the logi-thermal simulation system by means of a detailed model. With this model transient simulations mimicking the 'dual interface' conditions are performed and the simulated unit-step responses are turned into structure functions. The 'end' of the structure functions – beyond the simulated 'case' surface is discarded. On the other hand, thermal transient measurements with the package containing an appropriately chosen TTV are performed and measurement results are converted to structure functions. Such structure functions can be approximated by compact dynamic thermal models as shown in [25] From these the 'package internal' parts (the chip and its close environment) are discarded and the part representing the farther elements of the heat-flow path are added to the thermal simulation model as the wire bond models were added to the simulation model shown in [12]. This kind of thermal modeling can be used in the actual thermal simulation engines used in our logi-thermal simulation tools.

## 4. Conclusions

A new simulation principle aimed at the study of thermal effects in digital circuits, called logi-thermal simulation was introduced. With such simulation tools timing integrity problems caused by elevated and inhomogeneous chip surface temperatures can be detected. We have shown two approaches: the CellTherm tool aimed at logi-thermal simulation of standard cell designs is based on standard components and interfaces of IC design EDA tools. In another set of tools aimed at logi-thermal simulation through our proprietary logic simulation engine digital designs described on different abstraction levels can also be studied.

Both of our logi-thermal simulators are able to take the effect of the whole package into account by coupling compact models of the packaging, which can be partially created from measured or simulated thermal transient responses of real packages.

978-1-4673-1110-6/12 $31.00 © 2012 IEEE

## Acknowledgments

This work was partially financed by the 248603 THERMINATOR integrated project of the Framework 7 Program of the EU. The support of the Hungarian Government through the TÁMOP-4.2.1/B-09/1/KMR-2010-0002 project at the Budapest University of Technology and Economics is also acknowledged.

## References

[1] P. Landman, "High-level power estimation," International Symposium on Low Power Electronics and Design, pp. 29–35, 1996.

[2] M. Nemani and F. N. Najm, "Towards a high-level power estimation capability," IEEE Transactions on Computer-Aided Design of Integrated Circuits and Systems, vol. 15, no. 6, pp. 588–598, 1996.

[3] D. Crisu et al, "High-level energy estimation for ARM-based SOCs," Lecture Notes in Computer Science, vol. 3133/2004, no. 19-42, pp. 168–177, 2004.

[4] D. Chen et al, "High-level power estimation and low-power design space exploration for FPGAs," Design Automation Conference, pp. 529–534, 2007.

[5] J. Laurent et al, "High level power estimation based on a functional analysis for embedded DSP software," 2001.

[6] C. Yao et al, "Power and thermal constrained test scheduling under deep submicron technologies," IEEE Transactions on Computer-Aided Design of Integrated Circuits and Systems, vol. 30, no. 2, pp. 317–322, 2011.

[7] A. Bartolini et al, "A virtual platform environment for exploring power, thermal and reliability management control strategies in high-performance multicores," in GLSVLSI'10, Providence, Rhode Island, USA, May 2010.

[8] A. Bartolini et al, "A distributed and self-calibrating model-predictive controller for energy and thermal management of high-performance multicores," in DATE'11, 2011.

[9] G. Nagy et al, "Consideration of thermal effects in logic simulation", In: Proceedings of the 14th International Workshop on THERMal INvestigation of ICs and Systems (THERMINIC'08), 24-26 September 2008, Rome, Italy, pp. 229-234

[10] V. Székely et al, "Electro-thermal and logi-thermal simulation of VLSI designs", IEEE Transactions on Very Large Scale Integration (VLSI) Ssystems (ISSN: 1063-8210) 5:(3) pp. 258-269. (1997)

[11] M Rencz et al, "Electro-thermal simulation for the prediction of chip operation within the package", In: Proceedings of the 19th IEEE Semiconductor Thermal Measurement and Management Symposium (SEMI-THERM'03). San Jose, USA, 11-13 March 2003, pp. 168-175.

[12] A Poppe et al, "Electro-thermal and logi-thermal simulators aimed at the temperature-aware design of complex integrated circuits", In: Proceedings of the 24th IEEE Semiconductor Thermal Measurement and Management Symposium (SEMI-THERM'08). San Jose, USA, March 16-20 2008, pp. 69-77.

[13] A Timár et al, "Electro-thermal co-simulation of ICs with runtime back-annotation capability", In: Proceedings of the 16th International Workshop on THERMal INvestigation of ICs and Systems (THERMINIC'10). Barcelona, Spain, 6-8 October 2010, pp. 183-188.

[14] K. Torki and F. Ciontu, "Ic thermal map from digital and thermal simulations," in Proceedings of the 8th Therminic Workshop, Madrid, October 2002, pp. 303–308.

[15] S. Wünsche et al, "Electro-thermal circuit simulation using simulator coupling," IEEE Transactions on Very Large Scale Integration (VLSI) Systems, vol. 5, no. 3, September 1997.

[16] W. Huang et al, "Accurate pre-RTL temperature-aware design using a parameterized, geometric thermal model," IEEE Transactions on Computers, vol. 57, no. 8, August 2008.

[17] K. Sankaranarayanan et al, "A case for thermal-aware floorplanning at the microarchitectural level," Journal of Instruction-Level Parallelism, vol. 8, no. 1-16, 2005.

[18] W. Huang et al, "Hotspot: A compact thermal modeling methodology for early-stage vlsi design," IEEE Transactions on Very Large Scale Integration (VLSI) Systems, vol. 14, no. 5, May 2005

[19] W. Huang et al, "Compact thermal modeling for temperature-aware design," 41st Design Automation Conference (DAC), San Diego, CA, June 2004.

[20] M. Rencz et al, "Inclusion of RC compact models of packages into board level thermal simulation tools", In: Proceedings of the 18th IEEE Semiconductor Thermal Measurement and Management Symposium (SEMI-THERM'02). San Jose, CA, 11-14 March 2002, pp. 71-76.

[21] G. Nagy and A. Poppe, "A Novel Simulation Environment Enabling Multilevel Power Estimation of Digital Systems", In: Proceedings of the 17th International Workshop on THERMal INvestigation of ICs and Systems (THERMINIC'11), 27-29 September 2011, Paris, France

[22] Székely V: SUNRED a new thermal simulator and typical applications. 3rd International Workshop on THERMal INvestigation of ICs and Microstructures (THERMINIC'97), 21-23 September 1997, Cannes, France, pp. 229-234, pp. 84-90.

[23] L. Pohl et al, "Fast field solver for the simulation of large-area OLEDs", Microelectronics Journal, Volume 41, Issue 9, September 2010, pp 566-573

[24] K. Skadron et al, "A computer-architecture approach to thermal management in computer systems: Opportunities and challenges", In: Proceedings of the 5th International Conference on Thermal and Mechanical Simulation and Experiments in Microelectronics and Microsystems (EuroSimE'04), 10-12 May 2004, Bruxelles, Belgium, pp. 415-422.

[25] A. Vass-Várnai et al, "Measurement Based Compact Thermal Model Creation - Accurate Approach to Neglect Inaccurate TIM Conductivity Data", In: Proceedings of the 13th Electronics Packaging Technology Conference (EPCT 2011). 7-9 December 2011, Singapore, pp. 67-72. Paper A4.2.

[26] G. Perlaky and G. Farkas, "Thermal Transient Characterisation of Complex Circuits", In: Proceedings of the 14th International Workshop on THERMal INvestigation of ICs and Systems (THERMINIC'08), 24-26 September 2008, Rome, Italy, pp. 106-111

[27] G. Farkas et al "Dynamic Compact Models of Cooling Mounts for Fast Board Level Design", In: Proceedings of the 19th IEEE Semiconductor Thermal Measurement and Management Symposium (SEMI-THERM'03). 11-13 March 2003, San Jose, USA, pp. 255-262.

# An Innovative Passive Cooling Method for High Performance Light-emitting Diodes

Angie Fan[1], Richard Bonner[1], Stephen Sharratt[2] and Y. Sungtaek Ju[2]
[1]Advanced Cooling Technologies, Inc
1046 New Holland Ave, Lancaster, PA
angie.fan@1-act.com
[2]University of California
Los Angeles, CA

## Abstract

Thermal management challenges are becoming a major roadblock to the wide use of high-power LED lighting systems. Incremental improvements in conventional bulk metal heat sinks and thermal interface materials are projected to be insufficient to meet these challenges. Active cooling methods, such as forced air and pumped liquid cooling, may provide better performance but at the expense of higher cost and energy consumption. Passive phase change (liquid to vapor) cooling devices, such as heat pipes and thermosyphons, are well established in the electronics industry as a very effective and reliable way of removing excess waste heat at low thermal resistance. Successful application of heat pipes and thermosyphons in solid-state lighting (SSL) products will require adapting the technologies to the form-factor, material and cost requirements unique to SSL products. This paper describes a recent development effort that integrates a planar thermosyphon into a printed circuit board (PCB) for LED devices. The planar thermosyphon/PCB uses a dielectric fluid as the heat pipe working fluid, achieving significantly improved heat spreading performances over conventional PCBs. Analytical modeling showed a more than 50% thermal resistance reduction from typical metal core PCBs. A low temperature electroplating technique was also investigated to fabricate wick structures onto PCB surfaces to enhance the boiling heat transfer performance of the dielectric fluids. Test results showed that a boiling heat transfer coefficient of 20,000W/m²-K can be achieved with the 3M Novec fluid. In this paper, the preliminary study on heat transfer enhancement by using the PCB planar thermosyphon in single LED assembly was reported. Future development efforts will verify the design in practical applications, address manufacturing issues and improve the cost efficiency.

## Keywords

Two-phase passive cooling, PCB-based dielectric planar themosyphon, high power LEDs, boiling heat transfer enhancement, low temperature sintering

## 1. Introduction

According to the U.S. Department of Energy (DOE), solid-state lighting (SSL) technology has the potential to cut U.S. lighting energy usage by one-quarter and contribute significantly to our nation's climate change solutions. Compared with conventional white light sources such as incandescent, fluorescent, and metal halide lamps, light-emitting diodes (LEDs) provide significant benefits including compact size, long life, ease of maintenance, resistance to breakage and vibration, good performance in cold temperatures, reduced infrared or ultraviolet emissions, and instant-on performance. Table 1 shows the potential advantages as well as the challenges facing LED technology. Although the electrical power to visible light conversion efficiency of 20-30% represents significant improvement over the incandescent light sources, the 70-80% non-radiant heat dissipation poses a significant challenge to the thermal management of the device.

Cost competitiveness and quality have been identified by DOE as the two additional roadblocks in the commercialization path of the LED technology. Currently, LEDs cost 10 times more than incandescent lamps and 5 times more than compact fluorescent lamps (CFL). DOE's goal is to reduce the LED cost comparable to CFL's by 2015. Since a large portion of the energy in LED devices becomes waste heat and the LED junction temperature affects device's long-term reliability, developing thermally and cost effective thermal management methods plays a key role in improving LED's quality and cost competitiveness.

**Table 1:** Power Conversion for White Light Sources

|  | Incandescent | Fluorescent | Metal Halide | LED |
|---|---|---|---|---|
| **Visible Light** | 8% | 21% | 27% | 20-30% |
| **IR** | 73% | 37% | 17% | 0% |
| **UV** | 0% | 0% | 19% | 0% |
| **Total Radiant Energy** | 81% | 58% | 63% | 20-30% |
| **Non-Radiant Heat** | 19% | 42% | 37% | 70-80% |
| **Total Energy** | 100% | 100% | 100% | 100% |

High brightness LEDs (HBLEDs) are finding increasing usage in applications like LED lamps, display backlighting, and camera flash for cell phones. A typical high power LED chip has 1mm² surface area with a total power consumption of 1W. According to the 70-80% power to heat conversion rate in LEDs, the heat flux can be as high as 80 W/cm². By 2012, the heat flux will reach about 340 W/cm², which is 6-7 times higher than that of conventional CPU chips [1]. The high heat fluxes at the junction level, coupled with the dense packaging of many components into a small package, results in two thermal management challenges: temperature uniformity across multiple LED junctions and in-plane heat spreading at the heat sink and PCB package levels.

978-1-4673-1110-6/12 $31.00 © 2012 IEEE

To date, many heat dissipation solutions have been investigated for the thermal management of high-power LEDs, from the chip package level to the printed circuit board (PCB) level to the system level. The package-level thermal management research [1- 5, 11], which involves thermal material research, package design optimization such as 3D packaging design and LED array optimization, and theoretical simulations, is important to determine the packaging thermal resistance of LEDs as well as reduce the footprint. The board-level thermal management research [6-10] is mainly focused on solder material, bonding method improvement, and printed circuit board design optimization. On the system level [4, 12-20], fin-heat sinks with external active cooling is still the mainstream method in industry due to its high reliability and lowest cost. Aside from conventional fans, piezoelectric fans [3, 23] have gained increasing interest from industry. Two-phase passive cooling methods like heat pipes and vapor chambers [15] are becoming good options for emerging HBLEDs. Due to the very high flux heat dissipation requirements, active liquid cooling is widely studied [13-14, 18-19]. Other than active liquid cooling, some novel and advanced methods have also emerged, such as micro-channel coolers [24], electrohydrodynamic approaches [22], and thermoelectric cooling [17]. However, these strategies often involve complex design processes, reliability issues, cost issues, high power consumption, which are the main obstacles for their commercialization and utilization.

In this paper, an advanced high-power LED cooling technology targeted at reducing the thermal resistance on the board level is presented. This technology integrates a passive two-phase thermosyphon into a printed circuit board, which effectively converts the PCB into a high performance heat spreader. Although vapor chamber printed circuit boards and vapor chamber heat sinks have been investigated by other groups [9, 15], the novelty of the current concept is the direct bond between the PCB/heat spreader and LED devices that eliminates extra thermal interfaces as well as the need for electrical insulation between heat sink and PCB. Thus the new design can significantly enhance cooling that allows the LED to run at higher fluxes without degradation in performance, and improve heat spreading that is particularly effective in cooling arrays of high density chips.

## 2. Passive Heat Spreader (PHS) Printed Circuit Board (PCB)

The thermosyphon is a proven cooling technology with exceptional heat transfer performance. A traditional thermosyphon is a tubular metal structure that consists of an evaporator and a condenser section. A planar thermosyphon can be viewed as a highly efficient heat spreader. When subjected to heating by an electronic device attached to the evaporator, the working fluid inside the thermosyphon vaporizes and thereby limits temperature rise. The vapor condenses back to liquid at the condenser, which is cooled by an external heat sink. In the gravity field, the condensed liquid falls back to the evaporator in the form of a liquid film on the sides of the thermosyphon.

Figure 1(b) illustrates the design concept where the copper lid (at top), PCB and the copper thermal pad (at bottom) form the envelope of the thermosyphon. Compared with the typical surface mount LED in Figure 1(a), the thermosyphon design replaces the low thermally conductive dielectric layer by an enclosed vapor space filled with dielectric fluid. The copper thermal pad of the chip package can be directly soldered to the thermosyphon for minimal interface thermal resistance. It should be pointed out that one PCB thermosyphon heat spreader can have multiple, discrete copper thermal pads to accommodate arrays of LED devices. The dielectric working fluid for the thermosyphon provides the necessary electrical isolation to prevent short circuiting. This feature eliminates the need for a ceramic substrate separating the thermal pad from the electrical circuitry and eliminates the associated thermal resistance. In summary, the technology presented in this paper has the potential of improving thermal performance and simplifying the LED packaging, which will in turn result in cost savings.

**(a)**

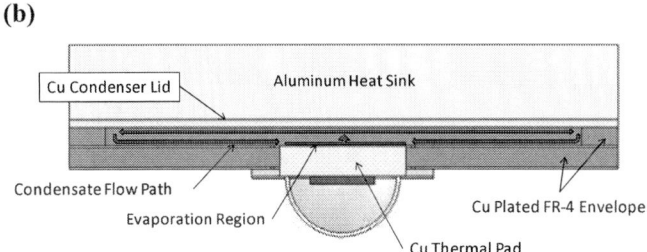

**(b)**

**Figure 1:** (a) Cross-sectional depiction of a typical surface mount LED mounted to an FR-4 PCB with open thermal vias (anode and cathode not shown). (b) Cross-sectional depiction of the PHS PCB for a single LED package. The LED thermal pad is integrated directly into the FR-4 envelope such that the thermal pad's top surface acts as the evaporator. A copper plated FR-4 frame is used to bond the FR-4 PCB and the copper lid (the condenser). Vapor and condensate flow paths are also shown within the vapor space formed by the FR-4 frame.

Thermosyphons are particularly suited for LED cooling since many artificial lighting applications are in the form of downlighting where the gravity aids in the condensate return inside the thermosyphon. For other orientations, a wick may be used to provide the capillary action needed to drive the liquid against the gravity. The use of a thin layer of wick over the evaporator area will also enhance the boiling heat transfer by providing extra nucleation sites and enhancing liquid supply to local high heat flux areas. Because the design involves a copper clad PCB board, fabricating and bonding the wick to the thermosyphon inner surfaces needs to occur at relatively low temperatures to avoid any damage to the PCB. A novel low temperature electroplating scheme was investigated, and the result is presented in this paper.

## 3. Results and Discussions

The prototype design was developed based on one of the high brightness LEDs - Cree XLamp® LEDs [21]. In this LED package, the typical dissipated heat fluxes range between 18 and 64W/cm² over an area of 4.25mm². MCPCBs and FR-4 PCBs with thermal vias as shown in Figure 2 are suggested by Cree to be used with this high power chip package. The PHS PCB was targeted to dramatically reduce the thermal resistances in the PCB and the associated TIM layers. Figure 3 shows the cross-sectional view of the prototype design. Table 2 ~ Table 4 listed the configurations for all three PCBs. The thermal performances of the prototype design as well as the other two PCBs were analyzed using commercial CFD software.

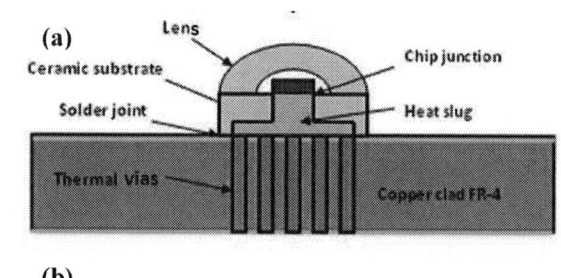

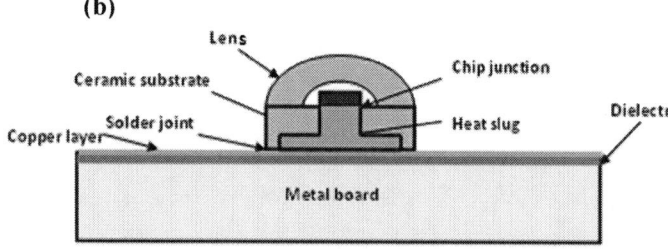

**Figure 2:** Diagrams of LED chip package mounted on (a) FR-4 PCB with thermal vias, and (b) metal core PCB

**Table 2:** FR-4 PCB with thermal vias configuration (PCB surface area 270mm²)

| Component | Thickness (µm) | Thermal conductivity (W/m-K) |
|---|---|---|
| Top layer copper | 70 | 398 |
| FR-4 | 1588 | 0.2 |
| Filled vias (SnAgCu) | 1588 | 59 |
| Bottom layer copper | 70 | 398 |
| Total | 1728 | -- |

**Table 3:** Metal core PCB configuration (PCB surface area 270mm²)

| Component | Thickness (µm) | Thermal conductivity (W/m-K) |
|---|---|---|
| Top Layer Copper | 70 | 398 |
| PCB dielectric | 100 | 2.2 |
| Al plate | 1588 | 150 |
| Total | 1758 | -- |

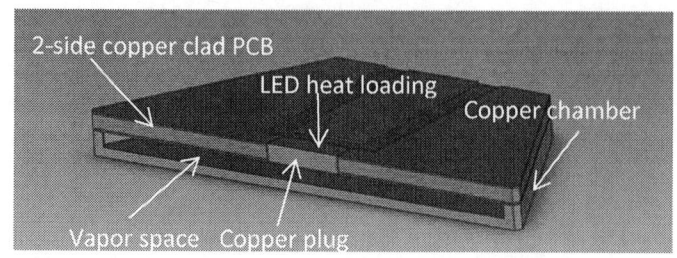

**Figure 3:** Cross-sectional view of the prototype design of the PCB planar thermosyphon for single LED. A copper plug is implemented in the middle of a two-side copper clad FR-4 PCB via a through hole to simulate the LED thermal pad and the FR-4 PCB is bonded to a copper chamber to form the vapor space which contains the dielectric working fluid.

**Table 4:** PHS PCB Configurations (PCB surface area 270mm²)

| Component | Thickness (µm) | Thermal conductivity (W/m-K) | Heat transfer coefficient (W/m²-K) |
|---|---|---|---|
| Copper plug (area 3.3mm x 1.65mm) | 728 | 398 | -- |
| Top layer copper | 70 | 398 | -- |
| FR-4 dielectric | 588 | 0.2 | -- |
| Bottom layer copper | 70 | 398 | -- |
| Evaporator surface | -- | -- | 20,000 |
| Vapor space | 600 | 100,000 | -- |
| Condenser surface | -- | -- | 10,000 |
| Copper chamber | 1000 | 398 | -- |
| Total | 1728 | -- | -- |

The modeling results are displayed in Figure 4. The total heat load $Q_{in}$ is 1W, and finned heat sink with forced air cooling is assumed on the back side. The total temperature gradient $\Delta T$ in the FR-4 PCB, the MCPCB, and the PHS PCB were calculated to be 26.6°C, 7.2°C and 3.7°C, respectively. Thermal resistance $R$ is defined as

$$R = \frac{\Delta T}{Q_{in}}$$

The thermal resistances of the FR-4 PCB, the MCPCB, and the PHS PCB were 26.6°C/W, 7.2°C/W, and 3.7°C/W. The thermal resistance of the PHS PCB is less than one-sixth of that of the FR-4 PCB and half of that of the MCPCB. The reduction in thermal resistance will be even more significant if the elimination of the solder joint between the LED package and the PCB is included in the calculation.

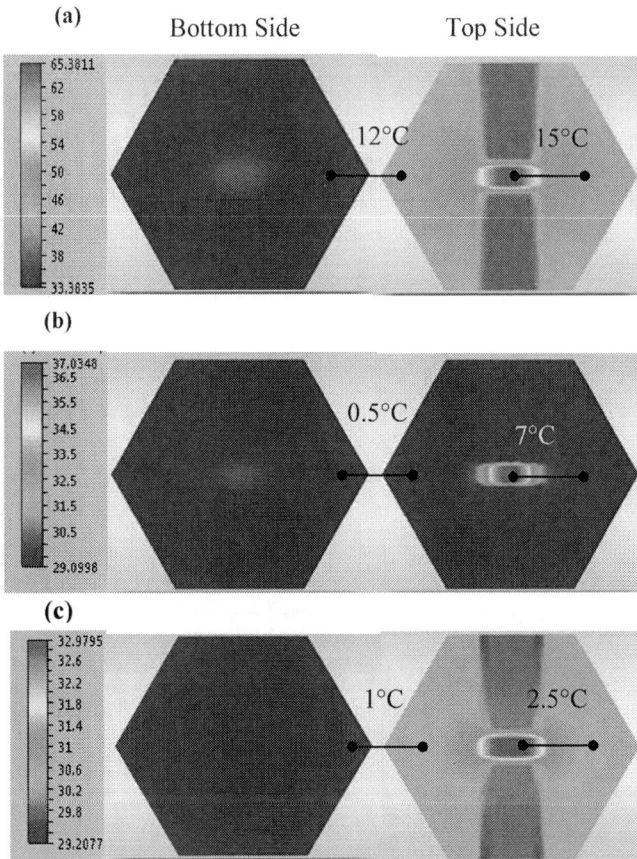

**(a)**

Bottom Side | Top Side

12°C | 15°C

**(b)**

0.5°C | 7°C

**(c)**

1°C | 2.5°C

**Figure 4:** CFD results of the temperature distribution over (a) FR-4 PCB, (b) MCPCB and (c) PHS PCB. Under the heat load of 1W, the temperature difference between the heat loading surface (top) and the bottom of the PCBs were 26.6°C, 7.2°C and 3.7°C, respectively. FR-4 PCB has poor thermal resistance in both heat spreading directions. MCPCB has good axial heat conduction but poor in-plane heat spreading. PHS PCB has low thermal resistance in both directions.

Materials selection for the prototypes was based on considerations of performance, compatibility, reliability, and cost. FR-4 PCB was selected as part of the thermosyphon envelope for its low cost and availability in LED systems. Oxygen free copper, a proven heat pipe envelope material, forms the other part of the thermosyphon envelope. Novec 7200 and 72DE were selected as the working fluids for their latent heat, vapor pressures, surface tension properties and their compatibility with other materials in the system. A permeation test was performed on two samples to examine the leakage rate and materials compatibility. In the first sample, a piece of copper clad FR-4 PCB was soldered onto a copper chamber to form a vacuum tight envelop. The second sample had the same setting as the first one except the copper clad layer was partially removed so that the working fluid met the FR-4 directly. Both samples were half filled with the Novec 7200. In 4 months, continuous weight loss was observed on the 2nd sample, but none on the 1st sample. The FR-4 is permeable, but hermetic sealing can be obtained by applying copper coating on the FR-4.

The CFD analysis shows that the evaporation heat transfer in the evaporator of the thermosyphon plays a critical role in the overall thermal performance. As discussed earlier, the use of a thin layer of wick over the evaporator area will enhance the evaporation heat transfer. In particular, sintered powder wicks have been demonstrated in many previous studies to provide high boiling/evaporation heat transfer coefficients at high heat flux conditions. Two approaches to wick fabrication were investigated: intense pulsed light (IPL) sintering and microfabrication electroplating. The IPL sintering technique yielded wicks with inconsistent quality. The electroplating technique, on the other hand, yielded successful results. Figure 5 shows the images of the top and angled views of a sample wick structure with porous posts under a scanning electron microscope (SEM).

**Figure 5:** SEM images of electroplated Cu posts on a PCB board. Left: top view. Right: angle view.

Two wick samples of different porosities were tested for their boiling heat transfer performance: a low-solid-fraction (.227) sample (LSF) and a high-solid-fraction (.463) sample (HSF). In both samples, the copper powder posts, 50 μm in diameter and 100 μm in height, were electrodeposited on a 500 μm-thick Si wafer. The pitch distance in LSF and HSF were 100 μm and 65 μm, respectively. The overall wick area was 2.2cm by 2.2cm. A 5 mm x 5 mm thin-film heater was used as the heat source. The samples were held vertically with the lower end dipped in the working fluid to allow the fluid to wick the evaporator region by capillary force. The experiments were performed in a vacuum-tight chamber with the saturation temperature at approximately 33°C for 72DE and 65°C for 7200.

The test results, as shown in Figure 6, indicate that the HSF wick exhibits higher heat transfer coefficients than the

LSF wick. For 3M dielectric fluid 72DE, although partial dry-out occurred at heat fluxes around 10W/cm$^2$, heat transfer coefficients as high as 10,000W/m$^2$ K was achieved with the HSF sample. In the tests with two different 3M dielectric fluids, heat transfer coefficients as high as 20,000W/m$^2$ K was obtained for the HSF sample in 72DE below 6W/cm$^2$.

**(a)**

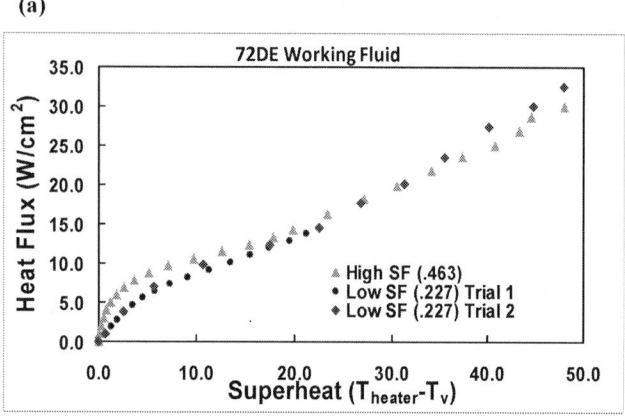

**(b)**

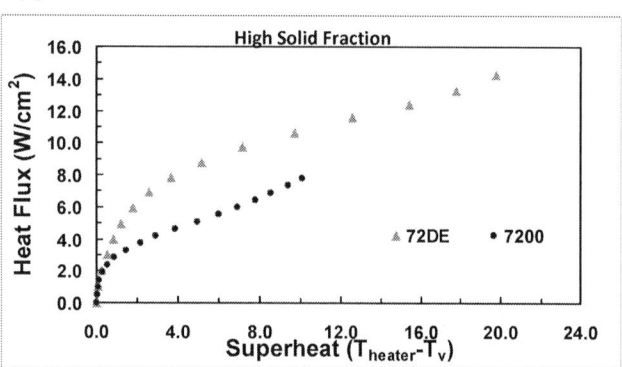

**Figure 6:** Experimental Results for Boiling Heat Transfer Tests. (a) Low solid fraction sample and high solid fraction sample were both tested with 3M Novec dielectric fluid 72DE. Partial dry-out started at heat fluxes around 10W/cm$^2$. (b) High sold fraction sample was tested with both 3M Novec dielectric fluids 72DE and 7200. Heat transfer coefficient as high as 20,000 W/m$^2$ K achieved for 72DE at heat fluxes up to 6W/cm$^2$.

## 4. Conclusions

The feasibility of a novel PCB based planar thermosyphon concept for high power LED cooling was demonstrated through numerical simulation and experimental study. This PHS PCB was made of a FR-4 PCB lid and a copper chamber, and used a dielectric working fluid. Although an initial permeation test showed the dielectric fluid permeated through the PCB wall, standard copper cladding solved the PCB permeation issue. Wick structures were applied to the evaporator of the thermosyphon to achieve the desired boiling heat transfer performance. Numerical simulations of a representative CREE LED package incorporating various thermal management methods, including FR-4 PCB with thermal vias, MCPCB and the new PHS PCB, were performed to identify the performance enhancements by the new concept. It was shown that the heat transfer performance

of the PHS PCB improved by 50% over the MCPCB and 86% over the FR-4 PCB. An experimental investigation on using various wicks for boiling heat transfer enhancement was conducted, which revealed that a boiling heat transfer coefficient of 20,000W/m$^2$ K can be achieved with 3M 72DE dielectric fluid and an advanced wick structure at heat fluxes up to 6W/cm$^2$. Various low-temperature wick fabrication techniques were also investigated  Further enhancing the phase change heat transfer will be studied by testing with different working fluids e.g. DI-water, and improving wick structures. Also, additional wick structures will be added to make the PHS independent of gravity.

**Acknowledgments**

This material is based upon work supported by the Department of Energy under Award Number DE-SC0004722.

**References**

1. Zhang, J., Niu, P., Gao, D., and Sun, L., "Research Progress on Packaging Thermal Management Techniques of High Power LED", Advanced Materials Research Vols. 347-353 (2012) pp 3989-3994

2. Horng, R. H., Hong, J. S., Tsai, L. T., Wuu, D. S., Chen, C. M., and Chen, C. J., "Optimized Thermal Management From a Chip to a Heat Sink for High-Power GaN-Based Light-Emitting Diodes", IEEE Trans on Electron Devices, Vol. 57, No. 9, Sep., 2010

3. Lau, J., Lee, R., Yuen, M., and Chan, P., "3D LED and IC wafer level packaging", Microelectronics International 27/2 (2010) 98–105

4. Qin, Y. X, and Hui S. Y. Ron, "Comparative Study on the Structural Designs of LED Devices and Systems Based on the General Photo-Electro-Thermal Theory", IEEE Trans on Power Electronics, Vol. 25, No. 2, Feb., 2010

5. Hui Yu, Jintang Shang, Chao Xu, Xinhu Luo, Jingdong Liu, Li Zhang, Chiming Lai "Chip-on-board (COB) Wafer Level Packaging of LEDs Using Silicon Substrates and Chemical Foaming Process(CFP)-made Glass-bubble Caps", 2011 International Conference on Electronic Packaging Technology & High Density Packaging

6. Wang, N., Hsu, A., Lim, A., Tan, J., Lin, C., Ru, H., Jiang, T., and Liao, D., "High Brightness LED Assembly using DPC substrate and SuperMCPCB", IEEE 4th International Microsystem Packaging Assembly and Circuits Technology Conference IMPACT, 2009

7. Chen, V., Oliver, G., Roberts, K., and Amey, D., "Benchmark Study of Metal Core Thermal Laminates", IEEE 5th International Microsystems Packaging Assembly and Circuits Technology Conference (IMPACT), 2010

8. Moon-Ho Lee; Tae Jin Lee; Hye Jin Lee; Young-Joo Kim, "Design and Fabrication of Metal PCB based on the Patterned Anodizing for Improving Thermal Dissipation of LED Lighting", IEEE 5th International Microsystems Packaging Assembly and Circuits Technology Conference (IMPACT), 2010

9. Hou, Fengze, Yang, Daoguo, Zhang, G.Q., Liu, Dongjing, "Research on heat dissipation of high heat flux multi-chip GaN-based white LED lamp", IEEE 12th

International Electronic Packaging Technology and High Density Packaging Conference (ICEPT-HDP), 2011

10. Bo-Hung Liou, Chih-Ming Chen, Ray-Hua Horng, Yi-Chen Chiang, Dong-Sing Wuu "Improvement of thermal management of high-power GaN-based light-emitting diodes" Article in press, Microelectronics Reliability, 2011

11. Dong Lu, Chenmin Liu, Xianxin Lang, Bo Wang, Zhiying Li, W. M. Peter Lee, S. W. Ricky Lee, "Enhancement of Thermal Conductivity of Die Attach Adhesives (DAAs) using Nanomaterials for High Brightness Light-Emitting Diode (HBLED)", IEEE 61st Electronic Components and Technology Conference (ECTC), 2011

12. Christensen, A., Graham, S., "Thermal effects in packaging high power light emitting diode arrays", Applied Thermal Engineering 29 (2009) 364 – 371

13. Yan Lai, Nicolas Cordero, Frank Barthel, Frank Tebbe, Jorg Kuhn, Robert Apfelbeck, Dagmar Wurtenberger, "Liquid cooling of bright LEDs for automotive applications", Applied Thermal Engineering 29 (2009) 1239–1244

14. Yueguang Deng, Jing Liu, "A liquid metal cooling system for the thermal management of high power LEDs", International Communications in Heat and Mass Transfer 37 (2010) 788–791

15. Luo, Xiaobing, Hu, Run, Guo, Tinghui, Zhu, Xiaolei, Chen, Wen, Mao, Zhangming, Liu, Sheng, "Low thermal resistance LED light source with vapor chamber coupled fin heat sink", Proceedings of 60th IEEE Electronic Components and Technology Conference (ECTC), 2010

16. K.C. Yung, H. Liem, H.S. Choy, W.K. Lun , "Thermal performance of high brightness LED array package on PCB", International Communications in Heat and Mass Transfer 37 (2010) 1266–1272

17. Junhui Li, Bangke Ma, Ruishan Wang, Lei Han, "Study on a cooling system based on thermoelectric cooler for thermal management of high-power LEDs", Microelectronics Reliability 51 (2011) 2210–2215

18. Maw-Tyan Sheen, Ming-Der Jean and Yu-Tsun Lai, "Application of Micro-Tube Water-Cooling Device for the Improvement of Heat Management in Mixed White Light Emitting Diode Modules", Advanced Materials Research Vols. 308-310 (2011) pp 2422-2427

19. Z.M. Wan, J.Liu, K.L.Su, X.H.Hu, S.S.M, "Flow and heat transfer in porous micro heat sink for thermal management of high power LEDs", Microelectronics Journal 42 (2011) 632–637

20. Kai Zhang, David G. W. Xia, Xiaohua Zhang, Haibo Fan, Zhaoli Gao, Matthew M. F. Yuen "Thermal Performance of LED Packages for Solid State Lighting with Novel Cooling Solutions", 2011 12th. Int. Conf. on Thermal, Mechanical and Multiphysics Simulation and Experiments in Microelectronics and Microsystems, EuroSimE 2011

21. Cree LED Lighting. Available from: <http://www.cree.com>.

22. S.W. Chau, C.H. Lin, C.H. Yeh, C. Yang, "Study on the cooling enhancement of LED heat sources via an electrohydrodynamic approach", 33rd Annual Conference of the IEEE Industrial Electronics Society, Taipei, 2007, pp. 2934–2937.

23. T. Acikalin, S.V. Garimella, J. Petroski, A. Raman, Optimal design of miniature piezoelectric fans for cooling light emitting diodes, 9th Intersociety Conference on Thermal and Thermomechanical Phenomena in Electronic Systems, Las Vegas, 2004, pp. 663–671.

24. L.L. Yuan, S. Liu, M.X. Chen, X.B. Luo, "Thermal analysis of high power LED array packaging with microchannel cooler", 7th International Conference on Electronics Packaging Technology, Shanghai, 2006, pp. 574–577.

# A Step Forward in Multi-domain Modeling of Power LEDs

## András Poppe[1,2]

[1]Budapest University of Technology and Economics, Department of Electron Devices
Magyar tudósok körútja 2, Bldg. Q, wing B, Budapest, H-1117 Hungary

[2]Mentor Graphics Mechanical Analysis MicReD Division
Gábor Dénes utca 2. Infopark Bldg. D, Budapest, H-1117 Hungary

[1] poppe@eet.bme.hu, [2] andras_poppe@mentor.com

## Abstract

Besides their electrical properties the optical parameters of LEDs also depend on junction temperature. For this reason thermal characterization and thermal management plays important role in case of power LEDs, necessitating tools both for physical measurements and simulation. This paper deals with the combined electrical, thermal and optical characterization and multi-domain modeling of power LEDs. The major target is to provide with a model architecture and set of model equations the parameters of which can be all measured using a combined thermal and radiometric and photometric test setup completed with a spectrometer. Current and temperature dependent modeling of the light output of white LEDs is also addressed.

## Keywords

Power LED, thermal characterization, LED multi-domain modeling

## 1. Introduction

Solid state light emitting device technology has made tremendous progress in the last decade. After the development of the and white Light Emitting Diodes (LEDs) [1], [2] the road led straight to the solid state lighting applications, thanks to drastic drop in "price of light". Therefore, today one may observe a tremendous change in the world-wide lighting industry. Solid-state lighting, especially high power / high brightness LEDs appeared on the market and with increasing energy conversion efficiency they are competing with traditional light sources. The development trend of LEDs (also known as Haitz's law [3]) resembles Moore's law of the conventional semiconductor industry. In case of LEDs, instead of the integration density, the light output (measured in lumens) per package is a metric of the development.

As shown in Fig.1, this metric follows an exponential trend: the emitted luminous flux per package doubles ever year. Another trend is the continuous price decrease measured in $/lumen (cost of generating 1 lumen of luminous flux): in 10 years this figure dropped by an order of magnitude. This allowed application of LEDs in the automotive industry (such as tail lights, head lights), in consumer electronics (e.g. back light units) as well as in real lighting application – both indoor and outdoor, such as street lighting.

As shown in Fig. 2, LEDs' operation involves 3 domains, which have strong mutual dependence: electrical thermal and optical characteristics of LEDs influence one another. The most often discussed aspect is how temperature influences the light output of an LED: energy conversion efficiency drops with increasing temperature. Lighting designers refer to this by estimating the "hot lumens" of LEDs at actual operating temperatures [5], using information provided on vendors' LED data sheets.

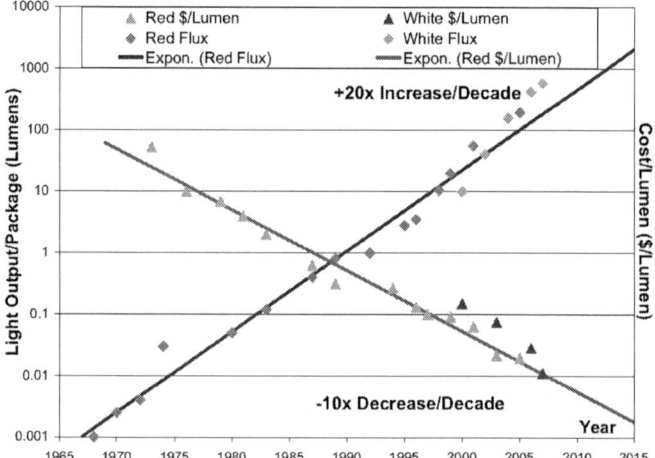

**Figure 1**: Development trend of LEDs: exponential growth of luminous flux/package and exponential decay of related costs (source: US DoE SSL-roadmap, March 2008 [4]).

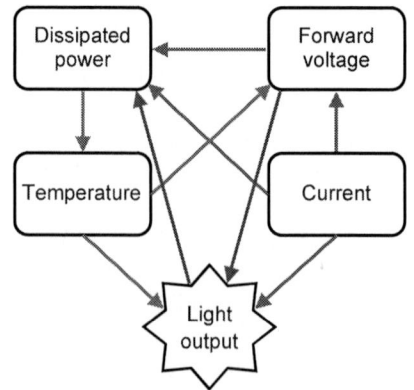

**Figure 2**: Mutual dependence of the parameters of the three operational domains of LEDs.

The aim of this paper is to draft an alternative to such spreadsheet based methods by suggesting a multi-domain LED model which includes model equations regarding the I-V characteristic as well as models for emission spectra and their temperature dependence together with a compact thermal network model of the LEDs' junction to ambient heat-flow path.

The motivation for this work comes from a Hungarian R&D consortium aiming the development of a LED based streetlighting luminaire family. The consortium had different goals in mind – one was the adaptive control of the luminaries which, for example adapts LEDs' operating point to environmental conditions. The major point is energy saving: in winter at lower temperatures energy conversion efficiency increases, thus less electrical energy is needed to achieve the illumination level prescribed for a given road category. How such a current reduction should be made can be well controlled based on a multi-domain LED model.

## 2. Prior work in LED characterization and LED modeling

Fig. 3 shows our the concept of physics based a multi-domain LED model [6], [7]. Up to now we concentrated our efforts on development of methodologies regarding compact thermal modeling of LED packages [6]. With the publication of the so called dual interface method in the JEDEC JESD51-14 standard [8] software support of generating compact dynamic thermal models of power semiconductor packages became commercially available, see Fig. 4. The 'bottom' node of the LED package is defined as the 'case' in JESD51-14 and the corresponding thermal resistance value can be determined by the new measurement standard from thermal transient measurement results of power LED packages. In their recent paper [9], Vass-Várnai et al describe such a compact model identification procedure together with validation against a detailed CFD model.

During the measurements the energy of the emitted light also has to be measured and considered in the heating power calculations [6], [7], by using a combined thermal and radiometric measurement setup. Such a system can be extended with a spectrometer, thus emission spectra and all other light output characteristics can be identified as function of temperature and operating current.

a)

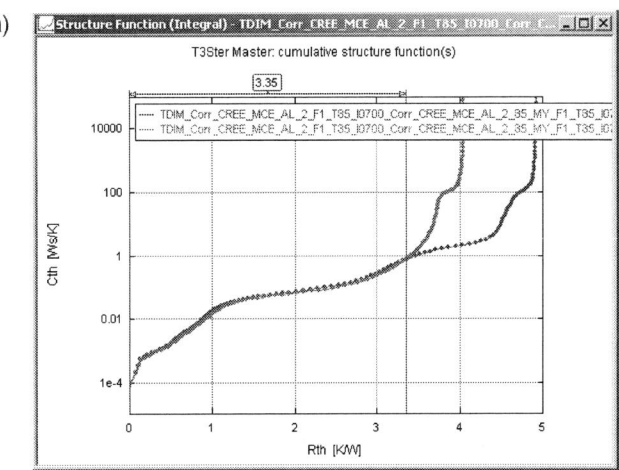

b)

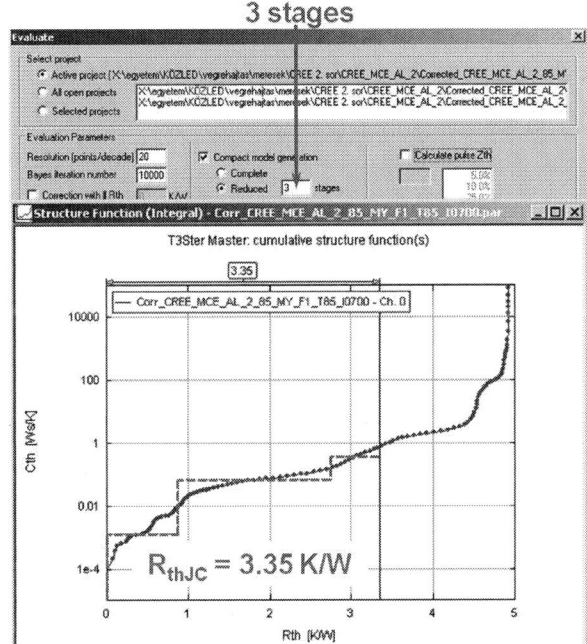

c)

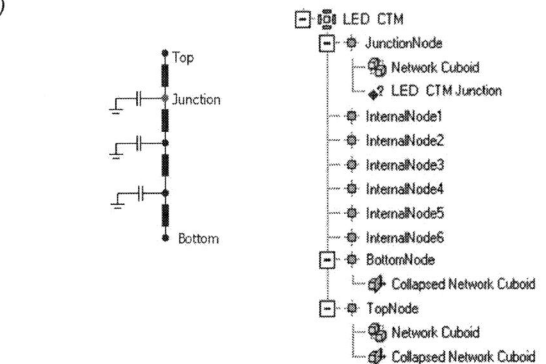

**Figure 4:** Dynamic compact thermal model generation of LED packages:

a) Finding the $R_{thJC}$ of the LED package by the application of the dual interface method defined in JEDEC JESD51-14 [8],
b) Generate a multiple RC-stage ladder model of the package,
c) In a commercial CFD tool the RC ladder is used to extend a JEDEC 2R model of an LED package to the transient domain. In the thermal impedance calculations the emitted optical power was calculated as suggested by [1] and [2].

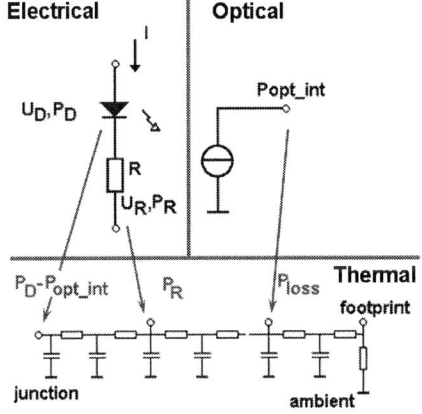

**Figure 3:** Concept of a multi-domain LED model [6], [7] showing details of the connection between the electrical and thermal sub-models.

In [10] besides considering the thermal effect of the series resistance we showed an electrical LED model. Schubert in his famous book about LEDs [11] provides details of different physical processes which take place in a LED chip and determine the electrical and light output characteristics of LEDs, but no comprehensive LED model is provided which could be used for simulation purposes.

Keppens in his PhD dissertation [12] experimentally confirmed the Schockly behavior of LEDs and investigated the details of the temperature dependence of LEDs' I-V characteristics. Our aim is to go further along this path: provide an electrical model completed with a thermal model and light emission model.

## 3. Light output modeling

Once we know the forward current and the temperature of an LED, we should be able to calculate the properties of its emitted light. Details of different mechanisms of white light generation were provided by Schubert [11]. Fig. 5 was created after Schubert's book – it shows how a multi-domain LED model looks like.

The ultimate goal is to be able to calculate emission spectra of white LEDs using the simplest possible models the parameters of which can be determined directly from optical, electrical and thermal measurements which are usual in LED testing.

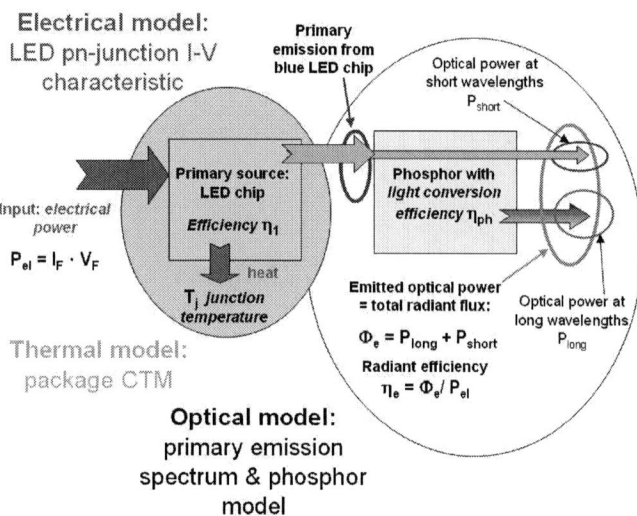

**Figure 5**: Components of a multi-domain LED model with some details of the light output modeling (after Schubert [11])

### 3.1. Modeling spectra of monochromatic LEDs

To start with, in our present work we concentrated on the calculation of emission spectra of monochromatic LEDs. Modeling emission spectra of monochromatic LEDs with current and/or temperature dependence was reported by e.g. Reifegerste et al [13] and later by Keppens et al [14].

We have chosen the model of Keppens as the basis of our work since it is more physics based than the model by Reifegerste. We extended Keppens' model with current dependence. In this extension we have split the LEDs' forward current into two portions: one portion is represented by the ideal diode equation – with a properly chosen ideality factor and the other portion is an effective recombination current which is directly related to light emission:

$$I_F = I_{id}(V_F) + I_{rec}(V_F) \qquad (1)$$

where $I_{id} = I_0 \cdot [\exp(V_F/nV_T)-1]$ is well known ideal diode equation with $V_T$ being the thermal voltage and $n$ the ideality factor, and $I_{rec}$ representing the recombination current associated with light output (current due to radiative recombination) and $V_F$ denotes the forward voltage.

From combined thermal and radiometric LED measurements (such as suggested in [1] and [2]) the total supplied electrical power of an LED is known together with the radiant flux (emitted optical power). On one hand:

$$P_{el} = I_F \cdot V_F = [I_{id}(V_F) + I_{rec}(V_F)] \cdot V_F \qquad (2)$$

On the other hand:

$$P_{el} = P_{heat} + P_{opt} \qquad (3)$$

where $P_{opt}$ is the measured $\Phi_e$ total radiant flux of the LED.

From these one can conlude, that the recombination current related to light output can be calculated from the measured emitted optical power:

$$I_{rec}(V_F) = P_{opt}/V_F = \Phi_e/V_F \qquad (4)$$

We call this current as effective current of an LED. This value can be easily determined from measurement data, such as shown in Fig. 6.

a)

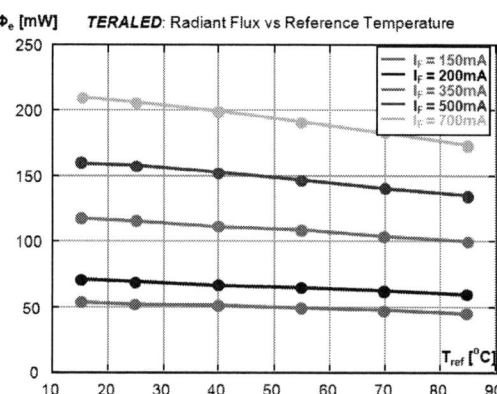

b)

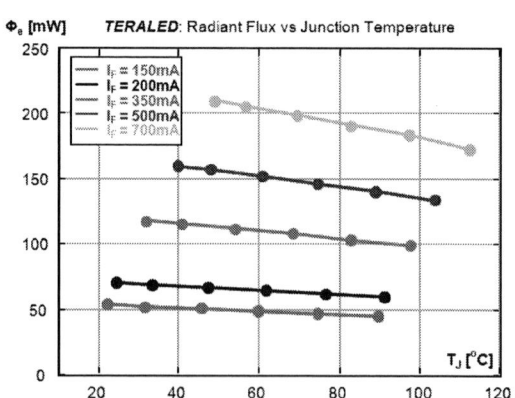

**Figure 6**: Measured total radiant flux values of an LED as function of forward current and
a) reference temperature, b) junction temperature

One has to note, that lighting designers are primarily interested in LEDs' light output at a given reference (ambient) temperature while for physical modeling the junction temperature should be used since all physical relationships

978-1-4673-1110-6/12 $31.00 © 2012 IEEE

describe the LED behavior as function of the junction temperature. The junction temperature of an LED can be related to the reference temperature as follows [7]:

$$T_J = R_{\text{th JA}} \cdot ( I_F \cdot V_F - \Phi_e) + T_{\text{ref}} \qquad (5)$$

We shall also use the junction temperature in our subsequent discussion. The effective current can be calculated from the actual energy conversion efficiency and the total forward current of the LED as well:

$$I_{eff} = I_{rec}(V_F) = I_F \cdot \eta \qquad (6)$$

wehere $\eta$ denotes the energy conversion efficiency.

In cooperation with prof. T. Rang's group at the Technical University of Tallinn (Estonia) extended the temperature dependent emission spectrum model of Keppens [14] with the dependence on the effective current [15]. The original set of model parameters published by Keppens [14] are used but these parameters now depend both on junction temperature and effective current. Figure 4 compares maesured and modelled emission spectra for blue LEDs where both current and temperature dependence was considered in the model. The ultimate spectrum model provides emitted optical power as function of light wavelength, forward current and junction temperature in the following format [15]:

$$\Phi_e(I_F, T_J, \lambda) = \left\{ \left( S_{1,ref} \, axp \left[ -[a + \theta_a \Delta I_{eff}] \left( \frac{c}{\lambda} - \frac{c}{\lambda_{p,ref}} + \gamma_p \Delta T_J + \theta_p \Delta I_{eff} \right) \right] + S_{2,ref} \, axp \left[ \frac{h \left( \frac{c}{\lambda} - \frac{c}{\lambda_{p,ref}} + \gamma_p \Delta T_J + \theta_p \Delta I_{eff} \right)}{k \left( \gamma_{T_c} \Delta T_J + \tau_{c,ref} + \theta_{T_c} \Delta I_{eff} \right)} \right]^{-1} + S_{3,ref} \, axp \left[ -\left( \frac{\frac{c}{\lambda} - \frac{c}{\lambda_{G,ref}} + \gamma_G \Delta T_J + \theta_G \Delta I_{eff}}{b + \theta_b \Delta I_{eff}} \right)^2 \right] \right) \right\} \times \exp \left( -\frac{\Delta T_J}{T_0} \right) \times \theta_{I_{eff}} I_{eff} \qquad (7)$$

where $c$ denotes the speed of light, $h$ is Planck's constant, $k$ is Boltzmann's constant, $\lambda$ is the wavelength $v_{p,ref}$, $S_{1,ref}$, $a$, $S_{2,ref}$, $T_{c,ref}$, $v_{G,ref}$, $S_{3,ref}$ and $b$ are determined from one reference spectrum at junction temperature $T_{J,ref}$ and $\Delta T_J = T_J - T_{J,ref}$.

For the determination of $\gamma p$, $\gamma G$, $c$ and $T_0$ a spectrum obtained at an additional temperature has to be available. Fitting parameters $\theta a$, $\theta p$, $\theta T c$, $\theta G$, $\theta b$ and $\theta I_{eff}$ can be determined using a spectrum obtained at an additional driving current and $\Delta I_{eff} = I_{eff} - I_{eff,ref}$ is the difference of the effective current corresponding to the actual spectrum and the reference spectrum. As a result, only a minimum of three measured spectra are needed to simulate the output spectrum of a LED.

One has to note that during parameter fitting the mismatch of the actual measured spectra and the modeled spectra (i.e. the fitting error) can be calculated in different ways. So far we calculated the color coordinates from the spectra and the distance between the loci in the color space obtained this was minimized. Another possible error metric could be calculated similarly to the definition of the f1 spectral mismatch error of photo detectors as described in the CIE 127-2007 document [16].

### 3.2. Modeling spectra of white LEDs

White light from LEDs is obtained by applying a thin layer of phosphor which absorbs most of the short wavelength primary emission of the blue LEDs and through photoluminescence converts the absorbed light to longer wavelength (see Fig. 5). Some of the primary emission of the LED chip gets through the phosphor layer, this way the emitted total flux of the LED covers almost the entire range of the visible spectrum of light. (Typically the phosphor layer is deposited on top of the LED chip therefore we assume that the temperature of the phosphor is equal to the LED's junction temperature.) The phosphor can be characterized by a single number, the $\eta_{ph}$ conversion efficiency introduced already in Fig. 5. This efficiency is

$$\eta_{ph} = \frac{\Phi_{e,white}}{\Phi_{e,blue}} \qquad (8)$$

where $\Phi_{e,white}$ denotes the emitted radiant flux of the phosphor converted white LED and $\Phi_{e,blue}$ is the total emitted flux of the blue LED chip.

This $\eta_{ph}$ conversion efficiency can be determined experimentally in a straightforward way. Usually the blue chips which are used in phosphor converted white LEDs are also sold by the LED vendors. Thus, the total emitted optical power of the blue LED and the white LED using the same blue chip can be measured. It is important, that in both measurements the reference temperatures of the LEDs are set such, that at the same constant forward current the junction temperature of both LEDs would be the same. (The junction temperatures could be determined either by using eq. (5) or by using the method proposed by Y. Zong and Y. Ohno from NIST [17].)

The conversion efficiency drops with increasing junction temperature. In case of the blue and white LED samples we measured, linear temperature dependence was found over a junction temperature range of about $\Delta T_J = 50$ °C.

If light output properties such as luminous flux, color coordinates, correlated color temperature or even color rendering index need to be predicted, the conversion efficiency of the phosphor alone would not be sufficient since from this and the modeled emission spectrum of the blue LED chip one cannot calculate the emission spectrum of phosphor

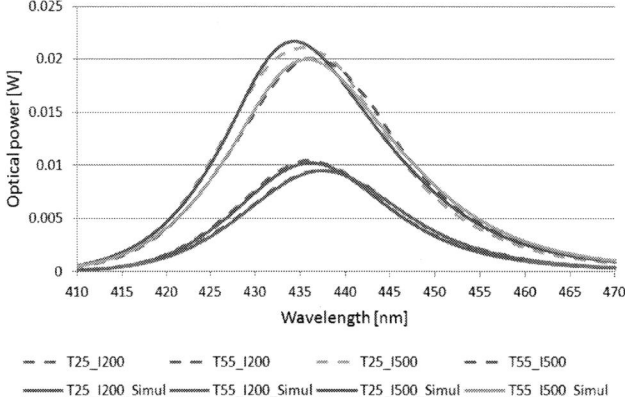

**Figure 7**: Measured and modeled emission spectra of a blue LED.

978-1-4673-1110-6/12 $31.00 © 2012 IEEE

converted white LED. Again, the workaround is provided by measuring emission spectra of both a blue LED and a white LED using the same blue chip at the same junction temperatures. For modeling purposes, if the total radiant flux values are already known, relative spectral power distributions (RSPD-s) of the LEDs are sufficient. A model for the net absorption and emission of the phosphor is obtained as follows:

$$C_{phosphor}(\lambda) = A \cdot \left[ RSPD_{white}(\lambda) - RSPD_{blue}(\lambda) \right] \qquad (9)$$

where $RSPD_{white}(\lambda)$ and $RSPD_{blue}(\lambda)$ denote the relative spectral power distributions of the white and the blue LEDs respectively, coefficient $A$ depends on the height of blue peaks of the absolute spectral power distributions of both LEDs. The value of $A$ scales with the junction temperature the same way as the $\eta_{ph}$ conversion efficiency. Fig. 8 shows an experimentally obtained plot corresponding to eq. (9).

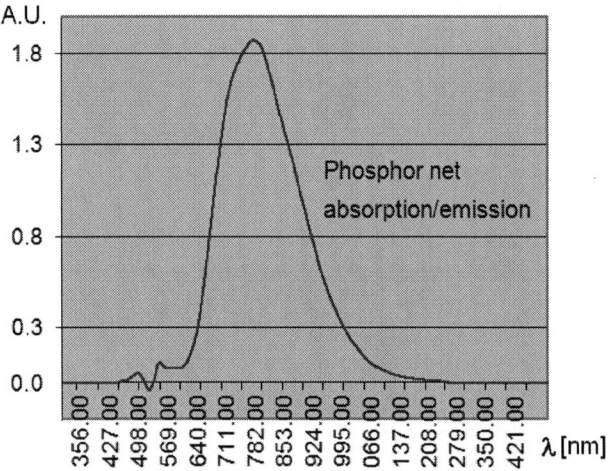

**Figure 8**: Net absorption/emission distribution of a yellow phosphor used in a phosphor converted white LED

The wavelength dependent function defined by eq. (9) can be represented by an array of a few hundred numbers, calculated from measured spectra.

## 4. The ultimate temperature and current dependent white LED model

A comprehensive model of a white LED includes the electro-thermal circuit model of the LED chip, the compact dynamic thermal of the LED package and the temperature and current dependent model of the light emission spectra.

Guidelines for the topology for an electro-thermal LED chip circuit model aimed at SPICE-like circuit simulations was provided in [6]. It is worth recalling, the besides describing the electrical operation of the LED's PN-junction, the model must be completed with a thermal node. The dissipation of the LED is calculated solely from the ideal diode current and any other current *not* associated with radiative recombination. This dissipation is forced into the thermal node of the model. (This thermal node must be terminated by the compact thermal model of the LED package.) Furthermore, the electro-thermal model should include the electro-thermal and thermo-electrical transconductances of the device – as indicated in Fig. 9.

This electro-thermal model together with the LED package compact thermal model would result in consistent total forward current, forward voltage and junction temperature values as well as in the corresponding value of the effective current responsible for light output.

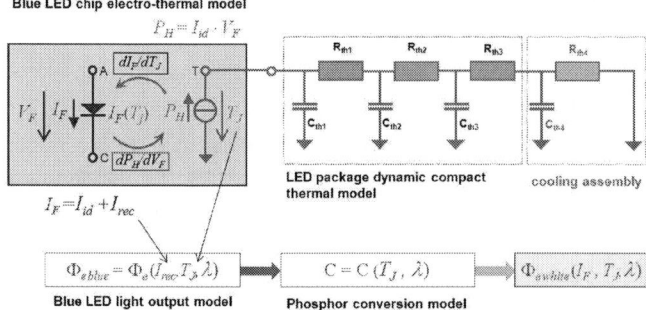

**Figure 9**: Structure of a comprehensive multi-domain LED model

This, together with the junction temperature provides input for the model of the primary emission of the blue LED chip. The light conversion model of the phosphor applied to this model spectrum provides the modeled spectral power distribution of the phosphor converted white LED.

## 5. Conclusions

The work reported in this paper was aimed at developing a forward current and temperature dependent model of light output of phosphor converted white LEDs. Such a model could provide lighting designers with exact values of light output properties of LEDs under different operating conditions. The paper recalled a very effective method for obtaining dynamic compact models of LED packages: results of thermal transient measurements can be converted into compact thermal models quickly and easily. Such package models can be used in CFD simulation tools or in form of SPICE netlists, can be connected to electro-thermal network models of LED's PN-junctions.

As a step forward in multi-domain modeling of LEDs, forward current and junction temperature dependent light emission spectrum models for monochromatic LEDs were also introduced. The introduced model was tested with high power red and blue LEDs. A simple principle of modeling emission spectra of phosphor converted white LEDs was also briefly described. By connecting these models light output of white LEDs can also be modeled as function of the operating conditions.

### Acknowledgments

This work was partially financed by the TECH_08-A4/2-2008-0168 KÖZLED project of the Hungarian National Development Agency. The support of the Hungarian Government through the TÁMOP-4.2.1/B-09/1/KMR-2010-0002 project at the Budapest University of Technology and Economics is also acknowledged.

The author thanks K. Paisnik for his work regarding spectrum measurements and modeling.

# References

[1] S. Nakamura, "Zn-doped InGaN growth and InGaN/AlGaN double-heterostructure blue-light-emitting diodes", Journal of Crystal Growth, Volume 145, Issues 1-4, 2 December 1994, Pages 911-917

[2] K. Bando, K. Sakano, Y. Noguchi, Y. Shimizu, "Development of high-bright and pure-white LED lamps", J. Light Visual Environ. 22, pp. 2-5, 1998.

[3] R. Haitz, F. Kish, J.Tsao, J. Nelson, "The Case for a National Research Program on Semiconductor Lighting", http://lighting.sandia.gov/lightingdocs/hpsnl_long.pdf, see also http://en.wikipedia.org/wiki/Haitz's_Law

[4] Solid-State Lighting R&D Multi-Year Program Plan FY'09-FY'14, http://apps1.eere.energy.gov/buildings/publications/pdfs/ssl/ssl_mypp2008_web.pdf

[5] C. Biber, "LED Light Emission as a Function of Thermal Conditions", In: Proceedings of the 24th IEEE Semiconductor Thermal Measurement and Management Symposium (SEMI-THERM'08), 16-20 March 2008, San Jose, USA, pp. 182-186

[6] A. Poppe, G. Farkas, V. Székely, Gy. Horváth, M. Rencz, "Multi-domain simulation and measurement of power LED-s and power LED assemblies", In: Proceedings of the 22nd IEEE Semiconductor Thermal Measurement and Management Symposium (SEMI-THERM'06). Dallas, USA, March 14-16 2006, pp. 191-198.

[7] A. Poppe, C. J. M. Lasance, "On the Standardization of Thermal Characterization of LEDs", In: Proceedings of the 25th IEEE Semiconductor Thermal Measurement and Management Symposium (SEMI-THERM'09). San Jose, USA, March 15-19 2009, pp. 151-158

[8] JEDEC Standard JESD51-14 "Transient Dual Interface Test Method for the Measurement of the Thermal Resistance Junction-To-Case of Semiconductor Devices with Heat Flow through a Single Path" http://www.jedec.org/sites/default/files/docs/JESD51-14.pdf

[9] A. Vass-Várnai et al, "Measurement Based Compact Thermal Model Creation - Accurate Approach to Neglect Inaccurate TIM Conductivity Data", In: Proceedings of the 13th Electronics Packaging Technology Conference (EPCT 2011). 7-9 December 2011, Singapore, pp. 67-72. Paper A4.2.

[10] G. Farkas, Q. van Voorst Vader, A. Poppe, Gy. Bognár, "Thermal Investigation of High Power Optical Devices by Transient Testing", IEEE Trans on Components and Packaging Technologies, Vol. 28, no. 1, (March 2005), pp. 45-50

[11] E. F. Schubert, Light-Emitting Diodes, 2nd ed. (Cambridge University Press, Cambridge, 2006)

[12] Arno KEPPENS, "Modelling and evaluation of high-power light-emitting diodes for general lighting", Katholieke Universiteit Leuven, Arenberg Doctoral School of Science, Engineering & Technology – Faculty of Engineering, ESAT/ELECTA, Kasteelpark Arenberg 10, B-3001 Leuven, Belgium, September, 2010, ISBN 978-94-6018-256-3

[13] F. Reifegerste and J. Lienig, "Modelling of the Temperature and Current Dependence of LED Spectra", J. Light & Vis. Env., Vol. 32, No. 3, pp.288-294 (2008)

[14] A. Keppens, W. Ryckaert, G. Deconinck, P. Hanselaer, "Modeling high power light-emitting diode spectra and their variation with junction temperature", J. Appl. Phys., Vol. 108, No. 4, pp.043104-043104-7 (2010)

[15] Kristo Paisnik "Multi-domain modeling of high power LEDs", Master's Thesis, Budapest University of Technology and Economics, Department of Electron Devices and Tallinn Uiversity of Technology, Thomas Johann Seebeck Department of Electronics, Budapest/Tallinn, 2011

[16] Technical Report, "Measurement of LEDs", CIE127-2007 (2007)

[17] Yuqin Zong and Yoshi Ohno: "New practical method for measurement of high-power LEDs", Proceedings of the CIE Expert Symposium 2008 on Advances in Photometry and Colorimetry: CIEx033-2008, pp.102-106 (2008)

# How Thermal Environment Affects OLEDs' Operational Characteristics

Zsolt Kohári, László Pohl, András Poppe

Budapest University of Technology and Economics, Department of Electron Devices
Magyar tudósok körútja 2, Bldg. Q, wing B, Budapest, H-1117, Hungary
<kohari∥pohl∥poppe>@eet.bme.hu

## Abstract

In recent years great effort has been put into development of organic light emitting diodes (OLEDs) worldwide. Among many concerns of developers is heat-removal from the thin film structure of the active layers of OLEDs which are typically realized on low thermal conductivity substrates such as glass or polymer foils. The other issue is to provide the OLEDs with transparent, yet high electrical conductivity electric power supply structure, therefore metallic shunting grids are added to the layer stack of OLEDs. These two major issues necessitate self-consistent electro-thermal simulation of large area OLEDs in which the temperature dependent I-V characteristics of the light emitting polymer layers are also considered. In this paper we discuss details of our 2.5D field-solver algorithm extended with such capabilities which was developed for the European funded Fast2Light project. The paper also presents a measurement and simulation example for a glass-based research OLED sample.

## Keywords

OLED, electro-thermal simulation

## 1. Introduction

Solid state light emitting device technology has made tremendous progress in the last decade. After the development of the inorganic blue and white Light Emitting Diodes (LEDs) [1], [2] the road led straight to the solid state lighting applications.

The latest developments achieved a lifetime of 62,000 hours for blue OLEDs of an operating brightness of 400 cd/m2 [3] thus, the development of lighting purpose OLED devices stepping from experimental phase to market phase. OLED application concepts [4] were announced and lighting experimental kits like Lumiblade [5] or Orbeos [6] are marketed already. Main targets of developments are to increase panel sizes and to try to use substrates other than glass (like flexible plastic foil), try to reduce manufacturing costs etc. These goals are also among the goals of a couple of European research projects like Fast2Light [7].

Since until recently there have been no reports on tools capable of simulating OLEDs' coupled thermal and electrical behavior, we started extending our former finite difference method based field solver algorithm SUNRED (SUccessive Node REDuction) [8], [9] with the ability of simulating coupled thermal and electrical problems with the emphasis on OLED structures [10], [11].

In their recent publication J. Park et al [12] describe coupled electro-thermal simulation of thin film OLED structures based on the drift-diffusion equation and Poisson's equation for the organic semiconductor material layers, coupled with the heat-flow equation (solved for the entire OLED stack). In their approach the coupling between the two sets of partial differential equations is realized by the temperature dependent carrier mobility models of holes and electrons used in the active layers of their OLED structure (hole transport layer, electron transport layer, LEP).

Our approach is somewhat different. In the basic SUNRED field solver algorithm [8], [9] the partial differential equation of heat conduction is treated numerically by the method of finite differences. The finite difference model of the thermal structure is numerically represented as a 3D model of discrete thermal resistances, and heat-flux sources. (In dynamic cases thermal capacitances are also replaced by their time discretized equivalent models composed of time-step dependent thermal resistances and heat-flux sources.) With the electro-thermal extension of the original solver code this "network" concept was maintained – the distributed electrical problem is also represented by a network of resistors: each resistor representing a segment of conducting material within a simulation grid cell.

In our first attempt of the electrical modeling of the light emitting polymer layer of OLEDs the values of electrical resistors in the FD model corresponding to the LEP layer were calculated from the differential resistance of the I-V characteristics [10].

Inspired by our prior studies regarding multi-domain modeling of LEDs [13] a compact, temperature dependent electrical model of the LEP of small area OLEDs was developed [14], resulting in a set of equations also suitable for application in a SPICE-like circuit simulator. Parameters of this model have been fitted to measured I-V characteristics of a glass based research OLED sample, following our OLED characterization methodology [10] also used in the Fast2Light project of the EU [7]. The concept of the electro-thermal model of LEP layers of OLEDs is illustrated by Fig. 1a.

This electro-thermal LEP model completed with a simple model of the OLEDs radiant efficiency was implemented in the SUNRED algorithm [15] and most recently the code was completed with a luminosity model as well [16]. Fig. 1b illustrates how such a LEP model connects the finite difference models of the thermal and electrical properties of the material layers of OLED structures. (In Fig. 1b the resistor network with solid red resistors represents the thermal model of a material layer, the dashed blue resistors belong to the FD model of the electrical properties of the same layer.)

In this paper first we briefly describe some details of this new version of the SUNRED algorithm and provide examples which highlight how local device temperature influences OLEDs' operational characteristics.

a)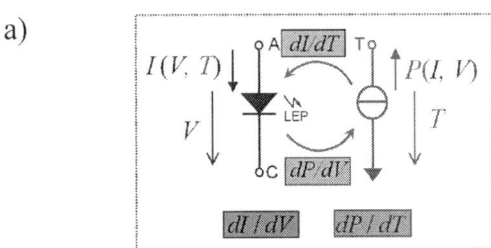

b)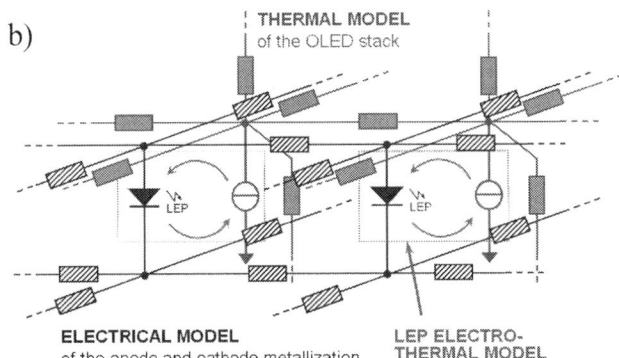

**Figure 1**: Electro-thermal modeling of OLED material stacks: a) topology of the compact electro-thermal model of active layers of OLEDs, b) finite difference network model of the electrical and thermal properties of OLED material layers.

## 2. LEP model in the electro-thermal field solver

### 2.1. The applied principles

In our earlier publication [14] we have demonstrated an OLED model described by a power function in the forward voltage region where the current of the device is expressed by the biasing voltage:

$$I_{OLED} = b_{LOW} \cdot V^{m_{LOW}} + b_{HIGH} \cdot V^{m_{HIGH}} \qquad (1)$$

Index *LOW* refers to the reverse direction, index *HIGH* refers to the forward direction of the OLED. In real applications the OLED is either turned on and gives light, operating in the *HIGH* range, or it is turned off. The bias domain where the *LOW* part is significant can be ignored because the device does not emit perceivable amount of light, so instead of Eq. (1) the following simplified equation can be used:

$$I_{OLED}(V,T) = b(T) \cdot V^{m(T)} \qquad (2)$$

The parameters *b* and *m* are temperature, structure and material dependent. This model fits very well the OLEDs' device characteristics that we measured.

The purpose of the electrical compact modeling work is to allow inclusion of an electro-thermal LEP model (illustrated by Fig. 1a) into the SUNRED field-solver algorithm with which temperature and potential distributions are calculated for the Fast2Light project. Fig. 1b illustrates this. The coupling between the electrical and thermal sides of the discretized distributed OLED numerical model is realized by the electro-thermal and thermo-electrical transconductances of the local LEP model denoted by *dI/dT* and *dP/dV* in Fig. 1a.

The LEP model described here briefly was implemented in the SUNRED algorithm used to solve thermal and electrical fields with the help of the finite differences method. The applicable fields are defined by partial differential equations

(PDE-s): Laplace-equation and Poisson's equation. OLED design requires a special version of the electro-thermal simulator where nonlinear; temperature dependent electrical characteristics can also be used in the simulation.

Electro-thermal fields in the steady-state (DC) case can be described by the following four partial differential equations:

$$\underline{j} = \sigma_e \underline{E} \qquad (3a)$$

$$\underline{p} = -\sigma \cdot grad\,T \qquad (3b)$$

Continuity equations:

$$div\,\underline{j} = 0 \qquad (3c)$$

$$div\,\underline{p} = \underline{j}\underline{E} \qquad (3d)$$

where *j* and *p* are the current and power density vectors, *E* is the electric field vector, *T* is the temperature, $\sigma_e$ and $\sigma$ denote the electric and thermal conductance. The $\sigma_e$ and $\sigma$ parameters are temperature-dependent. The original electro-thermal SUNRED model includes the Seebeck-/Peltier-effects and Joule-heating. OLED modeling requires Joule-heating only.

When the above PDE's are discretized for the numerical solution, the thermal and the electrical fields are mapped to coupled networks; the coupling (dissipation) is realized as controlled current sources (as indicated in Fig. 1). Local Joule-heating in the electrical side of the model is given by terms $V^2/R$; representing the power dissipated on a grid cell's *R* electrical resistance of the FDM model when *V* voltage drops on it. The Successive Node Reduction algorithm solves this resulting network for the given boundary conditions, electrical excitations and thermal loads, and determines all the voltages/temperatures and current densities/heat fluxes. The base algorithm [8], [9] is a direct solution method by its nature; there is no need for iteration but since both the electro-thermal coupling and the OLED equations are nonlinear, iteration is inevitable. This internal iteration however, requires much less steps than an inherently iterative solution method.

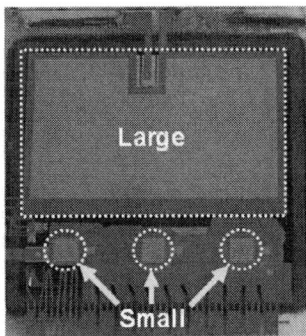

**Figure 2**: OLED samples realized in different size on the same substrate, which were used to create and validate our simulation models.

In the electro-thermal SUNRED algorithm there is an iteration loop between the electrical and thermal side of the model of the entire OLED structure to end up with electrical current, electrical potential, temperature and dissipation values which are consistent for the LEP layer. The electro-thermal model was established using measured characteristics of the smaller test devices (see Fig. 2) and the model was validated through simulation and measurement of the large OLED device realized on the same substrate.

978-1-4673-1110-6/12 $31.00 © 2012 IEEE          332

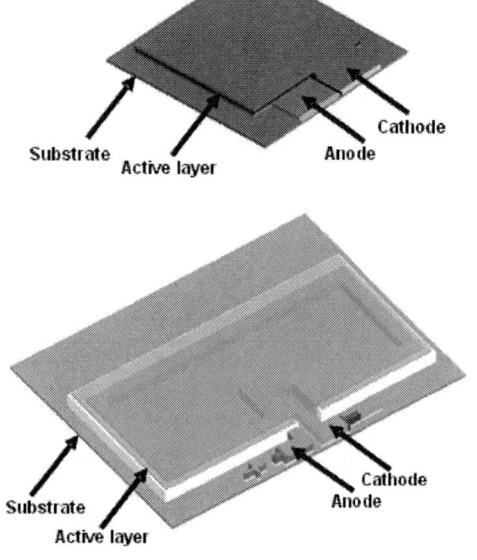

**Figure 3**: Simulation models of the small OLEDs (top) and the large OLED (bottom) devices used for model validation. The images are not to scale.

### 2.2. Model validation

Solution of model equation (2) provides I-V characteristics at any required temperature for the investigated device. These characteristics will change if the ambient (i.e. thermal boundary conditions) or the structure of the device changes, therefore for new conditions new simulation runs are needed. The quality of simulation results is determined by (i) the validity of the applied model equations and (ii) by the correctness of the applied material parameters and the set of other parameters used by the model equations (e.g. the set of $b$ and $m$ values). Steps of validation of our entire OLED electro-thermal modeling are as follows:

1. Measurement of the small OLEDs (see Fig. 2) and parameter fitting for the creation of the device model equation.
2. Building simulation model and performing electro-thermal simulation for the *small* OLEDs (Fig 3, top); comparing measurement and simulation results (to check correctness of the model geometry). Perfect matching of measured and simulated data is expected since model parameters were fitted using data of the small OLEDs
3. Measurement of the large OLED (Fig. 2, top).
4. Building simulation model and performing electro-thermal simulation for the *large* OLED (Fig. 3 bottom); comparing simulation and measurement results (Fig. 4).

Measured/simulated I-V characteristics are shown in Fig. 4.

As it can be seen in Fig. 4b, the electro-thermal simulation by our field solver extended for OLEDs resulted in a good fit for the large OLED device.

For the large OLED one would think that re-scaling the current of the smaller devices in proportion to the surface areas of the two devices would give correct results. By applying the scaling

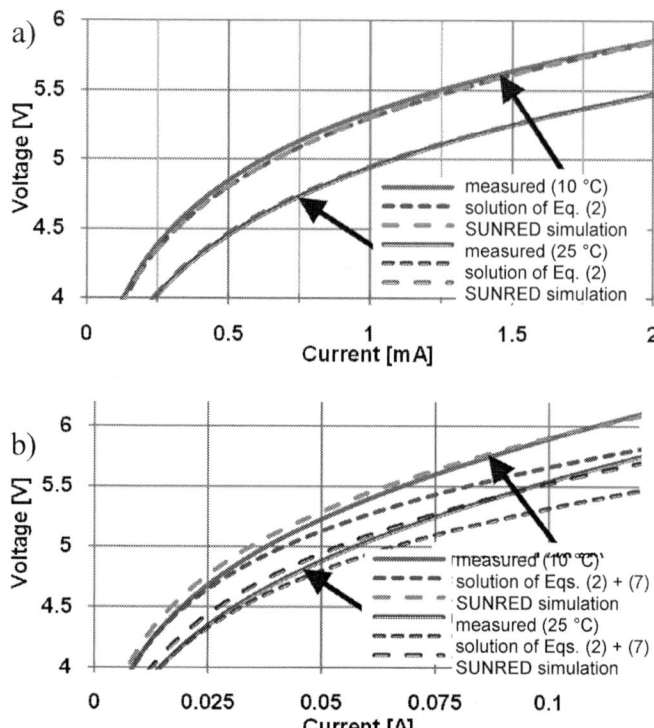

**Figure 4**: Measured and simulated OLED I-V characteristics at two different ambient temperature values: a) small OLEDs, b) large OLED

$$I_{large} = I_{small} \frac{A_{large}}{A_{small}} \qquad (7)$$

the fitting parameters of coefficient $b$ in formula (2) have to be re-evaluated. By doing so and solving the resulting formula however, the calculated I-V characteristic does not match the measured one as shown by the blue and purple dashed curves in Fig. 4b. This clearly demonstrates the need for the 2.5D numerical simulation of the electro-thermal behavior of OLEDs. Further details on the LEP model implementation, parameter identification and model validation are provided in [15] and in [17].

### 3. Prediction of the luminance distribution of large OLED panels

The ultimate goal is to design large area OLED panels with homogenous distribution of luminosity. Design of electrical shunting grids necessitated the development of our electro-thermal OLED simulation tool. The final test of a design is whether the required level of homogeneity of the luminosity is achieved in a given application. Therefore we performed certain parametric studies where design variables such as LEP energy conversion efficiency, shunting grid geometry etc. were considered. We investigated two simulation scenarios corresponding to two foreseen applications. In one case a vertically standing OLED panel was assumed, in a second case the same OLED panel was assumed to be attached to a horizontally placed cold-plate imitating the situation of an OLED luminaire attached to a concrete ceiling with practically infinite thermal capacitance (see Fig. 5). Note, that in the second case a relatively large interface resistance between the OLED panel and the ceiling was assumed.

The electro-thermal simulation results of the assumed OLED devices were post-processed using a local temperature, current and voltage dependent luminosity model [16] derived for a the same LEP material the parameters of which were applied during the electro-thermal simulation.

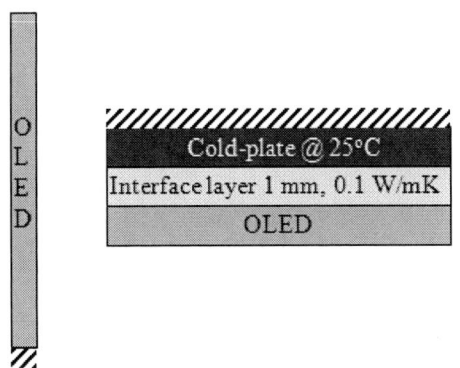

**Figure 5**: Arrangements of the foreseen OLED applications

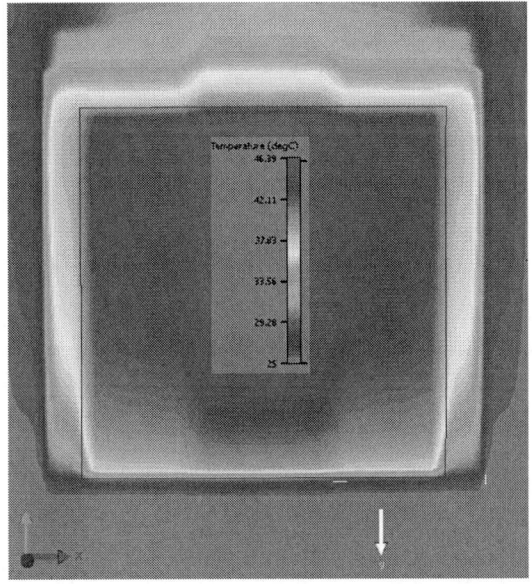

**Figure 6**: CFD simulation results of a vertically placed 12x12 cm OLED panel with 10% energy conversion efficiency and uniform dissipation (no electro-thermal interaction considered)

In both cases we expected the natural convection as a major heat-transfer mechanism towards the environment. For a correct thermal analysis of the outlined problems a full CFD solver with electro-thermal simulation capabilities (using our LEP model) would be needed. As no such code is available today and we did not want to extend our own simulation code with CFD capabilities, we applied an approximate solution.

We simulated a plate with similar geometry and surface properties as our assumed OLED device in a commercially available CFD tool (FloTHERM from Mentor Graphics) – see Fig. 6. From such simulations we derived local effective heat-transfer coefficient values. These values were then introduced as boundary condition in our conduction only mode electro-thermal solver.

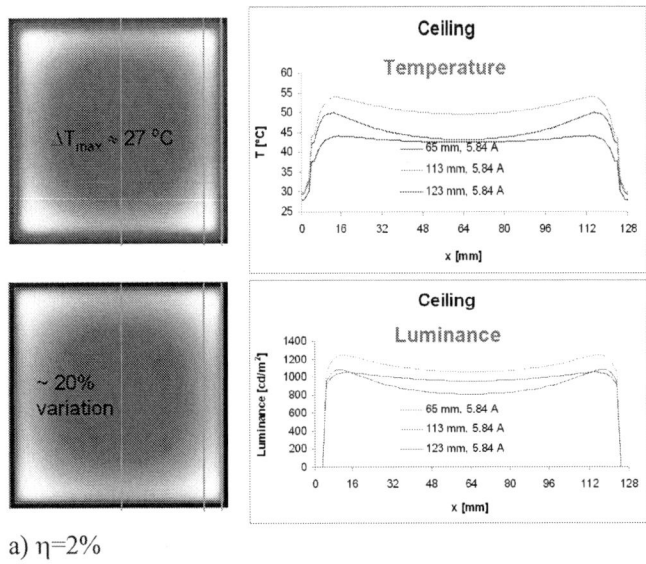

a) η=2%

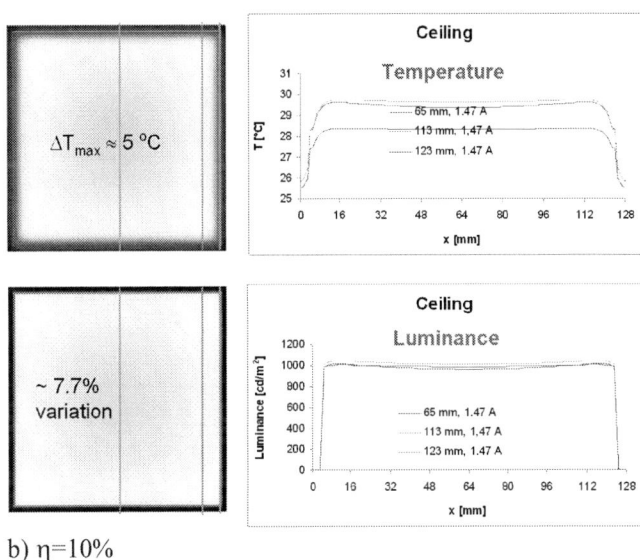

b) η=10%

**Figure 7**: Simulated temperature and luminance distributions of the horizontally placed lighting panel with assumed 2% and 10% LEP efficiency

Simulation results achieved this way are shown in Fig. 7 and 8. Beside the calculated temperature distribution the resulting luminance maps are also shown.

Comparing Fig. 7a and 7b suggest that higher LEP efficiencies help reduce the inhomogeneity of the luminance.The results shown in Fig.8. exhibit a good qualitative agreement with published results of M. Slawinski et al obtained for a 5x5 cm vertically placed OLED panel [18].

## 4. Validation by means of the analysis of dark spot

In the previous section simulation study of a large foil substrate based OLED panel was shown where direct validation was not possible since such OLED panels do not exist yet. Therefore, despite the fact that the electro-thermal field solver and the applied LEP models were jointly validated (see section 2.2) the question can be raised if one can trust the simulated luminance maps. To answer this question, the glass substrate based research OLED samples helped.

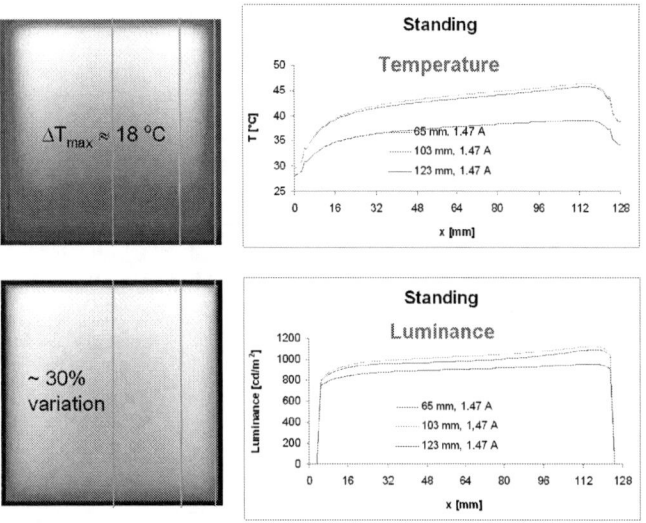

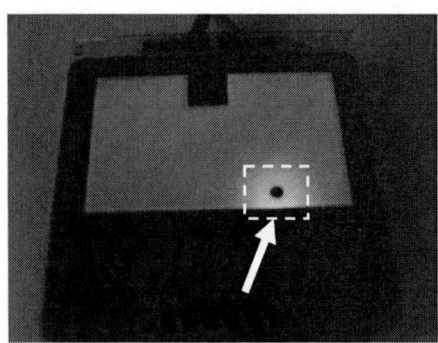

**Figure 8**: Simulated temperature and luminance distributions of the vertically placed lighting panel with an assumed 10% LEP efficiency

We had a sample in which a dark spot formed during long term measurements, see Fig. 9. There are dark spots which realize a medium impedance short across the LEP layer, between the anode and cathode electrodes. Thorough investigations proved that the dark spot in our sample device was of this kind.

**Figure 9**: Dark spot defect of an OLED resembling the image of a "total eclipse"

As the local potential difference between the anode and cathode is below the forward voltage of the light emitting polymer, In this case the small defect influences the light intensity and its distribution on the whole OLED surface. In the local environment of the defect there is no light. Farther away from the spot the light intensity rapidly increases, and reaches a maximum. Still farther away the light intensity is slowly decreasing and stabilizing at a level, which is less than that of the intact device. Though the center of the defect does not emit light, that is the hottest part of the device as shown by the IR (infrared) image (Fig. 10.). This high temperature is destructive to the organic material: the brightness of the device decreases until it gives no light anymore. IR

measurement provides surface temperatures of the glass substrate, the temperature of the dark spot in the LEP layer can be gained by simulation. Measured and simulated temperature profiles of the glass substrate are shown in Fig. 10.

a)

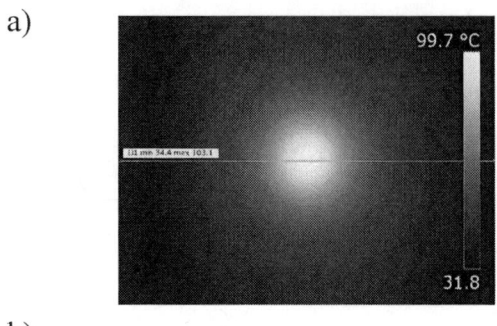

b)

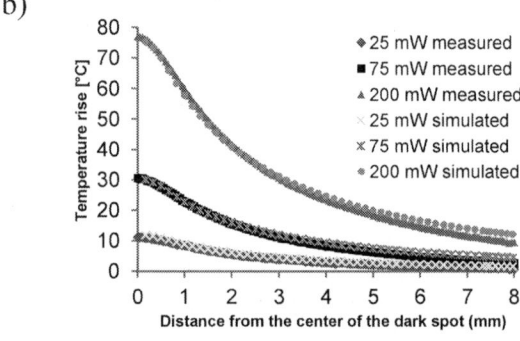

**Figure 10**: Measured and simulated temperature profile on the glass substrate around a dark spot failure of an OLED.    a) IR image, b) temperature profiles from IR measurements and simulations.

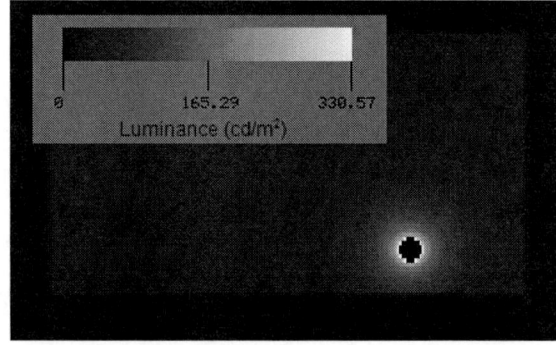

**Figure 11**: Simulated luminance distribution around a dark spot.

The simulated luminance distribution around the model of the dark spot failure can be seen in Fig. 11. This is also in good agreement with the physically observed image (Fig. 9), which suggests that our simulation model is correct. Analysis of the simulation results explains the cause of the phenomenon. Around the dark spots the vertical current density increases, this results in an increase of the intensity of the emitted light as it is proportional to the current density [19]. That is why the light emission increases around the dark spot. However, the higher current density results in higher temperature. Light emission of the LEP layer generally decreases with increasing temperature which partially compensates the light increasing effect of increased current.

## 5. Conclusions

The SUNRED field solver algorithm has been extended with new capabilities that allow self-consistent electro-thermal simulation of large area OLED devices.

By modeling electro-thermal interactions within OLEDs and precisely describing local heat generation in all electrically active layers of OLEDs accurate prediction of OLEDs' operational characteristics can be obtained. The developed electro-thermal OLED model was completed with empirical thermal boundary conditions and a current, voltage and temperature dependent luminance model. With these prediction of the effect of thermal properties of the operating environment on the luminosity distribution could be predicted for large are OLED panels. Results obtained by our simulation model shows good qualitative agreement with results of other teams [18]. Simulation and measurement results obtained for OLED defects have also validated our model [17].

## Acknowledgments

This work was partially financed by the ICT-2007.3.3/216641 Fast2Light project of the Framework 7 Program of the EU. The support of the Hungarian Government through the TÁMOP-4.2.1/B-09/1/KMR-2010-0002 project at the Budapest University of Technology and Economics is also acknowledged.

## References

[1] S. Nakamura, "Zn-doped InGaN growth and InGaN/AlGaN double-heterostructure blue-light-emitting diodes", Journal of Crystal Growth, Volume 145, Issues 1-4, 2 December 1994, Pages 911-917

[2] K. Bando, K. Sakano, Y. Noguchi, Y. Shimizu, "Development of high-bright and pure-white LED lamps", J. Light Visual Environ. 22, pp. 2-5, 1998.

[3] http://www.cdtltd.co.uk/press/archive_press_release_index/2007/602.asp, March 26, 2007

[4] http://www.worldarchitecturenews.com/index.php?fuseaction=wanappln.projectview&upload_id=12862

[5] http://www.lumiblade.com

[6] http://www.osram-os.com/appsos/ORBEOS/index.html

[7] http://www.fast2light.org

[8] V. Székely, "SUNRED a new thermal simulator and typical applications", In: Proc. of the 3rd International Workshop on THERMal INvestigations of ICs and Microstructures (THERMINIC'97). 21-23 September 1997, Cannes, France, pp. 84-90.

[9] Zs. Kohári, V. Székely, M. Rencz, A. Páhl, V. Dudek, B. Höfflinger, "Studies on the heat removal features of stacked SOI structures with a dedicated field solver program (SUNRED)", Microelectronics and Reliability, Volume 38, Issue 12, December 1998, pp 1881-1891

[10] A. Poppe, L. Pohl, E. Kollár, Zs. Kohári, H. Lifka, C. Tanase, "Methodology for thermal and electrical characterization of large area OLEDs", In: Proceedings of the 25th IEEE Semiconductor Thermal Measurement and Management Symposium (SEMI-THERM'09), San Jose, USA, 15-19 March 2009, pp. 38-44

[11] L. Pohl, Zs. Kohári, V. Székely, "Fast field solver for the simulation of large-area OLEDs", Microelectronics Journal, Volume 41, Issue 9, September 2010, pp 566-573

[12] J. Park, H. Ham, C. Park, "Heat transfer property of thin-film encapsulation for OLEDs", Organic Electronics, Volume 12, Issue 2, pp 227-233 (2011)

[13] A. Poppe, G. Farkas, V. Székely, Gy. Horváth, M. Rencz, "Multi-domain simulation and measurement of power LED-s and power LED assemblies", In: Proceedings of the 22nd IEEE Semiconductor Thermal Measurement and Management Symposium (SEMI-THERM'06). Dallas, USA, 14-16 March 2006, pp. 191-198.

[14] E. Kollár, I. Zólomy, A. Poppe, "Electro-thermal modeling of large-surface OLED", In: Proc. of the Symposium on Design, Test, Integration and Packaging of MEMS/MOEMS (DTIP'09). Rome, Italy, 1-3 April 2009, pp. 239-242.

[15] L. Pohl, E. Kollár, A. Poppe, "Nonlinear electro-thermal OLED model in SUNRED field simulator", In: Proceedings of the 16th International Workshop on THERMal INvestigation of ICs and Systems (THERMINIC'10), Barcelona, Spain, 6-8 October 2010, pp 149-153.

[16] László Pohl, Ernő Kollár, "Extension of the SUNRED algorithm for electrothermal simulation and its application in failure analysis of large area (organic) semiconductor devices", In: Proceedings of the 17th International Workshop on THERMal INvestigation of ICs and Systems (THERMINIC'11), 27-29 September 2011, Paris, France

[17] László Pohl, Ernő Kollár, András Poppe, Zsolt Kohári, "Nonlinear electro-thermal modeling and field-simulation of OLEDs for lighting applications I: algorithmic fundamentals", MICROELECTRONICS JOURNAL 43, (2011), DOI: 10.1016/j.mejo.2011.06.011

[18] M. Slawinski, D. Bertram, M. Heuken, H. Kalisch, A. Vescan, "Electrothermal characterization of large-area organic light-emitting diodes employing finite-element simulation", Organic Electronics 12: pp. 1399-1405 (2011), doi:10.1016/j.orgel.2011.05.010

[19] E. Kollár, I. Zólomy, A. Poppe: "Electro-thermal modeling of large-surface OLED", DTIP 2009. Rome, Italy 2009, pp. 239-242.

# Development of a Flexible Chip Infrared (IR) Thermal Imaging System for Product Qualification

Chenzhou Lian, Marc Knox, Kamal Sikka, Xiaojin Wei, and Alan J Weger

IBM Systems and Technology Group

2070 Route 52, Hopewell Jct, NY 12533

Email: clian@us.ibm.com

## Abstract

A flexible and efficient chip Infrared (IR) thermal imaging system was implemented on the product manufacturing test platform by collaboration with the burn-in/wafer test, systems, process, and failure analysis teams. A liquid cooling cell was successfully designed and tested. The imaging system was applied to investigate some wafer probe power/thermal issues for server high end products. Furthermore, we applied the method of Spatially-resolved Imaging of Microprocessor Power (SIMP) [1] to translate the thermal map into a power map. Finally, we propose a new concept of product thermal qualification as a supplement and potential alternative to the traditional thermal test vehicle (TTV) qualification.

## Keywords

Infrared (IR) thermal imaging, semiconductor, chip hot spot, thermal test vehicle (TTV), power map

## 1. Introduction

As driven by the increasing power consumption and decreasing feature size in modern microprocessor chips, the package power density and gradients have increased significantly. Thermal imaging of semiconductor die has become a more useful and informative tool to visualize thermal and power distributions on a die in real time. Besides, the effects of various workloads, patterns and conditions can be studied. By identifying chip hot spots, we could further address design, process, and defect related wafer and chip package issues. Thermal imaging capability by itself is not new or novel, but the novelty of this implementation lies in the attributes shown below. Most importantly, the system we developed is available to multiple users and for multiple products as soon as product burn-in (BI) hardware and Part Number Programs (PNPs) are available. This is the first time that we implement the IR imaging capability directly on top of an existing and fully supported manufacturing test platform and infrastructure, including hardware support (Burn In Boards (BIBs), sockets, tools), software support (all development & manufacturing infrastructure), Synergy of hardware development for new packages, and Test development support.

The core ideas and contributions of this study lie in the following:

• Develop a flexible and universal thermal imaging system directly on the production burn-in (BI) tool to evaluate chip spatial temperature and power distributions. A portable liquid cooling system for both the TTV and BI tool has been successfully designed and tested. This system can be used to evaluate responses to "fast" events such as power spikes. It can also be applied to validate electro-migration (EM) and Joule heating effects for both full chip scale and micro scale studies.

• Propose a new concept of chip product thermal qualification as a flexible/efficient supplement and potential alternative to traditional thermal test vehicle (TTV) qualification. This technology could reduce manpower, cost, and time, and only require the minimum customization and special resources to support. It allows us to study various non-uniform power patterns and conditions. In addition, it is executable for early silicon (new designs) and multiple products as soon as product BI hardware and PNPs are available.

## 2. Methodology and Experimental Setup

The imaging system developed in the project can be summarized in a work flow chart shown in Fig. 1. There are 3 steps - We start with the thermal test vehicle, proceed to the Product BI tool, and the final step is the realization of the product thermal qualification part. In Fig. 1 we distinguish the steps in 3 different colors. The thermal test vehicle part is the foundation and initiation of the whole program. The first goal is to design a liquid cooling cell assembly which could be used for both the TTV board and BI tool BIB. After the first design attempt, we measured the temperature map with an IR camera. We studied the flow patterns at different power levels and flow rates with flow in both directions. Then we checked whether the resulting flow is uniform and symmetric as expected. If not, we have to stay in Loop 1 and improve the design until we get the exact flow field that we want. Then, we can move to the next step of computational fluid dynamics (CFD) modeling. We compared the modeling results with the thermal map measurement to make sure the CFD model is correct with an error < 5%. If not, we have to stay in Loop 2 and improve the model. After that, we started to generate the uniform power map through the measured thermal map using the SIMP method [1]. We validated the total power to make sure the error is within 5%, so that we can get out of Loop 3 and move on.

For the Product BI tool part, we started with the product temperature measurement using an IR camera, and then we generated a non-uniform power map and validate the total power. After that, we then moved to the final goal of product thermal qualification. We generated the power map as the input for the FloTHERM thermal model to get TIM thermal impedance by matching the three thermal diode temperature readouts and qualify the products. Please also note that, the upper left parts above this dashed line are all the experimental testing steps, while the lower right parts below this dashed line are the CFD modeling steps. This study is a good integration of experimental and computational work.

978-1-4673-1110-6/12 $31.00 © 2012 IEEE

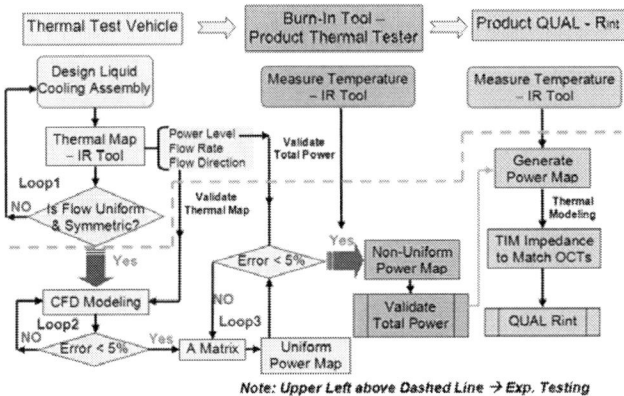

**Figure 1 Product Thermal Testing and Imaging Flow Chart**

## A. Production Burn-in Tool Modification

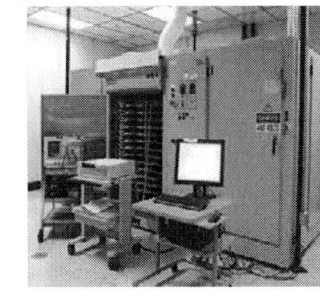

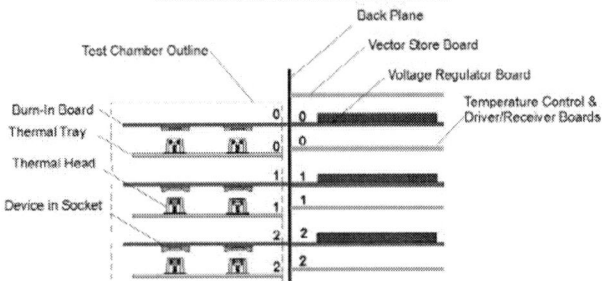

**Figure 2 Production burn-in tool and simplified side view of hardware system**

The production burn-in tool, details of which are shown in Fig. 2, is composed of a burn-in board with device in socket, a thermal tray with cooling thermal head, and temperature control & driver/receiver boards. To facilitate the thermal imaging capability, we first made the following modifications for the BIB (Fig. 3) and the final assembly is shown as a schematic plot in Fig. 4.

- Parts face up, thermal head removed and thermal tray modified
- Socket modified to accommodate liquid cooling approach to take IR images
- Chip thinned and coated for more accurate micro level temperature gradients
- Mold-compound or insulation layer recommended to cover the laminate surface to eliminate laminate conduction/convection effects
- Sapphire (IR transparent) window seated above the chip with cooling fluid flowing underneath
- Camera mounted above and looking down at chip

The hardware of the system includes a liquid recirculation system, a thermal test vehicle (TTV), a liquid cooling cell assembly, a single engineering version of BIB in the Product BI tool, and an IR camera with a 25 mm Len. The image resolution is around 100 um. The cooling liquid adopted is a flouro-carbon fluid, which is IR transparent and electrically insulating. We also noticed that imaging on small areas can be enhanced by thinning the chips. The silicon thickness between active circuits and the backside spreads the heat laterally thereby decreasing contrast because the camera can only captures the surface temperatures. Thinning allows more accurate micro level temperature gradients. Typically we recommend thinning for local area work and leaving the chip thick for full die imaging. However, thinning chips is also fairly costly and need to be performed in the Failure Analysis (FA) lab. In the preparation of the testing samples, we could also put a coating layer on the chip surface to enhance the image quality, especially for micron level thermal analysis. As we know, silicon is transparent to IR and underlying structures have various emissivities, therefore wiring and patterns are observed and may interfere the thermal image without the coating. Besides, coating can provide a consistent emissivity which can further enhances imaging.

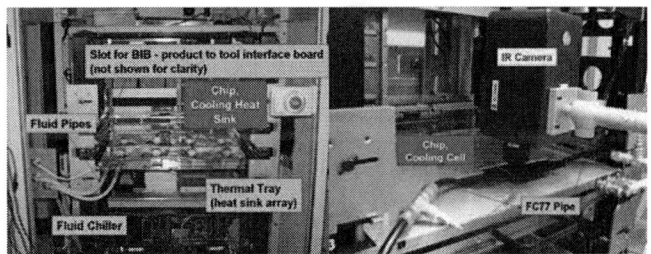

**Figure 3 Modification of the screener; Before (Left) and Final assembly (Right)**

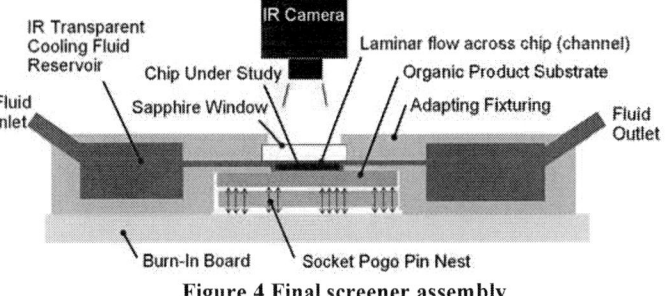

**Figure 4 Final screener assembly**

## B. Liquid cooling cell, recirculation system, and TTV

The liquid cooling cell is a critical and fundamental component in the whole system development. Fig. 5 shows the cooling cell and assembly on the Product BI tool board. Gaskets are located between parts to ensure good sealing and prevent leakage. The inner back-step design at the connection interface was implemented to prevent the inward squeezing of the gasket in the fluid flow path and avoid any interference with the flow. The cooling fluid goes over the chip top and the extensions of the cooling cell during tests. The liquid recirculation system includes a liquid reservoir, chiller, heat exchanger, flow meter, filter, and a switch between the liquid and compressed air loops.

We started with the experiments on a TTV. Fig. 6 shows the experimental setup and the heater/sensor layout for the

978-1-4673-1110-6/12 $31.00 © 2012 IEEE          338

TTV chip. There are total 8 heaters and 27 thermal sensors. The heaters were controlled by power supplies to obtain different power levels and patterns. The liquid cooling cell assembly was validated for various flow rates ranging from laminar to turbulent regime, both flow directions, and three types of power patterns. The generated power maps were successfully validated by both TTV power input and CFD modeling results.

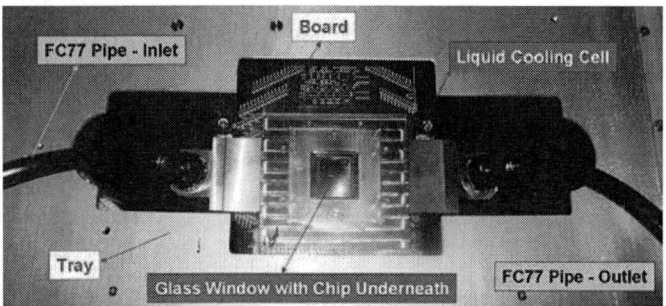

**Figure 5 Liquid cooling cell (top) and view when assembled on burn-in board (bottom)**

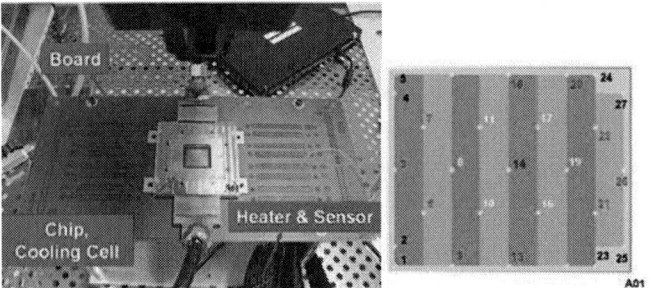

**Figure 6 Experiment Setup (Left) and schematic of TTV chip (Right)**

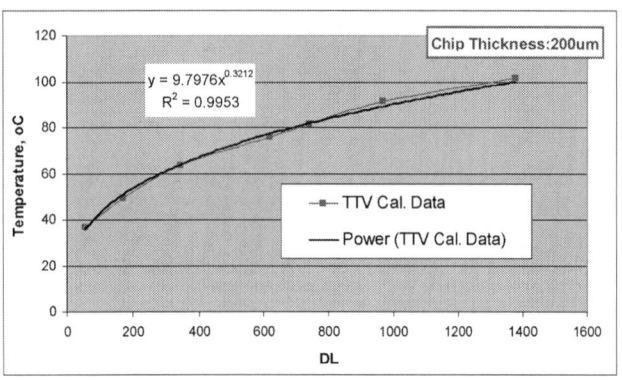

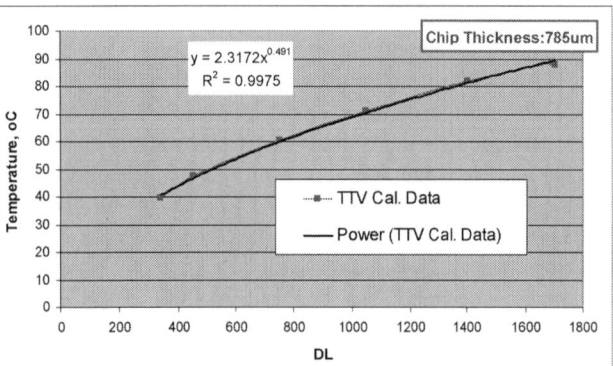

**Figure 7  IR Thermograph Calibration for two TTV chips - (Top: 200 um) and (Bottom: 785 um)**

## C. Experimental Measurements

Fig. 7 shows the IR thermograph calibration for two parts with different die thickness on TTV board. IR camera basically captured the digital level (DL) signals. Through the calibration between DL and known temperatures, we can obtain the temperature at the objective surface. The thermal sensors on the TTV and thermal diode readout on the product chip were adopted to provide temperatures for the calibration.

Figs. 8 and 9 summarize the thermal imaging for 200 and 785 um die thickness TTV chips. The flow rate is specified as 1.5 GPM corresponding to the Reynolds (Re) number ~ 3000. Heater layout for the chip is shown in Fig. 6. We tested the parts at three power levels and in both flow directions under conditions with three different heater controls.

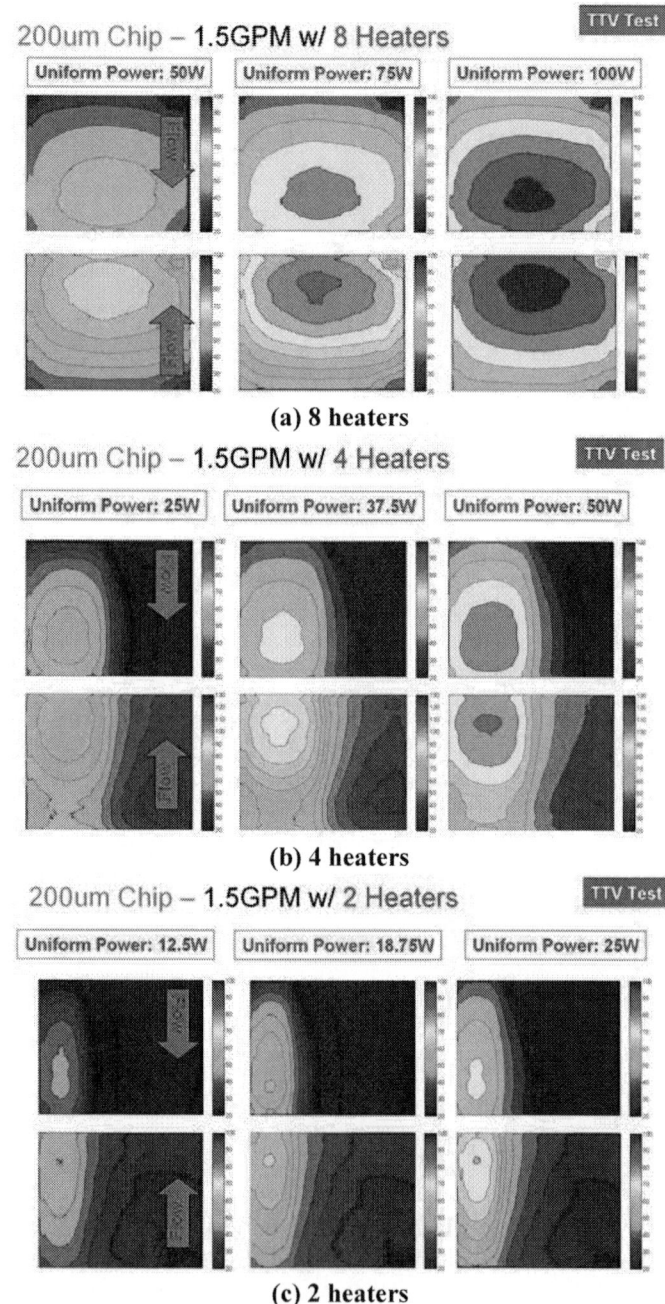

**Figure 8 IR thermal imaging – 200 um chip, 1.5GPM for various heater control**

978-1-4673-1110-6/12 $31.00 © 2012 IEEE          339

## 3. Computational modeling and validation

The computational modeling work was done in FloTHERM, a widely recognized commercial software in the microelectronic industry. Fig. 10 shows the geometry and configurations in the model. The total grid is around 1.35 million with the maximum aspect ratio of 20. The minimum grid sizes in x, y, and z are 5e-4, 5e-4, and 2e-5 m, respectively.

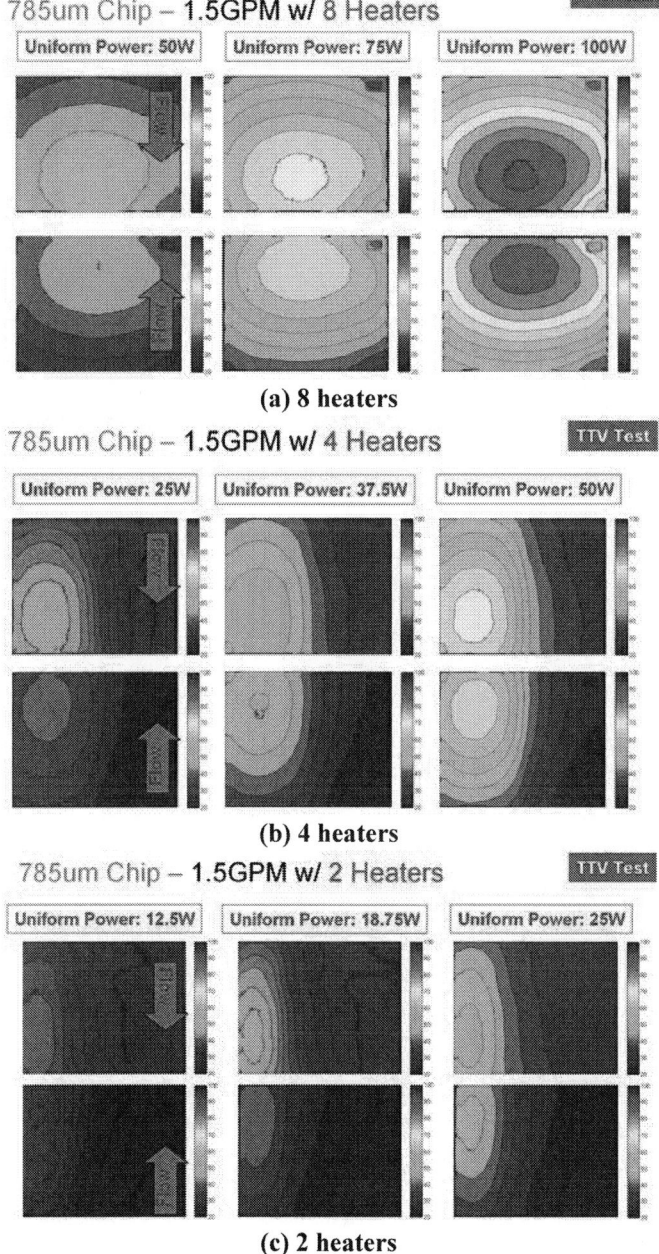

**(a) 8 heaters**

**(b) 4 heaters**

**(c) 2 heaters**

**Figure 9 IR thermal imaging – 785 um chip, 1.5GPM for various heater control**

Fig. 11 compares the modeling results with the experimental measurements for the chip top temperature distributions. The model predicted the hottest spot pretty well, but over predicted the temperature at leading edge. In terms of the temperature gradient across the chip in the flow direction, the modeling results match the testing data reasonably.

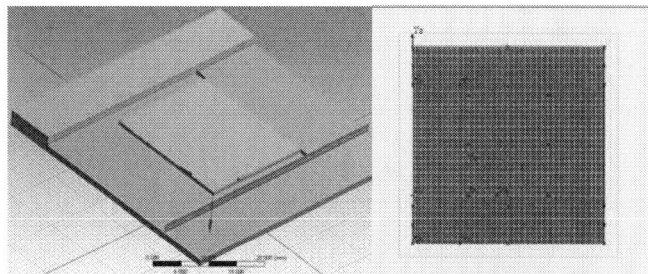

**Figure 10 Geometry and configurations in FloTHERM model**

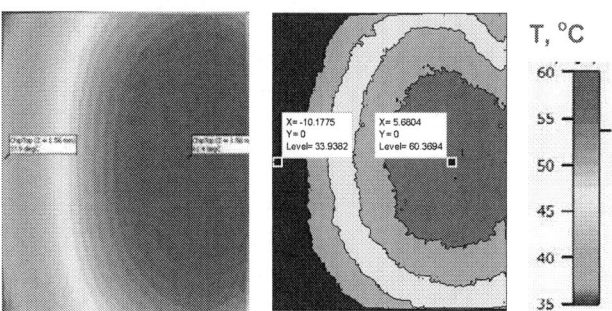

**Figure 11 Comparison between CFD modeling results (Left) and IR thermal image (Right)**

As described in Fig. 12, methodology adopted to generate the power map from thermal map is called Spatially-resolved Imaging of Microprocessor Power (SIMP) [1, 2]. The figure shows the chip grid size as n by m. Then we could defined the total modeling case number as n*m. The first case is shown on the left, and the last case is shown here on the right. The red spot represents the region where certain level of power is supplied. From the equation of convection, we defined A-matrix as the first equation in the figure. After having the n*m cases simulated, we can get the Q and T matrix and then obtain the A matrix. Using A matrix, we can translate between temperature and power map based on the second equation in Fig.12.

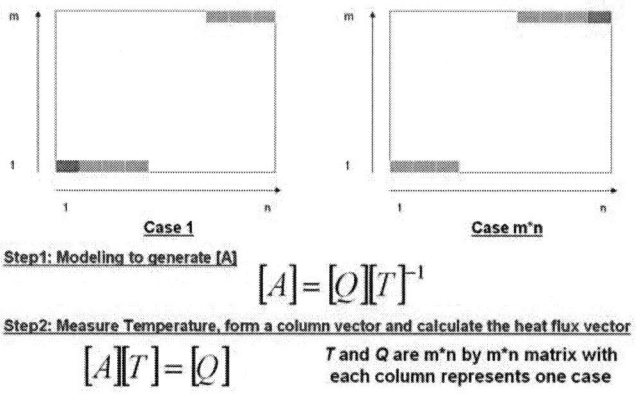

Step1: Modeling to generate [A]

$$[A] = [Q][T]^{-1}$$

Step2: Measure Temperature, form a column vector and calculate the heat flux vector

$$[A][T] = [Q]$$

T and Q are m*n by m*n matrix with each column represents one case

**Figure 12 Power map generation - Spatially-resolved Imaging of Microprocessor Power (SIMP)**

To validate the method, we proposed two study cases. As shown in Fig. 13, we supplied the power in the red square across the chip in study 1, and then we ran 10 cases in study 2, in which we supply only 1/10th of the region in Study 1. If the method is correct, the thermal map from Study 1 should be the same as the thermal map summation of the 10 cases in study 2. The SIMP method is based on the assumption that the heat transfer coefficients are the same at any locations for the case in Study 1 and any case in Study 2, and the thermal effects on the boundary layers are negligible.

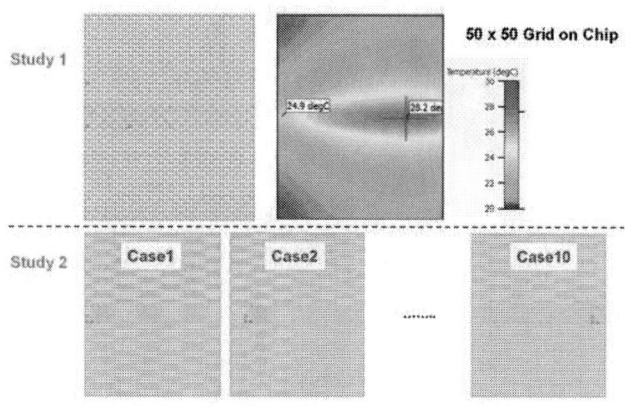

Note: Blue Square – Chip; Red Square – Power Supplied Region

**Figure 13 Description of two studies for thermal map comparison**

Fig. 14 shows the temperature distributions for 10 cases in Study 2. We observed the thermal dispersion for each case. The results are compared in Fig. 15. We studied the temperature profiles at 2 dash lines across the chip as shown in the figure. Results indicate that the assumption is acceptable and method is appropriate because these two lines are overlapped.

Before showing any power maps generated, we'd like to have some discussion on the relation between the supplied power and the resulting heat flux. Fig. 16 shows the heat flux contours at chip bottom and top for a case with uniform power supplied. We found that even for a uniform power map, the heat flux is not constant across the chip due to the flow heating effects. In the upstream, the fluid is colder, so that the heat flux is higher. While in the downstream, the cooling liquid has been heated in the upstream and the resulted hotter fluid causes the lower heat flux there. This phenomenon is more obvious at the chip top surface. The generated power map at the chip top surface is plotted in Fig. 17, which is quite comparable with the modeling results (right plot in Fig. 16).

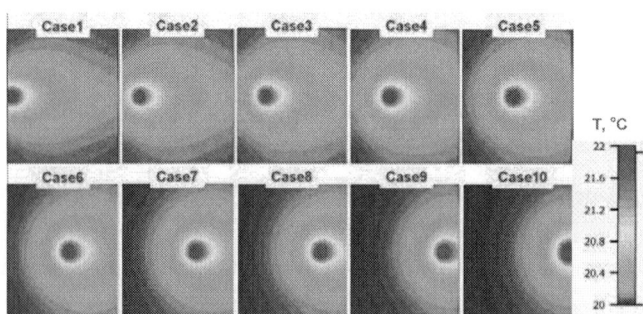

**Figure 14 Thermal maps for the cases in Study 2**

T, °C

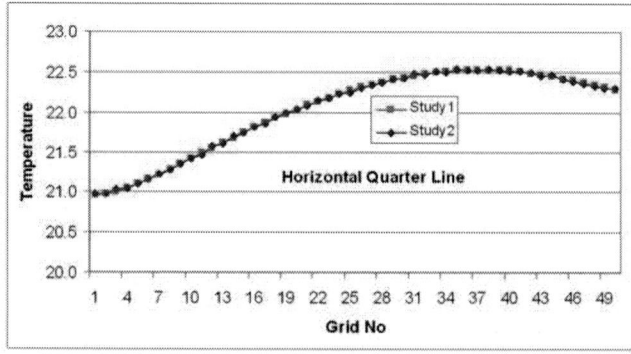

**Figure 15 Thermal map comparisons between Study 1 and 2**

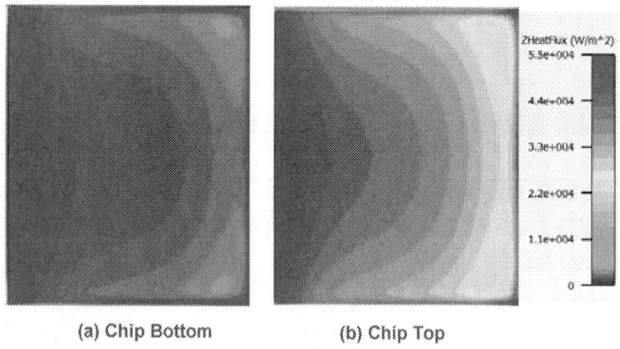

(a) Chip Bottom      (b) Chip Top

**Figure 16 Modeling power maps at chip bottom (Left) and chip top (Right)**

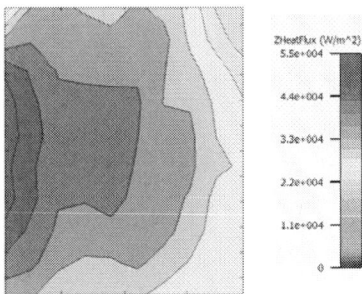

**Figure 17 Generated power map through thermal map – Chip top**

Fig. 18 shows the power map generated at the chip bottom for two 200 um die thickness TTV chips with 50 W total power uniformly distributed in all 8 heaters. The error of predicted total power is within 5%. The patterns and magnitude are quite comparable with the modeling results as shown in Fig. 16 (a).

The validation results for the cases with heater controlled are shown in Fig. 19. The power of each heater was kept the same as the previous case of 50W with all heaters on, while some of the heaters were turned off. The flow direction is from left to right. The power maps are shown in the most right plot in Fig. 19. Clearly, we observe that the power region boundary location is sitting in the middle of the die for the 4 heater (half of total 8) case, and at the quarter of the die for 2 heater (quarter of total 8) case.

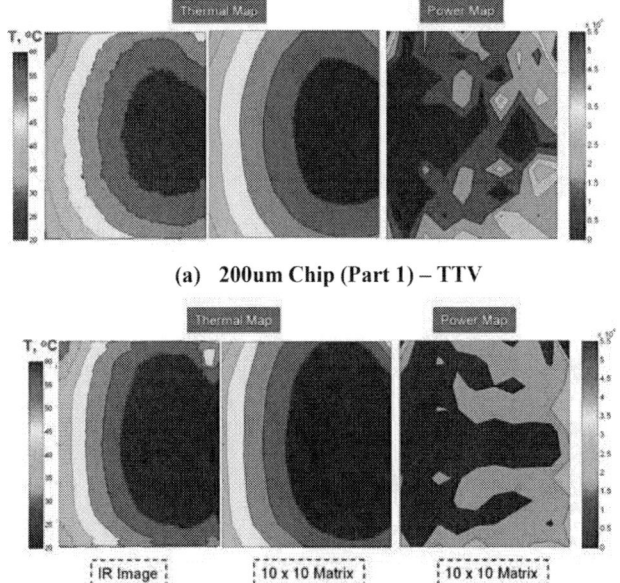

**Figure 18 Power map generation for two 200 um TTVs at Chip bottom – Uniform Power 50 W**

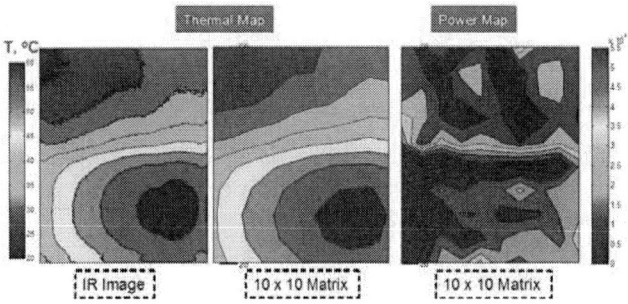

**(a) 200um Chip (Part 2) – TTV w/ 4 heaters**

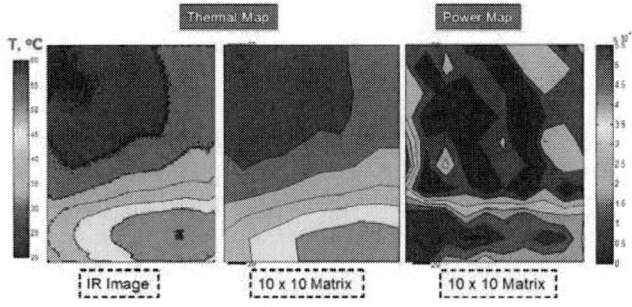

**(b) 200um Chip (Part 2) – TTV w/ 2 heaters**

**Figure 19 Power maps for a 200 um TTV at Chip bottom – Uniform Power 25 W; 4 heaters (Top) and 2 heaters (Bottom)**

In terms of the thermal imaging taken on the Product BI tool, we used a product chip as the example and conducted the hot spot measurements by controlling chip cores individually (Fig. 20). The measurement data were validated by the three On-Chip Thermal sensors (OCTs) readout in Table 1. Fig. 21 shows the non-uniform power map for the product chip, in which we are able to identify all the cores on the module precisely. This is not possible by just looking at the thermal image.

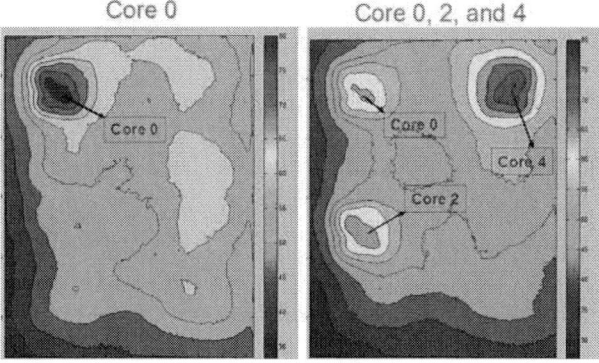

**Figure 20 Hot Spot Measurement – Control Core Individually**

978-1-4673-1110-6/12 $31.00 © 2012 IEEE

| Turn on Core 0; Ptotal = 85.3 W | OCTS SEN 0, °C | OCTS SEN 1, °C | OCTS SEN 2, °C |
|---|---|---|---|
| Sensor Readout | 84.0 | 50.7 | 58.5 |
| Extraction from IR Image | 83.0 | 51.1 | 59.6 |
| Error, % | 1.2 | 0.8 | 1.1 |

| Turn on Core 0, 2, and 4; Ptotal = 66.2 W | OCTS SEN 0, °C | OCTS SEN 1, °C | OCTS SEN 2, °C |
|---|---|---|---|
| Sensor Readout | 69.8 | 48.4 | 83 |
| Extraction from IR Image | 66.9 | 46.5 | 80.1 |
| Error, % | 4.2 | 3.9 | 3.5 |

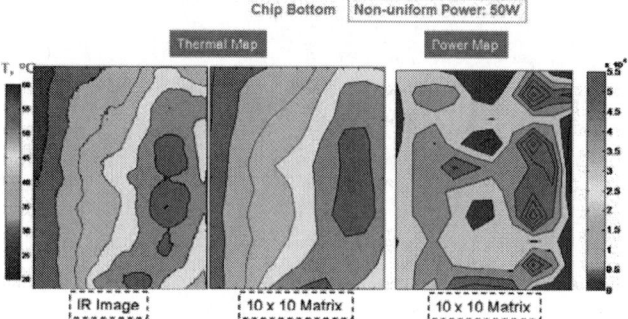

**Figure 21 Power map for a 200 um die thickness product chip –
Non-uniform Power 50 W**

## 4. Conclusions

A system was designed to visualize thermal map and predict power distributions on a chip in real time. The methodology developed has become an extremely useful and informative tool to identify chip hot spots. It has been widely used to address design, process, and defect related wafer and chip package issues. The present fast and universal thermal imaging capability was implemented directly on a fully supported manufacturing test platform and infrastructure. This work has been successfully applied to investigate wafer probe power/thermal issues for server high end products. Furthermore, a new concept of product thermal qualification was proposed as a supplement and potential alternative to the traditional thermal test vehicles (TTVs).

## Acknowledgments

The authors would like to thank Paul Aube and Paul Bodenweber for software/hardware support and useful inputs.

## References

1. US Patent: 7,167,806; H.F. Hamann, J.A. Lacey, M.P. O'Boyle, J.A. Wakil. A.J. Weger, Method and system for measuring temperature and power distribution of a device.

2. Hamann, H.F., Lacey, J., Weger, A., and Wakil, J. Spatially-resolved imaging of microprocessor power (SIMP): hotspots in microprocessors, Thermal and Thermomechanical Phenomena in Electronics Systems, 2006. ITHERM '06. The Tenth Intersociety Conference.

# Side-by-side Comparison between Infrared and Thermoreflectance Imaging using a Thermal Test Chip with Embedded Diode Temperature Sensors

Dustin Kendig[1]*, Kazuaki Yazawa[1], Amy Marconnet[2], Mehdi Asheghi[2] and Ali Shakouri[3]**,

[1]Microsanj LLC, 3287 Kifer Rd, Santa Clara, California 95051, USA

[2]Stanford Microscale Heat Transfer Lab, 450 Serra Mall Stanford, CA 94305–2004, USA

[3]Birck Nanotechnology Center, 1205 W State Street, West Lafayette, IN 47907-2057

*dustin@microsanj.com, **shakouri@purdue.edu

## Abstract

A side-by-side comparison between thermoreflectance imaging (TR) and infrared (IR) imaging is made using a specially designed thermal test chip with an embedded diode sensor array. IR thermal imaging is commonly used in industry. However, due to the infrared wavelength and the diffraction limit, IR has limited spatial resolution for chip level thermal characterization. In this paper we compare the spatial, thermal, and temporal resolutions of IR and TR methods and verify the results with integrated diode temperature sensors in the test chip. Thermoreflectance imaging showed higher spatial resolution, temporal resolution, and temperature accuracy on the metal heater. Infrared imaging showed to be less accurate on the metal without any coating to improve the emissivity. The TR measurement on the diode was within 1.7% of the diode reading, while the IR measurement was within 6%.

## Keywords

Thermoreflectance, infrared thermography, transient

## 1. Introduction

Temperature mapping of today's high-density electronic devices has been a grand challenge for chip packaging in electronics design. It is very important to design the chip layout to achieve better temperature uniformity in order to utilize silicon real-estate more efficiently and improve reliability. In this paper, we study the resolution and practical limitation of two widely used techniques, infrared and thermoreflectance imaging. [1-5]

## 2. Experiment Method

Infrared and thermoreflectance images were obtained for the thermal test chip (TTC) at high and low magnification. Both steady state and transient thermal images were obtained. Transient temperature change gives important information to understand the thermal structure of components in the chip as well as the external contacts.

### 2.1. Thermal Test Chip

We used a CMOS thermal test chip TTC-1002 with modular design, built for general characterization of the thermally induced phenomena by TEA (Figure 1). Each unit cell of the test chip includes two titanium heaters which cover 86% of the device area and 5 local temperature sensors at the center and various corners. Heaters in each unit cell are individually accessible. The TTC was wire bonded and packaged without any surface treatment for infrared imaging.

The thermoreflectance coefficient was calibrated for each material on the surface of the chip.

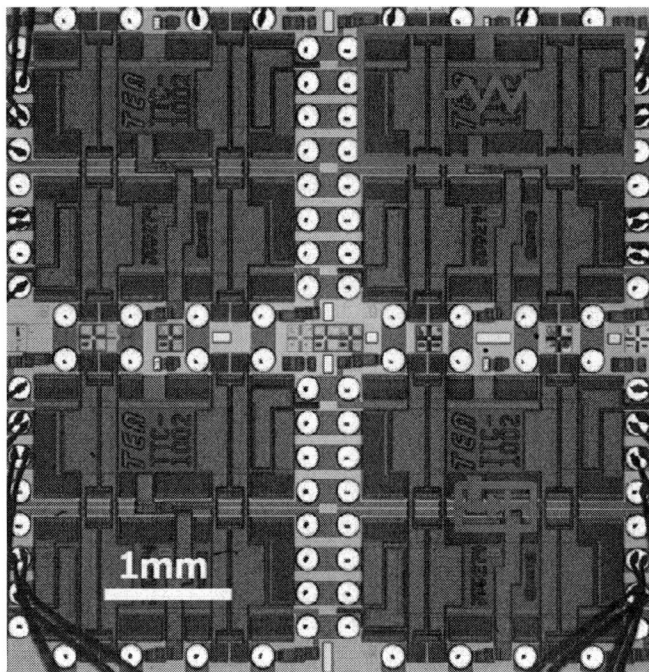

**Figure 1:** Optical Image of one TTC unit cell with highlighted (red) 7.6Ω heater and diode sensor

### 2.2. Infrared Imaging

Infrared imaging is a well known and well developed method that is based on blackbody radiation. All physical bodies emit electromagnetic radiation which is governed by Plank's Blackbody Law, which can be simplified to the Stefan Boltzman Law when integrated for all wavelengths. Objects at temperatures around 300K emit radiation in the infrared range which diffraction limits the spatial resolution of the thermal image to 3-10 microns.

How well an object emits this radiation is dependent on the emissivity of the material, which is between 0 and 1. The emissivity for metals and other reflective objects is low, while darker objects that absorb more light are much higher. For example, aluminum can have an emissivity of ~0.04 to 0.07 depending on roughness, while graphite has an emissivity of ~0.45. This difference in emissivity directly relates to the thermal signal coming from the device, and thus the signal to noise ratio (SNR) difference between aluminum and graphite would be ~10x different. Also, since the emissivity of a material is highly surface and material dependent, pixel-by-pixel calibration for each sample must be done for each new sample, even if it is the same material and manufacturing process. If the sample moves during the measurement, sample calibration must be redone. Sample movement such as thermal expansion at high magnifications also can be problematic for devices with many sharp features. For infrared imaging, it is common to coat the sample in a thin layer of material such as graphite to improve the emissivity and thus the signal. Often, surface coating is used in order to avoid feature by feature calibration and easy-to-use.

To improve SNR and temperature resolutions, lock-in thermography (LIT) techniques can be used to average out ambient noise and achieve temperature resolutions of μK with enough averaging. [6] At room temperature, radiation from other objects can give noisy data which reduces the detection limit of the system. To increase the possibility of detecting a small change in temperature, you can place the sample on a thermal stage and raise sample above room temperature. Since the radiation is proportional to the cube of the absolute temperature, a small change in device temperature will produce a much stronger signal at 50 or 60 C compared to room temperature. The ambient radiation of the detector is also an issue for the IR detector. This is remedied by using liquid nitrogen to cool the detector to temperatures where the radiation is negligible.

For measuring transient responses, IR cameras provide very crude resolution that is limited by the video frame rate. Single pixel IR detector can be used to achieve microsecond time resolution.

For this paper we used an older 256x256 pixel IR camera. Please note that there are newer cameras with 512x512 and 1024x1024 pixels, but this does not improve the ultimate spatial resolution of the system. Please also note that the IR system was not used with the lock-in thermography option and thus the temperature resolution/precision is larger. This however is negligible at the high temperatures the images were taken at and the accuracy of the temperature measurement was ultimately determined by the calibration for each material.

## 2.3. Thermoreflectance Imaging

Thermoreflectance imaging exploits the change in material reflectivity due to a change in temperature. A linear approximation of this relationship is often used when the temperature variation is small. This technique uses a probing light source to measure this change in reflected light rather than measuring the signal that is being emitted from the device. Because of this, the probing light can be pulsed to measure the temperature at specified time delays with regards to the biasing pulse. This can also be done at cryogenic temperatures since we are not limited by photons emitted by blackbody radiation. The amount that the reflectivity coefficient changes with temperature is called the thermoreflectance coefficient, and it is non-zero for most wavelengths, thus visible light can be used to measure the change in reflectance. This increases the spatial resolution of the thermal image by a full order of magnitude compared to IR imaging. This greater spatial resolution is important for

obtaining more accurate peak temperatures of the device under test.

We have adapted a differencing technique to obtain a full field, mega pixel thermal transient of devices. Using this technique we can obtain a series of images showing how the device heating propagates in time. This is different from the LIT technique that uses an excitation with 50% duty cycle and sine wave approximations of the thermal signal. Our current setup can obtain 100ns time resolution, and 800ps results have been obtained in university research with a pulsed laser. [7] The transient system works by opening the camera shutter and pulsing the light source. The pulsed light source samples the change in temperature of the device at a given delay with respect to the start of the excitation pulse. This thermal transient information is particularly useful as it can show the heat diffusion from microscale hot spots or features in the chip down to the thermal interface material and the package.

To determine the thermoreflectance coefficient for the TTC, we placed the sample on a thermoelectric cooler and modulated the temperature at low frequencies to insure uniform heating on the sample. We used a thermocouple to measure the temperature change of the stage. With this information we could determine the coefficient for each material on our test device.

## 3. Results and Discussion

We obtained thermal images of the TTC under different bias voltages and magnifications and used the integrated diode sensors as the reference to compare the thermal images to. Point by point emissivity calibration was used for the infrared images. For the TTC sample using a 530nm LED, we obtained a thermoreflectance coefficient of -2.1E-4 for the unpassivated metal heater (grey) and -2.2E-4 for the passivated (dark grey) material on the heater in Figure 1. The current thermoreflectance image processing software allows only two coefficients to be mapped to thermal image at a time, so the TR images in this paper will only be calibrated for the metal heaters.

### 3.1. Thermal Image Results

Thermal images of the TTC from the different measurements agreed well (Figure 2). Although the test device is designed to be uniform physically, a non-uniform temperature map was observed in all of the measurements. This temperature non-uniformity was measured to have 8% variation when looking at the diode temperature data. The infrared measurements showed a 15% temperature variation between the left and right side, and thermoreflectance measurements showed a 16% difference. The data shown here is a good example of why thermal imaging is important for device design and characterization. The roughness and other device features show in the thermoreflectance images more due to the high pixel count in Figure 2a. Small changes in reflectivity due to the passivation and roughness can be overcome by taking an average over a region of interest (ROI). For further data analysis, ROIs of the thermal images will be used to reduce error. In Figure 2b we can see in the

978-1-4673-1110-6/12 $31.00 © 2012 IEEE          345

thermal image where the differences in emissivity show device features similar to the TR images. The graphite coating in Figure 2d gives a uniform emissivity and removes any of the artifacts due to emissivity differences or surface roughness. This is ideal, but then we are slightly altering our device by putting this coating on. The TR image was taken at 29V, while the IR image was taken at 30V. Room temperature was added to TR images to make values absolute values in Celsius.

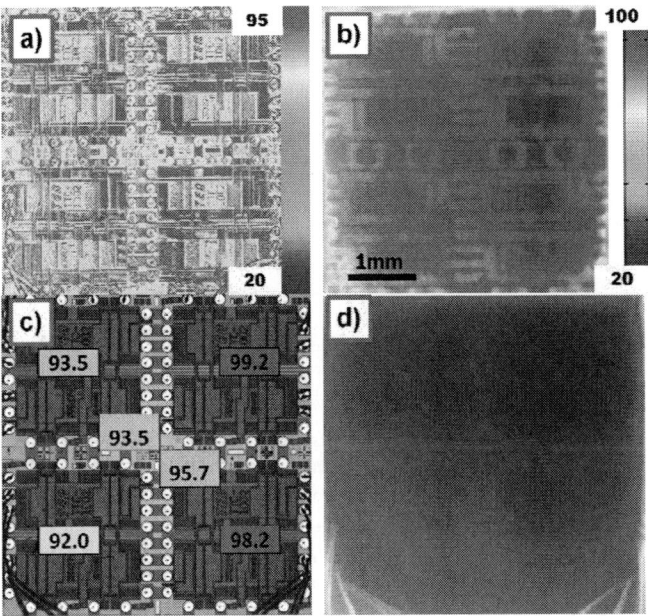

**Figure 2**: a) TR image calibrated for metal heaters only, b) IR image, c) diode temperature map, d) graphite coated IR thermal image

Thermal image sweeps were taken up to 30V to characterize the accuracy of the measurements. The 30V IR measurement was made at 5x magnification in the specified ROI. The measurements show good correlation overall, with the TR measurement within 1.7% of the diode and the IR measurement within 6%. Some of the error in the IR measurement could be due to the low emissivity of the metal or error due to thermal expansion and sample movement.

The two systems were also tested with a 120μm heater that has been fabricated on top of a microcooler with an insulating oxide layer in-between. This image pushes the spatial resolution limit of the IR system (3μm/pixel). These images also show the difference between a lock-in transient measurement and a DC measurement. With the DC IR thermal image, the heat has time to diffuse to throughout the device structures and substrate. The transient TR measurement is pulsed, thus not allowing the heat to completely propagate throughout the device. This is due to the diffusion length being proportional to $1/\sqrt{f}$. The faster you excite the device, the more localized the heating will be. The TR thermal image shows the 4μm heater lines clearly at 20x

magnification (600nm/pixel), however this still does not push the ultimate ~250nm limit of the system.

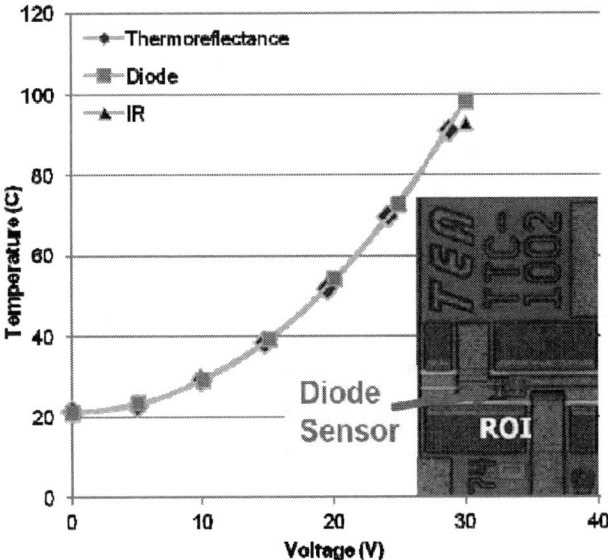

**Figure 3**: Temperature measurements of ROI next to diode

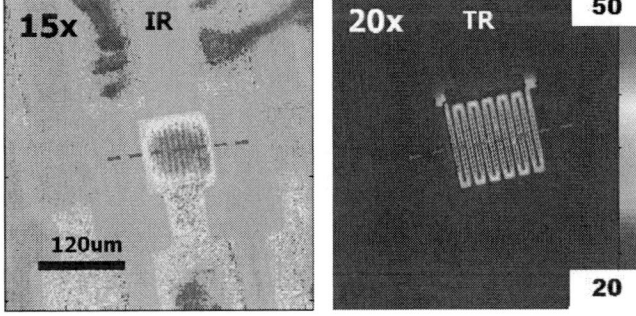

**Figure 4**: 2V DC IR and 3V/50 μs transient TR measurements of a microheater

### 3.2. Transient Data

Thermal transient data was obtained with the IR camera at 10Hz (full frame) and TR measurements were made up to 1MHz. IR measurements showed the slow turn on of the TTC which took about 500ms to reach 90% of the peak temperature. A transient TR image sweep was made at shorter time-scales to see how the heat propagated initially. Figure 5 shows that the heat from the heater blocks has not propagated to the substrate or interconnects 100μs after turning the TTC on. This data also shows that at these faster time scales the temperature between each heater is more uniform. There is no longer the 15% temperature gradient across the TTC. This hints that the non-uniformity is not a device issue, but a packaging/heat sinking issue which could be caused by how the die is attached. The thermal transient data in Figure 6 shows the sharp and fast thermal transient on top of the heater (μs regime), while the substrate thermal transient is much slower (ms regime). Heating under 10μs was negligible.

978-1-4673-1110-6/12 $31.00 © 2012 IEEE       346

**Figure 5**: Transient TR image @ 100µs and 60V

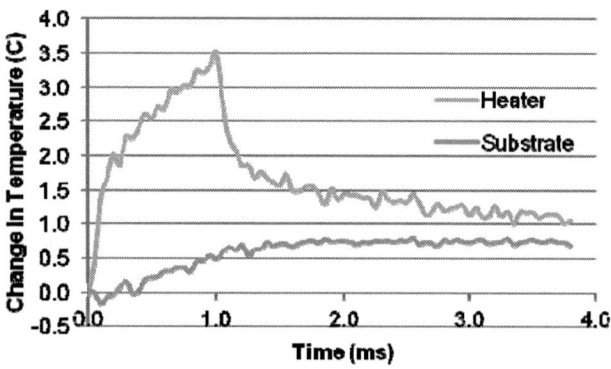

**Figure 6**: Thermal transient TR data of TTC at 30V, 1ms bias pulse, 10µs/ data point

## 4. Conclusions

We have shown data that compares infrared and thermoreflectance thermal images and verified results with a thermal test chip with integrated diode temperature sensors. These results showed a temperature gradient across the TTC with the IR data within 6% of the diode value and the TR data within 1.7%. Images of a microheater with 4µm heater lines showed the spatial limitations of IR. Transient TR image series were taken to view the thermal transient of the heater and substrate. These results showed the µs response time of the metal heater, while the substrate responded in the ms regime.

Both measurements had issues with thermal expansion and surface roughness at higher magnifications. The graphite coating on the sample solved this issue for the IR images and made a more uniform and smooth temperature map. Sources of error for the IR measurements could be due to low emissivity of the metal leading to a weaker signal or calibration discrepancies. Error in the thermoreflectance measurement is due to surface roughness/passivation non-

uniformities and SNR in the determination of the thermoreflectance coefficient.

## Acknowledgment

Authors acknowledge Professors Mehdi Asheghi-Roudheni and Ken Goodson at Stanford University, as well as Mr. Don Le and Mr. Joe Tran, for allowing us to use their infrared thermal imaging facility. Authors also thank Mr. Bernie Siegel of Thermal Engineering Associates for providing the thermal test chip and for valuable advice.

## References

1. M. Farzaneh et al., "CCD-based thermoreflectance microscopy: principles and applications", Journal of Physics D: Applied Physics 42, 143001 (20pp), 2009.
2. Lüerßen, D., Hudgings, J. A., Mayer, P. M., and Ram, R. J., "Nanoscale Thermoreflectance With 10mK Temperature Resolution Using Stochastic Resonance", 21st IEEE Semiconductor Thermal Measurement and Management Symposium, 2005.
3. Christofferson, J., Ezzahri, Y., Maize, K., Shakouri, A., "Transient Thermal Imaging of Pulsed-Operation Superlattice Micro-Refrigerators", 22nd IEEE Semiconductor Thermal Measurement and Management Symposium, pp. 45-49, 2009.
4. Christofferson, J, Shakouri, A., "Thermal measurements of active semiconductor micro-structures acquired through the substrate using near IR thermoreflectance", Microelectronics Journal, Vol. 35, (10), pp.791-796, 2004.
5. Komarov, P. L., Burzo, M. G., Raad, P. E., "A Thermoreflectance Thermography System for Measuring the Transient Surface Temperature Field of Activated Electronic Devices", 25th IEEE Semiconductor Thermal Measurement and Management Symposium, pp. 199-204, 2006.
6. Bauer, J., Breitenstein, O., Wagner, J. M., "Lock-in Thermography: a Versatile Tool for Failure Analysis of Solar Cells", Electronic Device Failure Analysis, Vol. 11, (3), pp. 6-12, 2009.
7. Christofferson, J., Yazawa, K., Shakouri, A., "Picosecond Transient Thermal Imaging Using a CCD Based Thermoreflectance System", 14th International Heat Transfer Conference (IHTC14)

# High Performance Thermal Interface Materials with Enhanced Reliability

Sihai Chen and Ning-Cheng Lee
Indium Corporation
34 Robinson Road, Clinton, NY 13323
schen@indium.com

## Abstract

Reliability has been a critical issue for thermal interface materials, especially when the materials are used for high power applications. In this paper, a series of reliability tests including baking, humidity chamber, temperature cycling, power cycling, and thermal shock have been conducted. Through comparison with conventional thermal pastes or phase change materials, a novel thermal paste with enhanced reliability is introduced. A mechanistic study has been performed to understand the reason for enhanced reliability. This paste possesses the following features and benefits:

- RoHS-compliant
- Room temperature storage
- No pump out or dry out
- Compliable to interfaces
- Minimal bond strength
- No need to cure
- No need to pre-reflow
- Good tack for component holding
- Easily printed or dispensed
- Direct replacement for grease
- Re-workable

This makes it a promising candidate for an advanced thermal dissipation solution in a variety of industrial applications.

## Keywords

Thermal interface materials, thermal paste, reliability, thermal cycling, temperature cycling, power cycling, humidity chamber test, thermal shock, baking test

## 1. Introduction

Thermal interface materials (TIMs) are a critical component placed in between heat generating and dissipating devices to facilitate the heat transfer. Conventionally, thermal greases have been widely available on the market because of their good thermal performance immediately after installation. However, after extended use, they can degrade significantly owing to "pump-out" and "dry-out" phenomena.[1-3] Due to the different coefficients of thermal expansion (CTE) between the heat generation (such as CPU die) and heat dissipation (such as heat sink) units, the powering up and down causes a relative motion between these units, resulting in "pumping" out of the paste from the interface gap. On the other hand, grease "drying out" occurs when the fillers separate from the organic matrix and/or the organics flow out at elevated temperature, resulting in delamination of the interface materials, therefore degrading the device reliability.

Many publications have tried to address the reliability issue of thermal interface materials. In one example,[4] a

composition comprised of a cured and cross-linked silicone-based gel is reported. It is found that these materials with proper mechanical properties could be used to avoid delamination. Another patent [5] describes a curable thermal interface material based on a silicone polymer matrix. However, in the 85°C 85% relative humidity chamber test, the thermal resistance increased nearly one order of magnitude after a little more than one month of treatment, indicating that the reliability of these materials is very poor.

Another approach to add the robustness of thermal interface materials is to form a polymer solder hybrid thermal interface material,[6] in which a solder with a low melting point is mixed with a composition containing polymer and a filler with a high melting temperature. The polymer is normally referred to as an epoxy or a siloxane- based organic, such as polydimethyl siloxane or poly (dimethyl diphenyl siloxane). These materials need a reflow process prior to real-time application, which increases the complexity and processing cost. Phase change materials are also applied to decrease the interface resistance;[7] however, due to the formation of the liquid phase, its reliability is doubtful, and it is easily pumped out, especially when the interface is vertically placed.

In summary, prior studies provide several means to improve the reliability of the thermal interface materials. Most of them [4-6] are using a silicone-based polymer as the main matrix. However, a silicone-based polymer normally has high permeability to both oxygen and water; therefore, it is not a preferred material suitable for highly reliable thermal interface materials.

In this paper, we would like to present a new thermal interface material called "Heat-Spring SC" (HS-SC).[8] This material is neither an epoxy adhesive nor a phase change material. A series of reliability test results shows that, as compared to two conventional thermal pastes (thermal pastes 1 and 2) and one phase change material (PCM 1), HS-SC is a more reliable material. Pastes 1 and 2 are both silicone-based materials. Paste 1 contains Al powders, while Paste 2 contains Ag and BN powders. PCM 1 mainly contains a low melting alloy (LMA).

## 2. Experimental Setup

A thermal test vehicle (TTV) instrument is used for sample tests. The details can be found elsewhere.[9] During the tests, thermal interface materials are placed in between a silicon die and a heat sink held in place by a spring-loaded clamping force. In the silicon die, electrically heated IC is instrumented with integral 4-wire resistance temperature detectors (RTDs). Since the gradients to the TIM surfaces are not directly measured, each TTV design is calibrated with a

978-1-4673-1110-6/12 $31.00 © 2012 IEEE

factor that includes the thermal resistance contribution of the thickness of the silicon die and the copper between the TIM and the thermocouple. The as-measured thermal resistance is called $R_j$ in this study. The real thermal resistance of the TIM is called $R_{TIM}$, which can be deduced from $R_j$ using a standard calibration curve.

For the same test, all the samples are placed in the same chamber to minimize experimental error.

## 3. Reliability Test Methods

The methods for reliability tests are shown in Table 1. For tests #1, 2, 3, and 5, samples are placed in the oven for a period of time, taken out, and cooled down to room temperature for measurement. For test #4, the experiments are conducted at room temperature. Occasionally, tests are paused to check how the pause will affect the $R_j$. No essential effects are observed. For the temperature cycling test, the profile of cycling is as follows: 25 min. from -55°C to 125°C, keeping at 125°C for 5 min., then 35 min. from 125°C to -55°C, keeping at -55°C for 5 min. For the power cycling test, switch-on power is kept at 50W, with 3 min. power up and 2 min. power down.

| # | Test methods | Cycle | Time (h) |
|---|---|---|---|
| 1 | 90°C Baking | | >2000 |
| 2 | 85°C, 85% RH | | >1800 |
| 3 | -55°C ~125°C | >1000 | |
| 4 | Power Cycling | >5000 | |
| 5 | 145°C, 30 min. | 4 | |

**Table 1.** Reliability test methods.

## 4. Comparison of $R_j$ Before Reliability Test

At least four measurements are conducted for each sample in order to understand how each paste performs before the reliability test. The graphic results and values are shown in Figure 1 and Table 2, respectively. It can be seen that Paste 2 gives the best performance in terms of low resistance and consistency. Paste 1 gives a similar performance as Paste 2. In contrast, PCM 1 displays a high mean thermal resistance and a large data variation.

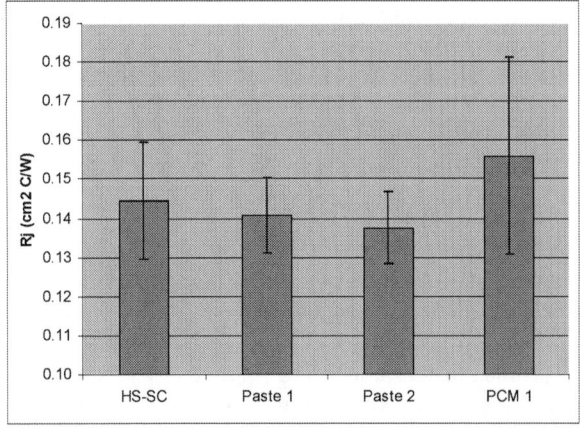

**Figure 1.** Thermal resistance of pastes before reliability test.

| | HS-SC | Paste 1 | Paste 2 | PCM 1 |
|---|---|---|---|---|
| Mean | 0.145 | 0.141 | 0.138 | 0.156 |
| StDev | 0.015 | 0.010 | 0.009 | 0.025 |
| StDev /Mean | 10.32% | 6.85% | 6.70% | 16.16% |

**Table 2.** Mean value and standard deviation of the thermal resistance of pastes before reliability test.

One very important point that we would like to stress is that the reliability of the paste is not related to the initial value of the thermal resistance measurement. In practice, people often favor a certain product once they have conducted several initial measurements of the thermal resistance. This is not right if you want a reliable product. As the experimental data shows later, Paste 2 shows bad reliability in our test results.

## 5. 90°C Baking Test

As shown in Figure 2, thermal resistance continuously increased for Pastes 1 and 2 as the aging time increased. For the PCM 1, it kept low and stable before 1000 h. However, it rapidly increased after further aging. The resistance of the HS-SC materials increased slightly in the early stage of aging, but became lower and stable through later aging steps in the test range, even over 2000 h. To further visualize the difference, average values of the thermal resistances of different pastes are shown in Figure 3.

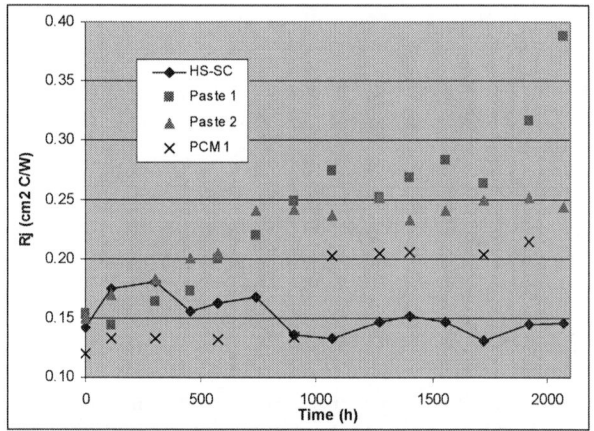

**Figure 2.** 90°C baking test results.

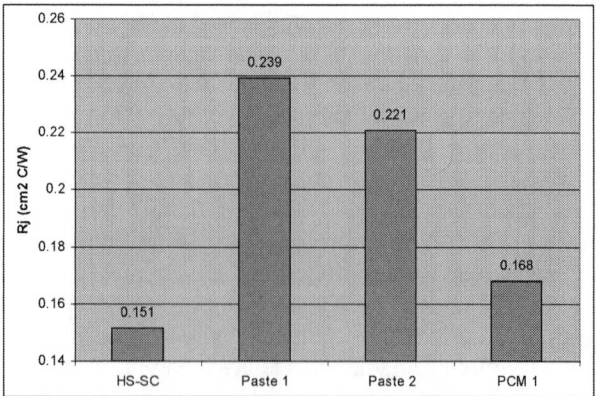

**Figure 3.** Average values of thermal resistance for different paste after 90°C baking tests.

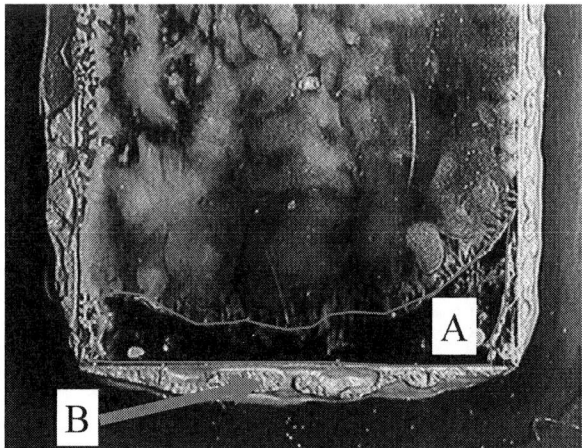

**Figure 4.** Photo showing the distribution of paste 1 on silicon substrate after 90°C baking tests.

In order to understand the underlying mechanism of the paste behavior after aging, an optical microscopic study is carried out by opening the device and observing the interfaces. Figure 4 shows Paste 1 on the silicon chip side of the device after the baking test. It is observed that little paste is left on the chip surface. In some areas (e.g., area A), there is almost no paste on it. However, more paste is found at the edge (area B), showing that "pumping out" is the main reason for performance degradation. A similar situation is also seen for Paste 2.

On the silicon or heat sink substrate, PCM 1 shows irregularly shaped metal materials, which are low melting point alloys. The surface of the alloy displays a black color and is glossless, indicating that the metal has been severely oxidized. In contrast, HS-SC paste does not show any of the above "pumping out" or oxidation phenomenon.

## 6. 85°C 85% RH Test

The test results and average thermal resistance values for different TIMs are shown in Figures 5 and 6, respectively. Paste 1 shows a sharp increase in thermal resistance after 460 h from 0.19 to 0.73 cm² C/W (Figure 5), indicating that some critical change occurred at this time. An interesting phenomenon is that the $R_j$ value decreases to 0.4 level before it increases again to about 0.6 cm² C/W after 1900 h. It is difficult to explain this decrease in resistance by "pumping out" or "drying out" mechanisms, since these two mechanisms will normally result in an increase in resistance. Optical microscopic study of the interface after the test gives us more insight. As displayed in Figure 7, channel-like structures are observed, implying that water vapor may have invaded the paste during the test. One possible explanation for the decrease of resistance may be as follows: since the test vehicle is taken out from the chamber for measurement, the vapor may condense so that the paste separated by vapor may recombine, recovering the contact between the interfaces.

A severe rise in thermal resistance is observed for Paste 2 and PCM 1 after 1400 h of aging (Figure 5). From the optical microscopy (Figure 8), a large dry area (e.g., area A) and even cracks (e.g., area B) are observed for Paste 2, indicating that "drying out" may account for the change. The metal surface of PCM 1 was completely oxidized resulting in complete failure of the device after 1400 h of aging. The HS-SC paste shows the least increase in thermal resistance even after 1900 h.

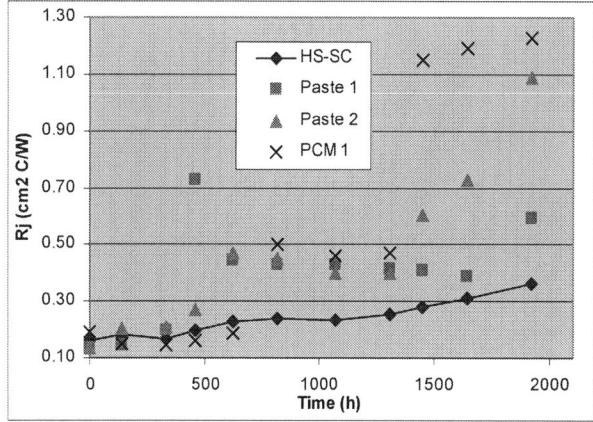

**Figure 5.** 85°C 85% relative humidity chamber test results.

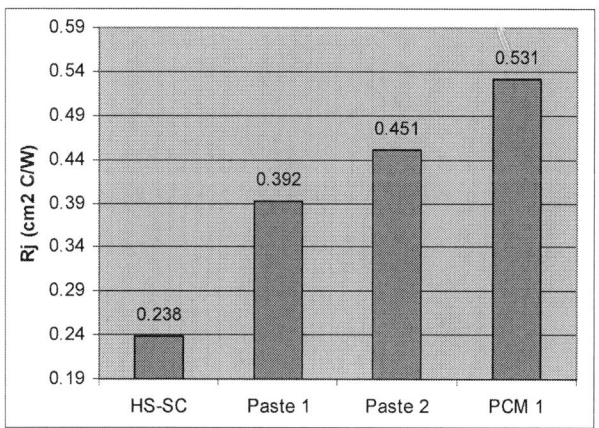

**Figure 6.** Average values of thermal resistance for different TIMs after 85-85 tests.

**Figure 7.** Photo showing the channel-like structures of paste 1 on silicon substrate after 85 - 85 tests.

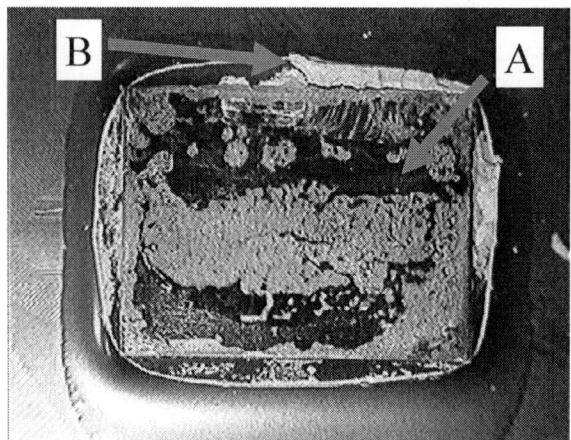

**Figure 8.** Photo showing the dry area of paste 2 on silicon substrate after 85 - 85 tests.

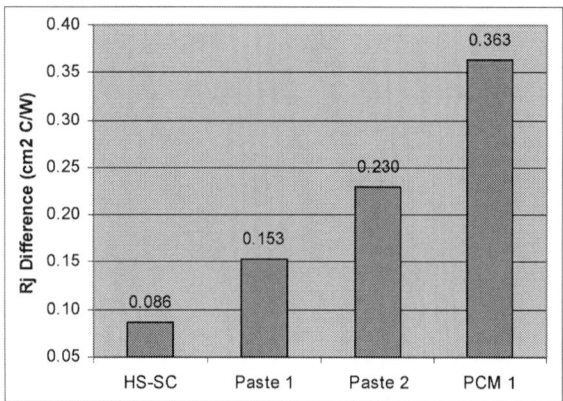

**Figure 9.** Average $R_j$ value difference between 85-85 and $90^{\circ}C$ tests.

The differences of average $R_j$ values between 85-85 and $90^{\circ}C$ test results are shown in Figure 9. If we regard the effect of the $5^{\circ}C$ temperature difference between these two test conditions as negligible, the data in Figure 9 will reflect the effect of humidity. One can see that humidity results in the increase of thermal resistance for all the samples (Figure 9). Among these, PCM 1 shows the biggest jump in resistance from average 0.168 to 0.531 cm$^2$ C/W (a difference of 0.363 cm$^2$ C/W), indicating that water is detrimental for this type of TIM. One explanation for this phenomenon may be that, because LMAs in the phase change materials are in the liquid state during operation, it is much more prone to be attacked by water. Pastes 1 and 2 give a moderate increase in thermal resistance at values of 0.153 and 0.23 cm$^2$ C/W, respectively. HS-SC kept at the lowest increase at a value of 0.086 cm$^2$ C/W.

## 7. Temperature Cycling Test

As displayed in Figures 10 and 11, temperature cycling data indicate that Paste 2 is the worst. Optical microscopic study reveals that the color of the paste has changed from grey to yellowish, implying that some essential change in chemistry of the composition may occur. At the same time, severe "pumping out" phenomenon was observed. Paste 1 and PCM

1 give moderate increase in thermal resistance. HS-SC shows the best performance overall.

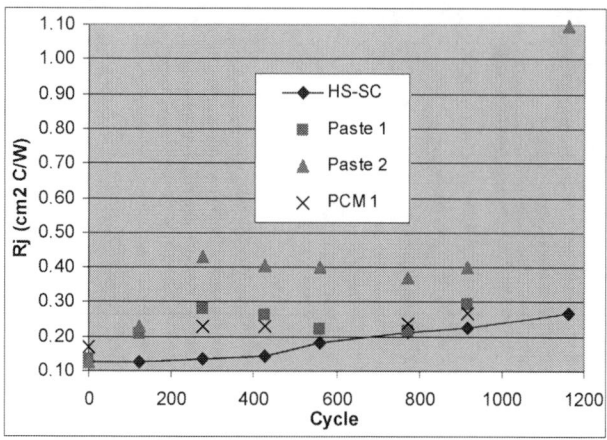

**Figure 10.** Temperature cycling test results.

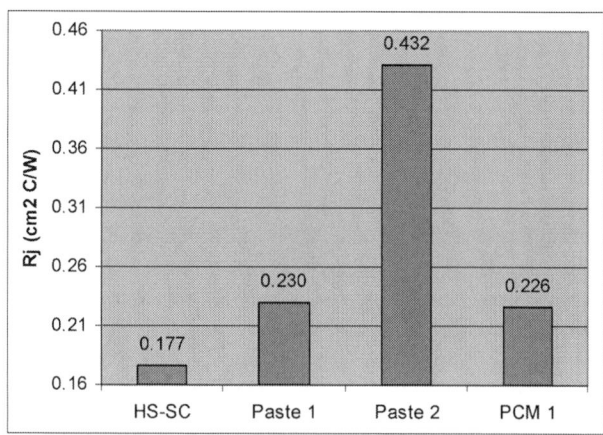

**Figure 11.** Average values of thermal resistance for different paste after temperature cycling tests.

## 8. Power Cycling Test

In order to mimic the real situation of computer operation, a power cycling test is conducted. As shown in Figure 12, Paste 2 gives the worst results. After 5000 h, its thermal resistance almost doubled, jumping to > 0.3 cm$^2$ C/W. Paste 1 ranked third place, PCM 1 ranked second, and HS-SC was the best. In order to clearly show the difference, thermal resistance values of 5000 cycles are averaged for each sample (Figure 13). One interesting phenomenon is that while other materials increased or kept almost constant in thermal resistance during the cycling course, HS-SC decreased in resistance as compared to the initial resistance value, which we attribute to the increased compliance of the materials with the interface.

The impact of $R_j$ value of the TIM on the temperature of the Si chip is big. If comparing Paste 2 with HS-SC materials, there is about three times the difference in thermal resistance at the end of 5000 cycles. At this time, junction temperature and heat sink temperature are 80.9 and $65.9^{\circ}C$, respectively for Paste 2, while they are 73.3 and $67.8^{\circ}C$ for HS-SC, respectively. This shows that the Si chip can be $7.6^{\circ}C$ cooler if HS-SC is used instead of Paste 2. Apparently, much heat has been transferred from Si chip to heat sink so that the heat sink temperature is $1.9^{\circ}C$ higher in the case of HS-SC.

978-1-4673-1110-6/12 $31.00 © 2012 IEEE

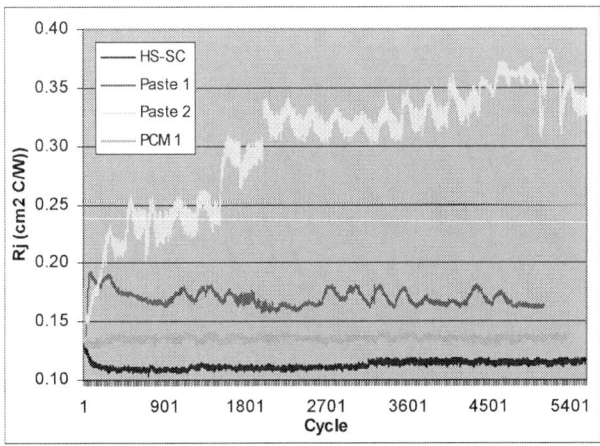

**Figure 12.** Power cycling test results.

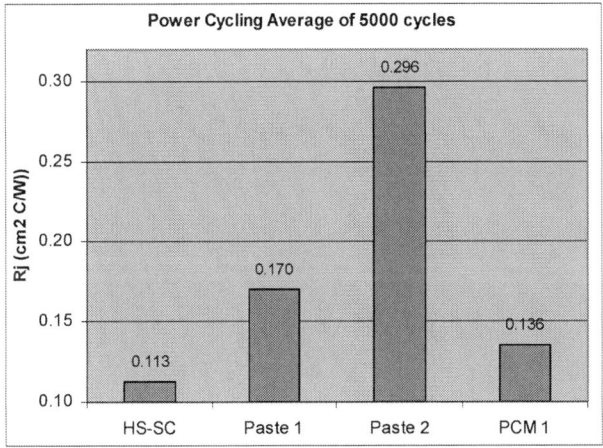

**Figure 13.** Average thermal resistance values of different samples.

## 9. 145°C Test

In order to test how high a temperature the HS-SC can survive, a thermal shock test is conducted. In order to protect the TTV device from dysfunction, we choose 145°C as the thermal shock test temperature. During the experiment, the TTV devices loaded with paste are placed in a 145°C oven for 30 min. and then taken out for measurement. We repeated this thermal shock process four times, and six samples were applied. The obtained $R_j$ values are listed in Table 3. It shows that before the thermal shock, $R_j$ is averaged at 0.146cm$^2$ C/W. After a 145°C treatment, it decreased to 0.130 cm$^2$ C/W level. Repeated experiments show the samples are very consistent in the test.

The decrease in thermal resistance is confirmed by tracing the measurement process as shown in Figure 14. This is a unique feature only observed for HS-SC paste. As we mentioned above, we attribute it to the compliance of the materials to the interface.

If converting the $R_j$ into thermal resistance of the TIM itself, we obtained the value of $R_{TIM}$ at 0.054 cm$^2$ C/W before test, at 0.04 cm$^2$ C/W after thermal shock tests.

| HS-SC | Before | 145C-30m-1st | 145c-30m-2nd | 145C-30m-3rd | 145c-30m-4th |
|---|---|---|---|---|---|
| 1# | 0.153 | 0.115 | 0.114 | 0.115 | 0.123 |
| 2# | 0.152 | 0.132 | 0.139 | 0.138 | 0.135 |
| 3# | 0.140 | 0.126 | 0.116 | 0.117 | 0.121 |
| 4# | 0.146 | 0.142 | 0.147 | 0.151 | 0.156 |
| 5# | 0.133 | 0.137 | 0.132 | 0.136 | 0.133 |
| 6# | 0.153 | 0.138 | 0.124 | 0.120 | 0.121 |
| $R_j$ (mean) | 0.146 | 0.132 | 0.129 | 0.129 | 0.132 |
| $R_j$ (StDev) | 0.008 | 0.010 | 0.013 | 0.014 | 0.013 |
| $R_{TIM}$(mean) | 0.054 | 0.041 | 0.039 | 0.040 | 0.041 |

**Table 3.** $R_j$ (mean and StDev) and $R_{TIM}$ (mean) values of the HS-SC paste before and after 145°C 30 min. treatment for six samples.

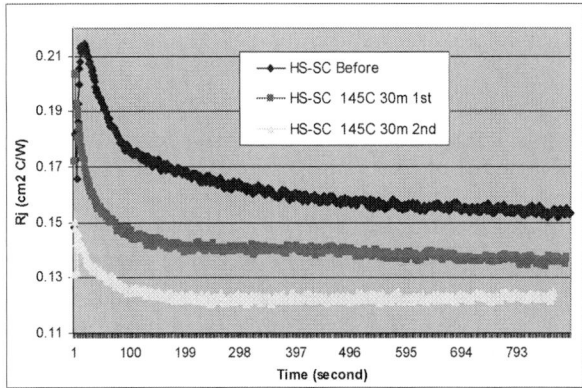

**Figure 14.** $R_j$ change before and after 145°C 30 min. treatment for one and two times.

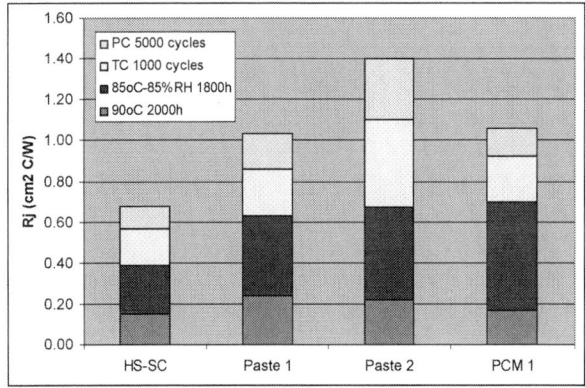

**Figure 15.** Stacked data of average thermal resistance values for different samples under HAST conditions.

## 10. Conclusions

As a summary, the $R_j$ values of different reliability test results for each material are stacked in Figure 15. The overall ranking based on reliability is: first (HS-SC); second (Paste 1); third (PCM 1); and fourth (Paste 2). Based on the data in Figure 1, the overall ranking before the reliability test is: first (Paste 2); second (Paste 1); third (1 HS-SC); and fourth (PCM). Comparing these ranking data, we conclude that a conventional thermal resistance test result without a highly accelerated stress test (HAST) cannot reflect the reliability of the thermal interface materials.

978-1-4673-1110-6/12 $31.00 © 2012 IEEE

Also, as shown in Figure 15, HS-SC proves to be the most reliable paste within the tested samples. In tests such as baking and thermal shock, its thermal resistance decreases gradually, which we attribute to its capability to comply with the interfaces.

Conventional silicone-based pastes normally suffer from performance degradation by "pumping out" and "drying out" mechanisms. In the case of a humidity test, water vapor may invade the paste, destroying its integrity.

Phase change materials are vulnerable to water attack due to its liquid state during operation.

**Acknowledgments**

The authors would like to thank R. N. Jarrett for help and discussion.

**References**

1. Viswanath, R., Wakharkar, V., Watwe, A., Lebonheur, V. "Thermal Performance Challenges from Silicon to Systems", Intel Technology Journal Q3, 2000.
2. Chiu, P-C., Chandran, B., Mello, M., Kelly, K., "An Accelerated Reliability Test Method to Predict Thermal Grease Pump-Out in Flip-Chip Applications", Electronic Components and Technology Conference, 2001.
3. Gowda, A., Esler, D., Nagarkar, K., Tonapi, S., "Reliability Testing of Thermal Greases", ElectronicsCooling, Volume 13, No. 4, pp. 10-16, 2007.
4. Matayabas, Jr. et al., "Electronic packages having good reliability comprising low modulus thermal interface materials" US Patent No. 6597575. Jul. 22, 2003.
5. Bhagwagar et al., "Thermal interface materials and methods for their preparation and use" US Patent No. 6791839. Sep. 24, 2004.
6. Koning et al., "Polymer solder hybrid" US Patent No. 6813153. Nov. 2, 2004.
7. Matayabas, Jr. et al., "Phase change thermal interface materials including polyester resin" US Patent No. 7408787. Aug. 5, 2008.
8. Heat-Spring® is the registered mark of Indium Corporation.
9. Jarrett, R. N., et al. "Comparison of test methods for high performance thermal interface materials", Proceeding of 23th IEEE SEMI-THERM symposium, 2007.

# Performance Improvements of Air-Cooled Thermal Tool with Advanced Technologies

Rahima K. Mohammed, Yi Xia, Ridvan A. Sahan, Ying-Feng Pang
Intel Corporation
MS: SC12-214, 3600 Juliette Lane, Santa Clara, CA-95054
Email: rahima.k.mohammed@intel.com

## Abstract

Air-cooled based thermal margining tool is simple, less costly and reliable cooling solution compared to liquid-cooled thermal tool. Air-cooled thermal tool is used by many Intel® groups for different purposes such as CPU, chipset or ASIC debug and validation. However, conventional air-cooled thermal margining tool do not provide sufficient cooling due to the increasing challenges in thermal management, such as form factor constraint, high power density and noise level limitation. High performance cooling technologies need to be investigated to meet the increasing demand for thermal performance. This paper introduces four advanced cooling strategies, their implementation and ongoing prototyping efforts to achieve performance improvement for air-cooled thermal tool design. In specific, this paper studies the optimization of heat sink design with different fin materials, different embedded heat pipe sizes and numbers, vapor chamber base, liquid chamber base, and combinations of different cooling technologies. The results show that copper fin with liquid chamber base and four $\phi$8mm heat pipes provide the best thermal performance among all the technologies studied.

## Keywords

Air cooling, thermal tool, high performance thermal technologies, heat pipe, vapor chamber, liquid chamber.

## Nomenclature

| | |
|---|---|
| AC-TT | Air-cooled thermal tool |
| Al | Aluminum |
| CFD | Computational fluid dynamics |
| CPU | Central processing unit |
| Cu | Copper |
| LC-TT | Liquid-cooled thermal tool |
| TDP | Thermal design power |
| TEC | Thermo-electric cooler |
| Tc | Case temperature |
| Tcp | Cold plate temperature |
| Tla | Local ambient temperature |
| TT | Thermal tool/head |
| TTV | Thermal test vehicle |

## 1. INTRODUCTION

Silicon debug, test and validation are essential for delivering world-class reliable products in the market. Temperature margining thermal tools are used for accelerating fault detection, identifying bugs, validating silicon, extracting thermal design power (TDP), reducing escapes and reducing time to market in the silicon debug, test, and validation environment. Many failure conditions can be accelerated through thermal stressing using either liquid-cooled TT (LC-TT) presented by Mohammed et al. [1] or air-cooled TTs (AC-TT) presented by Mohammed et al. [2]. Both liquid-cooled and air-cooled TTs can provide temperature margining capability by varying the case temperature for a certain range based on silicon TDP and TT's cooling capability. Compared to LC-TT, AC-TT is expected to provide a narrower range of margining capability due to the limitations of air cooling. If the AC-TT capability meets the validation customers' requirements, then the AC-TTs provide additional benefits on reduced cost, noise, elimination of chilled water, piping management, as well as messy tubing harnesses and cable management compared to the LC-TTs. This paper presents the efforts of improving the existing AC-TT as well as developing new AC-TT to widen the temperature margining capability and address the thermal management challenges caused by increased power and decreasing form factor. As shown in Figure 1, an AC-TT consists of a cold plate, a peltier device that function as a heat pump, a heat sink assembly with retention for cooling, a temperature controller for driving the temperature change to achieve the set point temperature, and a temperature sensor for providing feedback to the temperature controller. The demand for AC-TT heat sink performance is more critical than regular heat sink since it needs to dissipate the heat generated not only from silicon but also from the peltier device (2-3 times of the silicon power). Thus, this paper mainly focuses on the heat sink performance improvement. Various cooling technologies were investigated numerically and experimentally. In specific, this paper studies the heat sink with different fin materials, different embedded heat pipe sizes and numbers, vapor chamber base, liquid chamber base, and combinations of different cooling technologies.

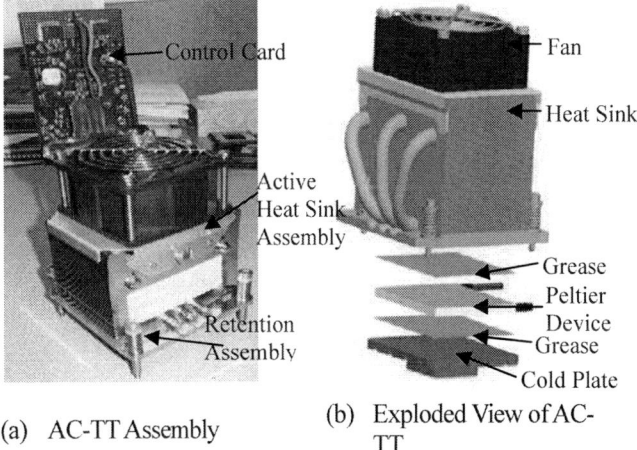

(a) AC-TT Assembly

(b) Exploded View of AC-TT

**Figure 1.** Elements of Air-Cooled Thermal Tool Design

978-1-4673-1110-6/12 $31.00 © 2012 IEEE

## 2. ADVANCED COOLING TECHNOLOGIES FOR AIR-COOLED THERMAL TOOL

Air-cooled thermal margining tool design has critical challenges due to form factor constraint, high power density and noise level limitation. Increasing power density and decreasing form factor requirement in every processor generation makes the thermal design more challenging. High performance cooling technologies need to be investigated and identified for air-cooled thermal tools to meet the increasing demand for thermal performance. This section is intended to provide an overview of the advanced cooling technologies investigated in this study to improve the thermal performance of the heat sink, widen the temperature margining capability and address the thermal management challenges.

### 2.1. Heat Pipe Technology

Heat pipes are widely used in electronics cooling applications [3]. As phase change heat transfer devices, heat pipes can effectively transfer heat over their length with a small temperature gradient. A heat pipe is an evacuated and sealed pipe which contains a small amount of working fluid and a wick structure. The working liquid evaporates from liquid to vapor at the hot end and condenses at the cold end while releasing its latent heat at the same time. The condensed liquid returns to the evaporator through the wick structure by capillary action. Heat pipes offer many advantages in their use and operation [4]. It has simple structures, no moving parts, no liquid leakage issue, does not need extra power, has low cost, is highly reliable and easy to maintain. The heat pipes take hot spots on the heat sink base and transport the heat across the heat sink base to the top side of fin for more efficient cooling. So, it not only enhances base-plate heat spreading but also boosts fin efficiencies as shown in Figure 2. In general, a heat sink with embedded heat pipes can offer thermal performance improvement of up to 20% when compared to a typical aluminum or copper base heat sink.

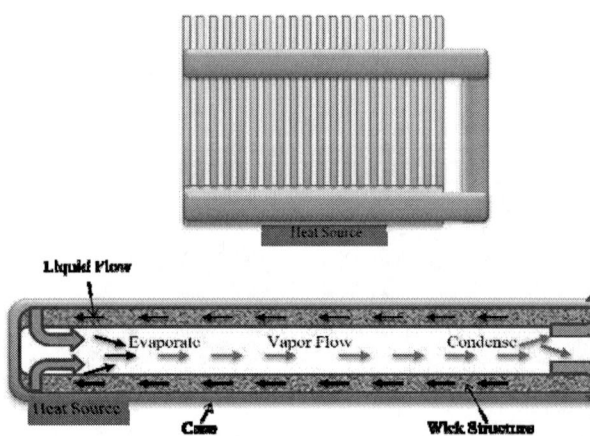

**Figure 2.** Heat Sink Design with Embedded Heat Pipe

Selection of the working fluid and heat pipe material is based on the operating temperature of the application. In computer application, the operating temperatures are normally between 50°C - 100°C. At this temperature range, copper and water are usually used to build heat pipe. Water is the best working fluid due to surface tension, latent heat, and vapor density. Its high latent heat of vaporization spreads more heat with less fluid flow. Water's high surface tension, when presented to a wick with small pore size generates a large capillary force. Its high thermal conductivity minimizes the ΔT associated with conduction through the wick. In addition to its thermodynamic properties, water is safe with no cost. A variety of wick structures have been produced as shown in Figure 3. Mesh screen and sintered powder wicks are heat pipe wick structures that are most often used for electrical cooling. Table 1 shows the typical parameters for embedded heat pipe heat sink. The embedded heat pipes need to be optimized by evaluating the size, number and routing location. Figure 4 shows the performance of heat pipes as a function of various diameters [5].

The use of heat pipe on AC-TT design is mainly for three purposes: (1) to aid heat spreading across the heat sink base, thereby effectively increasing the base thermal conductivity especially when aspect ratio of heat sink base size to heat source size is high; (2) to act as a primary heat conductive path for transmitting the heat from the source to a remote location where the heat can be managed due to space constraint; (3) to improve the effective conductivity and efficiency of a traditional heat sink.

**Figure 3.** Different Wick Structures for Heat Pipe

**Table 1.** Embedded Heat Pipe Heat Sink

| Typical Heat Pipe Size | 4mm, 6mm, 8mm, 10mm |
|---|---|
| Embedded Heat Pipe Options | Expanded, Soldered or Adhesively Bonded |
| Heat Pipe Material | Copper, Aluminum, Nickel, Stainless steel, Titanium, Monel, etc. |
| Working Fluid | Water, Methanol, Liquid Ammonia, Mercury, etc. |
| Wick Structure (Heat Dissipation Capability) | Axially Grooved Wicks ($< 40 W/cm^2$) Mesh Screen Wicks ($< 40 W/cm^2$) Sintered Powder Wicks ($< 250 W/cm^2$) Bi- dispersed Wicks ($< 1000 w/cm^2$) Porous Metal Wicks ($< 12 KW/cm^2$) |

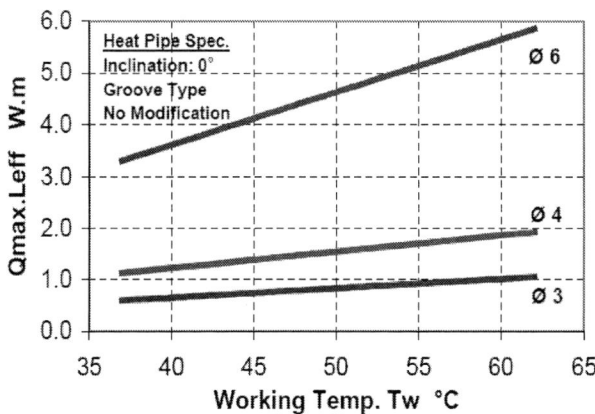

**Figure 4.** Performance of Heat Pipes As a Function of Various Diameters

### 2.2. Vapor Chamber

Vapor chamber is widely integrated with heat sink designs. Vapor chambers are typically referred as planar or flat heat pipes as shown in Figure 5. The principle for vapor chamber operation is similar to the heat pipe. They are both two-phase heat transfer devices. Unlike the cylindrical heat pipes that spread the heat axially, the vapor chamber spreads heat in two directions, allowing nearly isothermal heat dissipation for heat sink with vapor chamber base [6-7]. In addition, direct device attachment eliminates the thermal resistance associated with the interface between the heat pipe and the heat sink base. Vapor chamber heat sinks are lighter and more efficient than solid copper heat sink base. The temperature drop associated with the vapor spreading is negligible, providing effective means of spreading the heat from a concentrated source to a large surface. Thermal spreading resistance or the effective of thermal conductivity is significant at the base of traditional heat sinks when aspect ratio of the heat sink base size to heat source size is high. Vapor chamber can reduce the spreading resistance by providing an effective way to spread heat evenly across the base and remove hot spots. The heat sinks with vapor chamber base can offer thermal performance improvement up to 30% when compared to a typical aluminum or copper base spreaders.

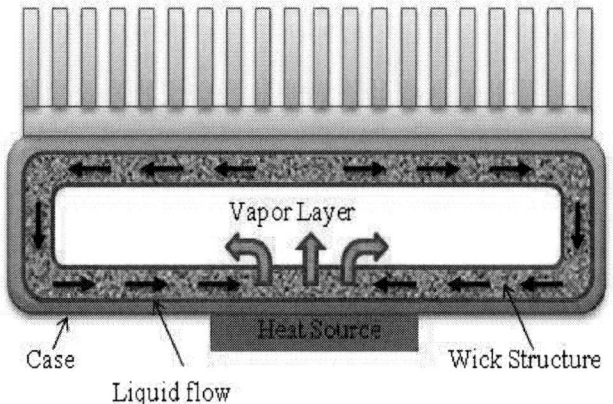

**Figure 5.** Vapor Chamber Based Heat Sink Design

### 2.3. Liquid Chamber

Liquid chamber is another two phase heat transfer device as shown in Figure 6. It can provide nearly uniform heat spreading at the heat sink base with low thermal resistance. Similar to the vapor chamber, liquid chamber is a flat sealed container with working liquid moving inside to transfer heat from heat source to cooler end. However, the principle of liquid chamber operation is different than the vapor chamber. First of all, there is no wick structure within the liquid chamber. Instead, it has special micro-porous coating surface near hot spot. The microporous coating has multi-layered microstructures that improves the heat carrying capacity by increasing the number of active nucleation sites in contact with liquid. The microporous surface makes the contact surface rough, increases the number of bubbles produced, allowing more heat to be carried away more quickly, leading to much greater cooling efficiency. Second, the heat is transmitted through pool boiling effect instead of evaporation [8]. Third, the liquid chamber is more efficient for high thermal design power applications since pool boiling becomes more prominent above 50°C. Vapor chamber does not have this limitation. It can function as long as there is temperature gradient existing among the chamber. Finally, the performance of liquid chamber heat sink is less impacted by the orientation change compared to heat pipe or vapor chamber which sometimes is constrained by the capillary limit. The difference between vapor chamber and liquid chamber are summarized in Table 2.

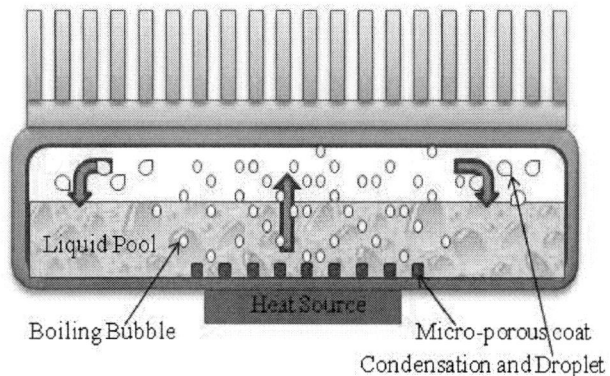

**Figure 6.** Liquid Chamber Based Heat Sink Design

Performance constraint of vapor chamber can be overcome by ample supply of liquid to the evaporator, coupled with boiling enhancement technology. Incipience overshoot is the excess temperature required to initiate boiling. After initiation, the bubble generation spreads quickly over the entire surface and the temperature drops dramatically. The critical heat flux is the maximum limit of heat dissipation achievable when entire surface is covered with a thin vapor blanket which causes the heat transfer rates drop again. Microporous coating is boiling enhancement surface used to reduce the incipience excursion and increase the critical heat flux (CHF). The heat transfer is caused by thin film evaporation inside the channels of the enhancement surface as well as the external convection induced by bubble agitation. The more superheat on the wall the more heat flux

978-1-4673-1110-6/12 $31.00 © 2012 IEEE

that the chamber can move. So heat sink performance can be improved by liquid chamber especially running at high temperature for high power application.

**Table 2.** Summary of the difference of vapor chamber and liquid chamber

| Vapor Chamber | Liquid Chamber |
|---|---|
| • Two phase heat transfer based on evaporation | • Two phase heat transfer based on pool boiling |
| • Latent heat only | • Latent heat plus micro convection |
| • Vacuum vessel with wicks structure | • Vacuum vessel with supporting/guiding ribs |
| • Heat source can be anywhere on Vapor chamber surface | • Heat source need to be located underneath of Microporous coating surface |
| • Small amount of liquid | • Need more liquid |
| • Liquid evaporate/condense when there is temperature gradient | • Liquid boils when temperature exceed boiling point |
| • Capillary action is slow down by gravity | • Insensitive to orientation |

Liquid chamber has simple internal structure as shown in Figure 7 [9] which eliminates performance and reliability problems that vapor chamber product often face. It can provide nearly uniform heat spreading at the heat sink base with low thermal resistance.

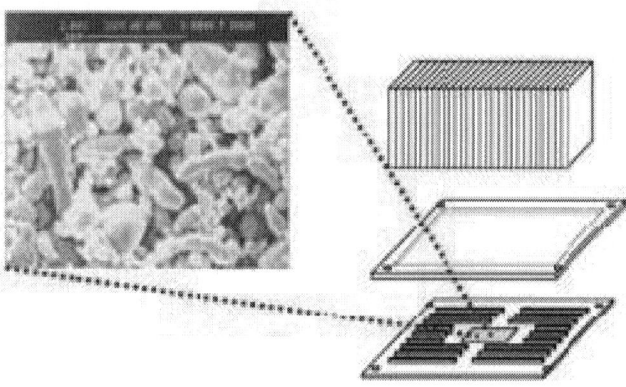

**Figure 7.** Liquid Chamber Heat Sink with Boiling Incipience Region

### 2.4. Combination of Heat Pipe with Vapor Chamber or Liquid Chamber

Combining heat pipe with vapor chamber or liquid chamber on a heat sink design is another advancement that can provide excellent heat transfer capability. Such hybrid heat sink design can provide superior cooling in a small form factor by utilizing the combination of two advanced cooling technologies. Figure 8(a) and 8(b) show the schematic of a thermal management device providing high cooling capability by spreading the heat horizontally through the vapor chamber

or liquid chamber heat sink base. Then, the heat is elevated vertically to the top side heat sink fins by the embedded heat pipes.

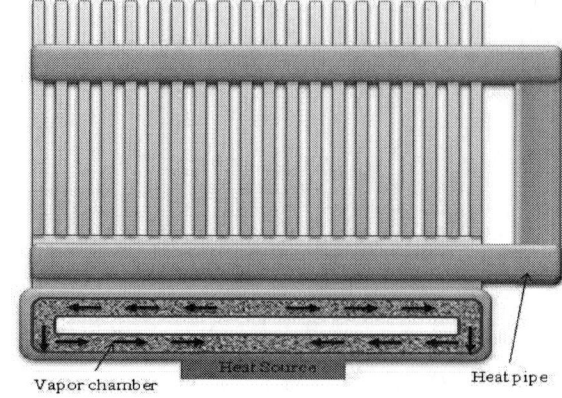

(a) Combination of Heat Pipe and Vapor Chamber

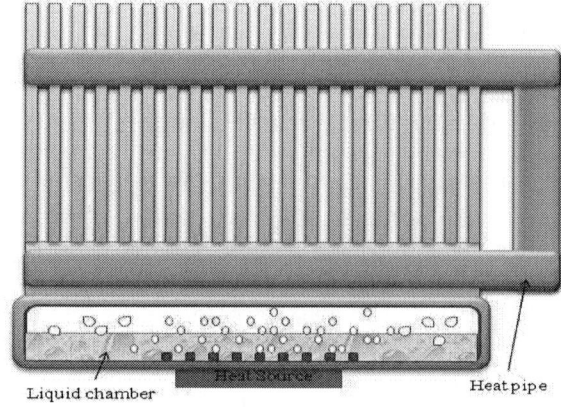

(b) Combination of Heat Pipe and Liquid Chamber

**Figure 8.** Combination of Heat Pipes and Vapor Chamber or Liquid Chamber Based Heat Sink Design

### 3. HEAT SINK DESIGNS

Many design factors were taken into consideration for AC-TT development process such as the keep out volume, silicon thermal design power, silicon size, cold plate, thermal interface material, heat sink, retention and thermo-electric cooler (TEC). In this study, four 2U heat sinks and two 1U heat sinks performance with different heat sink base material, fin material, heat pipe size and number of heat pipes were evaluated. Here, U refers to a unit of measurement (1.75") of the height of a rack-mounted device. Thus, a 1U product has a vertical measurement of 1.75"; 2U is 3.5". The performance evaluations of the heat sinks served as the fundamental guidance to the AC-TT designs. Table 3 summarizes the heat sink designs evaluated experimentally. The experiments were performed following the experimental procedure outlined by Mohammed et al. [10]. For both vapor and liquid chamber-based heat sink designs, the retention design needed upfront planning to prevent damage to the internal structure of both vapor and liquid chambers and provide the structural integrity of the chambers to accommodate the retention designs.

978-1-4673-1110-6/12 $31.00 © 2012 IEEE

**Table 3**. 1U and 2U Heat Sink Configurations Evaluated Experimentally

| Heat Sink Design | Base Material | Fin Material | Heat Pipe Size | Heat Pipe Qty | Heat Sink Height |
|---|---|---|---|---|---|
| 1 | Cu | Al | $\phi$8mm | 3 | 2U |
| 2 | Cu | Al | $\phi$6mm | 4 | 2U |
| 3 | Cu | Al | $\phi$8mm | 4 | 2U |
| 4 | Vapor Chamber | Cu | $\phi$8mm | 3 | 2U |
| 5 | Cu | Cu | n/a | n/a | 1U |
| 6 | Liquid Chamber | Cu | n/a | n/a | 1U |

## 4. AIR-COOLED THERMAL TOOL DESIGNS

The results from the heat sink experiments shown in Section 3 were then used as guidance for AC-TT performance studies. In this case, numerical simulations were carried out for feasibility study and performance optimization of the AC-TT.

Thermal design/analysis tools were employed: i) to build the computational fluid dynamics (CFD) models, ii) to study "what if" scenarios to perform parametric studies, iii) to predict the airflow distribution, and iv) to generate the thermal margining performance curves to successfully provide optimized thermal tool design for CPU, chipset, ASIC and memory [1-2]. Both detailed and compact heat sink models were used in thermal design simulations. TEC was modeled using the macro available in the software where all necessary inputs are provided by TEC vendor. These inputs include details of electric current being drawn, number of thermo-electric (TE) junctions, TE element height, TE element pitch, TE element area/height, ceramic based thickness, material properties of the TE element material and base material. The final model has high number of fins and each fin will require denser grid around them to resolve the flow field which increases total grid cells count. In order to reduce the total cell count as well as to achieve accurate results, CFD tool offers a methodology called non-conformal meshing. This method reduces the mesh cell count by at least 50% of the original grid count. In turn, this reduces the computational time required to perform the simulations and hence, cutting the thermal design cycle. Therefore, non-conformal meshing was used in all the optimization simulations to achieve local refinement around the critical components leading to accurate results with mesh independent solutions. For embedded heat pipes, vapor chamber and liquid chamber, lumped models were used with effective thermal conductivity obtained from heat sink experiments. For the studies presented in this paper, the CFD simulation results were used to guide the design for thermal tool performance improvements before prototyping.

Table 4 summarizes all the numerical simulations carried out to study nine different AC-TT designs. All designs are active heat sink with embedded heat pipes. In addition, the AC-TT designs were evaluated with the same cold plate, same fan, same TEC, and same local ambient temperature of 24°C at TDP of 150W except with various heat sink configurations.

**Table 4.** Air-Cooled Thermal Tool Designs

| AC-TT Design | Heat Sink Configuration | | | | |
|---|---|---|---|---|---|
| | Base Material | Fin Material | Heat Pipe Size | Heat Pipe Qty | Heat Sink Height |
| I | Cu | Al | $\phi$8mm | 3 | 2U |
| II | Cu | Al | $\phi$6mm | 4 | 2U |
| III | Cu | Al | $\phi$8mm | 4 | 2U |
| IV | Cu | Cu | $\phi$8mm | 3 | 2U |
| V | Cu | Al | $\phi$6mm $\phi$8mm | 2 2 | 2U |
| VI | Vapor Chamber | Al | $\phi$8mm | 3 | 2U |
| VII | Vapor Chamber | Cu | $\phi$8mm | 3 | 2U |
| VIII | Liquid Chamber | Al | $\phi$8mm | 3 | 2U |
| IX | Liquid Chamber | Al | $\phi$8mm | 4 | 2U |

The AC-TTs with embedded $\phi$8mm heat pipes shown in Figure 9 has copper base and aluminum fins. The AC-TT designs with vapor chamber and liquid chamber base are shown in Figure 10. The contact surface is flat for both designs. However, it is noticeable that the rest of the heat sink base surface of liquid chamber is forged with some bumpers for structural strength purpose.

(a) AC-TT with Three $\phi$8mm Heat Pipe

(b) AC-TT with Four $\phi$8mm Heat Pipe

**Figure 9.** AC-TT with Embedded Heat Pipes

 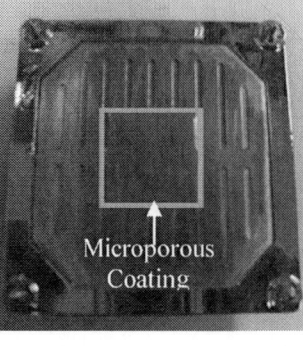

(a)  Vapor Chamber Base　　(b)  Liquid Chamber Base

**Figure 10.** AC-TT with Vapor Chamber and Liquid Chamber

# 5.  RESULT AND DISCUSSION

## 5.1 Heat Sink Performance

As shown in Table 2, heat sinks with different technologies were evaluated. The heat sink configurations were: i) 2U heat sink with solid copper base and Al fins, ii) 2U heat sink with different size and number of heat pipes, iii) 2U heat sink with vapor chamber base and Cu fins, and iv) 1U heat sink with liquid chamber heat sink base and solid Cu base and Cu fins. There is no embedded heat pipe in the 1U heat sink. The learning from heat sink experiments were then used to guide CPU AC-TT design.

Figure 11 compares the experimental data of the 2U heat sinks. It was found that having three $\phi$8mm heat pipes or four $\phi$6mm heat pipes in a 2U heat sink with Al fins design provides similar heat sink performance. Increasing the $\phi$8mm heat pipes from three to four could improve the heat sink with Al fins performance by 0.026°C/W or 3.9°C for a TDP of 150W at local ambient temperature of 24°C. In addition, replacing the copper base with vapor chamber base and Al fins with Cu fins could result in heat sink performance improvement of 0.041C/W. This provides ~6°C performance gain at TDP of 150W.

1U copper finned heat sink samples employing liquid chamber base as well as solid copper base were tested for performance comparison. Heat pipes are not embedded in this active heat sink. Figure 12 summarizes the experimental results of the 1U heat sinks. Both heat sinks were tested under the same ambient conditions using the same 80 mm X 25 mm fan providing the airflow required to cool the both heat sinks. Experimental results shown in Figure 12 indicate that as TDP increase from 60W to 240W, the liquid chamber based heat sink significantly performs better than the heat sink with copper base. The performance difference is up to 5.1°C at TDP of 240W. For liquid chamber base, the performance becomes more prominent at higher power since pool boiling becomes more efficient at higher temperature above 50°C.

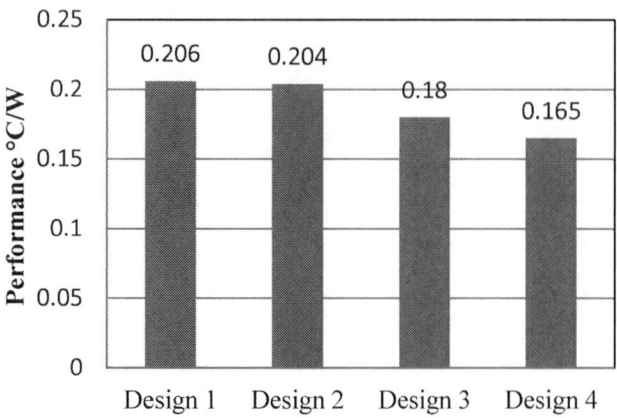

**Figure 11.** Experimental Data Comparison of 2U Heat Sink Performance

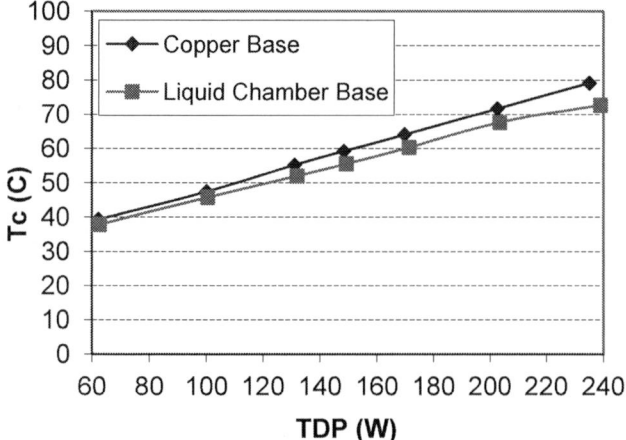

**Figure 12.** Experimental Data Comparison of 1U Heat Sink Performance

## 5.2 Air-Cooled Thermal Tool Performance

Nine different heat sink designs for AC-TT were numerically evaluated. The simulations were performed with package TDP of 150W and local ambient temperature of 24°C. The results shown in Figure 13 indicate that CPU AC-TT employing copper fins (Design IV) provides 3.3°C better performance than the CPU AC-TT using aluminum fins (Design I) at TDP of 150W. This is due to the fact that copper has almost twice higher thermal conductivity (387 W/m-K) than aluminum (205 W/m-K) which provides better heat transfer rate. Design I and Design II provide similar performances. This agrees well with the finding from the heat sink Design 1 and Design 2 experimental data shown in Figure 11. Similarly, the improvement on AC-TT with four $\phi$8mm heat pipe and copper base heat sink (Design III) provides 3.7°C improvement over the three $\phi$8mm heat pipe and copper base heat sink design (Design I). This also agrees well with the finding shown in Figure 11. AC-TT with combination of $\phi$6mm and $\phi$8mm heat pipes (Design V) only resulted in 1.6°C improvement over Design I.

978-1-4673-1110-6/12 $31.00 © 2012 IEEE

Replacing the copper heat sink base (Design I) with vapor chamber base (Design VI) provides 3.8% (1.4°C) improvement over Design I. An additional of 2.2°C improvement can be achieved by replacing the aluminum fins with copper fins (Design VII). On the other hand, replacing the copper heat sink base (Design I) with liquid chamber base (Design VIII) provides 2.8°C improvement. In addition, further improvement of as much as 5.1°C or 13.7% can be achieved by increasing the three $\phi$8mm heat pipe to four $\phi$8mm heat pipe (Design IX).

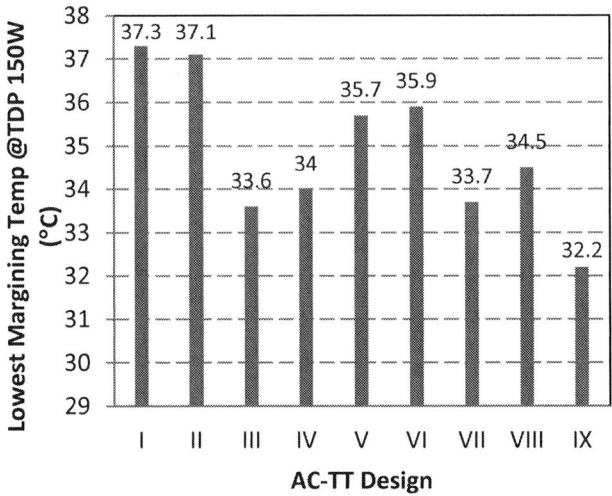

**Figure 13.** Simulation Results of the AC-TT with Various Heat Sink Configurations

In addition to the simulation results presented in Figure 13, experiments were also performed for Design I-IV. Experimental results show that Design IV provides ~3°C performance improvement compared to Design I with aluminum fins and solid copper base as shown in Figure 14. Comparison of experimental data of Design I, II and III for design option with three $\phi$8mm heat pipes, four $\phi$6mm heat pipes and four 8mm heat pipes are shown in Figure 15. We can observe that increasing the number of heat pipes from embedded three $\phi$8mm to embedded four $\phi$6mm did not give significant performance improvement while increasing the number of embedded $\phi$8mm heat pipes from three to four resulted in a performance improvement of 3.7°C or 9.9%. These experimental results are well in agreement with the simulation data presented in Figure 13. Experimental results of design option employing vapor chamber base, liquid chamber base and aluminum fins are not tested due to sample availability and the prototyping efforts are still ongoing.

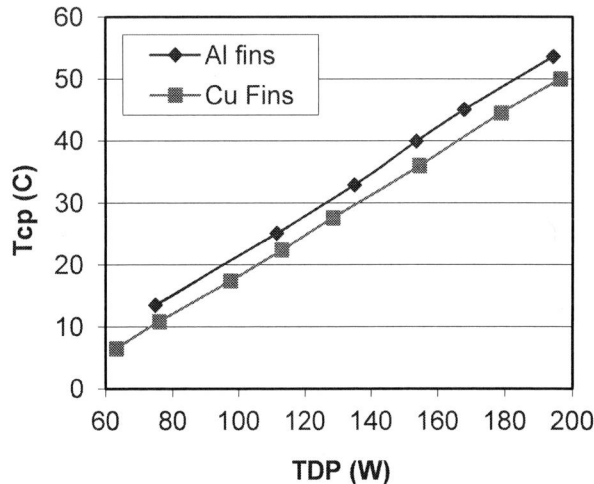

**Figure 14**. Experimental Results for CPU AC-TT Design I and Design IV (Tla=24°C)

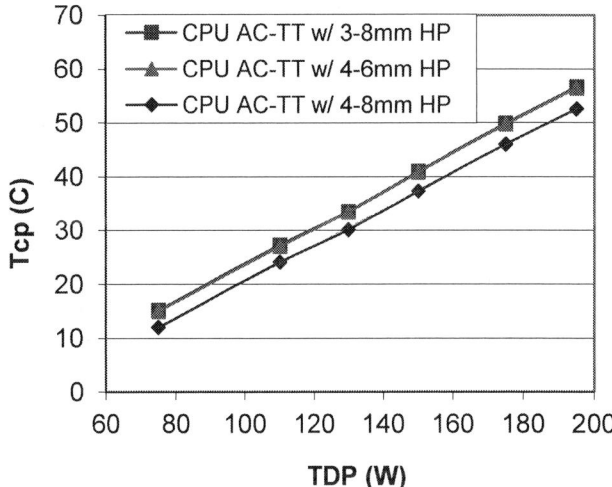

**Figure 15**. Experimental Results for AC-TT Design I-III Tested on CPU TTV (Tla=24°C)

## 6. SUMMARY

This paper presents the investigation of various advanced cooling technologies used for improving the thermal performance of AC-TTs. Results from numerical and experimental studies on the heat sink fin materials, vapor chamber heat sink base, liquid chamber heat sink base, heat sink with heat pipe, and combination of liquid chamber with heat pipe were summarized. The findings are:

- Heat sink performance found from the experimental results can be used to predict the AC-TT performance

- Increasing the number of embedded $\phi$8mm heat pipes from three to four resulted in 3.7°C performance improvement over three $\phi$8mm embedded heat pipes and copper base heat sink

- Combination of four $\phi$8mm embedded heat pipe and liquid chamber employed on heat sink base provide at

978-1-4673-1110-6/12 $31.00 © 2012 IEEE       360

least 5°C performance improvement over three $\phi$8mm embedded heat pipe and copper base heat sink

- The AC-TT performance is expected to improve about 3°C by replacing Al fin with Cu fin

As a result, Air-cooled thermal tools can be used for high temperature applications and reduce the dependency on Liquid-cooled thermal tools with the performance improvement achieved based on this study. The advanced cooling technologies and performance optimization strategies investigated in this paper can also benefit future generation of processor heat sink designs to meet the increasing thermal design power trends.

**Acknowledgments**

Authors would like to thank Ronaldo Albano and Victor Polyanko for their continuous support in the lab. The team also likes to extend their thanks to their group managers Adolfo Gascon, Floy Campbell and Ashok Kabadi to provide continuous support for the team's ongoing innovation efforts.

**References**

1. Mohammed, R. K., Sahan, R. A. and Prabhugoud, M., "Design Challenges of Thermal Margining Tools for Silicon Validation," 12th IEEE Intersociety Conference on Thermal and Thermo-mechanical Phenomena in Electronic Systems, ITHERM 2010.

2. Mohammed, R. K., Xia, Y., Sahan, R. A. and Pang, Y. F., "High Performance Air-Cooled Temperature Margining Tools for Silicon Validation," 26th IEEE SEMI-THERM Symposium, 2010.

3. Godet, C., Tantolin, C. and Zaghdoudi, M. C., "Use of Heat Pipes Cooling Systems in the Electronics Industry," Electronic Cooling, 2004 Nov.

4. Thayer, J., "Analysis of a Heat Pipe Assisted Heat Sink," Lancaster, PA: Thermacore International, Inc., 2005.

5. "Things to Consider When Designing With Heat Pipes," Tempe AZ, Enertron Inc., 2001.

6. Ahmed, M. etc., "Advanced Vapor Chamber for Thermal Management of High Performance Graphic Cards," International Forum on Heat Transfer (IFHT), September 2008, Tokyo.

7. Wuttijumnong, V, Nguyen, T. etc., "Overview Latest Technologies Using Heat Pipe and Vapor Chamber for Cooling of High Heat Generation Notebook Computer," 20th IEEE SEMI-THERM Symposium, 2004.

8. Murthy, S. S., "Thin Two-phase Heat Spreaders with Boiling Enhancement Microstructures for Thermal Management of Electronic Systems," PhD Dissertation University of Maryland, 2004.

9. Direct communication with vendor, Vapro Inc.

10. Mohammed, R. K., Xia, Y., Pang, Y. F, Prabhugoud, M and Sahan, R. A, "Experimental Techniques for Thermo-Mechanical Design in Silicon Validation Platforms", 25th IEEE SEMI-THERM Symposium, 2009.

# Sustainable Data Centers Powered by Renewable Energy

Levente J. Klein[1], Sergio Bermudez, Hans-Dieter Wehle*, Stephan Barabasi, Hendrik F. Hamann

IBM TJ Watson Research Center, Yorktown Heights, NY 10598

*IBM Systems &Technology Group, Böblingen, Germany, 71032

Email: [1] kleinl@us.ibm.com

## Abstract

The energy consumption of data centers (DCs) has dramatically increased in recent years, primarily due to the massive computing demands driven by communications, banking, online retail, and entertainment services. In today's data centers, the cooling and infrastructure operations require almost the same energy as the IT operations. The large energy consumption in data centers prompted government agencies, industries, professional organizations, and academic institutions to investigate sustainable growth paths. We discuss such scenarios based on current trends and projections and propose the required innovations to achieve a 10 fold increase in "performance per Watt" of IT operations. We discuss three possible technology components that would improve the operational performance of data centers: (1) integration of renewable energy sources (2) increasing energy efficiency through IT consolidation and workload optimization, and (3) multifunctional sensor networks for better cooling and infrastructure management. We discuss the key requirements and how these technologies can be combined to achieve a sustainable path.

## Keywords

Solar energy, air side economizer, energy forecasting, advanced control, IT scaling, data center

## 1. Introduction

In recent studies, it has been pointed out that the energy consumption in U.S. data centers is approximately 2% of the total U.S. electricity consumption and it increased 65% since 2000 [1,2]. These reports suggest that most of the energy-efficiency improvements that have resulted from new technologies and system design have been outpaced by the continued demand for more computing capacity.

The energy use of data centers is accelerated due to recent trends to expand information technology in every aspect of human life from work to entertainment, education, and health [1,3]. Furthermore, information technology will be absolutely critical for many emerging energy efficiency solutions such as smart grids, traffic guidance systems, building management systems, city operations and in general for smart planet applications where large amount of data could lead to better decision, planning, and emergency response [4].

Professional organization and the IT industry recently adopted various metrics to characterize the efficiency of data center operations. These metrics should enable comparisons of data centers located in different geographies and serving different business needs [5]. One such metrics is Data Center Infrastructure Efficiency (DCIE); it represents the ratio of the IT consumed power compared with the total power consumption. Furthermore, in response to energy consumption concerns, ASHRAE recently expanded the environmental operating envelope for data centers increasing the upper limit of temperature and relative humidity and implicitly allowing an increase of the DCIE [6].

While the energy consumption for industrial processes in the US is projected to not change until 2035, the data centers energy expansion is projected to be 500 billion kWh in 2035 [7]. This dramatic increase in energy consumption triggered the development of energy efficient technologies that could potentially mitigate the energy consumption. To bring data centers on a sustainable path, a holistic management is required, where new innovations from chip to data center levels are integrated and existing infrastructure operations and cooling are made more energy efficient.

Today's typical data centers may use 55 % of the power for IT processing, 16 % for power delivery (including power conditioning and distribution), and 29 % for the supporting facilities (mainly for cooling) [1,2]. The DCIE for today's typical data centers is 0.55 but could be highly improved using energy efficient technologies [8].

One possible path to improve the DCIE is a better integration and control of the electric grid, cooling strategies, data center building operations, and renewable energy solutions. While the above approaches can improve dramatically the DCIE metrics, other sustainability metrics like Carbon Usage Effectiveness (CUE) and Water Usage Effectiveness (WUE) should be optimized simultaneously [9]. Certainly one of the most straightforward ways to reduce CEU is the implementation of renewable energy solutions. A main challenge of renewable energy solutions is the intermittencies in generated power. It has been found that combining wind and solar energy could not eliminate the intermittencies and to mitigate power intermittencies the best approach is to use battery storage and weather forecasting. Currently, the most common approach is to use the electric grid as the storage to smoothen out intermittencies and power fluctuations. This is applicable as long as the power generated by renewable methods is a small fraction of the total energy portfolio of the electric utility companies.

In the table below we present a case study of how the operating conditions of data centers may evolve in the next 25 years based on a "business as usual" model and a sustainable data center based on current trends and projections [1,3]. For the table we assume that energy consumption per server for "business as usual" will increase 10% annually while the performance per watt will increase by 17% [10].

978-1-4673-1110-6/12 $31.00 © 2012 IEEE

| | 2011 | 2035 | Sustainable Data Center |
|---|---|---|---|
| IT Power | 55 % | 55 % | 90 % |
| Power Delivery | 16 % | 16 % | 4 % |
| Facilities | 29 % | 29 % | 6 % |
| DCIE | 0.55 | 0.55 →1.7 | 0.92 |
| Accumulated IT System Performance | 1 | 50 →6 | 290 |
| DC Power | 1 MW | 10 MW | 1 MW |
| Renewable Mix | <1 % | 20 % | 100 % |

Table 1 : Trends for data center operations for "business as usual" (BAU) and a sustainable data center.

For "business as usual", IT improvements in "performance per Watt" are expected to increase by approximately 50x for a given power level [10]. The 50x enhancement in "performance per Watt" will not be sufficient to meet the future computation needs which demand an increase of 500x (performance doubling every two years) [11]. Because only 50x is being offered by technology improvements in "business as usual", an additional improvement of factor 6 should be achieved in "performance per Watt" for the sustainable data center. For the sustainable data center more computation should be performed while energy consumption remains unchanged and this require both a fundamental IT system design and architectural changes [10]. Additionally, to achieve a DCIE of 0.9, require more power delivered for IT operations while cooling and infrastructure operations power is reduced. For the sustainable data centers the renewable power must be integrated in the operation of the infrastructure and ideally the fraction of power drawn from renewable energy sources would be 100%.

### 1. Sustainable data centers

Three technology components are highlighted to achieve a DCIE of 0.9 using energy efficiency innovations and renewable energy integration in data center operations. We discuss the sustainability from the perspective of energy efficiency, space utilization, reliability, and uptime of data centers.

#### a. Integration of renewable energy sources

Renewable energy integration is actively pursued by electrical utilities and by industries as a way to reduce dependency on fossil fuel and decrease carbon footprint. Solar and wind energy penetration is below 1 % in the US, however the Department of Energy is targeting a penetration of up to 15% by 2035 [12]. One challenge faced by renewable energy sources are their intermittent nature; solar energy is available during sunny days but not during cloudy days or night time. Same is true for wind energy, where energy is produced only on windy days.

While various strategies may be employed to compensate for the solar power fluctuations/intermittencies, the most straightforward one is the integration of battery systems to store excess energy when is available and to dispatch the energy when there is lack of it. Depending on the battery size, power compensation can range from minutes up to hours or days.

The electric grid can be used to compensate for power fluctuations as long as the renewable energy penetration is below 10% of the total available power but it becomes more challenging as penetration increase above 30% or even higher. For high level penetrations the demand has to be matched with supply. While this sound achievable, any data center requires uninterrupted operation 24/7 so battery management and scheduling should be integral part of solution.

To achieve this goal, three solutions must be implemented:

(1) Predicting availability of the solar radiation

(2) Dynamic power management to control peak and average power to maximizing performance per watt

(3) Dynamic workload and power management across the computational, communication and storage resources.

With advances in energy production metering, energy forecasting, and advanced control schemes, real time power availability can be integrated in infrastructure operation or matched with demand. For example, variable frequency drives of the Air Conditioning Units (ACU) units can be correlated with the availability of renewable energy, where ACU is operated at maximum speed while reducing the speed when power production drops. These precise and tight controls will require a reliable energy forecasting in addition to controlling the battery power dispatch and switching between various power sources. Currently the largest limitations are the availability and cost of battery systems and the accuracy of energy forecasting. In Fig 1 two possible connections of the renewable energy in the power distribution of data center are presented: the solar power could be connected (1) into the Power Distribution Units or the Uninterruptible Power Supply (UPS) backup battery systems or (2) directly to the Automatic Transfer Switch.

In the first case the energy is fed into the UPS or battery bank and requires the development of a charging and control circuit to direct excess generated energy into data center. In the second implementation, which is the most common approach, the photovoltaics output is converted to AC voltage and fed back into the main power line that supplies the data center. The first case (Fig 1a) is more desirable as it will maximize the contribution of the renewable energy integration by eliminating the losses associated with AC/DC conversion. This approach requires managing supply and demand and optimization and dynamic power management. In situations where the renewable energy is fed back to the grid (Fib 1b), the data center DCIE may not increase; to improve the DCIE the renewable energy sources should be directly integrated with the data center operation such that renewable energy reduces the total energy drawn from the electric grid.

978-1-4673-1110-6/12 $31.00 © 2012 IEEE

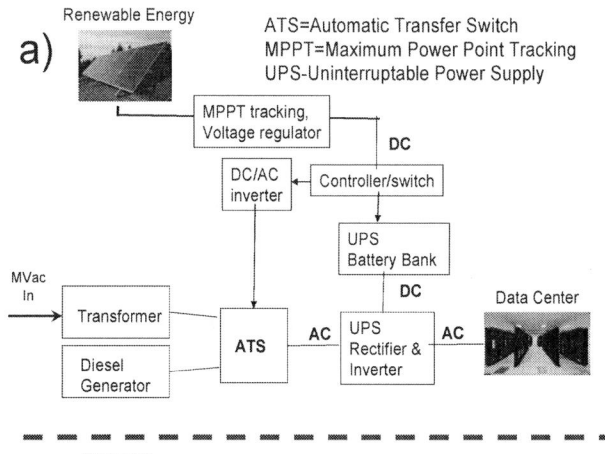

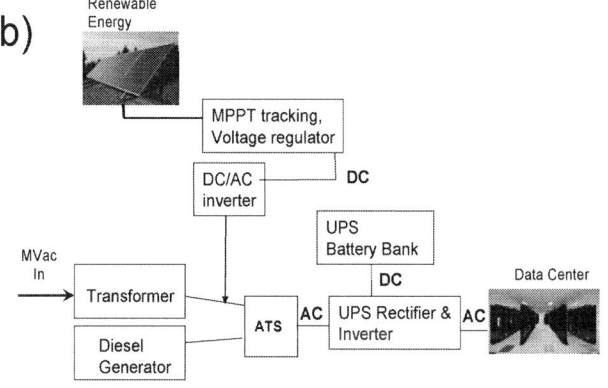

Fig 1: Integration of the renewable energy in UPS battery bank (a) or integration in the Automatic Transfer Switch (b).

For reliable renewable power management an energy forecasting technology has to be developed. While there are various components of solar and wind energy forecasting they are limited either in time span or in accuracy. The larger is the time horizon of forecasting, the more the accuracy is reduced. The highest accuracy can be achieved using a deterministic approach, where clouds detected by a ground base sensor can be tracked and their impact on the photovoltaic system can be predicted. Multiple forecasting technologies are under development (Fig. 2) to predict the solar radiation availability from minutes up to hours in advance. For short term forecasting the most common approaches are: (1) real time ground based sensor system and (2) satellite image processing. The local sensor could be a sky camera consisting of a hemispherical mirror with a camera positioned on the top. The image of the sky is reflected in the mirror and projected into the camera. Images are acquired every 30 sec and the clouds are delineated and their position on the sky is tracked from image to image. By knowing the position of the sun and projecting the cloud movement trajectory, the camera system can track the clouds over an area of a 1.5 mile radius around the location of the facility. The sky camera system has a very high spatial and temporal resolution with forecasting extending up to 30 minutes. For hours ahead forecasting satellite images from the Geo Stationary Satellite (GEOS) can be processed for cloud tracking. For the satellite images the spatial resolution of the image is around 40 km and

images are obtained every 30 min [13]. The satellite images track larger clouds extending over extended geographical areas. The integration of the two forecasting methods is under development, with the goal of matching the two models. Both forecasting methods require a self learning calibration to correlate pixel color intensities with the physical properties of the clouds. The physical properties of the clouds will determine the solar radiation that reaches the photovoltaic installations [14].

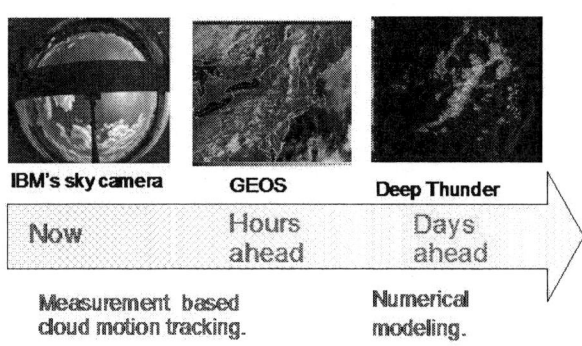

Fig 2 The time span of solar radiation forecasting using ground based sky camera system, satellite imaging system, and numerical weather forecasting. For day ahead forecasting all three methods have to be combined.

For forecasting spanning a few days in advance numerical weather models can be employed. One such weather model is Deep Thunder that is used for wind and solar energy forecasting [16]. The three forecasting methods can be utilized for managing and scheduling various electric load operations in the data center. The very short term forecasting is ideal for battery operation (like battery charging or power drop compensation to maintain a constant power level). The hourly forecast can be used for scheduling the load in the data center or to pre-cool the facility while maintaining the environmental parameters within ASHRAE guideline [2].

**b.  Increasing  energy  efficiency  through  IT consolidation and workload optimization**

Today's energy efficiency innovations are targeting better utilization of power distribution, cooling and dynamic allocation of the workload [15].

To achieve these goals, three solutions could be implemented:

1) Exploit virtualization to reduce the number of servers.

2) Use integrated approach for server consolidation.

3) Match IT workload allocations to the availability of renewable energy sources

Power utilization of IT servers in many data centers peaks during daytime when high computational loads are more likely to be scheduled. A typical three day power load on an IT server is shown in Figure 3. The maximum power consumption is around 2 pm for the three day period and remains relatively flat during nighttime and late afternoon. The maximum power production from a solar installation will

978-1-4673-1110-6/12 $31.00 © 2012 IEEE       364

peak at noon making solar integration with IT servers a feasible way to offset the high power demand of IT systems during daytime. We note that during daytime outdoor temperature increases making the indoor cooling demands to increase. Since there is an overlap between the maximum power demand from data centers and the maximum power production from solar panels, albeit phase shifted, solar power can be an effective way to match the maximum power demand. The IT load could potentially serve as load ballast for renewable energy production, thereby helping to drive down grid fluctuations due to the emergence of renewable energy resources.

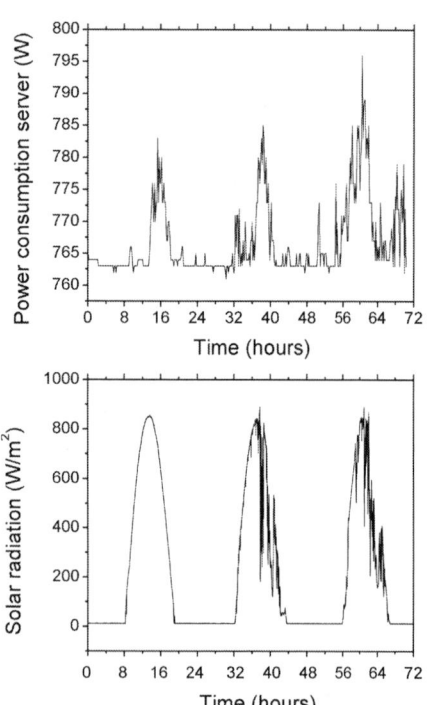

Fig 3: Power load on an IT server and the solar power generated during the same period where server power loads and solar power generation peaks to maximum value at the same time.

### c. Multifunctional sensor networks for better cooling and infrastructure management

Within the framework of energy efficient data centers, it is important to be able to reliable measure the environment of such spaces. There are monitoring networks of sensors that could be divided in wired or wireless technology, based on their communication channel. The use of wireless sensor networks for monitoring data centers is a more recent solution that has the benefits of installation simplicity (no burden of long cabling) as well as effortless maintenance (when moving equipment or reconfiguring the DC), compared to wired sensing solutions.

We developed a wireless sensing solution, called Low-power Mote Technology (LMT). LMT is an energy efficient platform that is easily configurable. An LMT mesh network consists of a set of motes and one gateway. An LMT mote is an autonomous device with a radio, a microcontroller, sensors, and power supply. One mote supports up to 20 sensors. The LMT manager keeps the mesh network working and publishes the raw data to IBM MMT application. LMT supports a variety of sensors and each mote in the network has a typical lifetime of 5 years. A mote can have multiple sensors attached to it, including temperature, relative humidity, differential pressure, and corrosion sensors among others.

Given the high resolution and real-time sensing provided by LMT and the use of analytics for modeling of MMT, this solution is ideal for monitoring and controlling mission-critical spaces in general, and DC in particular. LMT provides high spatio-temporal sensing resolution (real time data streams) which supports the accurate modeling of the DC environment (static and dynamic models of 3D temperature distribution and other analytics: like Humidity, Dew Point, Cooling Power, and Air flow). These analytics facilitate the uncovering of hot-spots or underutilized cooling zones. The wireless sensing solution uses a mesh network that is deployed across the whole data center; sensors are located at the air intake site of racks at various heights, under the plenum, at the inlet and outlet of the ACU units. The dense sensing is required in cases when ACU control is implemented to assure that every corner of the data centers is maintained at temperature within operating conditions.

The wireless sensors serve a double purpose, (1) to sense the environment and (2) to control ACU units based on local temperature and relative humidity data. The control algorithm turns on/off any ACU unit if sensors value exceeds an upper/lower threshold value that is set by operators. By managing ACU to better control indoor environmental parameters a reduction in energy usage of up to 10% of the total energy consumption can be achieved.

Recently, the air side economizer has been extensively discussed as a way to reduce energy consumption in data centers. The air side economizer enables using outside air for cooling when the temperature and relative humidity are within IT equipment operating specifications. ASHRAE recently extended the upper temperature levels to 27° C and the upper relative humidity levels to 80% for non condensing conditions [6].

While the benefits of the air side economizer have been documented, a potential downside is the unintentional introduction of gaseous and particulate contamination in the data centers along the outside air. To maintain reliability and uptime in data centers, air quality sensing should be implemented along with temperature and relative humidity sensing to assure that gaseous and particulate contamination is not exceeding ASHRAE recommendations [17]. Air quality sensor network can assure that corrosion rate levels are within acceptable range. Since the corrosion is a complex function of temperature, relative humidity, and contamination, a facility wide monitoring of air quality parameters is highly beneficial to prevent corrosion failures. To assure reliability of IT equipment and uptime for air side economized data centers, an ultra sensitive corrosion sensor was developed based on the resistive technique, where the silver and copper film thickness changes are monitored while the sensor is exposed to the data center environment. The sensor can sense a change in film thickness down to 1A; a single atomic layer. In places where air contamination may pose a risk, gaseous and particulate filtering should be integrated with air side economization. Pollution and air contamination have spatial and temporal variations and these variations should be considered when outside air may be used for cooling purposes [18]. Even with addition of air quality sensors and air filtration to maintain reliability and uptime of the DC, the benefits of air side economizer are significant as the energy spent on cooling can be reduced up to 50%.

An alternative way to store the renewable energy could be using the thermal mass of building. Knowing in advance the available power, the indoor air temperature can be adjusted to store thermal energy while maintaining its value within the safe operating zone of the data center. For example the data center can be overcooled when the renewable energy is available and let to warm up when the energy production is small. The building thermal envelope is used in this case as a "battery" to store the available renewable energy in the indoor air temperature. For building mass storage the renewable energy forecasting should be reliable on a very short to medium time scale to schedule ACU cooling

## 2. Discussion and Conclusion

Significant energy efficiency improvements will originate from integrating the various technology components of a modern IT facility. Such integration will enable a much more holistic management approach; this includes integrating the underlying IT technology, energy and thermal management, power delivery technologies, as well as cooling and facilities. Many of today's energy efficient technologies provide a clear pathway towards such integration. Innovation in technologies to improve IT performance and DCIE metrics should be integrated in DC operations to achieve a sustainable growth path.

### Acknowledgments

We acknowledge contributions from the world-wide MMT team including Andrew Stepanchuk, Alan Claassen, Dennis Manzer, Srinivas Yarlanki, Vanessa Lopez, Tom Keller, Michael Schappert, Fernando Marianno, and many more colleagues. We also thank Jon Lenchner, Jeff Kephart, Raja Das, Tom Sarasin, and Wayne Riley from the Tivoli/Maximo team. This work was partially supported by the Department of Energy (Grant Number DE-EE0002897).

### References

1) J. G. Koomey, "Estimating Total Power Consumption by Servers in the U.S. and the World", A report by the Lawrence Berkeley National Laboratory, February 15,2007.; ''Report to Congress on Server and Data Center Energy Efficiency,'' Public Law 109–431, United States Code 2008.

2) J.G. Koomey, "Growth in Data center electricity use 2005 to 2010" Oakland, CA: Analytics Press. August 1, 2011.

3) "Best Practices Guide for Energy-Efficient Data Center Design", 2011 available at http://www1.eere.energy.gov/femp/pdfs/eedatacenterbestpractices.pdf

4) Susanne Dirks, and Mary Keeling, "A vision of smarter cities", IBM Institute for Business Value, 2009 ftp://ftp.software.ibm.com/common/ssi/pm/xb/n/gbe03227usen/GBE03227USEN.PDF

5) J.R. Stanley,K.R. Brill, and J.Koomey, "Four metrics define Data Center "Greeness"", Uptime Institute, 2007 ;C. Belady,A. Rawson,J. Pflueger,T. Cader, The Green Grid Data Center Power Efficiency Metrics: PUE and DCiE, The Green Grid, 2008

6) ASHRAE TC 9.9, "2011 Thermal Guidelines for Data Processing Environments – Expanded Data Center Classes and Usage Guidance", Ashrae, 2011.

7) M.Iyengar, and R.Schmidt, "Energy Consumption of Information Technology Data Centers", Electronics Cooling 12, 2010.

8) H. F. Hamann, T. van Kessel, M. Iyengar, J.-Y. Chung, W. Hirt, M. Schappert, A. Claassen, J. Cook, W. Min, Y. Amemiya, V.López, "Uncovering Energy Efficiency Opportunities in Data Centers", IBM J. Res. & Dev. 53, 19 2009.

9) C.Belady, "Carbon Usage Effectiveness (CUE): A Green Grid Data Center Sustainability Metric", The Green Grid, 2010; M.Patterson, "Water Usage Effectiveness (WUE): A Green Grid Data Center Sustainability Metric", The Green Grid , 2010.

10) J. G. Koomey, C. Belady, M. Patterson, A. Santos, K. D Lange, "Assessing trends over time in performance, cost, and energy use for servers", Microsoft and Intel Corporation Report, 2009.

11) R. Bohn, J. Short, and C. Baru, "How Much Information?" 2010 Report on Enterprise Server Information. San Diego: Global Information Industry Center at the School of International Relations and Pacific Studies, UC San Diego, 2010

12) Sunshot Vision Study, DOE, 2012 http://www1.eere.energy.gov/solar/pdfs/47927.pdf

13) L. J. Klein, S. Bermudez Rodrigues, S. Nitta , R. Sandstrom, S. Guha, H.F. Hamann, "Optimization of photovoltaic power generation using a Measurement and Management Technology (MMT) platform", IMAPS 2010, 43rd International Symposium on Microelectronics, 2010.

14) D. King, J. Dudley and W. Boyson, "PVSIM: a simulation program for photovoltaic cells, modules and arrays," Proc. IEEE Photovoltaic Specialists Conference, 1996.

15) H. F. Hamann, M. Schappert, M. Iyengar, T. van Kessel, and A. Claassen, ''Methods and Techniques for Measuring and Improving Data Center Best Practices,'' Proc. 11th Intersociety Conference on Thermomechanical Phenomena

*in Electronic Systems,* Orlando, Florida, pp. 1146–1152, 2008.

16) L.A. Treinish, and A.P. Praino, Applications and Implementation of a Mesoscale Numerical Prediction and Visualization System, Proceedings of the 20th Conference on Weather Analysis and Forecasting/16th Conference on Numerical Weather Prediction, January 2004.

17) 2011 Gaseous and Particulate Contamination Guidelines For Data Centers, Ashrae White Paper, 2011.

18) L.J. Klein, P.J. Singh, M. Schappert. M. Griffel. H.F. Hamann, Corrosion management for data centers, Semiconductor Thermal Measurement and Management Symposium (SEMI-THERM), 2011 27th Annual IEEE, 21, 2011.

CURRAN ASSOCIATES INC.
proceedings
.com

9781467311106